Critical Values for Student's *t*

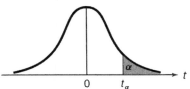

ν	$t_{.100}$	$t_{.050}$	$t_{.025}$	$t_{.010}$	$t_{.005}$	$t_{.001}$	$t_{.0005}$
1	3.078	6.314	12.706	31.821	63.657	318.31	636.62
2	1.886	2.920	4.303	6.965	9.925	22.326	31.598
3	1.638	2.353	3.182	4.541	5.841	10.213	12.924
4	1.533	2.132	2.776	3.747	4.604	7.173	8.610
5	1.476	2.015	2.571	3.365	4.032	5.893	6.869
6	1.440	1.943	2.447	3.143	3.707	5.208	5.959
7	1.415	1.895	2.365	2.998	3.499	4.785	5.408
8	1.397	1.860	2.306	2.896	3.355	4.501	5.041
9	1.383	1.833	2.262	2.821	3.250	4.297	4.781
10	1.372	1.812	2.228	2.764	3.169	4.144	4.587
11	1.363	1.796	2.201	2.718	3.106	4.025	4.437
12	1.356	1.782	2.179	2.681	3.055	3.930	4.318
13	1.350	1.771	2.160	2.650	3.012	3.852	4.221
14	1.345	1.761	2.145	2.624	2.977	3.787	4.140
15	1.341	1.753	2.131	2.602	2.947	3.733	4.073
16	1.337	1.746	2.120	2.583	2.921	3.686	4.015
17	1.333	1.740	2.110	2.567	2.898	3.646	3.965
18	1.330	1.734	2.101	2.552	2.878	3.610	3.922
19	1.328	1.729	2.093	2.539	2.861	3.579	3.883
20	1.325	1.725	2.086	2.528	2.845	3.552	3.850
21	1.323	1.721	2.080	2.518	2.831	3.527	3.819
22	1.321	1.717	2.074	2.508	2.819	3.505	3.792
23	1.319	1.714	2.069	2.500	2.807	3.485	3.767
24	1.318	1.711	2.064	2.492	2.797	3.467	3.745
25	1.316	1.708	2.060	2.485	2.787	3.450	3.725
26	1.315	1.706	2.056	2.479	2.779	3.435	3.707
27	1.314	1.703	2.052	2.473	2.771	3.421	3.690
28	1.313	1.701	2.048	2.467	2.763	3.408	3.674
29	1.311	1.699	2.045	2.462	2.756	3.396	3.659
30	1.310	1.697	2.042	2.457	2.750	3.385	3.646
40	1.303	1.684	2.021	2.423	2.704	3.307	3.551
60	1.296	1.671	2.000	2.390	2.660	3.232	3.460
120	1.289	1.658	1.980	2.358	2.617	3.160	3.373
∞	1.282	1.645	1.960	2.326	2.576	3.090	3.291

Source: This table is reproduced with the kind permission of the Trustees of Biometrika from E. S. Pearson and H. O. Hartley (eds.), *The Biometrika Tables for Statisticians*, Vol. 1, 3rd ed., *Biometrika*, 1996.

A SECOND COURSE IN STATISTICS: REGRESSION ANALYSIS

SIXTH EDITION

William Mendenhall
University of Florida

Terry Sincich
University of South Florida

PEARSON

Prentice Hall

PEARSON EDUCATION, INC.
Upper Saddle River, New Jersey 07458

Library of Congress Cataloging-in-Publication Data

Mendenhall, William.
  A second course in statistics : regression analysis / William Mendenhall, Terry Sincich.–
6th ed.
    p. cm.
  Includes bibliographical references and index.
  ISBN 0-13-022323-9
  1. Commercial statistics. 2. Statistics. 3. Regression analysis. I. Sincich, Terry. II. Title.

HF1017.M46 2003
519.5′36–dc21   2002044550

Acquisitions Editor: *George Lobell*
Editor in Chief: *Sally Yagan*
Vice-President/Director of Production and Manufacturing: *David W. Riccardi*
Production Editor: *Joan Wolk*
Manufacturing Buyer: *Lynda Castillo*
Manufacturing Manager: *Trudy Pisciotti*
Marketing Manager: *Halee Dinsey*
Marketing Assistant: *Rachel Beckman*
Director of Creative Services: *Paul Belfanti*
Art Editor: *Tom Benfatti*
Creative Director: *Carole Anson*
Interior Designer: *Bayani DeLeon*
Cover Art Director: *Jayne Conte*
Editorial Assistant: *Jennifer Brady*
Cover Image: *The Great Wall of China*: ©Underwood & Underwood/CORBIS
Art Studio: *Laserwords Private Limited*

© 2003, 1996 by Pearson Education, Inc.
Pearson Education, Inc.
Upper Saddle River, New Jersey 07458

Printed in the United States of America

10 9 8 7 6 5 4 3 2 1

ISBN 0-13-022323-9

Pearson Education LTD., *London*
Pearson Education Australia PTY, Limited, *Sydney*
Pearson Education Singapore, Pte. Ltd
Pearson Education North Asia Ltd., *Hong Kong*
Pearson Education Canada, Ltd., *Toronto*
Pearson Educación de Mexico, S.A. de C.V.
Pearson Education–Japan, *Tokyo*
Pearson Education Malaysia, Pte. Ltd.

CONTENTS

■ APPENDIX D   SAS for Windows Tutorial

■ APPENDIX E   SPSS for Windows Tutorial

■ APPENDIX F   MINITAB for Windows Tutorial

■ APPENDIX G   File Layouts for Large Data Sets

PREFACE

OVERVIEW

This text is designed for two types of statistics courses. The early chapters, combined with a selection of the case study chapters, are designed for use in the second half of a two-semester (or two-quarter) introductory statistics sequence for undergraduates with statistics or non-statistics majors. Or, the text can be used for a course in applied regression analysis for masters or Ph.D. students in other fields.

At first glance, these two uses for the text may seem inconsistent. How could a text be appropriate for both undergraduate and graduate students? The answer lies in the content. In contrast to a course in statistical theory, the level of mathematical knowledge required for an applied regression analysis course is minimal. Consequently, the difficulty encountered in learning the mechanics is much the same for both undergraduate and graduate students. The challenge is in the application–diagnosing practical problems, deciding on the appropriate linear model for a given situation, and knowing which inferential technique will answer the researcher's practical question. This *takes experience*, and it explains why a student with a non-statistics major can take an undergraduate course in applied regression analysis and still benefit from covering the same ground in a graduate course.

Introductory Statistics Course

It is difficult to identify the amount of material that should be included in the second semester of a two-semester sequence in introductory statistics. Optionally, a few lectures should be devoted to Chapter 1 (A Review of Basic Concepts) to make certain that all students possess a common background knowledge of the basic concepts covered in a first-semester (first-quarter) course. Chapter 2 (Introduction to Regression Analysis), Chapter 3 (Simple Linear Regression), Chapter 4 (Multiple Regression Models), Chapter 5 (Model Building), Chapter 6 (Variable Screening Methods), Chapter 7 (Some Regression Pitfalls), and Chapter 8 (Residual Analysis) provide the core for an applied regression analysis course. These chapters could be supplemented by the addition of Chapter 10 (Introduction to Time Series Modeling and Forecasting), Chapter 11 (Principles of Experimental Design), or Chapter 12 (The Analysis of Variance for Designed Experiments).

Applied Regression for Graduates

In our opinion, the quality of an applied graduate course is not measured by the number of topics covered or the amount of material memorized by the students. The measure is how well they can apply the techniques covered in the course to the solution of real problems encountered in their field of study. Consequently, we advocate moving on to new topics only after the students have demonstrated ability (through testing) to apply the techniques under discussion. In-class consulting sessions, where a case study is presented and the students have the opportunity to diagnose the problem and recommend an appropriate method of analysis, are very helpful in teaching applied regression analysis. This approach is particularly useful in helping students master the difficult topic of model selection and model building (Chapters 4–8) and relating questions about the model to real-world questions. The case study chapters (Chapters 13–17) illustrate the type of material that might be useful for this purpose.

A course in applied regression analysis for graduate students would start in the same manner as the undergraduate course, but would move more rapidly over the review material and would more than likely be supplemented by Appendix A (The Mechanics of a Multiple Regression Analysis), one of the statistical software Windows tutorials in Appendices D, E, or F (SAS, SPSS, or MINITAB), Chapter 9 (Special Topics in Regression), and other chapters selected by the instructor. As in the undergraduate course, we recommend the use of case studies and in-class consulting sessions to help students develop an ability to formulate appropriate statistical models and to interpret the results of their analyses.

FEATURES

1. **Readability.** We have purposely tried to make this a teaching (rather than a reference) text. Concepts are explained in a logical intuitive manner using worked examples.

2. **Emphasis on model building.** The formulation of an appropriate statistical model is fundamental to any regression analysis. This topic is treated in Chapters 4–8 and is emphasized throughout the text.

3. **Emphasis on developing regression skills.** In addition to teaching the basic concepts and methodology of regression analysis, this text stresses its use, as a tool, in solving applied problems. Consequently, a major objective of the text is to develop a skill in applying regression analysis to appropriate real-life situations.

4. **Numerous real data-based examples and exercises.** The text contains many worked examples that illustrate important aspects of model construction, data analysis, and the interpretation of results. Nearly every exercise is based on data and a problem extracted from a news article, magazine, or journal. Exercises are located at the ends of key sections and at the ends of chapters.

5. **Case study chapters.** The text contains five case study chapters, each of which addresses a real-life research problem. The student can see how regression analysis was used to answer the practical questions posed by the problem,

proceeding with the formulation of appropriate statistical models to the analysis and interpretation of sample data.

6. **Data sets.** The text contains four complete data sets that are associated with the case studies (Chapters 13–17). These can be used by instructors and students to practice model-building and data analyses.

7. **Extensive use of statistical software.** Tutorials on how to use any of three popular statistical software packages, SAS, SPSS, and MINITAB, are provided in Appendices D, E, and F, respectively. The printouts of the respective software packages are presented and discussed throughout the text.

NEW TO THE SIXTH EDITION

Although the scope and coverage remain the same, the sixth edition contains several substantial changes, additions, and enhancements:

1. **More computer printouts.** A SAS, SPSS, or MINITAB printout now accompanies every statistical technique presented, allowing the instructor to emphasize interpretations of the statistical results rather than the calculations required to obtain the results.

2. **Statistical software tutorials.** The Appendix now includes basic instructions on how to use the Windows versions of SAS, SPSS, and MINITAB. Step-by-step instructions and screen shots for each method presented in the text are shown.

3. **Describing qualitative data.** A new section (Sec. 1.3) on graphical and numerical methods of describing qualitative data has been added to Chapter 1.

4. **Paired comparisons for means.** New material on comparing two population means using a paired difference experiment is now included in Chapter 1 (Sec. 1.10).

5. **Reorganization of multiple regression models.** The multiple regression models presented in Chapter 4 have been reorganized according to order and complexity. First-order models are presented first, followed by interaction and second-order models.

6. **Model validation.** The section on external model validation (previously presented as a special topic in Chapter 9) has been moved to the model building chapter (Chapter 5). Several new examples are presented.

7. **Variable screening methods.** Stepwise regression and the all-possible-regressions-selection procedure are now included in a separate chapter (Chapter 6).

8. **Spline regression.** Spline regression methods are now discussed in the section on robust regression (Sec. 9.8) in Chapter 9: Special Topics.

9. **Case study 13: Residential property sale price data updated.** The data set for the case study on predicting sale prices of residential properties has been updated to reflect current economic trends.

Numerous less obvious changes in details have been made throughout the text in response to suggestions by current users of the earlier editions.

SUPPLEMENTS

The text is also accompanied by the following supplementary material:

1. **Student's solutions manual.** (by Mark Dummeldinger). A student's exercise solutions manual presents the full solutions to the odd exercises contained in the text.
2. **Instructor's solutions manual.** (by Mark Dummeldinger). The instructor's exercise solutions manual presents the full solutions to the other half (the even) exercises contained in the text. For adopters, the manual is complimentary from the publisher.
3. **Data CD.** The text is accompanied by a CD that contains files for all data sets marked with a CD icon in the text. These include data sets for text examples, exercises, and case studies. The data files are saved in ASCII format for easy importing into statistical software (SAS, SPSS, and MINITAB).

ACKNOWLEDGMENTS

We want to thank the many people who contributed time, advice, and other assistance to this project. We owe particular thanks to the many reviewers who provided suggestions and recommendations at the onset of the project and for the succeeding editions (including the 6th):

    Ruben Zamar, University of British Columbia

    Tom O'Gorman (Northern Illinois University)

    William Bridges, Jr. (Clemson University)

    Jeff Banfield, (Montana State University)

    Paul Maiste, (Johns Hopkins University)

    Mohammed Askalani, Mankato State University (Minnesota)

    Ken Boehm, Pacific Telesis (California)

    Andrew C. Brod, University of North Carolina at Greensboro

    James Daly, California State Polytechnic Institute at San Luis Obispo

    Assane Djeto, University of Nevada - Las Vegas

    Robert Elrod, Georgia State University

    James Ford, University of Delaware

    Carol Ghomi, University of Houston

    James Holstein, University of Missouri at Columbia

    Steve Hora, Texas Technological University

    K. G. Janardan, Eastern Michigan University

    Thomas Johnson, North Carolina State University

    Ann Kittler, Ryerson College (Toronto)

    James T. McClave, University of Florida

John Monahan, North Carolina State University
Kris Moore, Baylor University
Farrokh Nasri, Hofstra University
Robert Pavur, University of North Texas
P. V. Rao, University of Florida
Tom Rothrock, Info Tech, Inc.
Ray Twery, University of North Carolina at Charlotte
Joseph Van Matre, University of Alabama at Birmingham
William Weida, United States Air Force Academy
Dean Wichern, Texas A&M University
James Willis, Louisiana State University

We are particularly grateful to Charles Bond, Evan Anderson, Jim McClave, Herman Kelting, Ron Alderman, P.J. Taylor and Mike Jacob, who provided data sets and/or background information used in the case studies (Chapters 13–17).

William Mendenhall

Terry Sincich

A Review of Basic Concepts (Optional)

CONTENTS

OBJECTIVE

To review the basic concepts of statistics that are essential prerequisites to the study of regression analysis

Although we assume students have had a prerequisite introductory course in statistics, courses vary somewhat in content and in the manner in which they present statistical concepts. To be certain that we are starting with a common background, we will use this chapter to review some basic definitions and concepts. Coverage is optional.

1.1 Statistics and Data

According to *The Random House College Dictionary* (2001 ed.), statistics is "the science that deals with the collection, classification, analysis, and interpretation of numerical facts or data." In short, statistics is the **science of data**—a science that will enable you to be proficient data producers and efficient data users.

Definition 1.1
Statistics is the science of data. This involves collecting, classifying, summarizing, organizing, analyzing, and interpreting data.

Data are obtained by measuring some characteristic or property of the objects (usually people or things) of interest to us. These objects upon which the measurements (or observations) are made are called **experimental units**, and the properties being measured are called **variables** (since, in virtually all studies of interest, the property varies from one observation to another).

Definition 1.2
An **experimental unit** is an object (person or thing) upon which we collect data.

Definition 1.3
A **variable** is a characteristic (property) of the experimental unit with outcomes (data) that vary from one observation to the next.

All data (and consequently, the variables we measure) are either **quantitative** or **qualitative** in nature. Quantitative data are data that can be measured on a naturally occurring numerical scale. In general, qualitative data take values that are nonnumerical; they can only be classified into categories. The statistical tools that we use to analyze data depend on whether the data are quantitative or qualitative. Thus, it is important to be able to distinguish between the two types of data.

Definition 1.4
Quantitative data are observations measured on a naturally occurring numerical scale.

Definition 1.5
Nonnumerical data that can only be classified into one of a group of categories are said to be **qualitative data**.

EXAMPLE 1.1

Chemical and manufacturing plants often discharge toxic waste materials such as DDT into nearby rivers and streams. These toxins can adversely affect the plants and animals inhabiting the river and the riverbank. The U.S. Army Corps of Engineers conducted a study of fish in the Tennessee River (in Alabama) and its three tributary creeks: Flint Creek, Limestone Creek, and Spring Creek. A total of 144 fish were captured, and the following variables were measured for each:

1. River/creek where each fish was captured
2. Number of miles upstream where the fish was captured
3. Species (channel catfish, largemouth bass, or smallmouth buffalofish)
4. Length (centimeters)
5. Weight (grams)
6. DDT concentration (parts per million)

The data are saved in the **FISHDDT** file. Data for 10 of the 144 captured fish are shown in Table 1.1.

 FISHDDT

TABLE 1.1   Data Collected by U.S. Army Corps of Engineers

RIVER/CREEK	UPSTREAM	SPECIES	LENGTH	WEIGHT	DDT
FLINT	5	CHANNELCATFISH	42.5	732	10.00
FLINT	5	CHANNELCATFISH	44.0	795	16.00
SPRING	1	CHANNELCATFISH	44.5	1133	2.60
TENNESSEE	275	CHANNELCATFISH	48.0	986	8.40
TENNESSEE	275	CHANNELCATFISH	45.0	1023	15.00
TENNESSEE	280	SMALLMOUTHBUFF	49.0	1763	4.50
TENNESSEE	280	SMALLMOUTHBUFF	46.0	1459	4.20
TENNESSEE	285	LARGEMOUTHBASS	25.0	544	0.11
TENNESSEE	285	LARGEMOUTHBASS	23.0	393	0.22
TENNESSEE	285	LARGEMOUTHBASS	28.0	733	0.80

a. Identify the experimental units.

b. Classify each of the five variables measured as quantitative or qualitative.

Solution

a. Because the measurements are made for each fish captured in the Tennessee River and its tributaries, the experimental units are the 144 captured fish.

b. The variables upstream capture location, length, weight, and DDT concentration are quantitative because each is measured on a natural numerical scale: upstream in miles from the mouth of the river, length in centimeters, weight in grams, and DDT in parts per million. In contrast, river/creek and species cannot be measured quantitatively: They can only be classified into categories (e.g., channel catfish, largemouth bass, and smallmouth buffalofish for species). Consequently, data on river/creek and species are qualitative. ◆

EXERCISES

1.1. Colleges and universities are requiring an increasing amount of information about applicants before making acceptance and financial aid decisions. Classify each of the following types of data required on a college application as quantitative or qualitative.
   a. High school GPA
   b. High school class rank
   c. Applicant's score on the SAT or ACT
   d. Gender of applicant
   e. Parents' income
   f. Age of applicant

1.2. The data in the accompanying table were obtained from the *Model Year 2002 Fuel Economy Guide* for new automobiles.
   a. Identify the experimental units.

b. State whether each of the variables measured is quantitative or qualitative.

MODEL NAME	MFG	TRANS-MISSION TYPE	ENGINE SIZE (liters)	NUMBER OF CYLINDERS	EST. CITY MILEAGE (mpg)	EST. HIGHWAY MILEAGE (mpg)
NSX	Acura	Automatic	3.0	6	17	24
Golf	VW	Automatic	1.9	4	34	45
330CI	BMW	Manual	3.0	6	20	28
Escort	Ford	Automatic	2.0	4	25	33
Accord	Honda	Manual	2.3	4	25	32
Deville	Cadillac	Automatic	4.6	8	18	27

Source: Model Year 2002 Fuel Economy Guide, U.S. Dept. of Energy, U.S. Environmental Protection Agency.

1.3. The Cutter Consortium surveyed 154 U.S. companies to determine the extent of their involvement in electronic commerce (called *e-commerce*). Four of the questions they asked follow. (*Internet Week*, Septemeber 6, 1999.) For each question, determine the variable of interest and classify it as quantitative or qualitative.

  a. Do you have an overall e-commerce strategy?

  b. If you don't already have an e-commerce plan, when will you implement one?

  c. Are you delivering products over the Internet?

  d. What was your company's total revenue in the last fiscal year?

1.4. *The American Association of Nurse Anesthetists Journal* (Feb. 2000) published the results of a study on the use of herbal medicines before surgery. Each of 500 surgical patients was asked whether they used herbal or alternative medicines (e.g., garlic, ginkgo, kava, fish oil) against their doctor's advice before surgery. Surprisingly, 51% answered "yes."

  a. Identify the experimental unit for the study.

  b. Identify the variable measured for each experimental unit.

  c. Is the data collected quantitative or qualitative?

1.5. The *Journal of Performance of Constructed Facilities* (Feb. 1990) reported on the performance dimensions of water-distribution networks in the Philadelphia area. For one part of the study, the following data were collected for a sample of water pipe sections:

  1. Pipe diameter (inches)

  2. Pipe material

  3. Age (year of installation)

  4. Location

  5. Pipe length (feet)

  6. Stability of surrounding soil (unstable, moderately stable, or stable)

  7. Corrosiveness of surrounding soil (corrosive or noncorrosive)

  8. Internal pressure (pounds per square inch)

  9. Percentage of pipe under land cover

  10. Breakage rate (number of times pipe had to be repaired because of breakage)

  Identify the data as quantitative or qualitative.

1.2
Populations, Samples, and Random Sampling

When you examine a data set in the course of your study, you will be doing so because the data characterize a group of experimental units of interest to you. In statistics, the data set that is collected for all experimental units of interest is called a **population**. This data set, which is typically large, either exists in fact or is part of an ongoing operation and hence is conceptual. Some examples of statistical populations are given in Table 1.2.

> **Definition 1.6**
> A **population data set** is a collection (or set) of data measured on all experimental units of interest to you.

TABLE 1.2 Some Typical Populations

VARIABLE	EXPERIMENTAL UNITS	POPULATION DATA SET	TYPE
a. Starting salary of a graduating Ph.D. biologist	All Ph.D. biologists graduating this year	Set of starting salaries of all Ph.D. biologists who graduated this year	Existing
b. Breaking strength of water pipe in Philadelphia	All water pipe sections in Philadelphia	Set of breakage rates for all water pipe sections in Philadelphia	Existing
c. Quality of an item produced on an assembly line	All manufactured items	Set of quality measurements for all items manufactured over the recent past and in the future	Part existing, part conceptual
d. Sanitation inspection level of a cruise ship	All cruise ships	Set of sanitation inspection levels for all cruise ships	Existing

Many populations are too large to measure (because of time and cost); others cannot be measured because they are partly conceptual, such as the set of quality measurements (population c in Table 1.2). Thus, we are often required to select a subset of values from a population and to make **inferences** about the population based on information contained in a **sample**. This is one of the major objectives of modern statistics.

Definition 1.7

A **sample** is a subset of data selected from a population.

Definition 1.8

A **statistical inference** is an estimate, prediction, or some other generalization about a population based on information contained in a sample.

EXAMPLE 1.2

According to *USA Today* (Dec. 30, 1999), the average age of viewers of MSNBC cable television news programming is 50 years. Suppose a rival network executive hypothesizes that the average age of MSNBC news viewers is less than 50. To test her hypothesis, she samples 500 MSNBC news viewers and determines the age of each.

a. Describe the population.
b. Describe the variable of interest.
c. Describe the sample.
d. Describe the inference.

Solution

a. The population is the set of units of interest to the cable TV executive, which is the set of all MSNBC news viewers.
b. The age (in years) of each viewer is the variable of interest.
c. The sample must be a subset of the population. In this case, it is the 500 MSNBC viewers selected by the executive.
d. The inference of interest involves the *generalization* of the information contained in the sample of 500 viewers to the population of all MSNBC viewers. In particular, the executive wants to estimate the average age of the viewers in order to determine whether it is less than 50 years. She might accomplish this by calculating the average age in the sample and using the sample average to estimate the population average. ◆

Whenever we make an inference about a population using sample information, we introduce an element of uncertainty into our inference. Consequently, it is important to report the **reliability** of each inference we make. Typically, this is accomplished by using a probability statement that gives us a high level of confidence that the inference is true. In Example 1.2, we could support the inference about the average age of all MSNBC viewers by stating that the population average falls within 2 years of the calculated sample average with "95% confidence." (Throughout the text, we demonstrate how to obtain this measure of reliability—and its meaning—for each inference we make.)

Definition 1.9

A **measure of reliability** is a statement (usually quantified with a probability value) about the degree of uncertainty associated with a statistical inference.

The level of confidence we have in our inference, however, will depend on how **representative** our sample is of the population. Consequently, the sampling procedure plays an important role in statistical inference.

Definition 1.10

A **representative sample** exhibits characteristics typical of those possessed by the population.

The most common type of sampling procedure is one that gives every different sample of fixed size in the population an equal probability (chance) of selection. Such a sample—called a **random sample**—is likely to be representative of the population.

Definition 1.11

A **random sample** of n experimental units is one selected from the population in such a way that every different sample of size n has an equal probability (chance) of selection.

How can a random sample be generated? If the population is not too large, each observation may be recorded on a piece of paper and placed in a suitable container. After the collection of papers is thoroughly mixed, the researcher can remove n pieces of paper from the container; the elements named on these n pieces of paper are the ones to be included in the sample. Lottery officials utilize such a technique in generating the winning numbers for Florida's weekly 6/52 Lotto game. Fifty-two white Ping-Pong balls (the population), each identified from 1 to 52 in black numerals, are placed into a clear plastic drum and mixed by blowing air into the container. The Ping-Pong balls bounce at random until a total of six balls "pop" into a tube attached to the drum. The numbers on the six balls (the random sample) are the winning Lotto numbers.

This method of random sampling is fairly easy to implement if the population is relatively small. It is not feasible, however, when the population consists of a large number of observations. Since it is also very difficult to achieve a thorough mixing, the procedure only approximates random sampling. Most scientific studies, however, rely on computer software (with built-in random-number generators) to automatically generate the random sample. Almost all of the popular statistical software packages available (e.g., SAS, SPSS, MINITAB) have procedures for generating random samples.

EXERCISES

1.6. Refer to the *American Association of Nurse Anesthetists Journal* (Feb. 2000) study on the use of herbal medicines before surgery, Exercise 1.4. The 500 surgical patients that participated in the study were randomly selected from surgical patients at several metropolitan hospitals across the country.
  a. Do the 500 surgical patients represent a population or a sample? Explain.
  b. If your answer was sample in part a, is the sample likely to be representative of the population? If you answered population in part a, explain how to obtain a representative sample from the population.

1.7. Does a massage enable the muscles of tired athletes to recover from exertion faster than usual? To answer this question, researchers recruited eight amateur boxers to participate in an experiment (*British Journal of Sports Medicine*, April 2000). After a 10-minute workout in which each boxer threw 400 punches, half the boxers were given a 20-minute massage and half just rested for 20 minutes. Before returning to the ring for a second workout, the heart rate (beats per minute) and blood lactate level (micromoles) were recorded for each boxer. The researchers found no difference in the means of the two groups of boxers for either variable.
  a. Identify the experimental units of the study.
  b. Identify the variables measured and their type (quantitative or qualitative).
  c. What is the inference drawn from the analysis?
  d. Comment on whether this inference can be made about all athletes.

1.8. A Gallup Youth Poll was conducted to determine the topics that teenagers most want to discuss with their parents. The findings show that 46% would like more discussion about the family's financial situation, 37% would like to talk about school, and 30% would like to talk about religion. The survey was based on a national sampling of 505 teenagers, selected at random from all US. teenagers.
  a. Describe the sample.
  b. Describe the population from which the sample was selected.
  c. Is the sample representative of the population?

  d. What is the variable of interest?
  e. How is the inference expressed?
  f. Newspaper accounts of most polls usually give a *margin of error* (i.e., plus or minus 3%) for the survey result. What is the purpose of the margin of error and what is its interpretation?

1.9. *USA Today* (Aug. 14, 1995) reported on a study that suggests "frequently 'heading' the ball in soccer lowers players' IQs." A psychologist tested 60 male soccer players, ages 14–29, who played up to five times a week. Players who averaged 10 or more headers a game had an average IQ of 103, while players who headed one or fewer times per game had an average IQ of 112.
  a. Describe the population of interest to the psychologist.
  b. Identify the variables of interest.
  c. Identify the type (qualitative or quantitative) of the variables, part **b**.
  d. Describe the sample.
  e. What is the inference made by the psychologist?
  f. Discuss possible reasons why the inference, part **e**, may be misleading.

1.10. Are men or women more adept at remembering where they leave misplaced items (like car keys)? According to University of Florida psychology professor Robin West, women show greater competence in actually finding these objects (*Explore*, Fall 1998). Approximately 300 men and women from Gainesville, Florida, participated in a study in which each person placed 20 common objects in a 12-room "virtual" house represented on a computer screen. Thirty minutes later, the subjects were asked to recall where they put each of the objects. For each object, a recall variable was measured as "yes" or "no."
  a. Identify the population of interest to the psychology professor.
  b. Identify the sample.
  c. Does the study involve descriptive or inferential statistics? Explain.
  d. Are the variables measured in the study quantitative or qualitative?

1.3
Describing Qualitative Data

Consider a study of aphasia published in the *Journal of Communication Disorders* (Mar. 1995). Aphasia is the "impairment or loss of the faculty of using or understanding spoken or written language." Three types of aphasia have been identified by researchers: Broca's, conduction, and anomic. They wanted to determine whether one type of aphasia occurs more often than any other, and, if so, how often. Consequently, they measured aphasia type for a sample of 22 adult aphasiacs. Table 1.3 gives the type of aphasia diagnosed for each aphasiac in the sample.

For this study, the variable of interest, aphasia type, is qualitative in nature. Qualitative data are nonnumerical in nature; thus, the value of a qualitative variable can only be classified into categories called *classes*. The possible aphasia types—Broca's, conduction, and anomic—represent the classes for this qualitative variable. We can summarize such data numerically in two ways: (1) by computing the *class frequency*—the number of observations in the data set that fall into each class; or (2) by computing the *class relative frequency*—the proportion of the total number of observations falling into each class.

APHASIA

TABLE 1.3   Data on 22 Adult Aphasiacs

SUBJECT	TYPE OF APHASIA
1	Broca's
2	Anomic
3	Anomic
4	Conduction
5	Broca's
6	Conduction
7	Conduction
8	Anomic
9	Conduction
10	Anomic
11	Conduction
12	Broca's
13	Anomic
14	Broca's
15	Anomic
16	Anomic
17	Anomic
18	Conduction
19	Broca's
20	Anomic
21	Conduction
22	Anomic

Source: Li, E. C., Williams, S. E., and Volpe, R. D., "The effects of topic and listener familiarity of discourse variables in procedural and narrative discourse tasks." *Journal of Communication Disorders*, Vol. 28, No. 1, Mar. 1995, p. 44 (Table 1).

Definition 1.12

A **class** is one of the categories into which qualitative data can be classified.

Definition 1.13

The **class frequency** is the number of observations in the data set falling in a particular class.

Definition 1.14

The **class relative frequency** is the class frequency divided by the total number of observations in the data set, i.e.,

$$\text{class relative frequency} = \frac{\text{class frequency}}{n}$$

Examining Table 1.3, we observe that 5 aphasiacs in the study were diagnosed as suffering from Broca's aphasia, 7 from conduction aphasia, and 10 from anomic aphasia. These numbers—5, 7, and 10—represent the class frequencies for the three classes and are shown in the summary table, Table 1.4.

Table 1.4 also gives the relative frequency of each of the three aphasia classes. From Definition 1.14, we know that we calculate the relative frequency by dividing the class frequency by the total number of observations in the data set. Thus, the relative frequencies for the three types of aphasia are

$$\text{Broca's:} \quad \frac{5}{22} = .227$$

$$\text{Conduction:} \quad \frac{7}{22} = .318$$

$$\text{Anomic:} \quad \frac{10}{22} = .455$$

From these relative frequencies we observe that nearly half (45.5%) of the 22 subjects in the study are suffering from anomic aphasia.

TABLE 1.4 **Summary Table for Data on 22 Adult Aphasiacs**

CLASS	FREQUENCY	RELATIVE FREQUENCY
TYPE OF APHASIA	NUMBER OF SUBJECTS	PROPORTION
Broca's	5	.227
Conduction	7	.318
Anomic	10	.455
Totals	22	1.000

Although the summary table of Table 1.4 adequately describes the data of Table 1.3, we often want a graphical presentation as well. Figures 1.1 and 1.2 show two of the most widely used graphical methods for describing qualitative data—bar graphs and pie charts. Figure 1.1 shows the frequencies of aphasia types in a **bar graph** produced with SAS. Note that the height of the rectangle, or "bar," over each class is equal to the class frequency. (Optionally, the bar heights can be proportional to class relative frequencies.)

FIGURE 1.1

SAS bar graph for data on 22 aphasiacs

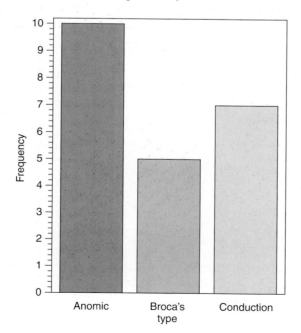

FIGURE 1.2
SPSS pie chart for data on 22
aphasiacs

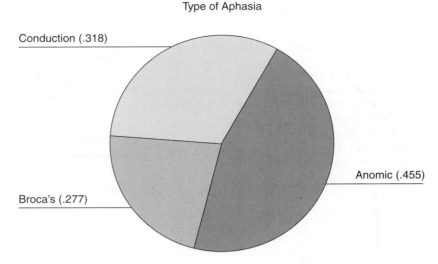

Type of Aphasia

Conduction (.318)

Anomic (.455)

Broca's (.277)

In contrast, Figure 1.2 shows the relative frequencies of the three types of aphasia in a **pie chart** generated with SPSS. Note that the pie is a circle (spanning 360°) and the size (angle) of the "pie slice" assigned to each class is proportional to the class relative frequency. For example, the slice assigned to anomic aphasia is 45.5% of 360°, or (.455)(360°) = 163.8°.

Step-by-step instructions and keystrokes for generating bar graphs and pie charts with SAS, SPSS, and MINITAB are provided in the Appendix.

EXERCISES

1.11. The International Rhino Federation estimates that there are 13,585 rhinoceroses living in the wild in Africa and Asia. A breakdown of the number of rhinos of each species is reported in the accompanying table.

RHINO SPECIES	POPULATION ESTIMATE
African Black	2,600
African White	8,465
(Asian) Sumatran	400
(Asian) Javan	70
(Asian) Indian	2,050
Total	13,585

Source: International Rhino Federation, July 1998.

a. Construct a relative frequency table for the data.
b. Display the relative frequencies in a bar graph.
c. What proportion of the 13,585 rhinos are African rhinos? Asian?

1.12. In *Psychology and Aging* (Dec. 2000), University of Michigan School of Public Health researchers studied the roles that elderly people feel are the most important to them in late life. The accompanying table summarizes the most salient roles identified by each in a national sample of 1,102 adults, 65 years or older.

MOST SALIENT ROLE	NUMBER
Spouse	424
Parent	269
Grandparent	148
Other relative	59
Friend	73
Homemaker	59
Provider	34
Volunteer, club, church member	36
Total	1,102

Source: Krause, N., and Shaw, B.A. "Role-specific feelings of control and mortality," *Psychology and Aging*, Vol. 15, No. 4, Dec. 2000 (Table 2).

a. Describe the qualitative variable summarized in the table. Give the categories associated with the variable.

b. Are the numbers in the table frequencies or relative frequencies?

c. Display the information in the table in a bar graph.

d. Which role is identified by the highest percentage of elderly adults? Interpret the relative frequency associated with this role.

1.13. The Moffitt Cancer Center at the University of South Florida treats over 25,000 patients a year. The graphic describes the types of cancer treated in Moffitt's patients during fiscal year 2000.

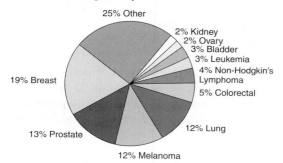

25% Other
2% Kidney
2% Ovary
3% Bladder
3% Leukemia
4% Non-Hodgkin's Lymphoma
5% Colorectal
19% Breast
13% Prostate
12% Lung
12% Melanoma

a. What type of graph is portrayed?

b. Which type of cancer is treated most often at Moffitt?

c. What percentage of Moffitt's patients are treated for melanoma, lymphoma, or leukemia?

1.14. Archaeologists excavating the ancient Greek settlement at Phylakopi classified the pottery found in trenches (*Chance*, Fall 2000). The accompanying table describes the collection of 837 pottery pieces uncovered in a particular layer at the excavation site. Construct and interpret a graph that will aid the archaeologists in understanding the distribution of the pottery types found at the site.

POT CATEGORY	NUMBER FOUND
Burnished	133
Monochrome	460
Slipped	55
Painted in curvilinear decoration	14
Painted in geometric decoration	165
Painted in naturalistic decoration	4
Cycladic white clay	4
Conical cup clay	2
Total	837

Source: Berg, I., and Bliedon, S. "The Pots of Phylakopi: Applying Statistical Techniques to Archaeology," Vol. 13, No. 4, Fall 2000.

 OILSPILL

TANKER	SPILLAGE (metric tons, thousands)	CAUSE
Atlantic Empress	257	Collision (C)
Castillo De Bellver	239	Fire/Explosion (FE)
Amoco Cadiz	221	Hull Failure (HF)
Odyssey	132	FE
Torrey Canyon	124	Grounding (G)
Sea Star	123	C
Hawaiian Patriot	101	HF
Independenta	95	C
Urquiola	91	G
Irenes Serenade	82	FE
Khark 5	76	FE
Nova	68	C
Wafra	62	G
Epic Colocotronis	58	G
Sinclair Petrolore	57	FE
Yuyo Maru No 10	42	C
Assimi	50	FE
Andros Patria	48	FE
World Glory	46	HF
British Ambassador	46	HF
Metula	45	G
Pericles G. C.	44	FE
Mandoil II	41	C
Jakob Maersk	41	G
Burmah Agate	41	C
J. Antonio Lavalleja	38	G
Napier	37	G
Exxon Valdez	36	G
Corinthos	36	C
Trader	36	HF
St. Peter	33	FE
Gino	32	C
Golden Drake	32	FE
Lonnis Angelicoussis	32	FE
Chryssi	32	HF
Irenes Challenge	31	HF
Argo Merchant	28	G
Heimvard	31	C
Pegasus	25	Unknown (U)
Pacocean	31	HF
Texaco Oklahoma	29	HF
Scorpio	31	G
Ellen Conway	31	G
Caribbean Sea	30	HF
Cretan Star	26	U
Grand Zenith	26	HF
Athenian Venture	26	FE
Venoil	26	C
Aragon	24	HF
Ocean Eagle	21	G

Source: Daidola, J. C. "Tanker structure behavior during collision and grounding." *Marine Technology*, Vol. 32, No. 1, Jan. 1995, p. 22 (Table 1).

1.15. Owing to several major ocean oil spills by tank vessels, Congress passed the 1990 Oil Pollution Act, which requires all tankers to be designed with thicker hulls. Further improvements in the structural design of a tank vessel have been proposed since then, each with the objective of reducing the likelihood of an oil spill and decreasing the amount of outflow in the event of a hull puncture. To aid in this development, *Marine Technology* (Jan. 1995) reported on the spillage amount and cause of puncture for 50 recent major oil spills from tankers and carriers. The data are reproduced in the table on p. 11.

    a. Use a graphical method to describe the cause of oil spillage for the 50 tankers.

    b. Does the graph, part **a**, suggest that any one cause is more likely to occur than any other? How is this information of value to the design engineers?

1.4
Describing Quantitative Data Graphically

A useful graphical method for describing quantitative data is provided by a relative frequency distribution. Like a bar graph for qualitative data, this type of graph shows the proportions of the total set of measurements that fall in various intervals on the scale of measurement. For example, Figure 1.3 shows the intelligence quotients (IQs) of identical twins. The area over a particular interval under a relative frequency distribution curve is proportional to the fraction of the total number of measurements that fall in that interval. In Figure 1.3, the fraction of the total number of identical twins with IQs that fall between 100 and 105 is proportional to the shaded area. **If we take the total area under the distribution curve as equal to 1, then the shaded area is equal to the fraction of IQs that fall between 100 and 105.**

FIGURE 1.3

Relative frequency distribution: IQs of identical twins

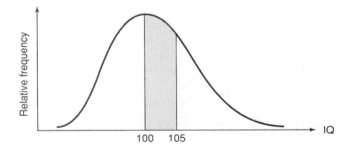

Throughout this text we will devote the quantitative variable measured by the symbol y. Observing a single value of y is equivalent to selecting a single measurement from the population. The probability that it will assume a value in an interval, say, a to b, is given by its relative frequency or **probability distribution**. The total area under a probability distribution curve is always assumed to equal 1. Hence, the probability that a measurement on y will fall in the interval between a and b is equal to the shaded area shown in Figure 1.4.

Since the theoretical probability distribution for a quantitative variable is usually unknown, we resort to obtaining a sample from the population: Our objective is to describe the sample and use this information to make inferences about the probability distribution of the population. **Stem-and-leaf plots** and **histograms** are two of the most popular graphical methods for describing quantitative data. Both display the frequency (or relative frequency) of observations that fall into specified intervals (or classes) of the variable's values.

For small data sets (say, 30 or fewer observations) with measurements with only a few digits, stem-and-leaf plots can be constructed easily by hand. Histograms, on

FIGURE 1.4

Probability distribution for a
quantitative variable

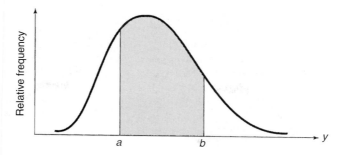

the other hand, are better suited to the description of larger data sets, and they
permit greater flexibility in the choice of classes. Both, however, can be generated
using the computer, as illustrated in the following examples.

EXAMPLE   1.3

The data in Table 1.5 represent sale prices (in thousands of dollars) for a random
sample of 25 residential properties sold in Tampa, Florida, in 1999. (These data were
extracted from the TAMSALES data file described in Appendix I. We analyze the
data more thoroughly in Case Study 12.) A MINITAB printout of a stem-and-leaf
plot for the 25 sale prices is shown in Figure 1.5. Interpret the figure.

**TABLE 1.5   Sale Prices for a Sample of
Properties from the TAMSALES file**

SALE PRICE (hundreds of dollars)				
66	59	106	50	63
89	129	74	82	84
71	95	72	57	76
109	77	68	101	65
42	36	148	94	112

Solution

In a stem-and-leaf plot, each measurement is partitioned into a stem and a leaf.
MINITAB has selected the last digit in the sale price to represent the leaf and the
preceding digits to represent the stem. For example, the value 148 (representing
a sale price of $148,000) is partitioned into a stem of 14 and a leaf of 8, as
illustrated here:

STEM	LEAF
14	8

The stems are listed in order in the second column of the plot, Figure 1.6, starting
with the smallest stem of 3 and ending with the largest stem of 14. The respective
leaves are then placed to the right in the appropriate stem row in increasing order.
For example, the stem row of 9 in Figure 1.5 has two leaves, 4 and 5, representing
the sale prices $94,000 and $95,000. Notice that the stem row of 7 (representing
sale prices in the $70,000s) has the most leaves (5). Thus, 5 of the 25 sale prices (or
20%) have values in the $70,000s. Notice also that 20 of the 25 sale prices (80%)
fall between $50,000 and $110,000. (That is, 20 of the sale prices have stems ranging

FIGURE 1.5

MINITAB stem-and-leaf display
for sale prices in Table 1.5

Stem-and-Leaf Display: salepric

```
Stem-and-leaf of salepric   N  = 25
Leaf Unit = 1.0

     1      3 6|
     2      4 2
     5      5 079
     9      6 3568
    (5)     7 12467
    11      8 249
     8      9 45
     6     10 169
     3     11 2
     2     12 9
     1     13
     1     14 8
```

from 5 to 10.) [*Note*: The first column in the MINITAB stem-and-leaf display gives
the cumulative number of measurements in the nearest "tail" of the distribution,
beginning with the stem row.]   ◆

EXAMPLE   1.4

Figure 1.6 is a SAS printout of a relative frequency histogram describing the sale
prices (in $ thousands) of over 1,000 residential properties stored in the TAMSALES
data file.

FIGURE 1.6

SAS histogram for the sale prices
(in $ thousands)

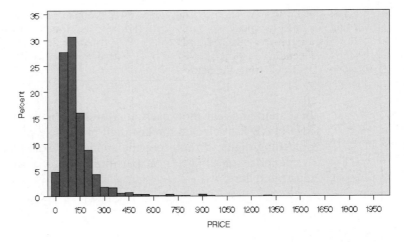

a. Interpret the graph.
b. Visually estimate the proportion of sale prices in the data set between $75,000
   and $175,000.

Solution

a. In constructing a histogram, the values of the sale prices are divided into intervals
   of equal length ($50,000), called **classes**. The midpoints of these classes are shown
   on the horizontal axis of Figure 1.6. For example, the class with a midpoint of 150

ranges from $125,000 to $175,000. The relative frequency (or percentage) of sale prices falling in each class interval is represented by the vertical bars over the class.

You can see from Figure 1.6 that the sale prices tend to pile up near $100,000; the class interval from $75,000 to $125,000 has the greatest relative frequency.

Figure 1.6 also shows a tendency for the data to tail out to the high side because of a few extremely large sale prices. Distributions of data with this feature are said to be skewed to the right, or **positively skewed**. (Similarly, distributions of data are skewed left, or **negatively skewed**, if they tend to tail out to the low side because of a few unusually small measurements.)

b. The interval $75,000 to $175,000 spans two sale price classes: 75–125 (midpoint of 100) and 125–175 (midpoint of 150). The proportion of sale prices between $75,000 and $175,000 is equal to the sum of the relative frequencies associated with these two classes. These two class relative frequencies are (approximately) .27 and .31, respectively. Consequently, the approximate proportion of sale prices between $75,000 and $175,000 is

$$(.27 + .31) = .58 = 58\%$$ ◆

The commands, menu selections, and keystrokes required to construct stem-and-leaf plots and histograms using SAS, SPSS, or MINITAB are provided in the Appendices.

EXERCISES

1.16. The United States Chess Federation (USCF) establishes a numerical rating for each competitive chess player. The USCF rating is a number between 0 and 4,000 that changes over time depending on the outcome of tournament games. The higher the rating, the better (more successful) the player. The graph describes the rating distribution of 27,563 players who were active competitors in 1997.
  a. What type of graph is displayed?
  b. Is the variable displayed on the graph quantitative or qualitative?
  c. What percentage of players has a USCF rating above 1,000?

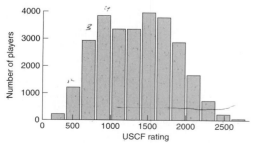

Source: United States Chess Federation, Jan. 1998.

1.17. A group of University of Virginia biologists studied nectivory (nectar drinking) in crab spiders to determine if adult males were feeding on nectar to prevent fluid loss (*Animal Behavior*, June 1995). Nine male spiders were weighed and then placed on the flowers of Queen Anne's lace. One hour later, the spiders were removed and reweighed. The evaporative fluid loss (in milligrams) of each of the nine male spiders is given in the table.

⬤ SPIDERS

MALE SPIDER	FLUID LOSS
A	.018
B	.020
C	.017
D	.024
E	.020
F	.024
G	.003
H	.001
I	.009

Source: Pollard, S. D., *et al.* "Why do male crab spiders drink nectar?" *Animal Behavior*, Vol. 49, No. 6, June 1995, p. 1445 (Table II).

a. Summarize the fluid losses of male crab spiders with a stem-and-leaf display.

b. Of the nine spiders, only three drank any nectar from the flowers of Queen Anne's lace. These three spiders are identified as G, H, and I in the table. Locate and circle these three fluid losses on the stem-and-leaf display. Does the pattern depicted in the graph give you any insight into whether feeding on flower nectar reduces evaporative fluid loss for male crab spiders? Explain.

BRAINPMI

POSTMORTEM INTERVALS FOR 22 HUMAN BRAIN SPECIMENS

5.5	14.5	6.0	5.5	5.3	5.8	11.0	6.1
7.0	14.5	10.4	4.6	4.3	7.2	10.5	6.5
3.3	7.0	4.1	6.2	10.4	4.9		

Source: Hayes. T. L., and Lewis, D. A. "Anatomical specialization of the anterior motor speech area: Hemispheric differences in magnopyramidal neurons." *Brain and Language*, Vol. 49, No. 3, June 1995, p. 292 (Table 1).

1.18. *Postmortem interval* (PMI) is defined as the elapsed time between death and an autopsy. Knowledge of PMI is considered essential when conducting medical research on human cadavers. The data in the table above are the PMIs of 22 human brain specimens obtained at autopsy in a recent study (*Brain and Language*, June 1995). Graphically describe the PMI data with a dot plot. Based on the plot, make a summary statement about the PMI of the 22 human brain specimens.

1.19. Data from a psychology experiment were reported and analyzed in *The American Statistician* (May 2001). Two samples of female students participated in the experiment. One sample consisted of 11 students known to suffer from the eating disorder bulimia; the other sample consisted of 14 students with normal eating habits. Each student completed a questionnaire from which a "fear of negative evaluation" (FNE) score was produced. (The higher the score, the greater the fear of negative evaluation.) The data are displayed in the table below.

a. Construct a dot plot or stem-and-leaf display for the FNE scores of all 25 female students.

b. Highlight the bulimic students on the graph, part **a.** Does it appear that bulimics tend to have a greater fear of negative evaluation? Explain.

c. Why is it important to attach a measure of reliability to the inference made in part **b**?

1.20. A Harris Corporation/University of Florida study was undertaken to determine whether a manufacturing process performed at a remote location could be established locally. Test devices (pilots) were set up at both the old and new locations, and voltage readings on the process were obtained. A "good" process was considered to be one with voltage readings of at least 9.2 volts (with larger readings better than smaller readings). The table contains voltage readings for 30 production runs at each location.

VOLTAGE

OLD LOCATION			NEW LOCATION		
9.98	10.12	9.84	9.19	10.01	8.82
10.26	10.05	10.15	9.63	8.82	8.65
10.05	9.80	10.02	10.10	9.43	8.51
10.29	10.15	9.80	9.70	10.03	9.14
10.03	10.00	9.73	10.09	9.85	9.75
8.05	9.87	10.01	9.60	9.27	8.78
10.55	9.55	9.98	10.05	8.83	9.35
10.26	9.95	8.72	10.12	9.39	9.54
9.97	9.70	8.80	9.49	9.48	9.36
9.87	8.72	9.84	9.37	9.64	8.68

Source: Harris Corporation, Melbourne, Fla.

a. Construct a relative frequency histogram for the voltage readings of the old process.

b. Construct a stem-and-leaf display for the voltage readings of the old process. Which of the two graphs in parts **a** and **b** is more informative?

c. Construct a frequency histogram for the voltage readings of the new process.

d. Compare the two graphs in parts **a** and **c**. (You may want to draw the two histograms on the same graph.) Does it appear that the manufacturing process can be established locally (i.e., is the new process as good as or better than the old)?

BULIMIA

Bulimic students:	21	13	10	20	25	19	16	21	24	13	14		
Normal students:	13	6	16	13	8	19	23	18	11	19	7	10	15	20

Source: Randles, R. H. "On Neutral Responses (Zeros) in the Sign Test and Ties in the Wilcoxon-Mann-Whitney Test." *The American Statistician*, Vol. 55, No. 2, May 2001 (Figure 3).

1.21. To minimize the potential for gastrointestinal disease outbreaks, all passenger cruise ships arriving at U.S. ports are subject to unannounced sanitation inspections. Ships are rated on a scale of 0 to 100 points, depending on how well they meet the Centers for Disease Control and Prevention sanitation standards. In general, the lower the score, the lower the level of sanitation. The table lists the sanitation inspection scores for 151 international cruise ships during 2001.

⊙ SHIPSANIT

SHIP NAME	SCORE	SHIP NAME	SCORE	SHIP NAME	SCORE
Grande Mariner	95	Crystal Symphony	93	Scotia Prince	92
Seabourn Sun	90	Statendam	94	Splendour of the Seas	90
Olympic Voyager	95	Texas Treasure	96	Silver Wind	96
Grandeur of the Seas	94	Elation	95	Clipper Adventurer	92
Jubilee	95	Delphin	99	Sea Bird	96
Club Med 2	94	Sea Princess	97	Grande Caribe	93
Costa Victoria	96	Melody	98	Spirit of Columbia	95
Wind Spirit	98	Crystal Harmony	96	Yorktown Clipper	90
Norwegian Majesty	94	Millennium	93	Hanseatic	97
Galaxy	98	Zenith	95	Pacific Sky	91
Costa Atlantica	100	Legacy	93	Big Red Boat II	94
Volendam	98	Vision of the Seas	92	Pacific Princess	88
Norwegian Dream	100	Westerdam	92	Big Red Boat III	92
Maasdam	94	C. Columbus	98	Oceanic	91
Mercury	96	Queen Elizabeth 2	92	Niagara Prince	88
Orient Venus	87	Fuji Maru	91	Sea Lion	92
Seven Seas Navigator	100	Pacific Venus	82	Regal Princess	93
Seabourn Pride	93	Universe Explorer	95	Dolphin IV	88
Infinity	97	The Emerald	91	Asuka	94
Seabourn Legend	97	Celebration	91	Enchanted Capri	93
Radisson Diamond	96	Inspiration	98	Horizon	92
Ocean Princess	97	Grand Princess	94	Rembrandt	94
Nordic Empress	98	Regal Empress	89	Enchanted Isle	89
Vistamar	83	Ocean Breeze	95	Maxim Gorky	90
Norwegian Wind	92	Palm Beach Princess	92	Enchanted Sun	89
Bolero	96	Regal Voyager	87	The Topaz	77
Oriana	100	Rotterdam	97	Costa Romantica	90
Nippon Maru	99	Dawn Princess	96	Royal Princess	94
Paradise	95	Noordam	95	Flamenco	95
Albatross	93	Enchantment of the Seas	99	Stella Solaris	93
Viking Serenade	92	Fascination	95	Contessa 1	93
Holiday	97	Caronia	95	Aegean I	52
Island Adventure	97	Arcadia	96	Silver Cloud	98
Carnival Victory	97	Seabourn Goddess II	94	Deutschland	91
Norway	93	Explorer of the Seas	91	Spirit of Ninety Eight	91
Imagination	94	Voyager of the Sea	92	Legend of the Seas	95
Majesty of the Seas	90	Europa	96	Astor	91
Crown Princess	96	Discovery Sun	95	Paul Gauguin	86
Arkona	90	Norwegian Sea	90	Norwegian Star	78
Sun Princess	98	Tropicale	86	Black Watch	86

(continued overleaf)

(continued)

Monarch of the Seas	93	Ryndam	96	Spirit of Alaska	93
Carnival Destiny	96	Amsterdam	93	Aida	90
Seabourn Goddess I	92	Fantasy	93	Spirit of Discovery	96
Zaandam	97	Silver Shadow	97	Spirit of Glacier Bay	95
Aurora	98	Century	93	Seabourn Spirit	92
Disney Wonder	99	Ecstasy	94	Triton	93
Carnival Triumph	98	Norwegian Sky	97	Costa Allegra	88
Sensation	90	Nantucket Clipper	88	Bremen	90
Victoria	94	Rhapsody of the Seas	94	Costa Classica	92
Sovereign of the Seas	98	Veendam	86		
Disney Magic	96	Le Levant	91		

Source: National Center for Environmental Health, Centers for Disease Control and Prevention, May 8, 2001.

a. A MINITAB stem-and-leaf display of the data is shown below. Identify the stems and leaves of the graph.

b. A score of 90 or higher at the time of inspection indicates the ship is providing an accepted standard of sanitation. Use the MINITAB graph to estimate the proportion of ships that have an accepted sanitation standard.

c. Locate the inspection score of 87 (Regal Voyager and Orient Venus) on the stem-and-leaf display.

Stem-and-Leaf Display: sanlevel

```
Stem-and-leaf of sanlevel   N  = 151
Leaf Unit = 1.0


        LO   52, 77, 78, 82, 83,

   11     8  666677
   19     8  88888999
   40     9  00000000000111111111
   72     9  222222222222223333333333333333
  (31)    9  4444444444444445555555555555555
   48     9  66666666666666677777777777777
   20     9  888888888889999
    4    10  0000
```

1.5
Describing Quantitative Data Numerically

Numerical descriptive measures provide a second (and often more powerful) method for describing a set of quantitative data. These measures, which locate the center of the data set and its spread, actually enable you to construct an approximate mental image of the distribution of the data set.

Note: Most of the formulas used to compute numerical descriptive measures require the summation of numbers. For instance, we may want to sum the observations in a data set, or we may want to square each observation and then sum the squared values. The symbol Σ (sigma) is used to denote a summation operation.

For example, suppose we denote the n sample measurements on a random variable y by the symbols $y_1, y_2, y_3, \ldots, y_n$. Then the sum of all n measurements in the sample is represented by the symbol

$$\sum_{i=1}^{n} y_i$$

This is read "summation y, y_1 to y_n" and is equal to the value

$$y_1 + y_2 + y_3 + \cdots + y_n$$

One of the most common measures of central tendency is the **mean**, or arithmetic average, of a data set. Thus, if we denote the sample measurements by the symbols $y_1, y_2, y_3, \ldots$, the sample mean is defined as follows:

Definition 1.15

The **mean** of a sample of n measurements $y_1, y_2, \ldots, y_n$ is

$$\bar{y} = \frac{\sum_{i=1}^{n} y_i}{n}$$

The mean of a population, or equivalently, the expected value of y, $E(y)$, is usually unknown in a practical situation (we will want to infer its value based on the sample data). Most texts use the symbol μ to denote the mean of a population. Thus, we will use the following notation:

Notation

Sample mean: $\bar{y}$
Population mean: $E(y) = \mu$

The spread or variation of a data set is measured by its **range**, its **variance**, or its **standard deviation**.

Definition 1.16

The **range** of a sample of n measurements $y_1, y_2, \ldots, y_n$ is the difference between the largest and smallest measurements in the sample.

EXAMPLE   1.5

If a sample consists of measurements 3, 1, 0, 4, 7, find the sample mean and the sample range.

Solution

The sample mean and range are

$$\bar{y} = \frac{\sum\limits_{i=1}^{n} y_i}{n} = \frac{15}{5} = 3$$

$$\text{Range} = 7 - 0 = 7$$

◆

The variance of a set of measurements is defined to be the average of the *squares of the deviations* of the measurements about their mean. Thus, the population variance, which is usually unknown in a practical situation, would be the mean or expected value of $(y - \mu)^2$, or $E[(y - \mu)^2]$. We use the symbol σ^2 to represent the variance of a population:

$$E[(y - \mu)^2] = \sigma^2$$

The quantity usually termed the **sample variance** is defined in the box.

Definition 1.17
The **variance** of a sample of n measurements $y_1, y_2, \ldots, y_n$ is defined to be

$$s^2 = \frac{\sum\limits_{i=1}^{n} (y_i - \bar{y})^2}{n - 1} = \frac{\sum\limits_{i=1}^{n} y_i^2 - n\bar{y}^2}{n - 1}$$

Note that the sum of squares of deviations in the sample variance is divided by $(n - 1)$, rather than n. Division by n produces estimates that tend to underestimate σ^2. Division by $(n - 1)$ corrects this problem.

EXAMPLE   1.6

Refer to Example 1.5. Calculate the sample variance for the sample 3, 1, 0, 4, 7.

Solution

We first calculate

$$\sum\limits_{i=1}^{n} (y_i - \bar{y})^2 = \sum\limits_{i=1}^{n} y_i^2 - n\bar{y}^2 = 75 - 5(3)^2 = 30$$

where $\bar{y} = 3$ from Example 1.4. Then

$$s^2 = \frac{\sum_{i=1}^{n}(y_i - \bar{y})^2}{n - 1} = \frac{30}{4} = 7.5$$

◆

The concept of a variance is important in theoretical statistics, but its square root, called a **standard deviation**, is the quantity most often used to describe data variation.

Definition 1.18

The **standard deviation** of a set of measurements is equal to the square root of their variance. Thus, the standard deviations of a sample and a population are

Sample standard deviation: s
Population standard deviation: σ

The standard deviation of a set of data takes on meaning in light of a theorem (Tchebysheff's theorem) and a rule of thumb.* Basically, they give us the following guidelines:

Guidelines for Interpreting a Standard Deviation

1. For *any* data set (population or sample), at least three-fourths of the measurements will lie within 2 standard deviations of their mean.
2. For *most* data sets of moderate size (say, 25 or more measurements) with a mound-shaped distribution, approximately 95% of the measurements will lie within 2 standard deviations of their mean.

EXAMPLE 1.7

Often, travelers who have no intention of showing up fail to cancel their hotel reservations in a timely manner. These travelers are known, in the parlance of the hospitality trade, as "no-shows." To protect against no-shows and late cancellations, hotels invariably overbook rooms. A study reported in the *Journal of Travel Research* examined the problems of overbooking rooms in the hotel industry. The data in Table 1.6, extracted from the study, represent daily numbers of late cancellations and no-shows for a random sample of 30 days at a large (500-room) hotel. Based on this sample, how many rooms, at minimum, should the hotel overbook each day?

---

*For a more complete discussion and a statement of Tchebysheff's theorem, see the references listed at the end of this chapter.

NOSHOWS

TABLE 1.6   Hotel No-Shows for a Sample of 30 Days

18	16	16	16	14	18	16	18	14	19
15	19	9	20	10	10	12	14	18	12
14	14	17	12	18	13	15	13	15	19

Source: Toh, R. S. "An inventory depletion overbooking model for the hotel industry." *Journal of Travel Research*, Vol. 23, No. 4. Spring 1985, p. 27. The *Journal of Travel Research* is published by the Travel and Tourism Research Association (TTRA) and the Business Research Division, University of Colorado at Boulder.

Solution

To answer this question, we need to know the range of values where most of the daily numbers of no-shows fall. We must compute $\bar{y}$ and s, and examine the shape of the relative frequency distribution for the data.

Figure 1.7 is a MINITAB printout that shows a stem-and-leaf display and descriptive statistics of the sample data. Notice from the stem-and-leaf display that the distribution of daily no-shows is mound-shaped, and only slightly skewed on the low (top) side of Figure 1.7. Thus, guideline 2 in the previous box should give a good estimate of the percentage of days that fall within 2 standard deviations of the mean.

FIGURE 1.7

MINITAB printout: Describing the no-show data, Example 1.6

Stem-and-Leaf Display: Noshows

```
Stem-and-leaf of Noshows   N  = 30
Leaf Unit = 0.10

    1      9  0
    3     10  00
    3     11
    6     12  000
    8     13  00
   13     14  00000
   (3)    15  000
   14     16  0000
   10     17  0
    9     18  00000
    4     19  000
    1     20  0
```

Descriptive Statistics: Noshows

Variable	N	Mean	Median	TrMean	StDev	SE Mean
Noshows	30	15.133	15.000	15.231	2.945	0.538

Variable	Minimum	Maximum	Q1	Q3
Noshows	9.000	20.000	13.000	18.000

The mean and standard deviation of the sample data, shaded on the MINITAB printout, are $\bar{y} = 15.133$ and $s = 2.945$. From guideline 2 in the box, we know that about 95% of the daily number of no-shows fall within 2 standard deviations of the mean, i.e., within the interval

$$\bar{y} \pm 2s = 15.133 \pm 2(2.945)$$

$$= 15.133 \pm 5.890$$

or between 9.243 no-shows and 21.023 no-shows. (If we count the number of measurements in this data set, we find that actually 29 out of 30, or 96.7%, fall in this interval.)

From this result, the large hotel can infer that there will be at least 9.243 (or, rounding up, 10) no-shows per day. Consequently, the hotel can overbook at least 10 rooms per day and still be highly confident that all reservations can be honored.◆

Numerical descriptive measures calculated from sample data are called **statistics**. Numerical descriptive measures of the population are called **parameters**. In a practical situation, we will not know the population relative frequency distribution (or equivalently, the population distribution for y). We will usually assume that it has unknown numerical descriptive measures, such as its mean μ and standard deviation σ, and by inferring (using **sample statistics**) the values of these parameters, we infer the nature of the population relative frequency distribution. Sometimes we will assume that we know the shape of the population relative frequency distribution and use this information to help us make our inferences. When we do this, we are postulating a model for the population relative frequency distribution, and we must keep in mind that the validity of the inference may depend on how well our model fits reality.

Definition 1.19

Numerical descriptive measures of a population are called **parameters**.

Definition 1.20

A **sample statistic** is a quantity calculated from the observations in a sample.

EXERCISES

1.22. Periodically, the Federal Trade Commission (FTC) ranks domestic cigarette brands according to tar, nicotine, and carbon monoxide content. The test results are obtained by using a sequential smoking machine to "smoke" cigarettes to a 23-millimeter butt length. The tar, nicotine, and carbon monoxide concentrations (rounded to the nearest milligram) in the residual "dry" particulate matter of the smoke are then measured. The SAS printouts on p. 24 describe the nicotine contents of the 500 cigarette brands recently tested by the FTC.

a. Examine the relative frequency histogram for nicotine content. Use the rule of thumb to describe the data set.

b. Locate $\bar{y}$ and s on the printout, then compute the interval $\bar{y} \pm 2s$.

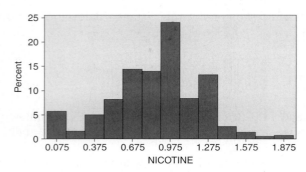

The MEANS Procedure

Analysis Variable : NICOTINE

N	Mean	Std Dev	Minimum	Maximum
500	0.8425000	0.3455250	0.0500000	1.9000000

c. Based on your answer to part **a**, estimate the percentage of cigarettes with nicotine contents in the interval formed in part **b**.

d. Use the information on the SAS histogram to determine the actual percentage of nicotine contents that fall within the interval formed in part **b**. Does your answer agree with your estimate of part **c**?

1.23. Refer to the Centers for Disease Control and Prevention study of sanitation levels for 151 international cruise ships, Exercise 1.21. A MINITAB printout of the descriptive statistics for the data is shown below. (Recall that sanitation scores range from 0 to 100.) Interpret the numerical descriptive measures of central tendency displayed on the printout.

Descriptive Statistics: sanlevel

Variable	N	Mean	Median	TrMean	StDev	SE Mean
sanlevel	151	93.113	94.000	93.585	5.184	0.422

Variable	Minimum	Maximum	Q1	Q3
sanlevel	52.000	100.000	91.000	96.000

WOMENPOWER

RANK	NAME	AGE	COMPANY	TITLE
1	Carly Fiorina	45	Hewlett-Packard	CEO
2	Heidi Miller	46	Citigroup	CFO
3	Mary Meeker	40	Morgan Stanley	Managing Director
4	Shelly Lazarus	52	Ogilvy & Mather	CEO
5	Meg Whitman	43	eBay	CEO
6	Debby Hopkins	44	Boeing	CFO
7	Marjorie Scardino	52	Pearson	CEO
8	Martha Stewart	58	Omnimedia	CEO
9	Nancy Peretsman	45	Allen & Co.	Ex.V.P.
10	Pat Russo	47	Lucent Technologies	Ex. V.P.
11	Patricia Dunn	46	Barclays Global Investors	Chairman
12	Abby Joseph Cohen	47	Goldman Sachs	Managing Director
13	Ann Livermore	41	Hewlett-Packard	CEO
14	Andrea Jung	41	Avon Products	COO
15	Sherry Lansing	55	Paramount Pictures	Chairman
16	Karen Katen	50	Pfizer	Ex.V.P.
17	Marilyn Carlson Nelson	60	Carlson Cos.	CEO
18	Judy McGrath	47	MTV & M2	President
19	Lois Juliber	50	Colgate-Palmolive	COO
20	Gerry Laybourne	52	Oxygen Media	CEO
21	Judith Estrin	44	Cisco Systems	Sr. V.P.
22	Cathleen Black	55	Hearst Magazines	President
23	Linda Sandford	46	IBM	General Manager
24	Ann Moore	49	Time Inc.	President
25	Jill Barad	48	Mattel	CEO
26	Oprah Winfrey	45	Harpo Entertainment	Chairman
27	Judy Lewent	50	Merck	Sr. V.P.
28	Joy Covey	36	Amazon.com	COO
29	Rebecca Mark	45	Azurix	CEO
30	Deborah Willingham	43	Microsoft	V.P.
31	Dina Dubion	46	Chase Manhattan	Ex. V.P.

(continued)

32	Patricia Woertz	46	Chevron	President
33	Lawton Fitt	46	Goldman Sachs	Man. Dir.
34	Ann Fudge	48	Kraft Foods	Ex. V.P.
35	Carolyn Ticknor	52	Hewlett-Packard	CEO
36	Dawn Lepore	45	Charles Schwab	CIO
37	Jeannine Rivet	51	United Healthcare	CEO
38	Jamie Gorelick	49	Fannie Mae	Vice Chairman
39	Jan Brandt	48	America Online	Mar. President
40	Bridget Macaskill	51	Oppenheimer Funds	CEO
41	Jeanne Jackson	48	Banana Republic	CEO
42	Cynthia Trudell	46	General Motors	V.P.
43	Nina DiSesa	53	McCann-Erickson	Chairman
44	Linda Wachner	53	Warnaco	Chairman
45	Darla Moore	45	Rainwater Inc.	President
46	Marion Sandler	68	Golden West	Co-CEO
47	Michelle Anthony	42	Sony Music	Ex. V.P.
48	Orit Gadlesh	48	Bain & Co.	Chairman
49	Charlotte Beers	64	J. Walter Thompson	Chairman
50	Abigail Johnson	37	Fidelity Investments	V.P.

Source: Fortune, Oct. 25, 1999.

1.24. *Fortune* (Oct. 25, 1999) published a list of the 50 most powerful women in America. The data on age (in years) and title of each of these 50 women (pp. 24–25) are stored in the WOMENPOWER file.

a. Find the mean and standard deviation of these 50 ages.

b. Give an interval that is highly likely to contain the age of a randomly selected woman from the list.

1.25. Three-way catalytic converters have been installed in new vehicles in order to reduce pollutants from motor vehicle exhaust emissions. However, these converters unintentionally increase the level of ammonia in the air. *Environmental Science & Technology* (Sept. 1, 2000) published a study on the ammonia levels near the exit ramp of a San Francisco highway tunnel. The data in the table represent daily ammonia concentrations (parts per million) on eight randomly selected days during afternoon drive-time in the summer of 1999.

🌀 **AMMONIA**

1.53	1.50	1.37	1.51	1.55	1.42	1.41	1.48

a. Find and interpret the mean daily ammonia level in air in the tunnel.

b. Find the standard deviation of the daily ammonia levels. Interpret the result.

c. Suppose the standard deviation of the ammonia levels during morning drive-time at the exit ramp is 1.45 ppm. Which time, morning or afternoon drive-time, has more variable ammonia levels?

1.26. Refer to the *Marine Technology* (Jan. 1995) data on spillage amounts (in thousands of metric tons) for 50 major oil spills, Exercise 1.15. An SPSS histogram for the 50 spillage amounts is shown below.

a. Interpret the histogram.

b. Descriptive statistics for the 50 spillage amounts are also shown on the SPSS histogram. Use this information to form an interval that can be used to predict the spillage amount for the next major oil spill.

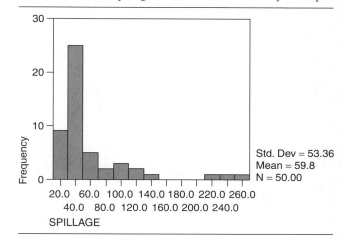

1.27. The National Education Longitudinal Survey (NELS) tracks a nationally representative sample of U.S. students from eighth grade through high school and college. Research published in *Chance* (Winter 2001) examined the Standardized Admission Test (SAT)

scores of 265 NELS students who paid a private tutor to help them improve their scores. The table summarizes the changes in both the SAT-Mathematics and SAT-Verbal scores for these students.

	SAT-MATH	SAT-VERBAL
Mean change in score	19	7
Standard deviation of score changes	65	49

a. Suppose one of the 265 students who paid a private tutor is selected at random. Give an interval that is likely to contain this student's change in the SAT-Math score.
b. Repeat part **a** for the SAT-Verbal score.
c. Suppose the selected student's score increased on one of the SAT tests by 140 points. Which test, the SAT-Math or SAT-Verbal, is the one most likely to have the 140-point increase? Explain.

1.6
The Normal Probability Distribution

One of the most commonly used models for a theoretical population relative frequency distribution for a quantitative variable is the **normal probability distribution**, as shown in Figure 1.8. The normal distribution is symmetric about its mean μ, and its spread is determined by the value of its standard deviation σ. Three normal curves with different means and standard deviations are shown in Figure 1.9.

Computing the area over an interval under the normal probability distribution can be a difficult task.* Consequently, we will use the computed areas listed in

FIGURE 1.8

A normal probability distribution

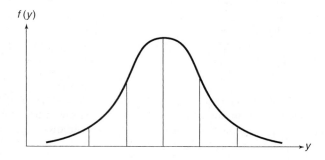

FIGURE 1.9

Several normal distributions with different means and standard deviations

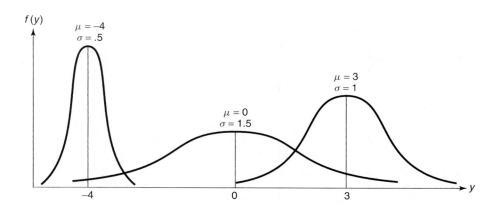

---

*Students with knowledge of calculus should note that the probability that y assumes a value in the interval $a < y < b$ is $P(a < y < b) = \int_a^b f(y)dy$, assuming the integral exists. The value of this definite integral can be obtained to any desired degree of accuracy by approximation procedures. For this reason, it is tabulated for the user.

TABLE 1.7   Reproduction of Part of Table 1 of Appendix C

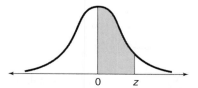

z	.00	.01	.02	.03	.04	.05	.06	.07	.08	.09
0	.0000	.0040	.0080	.0120	.0160	.0199	.0239	.0279	.0319	.0359
.1	.0398	.0438	.0478	.0517	.0557	.0596	.0636	.0675	.0714	.0753
.2	.0793	.0832	.0871	.0910	.0948	.0987	.1026	.1064	.1103	.1141
.3	.1179	.1217	.1255	.1293	.1331	.1368	.1406	.1443	.1480	.1517
.4	.1554	.1591	.1628	.1664	.1700	.1736	.1772	.1808	.1844	.1879
.5	.1915	.1950	.1985	.2019	.2054	.2088	.2123	.2157	.2190	.2224
.6	.2257	.2291	.2324	.2357	.2389	.2422	.2454	.2486	.2517	.2549
.7	.2580	.2611	.2642	.2673	.2704	.2734	.2764	.2794	.2823	.2852
.8	.2881	.2910	.2939	.2967	.2995	.3023	.3051	.3078	.3106	.3133
.9	.3159	.3186	.3212	.3238	.3264	.3289	.3315	.3340	.3365	.3389
1.0	.3413	.3438	.3461	.3485	.3508	.3531	.3554	.3577	.3599	.3621
1.1	.3643	.3665	.3686	.3708	.3729	.3749	.3770	.3790	.3810	.3830
1.2	.3849	.3869	.3888	.3907	.3925	.3944	.3962	.3980	.3997	.4015
1.3	.4032	.4049	.4066	.4082	.4099	.4115	.4131	.4147	.4162	.4177
1.4	.4192	.4207	.4222	.4236	.4251	.4265	.4279	.4292	.4306	.4319
1.5	.4332	.4345	.4357	.4370	.4382	.4394	.4406	.4418	.4429	.4441

Table 1 of Appendix C. A partial reproduction of this table is shown in Table 1.7. As you can see from the normal curve above the table, the entries give areas under the normal curve between the mean of the distribution and a standardized distance

$$z = \frac{y - \mu}{\sigma}$$

to the right of the mean. Note that z is the number of standard deviations σ between μ and y. The distribution of z, which has mean $\mu = 0$ and standard deviation $\sigma = 1$, is called a **standard normal distribution**.

EXAMPLE   1.8

Suppose y is a normal random variable with $\mu = 50$ and $\sigma = 15$. Find $P(30 < y < 70)$, the probability that y will fall within the interval $30 < y < 70$.

Solution

Refer to Figure 1.10. Note that $y = 30$ and $y = 70$ lie the same distance from the mean $\mu = 50$, with $y = 30$ below the mean and $y = 70$ above it. Then, because the normal curve is symmetric about the mean, the probability A_1 that y falls between $y = 30$ and $\mu = 50$ is equal to the probability A_2 that y falls between $\mu = 50$ and $y = 70$. The z score corresponding to $y = 70$ is

$$z = \frac{y - \mu}{\sigma} = \frac{70 - 50}{15} = 1.33$$

Therefore, the area between the mean $\mu = 50$ and the point $y = 70$ is given in Table 1 of Appendix C (and Table 1.7) at the intersection of the row corresponding

FIGURE 1.10

Normal probability distribution:
$\mu = 50, \sigma = 15$

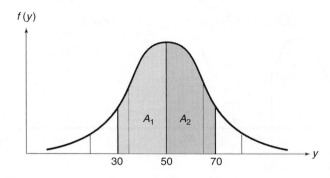

FIGURE 1.11

A distribution of z scores (a standard normal distribution)

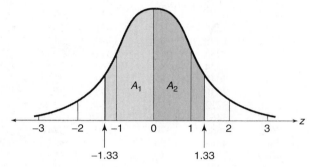

to $z = 1.3$ and the column corresponding to .03. This area (probability) is $A_2 = .4082$. Since $A_1 = A_2$, A_1 also equals .4082, and it follows that the probability that y falls in the interval $30 < y < 70$ is $P(30 < y < 70) = 2(.4082) = .8164$. The z scores corresponding to $y = 30 \ (z = -1.33)$ and $y = 70 \ (z = 1.33)$ are shown in Figure 1.11.

◆

EXAMPLE   1.9

Use Table 1 of Appendix C to determine the area to the right of the z score 1.64 for the standard normal distribution. That is, find $P(z \geq 1.64)$.

Solution

The probability that a normal random variable will fall more than 1.64 standard deviations to the right of its mean is indicated in Figure 1.12. Because the normal distribution is symmetric, half of the total probability (.5) lies to the right of the mean and half to the left. Therefore, the desired probability is

$$P(z \geq 1.64) = .5 - A$$

where A is the area between $\mu = 0$ and $z = 1.64$, as shown in the figure. Referring to Table 1, we find that the area A corresponding to $z = 1.64$ is .4495. So

$$P(z \geq 1.64) = .5 - A = .5 - .4495 = .0505$$

◆

We will not be making extensive use of the table of areas under the normal curve, but you should know some of the common tabulated areas. In particular, you should

FIGURE 1.12

Standard normal distribution:
$\mu = 0, \sigma = 1$

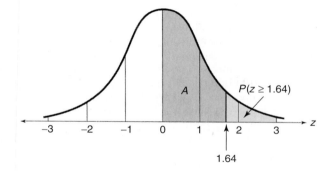

note that the area between $z = -2.0$ and $z = 2.0$, which gives the probability that y falls in the interval $\mu - 2\sigma < y < \mu + 2\sigma$, is .9544 and agrees with guideline 2 of Section 1.4.

EXERCISES

1.28. Use Table 1 of Appendix C to find each of the following:
   a. $P(-1 \leq z \leq 1)$ **b.** $P(-1.96 \leq z \leq 1.96)$
   c. $P(-1.645 \leq z \leq 1.645)$ **d.** $P(-3 \leq z \leq 3)$

1.29. Given that the random variable y has a normal probability distribution with mean 100 and variance 64, draw a sketch (i.e., graph) of the frequency function of y. Locate μ and the interval $\mu \pm 2\sigma$ on the graph. Find the following probabilities:
   a. $P(\mu - 2\sigma \leq y \leq \mu + 2\sigma)$ **b.** $P(y \geq 108)$
   c. $P(y \leq 92)$ **d.** $P(92 \leq y \leq 116)$
   e. $P(92 \leq y \leq 96)$ **f.** $P(76 \leq y \leq 124)$

1.30. Psychology students at Wittenberg University completed the Dental Anxiety Scale questionnaire (*Psychological Reports*, Aug. 1997). Scores on the scale range from 0 (no anxiety) to 20 (extreme anxiety). The mean score was 11 and the standard deviation was 3.5. Assume that the distribution of all scores on the Dental Anxiety Scale is normal with $\mu = 11$ and $\sigma = 3.5$.
   a. Suppose you score a 16 on the Dental Anxiety Scale. Find the z value for this score.
   b. Find the probability that someone scores between a 10 and a 15 on the Dental Anxiety Scale.
   c. Find the probability that someone scores above a 17 on the Dental Anxiety Scale.

1.31. The alkalinity level of water specimens collected from the Han River in Seoul, Korea, has a mean of 50 milligrams per liter and a standard deviation of 3.2 milligrams per liter (*Environmental Science & Engineering*, Sept. 1, 2000). Assume the distribution of alkalinity levels is approximately normal and find the probability that a water specimen collected from the river has an alkalinity level
   a. exceeding 45 milligrams per liter.
   b. below 55 milligrams per liter.
   c. between 51 and 52 milligrams per liter.

1.32. Refer to the *Chance* (Winter 2001) study of students who paid a private tutor to help them improve their Standardized Admission Test (SAT) scores, Exercise 1.27 (p. 25). The table summarizing the changes in both the SAT-Mathematics and SAT-Verbal scores for these students is reproduced here. Assume that both distributions of SAT score changes are approximately normal.

	SAT-MATH	SAT-VERBAL
Mean change in score	19	7
Standard deviation of score changes	65	49

   a. What is the probability that a student increases his or her score on the SAT-Math test by at least 50 points?
   b. What is the probability that a student increases his or her score on the SAT-Verbal test by at least 50 points?

1.33. A group of Florida State University psychologists examined the effects of alcohol on the reactions of people to a threat (*Journal of Abnormal Psychology*, Vol. 107, 1998). After obtaining a specified blood alcohol level, experimental subjects were placed in a room and threatened with electric shocks. Using sophisticated equipment to monitor the subjects' eye movements, the startle response (measured in milliseconds) was recorded for each subject. The mean and standard deviation of the startle responses were 37.9 and 12.4, respectively. Assume that the startle response y for a person with the specified blood alcohol level is approximately normally distributed.

a. Find the probability that y is between 40 and 50 milliseconds.

b. Find the probability that y is less than 30 milliseconds.

c. Give an interval for y, centered around 37.9 milliseconds, so that the probability that y falls in the interval is .95.

1.34. Based on data from the National Center for Health Statistics, N. Wetzel used the normal distribution to model the length of gestation for pregnant U.S. women (*Chance*, Spring 2001). Gestation length has a mean of 280 days with a standard deviation of 20 days.

a. Find the probability that gestation length is between 275.5 and 276.5 days. (This estimates the probability

that a women has her baby 4 days earlier than the "average" due date.)

b. Find the probability that gestation length is between 258.5 and 259.5 days. (This estimates the probability that a women has her baby 21 days earlier than the "average" due date.)

c. Find the probability that gestation length is between 254.5 and 255.5 days. (This estimates the probability that a women has her baby 25 days earlier than the "average" due date.)

d. The *Chance* article referenced a newspaper story about three sisters who all gave birth on the same day (March 11, 1998). Karralee had her baby 4 days early; Marrianne had her baby 21 days early; and Jennifer had her baby 25 days early. Use the results, parts **a–c**, to estimate the probability that three women have their babies 4, 21, and 25 days early, respectively. Assume the births are independent events. [Hint: If events A, B, and C are independent, then P(A and B and C) = P(A) × P(B) × P(C).]

1.7
Sampling Distributions and the Central Limit Theorem

Since we will use sample statistics to make inferences about population parameters, it is natural that we would want to know something about the reliability of the resulting inferences. For example, if we use a statistic to estimate the value of a population mean μ, we will want to know how close to μ our estimate is likely to fall. To answer this question, we need to know the probability distribution of the statistic.

The probability distribution for a statistic based on a random sample of n measurements could be generated in the following way. For purposes of illustration, we will suppose we are sampling from a population with $\mu = 10$ and $\sigma = 5$, the sample statistic is $\bar{y}$, and the sample size is $n = 25$. Draw a single random sample of 25 measurements from the population and suppose that $\bar{y} = 9.8$. Return the measurements to the population and try again. That is, draw another random sample of $n = 25$ measurements and see what you obtain for an outcome. Now, perhaps, $\bar{y} = 11.4$. Replace these measurements, draw another sample of $n = 25$ measurements, calculate $\bar{y}$, and so on. If this sampling process were repeated over and over again an infinitely large number of times, you would generate an infinitely large number of values of $\bar{y}$ that could be arranged in a relative frequency distribution. This distribution, which would appear as shown in Figure 1.13, is the probability distribution (or **sampling distribution**, as it is commonly called) of the statistic $\bar{y}$.

Definition 1.21

The **sampling distribution** of a sample statistic calculated from a sample of n measurements is the probability distribution of the statistic.

FIGURE 1.13

Sampling distribution for $\bar{y}$ based on a sample of $n = 25$ measurements

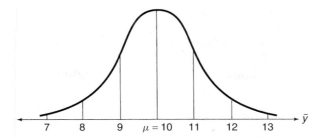

In actual practice, the sampling distribution of a statistic is obtained mathematically or by simulating the sampling on a computer using the procedure described previously.

If $\bar{y}$ has been calculated from a sample of $n = 25$ measurements selected from a population with mean $\mu = 10$ and standard deviation $\sigma = 5$, the sampling distribution shown in Figure 1.13 provides all the information you may wish to know about its behavior. For example, the probability that you will draw a sample of 25 measurements and obtain a value of $\bar{y}$ in the interval $9 \leq \bar{y} \leq 10$ will be the area under the sampling distribution over that interval.

Generally speaking, if we use a statistic to make an inference about a population parameter, we want its sampling distribution to center about the parameter (as is the case in Figure 1.13) and the standard deviation of the sampling distribution, called the **standard error of estimate**, to be as small as possible.

Two theorems provide information on the sampling distribution of a sample mean.

Theorem 1.1

If $y_1, y_2, \ldots, y_n$ represent a random sample of n measurements from a large (or infinite) population with mean μ and standard deviation σ, then, regardless of the form of the population relative frequency distribution, the mean and standard error of estimate of the sampling distribution of $\bar{y}$ will be

Mean: $E(\bar{y}) = \mu_{\bar{y}} = \mu$
Standard error of estimate: $\sigma_{\bar{y}} = \frac{\sigma}{\sqrt{n}}$

Theorem 1.2   The Central Limit Theorem

For large sample sizes, the mean $\bar{y}$ of a sample from a population with mean μ and standard deviation σ has a sampling distribution that is approximately normal, **regardless of the probability distribution of the sampled population**. The larger the sample size, the better will be the normal approximation to the sampling distribution of $\bar{y}$.

Theorems 1.1 and 1.2 together imply that for sufficiently large samples, the sampling distribution for the sample mean $\bar{y}$ will be approximately normal with mean μ and standard error $\sigma_{\bar{y}} = \sigma/\sqrt{n}$. The parameters μ and σ are the mean and standard deviation of the sampled population.

How large must the sample size n be so that the normal distribution provides a good approximation for the sampling distribution of $\bar{y}$? The answer depends on the shape of the distribution of the sampled population, as shown by Figure 1.14. Generally speaking, the greater the skewness of the sampled population distribution, the larger the sample size must be before the normal distribution is an adequate approximation for the sampling distribution of $\bar{y}$. For most sampled populations, sample sizes of $n \geq 30$ will suffice for the normal approximation to be reasonable. We will use the normal approximation for the sampling distribution of $\bar{y}$ when the sample size is at least 30.

FIGURE 1.14

Sampling distributions of $\bar{x}$ for different populations and different sample sizes

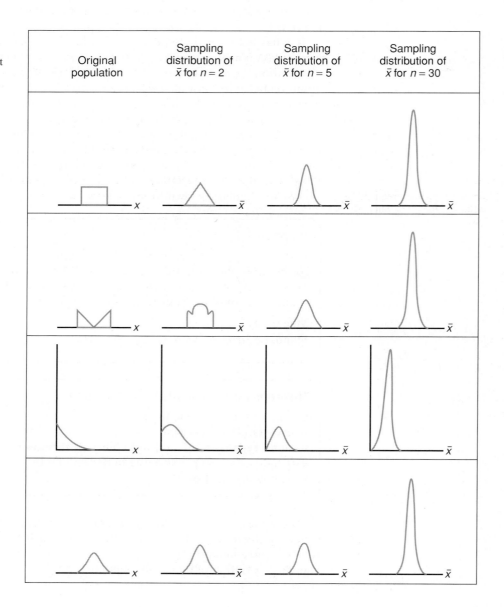

EXAMPLE   1.10

Suppose we have selected a random sample of $n = 25$ observations from a population with mean equal to 80 and standard deviation equal to 5. It is known that the population is not extremely skewed.

a. Sketch the relative frequency distributions for the population and for the sampling distribution of the sample mean, $\bar{y}$.

b. Find the probability that $\bar{y}$ will be larger than 82.

Solution

a. We do not know the exact shape of the population relative frequency distribution, but we do know that it should be centered about $\mu = 80$, its spread should be measured by $\sigma = 5$, and it is not highly skewed. One possibility is shown in Figure 1.15a. From the Central Limit Theorem, we know that the sampling distribution of $\bar{y}$ will be approximately normal since the sampled population distribution is not extremely skewed. We also know that the sampling distribution will have mean and standard deviation

$$\mu_{\bar{y}} = \mu = 80 \quad \text{and} \quad \sigma_{\bar{y}} = \frac{\sigma}{\sqrt{n}} = \frac{5}{\sqrt{25}} = 1$$

The sampling distribution of $\bar{x}$ is shown in Figure 1.15b.

FIGURE 1.15

A population relative frequency distribution and the sampling distribution for $\bar{x}$

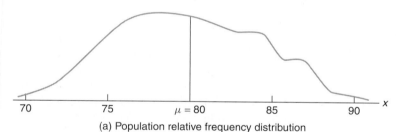

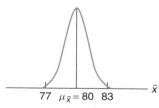

(b) Sampling distribution of $\bar{x}$

b. The probability that $\bar{y}$ will exceed 82 is equal to the highlighted area in Figure 1.15. To find this area, we need to find the z value corresponding to $\bar{y} = 82$. Recall that the standard normal random variable z is the difference between any normally distributed random variable and its mean, expressed in units of its standard deviation. Since $\bar{x}$ is a normally distributed random variable with mean $\mu_{\bar{y}} = \mu$ and standard deviation $\sigma_{\bar{y}} = \sigma/\sqrt{n}$, it follows that the standard normal z value corresponding to the sample mean, $\bar{x}$, is

$$z = \frac{\text{(Normal random variable)} - \text{(Mean)}}{\text{Standard Deviation}} = \frac{\bar{y} - \mu_{\bar{x}}}{\sigma_{\bar{x}}}$$

FIGURE 1.16

The sampling distribution of $\bar{y}$

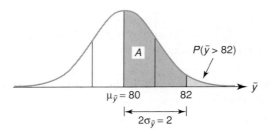

Therefore, for $\bar{y} = 82$, we have

$$z = \frac{\bar{y} - \mu_{\bar{y}}}{\sigma_{\bar{y}}} = \frac{82 - 80}{1} = 2$$

The area A in Figure 1.16 corresponding to $z = 2$ is given in the table of areas under the normal curve (see Table 1 of Appendix C) as .4772. Therefore, the tail area corresponding to the probability that $\bar{y}$ exceeds 82 is

$$P(\bar{y} > 82) = P(z > 2) = .5 - .4772 = .0228 \qquad \blacklozenge$$

The central limit theorem can also be used to justify the fact that the *sum* of the sample measurements possesses a sampling distribution that is approximately normal for large sample sizes. In fact, since many statistics are obtained by summing or averaging random quantities, the central limit theorem helps to explain why many statistics have mound-shaped (or approximately normal) sampling distributions.

As we proceed, we will encounter many different sample statistics, and we will need to know their sampling distributions to evaluate the reliability of each one for making inferences. These sampling distributions will be described as the need arises.

1.8 Estimating a Population Mean

We can make an inference about a population parameter in two ways:

1. Estimate its value.
2. Make a decision about its value (i.e., test a hypothesis about its value).

In this section, we will illustrate the concepts involved in estimation, using the estimation of a population mean as an example. Tests of hypotheses will be discussed in Section 1.9.

To estimate a population parameter, we choose a sample statistic that has two desirable properties: (1) a sampling distribution that centers about the parameter and (2) a small standard error. If the mean of the sampling distribution of a statistic equals the parameter we are estimating, we say that the statistic is an **unbiased estimator** of the parameter. If not, we say that it is **biased**.

In Section 1.7, we noted that the sampling distribution of the sample mean is approximately normally distributed for moderate to large sample sizes and that it possesses a mean μ and standard error $\sigma/\sqrt{n}$. Therefore, as shown in Figure 1.17, $\bar{y}$

is an unbiased estimator of the population mean μ, and the probability that $\bar{y}$ will fall within $1.96\sigma_{\bar{y}} = 1.96\sigma/\sqrt{n}$ of the true value of μ is approximately .95.[*]

FIGURE 1.17

Sampling distribution of $\bar{y}$

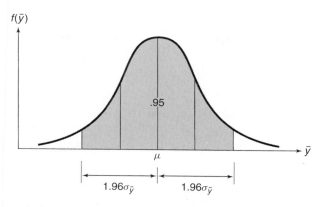

Since $\bar{y}$ will fall within $1.96\sigma_{\bar{y}}$ of μ approximately 95% of the time, it follows that the interval

$$\bar{y} - 1.96\sigma_{\bar{y}} \quad \text{to} \quad \bar{y} + 1.96\sigma_{\bar{y}}$$

will enclose μ approximately 95% of the time in repeated sampling. This interval is called a 95% **confidence interval**, and .95 is called the **confidence coefficient**.

Notice that μ is fixed and that the confidence interval changes from sample to sample. The probability that a confidence interval calculated using the formula

$$\bar{y} \pm 1.96\sigma_{\bar{y}}$$

will enclose μ is approximately .95. Thus, the confidence coefficient measures the confidence that we can place in a particular confidence interval.

Confidence intervals can be constructed using any desired confidence coefficient. For example, if we define $z_{\alpha/2}$ to be the value of a standard normal variable that places the area $\alpha/2$ in the right tail of the z distribution (see Figure 1.18), then a $100(1 - \alpha)\%$ confidence interval for μ is given in the box.

Large-Sample $100(1 - \alpha)\%$ Confidence Interval for μ

$$\bar{y} \pm z_{\alpha/2}\sigma_{\bar{y}}$$

where $z_{\alpha/2}$ is the z value with an area $\alpha/2$ to its right (see Figure 1.18) and $\sigma_{\bar{y}} = \sigma/\sqrt{n}$. The parameter σ is the standard deviation of the sampled population and n is the sample size. If σ is unknown, its value may be approximated by the sample standard deviation s. The approximation is valid for large samples (e.g., $n \geq 30$) only.

---

[*]Additionally, $\bar{y}$ has the smallest standard error among all unbiased estimators of μ. Consequently, we say that $\bar{y}$ is the **minimum variance unbiased estimator** (MVUE) for μ.

FIGURE 1.18

Locating $z_{\alpha/2}$ on the standard
normal curve

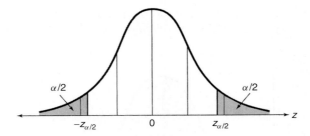

The confidence interval shown in the box is called a large-sample confidence interval because the sample size must be large enough to ensure approximate normality for the sampling distribution of $\bar{y}$. Also, and even more important, you will rarely, if ever, know the value of σ, so its value must be estimated using the sample standard deviation s. This approximation for σ will be adequate only when $n \geq 30$.

Typical confidence coefficients and corresponding values of $z_{\alpha/2}$ are shown in Table 1.8.

TABLE 1.8   **Commonly Used Values of $z_{\alpha/2}$**

CONFIDENCE COEFFICIENT $(1 - \alpha)$	α	$\alpha/2$	$z_{\alpha/2}$
.90	.10	.05	1.645
.95	.05	.025	1.96
.99	.01	.005	2.576

EXAMPLE 1.11

A fact long known but little understood is that twins, in their early years, tend to have lower intelligence quotients and pick up language more slowly than nontwins. Recently, psychologists have speculated that the slower intellectual growth of twins may be caused by benign parental neglect. Suppose we want to investigate this phenomenon. A random sample of $n = 50$ sets of 2 1/2-year-old twin boys is selected, and the total parental attention time given to each pair during 1 week is recorded. The data (in hours) are listed in Table 1.9. Estimate μ, the mean attention time given to all $2\frac{1}{2}$-year-old twin boys by their parents, using a 99% confidence interval. Interpret the interval in terms of the problem.

 ATTENTIMES

TABLE 1.9   **Attention Time for a Random Sample of $n = 50$ Sets of Twins**

20.7	14.0	16.7	20.7	22.5	48.2	12.1	7.7	2.9	22.2
23.5	20.3	6.4	34.0	1.3	44.5	39.6	23.8	35.6	20.0
10.9	43.1	7.1	14.3	46.0	21.9	23.4	17.5	29.4	9.6
44.1	36.4	13.8	0.8	24.3	1.1	9.3	19.3	3.4	14.6
15.7	32.5	46.6	19.1	10.6	36.9	6.7	27.9	5.4	14.0

Solution

The general form of the 99% confidence interval for a population mean is

$$\bar{y} \pm z_{\alpha/2}\sigma_{\bar{y}} = \bar{y} \pm z_{.01}\sigma_{\bar{y}}.$$

$$= \bar{y} \pm 2.575\left(\frac{\sigma}{\sqrt{n}}\right)$$

A SAS printout showing descriptive statistics for the sample of $n = 50$ attention times is displayed in Figure 1.19. The values of $\bar{y}$ and s, shaded on the printout, are $\bar{y} = 20.85$ and $s = 13.41$. Thus, for the 50 twins sampled, the 99% confidence interval is

$$20.85 \pm 2.575\left(\frac{\sigma}{\sqrt{50}}\right)$$

FIGURE 1.19

SAS descriptive statistics for $n = 50$ sample attention times

```
Sample Statistics for ATTIME

     N        Mean      Std. Dev.    Std. Error
-------------------------------------------------
    50        20.85      13.41         1.90



99% Confidence Interval for the Mean

          Lower Limit     Upper Limit
          -----------     -----------
             15.96           25.73
```

We do not know the value of σ (the standard deviation of the weekly attention time given to $2\frac{1}{2}$-year-old twin boys by their parents), so we use our best approximation, the sample standard deviation s. (Since the sample size, $n = 50$, is large, the approximation is valid.) Then the 99% confidence interval is

$$20.85 \pm 2.575\left(\frac{13.41}{\sqrt{50}}\right) = 20.85 \pm 4.88$$

or, from 15.97 to 25.73. That is, we can be 99% confident that the true mean weekly attention given to $2\frac{1}{2}$-year-old twin boys by their parents falls between 15.97 and 25.73 hours. [Note: This interval is also shown at the bottom of the SAS printout, Figure 1.19.] ◆

The large-sample method for making inferences about a population mean μ assumes that either σ is known or the sample size is large enough ($n \geq 30$) for the sample standard deviation s to be used as a good approximation to σ. The technique for finding a $100(1 - \alpha)\%$ confidence interval for a population mean μ for small sample sizes requires that the sampled population have a normal probability distribution. The formula, which is similar to the one for a large-sample confidence interval for μ, is

$$\bar{y} \pm t_{\alpha/2}s_{\bar{y}}$$

where $s_{\bar{y}} = s/\sqrt{n}$ is the estimated standard error of $\bar{y}$. The quantity $t_{\alpha/2}$ is directly analogous to the standard normal value $z_{\alpha/2}$ used in finding a large-sample confidence interval for μ except that it is an upper-tail t value obtained from a Student's t distribution. Thus, $t_{\alpha/2}$ is an upper-tail t value such that an area $\alpha/2$ lies to its right.

Like the standardized normal (z) distribution, a Student's t distribution is symmetric about the value $t = 0$, but it is more variable than a z distribution. The variability depends on the number of **degrees of freedom, df**, which in turn depends on the number of measurements available for estimating σ^2. The smaller the number of degrees of freedom, the greater will be the spread of the t distribution. For this application of a Student's t distribution, df $= n - 1$.* As the sample size increases (and df increases), the Student's t distribution looks more and more like a z distribution, and for $n \geq 30$, the two distributions will be nearly identical. A Student's t distribution based on df $= 4$ and a standard normal distribution are shown in Figure 1.20. Note the corresponding values of $z_{.025}$ and $t_{.025}$.

FIGURE 1.20

The $t_{.025}$ value in a t distribution with 4 df and the corresponding $z_{.025}$ value

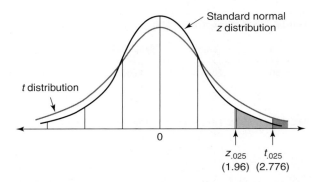

The upper-tail values of the Student's t distribution are given in Table 2 of Appendix C. An abbreviated version of the t table is presented in Table 1.10. To find the upper-tail t value based on 4 df that places .025 in the upper tail of the t distribution, we look in the row of the table corresponding to df $= 4$ and the column corresponding to $t_{.025}$. The t value is 2.776 and is shown in Figure 1.17.

The process of finding a small-sample confidence interval for μ is given in the next box.

Small-Sample Confidence Interval for μ

$$\bar{y} \pm t_{\alpha/2} s_{\bar{y}}$$

where $s_{\bar{y}} = s/\sqrt{n}$ and $t_{\alpha/2}$ is a t value based on $(n - 1)$ degrees of freedom, such that the probability that $t > t_{\alpha/2}$ is $\alpha/2$.

Assumptions: The relative frequency distribution of the sampled population is approximately normal.

---

*Think of df as the amount of information in the sample size n for estimating μ. We lose 1 df for estimating σ^2; hence df $= n - 1$.

TABLE 1.10 Reproduction of a Portion of Table 2 of Appendix C

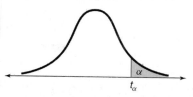

DEGREES OF FREEDOM	$t_{.100}$	$t_{.050}$	$t_{.025}$	$t_{.010}$	$t_{.005}$
1	3.078	6.314	12.706	31.821	63.657
2	1.886	2.920	4.303	6.965	9.925
3	1.638	2.353	3.182	4.541	5.841
4	1.533	2.132	2.776	3.747	4.604
5	1.476	2.015	2.571	3.365	4.032
6	1.440	1.943	2.447	3.143	3.707
7	1.415	1.895	2.365	2.998	3.499
8	1.397	1.860	2.306	2.896	3.355
9	1.383	1.833	2.262	2.821	3.250
10	1.372	1.812	2.228	2.764	3.169
11	1.363	1.796	2.201	2.718	3.106
12	1.356	1.782	2.179	2.681	3.055
13	1.350	1.771	2.160	2.650	3.012
14	1.345	1.761	2.145	2.624	2.977
15	1.341	1.753	2.131	2.602	2.947

EXAMPLE 1.12

The Geothermal Loop Experimental Facility, located in the Salton Sea in southern California, is a U.S. Department of Energy operation for studying the feasibility of generating electricity from the hot, highly saline water of the Salton Sea. Operating experience has shown that these brines leave silica scale deposits on metallic plant piping, causing excessive plant outages. Jacobsen et al. (*Journal of Testing and Evaluation*, Vol. 9, No. 2, Mar. 1981) have found that scaling can be reduced somewhat by adding chemical solutions to the brine. In one screening experiment, each of five antiscalants was added to an aliquot of brine, and the solutions were filtered. A silica determination (parts per million of silicon dioxide) was made on each filtered sample after a holding time of 24 hours, with the following results:

229 255 280 203 229

Estimate the mean amount of silicon dioxide present in the five antiscalant solutions. Use a 95% confidence interval.

Solution

The first step in constructing the confidence interval is to compute the mean, $\bar{y}$, and standard deviation, s, of the sample of five silicon dioxide amounts. These values, $\bar{y} = 239.2$ and $s = 29.3$, are provided in the MINITAB printout, Figure 1.21.

For a confidence coefficient of $1 - \alpha = .95$, we have $\alpha = .05$ and $\alpha/2 = .025$. Since the sample size is small ($n = 5$), our estimation technique requires the assumption that the amount of silicon dioxide present in an antiscalant solution has an approximately normal distribution (i.e., the sample of five silicon amounts is selected from a normal population).

FIGURE 1.21

MINITAB descriptive statistics
and confidence interval for
Example 1.12

One-Sample T: PPM

Variable	N	Mean	StDev	SE Mean	95.0% CI
PPM	5	239.2	29.3	13.1	(202.8, 275.6)

Substituting the values for $\bar{y}$, s, and n into the formula for a small-sample confidence interval for μ, we obtain

$$\bar{y} \pm t_{\alpha/2}(S_{\bar{y}}) = \bar{y} \pm t_{.025}\left(\frac{s}{\sqrt{n}}\right)$$

$$= 239.2 \pm t_{.025}\left(\frac{29.3}{\sqrt{5}}\right)$$

where $t_{.025}$ is the value corresponding to an upper-tail area of .025 in the Student's t distribution based on $(n-1) = 4$ degrees of freedom. From Table 2 of Appendix C, the required t value (shaded in Table 1.10) is $t_{.025} = 2.776$. Substituting this value yields

$$239.2 \pm t_{.025}\left(\frac{29.3}{\sqrt{5}}\right) = 239.2 \pm (2.776)\left(\frac{29.3}{\sqrt{5}}\right)$$

$$= 239.2 \pm 36.4$$

or 202.8 to 275.6 ppm.

Thus, if the distribution of silicon dioxide amounts is approximately normal, then we can be 95% confident that the interval (202.8. 275.6) encloses μ, the true mean amount of silicon dioxide present in an antiscalant solution. Remember, the 95% confidence level implies that if we were to employ our interval estimator on repeated occasions, 95% of the intervals constructed would capture μ.

The 95% confidence interval can also be obtained with statistical software. This interval is shaded on the MINITAB printout, Figure 1.20. You can see that the computer-generated interval is identical to our calculated one.   ◆

EXAMPLE   1.13

Suppose you want to reduce the width of the confidence interval obtained in Example 1.12. Specifically, you want to estimate the mean silicon dioxide content of an aliquot of brine correct to within 10 ppm with confidence coefficient approximately equal to .95. How many aliquots of brine would you have to include in your sample?

Solution

We will interpret the phrase, "correct to within 10 ppm ... equal to .95" to mean that we want half the width of a 95% confidence interval for μ to equal 10 ppm. That is, we want

$$t_{.025}\left(\frac{s}{\sqrt{n}}\right) = 10$$

To solve this equation for n, we need approximate values for $t_{.025}$ and s. Since we know from Example 1.10 that the confidence interval was wider than desired for $n = 5$, it is clear that our sample size must be larger than 5. Consequently, $t_{.025}$ will be very close to 2, and this value will provide a good approximation to $t_{.025}$. A good measure of the data variation is given by the standard deviation computed in Example 1.12. We substitute $t_{.025} \approx 2$ and $s \approx 29.3$ into the equation and solve for n:

$$t_{.025}\left(\frac{s}{\sqrt{n}}\right) = 10$$

$$2\left(\frac{29.3}{\sqrt{n}}\right) = 10$$

$$\sqrt{n} = 5.86$$

$$n = 34.3 \quad \text{or approximately } n = 34$$

Remember that this sample size is an approximate solution because we approximated the value of $t_{.025}$ and the value of s that might be computed from the prospective data. Nevertheless, $n = 34$ will be reasonably close to the sample size needed to estimate the mean silicon dioxide content correct to within 10 ppm. ◆

Important Note: Theoretically, the small-sample t procedure presented here requires that the sample data come from a population that is normally distributed. (See the assumption in the box, p. 38.) However, statisticians have found the one-sample t procedure to be **robust**, i.e., to yield valid results even when the data are nonnormal, as long as the population is not highly skewed.

EXERCISES

1.35. The table contains 50 random samples of random digits, $y = 0, 1, 2, 3, \ldots, 9$, where the probabilities corresponding to the values of y are given by the formula $p(y) = \frac{1}{10}$. Each sample contains $n = 6$ measurements.

EX1_35

SAMPLE	SAMPLE	SAMPLE	SAMPLE
8, 1, 8, 0, 6, 6	7, 6, 7, 0, 4, 3	4, 4, 5, 2, 6, 6	0, 8, 4, 7, 6, 9
7, 2, 1, 7, 2, 9	1, 0, 5, 9, 9, 6	2, 9, 3, 7, 1, 3	5, 6, 9, 4, 4, 2
7, 4, 5, 7, 7, 1	2, 4, 4, 7, 5, 6	5, 1, 9, 6, 9, 2	4, 2, 3, 7, 6, 3
8, 3, 6, 1, 8, 1	4, 6, 6, 5, 5, 6	8, 5, 1, 2, 3, 4	1, 2, 0, 6, 3, 3
0, 9, 8, 6, 2, 9	1, 5, 0, 6, 6, 5	2, 4, 5, 3, 4, 8	1, 1, 9, 0, 3, 2
0, 6, 8, 8, 3, 5	3, 3, 0, 4, 9, 6	1, 5, 6, 7, 8, 2	7, 8, 9, 2, 7, 0
7, 9, 5, 7, 7, 9	9, 3, 0, 7, 4, 1	3, 3, 8, 6, 0, 1	1, 1, 5, 0, 5, 1
7, 7, 6, 4, 4, 7	5, 3, 6, 4, 2, 0	3, 1, 4, 4, 9, 0	7, 7, 8, 7, 7, 6
1, 6, 5, 6, 4, 2	7, 1, 5, 0, 5, 8	9, 7, 7, 9, 8, 1	4, 9, 3, 7, 3, 9
9, 8, 6, 8, 6, 0	4, 4, 6, 2, 6, 2	6, 9, 2, 9, 8, 7	5, 5, 1, 1, 4, 0
3, 1, 6, 0, 0, 9	3, 1, 8, 8, 2, 1	6, 6, 8, 9, 6, 0	4, 2, 5, 7, 7, 9
0, 6, 8, 5, 2, 8	8, 9, 0, 6, 1, 7	3, 3, 4, 6, 7, 0	8, 3, 0, 6, 9, 7
8, 2, 4, 9, 4, 6	1, 3, 7, 3, 4, 3		

a. Use the 300 random digits to construct a relative frequency distribution for the data. This relative frequency distribution should approximate $p(y)$.

b. Calculate the mean of the 300 digits. This will give an accurate estimate of μ (the mean of the population) and should be very near to $E(y)$, which is 4.5.

c. Calculate s^2 for the 300 digits. This should be close to the variance of y, $\sigma^2 = 8.25$.

d. Calculate $\bar{y}$ for each of the 50 samples. Construct a relative frequency distribution for the sample means to see how close they lie to the mean of $\mu = 4.5$. Calculate the mean and standard deviation of the 50 means.

1.36. Refer to Exercise 1.35. To see the effect of sample size on the standard deviation of the sampling distribution of a statistic, combine pairs of samples (moving down the columns of the table) to obtain 25 samples of $n = 12$ measurements. Calculate the mean for each sample.

a. Construct a relative frequency distribution for the 25 means. Compare this with the distribution prepared for Exercise 1.35 that is based on samples of $n = 6$ digits.

b. Calculate the mean and standard deviation of the 25 means. Compare the standard deviation of this sampling distribution with the standard deviation of the sampling distribution in Exercise 1.35. What relationship would you expect to exist between the two standard deviations?

1.37. Let t_0 be a particular value of t. Use Table 2 of Appendix C to find t_0 values such that the following statements are true:

   a. $P(t \geq t_0) = .025$ where $n = 10$
   b. $P(t \geq t_0) = .01$ where $n = 5$
   c. $P(t \leq t_0) = .005$ where $n = 20$
   d. $P(t \leq t_0) = .05$ where $n = 12$

1.38. The Computer-Assisted Hypnosis Scale (CAHS) is designed to measure a person's susceptibility to hypnosis. In computer-assisted hypnosis, the computer serves as a facilitator of hypnosis by using digitized speech processing coupled with interactive involvement with the hypnotic subject. CAHS scores range from 0 (no susceptibility) to 12 (extremely high susceptibility). A study in *Psychological Assessment* (Mar. 1995) reported a mean CAHS score of 4.59 and a standard deviation of 2.95 for University of Tennessee undergraduates. Assume that $\mu = 4.29$ and $\sigma = 2.95$ for this population. Suppose a psychologist uses CAHS to test a random sample of 50 subjects.

   a. Would you expect to observe a sample mean CAHS score of $\bar{x} = 6$ or higher? Explain.

   b. Suppose the psychologist actually observes $\bar{x} = 6.2$. Based on your answer to part **a**, make an inference about the population from which the sample was selected.

1.39. Studies by neuroscientists at the Massachusetts Institute of Technology (MIT) reveal that melatonin, which is secreted by the pineal gland in the brain, functions naturally as a sleep-inducing hormone (*Tampa Tribune*, Mar. 1, 1994). Male volunteers were given various doses of melatonin or placebo; then they were placed in a dark room at midday and told to close their eyes and fall asleep on demand. Of interest to the MIT researchers is the time y (in minutes) required for each volunteer to fall asleep. With the placebo (i.e., no hormone), the researchers found that the mean time to fall asleep was 15 minutes. Assume that with the placebo treatment, $\mu = 15$ and $\sigma = 5$.

   a. Consider a random sample of $n = 20$ men who are given the sleep-inducing hormone, melatonin. Let $\bar{y}$ represent the mean time to fall asleep for this

sample. If the hormone is *not* effective in inducing sleep, describe the sampling distribution of $\bar{y}$.

   b. Refer to part a. Find $P(\bar{y} \leq 6)$.

   c. In the actual study, the mean time to fall asleep for the 20 volunteers was $\bar{y} = 5$. Use this result to make an inference about the true value of the μ for those taking the melatonin.

1.40. Refer to the *Chance* (Winter 2001) and National Education Longitudinal Survey (NELS) study of 265 students who paid a private tutor to help them improve their SAT scores, Exercise 1.27 (p. 25). The changes in both the SAT-Mathematics and SAT-Verbal scores for these students are reproduced in the table.

	SAT-MATH	SAT-VERBAL
Mean change in score	19	7
Standard deviation of score changes	65	49

   a. Construct and interpret a 95% confidence interval for the population mean change in SAT-Mathematics score for students who pay a private tutor.

   b. Repeat part **a** for the population mean change in SAT-Verbal score.

   c. Suppose the true population mean change in score on one of the SAT tests for all students who paid a private tutor is 15. Which of the two tests, SAT-Mathematics or SAT-Verbal, is most likely to have this mean change? Explain.

1.41. Animal behaviorists have discovered that the more domestic chickens peck at objects placed in their environment, the healthier the chickens seem to be. White string has been found to be a particularly attractive pecking stimulus. In one experiment, 72 chickens were exposed to a string stimulus. Instead of white string, blue-colored string was used. The number of pecks each chicken took at the blue string over a specified time interval was recorded. Summary statistics for the 72 chickens were: $\bar{y} = 1.13$ pecks, $s = 2.21$ pecks (*Applied Animal Behaviour Science*, Oct. 2000).

   a. Estimate the population mean number of pecks made by chickens pecking at blue string using a 99% confidence interval. Interpret the result.

   b. Previous research has shown that $\mu = 7.5$ pecks if chickens are exposed to white string. Based on the results, part **a**, is there evidence that chickens are more apt to peck at white string than blue string? Explain.

1.42. A group of Harvard University School of Public Health researchers studied the impact of cooking on the size of indoor air particles (*Environmental Science & Technology*, Sept. 1, 2000). The decay rate (measured as μ m/hour) for fine particles produced from oven cooking or toasting was recorded on six randomly selected days. These six measurements are

⬤ DECAY

.95	.83	1.20	.89	1.45	1.12

Source: Abt, E., *et al.* "Relative contribution of outdoor and indoor particle sources to indoor concentrations." *Environmental Science & Technology*, Vol. 34, No. 17, Sept. 1, 2000 (Table 3).

  a. Find and interpret a 95% confidence interval for the true average decay rate of fine particles produced from oven cooking or toasting.
  b. Explain what the phrase "95% confident" implies in the interpretation of part **a**.
  c. What must be true about the distribution of the population of decay rates for the inference to be valid?

1.43. In Exercise 1.18 (p. 16) you learned that postmortem interval (PMI) is the elapsed time between death and the performance of an autopsy on the cadaver. *Brain and Language* (June 1995) reported on the PMIs of 22 randomly selected human brain specimens obtained at autopsy. The data are reproduced in the table below.

⬤ BRAINPMI

5.5	14.5	6.0	5.5	5.3	5.8	11.0	6.4
7.0	14.5	10.4	4.6	4.3	7.2	10.5	6.5
3.3	7.0	4.1	6.2	10.4	4.9		

Source: Hayes, T.L., and Lewis, D.A. "Anatomical specialization of the anterior motor speech area: Hemispheric differences in magnopyramidal neurons." *Brain and Language*. Vol. 49, No. 3, June 1995, p. 292 (Table 1).

  a. Construct a 95% confidence interval for the true mean PMI of human brain specimens obtained at autopsy.
  b. Interpret the interval, part **a**.
  c. What assumption is required for the interval, part **a**, to be valid? Is this assumption satisfied? Explain.
  d. What is meant by the phrase "95% confidence"?

1.44. The "fear of negative evaluation" (FNE) scores for 11 bulimic female students and 14 normal female students, first presented in Exercise 1.19 (p. 16) are reproduced below. (Recall that the higher the score, the greater the fear of negative evaluation.)

⬤ BULIMIA

Bulimic										
students	21	13	10	20	25	19	16	21	24	13	14
Normal										
students	13	6	16	13	8	19	23	18	11	19	7	10	15	20

Source: Randles, R. H. "On neutral responses (zeros) in the sign test and ties in the Wilcoxon-Mann-Whitney Test." *The American Statistician*, Vol. 55, No. 2, May 2001 (Figure 3).

  a. Construct a 95% confidence interval for the mean FNE score of the population of bulimic female students. Interpret the result.
  b. Construct a 95% confidence interval for the mean FNE score of the population of normal female students. Interpret the result.
  c. What assumptions are required for the intervals of parts **a** and **b** to be statistically valid? Are these assumptions reasonably satisfied? Explain.

1.9
Testing a Hypothesis About a Population Mean

The procedure involved in testing a hypothesis about a population parameter can be illustrated with the procedure for a test concerning a population mean μ.

A statistical test of a hypothesis is composed of the following four parts:

1. A **null hypothesis**, denoted by the symbol H_0, which is the hypothesis that we postulate is true

2. An **alternative** (or **research**) **hypothesis**, denoted by the symbol H_a, which is counter to the null hypothesis and is what we want to support

3. A **test statistic**, calculated from the sample data, that functions as a decision maker
4. A **rejection region**, values of a test statistic for which we reject the null hypothesis and accept the alternative hypothesis

The test statistic for testing the null hypothesis that a population mean μ equals some specific value, say, μ_0, is the sample mean $\overline{y}$ or the standardized normal variable

$$z = \frac{\overline{y} - \mu_0}{\sigma_{\overline{y}}} \quad \text{where } \sigma_{\overline{y}} = \frac{\sigma}{\sqrt{n}}$$

The logic used to decide whether sample data *disagree* with this hypothesis can be seen in the sampling distribution of $\overline{y}$ shown in Figure 1.22. If the population mean μ is equal to μ_0 (i.e., if the null hypothesis is true), then the mean $\overline{y}$ calculated from a sample should fall, with high probability, within $2\sigma_{\overline{y}}$ of μ_0. If $\overline{y}$ falls too far away from μ_0, or if the standardized distance

$$z = \frac{\overline{y} - \mu_0}{\sigma_{\overline{y}}}$$

is too large, we conclude that the data disagree with our hypothesis, and we reject the null hypothesis.

FIGURE 1.22

The sampling distribution of $\overline{y}$ for $\mu = \mu_0$

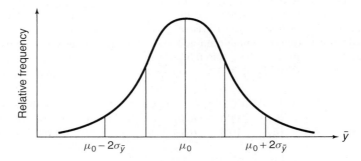

If we want to detect the alternative hypothesis that $\mu > \mu_0$, we locate the boundary of the rejection region in the upper tail of the z distribution, as shown in Figure 1.23a, at the point z_α. Similarly, to detect $\mu < \mu_0$, we place the rejection region in the lower tail of the z distribution, as shown in Figure 1.23b. These are called **one-tailed statistical tests**. To detect either $\mu > \mu_0$ or $\mu < \mu_0$—that is, $\mu \neq \mu_0$—we split α equally between the two tails of the z distribution and reject the null hypothesis if $z > z_{\alpha/2}$ or $z < -z_{\alpha/2}$, as shown in Figure 1.23c. This is called a **two-tailed statistical test**.

The z test, summarized in the next box, is called a *large-sample test* because we will rarely know σ and hence will need a sample size that is large enough so that the sample standard deviation s will provide a good approximation to σ. Normally, we recommend that the sample size be $n \geq 30$.

FIGURE 1.23

Location of the rejection region for various
alternative hypotheses

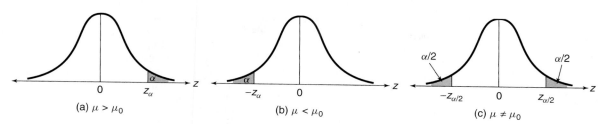

(a) $\mu > \mu_0$ (b) $\mu < \mu_0$ (c) $\mu \neq \mu_0$

Large-Sample ($n \geq 30$) Test of Hypothesis About μ

ONE-TAILED TEST	TWO-TAILED TEST
$H_0: \quad \mu = \mu_0$	$H_0: \quad \mu = \mu_0$
$H_a: \quad \mu < \mu_0$	$H_a: \quad \mu \neq \mu_0$
(or $H_a: \mu > \mu_0$)	

Test statistic: $z = \dfrac{\overline{y} - \mu_0}{\sigma_{\overline{y}}}$

Rejection region: $z < -z_\alpha$	*Rejection region*: $	z	> z_{\alpha/2}$
(or $z > z_\alpha$)	
where z_α is chosen so that	where $z_{\alpha/2}$ is chosen so that
$P(z > z_\alpha) = \alpha$	$P(z > z_{\alpha/2}) = \alpha/2$

We illustrate with an example.

EXAMPLE 1.14

Humerus bones from the same species of animal tend to have approximately the same length-to-width ratios. When fossils of humerus bones are discovered, archeologists can often determine the species of animal by examining the length-to-width ratios of the bones. It is known that species A has a mean ratio of 8.5. Suppose 41 fossils of humerus bones were unearthed at an archeological site in East Africa, where species A is believed to have flourished. (Assume that the unearthed bones were all from the same unknown species.) The length-to-width ratios of the bones were measured and are listed in Table 1.11. Do these data present sufficient evidence to indicate that the mean ratio of all bones of this species differs from 8.5? Use $\alpha = .05$.

Solution

Since we wish to determine whether $\mu \neq 8.5$, the elements of the test are

$$H_0: \quad \mu = 8.5$$
$$H_a: \quad \mu \neq 8.5$$

Test statistic: $z = \dfrac{\overline{y} - 8.5}{\sigma_{\overline{y}}} = \dfrac{\overline{y} - 8.5}{\sigma/\sqrt{n}} \approx \dfrac{\overline{y} - 8.5}{s/\sqrt{n}}$

Rejection region: $|z| > 1.96$ for $\alpha = .05$

BONES

TABLE 1.11   Length-to-Width Ratios of a Sample of Humerus Bones

10.73	9.57	6.66	9.89
8.89	9.29	9.35	8.17
9.07	9.94	8.86	8.93
9.20	8.07	9.93	8.80
10.33	8.37	8.91	10.02
9.98	6.85	11.77	8.38
9.84	8.52	10.48	11.67
9.59	8.87	10.39	8.30
8.48	6.23	9.39	9.17
8.71	9.41	9.17	12.00
			9.38

The data in Table 1.11 were analyzed using SPSS. The SPSS printout is displayed in Figure 1.24.

FIGURE 1.24

SPSS printout for Example 1.14

One-Sample Statistics

	N	Mean	Std. Deviation	Std. Error Mean
LWRATIO	41	9.2576	1.20357	.18797

One-Sample Test

	Test Value = 8.5					
					95% Confidence Interval of the Difference	
	t	df	Sig. (2-tailed)	Mean Difference	Lower	Upper
LWRATIO	4.030	40	.000	.7576	.3777	1.1375

Substituting the sample statistics $\bar{y} = 9.26$ and $s = 1.20$ (shown at the top of the SPSS printout) into the test statistic, we have

$$z \approx \frac{\bar{y} - 8.5}{s/\sqrt{n}} = \frac{9.26 - 8.5}{1.20/\sqrt{41}} = 4.03$$

The test statistic value is also shown (highlighted) at the bottom of the SPSS printout. Since the test statistic exceeds the critical value of 1.96, we can reject H_0 at $\alpha = .05$. The sample data provide sufficient evidence to conclude that the true mean length-to-width ratio of all humerus bones of this species differs from 8.5.

The *practical* implications of the result obtained in Example 1.14 remain to be seen. Perhaps the animal discovered at the archeological site is of some species other than A. Alternatively, the unearthed humeri may have larger than normal length-to-width ratios because they are the bones of specimens having unusual feeding habits for species A. **It is not always the case that a statistically significant result implies a practically significant result**. The researcher must retain his or her objectivity and judge the practical significance using, among other criteria, knowledge of the subject matter and the phenomenon under investigation. ◆

The reliability of a statistical test is measured by the probability of making an incorrect decision. For example, the probability of rejecting the null hypothesis and accepting the alternative hypothesis when the null hypothesis is true (called a **Type I error**) is α, the tail probability used in locating the rejection region. A second type of error could be made if we accepted the null hypothesis when, in fact, the alternative hypothesis is true (a **Type II error**). Thus, you never "accept" the null hypothesis unless you know the probability of making a Type II error. Since this probability (denoted by the symbol β) is often unknown, it is a common practice to defer judgment if a test statistic falls in the nonrejection region.

A small-sample test of the null hypothesis $\mu = \mu_0$ using a Student's t statistic is based on the assumption that the sample was randomly selected from a population with a normal relative frequency distribution. The test is conducted in exactly the same manner as the large-sample z test except that we use

$$t = \frac{\overline{y} - \mu_0}{s_{\overline{y}}} = \frac{\overline{y} - \mu_0}{s/\sqrt{n}}$$

as the test statistic and we locate the rejection region in the tail(s) of a Student's t distribution with df $= n - 1$. We summarize the technique for conducting a small-sample test of hypothesis about a population mean in the box.

Small-Sample Test of Hypothesis About μ

ONE-TAILED TEST	TWO-TAILED TEST
$H_0:\quad \mu = \mu_0$	$H_0:\quad \mu = \mu_0$
$H_a:\quad \mu < \mu_0$	$H_a:\quad \mu \neq \mu_0$
(or $H_a: \mu > \mu_0$)	

$$\text{Test statistic:} \quad t = \frac{\overline{y} - \mu_0}{s/\sqrt{n}}$$

Rejection region : $t < -t_\alpha$	Rejection region : $|t| > t_{\alpha/2}$
(or $t > t_\alpha$)	where t_α is based on $(n - 1)$ df
where t_α is based on $(n - 1)$ df	

Assumptions: The population from which the sample is drawn is approximately normal.

EXAMPLE 1.15

Scientists have labeled benzene, a chemical solvent commonly used to synthesize plastics, as a possible cancer-causing agent. Studies have shown that people who work with benzene more than 5 years have 20 times the incidence of leukemia than the general population. As a result, the federal government lowered the maximum allowable level of benzene in the workplace from 10 parts per million (ppm) to 1 ppm. Suppose a steel manufacturing plant, which exposes its workers to benzene daily, is under investigation by the Occupational Safety and Health Administration

BENZENE

TABLE 1.12   Benzene Content for 20 Air Samples

0.5	0.9	4.5	3.4	1.0
2.7	1.1	1.9	0.0	0.0
4.2	2.1	0.0	2.0	3.4
3.4	2.5	0.9	5.1	2.4

(OSHA). Twenty air samples, collected over a period of 1 month and examined for benzene content, yielded the data in Table 1.12.

Is the steel manufacturing plant in violation of the changed government standards? Test the hypothesis that the mean level of benzene at the steel manufacturing plant is greater than 1 ppm, using $\alpha = .05$.

Solution

OSHA wants to establish the research hypothesis that the mean level of benzene, μ, at the steel manufacturing plant exceeds 1 ppm. The elements of this small-sample one-tailed test are

$H_0: \quad \mu = 1$

$H_a: \quad \mu > 1$

$$\text{Test statistic}: t = \frac{\bar{y} - \mu_0}{s/\sqrt{n}}$$

Assumptions: The relative frequency distribution of the population of benzene levels for all air samples at the steel manufacturing plant is approximately normal.

Rejection region: For $\alpha = .05$ and df $= n - 1 = 19$, reject H_0 if $t > t_{.05} = 1.729$ (see Figure 1.25)

FIGURE 1.25

Rejection region for Example 1.15

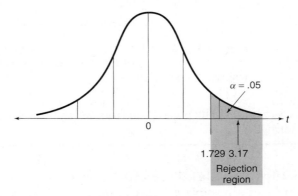

$\alpha = .05$

0

1.729 3.17

Rejection region

t

The SAS printout, Figure 1.25, gives summary statistics for the sample data. Substituting $\bar{y} = 2.1$ and $s = 1.55$ into the test statistic formula, we obtain:

$$t = \frac{\bar{y} - 1}{s/\sqrt{n}} = \frac{2.1 - 1}{1.55/\sqrt{20}} = 3.17$$

FIGURE 1.26

SAS output for testing benzene mean

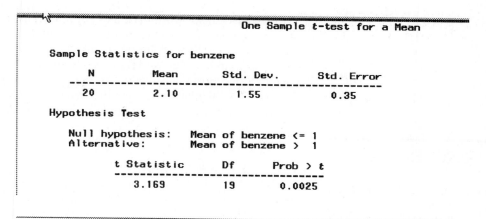

```
                                    One Sample t-test for a Mean

    Sample Statistics for benzene

         N          Mean         Std. Dev.        Std. Error
    --------------------------------------------------------------
        20           2.10          1.55             0.35

    Hypothesis Test

        Null hypothesis:    Mean of benzene <= 1
        Alternative:        Mean of benzene >  1

            t Statistic        Df         Prob > t
        ------------------------------------------
               3.169           19          0.0025
```

Since the calculated t falls in the rejection region, OSHA concludes that $\mu > 1$ ppm and the plant is in violation of the revised government standards. The reliability associated with this inference is $\alpha = .05$. This implies that if the testing procedure were applied repeatedly to random samples of data collected at the plant, OSHA would falsely reject H_0 for only 5% of the tests. Consequently, OSHA is highly confident (95% confident) that the plant is violating the new standards.   ◆

To conclude, we mention a second method that is useful in reporting the results of a statistical test. Some data analyzers indicate the degree to which the test statistic contradicts the null hypothesis (and hence supports the alternative hypothesis). This quantity, called the **observed significance level**, or **p-value**, of the test, is the probability of observing a value of the test statistic at least as contradictory to the null hypothesis as the observed value of the test statistic, *assuming the null hypothesis is true*. For example, suppose you conducted a large-sample z test to detect the alternative hypothesis $\mu > 100$, and the computed value of the test statistic was 2.12. The level of significance for this one-tailed test is the probability of observing a z value larger than 2.12 if $\mu = 100$, or

$$P(z > 2.12) = .0170 \text{ (obtained from Table 1, Appendix C)}$$

If the test were two-tailed (i.e., the alternative hypothesis were $\mu \neq 100$), then values more contradictory to the null hypothesis $\mu = 100$ would be z values greater than 2.12 or less than -2.12. The level of significance for this test would be $2(.0170) = .0340$.

Decisions about H_0 and H_a can be made by comparing the p-value of the test to your desired value of α, as shown in the box.

Reporting Test Results as p-Values: How to Decide Whether to Reject H_0

1. Choose the maximum value of α that you are willing to tolerate.
2. If the observed significance level (p-value) of the test is less than the maximum value of α, then reject the null hypothesis.

Most statistical software packages automatically compute the exact p-value of a test. For example, the p-value for the large sample test of $\mu = 8.5$ in Example 1.14 is highlighted on the SPSS printout, Figure 1.24 (p. 46). Since the two-tailed p-value, .000, is less than $\alpha = .05$, we reject H_0. Similarly, the SAS printout for the t test of Example 1.15 reports the p-value for the one-tailed test in Figure 1.26 (p. 49). This value, .0025, is highlighted under the **Prob** $> t$ column. This implies that we will reject H_0 for any α level that exceeds p-value $= .0025$.

EXERCISES

1.45. Define each of the following:
   a. H_0 **b.** H_a **c.** Type I error
   d. Type II error   **e.** α **f.** β

1.46. In hypothesis testing,
   a. who or what determines the size of the rejection region?
   b. does rejecting H_0 prove that the research hypothesis is correct?

1.47. For each of the following rejection regions, sketch the sampling distribution for z, indicate the location of the rejection region, and give the value of α:
   a. $z > 1.96$ **b.** $z > 1.645$ **c.** $z > 2.576$
   d. $z < -1.29$ **e.** $z < -1.645$ or   **f.** $z < -2.576$ or
                          $z > 1.645$ $z > 2.576$

1.48. *Science* (Jan. 1, 1999) reported that the mean listening time of 7-month-old infants exposed to a three-syllable sentence (e.g., "ga ti ti") is 9 seconds. Set up the null and alternative hypotheses for testing the claim.

1.49. According to a University of Florida wildlife ecology and conservation researcher, the average level of mercury uptake in wading birds in the Everglades has declined over the past several years (*UF News*, Dec. 15, 2000). In 1994, the average level was 15 parts per million.
   a. Give the null and alternative hypotheses for testing whether the average level in 2000 was less than 15 ppm.
   b. Describe a Type I error for this test.
   c. Describe a Type II error for this test.

1.50. The mean alkalinity level of water specimens collected from the Han River in Seoul, Korea, is 50 milligrams per liter (*Environmental Science & Engineering*, Sept. 1, 2000). Consider a random sample of 100 water specimens collected from a tributary of the Han River. Suppose the mean and standard deviation of the alkalinity levels for the sample are $\bar{y} = 67.8$ mpl and $s = 14.4$ mpl. Is there sufficient evidence (at $\alpha = .01$) to indicate that the population mean alkalinity level of water in the tributary exceeds 50 mpl?

1.51. The *Community Mental Health Journal* (Aug. 2000) presented the results of a survey of over 6,000 clients of the Department of Mental Health and Addiction Services (DMHAS) in Connecticut. One of the many variables measured for each mental health patient was frequency of social interaction (on a 5-point scale, where 1 = very infrequently, 3 = occasionally, and 5 = very frequently). The 6,681 clients who were evaluated had a mean social interaction score of 2.95 with a standard deviation of 1.10.
   a. Conduct a hypothesis test (at $\alpha = .01$) to determine if the true mean social interaction score of all Connecticut mental health patients differs from 3.
   b. Examine the results of the study from a practical view, then discuss why "statistical significance" does not always imply "practical significance."
   c. Because the variable of interest is measured on a 5-point scale, it is unlikely that the population of ratings will be normally distributed. Consequently, some analysts may perceive the test, part **a**, to be invalid and search for alternative methods of analysis. Defend or refute this position.

1.52. Radium-226 is a naturally occurring radioactive gas. Elevated levels of radium-226 in metropolitan Dade County (Florida) have been investigated (*Florida Scientist*, Summer/Autumn 1991). The data in the table are radium-226 levels (measured in picocuries per liter) for 26 soil specimens collected in southern Dade County. The Environmental Protection Agency (EPA) has set maximum exposure levels of radium-226 at 4.0 pCi/L. Use the information in the MINITAB printout (p. 51) to determine whether the mean radium-226 level of

RADIUM

1.46	.58	4.31	1.02	.17	2.92	.91	.43	.91
1.30	8.24	3.51	6.87	1.43	1.44	4.49	4.21	1.84
5.92	1.86	1.41	1.70	2.02	1.65	1.40	.75	

Source: Moore, H. E., and Gussow, D. G. "Radium and radon in Dade County ground-water and soil samples." *Florida Scientist*, Vol. 54, No. 3/4, Summer/Autumn, 1991, p. 155 (portion of Table 3).

soil specimens collected in southern Dade County is less than the EPA limit of 4.0 pCi/L. Use $\alpha = .10$.

MINITAB output for Exercise 1.52

```
One-Sample T: RadLevel. Test of mu = 4 vs mu not = 4

Variable        N       Mean     StDev    SE Mean
RadLevel        26      2.413    2.081    0.408

Variable             95.0% CI           T      P
RadLevel       (  1.573,   3.254)    -3.89  0.001
```

1.53. Research published in *Nature* (Aug. 27, 1998) revealed that people are more attracted to "feminized" faces, regardless of gender. In one experiment, 50 human subjects viewed both a Japanese female and Caucasian male face on a computer. Using special computer graphics, each subject could morph the faces (by making them more feminine or more masculine) until they attained the "most attractive" face. The level of feminization y (measured as a percentage) was measured.

a. For the Japanese female face, $\bar{y} = 10.2\%$ and $s = 31.3\%$. The researchers used this sample information to test the null hypothesis of a mean level of feminization equal to 0%. Verify that the test statistic is equal to 2.3.

b. Refer to part **a**. The researchers reported the p-value of the test as $p \approx .02$. Verify and interpret this result.

c. For the Caucasian male face, $\bar{y} = 15.0\%$ and $s = 25.1\%$. The researchers reported the test statistic (for the test of the null hypothesis stated in part **a**) as 4.23 with an associated p-value of approximately 0. Verify and interpret these results.

1.54. A study was conducted to evaluate the effectiveness of a new mosquito repellent designed by the U.S. Army to be applied as camouflage face paint (*Journal of the Mosquito Control Association*, June 1995). The repellent was applied to the forearms of five volunteers and then they were exposed to fifteen active mosquitos for a ten-hour period. The percentage of the forearm surface area protected from bites (called percent repellency) was calculated for each of the five volunteers. For one color of paint (loam), the following summary statistics were obtained:

$$\bar{y} = 83\% \qquad s = 15\%$$

a. The new repellent is considered effective if it provides a percent repellency of at least 95. Conduct a test to determine whether the mean repellency percentage of the new mosquito repellent is less than 95. Test using $\alpha = .10$.

b. What assumptions are required for the hypothesis test in part **a** to be valid?

1.55. "Hot Tamales" are chewy, cinnamon-flavored candies. A bulk vending machine is known to dispense, on average, 15 Hot Tamales per bag. *Chance* (Fall 2000) published an article on a classroom project in which students were required to purchase bags of Hot Tamales from the machine and count the number of candies per bag. One student group claimed they purchased five bags that had the following candy counts: 25, 23, 21, 21, and 20. There was some question as to whether the students had fabricated the data. Use a hypothesis test to gain insight into whether or not the data collected by the students are fabricated. Use a level of significance that gives the benefit of the doubt to the students.

1.10
Inferences About the Difference Between Two Population Means

The reasoning employed in constructing a confidence interval and performing a statistical test for comparing two population means is identical to that discussed in Sections 1.7 and 1.8. First, we present procedures that are based on the assumption that we have selected *independent* random samples from the two populations. The parameters and sample sizes for the two populations, the sample means, and the sample variances are shown in Table 1.13. The objective of the sampling is to make an inference about the difference $(\mu_1 - \mu_2)$ between the two population means.

Because the sampling distribution of the difference between the sample means $(\bar{y}_1 - \bar{y}_2)$ is approximately normal for large samples, the large-sample techniques are based on the standardized normal z statistic. Since the variances of the populations, σ_1^2 and σ_2^2, will rarely be known, we will estimate their values using s_1^2 and s_2^2.

To employ these large-sample techniques, we recommend that both samples sizes be large (i.e., each at least 30). The large-sample confidence interval and test are summarized in the boxes.

TABLE 1.13   **Two-Sample Notation**

	POPULATION	
	1	2
Sample size	n_1	n_2
Population mean	μ_1	μ_2
Population variance	σ_1^2	σ_2^2
Sample mean	$\bar{y}_1$	$\bar{y}_2$
Sample variance	s_1^2	s_2^2

Large-Sample Confidence Interval for $(\mu_1 - \mu_2)$: Independent Samples

$$(\bar{y}_1 - \bar{y}_2) \pm z_{\alpha/2} \sigma_{(\bar{y}_1 - \bar{y}_2)^*} = (\bar{y}_1 - \bar{y}_2) \pm z_{\alpha/2} \sqrt{\frac{\sigma_1^2}{n_1} + \frac{\sigma_2^2}{n_2}}$$

Assumptions: The two samples are randomly and independently selected from the two populations. The sample sizes, n_1 and n_2, are large enough so that $\bar{y}_1$ and $\bar{y}_2$ each have approximately normal sampling distributions and so that s_1^2 and s_2^2 provide good approximations to σ_1^2 and σ_2^2. This will be true if $n_1 \geq 30$ and $n_2 \geq 30$.

Large-Sample Test of Hypothesis About $(\mu_1 - \mu_2)$: Independent Samples

ONE-TAILED TEST	TWO-TAILED TEST
H_0: $(\mu_1 - \mu_2) = D_0$	H_0: $(\mu_1 - \mu_2) = D_0$
H_a: $(\mu_1 - \mu_2) < D_0$	H_a: $(\mu_1 - \mu_2) \neq D_0$
[or H_a: $(\mu_1 - \mu_2) > D_0$]	

where D_0 = Hypothesized difference between the means (this is often 0)

$$\text{Test statistic: } z = \frac{(\bar{y}_1 - \bar{y}_2) - D_0}{\sigma_{(\bar{y}_1 - \bar{y}_2)}}$$

where $\sigma_{(\bar{y}_1 - \bar{y}_2)} = \sqrt{\frac{\sigma_1^2}{n_1} + \frac{\sigma_2^2}{n_2}}$

Rejection region: $z < -z_\alpha$	Rejection region: $\lvert z \rvert > z_{\alpha/2}$
(or $z > z_\alpha$)	

Assumptions: Same as for the previous large-sample confidence interval.

---

*The symbol $\sigma_{(\bar{y}_1 - \bar{y}_2)}$ is used to denote the standard error of the distribution of $(\bar{y}_1 - \bar{y}_2)$.

EXAMPLE　1.16

A dietitian has developed a diet that is low in fats, carbohydrates, and cholesterol. Although the diet was initially intended to be used by people with heart disease, the dietitian wishes to examine the effect this diet has on the weights of obese people. Two random samples of 100 obese people each are selected, and one group of 100 is placed on the low-fat diet. The other 100 are placed on a diet that contains approximately the same quantity of food but is not as low in fats, carbohydrates, and cholesterol. For each person, the amount of weight lost (or gained) in a 3-week period is recorded. The data, saved in the DIETSTUDY file, are listed in Table 1.14. Form a 95% confidence interval for the difference between the population mean weight losses for the two diets. Interpret the result.

DIETSTUDY

TABLE 1.14　Diet Study Data, Example 1.16

WEIGHT LOSSES FOR LOW-FAT DIET

8	10	10	12	9	3	11	7	9	2
21	8	9	2	2	20	14	11	15	6
13	8	10	12	1	7	10	13	14	4
8	12	8	10	11	19	0	9	10	4
11	7	14	12	11	12	4	12	9	2
4	3	3	5	9	9	4	3	5	12
3	12	7	13	11	11	13	12	18	9
6	14	14	18	10	11	7	9	7	2
16	16	11	11	3	15	9	5	2	6
5	11	14	11	6	9	4	17	20	10

WEIGHT LOSSES FOR REGULAR DIET

6	6	5	5	2	6	10	3	9	11
14	4	10	13	3	8	8	13	9	3
4	12	6	11	12	9	8	5	8	7
6	2	6	8	5	7	16	18	6	8
13	1	9	8	12	10	6	1	0	13
11	2	8	16	14	4	6	5	12	9
11	6	3	9	9	14	2	10	4	13
8	1	1	4	9	4	1	1	5	6
14	0	7	12	9	5	9	12	7	9
8	9	8	10	5	8	0	3	4	8

Solution

Let μ_1 represent the mean of the conceptual population of weight losses for all obese people who could be placed on the low-fat diet. Let μ_2 be similarly defined for the other diet. We wish to form a confidence interval for $(\mu_1 - \mu_2)$.

Summary statistics for the diet data are displayed in the SPSS printout, Figure 1.27. Note that $\bar{y}_1 = 9.31$, $\bar{y}_2 = 7.40$, $s_1 = 4.67$, and $s_2 = 4.04$. Using these values and noting that $\alpha = .05$ and $z_{.025} = 1.96$, we find that the 95% confidence interval is:

$$(\bar{y}_1 - \bar{y}_2) \pm z_{.025}\sqrt{\frac{\sigma_1^2}{n_1} + \frac{\sigma_2^2}{n_2}} \approx$$

FIGURE 1.27

SPSS summary statistics for diet
study

Group Statistics

	DIET	N	Mean	Std. Deviation	Std. Error Mean
WTLOSS	REGULAR	100	7.40	4.035	.404
	LOW-FAT	100	9.31	4.668	.467

$$(9.31 - 7.40) \pm 1.96\sqrt{\frac{(4.67)^2}{100} + \frac{(4.04)^2}{100}} = 1.91 \pm (1.96)(.62) = 1.91 \pm 1.22$$

or (.69, 3.13). Using this estimation procedure over and over again for different samples, we know that approximately 95% of the confidence intervals formed in this manner will enclose the difference in population means ($\mu_1 - \mu_2$). Therefore, we are highly confident that the mean weight loss for the low-fat diet is between .69 and 3.13 pounds more than the mean weight loss for the other diet. With this information, the dietitian better understands the potential of the low-fat diet as a weight-reducing diet.   ◆

The small-sample statistical techniques used to compare μ_1 and μ_2 with independent samples are based on the assumptions that both populations have normal probability distributions and that the variation within the two populations is of the same magnitude, i.e., $\sigma_1^2 = \sigma_2^2$. When these assumptions are approximately satisfied, we can employ a Student's t statistic to find a confidence interval and test a hypothesis concerning ($\mu_1 - \mu_2$). The techniques are summarized in the following boxes.

Small-Sample Confidence Interval for ($\mu_1 - \mu_2$): Independent Samples

$$(\bar{y}_1 - \bar{y}_2) \pm t_{\alpha/2}\sqrt{s_p^2\left(\frac{1}{n_1} + \frac{1}{n_2}\right)}$$

where

$$s_p^2 = \frac{(n_1 - 1)s_1^2 + (n_2 - 1)s_2^2}{n_1 + n_2 - 2}$$

is a "pooled" estimate of the common population variance and $t_{\alpha/2}$ is based on $(n_1 + n_2 - 2)$ df.

Assumptions:

1. Both sampled populations have relative frequency distributions that are approximately normal.

2. The population variances are equal.

3. The samples are randomly and independently selected from the populations.

Small-Sample Test of Hypothesis About $(\mu_1 - \mu_2)$: Independent Samples

ONE-TAILED TEST

H_0: $(\mu_1 - \mu_2) = D_0$
H_a: $(\mu_1 - \mu_2) < D_0$
[or H_a: $(\mu_1 - \mu_2) > D_0$]

TWO-TAILED TEST

H_0: $(\mu_1 - \mu_2) = D_0$
H_a: $(\mu_1 - \mu_2) \neq D_0$

$$\text{Test statistic:} t = \frac{(\bar{y}_1 - \bar{y}_2) - D_0}{\sqrt{s_p^2 \left(\frac{1}{n_1} + \frac{1}{n_2}\right)}}$$

Rejection region: $t < -t_\alpha$
(or $t > t_\alpha$)

Rejection region: $|t| > t_{\alpha/2}$

where t_α is based on $(n_1 + n_2 - 2)$ df

Assumptions: Same as for the small-sample confidence interval for $(\mu_1 - \mu_2)$ in the previous box

---

EXAMPLE 1.17

Suppose you wish to compare a new method of teaching reading to "slow learners" to the current standard method. You decide to base this comparison on the results of a reading test given at the end of a learning period of 6 months. Of a random sample of 22 slow learners, 10 are taught by the new method and 12 are taught by the standard method. All 22 children are taught by qualified instructors under similar conditions for a 6-month period. The results of the reading test at the end of this period are given in Table 1.15.

READING

TABLE 1.15   **Reading Test Scores for Slow Learners**

NEW METHOD				STANDARD METHOD			
80	80	79	81	79	62	70	68
76	66	71	76	73	76	86	73
70	85			72	68	75	66

a. Use the data in the table to test whether the true mean test scores differ for the new method and the standard method. Use $\alpha = .05$.

b. What assumptions must be made in order that the estimate be valid?

Solution

a. For this experiment, let μ_1 and μ_2 represent the mean reading test scores of slow learners taught with the new and standard methods, respectively. Then, we want to test the following hypothesis:

H_0: $(\mu_1 - \mu_2) = 0$ (i.e., no difference in mean reading scores)

H_a: $(\mu_1 - \mu_2) \neq 0$ (i.e., $\mu_1 \neq \mu_2$)

To compute the test statistic, we need to obtain summary statistics (e.g., $\bar{y}$ and s) on reading test scores for each method. The data of Table 1.15 was entered into a computer, and SAS used to obtain these descriptive statistics. The SAS printout appears in Figure 1.28. Note that $\bar{y}_1 = 76.4$, $s_1 = 5.8348$, $\bar{y}_2 = 72.333$, and $s_2 = 6.3437$

FIGURE 1.28

SAS output for Example 1.17

Two Sample t-test for the Means of SCORE within METHOD

Sample Statistics

Group	N	Mean	Std. Dev.	Std. Error
NEW	10	76.4	5.8348	1.8451
STD	12	72.33333	6.3437	1.8313

Hypothesis Test

Null hypothesis: Mean 1 - Mean 2 = 0
Alternative: Mean 1 - Mean 2^ = 0

If Variances Are	t statistic	Df	Pr > t
Equal	1.552	20	0.1364
Not Equal	1.564	19.77	0.1336

95% Confidence Interval for the Difference between Two Means

Lower Limit	Upper Limit
−1.40	9.53

Next, we calculate the pooled estimate of variance:

$$s_{p}^{2} = \frac{(n_1 - 1)s_1^2 + (n_2 - 1)s_2^2}{n_1 + n_2 - 2}$$

$$= \frac{(10 - 1)(5.8348)^2 + (12 - 1)(6.3437)^2}{10 + 12 - 2} = 37.45$$

where s_{p}^{2} is based on $(n_1 + n_2 - 2) = (10 + 12 - 2) = 20$ degrees of freedom. Now, we compute the test statistic:

$$t = \frac{(\bar{y}_1 - \bar{y}_2) - D_0}{\sqrt{s_p^2 \left(\frac{1}{n_1} + \frac{1}{n_2} \right)}} \quad \frac{(76.4 - 72.33) - 0}{\sqrt{37.45 \left(\frac{1}{10} + \frac{1}{12} \right)}} = 1.55$$

The rejection region for this two-tailed test at $\alpha = .05$, based on 20 degrees of freedom, is

$$|t| > t_{.025} = 2.086$$

(See Figure 1.29)

FIGURE 1.29

Rejection region for Example 1.17

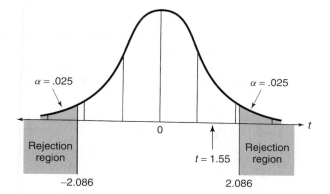

Since the computed value of t does not fall in the rejection region, we fail to reject H_0. There is insufficient evidence (at $\alpha = .05$) of a difference between the true mean test scores for the two reading methods.

This conclusion can also be obtained by using the p-value approach. Both the test statistic ($t = 1.552$) and p-value (.1364) are highlighted on the SAS printout, Figure 1.29. Since $\alpha = .05$ is less than the p-value, we fail to reject H_0.

b. To properly use the small-sample confidence interval, the following assumptions must be satisfied:

(1) The samples are randomly and independently selected from the populations of slow learners taught by the new method and the standard method.

(2) The test scores are normally distributed for both teaching methods.

(3) The variance of the test scores are the same for the two populations, that is $\sigma_1^2 = \sigma_2^2$.

The two-sample t statistic is a powerful tool for comparing population means when the necessary assumptions are satisfied. It has also been found that the two-sample t procedure is more robust against nonnormal data than the one-sample method. And, when the sample sizes are equal, the assumption of equal population variances can be relaxed. That is, when $n_1 = n_2$, σ_1^2 and σ_2^2 can be quite different and the test statistic will still have (approximately) a Student's t distribution. ◆

In Example 1.17, suppose it is possible to measure the slow learners' "reading IQs" *before* they are subjected to a teaching method. Eight pairs of slow learners with similar reading IQs are found, and one member of each pair is randomly assigned to the standard teaching method while the other is assigned to the new method. The data are given in Table 1.16. Do the data support the hypothesis that the population mean reading test score for slow learners taught by the new method is greater than the mean reading test score for those taught by the standard method?

Now, we want to test

$$H_0 : (\mu_1 - \mu_2) = 0$$

$$H_a : (\mu_1 - \mu_2) > 0$$

It appears that we could conduct this test using the t statistic for two independent samples, as in Example 1.17. However, *the independent samples t-test is not a valid procedure to use with this set of data*. Why?

⬡ PAIREDSCORES

TABLE 1.16   **Reading Test Scores for Eight Pairs of Slow Learners**

PAIR	NEW METHOD (1)	STANDARD METHOD (2)
1	77	72
2	74	68
3	82	76
4	73	68
5	87	84
6	69	68
7	66	61
8	80	76

TABLE 1.17   **Differences in Reading Test Scores**

PAIR	NEW METHOD	STANDARD METHOD	DIFFERENCE (NEW METHOD – STANDARD METHOD)
1	77	72	5
2	74	68	6
3	82	76	6
4	73	68	5
5	87	84	3
6	69	68	1
7	66	61	5
8	80	76	4

The t-test is inappropriate because the assumption of independent samples is invalid. We have randomly chosen *pairs of test scores*, and thus, once we have chosen the sample for the new method, we have *not* independently chosen the sample for the standard method. The dependence between observations within pairs can be seen by examining the pairs of test scores, which tend to rise and fall together as we go from pair to pair. This pattern provides strong visual evidence of a violation of the assumption of independence required for the two-sample t-test used in Example 1.17.

We now consider a valid method of analyzing the data of Table 1.16. In Table 1.17 we add the column of differences between the test scores of the pairs of slow learners. We can regard these differences in test scores as a random sample of differences for all pairs (matched on reading IQ) of slow learners, past and present. Then we can use this sample to make inferences about the mean of the population of differences, μ_d, which is equal to the difference $(\mu_1 - \mu_2)$. That is, the mean of the population (and sample) of differences equals the difference between the population (and sample) means. Thus, our test becomes

$$H_0: \mu_d = 0 \quad (\mu_1 - \mu_2 = 0)$$

$$H_a: \mu_d > 0 \quad (\mu_1 - \mu_2 > 0)$$

The test statistic is a one-sample t (Section 1.9), since we are now analyzing a single sample of differences for small n:

$$\text{Test statistic: } t = \frac{\bar{y}_d - 0}{s_d/\sqrt{n_d}}$$

where $\bar{y}_d$ = Sample mean difference
s_d = Sample standard deviation of differences
n_d = Number of differences = Number of pairs

Assumptions: The population of differences in test scores is approximately normally distributed. The sample differences are randomly selected from the population differences. [*Note*: We do not need to make the assumption that $\sigma_1^2 = \sigma_2^2$.]

Rejection region: At significance level $\alpha = .05$, we will reject H_0: if $t > t_{.05}$, where $t_{.05}$ is based on $(n_d - 1)$ degrees of freedom.

Referring to Table in Appendix, we find the t value corresponding to $\alpha = .05$ and $n_d - 1 = 8 - 1 = 7$ df to be $t_{.05} = 1.895$. Then we will reject the null hypothesis if $t > 1.895$ (see Figure 1.30). Note that the number of degrees of freedom decreases from $n_1 + n_2 - 2 = 14$ to 7 when we use the paired difference experiment rather than the two independent random samples design.

FIGURE 1.30

Rejection region for analysis of data in Table 1.17

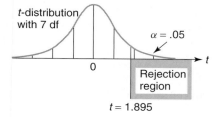

Summary statistics for the $n = 8$ differences are shown in the MINITAB printout, Figure 1.31. Note that $\bar{y}_d = 4.375$ and $s_d = 1.685$. Substituting these values into the formula for the test statistic, we have

$$t = \frac{\bar{y}_d - 0}{s_d/\sqrt{n_d}} = \frac{4.375}{1.685/\sqrt{8}} = 7.34$$

Because this value of t falls in the rejection region, we conclude (at $\alpha = .05$) that the population mean test score for slow learners taught by the new method exceeds the population mean score for those taught by the standard method. We can reach the same conclusion by noting that the p-value of the test, highlighted in Figure 1.31, is much smaller than $\alpha = .05$.

This kind of experiment, in which observations are paired and the differences are analyzed, is called a **paired difference experiment**. The hypothesis-testing procedures and the method of forming confidence intervals for the difference between two means using a paired difference experiment are summarized in the next two boxes for both large and small n.

FIGURE 1.31
MINITAB paired difference
analysis of data in Table 1.17

```
Paired T for NEW - STANDARD

                 N      Mean     StDev    SE Mean
NEW              8     76.00      6.93      2.45
STANDARD         8     71.63      7.01      2.48
Difference       8     4.375     1.685     0.596

95% CI for mean difference: (2.965, 5.785)
T-Test of mean difference = 0 (vs not = 0): T-Value = 7.34   P-Value = 0.000
```

Paired Difference Confidence Interval for $\mu_d = \mu_1 - \mu_2$

LARGE SAMPLE

$$\overline{y}_d \pm z_{\alpha/2} \frac{\sigma_d}{\sqrt{n_d}} \approx \overline{y}_d \pm z_{\alpha/2} \frac{s_d}{\sqrt{n_d}}$$

Assumption: The sample differences are randomly selected from the population of differences.

SMALL SAMPLE

$$\overline{y}_d \pm t_{\alpha/2} \frac{s_d}{\sqrt{n_d}}$$

where $t_{\alpha/2}$ is based on $(n_d - 1)$ degrees of freedom

Assumptions:

1. The relative frequency distribution of the population of differences is normal.
2. The sample differences are randomly selected from the population of differences.

Paired Difference Test of Hypothesis for $\mu_D = \mu_1 - \mu_2$

ONE-TAILED TEST	TWO-TAILED TEST
$H_0: \mu_d = D_0$	$H_0: \mu_d = D_0$
$H_a: \mu_d < D_0$	$H_a: \mu_d \neq D_0$
[or $H_a: \mu_d > D_0$]	

LARGE SAMPLE

$$\text{Test statistic}: z = \frac{\overline{y}_d - D_0}{\sigma_d/\sqrt{n_d}} \approx \frac{\overline{y}_d - D_0}{s_d/\sqrt{n_d}}$$

Rejection region: $z < -z_\alpha$ Rejection region: $|z| > z_{\alpha/2}$
[or $z > z_\alpha$ when $H_a: \mu_d > D_0$]

Assumption: The differences are randomly selected from the population of differences.

SMALL SAMPLE

$$\text{Test statistic: } t = \frac{\overline{y}_\text{d} - D_0}{s_\text{d}/\sqrt{n_\text{d}}}$$

Rejection region: $t < -t_\alpha$ Rejection region: $|t| > t_{\alpha/2}$
[or $t > t_\alpha$ when $H_\text{a}: \mu_\text{d} > D_0$]

where t_α and $t_{\alpha/2}$ are based on $(n_\text{d} - 1)$ degrees of freedom

Assumptions:

1. The relative frequency distribution of the population of differences is normal.
2. The differences are randomly selected from the population of differences.

EXERCISES

1.56. To use the t statistic to test for differences between the means of two populations based on independent samples, what assumptions must be made about the two sampled populations? What assumptions must be made about the two samples?

1.57. Describe the sampling distribution of $(\overline{y}_1 - \overline{y}_2)$.

1.58. Are children who repeat a grade in elementary school shorter on average, than their peers? To answer this question, researchers compared the heights of Australian school children who repeated a grade to those

who did not (*The Archives of Disease in Childhood*, Apr. 2000). All height measurements were standardized using z-scores. A summary of the results, by gender, is shown in the table.

a. Conduct a test of hypothesis to determine whether the average height of Australian boys who repeated a grade is less than the average height of boys who never repeated. Use $\alpha = .05$.

b. Repeat part **a** for Australian girls.

c. Summarize the results of the hypothesis tests in the words of the problem.

1.59. The "fear of negative evaluation" (FNE) scores for 11 female students known to suffer from the eating disorder bulimia and 14 female students with normal eating habits, first presented in Exercise 1.19 (p. 16), are reproduced below. (Recall that the higher the score, the greater the fear of negative evaluation.)

a. Find a 95% confidence interval for the difference between the population means of the FNE scores for bulimic and normal female students. Interpret the result.

	NEVER REPEATED	REPEATED A GRADE
Boys	$n = 1{,}349$	$n = 86$
	$\overline{x} = .30$	$\overline{x} = -.04$
	$s = .97$	$s = 1.17$
Girls	$n = 1{,}366$	$n = 43$
	$\overline{x} = .22$	$\overline{x} = .26$
	$s = 1.04$	$s = .94$

Source: Wake, M., Coghlan, D., and Hesketh, K. "Does height influence progression through primary school grades?" *The Archives of Disease in Childhood*, Vol. 82, Apr. 2000 (Table 3).

BULIMIA

Bulimic students	21	13	10	20	25	19	16	21	24	13	14		
Normal students	13	6	16	13	8	19	23	18	11	19	7	10	15	20

b. What assumptions are required for the interval of part **a** to be statistically valid? Are these assumptions reasonably satisfied? Explain.

1.60. The *Journal of Agricultural, Biological, and Environmental Statistics* (Sept. 2000) reported on an impact study of a tanker oil spill on the seabird population in Alaska. For each of 96 shoreline locations (called transects), the number of seabirds found, the length (in kilometers) of the transect, and whether or not the transect was in an oiled area were recorded. (The data are saved in the EVOS file.) Observed seabird density is defined as the observed count divided by the length of the transect. A comparison of the mean densities of oiled and unoiled transects is displayed in the following MINITAB printout. Use this information to make an inference about the difference in the population mean seabird densities of oiled and unoiled transects.

 EVOS

Two-Sample T-Test and CI: density, oil

```
Two-sample T for density

oil      N     Mean    StDev   SE Mean
no      36     3.27     6.70     1.1
yes     60     3.50     5.97     0.77

Difference = mu (no ) - mu (yes)
Estimate for difference:  -0.22
95% CI for difference: (-2.93, 2.49)
T-Test of difference = 0 (vs not =): T-Value = -0.16  P-Value = 0.871  DF = 67
```

1.61. Teachers Involve Parents in Schoolwork (TIPS) is an interactive homework process designed to improve the quality of homework assignments for elementary, middle, and high school students. TIPS homework assignments require students to conduct interactions with family partners (parents, guardians, etc.) while completing the homework. Frances Van Voorhis (Johns Hopkins University) conducted a study to investigate the effects of TIPS in science, mathematics, and language arts homework assignments (April, 2001). Each in a sample of 128 middle school students was assigned to complete TIPS homework assignments, while 98 students in a second sample were assigned traditional, non-interactive, homework assignments (called ATIPS). At the end of the study, all students reported on the level of family involvement in their homework on a 4-point scale (0 = never, 1 = rarely, 2 = sometimes, 3 = frequently, 4 = always). Three scores were recorded for each student: one for science homework, one for math homework, and one for language arts homework. The data for the study are saved in the HWSTUDY file. (The first 5 and last 5 observations in the data set are listed in the next table)

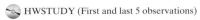 HWSTUDY (First and last 5 observations)

HOMEWORK CONDITION	SCIENCE	MATH	LANGUAGE
ATIPS	1	0	0
ATIPS	0	1	1
ATIPS	0	1	0
ATIPS	1	2	0
ATIPS	1	1	2
TIPS	2	3	2
TIPS	1	4	2
TIPS	2	4	2
TIPS	4	0	3
TIPS	2	0	1

Source: Van Voorhis, F. L., "Teachers' use of interactive homework and its effects on family involvement and science achievement of middle grade students." Paper presented at the annual meeting of the American Educational Research Association, Seattle, April 2001.

a. Conduct an analysis to compare the mean level of family involvement in science homework assignments of TIPS and ATIPS students. Use $\alpha = .05$. Make a practical conclusion.

b. Repeat part **a** for mathematics homework assignments.

c. Repeat part **a** for language arts homework assignments.

d. What assumptions are necessary for the inferences of parts **a–c** to be valid? Are they reasonably satisfied?

1.62. Refer to the Harris Corporation/University of Florida study to determine whether a manufacturing processperformed at a remote location could be established locally, Exercise 1.20 (p. 16). Test devices (pilots) were set up at both the old and new locations, and voltage readings on 30 production runs at each location were obtained. The data are reproduced in the table below. Descriptive statistics are displayed in the SAS printout on p. 63. [*Note*: Larger voltage readings are better than smaller voltage readings.]

VOLTAGE

OLD LOCATION			NEW LOCATION		
9.98	10.12	9.84	9.19	10.01	8.82
10.26	10.05	10.15	9.63	8.82	8.65
10.05	9.80	10.02	10.10	9.43	8.51
10.29	10.15	9.80	9.70	10.03	9.14
10.03	10.00	9.73	10.09	9.85	9.75
8.05	9.87	10.01	9.60	9.27	8.78
10.55	9.55	9.98	10.05	8.83	9.35
10.26	9.95	8.72	10.12	9.39	9.54
9.97	9.70	8.80	9.49	9.48	9.36
9.87	8.72	9.84	9.37	9.64	8.68

Source: Harris Corporation, Melbourne, Fla.

SAS output for Exercise 1.62.

```
--------------------------------- location=NEW ---------------------------------.
                              The MEANS Procedure
                           Analysis Variable : voltage
     N          Mean          Std Dev        Minimum        Maximum
    30       9.4223333       0.4788757      8.5100000     10.1200000


--------------------------------- location=OLD ---------------------------------.
                           Analysis Variable : voltage
     N          Mean          Std Dev        Minimum        Maximum
    30       9.8036667       0.5409155      8.0500000     10.5500000
```

a. Compare the mean voltage readings at the two locations using a 90% confidence interval.

b. Based on the interval, part **a**, does it appear that the manufacturing process can be established locally?

1.63. Does winning an Academy of Motion Picture Arts and Sciences award lead to long-term mortality for movie actors? In an article in the *Annals of Internal Medicine* (May 15, 2001), researchers identified 762 Academy Award winners and matched each one with another actor of the same sex who was in the same winning film and was born in the same era. The life expectancy (age) of each pair of actors was compared.

a. Explain why the data should be analyzed as a paired difference experiment.

b. Set up the null hypothesis for a test to compare the mean life expectancies of Academy Award winners and non-winners.

c. The sample mean life expectancies of Academy Award winners and non-winners were reported as 79.7 years and 75.8 years, respectively. The *p*-value for comparing the two population means was reported as $p = .003$. Interpret this value in the context of the problem.

1.64. Researchers at the University of South Alabama compared the attitudes of male college students toward their fathers with their attitudes toward their mothers (*Journal of Genetic Psychology*, March 1998). Each of a sample of 13 males was asked to complete the following statement about each of their parents: My relationship with my father (mother) can best be described as: (1) Awful, (2) Poor, (3) Average, (4) Good, or (5) Great. The following data were obtained:

FMATTITUDES

STUDENT	ATTITUDE TOWARD FATHER	ATTITUDE TOWARD MOTHER
1	2	3
2	5	5
3	4	3
4	4	5
5	3	4
6	5	4
7	4	5
8	2	4
9	4	5
10	5	4
11	4	5
12	5	4
13	3	3

Source: Adapted from Vitulli, W.F., and Richardson, D.K. "College student's attitudes toward relationships with parents: A five-year comparative analysis." *Journal of Genetic Psychology*, Vol. 159, No. 1, (March 1998), pp. 45–52.

a. Specify the appropriate hypotheses for testing whether male students' attitudes toward their fathers differ from their attitudes toward their mothers, on average.

b. Conduct the test of part **a** at $\alpha = .05$. Interpret the results in the context of the problem.

1.65. Executives of an industrial plant want to determine which of two types of power—gas or electric—will produce more useful energy at the lower cost. One measure of economical energy production, called the plant investment per delivered quad, is calculated by taking the amount of money (in dollars) invested in the particular utility by the plant, and dividing by the delivered amount of energy (in quadrillion British thermal units). The smaller this ratio, the less an industrial plant pays for its delivered energy. Random samples of 11 plants using electric utilities and 16 plants using gas utilities were taken, and the plant investment/quad was calculated for each. The data are listed in the table, followed by a MINITAB printout of the analysis of the data. Do these data provide sufficient evidence at the $\alpha = .05$ level of significance to indicate a difference in the average investment/quad between the plants using gas and those using electric utilities? What assumptions are required for the procedure you used to be valid?

POWER

ELECTRIC				GAS			
204.15	.57	62.76	89.72	.78	16.66	74.94	.01
	.35	85.46	.78	.65	.54	23.59	88.79	.64
44.38	9.28	78.60		.82	91.84	7.20	66.64
				.74	64.67	165.60	.36

```
Two-Sample T-Test and CI: electric, gas

Two-sample T for electric vs gas

            N     Mean   StDev   SE Mean
electric   11     52.4    62.4     19
gas        16     37.7    49.0     12

Difference = mu electric - mu gas
Estimate for difference:  14.7
95% CI for difference: (-29.5, 58.9)
T-Test of difference = 0 (vs not =): T-Value = 0.68  P-Value = 0.500  DF = 25
Both use Pooled StDev = 54.8
```

1.11 Comparing Two Population Variances

Suppose you want to use the two-sample t statistic to compare the mean productivity of two paper mills. However, you are concerned that the assumption of equal variances of the productivity for the two plants may be unrealistic. It would be helpful to have a statistical procedure to check the validity of this assumption.

The common statistical procedure for comparing population variances σ_1^2 and σ_2^2 is to make an inference about the ratio, σ_1^2/σ_2^2, using the ratio of the sample variances, s_1^2/s_2^2. Thus, we will attempt to support the research hypothesis that the ratio σ_1^2/σ_2^2 differs from 1 (i.e., the variances are unequal) by testing the null hypothesis that the ratio equals 1 (i.e., the variances are equal).

$$H_0: \frac{\sigma_1^2}{\sigma_2^2} = 1 \ (\sigma_1^2 = \sigma_2^2)$$

$$H_a: \frac{\sigma_1^2}{\sigma_2^2} \neq 1 \ (\sigma_1^2 \neq \sigma_2^2)$$

We will use the test statistic

$$F = \frac{s_1^2}{s_2^2}$$

To establish a rejection region for the test statistic, we need to know how s_1^2/s_2^2 is distributed in repeated sampling. That is, we need to know the sampling distribution of s_1^2/s_2^2. As you will subsequently see, the sampling distribution of s_1^2/s_2^2 depends on two of the assumptions already required for the t test, as follows:

1. The two sampled populations are normally distributed.

2. The samples are randomly and independently selected from their respective populations.

When these assumptions are satisfied and when the null hypothesis is true (i.e., $\sigma_1^2 = \sigma_2^2$), the sampling distribution of s_1^2/s_2^2 is an **F distribution** with $(n_1 - 1)$ df and $(n_2 - 1)$ df, respectively. The shape of the F distribution depends on the degrees of freedom associated with s_1^2 and s_2^2, i.e., $(n_1 - 1)$ and $(n_2 - 1)$. An F distribution with 7 and 9 df is shown in Figure 1.32. As you can see, the distribution is skewed to the right.

FIGURE 1.32

An F distribution with 7 and 9 df

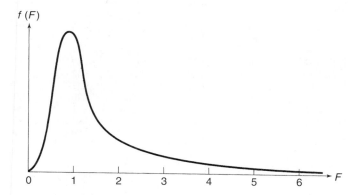

When the population variances are unequal, we expect the ratio of the sample variances, $F = s_1^2/s_2^2$, to be either very large or very small. Therefore, we will need to find F values corresponding to the tail areas of the F distribution to establish the rejection region for our test of hypothesis. The upper-tail F values can be found in Tables 3, 4, 5, and 6 of Appendix C. Table 4 is partially reproduced in Table 1.18. It gives F values that correspond to $\alpha = .05$ upper-tail areas for different degrees of freedom. The columns of the tables correspond to various degrees of freedom for the numerator sample variance s_1^2, whereas the rows correspond to the degrees of freedom for the denominator sample variance s_2^2.

Thus, if the numerator degrees of freedom is 7 and the denominator degrees of freedom is 9, we look in the seventh column and ninth row to find $F_{.05} = 3.29$. As shown in Figure 1.33, $\alpha = .05$ is the tail area to the right of 3.29 in the F distribution with 7 and 9 df. That is, if $\sigma_1^2 = \sigma_2^2$, the probability that the F statistic will exceed 3.29 is $\alpha = .05$.

Suppose we want to compare the variability in production for two paper mills and we have obtained the following results:

SAMPLE 1	SAMPLE 2
$n_1 = 13$ days	$n_2 = 18$ days
$\bar{y}_1 = 26.3$ production units	$\bar{y}_2 = 19.7$ production units
$s_1 = 8.2$ production units	$s_2 = 4.7$ production units

TABLE 1.18   Reproduction of Part of Table 4 of Appendix C: $\alpha = .05$

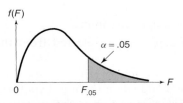

ν_2	NUMERATOR DEGREES OF FREEDOM								
	1	2	3	4	5	6	7	8	9
1	161.4	199.5	215.7	224.6	230.2	234.0	236.8	238.9	240.5
2	18.51	19.00	19.16	19.25	19.30	19.33	19.35	19.37	19.38
3	10.13	9.55	9.28	9.12	9.01	8.94	8.89	8.85	8.81
4	7.71	6.94	6.59	6.39	6.26	6.16	6.09	6.04	6.00
5	6.61	5.79	5.41	5.19	5.05	4.95	4.88	4.82	4.77
6	5.99	5.14	4.76	4.53	4.39	4.28	4.21	4.15	4.10
7	5.59	4.74	4.35	4.12	3.97	3.87	3.79	3.73	3.68
8	5.32	4.46	4.07	3.84	3.69	3.58	3.50	3.44	3.39
9	5.12	4.26	3.86	3.63	3.48	3.37	3.29	3.23	3.18
10	4.96	4.10	3.71	3.48	3.33	3.22	3.14	3.07	3.02
11	4.84	3.98	3.59	3.36	3.20	3.09	3.01	2.95	2.90
12	4.75	3.89	3.49	3.25	3.11	3.00	2.91	2.85	2.80
13	4.67	3.81	3.41	3.18	3.03	2.92	2.83	2.77	2.71
14	4.60	3.74	3.34	3.11	2.96	2.85	2.76	2.70	2.65

The leftmost label column reads ν_1 above ν_2, and the side label reads DENOMINATOR DEGREES OF FREEDOM.

FIGURE 1.33
An F distribution for 7 and 9 df:
$\alpha = .05$

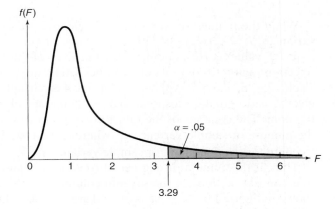

To form the rejection region for a two-tailed F test, we want to make certain that the upper tail is used, because only the upper-tail values of F are shown in Tables 3, 4, 5, and 6. To accomplish this, **we will always place the larger sample variance in the numerator of the F test**. This doubles the tabulated value for α, since we double the probability that the F ratio will fall in the upper tail by always placing the larger sample variance in the numerator. In effect, we make the test two-tailed by putting the larger variance in the numerator rather than establishing rejection regions in both tails.

Thus, for our production example, we have a numerator s_1^2 with df $= n_1 - 1 = 12$ and a denominator s_2^2 with df $= n_2 - 1 = 17$. Therefore, the test statistic will be

$$F = \frac{\text{Larger sample variance}}{\text{Smaller sample variance}} = \frac{s_1^2}{s_2^2}$$

and we will reject $H_0: \sigma_1^2 = \sigma_2^2$ for $\alpha = .10$ if the calculated value of F exceeds the tabulated value:

$$F_{.05} = 2.38 \text{ (see Figure 1.34)}$$

Now, what do the data tell us? We calculate

$$F = \frac{s_1^2}{s_2^2} = \frac{(8.2)^2}{(4.7)^2} = 3.04$$

and compare it to the rejection region shown in Figure 1.34. Since the calculated F value, 3.04, falls in the rejection region, the data provide sufficient evidence to indicate that the population variances differ. Consequently, we would be reluctant to use the two-sample t statistic to compare the population means, since the assumption of equal population variances is apparently untrue.

What would you have concluded if the value of F calculated from the samples had not fallen in the rejection region? Would you conclude that the null hypothesis of equal variances is true? No, because then you risk the possibility of a Type II error (accepting H_0 when H_a is true) without knowing the probability of this error (the probability of accepting $H_0: \sigma_1^2 = \sigma_2^2$ when it is false). Since we will not consider the calculation of β for specific alternatives in this text, when the F statistic does not fall in the rejection region, we simply conclude that **insufficient sample evidence exists to refute the null hypothesis that $\sigma_1^2 = \sigma_2^2$.**

The F test for equal population variances is summarized in the box.

FIGURE 1.34

Rejection region for production example F distribution

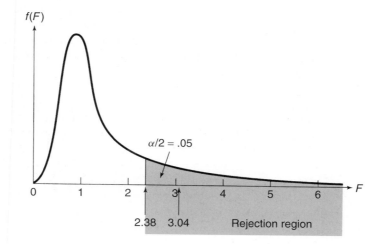

F-Test for Equal Population Variances*

ONE-TAILED TEST

$H_0: \sigma_1^2 = \sigma_2^2$
$H_a: \sigma_1^2 < \sigma_2^2$
  (or $H_a: \sigma_1^2 > \sigma_2^2$)

Test statistic:

$$F = \frac{s_2^2}{s_1^2}$$

$$\left(\text{or } F = \frac{s_1^2}{s_2^2} \text{ when } H_a: \sigma_1^2 > \sigma_2^2 \right)$$

Rejection region:

$F > F_\alpha$

TWO-TAILED TEST

$H_0: \sigma_1^2 = \sigma_2^2$
$H_a: \sigma_1^2 \neq \sigma_2^2$

Test statistic:

$$F = \frac{\text{Larger sample variance}}{\text{Smaller sample variance}}$$

$$= \frac{s_1^2}{s_2^2} \text{ when } s_1^2 > s_2^2$$

$$\left(\text{or } \frac{s_2^2}{s_1^2} \text{ when } s_2^2 > s_1^2 \right)$$

Rejection region:

$F > F_{\alpha/2}$

where F_α and $F_{\alpha/2}$ are based on $\nu_1 =$ numerator degrees of freedom and $\nu_2 =$ denominator degrees of freedom; ν_1 and ν_2 are the degrees of freedom for the numerator and denominator sample variances, respectively.

Assumptions:

1. Both sampled populations are normally distributed.
2. The samples are random and independent.

EXAMPLE   1.18

In Example 1.17 we used the two-sample t statistic to compare the mean reading scores of two groups of slow learners who had been taught to read using two different methods. The data are repeated in Table 1.19 for convenience. The use of the t statistic was based on the assumption that the population variances of the test scores were equal for the two methods. Conduct a test of hypothesis to check this assumption at $\alpha = .10$.

---

*Although a test of a hypothesis of equality of variances is the most common application of the F-test, it can also be used to test a hypothesis that the ratio between the population variances is equal to some specified value, $H_0: \sigma_1^2/\sigma_2^2 = k$. The test is conducted in exactly the same way as specified in the box, except that we use the test statistic

$$F = \left(\frac{s_1^2}{s_2^2} \right) \left(\frac{1}{k} \right)$$

READING

TABLE 1.19 Reading Test Scores for Slow Learners

NEW METHOD				STANDARD METHOD			
80	80	79	81	79	62	70	68
76	66	71	76	73	76	86	73
70	85			72	68	75	66

FIGURE 1.35

SAS F-test for the data in Table 1.19

```
Two Sample Test for Variances of SCORE within METHOD

Sample Statistics

    METHOD
    Group          N       Mean      Std. Dev.    Variance
    -----------------------------------------------------------
    NEW            10       76.4        5.8348      34.04444
    STD            12     72.33333      6.3437      40.24242

Hypothesis Test

    Null hypothesis:      Variance 1 / Variance 2 =  1
    Alternative:          Variance 1 / Variance 2 ^= 1

                  - Degrees of Freedom -
        F           Numer.    Denom.              Pr > F
    -----------------------------------------------------------
       0.85            9        11                0.8148
```

Solution

We want to test

$$H_0: \frac{\sigma_1^2}{\sigma_2^2} = 1 \text{ (i.e., } \sigma_1^2 = \sigma_2^2)$$

$$H_a: \frac{\sigma_1^2}{\sigma_2^2} \neq 1 \text{ (i.e., } \sigma_1^2 \neq \sigma_2^2)$$

The data were entered into SAS, and the SAS printout shown in Figure 1.35 was obtained. Both the test statistic, $F = .85$, and two-tailed p-value, .8148, are highlighted on the printout. Since $\alpha = .10$ is less than the p-value, we do not reject the null hypothesis that the population variances of the reading test scores are equal. ◆

The previous examples demonstrate how to conduct a two-tailed F test when the alternative hypothesis is $H_a: \sigma_1^2 \neq \sigma_2^2$. One-tailed tests for determining whether one population variance is larger than another population variance (i.e., $H_a: \sigma_1^2 > \sigma_2^2$) are conducted similarly. However, the α value no longer needs to be doubled since the area of rejection lies only in the upper (or lower) tail area of the F distribution. The procedure for conducting an upper-tailed F test is outlined in the previous box. Whenever you conduct a one-tailed F test, be sure to write H_a in the form of an upper-tailed test. This can be accomplished by numbering the populations so that the variance hypothesized to be larger in H_a is associated with population 1 and the hypothesized smaller variance is associated with population 2.

Important: As a final comment, we note that (unlike the small-sample t procedure for means) the F test for comparing variances is not very robust against nonnormal data. Consequently, with nonnormal data it is difficult to determine whether a significant F value implies that the population variances differ or is simply due to the fact that the populations are not normally distributed.

1.66. Use Tables 3, 4, 5, and 6 of Appendix C to find F_α for α, numerator df, and denominator df equal to:
  a. .05, 8, 7   **b.** .01, 15, 20
  c. .025, 12, 5   **d.** .01, 5, 25
  e. .10, 5, 10   **f.** .05, 20, 9

1.67. Refer to Exercise 1.59 (p. 61). The "fear of negative evaluation" (FNE) scores for the 11 bulimic females and 14 females with normal eating habits are reproduced in the table. The confidence interval you constructed in Exercise 1.59 requires that the variance of the FNE scores of bulimic females is equal to the variance of the FNE scores of normal females. Conduct a test (at $\alpha = .05$) to determine the validity of this assumption.

⬤ BULIMIA

Bulimic
  students 21 13 10 20 25 19 16 21 24 13 14
Normal
  students 13   6 16 13   8 19 23 18 11 19   7 10 15 20

Source: Randles, R. H. "On neutral responses (zeros) in the sign test and ties in the Wilcoxon-Mann-Whitney test." *The American Statistician*, Vol. 55, No. 2, May 2001 (Figure 3).

1.68. Tests of product quality using human inspectors can lead to serious inspection error problems (*Journal of Quality Technology*, Apr. 1986). To evaluate the performance of inspectors in a new company, a quality manager had a sample of 12 novice inspectors evaluate 200 finished products. The same 200 items were evaluated by 12 experienced inspectors. The quality of each item—whether defective or nondefective—was known to the manager. The next table lists the number of inspection errors (classifying a defective item as nondefective or vice versa) made by each inspector. A SAS printout comparing the two types of inspectors is shown in the next column.
  a. Prior to conducting this experiment, the manager believed the variance in inspection errors was lower for experienced inspectors than for novice inspectors. Do the sample data support her belief? Test using $\alpha = .05$.

⬤ INSPECT

NOVICE INSPECTORS				EXPERIENCED INSPECTORS			
30	35	26	40	31	15	25	19
36	20	45	31	28	17	19	18
33	29	21	48	24	10	20	21

  b. What is the appropriate *p*-value of the test you conducted in part **a**?

```
Two Sample Test for Variances of ERRORS within INSPECT

Sample Statistics

  INSPECT
  Group        N      Mean     Std. Dev.    Variance
  --------------------------------------------------
  1NOVICE     12   32.83333    8.6427       74.69697
  2EXPER      12   20.58333    5.7439       32.99242

Hypothesis Test

  Null hypothesis:        Variance 1 / Variance 2 <= 1
  Alternative:            Variance 1 / Variance 2 >  1

               - Degrees of Freedom -
       F          Numer.     Denom.              Pr > F
  --------------------------------------------------
      2.26          11         11               0.0955
```

1.69. Wet samplers are standard devices used to measure the chemical composition of precipitation. The accuracy of the wet deposition readings, however, may depend on the number of samplers stationed in the field. Experimenters in The Netherlands collected wet deposition measurements using anywhere from one to eight identical wet samplers (*Atmospheric Environment*, Vol. 24A, 1990). For each sampler (or sampler combination), data were collected every 24 hours for an entire year; thus, 365 readings were collected per sampler (or sampler combination). When one wet sampler was used, the standard deviation of the hydrogen readings (measured as percentage relative to the average reading from all eight samplers) was 6.3%. When three wet samplers were used, the standard deviation of the hydrogen readings (measured as percentage relative to the average reading from all eight samplers) was 2.6%. Conduct a test to compare the variation in hydrogen readings for the two sampling schemes (i.e., one wet sampler versus three wet samplers). Test using $\alpha = .05$.

1.70. Following the Persian Gulf War, the Pentagon changed its logistics processes to be more corporate-like. The extravagant "just-in-case" mentality was replaced with "just-in-time" systems. Emulating Federal Express and United Parcel Service, deliveries from factories to foxholes are now expedited using bar codes, laser cards, radio tags, and databases to track supplies. The table contains order-to-delivery times (in days) for a sample of shipments from the U.S. to the Persian Gulf in 1991 and a sample of shipments to Bosnia in 1995.

a. Determine whether the variances in order-to-delivery times for Persian Gulf and Bosnia shipments are equal. Use $\alpha = .05$.

b. Given your answer to part **a**, is it appropriate to construct a confidence interval for the difference between the mean order-to-delivery times? Explain.

⊘ ORDTIMES

PERSIAN GULF (1991)	BOSNIA (1995)
28.0	15.1
20.0	6.4
26.5	5.0
10.6	11.4
9.1	6.5
35.2	6.5
29.1	3.0
41.2	7.0
27.5	5.5

Source: Adapted from Crock, "The Pentagon goes to B-school." *Business Week*, Dec. 11, 1995, p. 98.

Summary

The preceding sections summarize many of the basic concepts and methods presented in an introductory statistics course. We presented the concepts of a **population, random sampling**, and the ultimate objective of most statistical investigations, **making an inference about a population based on information contained in a sample**. Because inference implies description, we first considered methods for describing a set of data—two graphical methods for qualitative data (**bar graph** and **pie chart**), two graphical methods for quantitative data (**relative frequency histogram** and **stem-and-leaf plot**) and **numerical descriptive methods** that provide measures of centrality and variability for a quantitative data set.

We noted that quantities computed from sample data—**statistics**—are used to estimate population numerical descriptive measures—**parameters**—and to make decisions about their values. To evaluate the properties of these statistics, we need to know the probabilities that they will assume specific sets of values in repeated sampling; that is, we need to know their **probability sampling distributions**. If we know the sampling distribution for a statistic, we can make probabilistic statements that measure the **reliability** of the statistic when it is used as an estimator or as the basis of a decision.

Finally, we summarized the basic concepts involved in **interval estimation** and **tests of hypotheses**. In particular, we presented **large- and small-sample confidence intervals** and **statistical tests for making inferences about a single population mean** and for **comparing two population means or variances based on independent random sampling**.

To aid in the formulation of confidence intervals and test statistics, we present two summary tables. Table 1.20 contains a list of parameters and their corresponding estimators and standard errors. Once you have identified the parameter of interest in Table 1.20, use Table 1.21 to formulate confidence intervals and test statistics.

TABLE 1.20   Some Population Parameters and Corresponding Estimators and Standard Errors

PARAMETER (θ)	ESTIMATOR ($\hat{\theta}$)	STANDARD ERROR ($\sigma_{\hat{\theta}}$)	ESTIMATE OF STANDARD ERROR ($S_{\hat{\theta}}$)
μ Mean (average)	$\bar{y}$	$\dfrac{\sigma}{\sqrt{n}}$	$\dfrac{s}{\sqrt{n}}$
$\mu_1 - \mu_2$ Difference between means (averages), independent samples	$\bar{y}_1 - \bar{y}_2$	$\sqrt{\dfrac{\sigma_1^2}{n_1} + \dfrac{\sigma_2^2}{n_2}}$	$\sqrt{\dfrac{s_1^2}{n_1} + \dfrac{s_2^2}{n_2}},\ n_1 \geq 30, n_2 \geq 30$ $\sqrt{s_p^2 \left(\dfrac{1}{n_1} + \dfrac{1}{n_2}\right)}$, either $n_1 < 30$ or $n_2 < 30$ where $s_p^2 = \dfrac{(n_1-1)s_1^2+(n_2-1)s_2^2}{n_1+n_2-2}$
$\mu_d = \mu_1 - \mu_2$, Difference between means, paired samples	$\bar{y}_d$	$\sigma_d/\sqrt{n}$	$s_d/\sqrt{n}$
$\dfrac{\sigma_1^2}{\sigma_2^2}$ Ratio of variances	$\dfrac{s_1^2}{s_2^2}$	(not necessary)	(not necessary)

TABLE 1.21   Formulation of Confidence Intervals for a Population Parameter θ and Test Statistics for $H_0: \theta = \theta_0$, where $\theta = \mu$ or $(\mu_1 - \mu_2)$

SAMPLE SIZE	CONFIDENCE INTERVAL	TEST STATISTIC
Large	$\hat{\theta} \pm z_{\alpha/2}s_{\hat{\theta}}$	$z = \dfrac{\hat{\theta}-\theta_0}{s_{\hat{\theta}}}$
Small	$\hat{\theta} \pm t_{\alpha/2}s_{\hat{\theta}}$	$t = \dfrac{\hat{\theta}-\theta_0}{s_{\hat{\theta}}}$

Note: The test statistic for testing $H_0: \sigma_1^2/\sigma_2^2 = 1$ is $F = s_1^2/s_2^2$ (see the box on page 68).

SUPPLEMENTARY EXERCISES

1.71. For each of the following data sets, compute $\bar{y}$, s, and s^2.
   a. 11, 2, 2, 1, 9   **b.** 22, 9, 21, 15
   c. 1, 0, 1, 10, 11, 11, 0   **d.** 4, 4, 4, 4

1.72. Tchebysheff's theorem states that at least $1 - (1/K^2)$ of a set of measurements will lie within K standard deviations of the mean of the data set. Use Tchebysheff's theorem to find the fraction of a set of measurements that will lie within:
   a. 2 standard deviations of the mean ($K = 2$)
   b. 3 standard deviations of the mean
   c. 1.5 standard deviations of the mean

1.73. Suppose the random variable y has mean $\mu = 30$ and standard deviation $\sigma = 5$. How many standard deviations away from the mean of y is each of the following y values?
   a. $y = 10$ **b.** $y = 32.5$ **c.** $y = 30$ **d.** $y = 60$

1.74. Use Table 1 of Appendix C to find each of the following:
   a. $P(z \geq 2)$ **b.** $P(z \leq -2)$ **c.** $P(z \geq -1.96)$
   d. $P(z \geq 0)$ **e.** $P(z \leq -.5)$ **f.** $P(z \leq -1.96)$

1.75. "Deep hole" drilling is a family of drilling processes used when the ratio of hole depth to hole diameter exceeds 10. Successful deep hole drilling depends on the satisfactory discharge of the drill chip. An experiment was conducted to investigate the performance of deep hole drilling when chip congestion exists (*Journal of Engineering for Industry*, May 1993). Some important variables in the drilling process are described here. Identify the data type for each variable.

a. Chip discharge rate (number of chips discarded per minute)

b. Drilling depth (millimeters)

c. Oil velocity (millimeters per second)

d. Type of drilling (single-edge, BTA, or ejector)

e. Quality of hole surface

1.76. Audiologists have recently developed a rehabilitation program for hearing-impaired patients in a Canadian home for senior citizens (*Journal of the Academy of Rehabilitative Audiology*, 1994). Each of the 30 residents of the home were diagnosed for degree and type of sensorineural hearing loss, coded as follows: 1 = hear within normal limits, 2 = high-frequency hearing loss, 3 = mild loss, 4 = mild-to-moderate loss, 5 = moderate loss, 6 = moderate-to-severe loss, and 7 = severe-to-profound loss. The data are listed in the accompanying table. Use a graph to portray the results. Which type of hearing loss appears to be the most prevalent among nursing home residents?

HEARLOSS

6	7	1	1	2	6	4	6	4	2	5	2	5
1	5	4	6	6	5	5	5	2	5	3	6	4
6	6	4	2									

Source: Jennings, M. B., and Head, B. G. "Development of an ecological audiologic rehabilitation program in a home-for-the-aged." *Journal of the Academy of Rehabilitative Audiology*, Vol. 27, 1994, p. 77 (Table 1).

1.77. *Choice* magazine provides new-book reviews each issue. A random sample of 375 *Choice* book reviews in American history, geography, and area studies was selected and the "overall opinion" of the book stated in each review was ascertained (*Library Acquisitions: Practice and Theory*, Vol. 19, 1995). Overall opinion was coded as follows: 1 = would not recommend, 2 = cautious or very little recommendation, 3 = little or no preference, 4 = favorable/recommended, 5 = outstanding/significant contribution. A summary of the data is provided in the accompanying bar graph.

a. Find the opinion that occurred most often. What proportion of the books reviewed had this opinion?

b. Do you agree with the following statement extracted from the study: "A majority (more than 75%) of books reviewed are evaluated favorably and recommended for purchase."?

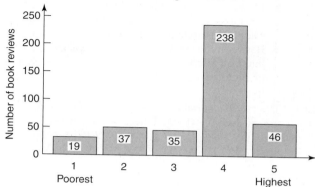

Source: Carlo, P.W., and Natowitz, A. "*Choice* book reviews in American history, geography, and area studies: An analysis for 1988–1993." *Library Acquisitions: Practice & Theory*, Vol. 19, No. 2, 1995, p. 159 (Figure 1).

1.78. Refer to the data on process voltage readings at two locations, Exercises 1.20 and 1.62. Use SAS printout for Exercise 1.62 (p. 63) and the rule of thumb to compare the voltage reading distributions for the two locations.

1.79. Saturn has five satellites that rotate around the planet. *Astronomy* (Aug. 1995) lists 19 different events involving eclipses or occults of Saturnian satellites during the month of August. For each event, the percent of light lost by the eclipsed or occulted satellite at midevent is recorded in the table.

SATURN

DATE	EVENT	LIGHT LOSS (%)
Aug. 2	Eclipse	65
4	Eclipse	61
5	Occult	1
6	Eclipse	56
8	Eclipse	46
8	Occult	2
9	Occult	9
11	Occult	5
12	Occult	39
14	Occult	1
14	Eclipse	100
15	Occult	5
15	Occult	4
16	Occult	13
20	Occult	11
23	Occult	3
23	Occult	20
25	Occult	20
28	Occult	12

Source: *Astronomy* magazine, Aug. 1995, p. 60.

a. Construct a stem-and-leaf display for light loss percentage of the 19 events.

b. Locate on the stem-and-leaf plot, part **a**, the light losses associated with eclipses of Saturnian satellites. (Circle the light losses on the plot.)

c. Based on the marked stem-and-leaf display, part **b**, make an inference about which event type (eclipse or occult) is more likely to lead to a greater light loss.

1.80. The random variable y has a normal distribution with $\mu = 80$ and $\sigma = 10$. Find the following probabilities:

a. $P(y \leq 75)$ **b.** $P(y \geq 90)$ **c.** $P(60 \leq y \leq 70)$

d. $P(y \geq 75)$ **e.** $P(y = 75)$ **f.** $P(y \leq 105)$

1.81. Beginning in 1991, the U.S. Department of Education began taking corrective and punitive actions against colleges and universities with high student-loan default rates. Those schools with default rates above 60% face suspension from the government's massive student-loan program, whereas schools with default rates between 40% and 60% are mandated to reduce their default rates by 5% a year or face a similar penalty (*Tampa Tribune*, June 21, 1989). A list of 66 colleges and universities in Florida with their student-loan default rate is provided in the table. An SPSS printout giving descriptive statistics for the data set is shown on p. 75.

⊙ STUDLOAN

COLLEGE/UNIVERSITY	DEFAULT RATE	COLLEGE/UNIVERSITY	DEFAULT RATE
Florida College of Business	76.2	Brevard CC	9.4
Ft. Lauderdale College	48.5	College of Boca Raton	9.1
Florida Career College	48.3	Florida International Univ.	8.7
United College	46.8	Santa Fe CC	8.6
Florida Memorial College	46.2	Edison CC	8.5
Bethune Cookman College	43.0	Palm Beach Junior College	8.0
Edward Waters College	38.3	Eckerd College	7.9
Florida College of Medical		University of Tampa	7.6
and Dental Careers	32.6	Lakeland College of Business	7.2
International Fine Arts College	26.5	Pensacola Junior College	6.8
Tampa College	23.9	University of Miami	6.7
Miami Technical College	23.3	Florida Institute of Technology	6.7
Tallahassee CC	20.6	University of West Florida	6.3
Charron Williams College	20.2	Palm Beach Atlantic College	6.0
Florida CC	19.1	University of Central Florida	5.7
Miami-Dade CC	19.0	Seminole CC	5.6
Broward CC	18.4	Polk CC	5.6
Daytona Beach CC	16.9	Phillips Junior College	5.6
Lake Sumter CC	16.7	Nova University	5.5
Florida Technical College	16.6	Rollins College	5.5
Florida A&M University	15.8	St. Leo College	5.5
Prospect Hall College	15.1	Gulf Coast CC	5.4
Hillsborough CC	14.4	Southern College	5.3
Pasco-Hernando CC	13.5	Flagler College	4.7
Orlando College	13.5	Florida Atlantic University	4.4
Jones College	13.1	University of South Florida	4.2
Webber College	11.8	Manatee Junior College	4.1
Warner Southern College	11.8	Florida State University	4.0
Central Florida CC	11.8	University of North Florida	3.9
Indian River CC	11.8	Barry University	3.1
St. Petersburg CC	11.3	University of Florida	3.1
Valencia CC	10.8	Stetson University	2.9
Florida Southern College	10.3	Jacksonville University	1.5
Lake City CC	9.8		

SPSS output for Exercise 1.81

Descriptive Statistics

	N	Minimum	Maximum	Mean	Std. Deviation
DEFRATE	66	1.50	76.20	14.6818	14.14121
Valid N (listwise)	66				

a. Locate the mean default rate on the printout.

b. Locate the variance and standard deviation of the default rates on the printout.

c. What proportion of measurements would you expect to find within 2 standard deviations of the mean?

d. Determine the proportion of measurements (default rates) that actually fall within the interval of part **c.** Compare this result with your answer to part **c.**

e. Suppose the college with the highest default rate (Florida College of Business—76.2%) was omitted from the analysis. Would you expect the mean to increase or decrease? Would you expect the standard deviation to increase or decrease?

f. Calculate the mean and standard deviation for the data set with Florida College of Business excluded. Compare these results with your answer to part **e.**

g. Answer parts **c** and **d** using the recalculated mean and standard deviation. This problem illustrates the dramatic effect a single observation can have on an analysis.

1.82. Paleomagnetic studies of Canadian volcanic rock strata known as the Carmacks Group have revealed that the northward displacement of the rock layers has an approximately normal distribution with standard deviation of 500 km (*Canadian Journal of Earth Sciences*, Vol. 27, 1990). One group of researchers estimated the mean displacement at 1,500 km, whereas a second group estimated the mean at 1,200 km.

a. Assuming the mean is 1,500 km, what is the probability of a northward displacement of less than 500 km?

b. Assuming the mean is 1,200 km, what is the probability of a northward displacement of less than 500 km?

c. If, in fact, the northward displacement is less than 500 km, which is the more plausible mean, 1,200 or 1,500 km?

1.83. The physical fitness of a patient is often measured by the patient's maximum oxygen uptake (recorded in milliliters per kilogram, ml/kg). The mean maximum oxygen uptake for cardiac patients who regularly participate in sports or exercise programs was found to be 24.1 with a standard deviation of 6.30 (*Adapted*

Physical Activity Quarterly, Oct. 1997). Assume this distribution is approximately normal.

a. What is the probability that a cardiac patient who regularly participates in sports has a maximum oxygen uptake of at least 20 ml/kg?

b. What is the probability that a cardiac patient who regularly exercises has a maximum oxygen uptake of 10.5 ml/kg or lower?

c. Consider a cardiac patient with a maximum oxygen uptake of 10.5. Is it likely that this patient participates regularly in sports or exercise programs? Explain.

1.84. Interpersonal violence (e.g., rape) generally leads to psychological stress for the victim. *Clinical Psychology Review* (Vol. 15, 1995) reported on the results of all recently published studies of the relationship between interpersonal violence and psychological stress. The distribution of the time elapsed between the violent incident and the initial sign of stress has a mean of 5.1 years and a standard deviation of 6.1 years. Consider a random sample of $n = 150$ victims of interpersonal violence. Let $\bar{y}$ represent the mean time elapsed between the violent act and the first sign of stress for the sampled victims.

a. Give the mean and standard deviation of the sampling distribution of $\bar{y}$.

b. Will the sampling distribution of $\bar{y}$ be approximately normal? Explain.

c. Find $P(\bar{y} > 5.5)$.

d. Find $P(4 < \bar{y} < 5)$.

1.85. *The Australian Journal of Zoology* (Vol. 43, 1995) reported on a study of the diets and water requirements of spinifex pigeons. Sixteen pigeons were captured in the desert and the crop (i.e., stomach) contents of each examined. The accompanying table reports the weight (in grams) of dry seed in the crop of each pigeon. Use the SAS printout on p. 76 to find a 99% confidence interval for the average weight of dry seeds in the crops of spinifex pigeons inhabiting the Western Australian desert. Interpret the result.

PIGEONS

.457	3.751	.238	2.967	2.509	1.384	1.454	.818
.335	1.436	1.603	1.309	.201	.530	2.144	.834

Source: Excerpted from Williams, J.B., Bradshaw, D., and Schmidt, L. "Field metabolism and water requirements of spinifex pigeons (*Geophaps plumifera*) in Western Australia." *Australian Journal of Zoology*, Vol. 43, No. 1, 1995, p. 7 (Table 2).

SAS output for Exercise 1.85

Sample Statistics for SEEDWT

N	Mean	Std. Dev.	Std. Error
16	1.37	1.03	0.26

99 % Confidence Interval for the Mean

Lower Limit: 0.61
Upper Limit: 2.13

1.86. The Occupational Safety and Health Administration (OSHA) conducted a study to evaluate the level of exposure of workers to the dioxin TCDD. The distribution of TCDD levels in parts per trillion (ppt) of production workers at a Newark, New Jersey, chemical plant had a mean of 293 ppt and a standard deviation 847 ppt (*Chemosphere*, Vol. 20, 1990). A graph of the distribution is shown here. In a random sample of $n = 50$ workers selected at the New Jersey plant, let $\bar{y}$ represent the sample mean TCDD level.

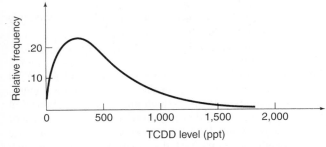

a. Find the mean and standard deviation of the sampling distribution of $\bar{y}$.
b. Draw a sketch of the sampling distribution of $\bar{y}$. Locate the mean on the graph.
c. Find the probability that $\bar{y}$ exceeds 550 ppt.

1.87. Many Vietnam veterans have dangerously high levels of the dioxin TCDD (2,3,7,8-tetrachlorodibenzo-*p*-dioxin) in blood and fat tissue as a result of their exposure to the defoliant Agent Orange. A study published in *Chemosphere* (Vol. 20, 1990) reported on the TCDD levels of 20 Massachusetts Vietnam veterans who were possibly exposed to Agent Orange. The amounts of TCDD (measured in parts per trillion) in blood plasma drawn from each veteran are shown in the table.

⬤ TCDD

VETERAN	TCDD LEVELS IN PLASMA	VETERAN	TCDD LEVELS IN PLASMA
1	2.5	11	6.9
2	3.1	12	3.3
3	2.1	13	4.6
4	3.5	14	1.6
5	3.1	15	7.2
6	1.8	16	1.8
7	6.0	17	20.0
8	3.0	18	2.0
9	36.0	19	2.5
10	4.7	20	4.1

Source: Schecter. A., *et al.* "Partitioning of 2,3,7,8-chlorinated dibenzo-*p*-dioxins and dibenzofurans between adipose tissue and plasma lipid of 20 Massachusetts Vietnam veterans." *Chemosphere*, Vol. 20. Nos. 7–9, 1990, pp. 954–955 (Table 1).

a. Construct a 90% confidence interval for the true mean TCDD level in the plasma of all Vietnam veterans exposed to Agent Orange.
b. Interpret the interval, part a.
c. What assumption is required for the interval estimation procedure to be valid?

1.88. The Cleveland Casting Plant produces iron automotive castings for Ford Motor Company. When the process is stable, the target pouring temperature of the molten iron is 2,550 degrees (*Quality Engineering*, Vol. 7, 1995). The pouring temperatures (in degrees Fahrenheit) for a random sample of 10 crankshafts produced at the plant are listed in the table on p. 77. Conduct a test to determine whether the true mean pouring temperature differs from the target setting. Test using $\alpha = .01$.

IRONTEMP

2,543	2,541	2,544	2,620	2,560	2,559	2,562
2,553	2,552	2,553				

Source: Price, B., & Barth, B. "A structural model relating process inputs and final product characteristics." *Quality Engineering*, Vol. 7, No. 4, 1995, p. 696 (Table 2).

1.89. *Genetical Research* (June 1995) published a study of the mating habits of hermaphroditic snails. The mating habits of the snails were identified as either self-fertilizing or cross-fertilizing. The effective population sizes of the two groups were compared. The data for the study are summarized in the table. Geneticists are interested in comparing the variation in population size of the two types of mating systems. Conduct this analysis for the researcher. Interpret the result.

	EFFECTIVE POPULATION SIZE		
SNAIL MATING SYSTEM	SAMPLE SIZE	MEAN	STANDARD DEVIATION
Cross-fertilizing	17	4,894	1,932
Self-fertilizing	5	4,133	1,890

Source: Jarne, P. "Mating system, bottlenecks, and genetic polymorphism in hermaphroditic animals." *Genetical Research*, Vol. 65, No. 3, June 1995, p. 197 (Table 4). Copyright 1995 Genetical Research, Cambridge University Press.

1.90. One of the most feared predators in the ocean is the great white shark. Although it is known that the great white shark grows to a mean length of 21 feet, a marine biologist believes that great white sharks off the Bermuda coast grow much longer because of unusual feeding habits. To test this claim, researchers plan to capture a number of full-grown great white sharks off the Bermuda coast, measure them, then set them free. However, because capturing sharks is difficult, costly, and very dangerous, only three are sampled. Their lengths are 24, 20, and 22 feet.
   a. Do the data provide sufficient evidence to support the marine biologist's claim? Test at significance level $\alpha = .05$.
   b. What assumptions are required for the hypothesis test of part **a** to be valid? Do you think these assumptions are likely to be satisfied in this particular sampling situation?

1.91. An investigation of ethnic differences in reports of pain perception was presented at the annual meeting of the American Psychosomatic Society (March 2001). A sample of 55 blacks and 159 whites participated in the study. Subjects rated (on a 13-point scale) the intensity and unpleasantness of pain felt when a bag of ice was placed on their foreheads for two minutes. (Higher ratings correspond to higher pain intensity.) A summary of the results is provided in the accompanying table.

	BLACKS	WHITES
Sample size	55	159
Mean pain intensity	8.2	6.9

   a. Why is it dangerous to draw a statistical inference from the summarized data? Explain.
   b. Give values of the missing sample standard deviations that would lead you to conclude (at $\alpha = .05$) that blacks, on average, have a higher pain intensity rating than whites.
   c. Give values of the missing sample standard deviations that would lead you to an inconclusive decision (at $\alpha = .05$) regarding whether blacks or whites have a higher mean intensity rating.

1.92. When searching for an item (e.g., a roadside traffic sign, a lost earring, or a tumor in a mammogram), common sense dictates that you will not re-examine items previously rejected. However, researchers at Harvard Medical School found that a visual search has no memory (*Nature*, Aug. 6, 1998). In their experiment, nine subjects searched for the letter "T" mixed among several letters "L." Each subject conducted the search under two conditions: random and static. In the random condition, the location of the letters were changed every 111 milliseconds; in the static condition, the location of the letters remained unchanged. In each trial, the reaction time (i.e., the amount of time it took the subject to locate the target letter) was recorded in milliseconds.
   a. One goal of the research is to compare the mean reaction times of subjects in the two experimental conditions. Explain why the data should be analyzed as a paired-difference experiment.
   b. If a visual search has no memory, then the main reaction times in the two conditions will not differ. Specify H_0 and H_a for testing the "no memory" theory.
   c. The test statistic was calculated as $t = 1.52$ with p-value $= .15$. Make the appropriate conclusion.

1.93. On average, do mates outperform females in mathematics? To answer this question, psychologists at the University of Minnesota compared the scores of male and female eighth-grade students who took a basic skills mathematics achievement test (*American Educational Research Journal*, Fall 1998). One form of the test consisted of 68 multiple-choice questions. A summary of the test scores is displayed in the table.

	MALES	FEMALES
Sample size	1,764	1,739
Mean	48.9	48.4
Standard deviation	12.96	11.85

Source: Bielinski, J., and Davison, M. L. "Gender differences by item difficulty interactions in multiple-choice mathematics items." *American Educational Research Journal*, Vol. 35, No. 3, Fall 1998, p. 464 (Table 1).

a. Is there evidence of a difference between the true mean mathematics test scores of male and female eighth-graders?

b. Use a 90% confidence interval to estimate the true difference in mean test scores between males and females. Does the confidence interval support the result of the test you conducted in part **a**?

c. What assumptions about the distributions of the populations of test scores are necessary to ensure the validity of the inferences you made in parts **a** and **b**?

d. What is the observed significance level of the test you conducted in part **a**?

e. The researchers hypothesized that the distribution of test scores for males is more variable than the distribution for females. Test this claim at $\alpha = .05$.

1.94. *Scram* is the term used by nuclear engineers to describe a rapid emergency shutdown of a nuclear reactor. The nuclear industry has made a concerted effort to significantly reduce the number of unplanned scrams. The accompanying table gives the number of scrams at each of 56 U.S. nuclear reactor units in a recent year. A MINITAB printout showing both a graphical and numerical description of the data follows.

⊙ SCRAMS

1	0	3	1	4	2	10	6	5	2	0	3	1	5
4	2	7	12	0	3	8	2	0	9	3	3	4	7
2	4	5	3	2	7	13	4	2	3	3	7	0	9
4	3	5	2	7	8	5	2	4	3	4	0	1	7

Descriptive Statistics: SCRAMS

Variable	N	Mean	Median	TrMean	StDev	SE Mean
SCRAMS	56	4.036	3.000	3.820	3.027	0.404

Variable	Minimum	Maximum	Q1	Q3
SCRAMS	0.000	13.000	2.000	5.750

a. Fully interpret the results.

b. Would you expect to observe a nuclear reactor in the future with 11 unplanned scrams? Explain.

c. Suppose the data for nuclear reactors with 12 and 13 scrams were omitted from the analysis. Would you expect $\bar{y}$ to increase or decrease? Would you expect s to increase or decrease?

d. Recalculate $\bar{y}$ and s, excluding the observations 12 and 13. Compare these results with your answer to part **c**.

Stem-and-Leaf Display: SCRAMS

```
Stem-and-leaf of SCRAMS    N  = 56
Leaf Unit = 0.10


     6     0  000000
    10     1  0000
    19     2  000000000
   (10)    3  0000000000
    27     4  00000000
    19     5  00000
    14     6  0
    13     7  000000
     7     8  00
     5     9  00
     3    10  0
     2    11
     2    12  0
     1    13  0
```

1.95. A *homophone* is a word whose pronunciation is the same as that of another word having a different meaning and spelling (e.g., *nun* and *none*, *doe* and *dough*, etc.). *Brain and Language* (Apr. 1995) reported on a study of homophone spelling in patients with Alzheimer's disease. Twenty Alzheimer's patients were asked to spell 24 homophone pairs given in random order, then the number of homophone confusions (e.g., spelling *doe* given the context, *bake bread dough*) was recorded for each patient. One year later, the same test was given to the same patients. The data for the study are provided in the table on p. 79. The researchers posed the following question: "Do Alzheimer's patients show a significant increase in mean homophone confusion errors over time?" Perform an analysis of the data to answer the researchers' question. Use the relevant information in the SAS printout. What assumptions are necessary for the procedure used to be valid? Are they satisfied?

HOMOPHONE

PATIENT	TIME 1	TIME 2
1	5	5
2	1	3
3	0	0
4	1	1
5	0	1
6	2	1
7	5	6
8	1	2
9	0	9
10	5	8
11	7	10
12	0	3
13	3	9
14	5	8
15	7	12
16	10	16
17	5	5
18	6	3
19	9	6
20	11	8

Source: Neils, J., Roeltgen, D. P., and Constantinidou, F. "Decline in homophone spelling associated with loss of semantic influence on spelling in Alzheimer's disease." *Brain and Language*, Vol. 49, No. 1, Apr. 1995, p. 36 (Table 3).

SAS output for Exercise 1.95

```
Two Sample Paired t-test for the Means of TIME1 and TIME2

Sample Statistics

    Group       N      Mean     Std. Dev.    Std. Error
    ------------------------------------------------------
    TIME1       20     4.15      3.4985        0.7823
    TIME2       20     5.8       4.2128        0.942

Hypothesis Test

    Null hypothesis:      Mean of (TIME1 - TIME2) => 0
    Alternative:          Mean of (TIME1 - TIME2) <  0

        t Statistic       Df        Prob > t
    --------------------------------------------
          -2.306          19         0.0163

95% Confidence Interval for the Difference between Two Paired Means

        Lower Limit        Upper Limit
        -----------        -----------
          -3.15              -0.15
```

REFERENCES

FREEDMAN, D., PISANI, R., and PURVES, R. *Statistics*. New York: W. W. Norton and Co., 1978.

MCCLAVE, J. T., and SINCICH, T. *A First Course in Statistics*, 8th ed. Upper Saddle River, N.J.: Prentice Hall, 2002.

MENDENHALL, W., BEAVER, R. J., and BEAVER, B. M. *Introduction to Probability and Statistics*, 10th ed. N. Scituate: Duxbury, 1999.

TANUR, J. M., MOSTELLER, F., KRUSKAL, W. H., LINK, R. F., PIETERS, R. S., and RISING, G. R. (eds.). *Statistics: A Guide to the Unknown*. San Francisco: Holden-Day, 1989.

TUKEY, J. *Exploratory Data Analysis*. Reading, Mass.: Addison-Wesley, 1977.

Introduction to Regression Analysis

CONTENTS

OBJECTIVE

To explain the concept of a statistical model; to describe applications of regression

Many applications of inferential statistics are much more complex than the methods presented in Chapter 1. Often, you will want to use sample data to investigate the relationships among a group of variables, ultimately to create a model for some variable (e.g., IQ, grade point average, etc.) that can be used to predict its value in the future. The process of finding a mathematical model (an equation) that best fits the data is part of a statistical technique known as **regression analysis**.

2.1 Modeling a Response

Suppose the dean of students at a university wants to predict the grade point average (GPA) of all students at the end of their freshman year. One way to do this is to select a random sample of freshmen during the past year, note the GPA y of each, and then use these GPAs to estimate the true mean GPA of all freshmen. The dean could then predict the GPA of each freshman using this estimated GPA.

Predicting the GPA of every freshman by the mean GPA is tantamount to using the mean GPA as a **model** for the true GPA of each freshman enrolled at the university.

In regression, the variable y to be modeled is called the **dependent** (or **response**) **variable** and its true mean (or **expected value**) is denoted $E(y)$. In this example,

$$y = \text{GPA of a student at the end of his or her freshman year}$$

$$E(y) = \text{Mean GPA of all freshmen}$$

Definition 2.1

The variable to be predicted (or modeled), y, is called the **dependent** (or **response**) **variable**.

The dean knows that the actual value of y for a particular student will depend on IQ, SAT score, major, and many other factors. Consequently, the real GPAs for all freshmen may have the distribution shown in Figure 2.1. Thus, the dean is modeling the first-year GPA y for a particular student by stating that y is equal to the mean GPA $E(y)$ of all freshmen plus or minus some random amount, which is unknown to the dean; that is,

$$y = E(y) + \text{Random error}$$

Since the dean does not know the value of the random error for a particular student, one strategy would be to predict the freshman's GPA with the estimate of the mean GPA $E(y)$.

FIGURE 2.1

Distribution of freshman GPAs

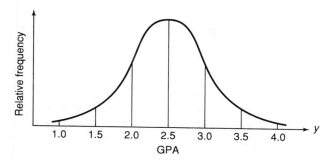

This model is called a **probabilistic model** for y. The adjective *probabilistic* comes from the fact that, when certain assumptions about the model are satisfied, we can make a probability statement about the magnitude of the deviation between y and $E(y)$. For example, if y is normally distributed with mean 2.5 grade points and standard deviation .5 grade point (as shown in Figure 2.1), then the probability that y will fall within 2 standard deviations (i.e., 1 grade point) of its mean is .95. The probabilistic model shown in the box is the foundation of all models considered in this text.

In practice, we will need to use sample data to estimate the parameters of the probabilistic model—namely, the mean $E(y)$ and the random error ε. In Chapter 3, we will learn a standard assumption in regression: the mean error is 0. Based on this assumption, our best estimate of ε is 0. Thus, we need only estimate $E(y)$.

General Form of Probabilistic Model in Regression

$$y = E(y) + \varepsilon$$

where y = Dependent variable
           $E(y)$ = Mean (or expected) value of y
            ε = Unexplainable, or random, error

The simplest method of estimating $E(y)$ is to use the technique of Section 1.8. For example, the dean could select a random sample of freshmen students during the past year and record the GPA y of each. The sample mean $\bar{y}$ could be used as an estimate of the true mean GPA $E(y)$. If we denote the predicted value of y as $\hat{y}$, the prediction equation for the simple model is

$$\hat{y} = \bar{y}$$

Therefore, with this simple model, the sample mean GPA $\bar{y}$ is used to predict the true GPA y of any student at the end of his or her freshman year.

Unfortunately, this simple model does not take into consideration a number of variables, called **independent variables**,* that are highly related to a freshman's GPA. Logically, a more accurate model can be obtained by using the independent variables (e.g., IQ, SAT score, major, etc.) to estimate $E(y)$. The process of finding the mathematical model that relates y to a set of independent variables and best fits the data is part of the process known as **regression analysis**.

Definition 2.2
The variables used to predict (or model) y are called **independent variables** and are denoted by the symbols x_1, x_2, x_3, etc.

For example, suppose the dean decided to relate freshman GPA y to a single independent variable x, defined as the student's SAT score. The dean might select a random sample of freshmen, record y and x for each, and then plot them on a graph as shown in Figure 2.2. Finding the equation of the smooth curve that best fits the data points is part of a regression analysis. Once obtained, this equation (a graph of which is superimposed on the data points in Figure 2.2) provides a model for estimating the mean GPA for freshmen with any specific SAT score. The dean can use the model to predict the GPA of any freshman as long as the SAT score for that freshman is known. As you can see from Figure 2.2, the model would also predict with some error (most of the points do not lie exactly on the curve), but the error of prediction will be much less than the error obtained using the model represented in Figure 2.1. As shown in Figure 2.1, a good estimate of GPA for a freshman student would be a value near the center of the distribution, say, the mean.

---

*The word *independent* should not be interpreted in a probabilistic sense. The phrase *independent variable* is used in regression analysis to refer to a predictor variable for the response y.

Since this prediction does not take SAT score into account, the error of prediction will be larger than the error of prediction for the model of Figure 2.2. Consequently, we would state that the model utilizing information provided by the independent variable, SAT score, is superior to the model represented in Figure 2.1.

FIGURE 2.2

Relating GPA of a freshman to SAT score

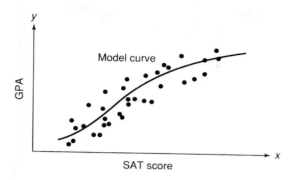

2.2

Overview of Regression Analysis

Regression analysis is a branch of statistical methodology concerned with relating a response y to a set of independent, or predictor, variables $x_1, x_2, \ldots, x_k$. The goal is to build a good model—a prediction equation relating y to the independent variables—that will enable us to predict y for given values of $x_1, x_2, \ldots, x_k$, and to do so with a small error of prediction. When using the model to predict y for a particular set of values of $x_1, x_2, \ldots, x_k$, we will want a measure of the reliability of our prediction. That is, we will want to know how large the error of prediction might be. All these elements are parts of a regression analysis, and the resulting prediction equation is often called a **regression model**.

For example, a property appraiser might like to relate percentage price increase y of residential properties to the two quantitative independent variables x_1, square footage of heated space, and x_2, lot size. This model could be represented by a **response surface** (see Figure 2.3) that traces the mean percentage price increase $E(y)$ for various combinations of x_1 and x_2. To predict the percentage price increase y for a given residential property with $x_1 = 2,000$ square feet of heated space and lot size $x_2 = .7$ acre, you would locate the point $x_1 = 2,000$, $x_2 = .7$ on the x_1, x_2-plane (see Figure 2.3). The height of the surface above that point gives the mean percentage increase in price $E(y)$, and this is a reasonable value to use to predict the percentage price increase for a property with $x_1 = 2,000$ and $x_2 = .7$.

The response surface is a convenient method for modeling a response y that is a function of two quantitative independent variables, x_1 and x_2. The mathematical equivalent of the response surface shown in Figure 2.3 might be given by the deterministic model

$$E(y) = \beta_0 + \beta_1 x_1 + \beta_2 x_2 + \beta_3 x_1 x_2 + \beta_4 x_1^2 + \beta_5 x_2^2$$

where $E(y)$ is the mean percentage price increase for a set of values x_1 and x_2, and $\beta_0, \beta_1, \ldots, \beta_5$ are constants (or weights) with values that would have to be estimated from the sample data. Note that the model for $E(y)$ is deterministic because, if the

FIGURE 2.3

Mean percentage price increase as
a function of heated square
footage, x_1, and lot size, x_2

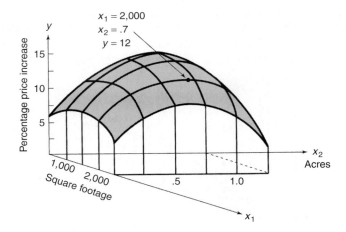

constants $\beta_0, \beta_1, \ldots, \beta_5$ are known, the values of x_1 and x_2 determine exactly the value of $E(y)$.

Replacing $E(y)$ with $\beta_0 + \beta_1 x_1 + \beta_2 x_2 + \beta_3 x_1 x_2 + \beta_4 x_1^2 + \beta_5 x_2^2$ in the probabilistic model for y, we obtain the full equation for y:

$$y = \beta_0 + \beta_1 x_1 + \beta_2 x_2 + \beta_3 x_1 x_2 + \beta_4 x_1^2 + \beta_5 x_2^2 + \varepsilon$$

Now the property appraiser would obtain a sample of residential properties and record square footage, x_1, and lot size, x_2, in addition to percentage increase y in assessed value (see Section 2.4). Subjecting the sample data to a regression analysis will yield estimates of the model parameters and enable the appraiser to predict percentage increase y for a particular property. The prediction equation takes the form

$$\hat{y} = \hat{\beta}_0 + \hat{\beta}_1 x_1 + \hat{\beta}_2 x_2 + \hat{\beta}_3 x_1 x_2 + \hat{\beta}_4 x_1^2 + \hat{\beta}_5 x_2^2$$

where $\hat{y}$ is the predicted value of y, and $\hat{\beta}_0, \hat{\beta}_1, \ldots, \hat{\beta}_5$ are estimates of the model parameters.

In practice, the appraiser would construct a deterministic model for $E(y)$ that takes into account other quantitative variables, as well as qualitative independent variables, such as location and type of construction. In the following chapters, we will show how to construct a model relating a response to both quantitative and qualitative independent variables, and we will fit the model to a set of sample data using a regression analysis.

The preceding description of regression analysis is oversimplified, but it provides a preliminary view of the methodology that is the subject of this text. In addition to predicting y for specific values of $x_1, x_2, \ldots, x_k$, a regression model can be used to estimate the mean value of y for given values of $x_1, x_2, \ldots, x_k$ and to answer other questions concerning the relationship between y and one or more of the independent variables. The practical values attached to these inferences will be illustrated by examples in the following chapters.

We conclude this section with a summary of the major steps involved in a regression analysis.

Regression Modeling: Six-Step Procedure

1. Hypothesize the form of the model for $E(y)$.
2. Collect the sample data.
3. Use the sample data to estimate unknown parameters in the model.
4. Specify the probability distribution of the random error term, and estimate any unknown parameters of this distribution.
5. Statistically check the usefulness of the model.
6. When satisfied that the model is useful, use it for prediction, estimation, and so on.

2.3 Regression Applications

Regression analysis of data is a very powerful statistical tool. It provides a technique for building a statistical predictor of a response and enables you to place a bound (an approximate upper limit) on your error of prediction. For example, suppose you manage a construction company and you would like to predict the profit y per construction job as a function of a set of independent variables $x_1, x_2, \ldots , x_k$. If you could find the right combination of independent variables and could postulate a reasonable mathematical equation to relate y to these variables, you could possibly deduce which of the independent variables were causally related to profit per job and then control these variables to achieve a higher company profit. In addition, you could use the forecasts in corporate planning. The following examples illustrate a few of the many successful applications of regression analysis to real-world problems.

EXAMPLE 2.1

Education The Standardized Admission Test (SAT) scores of 3,492 high school and college students, some of whom paid a private tutor in an effort to obtain a higher score, were analyzed in *Chance* (Winter, 2001). Multiple regression was used to successfully estimate the effect of coaching on the SAT-Mathematics score, y. The independent variables included in the model were scores on PSAT, whether the student was coached, student ethnicity, socioeconomic status, overall high school GPA, number of mathematics courses taken in high school, and overall GPA for the math courses. ◆

EXAMPLE 2.2

Psychology Where do you look when you are listening to someone speak? Researchers used regression to discover that listeners tend to gaze at the eyes or mouth of the speaker. In a study published in *Perception & Psychophysics* (Aug. 1998), subjects watched a videotape of a speaker giving a series of short monologues at a social gathering. The level of background noise (multilingual voices and music) was varied during the listening sessions. The response variable of interest was the proportion y of times the subject's eyes fixated on the speaker's mouth (determined using an infrared corneal detection system). ◆

EXAMPLE 2.3

Engineering A multiple regression model was applied to motor vehicle toxic emissions data collected between 1984 and 1999 in Mexico City (*Environmental Science &*

Engineering, Sept. 1, 2000). The percentage y of motor vehicles without catalytic converters was modeled as a quadratic function of year x. The researchers used the model to estimate "that just after the year 2021 the fleet of cars with catalytic converters will completely disappear." ◆

EXAMPLE 2.4

Management   To reward their executives appropriately, many large corporations receive advice from consulting firms (e.g., Towers, Perrin, Forster, & Crosby) regarding the amount of compensation that each executive should receive. To provide this advice, the consulting firm collects information on the compensation y received by a large number of corporate executives. For each of these executives, the firm records the values of many independent variables, some of which are the following:

1. Experience (years)
2. College education (years)
3. Number of employees supervised
4. Corporate assets (dollars)
5. Age of the executive
6. Whether the executive is on the company's board of directors (1 if yes; 0 if no)
7. Whether the executive has international responsibility (1 if yes; 0 if no)

The consulting firm then uses a regression analysis to build a good prediction equation for y, an executive's annual compensation, as a function of the independent variables listed here. If it is successful, the company can sell its services to both participating and nonparticipating corporations, providing them with reasonable compensation projections for their executives. ◆

EXAMPLE 2.5

Mental Health   The degree to which clients of the Department of Mental Health and Addiction Services in Connecticut adjust to their surrounding community was investigated in the *Community Mental Health Journal* (Aug. 2000). Multiple regression analysis was used to model the dependent variable, community adjustment y (measured quantitatively, where lower scores indicate better adjustment). The model contained a total of 21 independent variables categorized as follows: demographic (four variables), diagnostic (seven variables), treatment (four variables), and community (six variables). ◆

2.4 Collecting the Data for Regression

Recall from Section 2.2 that the initial step in regression analysis is to hypothesize a deterministic model for the mean response, $E(y)$, as a function of one or more independent variables. Once a model for $E(y)$ has been hypothesized, the next step is to collect the sample data that will be used to estimate the unknown model parameters (β's). This entails collecting observations on both the response y and the independent variables, $x_1, x_2, \ldots, x_k$, for each experimental unit in the sample. Thus, a sample to be analyzed by regression includes observations on several variables $(y, x_1, x_2, \ldots, x_k)$, not just a single variable.

The data for regression can be of two types: **observational** or **experimental**. Observational data are obtained if no attempt is made to control the values of the

TABLE 2.1   **Observational Data for Five Executives**

	EXECUTIVE				
	1	2	3	4	5
Annual compensation, y ($)	85,420	61,333	107,500	59,225	98,400
Experience, x_1 (years)	8	2	7	3	11
College education, x_2 (years)	4	8	6	7	2
No. of employees supervised, x_3	13	6	24	9	4
Corporate assets, x_4 (millions)	1.60	0.25	3.14	0.10	2.22
Age, x_5 (years)	42	30	53	36	51
Board of directors, x_6	0	0	1	0	1
International responsibility, x_7	1	0	1	0	0

independent variables (x's). For example, suppose you want to relate an executive's compensation y to the set of predictors listed in Example 2.4. One way to obtain the data for regression is to select a random sample of $n = 100$ executives and record the value of y and the values of each of the predictor variables. The data for the first five executives in the sample are displayed in Table 2.1. Note that in this example, the x values, such as experience, college education, number of employees supervised, etc., for each executive are not specified in advance of observing salary y; that is, the x values were uncontrolled. Therefore, the sample data are observational.

Definition 2.3

If the values of the independent variables (x's) in regression are uncontrolled (i.e., not set in advance before the value of y is observed) but are measured without error, the data are **observational**.

How large a sample should be selected when regression is applied to observational data? In Section 1.8, we learned that when estimating a population mean, the sample size n will depend on (1) the (estimated) population standard deviation, (2) the confidence level, and (3) the desired half-width of the confidence interval used to estimate the mean. Because regression involves estimation of the mean response, $E(y)$, the sample size will depend on these three factors. The problem, however, is not as straightforward as that in Section 1.8, since $E(y)$ is modeled as a function of a set of independent variables, and the additional parameters in the model (i.e., the β's) must also be estimated. In regression, the sample size should be large enough so that the β's are both estimable and testable. This will not occur unless n is at least as large as the number of β parameters included in the model for $E(y)$. To ensure a sufficiently large sample, a good rule of thumb is to select n greater than or equal to 10 times the number of β parameters in the model.

For example, suppose the consulting firm wants to use the following model for annual compensation, y, of a corporate executive:

$$E(y) = \beta_0 + \beta_1 x_1 + \beta_2 x_2 + \cdots + \beta_7 x_7$$

where $x_1, x_2, \ldots, x_7$ are defined in Example 2.2. Excluding β_0, there are seven β parameters in the model; thus, the firm should include at least $10 \times 7 = 70$ corporate executives in its sample.

The second type of data in regression, experimental data, are generated by designed experiments where the values of the independent variables are set in advance (i.e., controlled) before the value of y is observed. For example, if a production supervisor wants to investigate the effect of two quantitative independent variables, say, temperature x_1 and pressure x_2, on the purity of batches of a chemical, the supervisor might decide to employ three values of temperature (100°C, 125°C, and 150°C) and three values of pressure (50, 60, and 70 pounds per square inch) and to produce and measure the impurity y in one batch of chemical for each of the $3 \times 3 = 9$ temperature–pressure combinations (see Table 2.2). For this experiment, the settings of the independent variables are controlled, in contrast to the uncontrolled nature of observational data in the real estate sales example.

Definition 2.4

If the values of the independent variables (x's) in regression are controlled using a designed experiment (i.e., set in advance before the value of y is observed), the data are **experimental**.

TABLE 2.2   **Experimental Data**

TEMPERATURE, x_1	PRESSURE, x_2	IMPURITY, y
100	50	2.7
	60	2.4
	70	2.9
125	50	2.6
	60	3.1
	70	3.0
150	50	1.5
	60	1.9
	70	2.2

In many studies, it is usually not possible to control the values of the x's; consequently, most data collected for regression applications are observational. (Consider the regression analysis in Example 2.2. Clearly, it is impossible or impractical to control the values of the independent variables.) Therefore, you may want to know why we distinguish between the two types of data. We will learn (Chapter 7) that inferences made from regression studies based on observational data have more limitations than those based on experimental data. In particular, we will learn that establishing a cause-and-effect relationship between variables is much more difficult with observational data than with experimental data.

The majority of the examples and exercises in Chapters 3–10 are based on observational data. In Chapters 11–12, we describe regression analyses based on data collected from a designed experiment.

Summary

Psychologists, sociologists, engineers, managers, medical researchers, physicists, chemists, and others strive for a better understanding of the phenomena that affect the variables of interest in their field of study. To achieve this understanding, they seek the assistance of mathematical models that relate the mean value of a **response** (e.g., profit) to various **independent variables** (e.g., advertising budget, size of inventory, etc.). Since even a perfect mathematical description of this relationship will still predict the response with error, a random component is included in the model to account for the many other variables that have been purposely or inadvertently excluded.

The mathematical relationship that we have described forms a model for the relative frequency distribution of the population of response measurements that would be generated when the process is in a specific state. For example, a model might represent a relative frequency distribution of the population of monthly profits that a business might generate, now and in the immediate future, when the business is operating with a \$1,000,000 inventory and a \$500,000 annual advertising budget. Estimating the unknown parameters for this population, i.e., the unknown parameters in the model, and using the model to make predictions with known reliability, is the objective of a **regression analysis**.

The sample data for regression can be either **observational** (in which the values of the x's are uncontrolled) or **experimental** (in which the x's are set in advance of observing y). For practical reasons, most regression applications in business are based on observational data.

Simple Linear Regression

OBJECTIVE

To present the basic concepts of regression analysis based on a simple linear relation between a response y and a single predictor variable x

3.1 Introduction

As noted in Chapter 2, much research is devoted to the topic of **modeling**, i.e., trying to describe how variables are related. For example, a physician might be interested in modeling the relationship between the level of carboxyhemoglobin and the oxygen pressure in the blood of smokers. An advertising agency might want to know the relationship between a firm's sales revenue and the amount spent on advertising. And a psychologist may be interested in relating a child's age to the child's performance on a vocabulary test.

The simplest graphical model for relating a response variable y to a single independent variable x is a straight line. In this chapter, we will discuss **simple linear (straight-line) models**, and we will show how to fit them to a set of data points using the **method of least squares**. We will then show how to judge whether a relationship exists between y and x, and how to use the model either to estimate $E(y)$, the mean value of y, or to predict a future value of y for a given value of x. The totality of these methods is called a **simple linear regression analysis**.

Most models for response variables are much more complicated than implied by a straight-line relationship. Nevertheless, the methods of this chapter are very useful, and they set the stage for the formulation and fitting of more complex models in succeeding chapters. Thus, this chapter will provide an intuitive justification for the techniques employed in a regression analysis, and it will identify most of the types of inferences that we will want to make using a **multiple regression analysis** later in this book.

3.2
The Straight-Line Probabilistic Model

An important consideration in merchandising a product is the amount of money spent on advertising. Suppose you want to model the monthly sales revenue y of an appliance store as a function of the monthly advertising expenditure x. The first question to be answered is this: Do you think an exact (deterministic) relationship exists between these two variables? That is, can the exact value of sales revenue be predicted if the advertising expenditure is specified? We think you will agree that this is not possible for several reasons. Sales depend on many variables other than advertising expenditure—for example, time of year, state of the general economy, inventory, and price structure. However, even if many variables are included in the model (the topic of Chapter 4), it is still unlikely that we can predict the monthly sales *exactly*. There will almost certainly be some variation in sales due strictly to **random phenomena** that cannot be modeled or explained.

Consequently, we need to propose a probabilistic model for sales revenue that accounts for this random variation:

$$y = E(y) + \varepsilon$$

The random error component, ε, represents all unexplained variations in sales caused by important but omitted variables or by unexplainable random phenomena.

As you will subsequently see, the random error ε will play an important role in testing hypotheses or finding confidence intervals for the deterministic portion of the model; it will also enable us to estimate the magnitude of the error of prediction when the model is used to predict some value of y to be observed in the future.

We begin with the simplest of probabilistic models—a **first-order linear model**[*] that graphs as a straight line. The elements of the straight-line model are summarized in the box.

A First-Order (Straight-Line) Model

$y = \beta_0 + \beta_1 x + \varepsilon$

where

$y =$ **Dependent** variable (variable to be modeled—sometimes called the **response** variable)

$x =$ Independent variable (variable used as a **predictor** of y)

$E(y) = \beta_0 + \beta_1 x =$ Deterministic component

$\varepsilon =$ (epsilon) $=$ Random error component

$\beta_0 =$ (beta zero) $=$ **y-intercept** of the line, i.e., point at which the line intercepts or cuts through the y-axis (see Figure 3.1)

$\beta_1 =$ (beta one) $=$ **Slope** of the line, i.e., amount of increase (or decrease) in the mean of y for every 1-unit increase in x (see Figure 3.1)

---

[*]A general definition of the expression *first-order* is given in Section 5.3.

FIGURE 3.1

The straight-line model

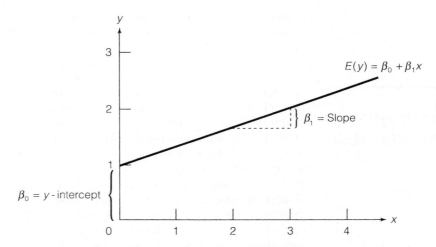

In Section 3.4, we make the standard assumption that the average of the random errors is zero, i.e., $E(\varepsilon) = 0$. Then the deterministic component of the straight-line probabilistic model represents the line of means $E(y) = \beta_0 + \beta_1 x$. Note that we use Greek symbols β_0 and β_1 to represent the y-intercept and slope of the line. They are population parameters with numerical values that will be known only if we have access to the entire population of (x, y) measurements.

Recall from Section 2.2 that it is helpful to think of regression modeling as a six-step procedure:

Steps in Regression Analysis

STEP 1   Hypothesize the form of the model for $E(y)$.

STEP 2   Collect the sample data.

STEP 3   Use the sample data to estimate unknown parameters in the model.

STEP 4   Specify the probability distribution of the random error term, and estimate any unknown parameters of this distribution.

STEP 5   Statistically check the usefulness of the model.

STEP 6   When satisfied that the model is useful, use it for prediction, estimation, and so on.

In this chapter, we will skip step 1 and deal only with the straight-line model. In Chapters 4 and 5, we will discuss how to build more complex models.

EXERCISES

3.1. In each case, graph the line that passes through the points.

  a. $(0, 2)$ and $(2, 6)$ **b.** $(0, 4)$ and $(2, 6)$

  c. $(0, -2)$ and $(-1, -6)$ **d.** $(0, -4)$ and $(3, -7)$

3.2. The equation for a straight line (deterministic) is

$$y = \beta_0 + \beta_1 x$$

If the line passes through the point $(0, 1)$, then $x = 0, y = 1$ must satisfy the equation. That is,

$$1 = \beta_0 + \beta_1(0)$$

Similarly, if the line passes through the point $(2, 3)$, then $x = 2, y = 3$ must satisfy the equation:

$$3 = \beta_0 + \beta_1(2)$$

Use these two equations to solve for β_0 and β_1, and find the equation of the line that passes through the points $(0, 1)$ and $(2, 3)$.

3.3. Find the equations of the lines passing through the four sets of points given in Exercise 3.1.

3.4. Plot the following lines:
   a. $y = 3 + 2x$ **b.** $y = 1 + x$ **c.** $y = -2 + 3x$
   d. $y = 5x$ **e.** $y = 4 - 2x$

3.5. Give the slope and y-intercept for each of the lines defined in Exercise 3.4.

3.3
Fitting the Model: The Method of Least Squares

Suppose an appliance store conducts a 5-month experiment to determine the effect of advertising on sales revenue. The results are shown in Table 3.1. (The number of measurements is small, and the measurements themselves are unrealistically simple to avoid arithmetic confusion in this initial example.) The straight-line model is hypothesized to relate sales revenue y to advertising expenditure x. That is,

$$y = \beta_0 + \beta_1 x + \varepsilon$$

The question is this: How can we best use the information in the sample of five observations in Table 3.1 to estimate the unknown y-intercept β_0 and slope β_1?

ADSALES

TABLE 3.1   **Appliance Store Data**

MONTH	ADVERTISING EXPENDITURE x, hundreds of dollars	SALES REVENUE y, thousands of dollars
1	1	1
2	2	1
3	3	2
4	4	2
5	5	4

To gain some information on the approximate values of these parameters, it is helpful to plot the sample data. Such a graph, called a **scatterplot**, locates each of the five data points on a graph, as in Figure 3.2. Note that the scatterplot suggests a general tendency for y to increase as x increases. If you place a ruler on the scatterplot, you will see that a line may be drawn through three of the five points, as shown in Figure 3.3. To obtain the equation of this visually fitted line, notice that the line intersects the y-axis at $y = -1$, so the y-intercept is -1. Also, y increases exactly 1 unit for every 1-unit increase in x, indicating that the slope is $+1$. Therefore, the equation is

$$\tilde{y} = -1 + 1(x) = -1 + x$$

where $\tilde{y}$ is used to denote the predictor of y based on the visually fitted model.

FIGURE 3.2

Scatterplot for data in Table 3.1

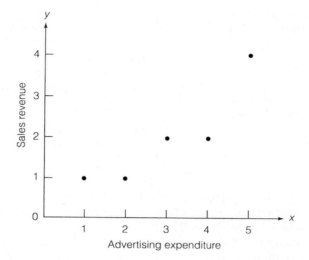

FIGURE 3.3

Visual straight-line fit to data in Table 3.1

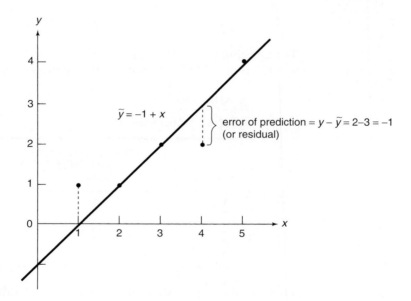

One way to decide quantitatively how well a straight line fits a set of data is to determine the extent to which the data points deviate from the line. For example, to evaluate the visually fitted model in Figure 3.3, we calculate the magnitude of the **deviations**, i.e., the differences between the observed and the predicted values of y. These deviations, or **errors of prediction** are the vertical distances between observed and predicted values of y (see Figure 3.3). The observed and predicted values of y, their differences, and their squared differences are shown in Table 3.2. Note that the **sum of the errors (SE)** equals 0 and the **sum of squares of the errors (SSE)**, which gives greater emphasis to large deviations of the points from the line, is equal to 2.

By shifting the ruler around the graph, we can find many lines for which the sum of the errors is equal to 0, but it can be shown that there is one (and only one) line

TABLE 3.2   Comparing Observed and Predicted Values for the Visual Model

x	y	$\tilde{y} = -1 + x$	$(y - \tilde{y})$	$(y - \tilde{y})^2$
1	1	0	$(1 - 0) = 1$	1
2	1	1	$(1 - 1) = 0$	0
3	2	2	$(2 - 2) = 0$	0
4	2	3	$(2 - 3) = -1$	1
5	4	4	$(4 - 4) = 0$	0
			Sum of errors (SE) $= 0$	Sum of squared errors (SSE) $= 2$

for which the SSE is a *minimum*. This line is called the **least squares line, regression line**, or **least squares prediction equation**.

To find the least squares line for a set of data, assume that we have a sample of n data points that can be identified by corresponding values of x and y, say, $(x_1, y_1), (x_2, y_2), \ldots, (x_n, y_n)$. For example, the $n = 5$ data points shown in Table 3.2 are $(1, 1), (2, 1), (3, 2), (4, 2),$ and $(5, 4)$. The straight-line model for the response y in terms of x is

$$y = \beta_0 + \beta_1 x + \varepsilon$$

The line of means is

$$E(y) = \beta_0 + \beta_1 x$$

and the fitted line, which we hope to find, is represented as

$$\hat{y} = \hat{\beta}_0 + \hat{\beta}_1 x$$

The "hats" can be read as "estimator of." Thus, $\hat{y}$ is an estimator of the mean value of y, $E(y)$, and a predictor of some future value of y; and $\hat{\beta}_0$ and $\hat{\beta}_1$ are estimators of β_0 and β_1, respectively.

For a given data point, say, (x_i, y_i), the observed value of y is y_i and the predicted value of y is obtained by substituting x_i into the prediction equation:

$$\hat{y}_i = \hat{\beta}_0 + \hat{\beta}_1 x_i$$

The deviation of the ith value of y from its predicted value, called the **ith residual**, is

$$(y_i - \hat{y}_i) = [y_i - (\hat{\beta}_0 + \hat{\beta}_1 x_i)]$$

Then the sum of squares of the deviations of the y values about their predicted values (i.e., the **sum of squares of residuals**) for all of the n data points is

$$\text{SSE} = \sum_{i=1}^{n} [y_i - (\hat{\beta}_0 + \hat{\beta}_1 x_i)]^2$$

The quantities $\hat{\beta}_0$ and $\hat{\beta}_1$ that make the SSE a minimum are called the **least squares estimates** of the population parameters β_0 and β_1, and the prediction equation $\hat{y} = \hat{\beta}_0 + \hat{\beta}_1 x$ is called the **least squares line**.

Definition 3.1

The **least squares line** is one that satisfies the following two properties:

1. $\text{SE} = \sum(y_i - \hat{y}_i) = 0$; i.e., the sum of the residuals is 0.
2. $\text{SSE} = \sum(y_i - \hat{y}_i)^2$; i.e., the sum of squared errors, is smaller than for any other straight-line model with $\text{SE} = 0$.

The values of $\hat{\beta}_0$ and $\hat{\beta}_1$ that minimize the SSE are given by the formulas in the box.*

Formulas for the Least Squares Estimates

$$\text{Slope}: \hat{\beta}_1 = \frac{\text{SS}_{xy}}{\text{SS}_{xx}}$$

$$\text{y-intercept}: \hat{\beta}_0 = \overline{y} - \hat{\beta}_1 \overline{x}$$

where

$$\text{SS}_{xy} = \sum_{i=1}^{n}(x_i - \overline{x})(y_i - \overline{y}) = \sum_{i=1}^{n} x_i y_i - n\overline{x}\,\overline{y}$$

$$\text{SS}_{xx} = \sum_{i=1}^{n}(x_i - \overline{x})^2 = \sum_{i=1}^{n} x_i^2 - n(\overline{x})^2$$

$$n = \text{Sample size}$$

Preliminary computations for finding the least squares line for the advertising–sales example are given in Table 3.3. We can now calculate.[†]

$$\text{SS}_{xy} = \sum x_i y_i - n\overline{x}\,\overline{y} = 37 - 5(3)(2) = 37 - 30 = 7$$

$$\text{SS}_{xx} = \sum x_i^2 - n(\overline{x})^2 = 55 - 5(3)^2 = 55 - 45 = 10$$

---

*Students who are familiar with calculus should note that the values of β_0 and β_1 that minimize $\text{SSE} = \sum(y_i - \hat{y}_i)^2$ are obtained by setting the two partial derivatives $\partial\text{SSE}/\partial\beta_0$ and $\partial\text{SSE}/\partial\beta_1$ equal to 0. The solutions to these two equations yield the formulas shown in the box. (The complete derivation is provided in Appendix A.) Furthermore, we denote the *sample* solutions to the equations by $\hat{\beta}_0$ and $\hat{\beta}_1$, whereas the "∧" (hat) denotes that these are sample estimates of the true population intercept β_0 and slope β_1.

[†]Since summations will be used extensively from this point on, we will omit the limits on $\sum$ when the summation includes all the measurements in the sample; i.e., when the summation is $\sum_{i=1}^{n}$, we will write $\sum$.

TABLE 3.3 **Preliminary Computations for the Advertising–Sales Example**

x_i	y_i	x_i^2	$x_i y_i$
1	1	1	1
2	1	4	2
3	2	9	6
4	2	16	8
5	4	25	20
Totals: $\sum x_i = 15$	$\sum y_i = 10$	$\sum x_i^2 = 55$	$\sum x_i y_i = 37$
Means: $\bar{x} = 3$	$\bar{y} = 2$		

Then, the slope of the least squares line is

$$\hat{\beta}_1 = \frac{SS_{xy}}{SS_{xx}} = \frac{7}{10} = .7$$

and the y-intercept is

$$\hat{\beta}_0 = \bar{y} - \hat{\beta}_1 \bar{x}$$
$$= 2 - (.7)(3) = 2 - 2.1 = -.1$$

The least squares line is then

$$\hat{y} = \hat{\beta}_0 + \hat{\beta}_1 x = -.1 + .7x$$

The graph of this line is shown in Figure 3.4.

FIGURE 3.4

Plot of the least squares line $\hat{y} = -.1 + .7x$

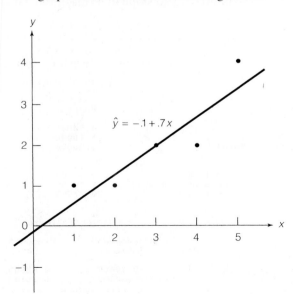

The observed and predicted values of y, the deviations of the y values about their predicted values, and the squares of these deviations are shown in Table 3.4. Note that the sum of squares of the deviations, SSE, is 1.10, and (as we would expect) this is less than the SSE = 2.0 obtained in Table 3.2 for the visually fitted line.

TABLE 3.4   Comparing Observed and Predicted Values for the Least Squares Model

x	y	$\hat{y} = -.1 + .7x$	$(y - \hat{y})$	$(y - \hat{y})^2$
1	1	.6	$(1 - .6) = \quad .4$	.16
2	1	1.3	$(1 - 1.3) = -.3$	.09
3	2	2.0	$(2 - 2.0) = \quad 0$	.00
4	2	2.7	$(2 - 2.7) = -.7$	.49
5	4	3.4	$(4 - 3.4) = \quad .6$	.36
			Sum of errors (SE) = 0	SSE = 1.10

The calculations required to obtain $\hat{\beta}_0$, $\hat{\beta}_1$, and SSE in simple linear regression, although straightforward, can become rather tedious. Even with the use of a calculator, the process is laborious and susceptible to error, especially when the sample size is large. Fortunately, the use of statistical computer software can significantly reduce the labor involved in regression calculations. The SAS, SPSS, and MINITAB outputs for the simple linear regression of the data in Table 3.1 are displayed in Figure 3.5a–c. The values of $\hat{\beta}_0$ and $\hat{\beta}_1$ are highlighted on the printouts. These values, $\hat{\beta}_0 = -.1$ and $\hat{\beta}_1 = .7$, agree exactly with our hand-calculated values. The value of SSE = 1.10 is also highlighted on the printouts.

Whether you use a calculator or a computer, it is important that you be able to interpret the intercept and slope in terms of the data being utilized to fit the model.

FIGURE 3.5

SAS printout for advertising-sales regression

The REG Procedure
Dependent Variable: SALES_Y

Analysis of Variance

Source	DF	Sum of Squares	Mean Square	F Value	Pr > F
Model	1	4.90000	4.90000	13.36	0.0354
Error	3	1.10000	0.36667		
Corrected Total	4	6.00000			

Root MSE	0.60553	R-Square	0.8167	
Dependent Mean	2.00000	Adj R-Sq	0.7556	
Coeff Var	30.27650			

Parameter Estimates

Variable	DF	Parameter Estimate	Standard Error	t Value	Pr > |t|
Intercept	1	-0.10000	0.63509	-0.16	0.8849
ADV_X	1	0.70000	0.19149	3.66	0.0354

FIGURE 3.5a

SPSS printout for advertising-sales regression

Model Summary

Model	R	R Square	Adjusted R Square	Std. Error of the Estimate
1	.904[a]	.817	.756	.606

a. Predictors: (Constant), ADV_X

ANOVA[b]

Model		Sum of Squares	df	Mean Square	F	Sig.
1	Regression	4.900	1	4.900	13.364	.035[a]
	Residual	1.100	3	.367		
	Total	6.000	4			

a. Predictors: (Constant), ADV_X

b. Dependent Variable: SALES_Y

Coefficients[a]

Model		Unstandardized Coefficients		Standardized Coefficients	t	Sig.
		B	Std. Error	Beta		
1	(Constant)	-1.00E-01	.635		-.157	.885
	ADV_X	.700	.191	.904	3.656	.035

a. Dependent Variable: SALES_Y

FIGURE 3.5c

MINITAB printout for advertising-sales regression

```
The regression equation is
SALES_Y = - 0.100 + 0.700 ADV_X

Predictor       Coef      SE Coef        T        P
Constant      -0.1000      0.6351     -0.16    0.885
ADV_X          0.7000      0.1915      3.66    0.035

S = 0.6055      R-Sq = 81.7%     R-Sq(adj) = 75.6%

Analysis of Variance

Source          DF         SS          MS        F        P
Regression       1      4.9000      4.9000    13.36    0.035
Residual Error   3      1.1000      0.3667
Total            4      6.0000
```

In the advertising-sales example, our interpretation of the least squares slope, $\hat{\beta}_1 = .7$, is that the mean of sales revenue y will increase .7 unit for every 1-unit increase in advertising expenditure x. Since y is measured in units of $1,000 and x in units of $100, our interpretation is that mean monthly sales revenue increases $700 for every $100 increase in monthly advertising expenditure. (We will attach a measure of reliability to this inference in Section 3.6.)

The least squares intercept, $\hat{\beta}_0 = -.1$, is our estimate of mean sales revenue y when advertising expenditure is set at $x = \$0$. Since sales revenue can never be negative, why does such a nonsensical result occur? The reason is that we are attempting to use the least squares model to predict y for a value of x ($x = 0$) that is outside the range of the sample data and therefore impractical. (We have more to say about predicting outside the range of the sample data—called **extrapolation**—in Section 3.9.) Consequently, $\hat{\beta}_0$ will not always have a practical interpretation. Only when $x = 0$ is within the range of the x values in the sample and is a practical value will $\hat{\beta}_0$ have a meaningful interpretation.

Even when the interpretations of the estimated parameters are meaningful, we need to remember that they are only estimates based on the sample. As such, their values will typically change in repeated sampling. How much confidence do we have that the estimated slope, $\hat{\beta}_1$, accurately approximates the true slope, β_1? This requires statistical inference, in the form of confidence intervals and tests of hypotheses, which we address in Section 3.6.

To summarize, we have defined the best-fitting straight line to be the one that satisfies the least squares criterion; that is, the sum of the squared errors will be smaller than for any other straight-line model. This line is called the **least squares line**, and its equation is called the **least squares prediction equation**. In subsequent sections, we show how to make statistical inferences about the model.

EXERCISES

3.6. Use the method of least squares to fit a straight line to these six data points:

⊙ EX3_6

x	1	2	3	4	5	6
y	1	2	2	3	5	5

a. What are the least squares estimates of β_0 and β_1?
b. Plot the data points and graph the least squares line on the scatterplot.

3.7. Use the method of least squares to fit a straight line to these five data points:

⊙ EX3_7

x	−2	−1	0	1	2
y	4	3	3	1	−1

a. What are the least squares estimates of β_0 and β_1?
b. Plot the data points and graph the least squares line on the scatterplot.

3.8. Real estate investors, home buyers, and home owners often use the appraised value of a property as a basis for predicting sale price. Data on sale prices and total appraised values of 92 residential properties sold in 1999 in an upscale Tampa, Florida, neighborhood named Tampa Palms are saved in the TAMPALMS file. The first five and last five observations of the data set are listed in the accompanying table.

⊙ TAMPALMS

PROPERTY	APPRAISED VALUE	SALE PRICE
1	$ 170,432	$ 180,000
2	212,827	245,100
3	68,130	85,400
4	65,505	87,900
5	68,655	84,200
⋮	⋮	⋮
88	195,862	244,000
89	176,850	219,000
90	95,718	132,000
91	137,108	156,900
92	183,704	263,000

Source: Hillsborough County (Florida) Property Appraiser's Office.

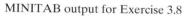

MINITAB output for Exercise 3.8

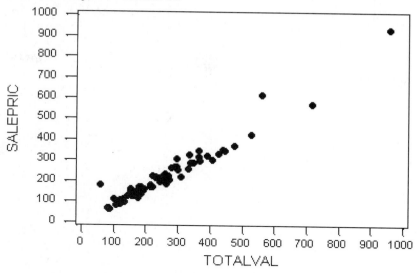

```
The regression equation is
SALEPRIC = - 7.50 + 0.885 TOTALVAL

Predictor         Coef     SE Coef         T         P
Constant        -7.496       6.148     -1.22     0.226
TOTALVAL       0.88455     0.02242     39.45     0.000

S = 29.83        R-Sq = 94.5%      R-Sq(adj) = 94.5%

Analysis of Variance

Source           DF          SS          MS         F         P
Regression        1     1384802     1384802   1556.48     0.000
Residual Error   90       80073         890
Total            91     1464876
```

a. Propose a straight-line model to relate the appraised property value x to the sale price y for residential properties in this neighborhood.

b. A MINITAB scatterplot of the data is shown above. [*Note*: Both sale price and total appraised value are shown in thousands of dollars.] Does it appear that a straight-line model will be an appropriate fit to the data?

c. A MINITAB simple linear regression printout is also shown above. Find the equation of the best-fitting line through the data on the printout.

d. Interpret the y-intercept of the least squares line. Does it have a practical meaning for this application? Explain.

e. Interpret the slope of the least squares line. Over what range of x is the interpretation meaningful?

f. Use the least squares model to estimate the mean sale price of a property appraised at $300,000.

3.9. In Denver, Colorado, environmentalists have discovered a link between high arsenic levels in soil and a crabgrass killer used in the 1950s and 1960s (*Environmental Science & Technology*, Sept. 1 2000). The recent discovery was based, in part, on the scatterplots shown on p. 102. The graphs plot the level of the metals cadmium and arsenic, respectively, against the distance from a former smelter plant for samples of soil taken from Denver residential properties.

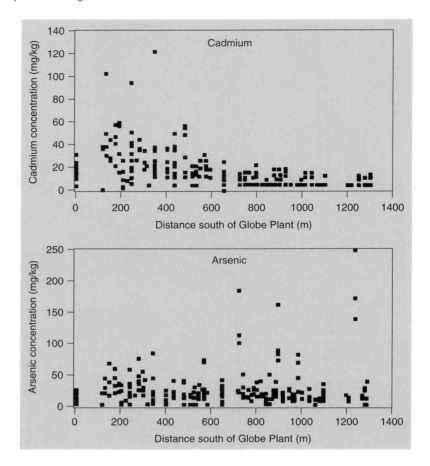

a. Normally, the metal level in soil decreases as distance from the source (e.g., a smelter plant) increases. Propose a straight-line model relating metal level y to distance from the plant x. Based on the theory, would you expect the slope of the line to be positive or negative?

b. Examine the scatterplot for cadmium. Does the plot support the theory, part **a**?

c. Examine the scatterplot for arsenic. Does the plot support the theory, part **a**? (*Note:* This finding led investigators to discover the link between high arsenic levels and the use of the crabgrass killer.)

3.10. In *Chance* (Winter, 2000), statistician Howard Wainer and two students compared men's and women's winning times in the Boston Marathon. One of the graphs used to illustrate gender differences is reproduced on p. 103. The scatterplot graphs the winning times (in minutes) against year in which the race was run. Men's times are represented by solid dots and women's times by open circles.

a. Consider only the winning times for men. Is there evidence of a linear trend? If so, propose a straight-line model for predicting winning time (y) based on year (x). Would you expect the slope of this line to be positive or negative?

b. Repeat part **b** for women's times.

c. Which slope, men's or women's, will be greater in absolute value?

d. Would you recommend using the straight-line models to predict the winning time in the 2020 Boston Marathon? Why or why not?

3.11. Modern warehouses use computerized and automated guided vehicles for materials handling. Consequently, the physical layout of the warehouse must be carefully designed to prevent vehicle congestion and optimize

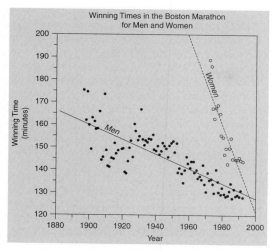

response time. Optimal design of an automated warehouse was studied in the *Journal of Engineering for Industry* (Aug. 1993). The layout assumes that vehicles do not block each other when they travel within the warehouse, i.e., that there is no congestion. The validity of this assumption was checked by simulating (on a computer) warehouse operations. In each simulation, the number of vehicles was varied and the congestion time (total time one vehicle blocked another) was recorded. The data are shown in the accompanying table. Of interest to the researchers is the relationship between congestion time (y) and number of vehicles (x).

WAREHOUSE

NUMBER OF VEHICLES	CONGESTION TIME, minutes	NUMBER OF VEHICLES	CONGESTION TIME, minutes
1	0	9	.02
2	0	10	.04
3	.02	11	.04
4	.01	12	.04
5	.01	13	.03
6	.01	14	.04
7	.03	15	.05
8	.03		

Source: Pandit, R., and U. S. Palekar. "Response time considerations for optimal warehouse layout design." *Journal of Engineering for Industry*, Transactions of the ASME, Vol. 115, Aug. 1993, p. 326 (Table 2).

a. Construct a scatterplot for the data.
b. Find the least squares line relating number of vehicles (x) to congestion time (y).
c. Plot the least squares line on the graph, part **a**.
d. Interpret the values of $\hat{\beta}_0$ and $\hat{\beta}_1$.

3.12. In *Brain and Behavior Evolution* (Apr. 2000), Zoologists conducted a study of the feeding behavior of blackbream fish. The zoologists recorded the number of aggressive strikes of two blackbream fish feeding at the bottom of an aquarium in the 10-minute period following the addition of food. The table listing the weekly number of strikes and age of the fish (in days) is reproduced here.

BLACKBREAM

WEEK	NUMBER OF STRIKES	AGE OF FISH (days)
1	85	120
2	63	136
3	34	150
4	39	155
5	58	162
6	35	169
7	57	178
8	12	184
9	15	190

Source: Shand, J., *et al.* "Variability in the location of the retinal ganglion cell area centralis is correlated with ontogenetic changes in feeding behavior in the Blackbream, Acanthopagrus 'butcher'." *Brain and Behavior*, Vol. 55, No. 4, Apr. 2000 (Figure H).

a. Write the equation of a straight-line model relating number of strikes (y) to age of fish (x).
b. Fit the model to the data using the method of least squares and give the least squares prediction equation.
c. Give a practical interpretation of the value of $\hat{\beta}_0$, if possible.
d. Give a practical interpretation of the value of $\hat{\beta}_1$, if possible.

3.13. The quality of the orange juice produced by a manufacturer (e.g., Minute Maid, Tropicana) is constantly monitored. There are numerous sensory and chemical components that combine to make the best tasting orange juice. For example, one manufacturer has developed a quantitative index of the "sweetness" of orange juice. (The higher the index, the sweeter the juice.) Is there a relationship between the sweetness index and a chemical measure such as the amount of water soluble pectin (parts per million) in the orange juice? Data collected on these two variables for 24 production runs at a juice manufacturing plant are shown in the table on p. 104. Suppose a manufacturer wants to use simple linear regression to predict the sweetness (y) from the amount of pectin (x).

a. Find the least squares line for the data.
b. Interpret $\hat{\beta}_0$ and $\hat{\beta}_1$ in the words of the problem.

c. Predict the sweetness index if amount of pectin in the orange juice is 300 ppm. [*Note*: A measure of reliability of such a prediction is discussed in Section 3.9.]

OJUICE

RUN	SWEETNESS INDEX	PECTIN (*ppm*)
1	5.2	220
2	5.5	227
3	6.0	259
4	5.9	210
5	5.8	224
6	6.0	215
7	5.8	231
8	5.6	268
9	5.6	239
10	5.9	212
11	5.4	410
12	5.6	256
13	5.8	306
14	5.5	259
15	5.3	284
16	5.3	383
17	5.7	271
18	5.5	264
19	5.7	227
20	5.3	263
21	5.9	232
22	5.8	220
23	5.8	246
24	5.9	241

Note: The data in the table are authentic. For confidentiality reasons, the manufacturer cannot be disclosed.

3.14. Two species of predatory birds, collard flycatchers and tits, compete for nest holes during breeding season on the island of Gotland, Sweden. Frequently, dead fly-catchers are found in nest boxes occupied by tits. A field study examined whether the risk of mortality to flycatchers is related to the degree of competition between the two bird species for nest sites (*The Condor*, May 1995). The next table gives data on the number y of flycatchers killed at each of 14 discrete locations (plots) on the island as well as the nest box tit occupancy x (that is, the percentage of nest boxes occupied by tits) at each plot. SAS was used to conduct a simple linear regression analysis for the model, $E(y) = \beta_0 + \beta_1 x$. The printout is shown below.

CONDOR2

PLOT	NUMBER OF FLYCATCHERS KILLED y	NEST BOX TIT OCCUPANCY x (%)
1	0	24
2	0	33
3	0	34
4	0	43
5	0	50
6	1	35
7	1	35
8	1	38
9	1	40
10	2	31
11	2	43
12	3	55
13	4	57
14	5	64

Source: Merila, J., and Wiggins, D. A. "Interspecific competition for nest holes causes adult mortality in the collard flycatcher." *The Condor*, Vol. 97, No. 2, May 1995, p. 449 (Figure 2), Cooper Ornithological Society.

Dependent Variable: NOKILLED

Analysis of Variance

Source	DF	Sum of Squares	Mean Square	F Value	Pr > F
Model	1	19.11669	19.11669	16.03	0.0018
Error	12	14.31188	1.19266		
Corrected Total	13	33.42857			

Root MSE	1.09209	R-Square	0.5719	
Dependent Mean	1.42857	Adj R-Sq	0.5362	
Coeff Var	76.44618			

Parameter Estimates

Variable	DF	Parameter Estimate	Standard Error	t Value	Pr > |t|
Intercept	1	-3.04686	1.15533	-2.64	0.0217
TITPCT	1	0.10766	0.02689	4.00	0.0018

a. Graph the data in a scatterplot. Does the frequency of flycatcher casualties per plot appear to increase linearly with increasing proportion of nest boxes occupied by tits?

b. Find the estimates of β_0 and β_1 in the SAS printout. Interpret their values.

3.15. The *Journal of Experimental Psychology-Applied* (June 2000) published a study in which the "name game" was used to help groups of students learn the names of other students in the group. The "name game" requires the first student in the group to state his/her full name, the second student to say his/her name and the name of the first student, the third student to say his/her name and the names of the first two students, etc. After making their introductions, the students listened to a seminar speaker for 30 minutes. At the end of the seminar, all students were asked to remember the full name of each of the other students in their group and the researchers measured the proportion of names recalled for each. One goal of the study was to investigate the linear trend between y = recall proportion and x = position (order) of the student during the game. The data (simulated based on summary statistics provided in the research article) for 144 students in the first eight positions are saved

in the NAMEGAME2 file. The first five and last five observations in the data set are listed in the table. [*Note*: Since the student in position 1 actually must recall the names of all the other students, he or she is assigned the position number 9 in the data set.] Use the method of least squares to estimate the line, $E(y) = \beta_0 + \beta_1 x$. Interpret the β estimates in the words of the problem.

NAMEGAME2

POSITION	RECALL
2	0.04
2	0.37
2	1.00
2	0.99
2	0.79
⋮	⋮
9	0.72
9	0.88
9	0.46
9	0.54
9	0.99

Source: Morris, P.E., and Fritz, C.O. "The name game: Using retrieval practice to improve the learning of names." *Journal of Experimental Psychology-Applied*, Vol. 6, No. 2, June 2000 (data simulated from Figure 2).

3.4
Model Assumptions

In the advertising–sales example presented in Section 3.3, we assumed that the probabilistic model relating the firm's sales revenue y to advertising dollars x is

$$y = \beta_0 + \beta_1 x + \varepsilon$$

Recall that the least squares estimate of the deterministic component of the model $\beta_0 + \beta_1 x$ is

$$\hat{y} = \hat{\beta}_0 + \hat{\beta}_1 x = -.1 + .7x$$

Now we turn our attention to the random component ε of the probabilistic model and its relation to the errors of estimating β_0 and β_1. In particular, we will see how the probability distribution of ε determines how well the model describes the true relationship between the dependent variable y and the independent variable x.

We will make four basic assumptions about the general form of the probability distribution of ε:

Assumption 1   The mean of the probability distribution of ε is 0. That is, the average of the errors over an infinitely long series of experiments is 0 for each setting of the independent variable x. This assumption implies that the mean value of y, $E(y)$, for a given value of x is $E(y) = \beta_0 + \beta_1 x$.

Assumption 2   The variance of the probability distribution of ε is constant for all settings of the independent variable x. For our straight-line model, this assumption means that the variance of ε is equal to a constant, say, σ^2, for all values of x.

Assumption 3   The probability distribution of ε is normal.

Assumption 4　The errors associated with any two different observations are independent. That is, the error associated with one value of y has no effect on the errors associated with other y values.

The implications of the first three assumptions can be seen in Figure 3.6, which shows distributions of errors for three particular values of x, namely, x_1, x_2, and x_3. Note that the relative frequency distributions of the errors are normal, with a mean of 0, and a constant variance σ^2 (all the distributions shown have the same amount of spread or variability). A point that lies on the straight line shown in Figure 3.6 represents the mean value of y for a given value of x. We will denote this mean value as $E(y)$. Then, the line of means is given by the equation

$$E(y) = \beta_0 + \beta_1 x$$

FIGURE 3.6

The probability distribution of ε

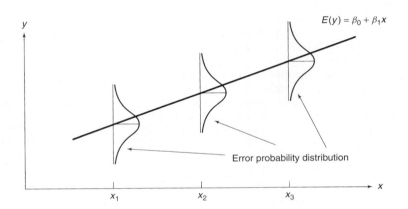

These assumptions make it possible for us to develop measures of reliability for the least squares estimators and to develop hypothesis tests for examining the utility of the least squares line. Various techniques exist for checking the validity of these assumptions, and there are remedies to be applied when the assumptions appear to be invalid. We discuss these techniques in detail in Chapter 8. In actual practice, the assumptions need not hold exactly for least squares estimators and test statistics (to be described subsequently) to possess the measures of reliability that we would expect from a regression analysis. The assumptions will be satisfied adequately for many applications encountered in the real world.

3.5
An Estimator of σ^2

It seems reasonable to assume that the greater the variability of the random error ε (which is measured by its variance σ^2), the greater will be the errors in the estimation of the model parameters β_0 and β_1, and in the error of prediction when $\hat{y}$ is used to predict y for some value of x. Consequently, you should not be surprised, as we proceed through this chapter, to find that σ^2 appears in the formulas for all confidence intervals and test statistics that we use.

In most practical situations, σ^2 will be unknown, and we must use the data to estimate its value. The best (proof omitted) estimate of σ^2 is s^2, which is obtained by dividing the sum of squares of residuals

$$\text{SSE} = \sum (y_i - \hat{y}_i)^2$$

by the number of degrees of freedom (df) associated with this quantity. We use 2 df to estimate the y-intercept and slope in the straight-line model, leaving $(n - 2)$ df for the error variance estimation (see the formulas in the box).

Estimation of σ^2 and σ for the Straight-Line (First-Order) Model

$$s^2 = \frac{\text{SSE}}{\text{Degrees of freedom for error}} = \frac{\text{SSE}}{n - 2}, \quad s = \sqrt{s^2}$$

where

$$\text{SSE} = \sum (y_i - \hat{y}_i)^2$$

$$= \text{SS}_{yy} - \hat{\beta}_1 \text{SS}_{xy} \text{ (calculation formula)}$$

$$\text{SS}_{yy} = \sum (y_i - \overline{y})^2 = \sum y_i^2 - n(\overline{y})^2$$

We refer to s as the **estimated standard error of the regression model**.

Warning: When performing these calculations, you may be tempted to round the calculated values of SS_{yy}, $\hat{\beta}_1$, and SS_{xy}. Be certain to carry at least six significant figures for each of these quantities to avoid substantial errors in the calculation of the SSE.

In the advertising–sales example, we previously calculated SSE = 1.10 for the least squares line $\hat{y} = -.1 + .7x$. Recalling that there were $n = 5$ data points, we have $n - 2 = 5 - 2 = 3$ df for estimating σ^2. Thus,

$$s^2 = \frac{\text{SSE}}{n - 2} = \frac{1.10}{3} = .367$$

is the estimated variance, and

$$s = \sqrt{.367} = .61$$

is the estimated standard deviation of ε.

The values of s^2 and s can also be obtained from a simple linear regression printout. The SAS printout for the advertising-sales example is reproduced in Figure 3.7. The value of s^2 is highlighted on the printout (in the **Mean Square** column in the row labeled **Error**). The value, $s^2 = .36667$, rounded to three decimal places, agrees with the one calculated using the formulas. The value of s is also highlighted in Figure 3.7 (to the right of the heading **Root MSE**). This value, $s = .60553$, agrees (except for rounding) with the calculated value.

You may be able to obtain an intuitive feeling for s by recalling the interpretation given to a standard deviation in Chapter 1 and remembering that the least squares line estimates the mean value of y for a given value of x. Since s measures the spread of the distribution of y values about the least squares line and these errors

FIGURE 3.7

SAS printout for advertising-sales regression

The REG Procedure
Dependent Variable: SALES_Y

Analysis of Variance

Source	DF	Sum of Squares	Mean Square	F Value	Pr > F
Model	1	4.90000	4.90000	13.36	0.0354
Error	3	1.10000	0.36667		
Corrected Total	4	6.00000			

Root MSE	0.60553	R-Square	0.8167
Dependent Mean	2.00000	Adj R-Sq	0.7556
Coeff Var	30.27650		

Parameter Estimates

Variable	DF	Parameter Estimate	Standard Error	t Value	Pr > |t|
Intercept	1	-0.10000	0.63509	-0.16	0.8849
ADV_X	1	0.70000	0.19149	3.66	0.0354

are assumed to be normally distributed, we should not be surprised to find that most (about 95%) of the observations lie within $2s$ or $2(.61) = 1.22$ of the least squares line. In the words of the problem, most of the monthly sales revenue values fall within \$1,220 of their respective predicted values using the least squares line. For this simple example (only five data points), all five monthly sales revenues fall within \$1,220 of the least squares line.

Interpretation of s, the Estimated Standard Deviation of ε

We expect most (approximately 95%) of the observed y values to lie within $2s$ of their respective least squares predicted values, $\hat{y}$.

How can we use the magnitude of s to judge the utility of the least squares prediction equation? Or stated another way, when is the value of s too large for the least squares prediction equation to yield useful predicted y-values? A good approach is to utilize your substantive knowledge of the variables and associated data. In the advertising-sales example, an error of prediction of \$1,220 is probably acceptable if monthly sales revenue values are relatively large (e.g., \$100,000). On the other hand, an error of \$1,220 is undesirable if monthly sales revenues are small (e.g., \$1,000 to \$5,000). A number that will aid in this decision is the **coefficient of variation (CV)**.

Definition 3.2

The **coefficient of variation** is the ratio of the estimated standard deviation of ε to the sample mean of the dependent variable, $\bar{y}$, measured as a percentage:

$$\text{C.V.} = 100(s/\bar{y})$$

The value of CV for the advertising-sales example, highlighted on Figure 3.7, is CV = 30.3. This implies that the value of s for the least squares line is 30% of the value of the sample mean sales revenue, $\bar{y}$. As a rule of thumb, most regression analysts desire regression models with CV values of 10% or smaller (i.e., models with a value of s that is only 10% of the mean of the dependent variable). Models with this characteristic usually lead to accurate predictions. The value of s for the advertising-sales regression is probably too large to consider using the least squares line in practice.

In the remaining sections of this chapter, the value of s will be utilized in tests of model adequacy, in evaluating model parameters, and in providing measures of reliability for future predictions.

EXERCISES

3.16. Suppose you fit a least squares line to nine data points and calculate SSE = .219.
  a. Find s^2, the estimator of the variance σ^2 of the random error term ε.
  b. Calculate s and interpret the result.

3.17. Find SSE, s^2, and s for the least squares lines in the following exercises. Interpret the value of s.
  a. Exercise 3.7   **b.** Exercise 3.11
  c. Exercise 3.13   **d.** Exercise 3.15

3.18. Find SSE, s^2, and s for the least squares lines in the following exercises. Interpret the value of s.
  a. Exercise 3.6   **b.** Exercise 3.8
  c. Exercise 3.10   **d.** Exercise 3.12

3.19. *Statistical Bulletin* (Oct.–Dec. 1999) reported the average hospital charge and the average length of hospital stay for patients undergoing radical prostatectomies in a sample of 12 states. The data are listed in the accompanying table.

◉ HOSPITAL

STATE	AVERAGE HOSPITAL CHARGE ($)	AVERAGE LENGTH OF STAY (days)
Massachusetts	11,680	3.64
New Jersey	11,630	4.20
Pennsylvania	9,850	3.84
Minnesota	9,950	3.11
Indiana	8,490	3.86
Michigan	9,020	3.54
Florida	13,820	4.08
Georgia	8,440	3.57
Tennessee	8,790	3.80
Texas	10,400	3.52
Arizona	12,860	3.77
California	16,740	3.78

Source: Statistical Bulletin, Vol. 80, No. 4, Oct.–Dec. 1999, p. 13.

  a. Construct a scatterplot for the data.
  b. Use the method of least squares to model the relationship between average hospital charge (y) and length of hospital stay (x).
  c. Find the estimated standard error of the regression model and interpret its value in the context of the problem.
  d. For a hospital stay of length $x = 4$ days, find $\hat{y} \pm 2s$.
  e. What fraction of the states in the sample have average hospital charges within $\pm 2s$ of the least squares line?

3.20. A study was conducted to model the thermal performance of integral-fin tubes used in the refrigeration and process industries (*Journal of Heat Transfer,* Aug. 1990). Twenty-four specially manufactured integral-fin tubes with rectangular fins made of copper were used in the experiment. Vapor was released downward into each tube and the vapor-side heat transfer coefficient (based upon the outside surface area of the tube) was measured. The dependent variable for the study is the heat transfer enhancement ratio, y, defined as the ratio of the vapor-side coefficient of the fin tube to the vapor-side coefficient of a smooth tube evaluated at the same temperature. Theoretically, heat transfer will be related to the area at the top of the tube that is "unflooded" by condensation of the vapor. The data in the table (p. 110) are the unflooded area ratio (x) and heat transfer enhancement (y) values recorded for the 24 integral-fin tubes.
  a. Fit a least squares line to the data.
  b. Plot the data and graph the least squares line as a check on your calculations.
  c. Calculate SSE and s^2.
  d. Calculate s and interpret its value.

HEAT

UNFLOODED AREA RATIO, x	HEAT TRANSFER ENHANCEMENT, y	UNFLOODED AREA RATIO, x	HEAT TRANSFER ENHANCEMENT, y
1.93	4.4	2.00	5.2
1.95	5.3	1.77	4.7
1.78	4.5	1.62	4.2
1.64	4.5	2.77	6.0
1.54	3.7	2.47	5.8
1.32	2.8	2.24	5.2
2.12	6.1	1.32	3.5
1.88	4.9	1.26	3.2
1.70	4.9	1.21	2.9
1.58	4.1	2.26	5.3
2.47	7.0	2.04	5.1
2.37	6.7	1.88	4.6

Source: Marto, P. J., et al. "An experimental study of R-113 film condensation on horizontal integral-fin tubes." *Journal of Heat Transfer*, Vol. 112, Aug. 1990, p. 763 (Table 2).

3.6
Assessing the Utility of the Model: Making Inferences About the Slope β_1

Refer to the advertising-sales data of Table 3.1 and suppose that the appliance store's sales revenue is *completely unrelated* to the advertising expenditure. What could be said about the values of β_0 and β_1 in the hypothesized probabilistic model

$$y = \beta_0 + \beta_1 x + \varepsilon$$

if x contributes no information for the prediction of y? The implication is that the mean of y, i.e., the deterministic part of the model $E(y) = \beta_0 + \beta_1 x$, does not change as x changes. Regardless of the value of x, you always predict the same value of y. In the straight-line model, this means that the true slope, β_1, is equal to 0 (see Figure 3.8.). Therefore, to test the null hypothesis that x contributes no information for the prediction of y against the alternative hypothesis that these variables are linearly related with a slope differing from 0, we test

$$H_0: \beta_1 = 0$$

$$H_a: \beta_1 \neq 0$$

If the data support the alternative hypothesis, we will conclude that x does contribute information for the prediction of y using the straight-line model [although the true relationship between $E(y)$ and x could be more complex than a straight line]. Thus, to some extent, this is a test of the utility of the hypothesized model.

FIGURE 3.8

Graphing the model with $\beta_1 = 0: y = \beta_0 + \varepsilon$

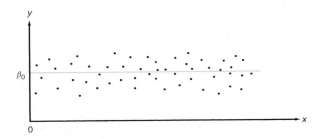

   The appropriate test statistic is found by considering the sampling distribution of $\hat{\beta}_1$, the least squares estimator of the slope β_1.

Sampling Distribution of $\hat{\beta}_1$

If we make the four assumptions about ε (see Section 3.4), then the sampling distribution of $\hat{\beta}_1$, the least squares estimator of the slope, will be a normal distribution with mean β_1 (the true slope) and standard deviation

$$\sigma_{\hat{\beta}_1} = \frac{\sigma}{\sqrt{SS_{xx}}} \quad \text{(See Figure 3.9.)}$$

FIGURE 3.9

Sampling distribution of $\hat{\beta}_1$

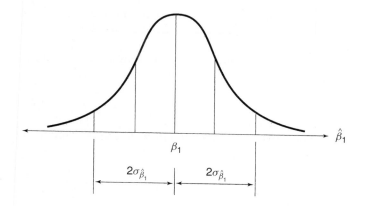

   Since σ will usually be unknown, the appropriate test statistic will generally be a Student's t statistic formed as follows:

$$t = \frac{\hat{\beta}_1 - \text{Hypothesized value of } \beta_1}{s_{\hat{\beta}_1}}$$

$$= \frac{\hat{\beta}_1 - 0}{s/\sqrt{SS_{xx}}}$$

where $s_{\hat{\beta}_1} = \dfrac{s}{\sqrt{SS_{xx}}}$

   Note that we have substituted the estimator s for σ, and then formed $s_{\hat{\beta}_1}$ by dividing s by $\sqrt{SS_{xx}}$. The number of degrees of freedom associated with this t statistic is the same as the number of degrees of freedom associated with s. Recall that this will be $(n-2)$ df when the hypothesized model is a straight line (see Section 3.5).

   The test of the utility of the model is summarized in the next box.

A Test of Model Utility: Simple Linear Regression

ONE-TAILED TEST   TWO-TAILED TEST

$H_0: \beta_1 = 0$ $H_0: \beta_1 = 0$

$H_a: \beta_1 < 0$ $H_a: \beta_1 \neq 0$
  (or $H_a: \beta_1 > 0$)

$$\text{Test statistic}: t = \frac{\hat{\beta}_1}{s_{\hat{\beta}_1}} = \frac{\hat{\beta}_1}{s/\sqrt{\text{SS}_{xx}}}$$

Rejection region: $t < -t_\alpha$ Rejection region: $|t| > t_{\alpha/2}$
  (or $t > t_\alpha$)
where t_α is based on $(n - 2)\,df$ where $t_{\alpha/2}$ is based on $(n - 2)$ df

Assumptions: The four assumptions about ε listed in Section 3.4.

For the advertising–sales example, we will choose $\alpha = .05$ and, since $n = 5$, df $= (n - 2) = 5 - 2 = 3$. Then the rejection region for the two-tailed test is

$$|t| > t_{.025} = 3.182$$

We previously calculated $\hat{\beta}_1 = .7$, $s = .61$, and $\text{SS}_{xx} = 10$. Thus,

$$t = \frac{\hat{\beta}_1}{s/\sqrt{\text{SS}_{xx}}} = \frac{.7}{.61/\sqrt{10}} = \frac{.7}{.19} = 3.7$$

Since this calculated t value falls in the upper-tail rejection region (see Figure 3.10), we reject the null hypothesis and conclude that the slope β_1 is not 0. The sample evidence indicates that advertising expenditure x contributes information for the prediction of sales revenue y using a linear model.

FIGURE 3.10

Rejection region and calculated t value for testing whether the slope $\beta_1 = 0$

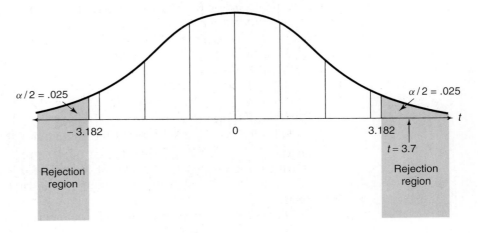

We can reach the same conclusion by using the observed significance level (p-value) of the test obtained from a computer printout. The SAS printout for the advertising–sales example is reproduced in Figure 3.11. The test statistic and two-tailed p-value are highlighted on the printout. Since p-value $= .0354$ is smaller than $\alpha = .05$, we will reject H_0.

FIGURE 3.11

SAS printout for advertising-sales regression

The REG Procedure
Model: MODEL1
Dependent Variable: SALES_Y

Analysis of Variance

Source	DF	Sum of Squares	Mean Square	F Value	Pr > F
Model	1	4.90000	4.90000	13.36	0.0354
Error	3	1.10000	0.36667		
Corrected Total	4	6.00000			

Root MSE	0.60553	R-Square	0.8167	
Dependent Mean	2.00000	Adj R-Sq	0.7556	
Coeff Var	30.27650			

Parameter Estimates

Variable	DF	Parameter Estimate	Standard Error	t Value	Pr > |t|	95% Confidence Limits
Intercept	1	-0.10000	0.63509	-0.16	0.8849	-2.12112 1.92112
ADV_X	1	0.70000	0.19149	3.66	0.0354	0.09061 1.30939

What conclusion can be drawn if the calculated t value does not fall in the rejection region? We know from previous discussions of the philosophy of hypothesis testing that such a t value does *not* lead us to accept the null hypothesis. That is, we do not conclude that $\beta_1 = 0$. Additional data might indicate that β_1 differs from 0, or a more complex relationship may exist between x and y, requiring the fitting of a model other than the straight-line model. We will discuss several such models in Chapter 4.

Another way to make inferences about the slope β_1 is to estimate it using a confidence interval. This interval is formed as shown in the next box.

A $100(1 - \alpha)\%$ Confidence Interval for the Simple Linear Regression Slope β_1

$$\hat{\beta}_1 \pm (t_{\alpha/2})s_{\hat{\beta}_1} \quad \text{where } s_{\hat{\beta}_1} = \frac{s}{\sqrt{\text{SS}_{xx}}}$$

and $t_{\alpha/2}$ is based on $(n - 2)$ df

For the advertising–sales example, a 95% confidence interval for the slope β_1 is

$$\hat{\beta}_1 \pm (t_{.025})s_{\hat{\beta}_1} = .7 \pm (3.182)\left(\frac{s}{\sqrt{\text{SS}_{xx}}}\right)$$

$$= .7 \pm (3.182)\left(\frac{.61}{\sqrt{10}}\right) = .7 \pm .61 = (.09, 1.31)$$

This 95% confidence interval for the slope parameter β_1 is also shown (highlighted) at the bottom of the SAS printout, Figure 3.11.

Remembering that y is recorded in units of \$1,000 and x in units of \$100, we can say, with 95% confidence, that the mean monthly sales revenue will increase between \$90 and \$1,310 for every \$100 increase in monthly advertising expenditure.

Since all the values in this interval are positive, it appears that β_1 is positive and that the mean of y, $E(y)$, increases as x increases. However, the rather large width of the confidence interval reflects the small number of data points (and, consequently, a lack of information) in the experiment. We would expect a narrower interval if the sample size were increased.

EXERCISES

3.21. Do the data provide sufficient evidence to indicate that β_1 differs from 0 for the least squares analyses in the following exercises? Use $\alpha = .05$.
   a. Exercise 3.6   **b.** Exercise 3.7

3.22. Refer to the data on sale prices and total appraised values of 92 residential properties in an upscale Tampa, Florida, neighborhood, Exercise 3.8 (p. 100). A SAS simple linear regression printout for the analysis is given at the bottom of the page.
   a. Use the printout to determine whether there is a positive linear relationship between appraised property value x and sale price y for residential properties sold in this neighborhood. That is, determine if there is sufficient evidence (at $\alpha = .01$) to indicate that β_1, the slope of the straight-line model, is positive.

   b. Find a 95% confidence interval for the slope, β_1, on the printout. Interpret the result practically.

   c. What can be done to obtain a narrower confidence interval in part **b**?

3.23. Refer to Exercise 3.13 (p. 103) and the simple linear regression relating the sweetness index (y) of an orange juice sample to the amount of water soluble pectin (x) in the juice. Find a 90% confidence interval for the true slope of the line. Interpret the result.

3.24. Refer to Exercise 3.14 (p. 104) and the simple linear regression relating number of fly-catchers killed y to nest box tit occupancy x.
   a. Refer to the SAS printout (p. 104) and test whether y is positively linearly related to x. Use $\alpha = .01$.

SAS Simple Linear Regression Output for Exercise 3.22

Dependent Variable: SALEPRIC

Analysis of Variance

Source	DF	Sum of Squares	Mean Square	F Value	Pr > F
Model	1	1384802	1384802	1556.48	<.0001
Error	90	80073	889.70234		
Corrected Total	91	1464876			

Root MSE		29.82788	R-Square	0.9453	
Dependent Mean		201.75036	Adj R-Sq	0.9447	
Coeff Var		14.78455			

Parameter Estimates

Variable	DF	Parameter Estimate	Standard Error	t Value	Pr > |t|	95% Confidence Limits	
Intercept	1	-7.49603	6.14824	-1.22	0.2259	-19.71058	4.71853
TOTALVAL	1	0.88455	0.02242	39.45	<.0001	0.84000	0.92909

b. Construct a 99% confidence interval for β_1. Practically interpret the result.

3.25. How do eye and head movements relate to body movements when reacting to a visual stimulus? Scientists at the California Institute of Technology designed an experiment to answer this question and reported their results in *Nature* (Aug. 1998). Adult male rhesus monkeys were exposed to a visual stimulus (i.e., a panel of light-emitting diodes) and their eye, head, and body movements were electronically recorded. In one variation of the experiment, two variables were measured: active head movement (x, percent per degree) and body plus head rotation (y, percent per degree). The data for $n = 39$ trials were subjected to a simple linear regression analysis, with the following results: $\hat{\beta}_1 = .88$, $s_{\hat{\beta}_1} = .14$

a. Conduct a test to determine whether the two variables, active head movement x and body plus head rotation y are positively linearly related. Use $\alpha = .05$.

b. Construct and interpret a 90% confidence interval for β_1.

c. The scientists want to know if the true slope of the line differs significantly from 1. Based on your answer to part **b**, make the appropriate inference.

3.26. Refer to the *Journal of Heat Transfer* study of the straight-line relationship between heat transfer enhancement (y) and unflooded area ratio (x), Exercise 3.20 (p. 109). Construct a 95% confidence interval for β_1, the slope of the line. Interpret the result.

3.27. The *British Journal of Sports Medicine* (Apr. 2000) published *a* study of the effect of massage on boxing performance. Two variables measured on the boxers were blood lactate concentration (mM) and the boxer's perceived recovery (28-point scale). Based on information provided in the article, the data in the next table were obtained for 16 five-round boxing performances, where a massage was given to the boxer between rounds. Conduct a test to determine whether blood lactate level (y) is linearly related to perceived recovery (x). Use $\alpha = .10$.

3.28. The U.S. Department of Agriculture has developed and adopted the Universal Soil Loss Equation (USLE) for predicting water erosion of soils. In geographic areas where runoff from melting snow is common, calculating the USLE requires an accurate estimate of snowmelt runoff erosion. An article in the *Journal of Soil and Water Conservation* (Mar.–Apr. 1995) used simple linear regression to develop a snowmelt erosion

⬤ BOXING2

BLOOD LACTATE LEVEL	PERCEIVED RECOVERY
3.8	7
4.2	7
4.8	11
4.1	12
5.0	12
5.3	12
4.2	13
2.4	17
3.7	17
5.3	17
5.8	18
6.0	18
5.9	21
6.3	21
5.5	20
6.5	24

Source: Hemmings, B., Smith, M., Graydon, J., and Dyson, R. "Effects of massage on physiological restoration, perceived recovery, and repeated sports performance." *British Journal of Sports Medicine*, Vol. 34, No. 2, Apr. 2000 (data adapted from Figure 3).

index. Data for 54 climatological stations in Canada were used to model the McCool winter-adjusted rainfall erosivity index, y, as a straight-line function of the once-in-5-year snowmelt runoff amount, x (measured in millimeters).

a. The data points are plotted in the graph shown on p. 116. Is there visual evidence of a linear trend?

b. The data for seven stations were removed from the analysis due to lack of snowfall during the study period. Why is this strategy advisable?

c. The simple linear regression on the remaining $n = 47$ data points yielded the following results: $\hat{y} = -6.72 + 1.39x$, $s_{\hat{\beta}_1} = .06$. Use this information to construct a 90% confidence interval for β_1.

d. Interpret the interval, part **c**.

⬤ NAMEGAME2

3.29. Refer to the *Journal of Experimental Psychology-Applied* (June 2000) name retrieval study, Exercise 3.15 (p. 105). Recall that the goal of the study was to investigate the linear trend between proportion of names recalled (y) and position (order) of the student (x) during the "name game." Is there sufficient evidence (at $\alpha = .01$) of a linear trend? Answer the question by analyzing the data for 144 students saved in the NAMEGAME2 file.

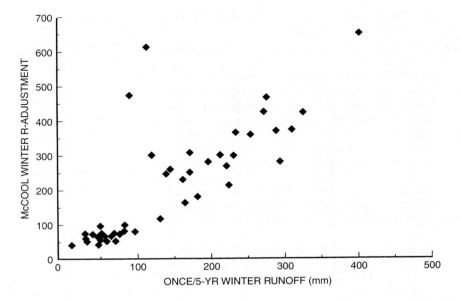

3.30. One of the most difficult tasks of developing and managing a global portfolio is assessing the risks of potential foreign investments. Duke University researcher C. R. Henry collaborated with two First Chicago Investment Management Company directors to examine the use of country credit ratings as a means of evaluating foreign investments (*Journal of Portfolio Management*, Winter 1995). To be effective, such a measure should help explain and predict the volatility of the foreign market in question. The researchers analyzed data on annualized risk (y) and average credit rating (x) for the 40 countries shown in the table on p. 117. A MINITAB printout for a simple linear regression analysis conducted on the data is shown below.

a. Locate the least squares estimates of β_0 and β_1 on the printout.
b. Graph the data in a scatterplot, then sketch the least squares line on the graph.
c. Do the data provide sufficient evidence to conclude that credit rating (x) contributes information for the prediction of annualized risk (y)?
d. Use the plot, part **b**, to locate any unusual data points (outliers).
e. Eliminate the outlier(s), part **d**, from the data set and rerun the simple linear regression analysis. Note any dramatic changes in the results.

MINITAB Output for Exercise 3.30

```
The regression equation is
RISK = 57.8 - 0.400 RATING

Predictor      Coef     SE Coef         T        P
Constant     57.755       6.128      9.43    0.000
RATING      -0.39961     0.09152     -4.37    0.000

S = 12.68      R-Sq = 33.4%     R-Sq(adj) = 31.7%

Analysis of Variance

Source          DF          SS          MS        F        P
Regression       1      3064.4      3064.4    19.07    0.000
Residual Error  38      6107.5       160.7
Total           39      9171.9
```

GLOBRISK

COUNTRY	ANNUALIZED RISK (%)	AVERAGE CREDIT RATING
Argentina	87.0	31.8
Australia	26.9	78.2
Austria	26.3	83.8
Belgium	22.0	78.4
Brazil	64.8	36.2
Canada	19.2	87.1
Chile	31.6	38.6
Colombia	31.5	44.4
Denmark	20.6	72.6
Finland	26.1	76.0
France	23.8	85.3
Germany	23.0	93.4
Greece	39.6	51.9
Hong Kong	34.3	69.6
India	30.0	46.6
Ireland	23.4	66.4
Italy	28.0	75.5
Japan	25.7	94.5
Jordan	17.6	33.6
Korea	30.7	62.2
Malaysia	26.7	64.4
Mexico	46.3	43.3

(Continued)

COUNTRY	ANNUALIZED RISK (%)	AVERAGE CREDIT RATING
Netherlands	18.5	87.6
New Zealand	26.3	68.9
Nigeria	41.4	30.6
Norway	28.3	83.0
Pakistan	24.4	26.4
Philippines	38.4	29.6
Portugal	47.5	56.7
Singapore	26.4	77.6
Spain	24.8	70.8
Sweden	24.5	79.5
Switzerland	19.6	94.7
Taiwan	53.7	72.9
Thailand	27.0	55.8
Turkey	74.1	32.6
United Kingdom	21.8	87.6
United States	15.4	93.4
Venezuela	46.0	45.0
Zimbabwe	35.6	24.5

Source: Erb, C. B., Harvey, C. R., and Viskanta, T. E. "Country risk and global equity selection." *Journal of Portfolio Management*, Vol. 21, No. 2, Winter 1995, p. 76.

3.7
The Coefficient of Correlation

The claim is often made that the crime rate and the unemployment rate are "highly correlated." Another popular belief is that IQ and academic performance are "correlated." Some people even believe that the Dow Jones Industrial Average and the lengths of fashionable skirts are "correlated." Thus, the term *correlation* implies a relationship or "association" between two variables.

The **Pearson product moment correlation coefficient r**, defined in the box, provides a quantitative measure of the strength of the linear relationship between x and y, just as does the least squares slope $\hat{\beta}_1$. However, unlike the slope, the correlation coefficient r is *scaleless*. The value of r is always between -1 and $+1$, regardless of the units of measurement used for the variables x and y.

Definition 3.3

The **Pearson product moment coefficient of correlation** r is a measure of the strength of the *linear* relationship between two variables x and y. It is computed (for a sample of n measurements on x and y) as follows:

$$r = \frac{SS_{xy}}{\sqrt{SS_{xx} \, SS_{yy}}}$$

Note that r is computed using the same quantities used in fitting the least squares line. Since both r and $\hat{\beta}_1$ provide information about the utility of the model, it is not surprising that there is a similarity in their computational formulas. In particular, note that SS_{xy} appears in the numerators of both expressions and, since both denominators are always positive, r and $\hat{\beta}_1$ will always be of the same sign (either both positive or both negative).

A value of r near or equal to 0 implies little or no linear relationship between y and x. In contrast, the closer r is to 1 or -1, the stronger the linear relationship between y and x. And, if $r = 1$ or $r = -1$, all the points fall exactly on the least squares line. Positive values of r imply that y increases as x increases; negative values imply that y decreases as x increases. Each of these situations is portrayed in Figure 3.12.

FIGURE 3.12
Values of r and their implications

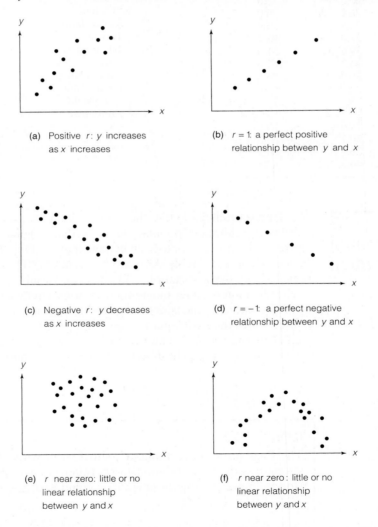

(a) Positive r: y increases as x increases

(b) $r = 1$: a perfect positive relationship between y and x

(c) Negative r: y decreases as x increases

(d) $r = -1$: a perfect negative relationship between y and x

(e) r near zero: little or no linear relationship between y and x

(f) r near zero: little or no linear relationship between y and x

We demonstrate how to calculate the coefficient of correlation r using the data in Table 3.1 for the advertising-sales example. The quantities needed to calculate r are SS_{xy}, SS_{xx}, and SS_{yy}. The first two quantities have been calculated previously and are repeated here for convenience:

$$SS_{xy} = 7, SS_{xx} = 10, SS_{yy} = \sum y^2 - \frac{\left(\sum y\right)^2}{n}$$

$$= 26 - \frac{(10)^2}{5} = 26 - 20 = 6$$

We now find the coefficient of correlation:

$$r = \frac{SS_{xy}}{\sqrt{SS_{xx}SS_{yy}}} = \frac{7}{\sqrt{(10)(6)}} = \frac{7}{\sqrt{60}} = .904$$

The fact that r is positive and near 1 in value indicates that monthly sales revenue y tends to increase as advertising expenditures x increases—*for this sample of five months.* This is the same conclusion we reached when we found the calculated value of the least squares slope to be positive.

EXAMPLE 3.1

Legalized gambling is available on several riverboat casinos operated by a city in Mississippi. The mayor of the city wants to know the correlation between the number of casino employees and yearly crime rate. The records for the past 10 years are examined, and the results listed in Table 3.5 are obtained. Find and interpret the coefficient of correlation r for the data.

 CASINO

TABLE 3.5   **Data on Casino Employees and Crime Rate, Example 3.1**

YEAR	NUMBER OF CASINO EMPLOYEES x (thousands)	CRIME RATE y (number of crimes per 1,000 population)
1993	15	1.35
1994	18	1.63
1995	24	2.33
1996	22	2.41
1997	25	2.63
1998	29	2.93
1999	30	3.41
2000	32	3.26
2001	35	3.63
2002	38	4.15

Solution

Rather than use the computing formula given in Definition 3.3, we resort to a statistical software package. The data of Table 3.1 were entered into a computer and MINITAB was used to compute r. The MINITAB printout is shown in Figure 3.13

The coefficient of correlation, highlighted on the printout, is $r = .987$. Thus, the size of the casino workforce and crime rate in this city are very highly correlated—at least over the past 10 years. The implication is that a strong positive linear relationship exists between these variables (see Figure 3.14). We must be careful, however, not to jump to any unwarranted conclusions. For instance, the mayor may be tempted to conclude that hiring more casino workers next year will increase the crime rate—that is, that there is a *causal relationship* between the two

FIGURE 3.13

MINITAB correlation printout for Example 3.1

Correlations: EMPLOY, CRIME

Pearson correlation of EMPLOY and CRIME = 0.987
P-Value = 0.000

FIGURE 3.14

MINITAB scatterplot for Example 3.1

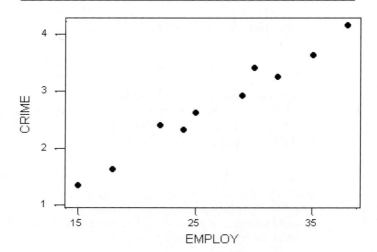

variables. However, high correlation does not imply causality. The fact is, many things have probably contributed both to the increase in the casino workforce and to the increase in crime rate. The city's tourist trade has undoubtedly grown since legalizing riverboat casinos and it is likely that the casinos have expanded both in services offered and in number. *We cannot infer a causal relationship on the basis of high sample correlation. When a high correlation is observed in the sample data, the only safe conclusion is that a linear trend may exist between x and y.* Another variable, such as the increase in tourism, may be the underlying cause of the high correlation between *x* and *y*. ◆

Warning

High correlation does *not* imply causality. If a large positive or negative value of the sample correlation coefficient *r* is observed, it is incorrect to conclude that a change in *x* causes a change in *y*. The only valid conclusion is that *a linear trend may exist* between *x* and *y*.

Keep in mind that the correlation coefficient *r* measures the correlation between *x* values and *y* values in the sample, and that a similar linear coefficient of correlation exists for the population from which the data points were selected. The **population correlation coefficient** is denoted by the symbol ρ (rho). As you might expect, ρ is estimated by the corresponding sample statistic, *r*. Or, rather than estimating ρ, we might want to test

$$H_0: \rho = 0$$

against

$$H_a: \rho \neq 0$$

That is, we might want to test the hypothesis that x contributes no information for the prediction of y, using the straight-line model against the alternative that the two variables are at least linearly related. However, we have already performed this identical test in Section 3.6 when we tested $H_0: \beta_1 = 0$ against $H_a: \beta_1 \neq 0$.

It can be shown (proof omitted) that $r = \hat{\beta}_1 \sqrt{SS_{xx}/SS_{yy}}$. Thus, $\hat{\beta}_1 = 0$ implies $r = 0$, and vice versa. Consequently, the null hypothesis $H_0: \rho = 0$ is equivalent to the hypothesis $H_0: \beta_1 = 0$. When we tested the null hypothesis $H_0: \beta_1 = 0$ in connection with the previous example, the data led to a rejection of the null hypothesis for $\alpha = .05$. This implies that the null hypothesis of a zero linear correlation between the two variables, crime rate and number of employees, can also be rejected at $\alpha = .05$. The only real difference between the least squares slope $\hat{\beta}_1$ and the coefficient of correlation r is the measurement scale.* Therefore, the information they provide about the utility of the least squares model is to some extent redundant. Furthermore, the slope $\hat{\beta}_1$ gives us additional information on the amount of increase (or decrease) in y for every 1-unit increase in x. For this reason, we recommend using the slope to make inferences about the existence of a positive or negative linear relationship between two variables.

For those who prefer to test for a linear relationship between two variables using the coefficient of correlation r, we outline the procedure in the following box.

Test of Hypothesis for Linear Correlation

ONE-TAILED TEST   TWO-TAILED TEST

$H_0: \rho = 0$ $H_0: \rho = 0$

$H_a: \rho > 0$ $H_a: \rho \neq 0$
  (or $H_a: \rho < 0$)

$$\text{Test statistic}: t = \frac{r\sqrt{n-2}}{\sqrt{1-r^2}}$$

Rejection region: $t > t_\alpha$ Rejection region: $|t| > t_{\alpha/2}$
  (or $t < -t_\alpha$)

where the distribution of t depends on $(n-2)$ df.

Assumption: The sample of (x, y) values is randomly selected from a normal population.

The next example illustrates how the correlation coefficient r may be a misleading measure of the strength of the association between x and y in situations where the true relationship is nonlinear.

---

*The estimated slope, $\hat{\beta}_1$, is measured in the same units as y. However, the correlation coefficient r is independent of scale.

EXAMPLE   3.2

Underinflated or overinflated tires can increase tire wear and decrease gas mileage. A manufacturer of a new tire tested the tire for wear at different pressures with the results shown in Table 3.6. Calculate the coefficient of correlation r for the data. Interpret the result.

 TIRES

TABLE 3.6   **Data for Example 3.2**

PRESSURE	MILEAGE
x, pounds per sq. inch	y, thousands
30	29.5
30	30.2
31	32.1
31	34.5
32	36.3
32	35.0
33	38.2
33	37.6
34	37.7
34	36.1
35	33.6
35	34.2
36	26.8
36	27.4

Solution

Again, we use a computer to find the value of r. An SPSS printout of the correlation analysis is shown in Figure 3.15. The value of r, shaded on the printout, is $r = -.114$. This relatively small value for r describes a weak linear relationship between pressure (x) and mileage (y).

The p-value for testing $H_0: p = 0$ against $H_a: e \neq 0$, $p = .699$, is also shaded on the SPSS printout. This value indicates that there is no evidence of a linear correlation in the population at $\alpha = .05$. The manufacturer, however, would be remiss in concluding that tire pressure has little or no impact on wear of the tire. On the contrary, the relationship between pressure and wear is fairly strong, as the SPSS scatterplot in Figure 3.16 illustrates. Note that the relationship is not linear, but curvilinear; the underinflated tires (low pressure values) and overinflated tires (high pressure values) *both* lead to low mileage.   ◆

A statistic related to the coefficient of correlation is defined and discussed in the next section.

FIGURE 3.15

SPSS correlation analysis of tire data

Correlations

		PRESSURE	MILEAGE
PRESSURE	Pearson Correlation	1	-.114
	Sig. (2-tailed)	.	.699
	N	14	14
MILEAGE	Pearson Correlation	-.114	1
	Sig. (2-tailed)	.699	.
	N	14	14

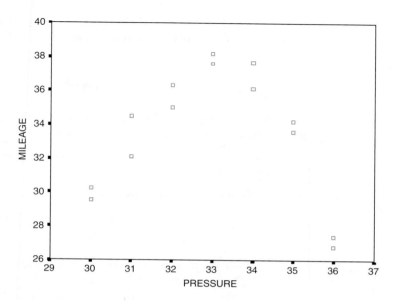

FIGURE 3.16

SPSS scatterplot of tire data

3.8

The Coefficient of Determination

Another way to measure the utility of the regression model is to quantify the contribution of x in predicting y. To do this, we compute how much the errors of prediction of y were reduced by using the information provided by x.

To illustrate, suppose a sample of data produces the scatterplot shown in Figure 3.17a. If we assume that x contributes no information for the prediction of y, the best prediction for a value of y is the sample mean $\overline{y}$, which graphs as the horizontal line shown in Figure 3.17b. The vertical line segments in Figure 3.17b are the deviations of the points about the mean $\overline{y}$. Note that the sum of squares of deviations for the model $\hat{y} = \overline{y}$ is

$$SS_{yy} = \sum (y_i - \overline{y})^2$$

Now suppose you fit a least squares line to the same set of data and locate the deviations of the points about the line as shown in Figure 3.17c. Compare the deviations about the prediction lines in Figure 3.17b and 3.17c. You can see that:

1. If x contributes little or no information for the prediction of y, the sums of squares of deviations for the two lines

$$SS_{yy} = \sum (y_i - \overline{y})^2 \quad \text{and} \quad SSE = \sum (y_i - \hat{y}_i)^2$$

will be nearly equal.

2. If x does contribute information for the prediction of y, then SSE will be smaller than SS_{yy}. In fact, if all the points fall on the least squares line, then SSE $= 0$.

A convenient way of measuring how well the least squares equation $\hat{y} = \hat{\beta}_0 + \hat{\beta}_1 x$ performs as a predictor of y is to compute the reduction in the sum of squares of deviations that can be attributed to x, expressed as a proportion of SS_{yy}. This quantity, called the **coefficient of determination** (and denoted), is

$$r^2 = \frac{SS_{yy} - SSE}{SS_{yy}}$$

FIGURE 3.17

A comparison of the sum of squares of deviations for two models

(a) Scatterplot of data

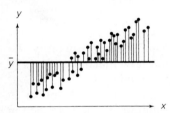

(b) Assumption: x contributes no
     information for predicting y;
     $\hat{y} = \bar{y}$

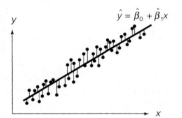

(c) Assumption: x contributes
     information for predicting y;
     $\hat{y} = \hat{\beta}_0 + \hat{\beta}_1 x$

In simple linear regression, it can be shown that this quantity is equal to the square of the simple linear coefficient of correlation r.

Definition 3.4

The **coefficient of determination** is

$$r^2 = \frac{SS_{yy} - SSE}{SS_{yy}} = 1 - \frac{SSE}{SS_{yy}}$$

It represents the proportion of the sum of squares of deviations of the y values about their mean that can be attributed to a linear relationship between y and x. (In simple linear regression, it may also be computed as the square of the coefficient of correlation r.)

Note that r^2 is always between 0 and 1, because $SSE \leq SS_{yy}$. Thus, an r^2 of .60 means that the sum of squares of deviations of the y values about their predicted values has been reduced 60% by using the least squares equation $\hat{y}$, instead of $\bar{y}$, to predict y.

A more practical interpretation of r^2 is derived as follows. If we let SS_{yy} represent the "total sample variability" of the y-values around the mean, and let SSE represent the "unexplained sample variability" after fitting the least squares line, $\hat{y}$, then $(SS_{yy} - SSE)$ is the "explained sample variability" of the y-values attributable to the linear relationship with x. Therefore, a verbal description of r^2 is:

$$r^2 = \frac{(SS_{yy} - SSE)}{SS_{yy}} = \frac{\text{Explained sample variability}}{\text{Total sample variability}}$$

= Proportion of total sample variability of the y-values explained by the linear relationship between y and x.

Practical Interpretation of the Coefficient of Determination, r^2

About $100(r^2)\%$ of the sample variation in y (measured by the total sum of squares of deviations of the sample y values about their mean $\bar{y}$) can be explained by (or attributed to) using x to predict y in the straight-line model.

EXAMPLE 3.3

Calculate the coefficient of determination for the advertising-sales example. The data are repeated in Table 3.7.

TABLE 3.7

ADVERTISING EXPENDITURE x, hundreds of dollars	SALES REVENUE y, thousands of dollars
1	1
2	1
3	2
4	2
5	4

Solution

We first calculate

$$SS_{yy} = \sum y_i^2 - n\bar{y}^2 = 26 - 5(2)^2$$
$$= 26 - 20 = 6$$

From previous calculations,

$$SSE = \sum (y_i - \hat{y}_i)^2 = 1.10$$

Then, the coefficient of determination is given by

$$r^2 = \frac{SS_{yy} - SSE}{SS_{yy}} = \frac{6.0 - 1.1}{6.0} = \frac{4.9}{6.0}$$
$$= .817$$

◆

This value is also shown (highlighted) on the SPSS printout, Figure 3.18. Our interpretation is: About 82% of the sample variation in sales revenues values can be "explained" by using monthly advertising expenditure x to predict sales revenue y with the least squares line

$$\hat{y} = -.1 + .7x$$

FIGURE 3.18

Portion of SPSS printout for advertising-sales regression

Model Summary

Model	R	R Square	Adjusted R Square	Std. Error of the Estimate
1	.904[a]	.817	.756	.606

a. Predictors: (Constant), ADV_X

In situations where a straight-line regression model is found to be a statistically adequate predictor of y, the value of r^2 can help guide the regression analyst in the search for better, more useful models. For example, design engineers used a simple linear model to relate cost of mechanical work in construction (heating, ventilating, and plumbing) to floor area. Based on the data associated with 26 factory and warehouse buildings, the least squares prediction equation given in Figure 3.19 was found. It was concluded that floor area and mechanical cost are linearly related, since the t statistic (for testing $H_0: \beta_1 = 0$) was found to equal 3.61, which is significant with an α as small as .002.* Thus, floor area should be useful when predicting the mechanical cost of a factory or warehouse. However, the value of the coefficient of determination r^2 was found to be .35. This tells us that only 35% of the variation among mechanical costs is accounted for by the differences in floor areas. This relatively small r^2 value led the engineers to include other independent variables (e.g., volume, amount of glass) in the model to account for a significant portion of the remaining 65% of the variation in mechanical cost not explained by floor area. In the next chapter, we discuss this important aspect of relating a response to more than one independent variable.

FIGURE 3.19

Simple linear model relating cost to floor area

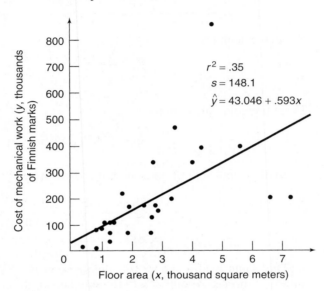

EXERCISES

3.31. Find the correlation coefficient and the coefficient of determination for the sample data of each of the following exercises. Interpret your results.
  a. Exercise 3.6   **b.** Exercise 3.7

3.32. Describe the slope of the least squares line if
  a. $r = .7$ **b.** $r = -.7$
  c. $r = 0$ **d.** $r^2 = .64$

3.33. Do you believe that the grade point average of a college student is correlated with the student's intelligence quotient (IQ)? If so, will the correlation be positive or negative? Explain.

3.34. Research by law enforcement agencies has shown that the crime rate is correlated with the U.S. population.

Would you expect the correlation to be positive or negative? Explain.

3.35. Give an example of two variables in your field of study that are
  a. positively correlated   **b.** negatively correlated

⊙ TAMPALMS

3.36. Refer to the data on sale prices and total appraised values of 92 residential properties recently sold in an upscale Tampa, Florida, neighborhood, Exercise 3.8 (p. 100). The MINITAB simple linear regression printout relating sale price (y) to appraised property value (x) is reproduced at the top of page 127, followed by a MINITAB correlation printout.

---

*Crandall, J. S., and Cedercreutz, M. "Preliminary cost estimates for mechanical work." *Building Systems Design*, Oct.–Nov. 1976, Vol. 73, pp. 35–51.

MINITAB Simple Linear Regression Output for Exercise 3.36

```
The regression equation is
SALEPRIC = - 7.50 + 0.885 TOTALVAL

Predictor          Coef      SE Coef          T          P
Constant         -7.496        6.148      -1.22      0.226
TOTALVAL        0.88455      0.02242      39.45      0.000

S = 29.83         R-Sq = 94.5%      R-Sq(adj) = 94.5%

Analysis of Variance

Source            DF           SS          MS          F          P
Regression         1      1384802     1384802    1556.48      0.000
Residual Error    90        80073         890
Total             91      1464876
```

Correlations: SALEPRIC, TOTALVAL

```
Pearson correlation of SALEPRIC and TOTALVAL = 0.972
P-Value = 0.000
```

a. Find the coefficient of correlation between appraised property value and sale price on the printout. Interpret this value.

b. Find the coefficient of determination between appraised property value and sale price on the printout. Interpret this value.

3.37. *Perception & Psychophysics* (July 1998) reported on a study of how people view the 3-dimensional objects projected onto a rotating 2-dimensional image. Each in a sample of 25 university students viewed various depth-rotated objects (e.g., hairbrush, duck, shoe) until they recognized the object. The recognition exposure time—that is, the minimum time (in milliseconds) required for the subject to recognize the object—was recorded for each. In addition, each subject rated the "goodness of view" of the object on a numerical scale, where lower scale values correspond to better views. The next table gives the correlation coefficient, r, between recognition exposure time and goodness of view for several different rotated objects,

OBJECT	r	t
Piano	.447	2.40
Bench	−.057	.27
Motorbike	.619	3.78
Armchair	.294	1.47
Teapot	.949	14.50

a. Interpret the value of r for each object.

b. Calculate and interpret the value of r^2 for each object.

c. The table also includes the t-value for testing the null hypothesis of no correlation (i.e., for testing $H_0: \beta_1 = 0$). Interpret these results.

3.38. Many high school students experience "math anxiety." Does such an attitude carry over to learning computer skills? A researcher at Duquesne University investigated this question and published her results in *Educational Technology* (May–June 1995). A sample of high school students—902 boys and 828 girls—from public schools in Pittsburgh, Pennsylvania, participated in the study. Using 5-point Likert scales, where 1 = "strongly disagree" and 5 = "strongly agree," the researcher measured the students' interest and confidence in both mathematics and computers.

a. For boys, math confidence and computer interest were correlated at $r = .14$. Fully interpret this result.

b. For girls, math confidence and computer interest were correlated at $r = .33$. Fully interpret this result.

3.39. In cotherapy two or more therapists lead a group. An article in the *American Journal of Dance Therapy* (Spring/Summer 1995) examined the use of cotherapy in dance/movement therapy. Two of several variables measured on each of a sample of 136 professional dance/movement therapists were years of formal training x and reported success rate y (measured as a percentage) of coleading dance/movement therapy groups.

a. Propose a linear model relating y to x.

b. The researcher hypothesized that dance/movement therapists with more years in formal dance training will report higher perceived success rates in cotherapy relationships. State the hypothesis in terms of the parameter of the model, part a.

c. The correlation coefficient for the sample data was reported as $r = -.26$. Interpret this result.

d. Does the value of r in part c support the hypothesis in part b? Test using $\alpha = .05$.

NAMEGAME2

3.40. Refer to the *Journal of Experimental Psychology-Applied* (June 2000) name retrieval study, Exercises 3.15 (p. 105) and 3.29 (p. 115). Find and interpret the values of r and r^2 for the simple linear regression relating the proportion of names recalled (y) and position (order) of the student (x) during the "name game."

BOXING2

3.41. Refer to the *British Journal of Sports Medicine* (April 2000) study of the effect of massage on boxing performance, Exercise 3.27 (p. 115). Find and interpret the values of r and r^2 for the simple linear regression relating the blood lactate concentration and the boxer's perceived recovery.

3.42. Botanists at the University of Toronto conducted a series of experiments to investigate the feeding habits of baby snow geese (*Journal of Applied Ecology*, Vol. 32, 1995). Goslings were deprived of food until their guts were empty, then were allowed to feed for 6 hours on a diet of plants or Purina Duck Chow. For each feeding trial, the change in the weight of the gosling after 2.5 hours was recorded as a percentage of initial weight. Two other variables recorded were digestion efficiency (measured as a percentage) and amount of acid-detergent fibre in the digestive tract (also measured as a percentage). The data for 42 feeding trials are listed in the table below.

a. The botanists were interested in the correlation between weight change (y) and digestion efficiency (x). Plot the data for these two variables in a scatterplot. Do you observe a trend?

b. Find the coefficient of correlation relating weight change y to digestion efficiency x. Interpret this value.

c. Conduct a test to determine whether weight change y is correlated with a digestion efficiency x. Use $\alpha = .01$.

d. Repeat parts b and c, but exclude the data for trials that used duck chow from the analysis. What do you conclude?

e. The botanists were also interested in the correlation between digestion efficiency y and acid-detergent fibre x. Repeat parts a–d for these two variables.

SNOWGEESE

Feeding Trial	DIET	WEIGHT CHANGE (%)	DIGESTION EFFICIENCY (%)	Acid-Detergent Fibre (%)
1	Plants	−6	0	28.5
2	Plants	−5	2.5	27.5
3	Plants	−4.5	5	27.5
4	Plants	0	0	32.5
5	Plants	2	0	32
6	Plants	3.5	1	30
7	Plants	−2	2.5	34
8	Plants	−2.5	10	36.5
9	Plants	−3.5	20	28.5
10	Plants	−2.5	12.5	29
11	Plants	−3	28	28
12	Plants	−8.5	30	28
13	Plants	−3.5	18	30
14	Plants	−3	15	31
15	Plants	−2.5	17.5	30
16	Plants	−.5	18	22

(continued)

Feeding Trial	DIET	WEIGHT CHANGE (%)	DIGESTION EFFICIENCY (%)	ACID-DETERGENT FIBRE (%)
17	Plants	0	23	22.5
18	Plants	1	20	24
19	Plants	2	15	23
20	Plants	6	31	21
21	Plants	2	15	24
22	Plants	2	21	23
23	Plants	2.5	30	22.5
24	Plants	2.5	33	23
25	Plants	0	27.5	30.5
26	Plants	.5	29	31
27	Plants	−1	32.5	30
28	Plants	−3	42	24
29	Plants	−2.5	39	25
30	Plants	−2	35.5	25
31	Plants	.5	39	20
32	Plants	5.5	39	18.5
33	Plants	7.5	50	15
34	Duck Chow	0	62.5	8
35	Duck Chow	0	63	8
36	Duck Chow	2	69	7
37	Duck Chow	8	42.5	7.5
38	Duck Chow	9	59	8.5
39	Duck Chow	12	52.5	8
40	Duck Chow	8.5	75	6
41	Duck Chow	10.5	72.5	6.5
42	Duck Chow	14	69	7

Source: Gadallah, F. L., and Jefferies, R. L. "Forage quality in brood rearing areas of the lesser snow goose and the growth of captive goslings." *Journal of Applied Biology*, Vol. 32, No. 2, 1995, pp. 281–282 (adapted from Figures 2 and 3).

3.43. A study published in *Psychosomatic Medicine* (Mar./Apr. 2001) explored the relationship between reported severity of pain and actual pain tolerance in 337 patients who suffer from chronic pain. Each patient reported his/her severity of chronic pain on a 7-point scale (1 = no pain, 7 = extreme pain). To obtain a pain tolerance level, a tourniquet was applied to the arm of each patient and twisted. The maximum pain level tolerated was measured on a quantitative scale.

a. According to the researchers, "correlational analysis revealed a small but significant inverse relationship between [actual] pain tolerance and the reported severity of chronic pain." Based on this statement, is the value of r for the 337 patients positive or negative?

b. Suppose that the result reported in part **a** is significant at $\alpha = .05$. Find the approximate value of r for the sample of 337 patients. [*Hint*: Use the formula $t = r\sqrt{(n-2)}/\sqrt{(1-r^2)}$.]

3.9
Using the Model for Estimation and Prediction

If we are satisfied that a useful model has been found to describe the relationship between sales revenue and advertising, we are ready to accomplish the original objectives for building the model: using it to estimate or to predict sales on the basis of advertising dollars spent.

The most common uses of a probabilistic model can be divided into two categories. The first is the use of the model for **estimating the mean value of** y, $E(y)$, **for a specific value of** x. For our example, we may want to estimate the mean sales revenue for *all* months during which \$400 ($x = 4$) is spent on advertising. The second use of the model entails **predicting a particular** y **value for a given** x. That is, if we decide to spend \$400 next month, we want to predict the firm's sales revenue for that month.

In the case of estimating a mean value of y, we are attempting to estimate the mean result of a very large number of experiments at the given x value. In the second case, we are trying to predict the outcome of a single experiment at the given x value. In which of these model uses do you expect to have more success; i.e., which value—the mean or individual value of y—can we estimate (or predict) with more accuracy?

Before answering this question, we first consider the problem of choosing an estimator (or predictor) of the mean (or individual) y value. We will use the least squares model

$$\hat{y} = \hat{\beta}_0 + \hat{\beta}_1 x$$

both to estimate the mean value of y and to predict a particular value of y for a given value of x. For our example, we found

$$\hat{y} = -.1 + .7x$$

so that the estimated mean value of sales revenue for all months when $x = 4$ (advertising = \$400) is

$$\hat{y} = -.1 + .7(4) = 2.7$$

or \$2,700 (the units of y are thousands of dollars). The identical value is used to predict the y value when $x = 4$. That is, both the estimated mean value and the predicted value of y equal $\hat{y} = 2.7$ when $x = 4$, as shown in Figure 3.20.

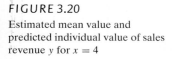

FIGURE 3.20

Estimated mean value and predicted individual value of sales revenue y for $x = 4$

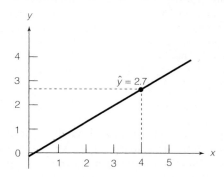

The difference in these two model uses lies in the relative accuracy of the estimate and the prediction. These accuracies are best measured by the repeated sampling errors of the least squares line when it is used as an estimator and as a predictor, respectively. These errors are given in the box.

Sampling Errors for the Estimator of the Mean of y and the Predictor of an Individual y for $x = x_p$

1. The standard deviation of the sampling distribution of the estimator $\hat{y}$ of the mean value of y at a particular value of x, say, x_p, is

$$\sigma_{\hat{y}} = \sigma \sqrt{\frac{1}{n} + \frac{(x_p - \bar{x})^2}{SS_{xx}}}$$

where σ is the standard deviation of the random error ε. We refer to $\sigma_{\hat{y}}$ as the *standard error* of $\hat{y}$.

2. The standard deviation of the prediction error for the predictor $\hat{y}$ of an individual y value for $x = x_p$ is

$$\sigma_{(y-\hat{y})} = \sigma \sqrt{1 + \frac{1}{n} + \frac{(x_p - \bar{x})^2}{SS_{xx}}}$$

where σ is the standard deviation of the random error ε. We refer to $\sigma_{(y-\hat{y})}$ as the *standard error of prediction*.

The true value of σ will rarely be known. Thus, we estimate σ by s and calculate the estimation and prediction intervals as shown in the next two boxes. The procedure is demonstrated in Example 3.4.

A $100(1 - \alpha)\%$ Confidence Interval for the Mean Value of y for $x = x_p$

$$\hat{y} \pm t_{\alpha/2}(\text{Estimated standard deviation of} \hat{y})$$

or

$$\hat{y} \pm (t_{\alpha/2})s \sqrt{\frac{1}{n} + \frac{(x_p - \bar{x})^2}{SS_{xx}}}$$

where $t_{\alpha/2}$ is based on $(n - 2)$df

A $100(1 - \alpha)\%$ Prediction Interval for an Individual y for $x = x_p$

$$\hat{y} \pm t_{\alpha/2}[\text{Estimated standard deviation of} (y - \hat{y})]$$

or

$$\hat{y} \pm (t_{\alpha/2})s \sqrt{1 + \frac{1}{n} + \frac{(x_p - \bar{x})^2}{SS_{xx}}}$$

where $t_{\alpha/2}$ is based on $(n - 2)$ df

EXAMPLE   3.4

Find a 95% confidence interval for mean monthly sales when the appliance store spends $400 on advertising.

Solution

For a $400 advertising expenditure, $x_p = 4$ and, since $n = 5$, df $= n - 2 = 3$. Then the confidence interval for the mean value of y is

$$\hat{y} \pm (t_{\alpha/2})s\sqrt{\frac{1}{n} + \frac{(x_p - \bar{x})^2}{SS_{xx}}}$$

or

$$\hat{y} \pm (t_{.025})s\sqrt{\frac{1}{5} + \frac{(4 - \bar{x})^2}{SS_{xx}}}$$

Recall that $\hat{y} = 2.7$, $s = .61$, $\bar{x} = 3$, and $SS_{xx} = 10$. From Table 2 of Appendix C, $t_{.025} = 3.182$. Thus, we have

$$2.7 \pm (3.182)(.61)\sqrt{\frac{1}{5} + \frac{(4 - 3)^2}{10}} = 2.7 \pm (3.182)(.61)(.55)$$

$$= 2.7 \pm 1.1 = (1.6, 3.8)$$

We estimate, with 95% confidence, that the interval from $1,600 to $3,800 encloses the mean sales revenue for all months when the store spends $400 on advertising. Note that we used a small amount of data for purposes of illustration in fitting the least squares line and that the width of the interval could be decreased by using a larger number of data points.   ◆

EXAMPLE   3.5

Predict the monthly sales for next month if a $400 expenditure is to be made on advertising. Use a 95% prediction interval.

Solution

To predict the sales for a particular month for which $x_p = 4$, we calculate the 95% prediction interval as

$$\hat{y} \pm (t_{\alpha/2})s\sqrt{1 + \frac{1}{n} + \frac{(x_p - \bar{x})^2}{SS_{xx}}} = 2.7 \pm (3.182)(.61)\sqrt{1 + \frac{1}{5} + \frac{(4 - 3)^2}{10}}$$

$$= 2.7 \pm (3.182)(.61)(1.14) = 2.7 \pm 2.2 = (.5, 4.9)$$

Therefore, with 95% confidence we predict that the sales next month (i.e., a month where we spend $400 in advertising) will fall in the interval from $500 to $4,900. As in the case of the confidence interval for the mean value of y, the prediction interval for y is quite large. This is because we have chosen a simple example (only five data points) to fit the least squares line. The width of the prediction interval could be reduced by using a larger number of data points.   ◆

Both the confidence interval for $E(y)$ and the prediction interval for y can be obtained using statistical software. Figures 3.21 and 3.22 are SAS printouts showing confidence intervals and prediction intervals, respectively, for the advertising-sales example. These intervals (highlighted on the printouts) agree, except for rounding, with our calculated intervals.

FIGURE 3.21

SAS printout showing 95% confidence intervals for $E(y)$

§

Dependent Variable: SALES_Y

Output Statistics

Obs	ADV_X	Dep Var SALES_Y	Predicted Value	Std Error Mean Predict	95% CL Mean	
1	1	1.0000	0.6000	0.4690	-0.8927	2.0927
2	2	1.0000	1.3000	0.3317	0.2445	2.3555
3	3	2.0000	2.0000	0.2708	1.1382	2.8618
4	4	2.0000	2.7000	0.3317	1.6445	3.7555
5	5	4.0000	3.4000	0.4690	1.9073	4.8927

FIGURE 3.22

SAS printout showing 95% prediction intervals for y

Dependent Variable: SALES_Y

Output Statistics

Obs	ADV_X	Dep Var SALES_Y	Predicted Value	Std Error Mean Predict	95% CL Predict	
1	1	1.0000	0.6000	0.4690	-1.8376	3.0376
2	2	1.0000	1.3000	0.3317	-0.8972	3.4972
3	3	2.0000	2.0000	0.2708	-0.1110	4.1110
4	4	2.0000	2.7000	0.3317	0.5028	4.8972
5	5	4.0000	3.4000	0.4690	0.9624	5.8376

Note that the confidence interval in Example 3.4 is wider than the prediction interval in Example 3.5. Will this always be true? The answer is "yes." The error in estimating the mean value of y, $E(y)$, for a given value of x, say, x_p, is the distance between the least squares line and the true line of means, $E(y) = \beta_0 + \beta_1 x$. This error, $[\hat{y} - E(y)]$, is shown in Figure 3.23. In contrast, the error $(y_p - \hat{y})$ in predicting some future value of y is the sum of two errors—the error of estimating the mean of y, $E(y)$, shown in Figure 3.23, plus the random error that is a component of the value of y to be predicted (see Figure 3.24). Consequently, the error of predicting a particular value of y will always be larger than the error of estimating the mean value of y for a particular value of x. Note from their formulas that both the error of estimation and the error of prediction take their smallest values when $x_p = \bar{x}$. The farther x lies from $\bar{x}$, the larger will be the errors of estimation and prediction. You can see why this is true by noting the deviations for different values of x between the line of means $E(y) = \beta_0 + \beta_1 x$ and the predicted line $\hat{y} = \hat{\beta}_0 + \hat{\beta}_1 x$ shown in Figure 3.24. The deviation is larger at the extremities of the interval where the largest and smallest values of x in the data set occur.

A graph showing both the confidence limits for $E(y)$ and the prediction limits for y over the entire range of the advertising expenditure (x) values is displayed in Figure 3.25. You can see that the confidence interval is always narrower than the prediction interval, and that they are both narrowest at the mean $\bar{x}$, increasing steadily as the distance $|x - \bar{x}|$ increases. In fact, when x is selected far enough away from $\bar{x}$ so that it falls outside the range of the sample data, it is dangerous to make any inferences about $E(y)$ or y.

FIGURE 3.23

Error of estimating the mean value of y for a given value of x

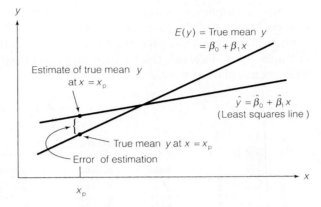

FIGURE 3.24

Error of predicting a future value of y for a given value of x

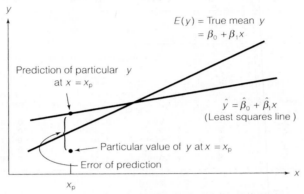

Caution

Using the least squares prediction equation to estimate the mean value of y or to predict a particular value of y for values of x that fall outside the range of the values of x contained in your sample data may lead to errors of estimation or prediction that are much larger than expected. Although the least squares model may provide a very good fit to the data over the range of x values contained in the sample, it could give a poor representation of the true model for values of x outside this region.

The confidence interval width grows smaller as n is increased; thus, in theory, you can obtain as precise an estimate of the mean value of y as desired (at any given x) by selecting a large enough sample. The prediction interval for a new value of y also grows smaller as n increases, but there is a lower limit on its width. If you examine the formula for the prediction interval, you will see that the interval can get no smaller than $\hat{y} \pm z_{\alpha/2}\sigma$.* Thus, the only way to obtain more accurate predictions for new values of y is to reduce the standard deviation of the regression model, σ. This can be accomplished only by improving the model, either by using a curvilinear

---

*The result follows from the facts that, for large n, $t_{\alpha/2} \approx z_{\alpha/2}$, $s \approx \sigma$, and the last two terms under the radical in the standard error of the predictor are approximately 0.

FIGURE 3.25

Comparison of widths of 95% confidence and prediction intervals

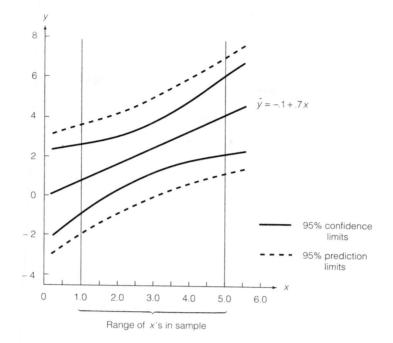

$\hat{y} = -.1 + .7x$

— 95% confidence limits

- - - 95% prediction limits

Range of x's in sample

(rather than linear) relationship with x or by adding new independent variables to the model, or both. Methods of improving the model are discussed in Chapter 4.

EXERCISES

3.44. **a.** Explain why for a particular x value, the prediction interval for an individual y value will always be wider than the confidence interval for a mean value of y.

b. Explain why the confidence interval for the mean value of y for a particular x value, say, x_p, gets wider the farther x_p is from $\bar{x}$. What are the implications of this phenomenon for estimation and prediction?

3.45. A simple linear regression analysis for $n = 20$ data points produced the following results:

$$\hat{y} = 2.1 + 3.4x \quad SS_{xx} = 4.77$$
$$\bar{x} = 2.5 \quad SS_{yy} = 59.21$$
$$\bar{y} = 10.6 \quad SS_{xy} = 16.22$$

a. Find SSE and s^2.

b. Find a 95% confidence interval for $E(y)$ when $x = 2.5$. Interpret this interval.

c. Find a 95% confidence interval for $E(y)$ when $x = 2.0$. Interpret this interval.

d. Find a 95% confidence interval for $E(y)$ when $x = 3.0$. Interpret this interval.

e. Examine the widths of the confidence intervals obtained in parts b, c, and d. What happens to the width of the confidence interval for $E(y)$ as the value of x moves away from the value of $\bar{x}$?

f. Find a 95% prediction interval for a value of y to be observed in the future when $x = 3.0$. Interpret its value.

3.46. Refer to the data on sale prices and total appraised values of 92 residential properties in an upscale Tampa, Florida, neighborhood, Exercise 3.8 (p. 100).

a. In Exercise 3.8, you determined that appraised property value x and sale price y are positively related for homes sold in the Tampa Palms subdivision. Does this result guarantee that appraised value will yield accurate predictions of sale price? Explain.

b. MINITAB was used to predict the sale price of a residential property in the subdivision with a total appraised value of $300,000. Locate a 95% prediction interval for the sale price of this property on the printout (p. 136) and interpret the result.

c. Locate a 95% confidence interval for $E(y)$ on the printout and interpret the result.

MINITAB Output for Exercise 3.46

```
Predicted Values for New Observations

New Obs      Fit     SE Fit          95.0% CI              95.0% PI
1         257.87       3.42    ( 251.07,  264.66)   ( 198.22,  317.51)


Values of Predictors for New Observations

New Obs   TOTALVAL
1             300
```

WAREHOUSE

3.47. Refer to Exercise 3.11 (p. 102). Consider an automated warehouse that operates $x = 10$ vehicles.

a. Find a 90% prediction interval for the congestion time y at this warehouse. Interpret the result.

b. Find a 90% confidence interval for the mean congestion time, $E(y)$, at all warehouses with $x = 10$ vehicles. Interpret the result.

c. Compare and comment on the sizes of the intervals in parts **a** and **b**.

d. Could you reduce the size of either or both intervals by increasing your sample size? Explain.

3.48. Refer to the simple linear regression of sweetness index y and amount of pectin x for $n = 24$ orange juice samples, Exercise 3.13 (p. 103). The SPSS printout of the analysis is shown at right. A 90% confidence interval for the mean sweetness index, $E(y)$, for each value of x is shown on the SPSS spreadsheet. Select an observation and interpret this interval.

3.49. Refer to the simple linear regression of number of fly-catchers killed y and nest box tit occupancy x for $n = 14$ nest sites, Exercise 3.14 (p. 104).

a. A 95% prediction interval for y when $x = 64$ is shown at the bottom of the SAS printout on p. 137. Interpret this interval.

b. How would the width of a 95% confidence interval for $E(y)$ when $x = 64$ compare to the interval, part **a**?

c. Would you recommend using the model to predict the number of flycatchers killed at a site with a nest box tit occupancy of 15%? Explain.

Model Summary[b]

Model	R	R Square	Adjusted R Square	Std. Error of the Estimate
1	.478[a]	.229	.194	.2150

a. Predictors: (Constant), PECTIN
b. Dependent Variable: SWEET

ANOVA[b]

Model		Sum of Squares	df	Mean Square	F	Sig.
1	Regression	.301	1	.301	6.520	.018[a]
	Residual	1.017	22	.046		
	Total	1.318	23			

a. Predictors: (Constant), PECTIN
b. Dependent Variable: SWEET

Coefficients[a]

Model		Unstandardized Coefficients		Standardized Coefficients	t	Sig.
		B	Std. Error	Beta		
1	(Constant)	6.252	.237		26.422	.000
	PECTIN	-2.31E-03	.001	-.478	-2.554	.018

a. Dependent Variable: SWEET

	run	sweet	pectin	lower90m	upper90m
1	1	5.2	220	5.64898	5.83848
2	2	5.5	227	5.63898	5.81613
3	3	6.0	259	5.57819	5.72904
4	4	5.9	210	5.66194	5.87173
5	5	5.8	224	5.64337	5.82560
6	6	6.0	215	5.65564	5.85493
7	7	5.8	231	5.63284	5.80379
8	8	5.6	268	5.55553	5.71011
9	9	5.6	239	5.61947	5.78019
10	10	5.9	212	5.65946	5.86497
11	11	5.4	410	5.05526	5.55416
12	12	5.6	256	5.58517	5.73592
13	13	5.8	306	5.43785	5.65219
14	14	5.5	259	5.57819	5.72904
15	15	5.3	284	5.50957	5.68213
16	16	5.3	383	5.15725	5.57694
17	17	5.7	271	5.54743	5.70434
18	18	5.5	264	5.56591	5.71821
19	19	5.7	227	5.63898	5.81613
20	20	5.3	263	5.56843	5.72031
21	21	5.9	232	5.63125	5.80075
22	22	5.8	220	5.64898	5.83848
23	23	5.8	246	5.60640	5.76091
24	24	5.9	241	5.61587	5.77454

SAS Output for Exercise 3.49

			Output Statistics				
Obs	TITPCT	Dep Var NOKILLED	Predicted Value	Std Error Mean Predict	95% CL Predict		Residual
1	24	0	-0.4631	0.5554	-3.1326	2.2064	0.4631
2	33	0	0.5058	0.3719	-2.0078	3.0194	-0.5058
3	34	0	0.6135	0.3559	-1.8891	3.1161	-0.6135
4	43	0	1.5824	0.2944	-0.8820	4.0468	-1.5824
5	50	0	2.3360	0.3695	-0.1760	4.8479	-2.3360
6	35	1.0000	0.7211	0.3412	-1.7718	3.2140	0.2789
7	35	1.0000	0.7211	0.3412	-1.7718	3.2140	0.2789
8	38	1.0000	1.0441	0.3073	-1.4278	3.5159	-0.0441
9	40	1.0000	1.2594	0.2949	-1.2053	3.7241	-0.2594
10	31	2.0000	0.2905	0.4074	-2.2492	2.8301	1.7095
11	43	2.0000	1.5824	0.2944	-0.8820	4.0468	0.4176
12	55	3.0000	2.8742	0.4643	0.2887	5.4598	0.1258
13	57	4.0000	3.0896	0.5073	0.4659	5.7132	0.9104
14	64	5.0000	3.8431	0.6700	1.0516	6.6347	1.1569

NAMEGAME2

3.50. Refer to the *Journal of Experimental Psychology-Applied* (June 2000) name retrieval study, Exercise 3.15 (p. 105).

a. Find a 99% confidence interval for the mean recall proportion for students in the fifth position during the "name game." Interpret the result.

b. Find a 99% prediction interval for the recall proportion of a particular student in the fifth position during the "name game." Interpret the result.

c. Compare the two intervals, parts **a** and **b**. Which interval is wider? Will this always be the case? Explain.

3.51. The Sasakawa Sports Foundation conducted a national survey to assess the physical activity patterns of Japanese adults. The table below lists the frequency (average number of days in the past year) and duration of time (average number of minutes per single activity) Japanese adults spent participating in a sample of 11 sports activities.

JAPANSPORTS

ACTIVITY	FREQUENCY x (days/year)	DURATION y (minutes)
Jogging	135	43
Cycling	68	99
Aerobics	44	61
Swimming	39	60
Volleyball	30	80
Tennis	21	100
Softball	16	91
Baseball	19	127
Skating	7	115
Skiing	10	249
Golf	5	262

Source: J. Bennett, ed. *Statistics in Sport*. London: Arnold, 1998 (adapted from Figure 11.6).

a. Write the equation of a straight-line model relating duration (y) to frequency (x).

b. Find the least squares prediction equation.

c. Is there evidence of a linear relationship between y and x? Test using $\alpha = .05$.

d. Use the least squares line to predict the duration of time Japanese adults participate in a sport that they play 25 times a year. Form a 95% confidence interval around the prediction and interpret the result.

3.52. In forestry, the diameter of a tree at breast height (which is fairly easy to measure) is used to predict the height of the tree (a difficult measurement to obtain). Silviculturists working in British Columbia's boreal forest conducted a series of spacing trials to predict the heights of several species of trees. The data in the accompanying table are the breast height diameters (in centimeters) and heights (in meters) for a sample of 36 white spruce trees.

a. Construct a scatterplot for the data.

b. Assuming the relationship between the variables is best described by a straight line, use the method of least squares to estimate the y-intercept and slope of the line.

WHITESPRUCE

BREAST HEIGHT DIAMETER x, cm	HEIGHT y, m	BREAST HEIGHT DIAMETER x, cm	HEIGHT y, m
18.9	20.0	16.6	18.8
15.5	16.8	15.5	16.9
19.4	20.2	13.7	16.3
20.0	20.0	27.5	21.4
29.8	20.2	20.3	19.2
19.8	18.0	22.9	19.8
20.3	17.8	14.1	18.5
20.0	19.2	10.1	12.1
22.0	22.3	5.8	8.0

(continued overleaf)

(*Continued*)

BREAST HEIGHT DIAMETER x, cm	HEIGHT y, m	BREAST HEIGHT DIAMETER x, cm	HEIGHT y, m
23.6	18.9	20.7	17.4
14.8	13.3	17.8	18.4
22.7	20.6	11.4	17.3
18.5	19.0	14.4	16.6
21.5	19.2	13.4	12.9
14.8	16.1	17.8	17.5
17.7	19.9	20.7	19.4
21.0	20.4	13.3	15.5
15.9	17.6	22.9	19.2

Source: Scholz, H., Northern Lights College, British Columbia.

c. Plot the least squares line on your scatterplot.
d. Do the data provide sufficient evidence to indicate that the breast height diameter x contributes information for the prediction of tree height y? Test using $\alpha = .05$.
e. Use your least squares line to find a 90% confidence interval for the average height of white spruce trees with a breast height diameter of 20 cm. Interpret the interval.

3.10 A Complete Example

In the previous sections, we have presented the basic elements necessary to fit and use a straight-line regression model. In this section, we will assemble these elements by applying them in an example with the aid of computer software.

Suppose a fire safety inspector wants to relate the amount of fire damage in major residential fires to the distance between the residence and the nearest fire station. The study is to be conducted in a large suburb of a major city; a sample of 15 recent fires in this suburb is selected.

STEP 1   First, we hypothesize a model to relate fire damage y to the distance x from the nearest fire station. We will hypothesize a straight-line probabilistic model:

$$y = \beta_0 + \beta_1 x + \varepsilon$$

STEP 2   Second, we collect the (x, y) values for each of the $n = 15$ experimental units (residential fires) in the sample. The amount of damage y and the distance x between the fire and the nearest fire station are recorded for each fire, as listed in Table 3.8.

STEP 3   Next, we enter the data of Table 3.8 into a computer and use statistical software to estimate the unknown parameters in the deterministic component of the hypothesized model. The SAS printout for the simple linear regression analysis is shown in Figure 3.26.

The least squares estimates of β_0 and β_1, highlighted on the printout, are

$$\hat{\beta}_0 = 10.27793, \quad \hat{\beta}_1 = 4.91933$$

Thus, the least squares equation is (after rounding)

$$\hat{y} = 10.28 + 4.92x$$

This prediction equation is shown on the scatterplot, Figure 3.27.

FIREDAM

TABLE 3.8 Fire Damage Data

DISTANCE FROM FIRE STATION x, miles	FIRE DAMAGE y, thousands of dollars
3.4	26.2
1.8	17.8
4.6	31.3
2.3	23.1
3.1	27.5
5.5	36.0
.7	14.1
3.0	22.3
2.6	19.6
4.3	31.3
2.1	24.0
1.1	17.3
6.1	43.2
4.8	36.4
3.8	26.1

The least squares estimate of the slope, $\hat{\beta}_1 = 4.92$, implies that the estimated mean damage increases by \$4,920 for each additional mile from the fire station. This interpretation is valid over the range of x, or from .7 to 6.1 miles from the station. The estimated y-intercept, $\hat{\beta}_0 = 10.28$, has the interpretation that a fire 0 miles from the fire station has an estimated mean damage of \$10,280. Although this would seem to apply to the fire station itself, remember that the y-intercept is meaningfully interpretable only if $x = 0$ is within the sampled range of the independent variable. Since $x = 0$ is outside the range, $\hat{\beta}_0$ has no practical interpretation.

STEP 4 Now, we specify the probability distribution of the random error component ε. The assumptions about the distribution will be identical to those listed in Section 3.4:

(1) $E(\varepsilon) = 0$
(2) $\text{Var}(\varepsilon) = \sigma^2$ is constant for all x-values
(3) ε has a normal distribution
(4) ε's are independent

Although we know that these assumptions are not completely satisfied (they rarely are for any practical problem), we are willing to assume they are approximately satisfied for this example. The estimate of σ^2, shaded on the printout is

$$s^2 = 5.36546$$

(This value is also called **mean square for error**, or **MSE**.)

FIGURE 3.26

SAS printout for fire damage
linear regression

Dependent Variable: DAMAGE

Analysis of Variance

Source	DF	Sum of Squares	Mean Square	F Value	Pr > F
Model	1	841.76636	841.76636	156.89	<.0001
Error	13	69.75098	5.36546		
Corrected Total	14	911.51733			

Root MSE	2.31635	R-Square	0.9235
Dependent Mean	26.41333	Adj R-Sq	0.9176
Coeff Var	8.76961		

Parameter Estimates

Variable	DF	Parameter Estimate	Standard Error	t Value	Pr > |t|	95% Confidence Limits	
Intercept	1	10.27793	1.42028	7.24	<.0001	7.20960	13.34625
DISTANCE	1	4.91933	0.39275	12.53	<.0001	4.07085	5.76781

Output Statistics

Obs	DISTANCE	Dep Var DAMAGE	Predicted Value	Std Error Mean Predict	95% CL Predict		Residual
1	3.4	26.2000	27.0037	0.5999	21.8344	32.1729	-0.8037
2	1.8	17.8000	19.1327	0.8340	13.8141	24.4514	-1.3327
3	4.6	31.3000	32.9068	0.7915	27.6186	38.1951	-1.6068
4	2.3	23.1000	21.5924	0.7112	16.3577	26.8271	1.5076
5	3.1	27.5000	25.5279	0.6022	20.3573	30.6984	1.9721
6	5.5	36.0000	37.3342	1.0573	31.8334	42.8351	-1.3342
7	0.7	14.1000	13.7215	1.1766	8.1087	19.3342	0.3785
8	3.0	22.3000	25.0359	0.6081	19.8622	30.2097	-2.7359
9	2.6	19.6000	23.0682	0.6550	17.8678	28.2686	-3.4682
10	4.3	31.3000	31.4311	0.7198	26.1908	36.6713	-0.1311
11	2.1	24.0000	20.6085	0.7566	15.3442	25.8729	3.3915
12	1.1	17.3000	15.6892	1.0444	10.1999	21.1785	1.6108
13	6.1	43.2000	40.2858	1.2587	34.5906	45.9811	2.9142
14	4.8	36.4000	33.8907	0.8450	28.5640	39.2175	2.5093
15	3.8	26.1000	28.9714	0.6320	23.7843	34.1585	-2.8714
16	3.5	.	27.4956	0.6043	22.3239	32.6672	.

Sum of Residuals		0
Sum of Squared Residuals		69.75098
Predicted Residual SS (PRESS)		93.21169

FIGURE 3.27

Least squares model for the fire
damage data

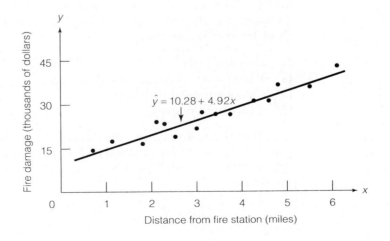

The estimated standard deviation of ε, also highlighted on the printout, is

$$s = 2.31635$$

The value of s implies that most of the observed fire damage (y) values will fall within approximately $2s = 4.64$ thousand dollars of their respective predicted values when using the least squares line.

STEP 5   We can now check the utility of the hypothesized model, that is, whether x really contributes information for the prediction of y using the straight-line model.

(a) *Test of model utility*: First, test the null hypothesis that the slope β_1 is 0, i.e., that there is no linear relationship between fire damage and the distance from the nearest fire station, against the alternative that x and y are positively linearly related, at $\alpha = .05$. The null and alternative hypotheses are:

$$H_0: \beta_1 = 0$$

$$H_a: \beta_1 > 0$$

The value of the test statistic highlighted on the printout is $t = 12.53$, and the two-tailed p-value of the test also highlighted is .0001. Thus, the p-value for our 1-tailed test is

$$p = \frac{.0001}{2} = .00005$$

Since $\alpha = .05$ exceeds this small p-value, there is sufficient evidence to reject H_0 and conclude that distance between the fire and the fire station contributes information for the prediction of fire damage and that mean fire damage increases as the distance increases.

(b) *Confidence interval for slope*: We gain additional information about the relationship by forming a confidence interval for the slope β_1. A 95% confidence interval for β (highlighted on the printout) is (4.07085, 5.76781).

We are 95% confident that the interval from \$4,071 to \$5,768 encloses the mean increase (β_1) in fire damage per additional mile distance from the fire station.

(c) *Numerical descriptive measures of model adequacy*: The coefficient of determination (highlighted on the printout) is

$$r^2 = .9235$$

This value implies that about 92% of the sample variation in fire damage (y) is explained by the distance x between the fire and the fire station in a straight-line model.

The coefficient of correlation r, which measures the strength of the linear relationship between y and x, is not shown on Figure 3.26. Using the facts that $r = \sqrt{r^2}$ in simple linear regression and that r and $\hat{\beta}_1$ have the same sign, we find

$$r = +\sqrt{r^2} = \sqrt{.9235} = .96$$

The high correlation confirms our conclusion that β_1 differs from 0; it appears that fire damage and distance from the fire station are linearly correlated.

The results of the test for β_1, the high value of r^2, and the relatively small $2s$ value (STEP 4), all point to a strong linear relationship between x and y.

STEP 6   We are now prepared to use the least squares model. Suppose the insurance company wants to predict the fire damage if a major residential fire were to occur 3.5 miles from the nearest fire station, i.e., $x_p = 3.5$. The predicted value, shaded at the bottom of the SAS printout, is $\hat{y} = 27.4956$, while the corresponding 95% prediction interval (also highlighted) is (22.3239, 32.6672). Therefore, we predict (with 95% confidence) that the fire damage for a major residential fire 3.5 miles from the nearest fire station will fall between $22,324 and $32,667.

Caution: We would not use this prediction model to make predictions for homes less than .7 mile or more than 6.1 miles from the nearest fire station. A look at the data in Table 3.8 reveals that all the x values fall between .7 and 6.1. Recall from Section 3.9 that it is dangerous to use the model to make predictions outside the region in which the sample data fall. A straight line might not provide a good model for the relationship between the mean value of y and the value of x when stretched over a wider range of x values.

[*Note*: Details on how to perform a simple linear regression analysis on the computer using each of three statistical software packages, SAS, SPSS, and MINITAB are provided in the Appendix.]

EXERCISES

3.53. *The Journal of Information Systems* (Spring 1992) published a study of a computerized intrusion-detection system. The input-output (I/O) units and the central processing unit (CPU) time (in seconds) utilized by a sample of 44 system users were recorded. The data for both variables are listed on p. 144 (CPUIO). A simple linear regression analysis relating CPU times (y) to I/O units (x) was conducted on the data. The results are shown in the SAS printout on p. 143. Conduct a complete simple linear regression analysis.

3.54. Two processes for hydraulic drilling of rock are dry drilling and wet drilling. In a dry hole, compressed air is forced down the drill rods to flush the cuttings and drive the hammer; in a wet hole, water is forced down. An experiment was conducted to determine whether the time y it takes to dry drill a distance of 5 feet in rock increases with depth x. The results (extracted from *The American Statistician* Feb. 1991) for one portion of the experiment are shown in the table on p. 144 (DRILLROCK).

SAS Output for Exercise 3.53

Dependent Variable: CPU

Analysis of Variance

Source	DF	Sum of Squares	Mean Square	F Value	Pr > F
Model	1	937.54390	937.54390	2.97	0.0919
Error	42	13236	315.15372		
Corrected Total	43	14174			

Root MSE		17.75257	R-Square	0.0661	
Dependent Mean		32.00000	Adj R-Sq	0.0439	
Coeff Var		55.47678			

Parameter Estimates

Variable	DF	Parameter Estimate	Standard Error	t Value	Pr > |t|	95% Confidence Limits	
Intercept	1	28.89432	3.22565	8.96	<.0001	22.38470	35.40395
IOUNITS	1	0.90497	0.52468	1.72	0.0919	-0.15389	1.96382

Predictions

Obs	CPU	IOUNITS	Predicted CPU	Lower prediction limit of CPU	Upper prediction limit of CPU
1	54	15	42.4688	29.0819	55.8557
2	55	5	33.4191	27.7687	39.0696
3	15	2	30.7043	25.0945	36.3140
4	41	17	44.2787	28.9303	59.6271
5	54	4	32.5142	27.0798	37.9486

MINITAB Output for Exercise 3.54

The regression equation is
TIME = 4.79 + 0.0144 DEPTH

Predictor	Coef	SE Coef	T	P
Constant	4.7896	0.6663	7.19	0.000
DEPTH	0.014388	0.002847	5.05	0.000

S = 1.432   R-Sq = 63.0%   R-Sq(adj) = 60.5%

Analysis of Variance

Source	DF	SS	MS	F	P
Regression	1	52.378	52.378	25.54	0.000
Residual Error	15	30.768	2.051		
Total	16	83.146			

Predicted Values for New Observations

New Obs	Fit	SE Fit	95.0% CI	95.0% PI
1	7.667	0.347	(6.927, 8.408)	(4.526, 10.808)

Values of Predictors for New Observations

New Obs	DEPTH
1	200

CPUIO

CPU TIME	I/O UNITS	CPU TIME	I/O UNITS	CPU TIME	I/O UNITS
54	15	55	5	15	2
41	17	54	4	27	3
28	1	37	1	28	0
32	0	18	0	19	0
17	0	53	0	21	0
23	20	42	9	20	0
19	0	40	0	13	1
20	6	28	1	27	3
30	1	52	0	102	6
14	0	19	0	54	0
26	0	35	1	62	14
19	0	46	7	59	0
23	2	38	9	23	4
16	0	15	0	16	0
13	9	15	10		

Source: O'Leary, D. E. "Intrusion-detection systems." *Journal of Information Systems*, Spring 1992, p. 68 (Table 2).

Use the MINITAB printout on p. 143 above to conduct a complete simple linear regression analysis of the data.

3.55. Beanie Babies are toy stuffed animals that have become valuable collector's items. *Beanie World Magazine* provided the information on 50 Beanie Babies shown in the table below.

DRILLROCK

DEPTH AT WHICH DRILLING BEGINS x, feet	TIME TO DRILL 5 FEET y, minutes
0	4.90
25	7.41
50	6.19
75	5.57
100	5.17
125	6.89
150	7.05
175	7.11
200	6.19
225	8.28
250	4.84
275	8.29
300	8.91
325	8.54
350	11.79
375	12.12
395	11.02

Source: Penner, R., and Watts, D. G. "Mining information." *The American Statistician*, Vol. 45, No. 1, Feb. 1991, p. 6 (Table 1).

Can age of a Beanie Baby be used to accurately predict its market value? Answer this question by conducting a complete simple linear regression analysis on the data.

BEANIE

NAME	AGE (months) AS OF SEPT. 1998	RETIRED (R)/ CURRENT (C)	VALUE ($)
1. Ally the Alligator	52	R	55.00
2. Batty the Bat	12	C	12.00
3. Bongo the Brown Monkey	28	R	40.00
4. Blackie the Bear	52	C	10.00
5. Bucky the Beaver	40	R	45.00
6. Bumble the Bee	28	R	600.00
7. Crunch the Shark	21	C	10.00
8. Congo the Gorilla	28	C	10.00
9. Derby the Coarse Mane Horse	28	R	30.00
10. Digger the Red Crab	40	R	150.00
11. Echo the Dolphin	17	R	20.00
12. Fetch the Golden Retriever	5	C	15.00
13. Early the Robin	5	C	20.00
14. Flip the White Cat	28	R	40.00
15. Garcia the Teddy	28	R	200.00
16. Happy the Hippo	52	R	20.00
17. Grunt the Razorback	28	R	175.00
18. Gigi the Poodle	5	C	15.00
19. Goldie the Goldfish	52	R	45.00

(*continued*)

⬤ BEANIE

NAME	AGE (months) AS OF SEPT. 1998	RETIRED (R)/ CURRENT (C)	VALUE ($)
20. Iggy the Iguana	10	C	10.00
21. Inch the Inchworm	28	R	20.00
22. Jake the Mallard Duck	5	C	20.00
23. Kiwi the Toucan	40	R	165.00
24. Kuku the Cockatoo	5	C	20.00
25. Mystic the Unicorn	11	R	45.00
26. Mel the Koala Bear	21	C	10.00
27. Nanook the Husky	17	C	15.00
28. Nuts the Squirrel	21	C	10.00
29. Peace the Tie Dyed Teddy	17	C	25.00
30. Patty the Platypus	64	R	800.00
31. Quacker the Duck	40	R	15.00
32. Puffer the Penguin	10	C	15.00
33. Princess the Bear	12	C	65.00
34. Scottie the Scottie	28	R	28.00
35. Rover the Dog	28	R	15.00
36. Rex the Tyrannosaurus	40	R	825.00
37. Sly the Fox	28	C	10.00
38. Slither the Snake	52	R	1,900.00
39. Skip the Siamese Cat	21	C	10.00
40. Splash the Orca Whale	52	R	150.00
41. Spooky the Ghost	28	R	40.00
42. Snowball the Snowman	12	R	40.00
43. Stinger the Scorpion	5	C	15.00
44. Spot the Dog	52	R	65.00
45. Tank the Armadillo	28	R	85.00
46. Stripes the Tiger (Gold/Black)	40	R	400.00
47. Teddy the 1997 Holiday Bear	12	R	50.00
48. Tuffy the Terrier	17	C	10.00
49. Tracker the Basset Hound	5	C	15.00
50. Zip the Black Cat	28	R	40.00

Source: Beanie World Magazine, Sept. 1998.

3.11
Regression Through the Origin (Optional)

In practice, we occasionally know in advance that the true line of means $E(y)$ passes through the point $(x = 0, y = 0)$, called the **origin**. For example, a chain of convenience stores may be interested in modeling sales y of a new diet soft drink as a linear function of amount x of the new product in stock for a sample of stores. Or, a medical researcher may be interested in the linear relationship between dosage x of a drug for cancer patients and increase y in pulse rate of the patient 1 minute after taking the drug. In both cases, it is known that the regression line must pass through the origin. The convenience store chain knows that if one of its stores chooses not to stock the new diet soft drink, it will have zero sales of the new product. Likewise, if the cancer patient takes no dosage of the drug, the theoretical increase in pulse rate 1 minute later will be 0.

For situations in which we know that the regression line passes through the origin, the y-intercept is $\beta_0 = 0$ and the probabilistic straight-line model takes the form

$$y = \beta_1 x + \varepsilon$$

When the regression line passes through the origin, the formula for the least squares estimate of the slope β_1 differs from the formula given in Section 3.3. Several other formulas required to perform the regression analysis are also different. These new computing formulas are provided in the following box.

Formulas for Regression Through the Origin: $y = \beta_1 x + \varepsilon$

Least squares slope:

$$\hat{\beta}_1 = \frac{\sum x_i y_i}{\sum x_i^2}$$

Estimate of σ^2:

$$s^2 = \frac{\text{SSE}}{n-1}, \quad \text{where SSE} = \sum y_i^2 - \hat{\beta}_1 \sum x_i y_i$$

Estimate of $\sigma_{\hat{\beta}_1}$:

$$s_{\hat{\beta}_1} = \frac{s}{\sqrt{\sum x_i^2}}$$

Estimate of $\sigma_{\hat{y}}$ for estimating $E(y)$ when $x = x_p$:

$$s_{\hat{y}} = s \left(\frac{x_p}{\sqrt{\sum x_i^2}} \right)$$

Estimate of $\sigma_{(y-\hat{y})}$ for predicting y when $x = x_p$:

$$s_{(y-\hat{y})} = s \sqrt{1 + \frac{x_p^2}{\sum x_i^2}}$$

Note that the denominator of s^2 is $n - 1$, not $n - 2$ as in the previous sections. This is because we need to estimate only a single parameter β_1 rather than both β_0 and β_1. Consequently, we have one additional degree of freedom for estimating σ^2, the variance of ε. Tests and confidence intervals for β_1 are carried out exactly as outlined in the previous sections, except that the t distribution is based on $(n - 1)$ df. The test statistic and confidence intervals are given in the next box.

EXAMPLE   3.6

As part of a computer system performance evaluation, a systems manager is interested in predicting the response time for computer terminals. *Terminal response time* is defined as the length of time (in seconds) it takes the computer to respond to a command sent from a computer terminal by pressing one of the terminal's program function keys. Although many variables influence terminal response time, the systems manager will model the response time y as a function of the number x of simultaneous users (i.e., the number of users who are accessing the computer's

Tests and Confidence Intervals for Regression Through the Origin

Test statistic for H_0: $\beta_1 = 0$:

$$t = \frac{\hat{\beta}_1 - 0}{s_{\hat{\beta}_1}} = \frac{\hat{\beta}_1}{s / \sqrt{\sum x_i^2}}$$

$100(1 - \alpha)\%$ *confidence interval for β_1:*

$$\hat{\beta}_1 \pm (t_{\alpha/2}) s_{\hat{\beta}_1} = \hat{\beta}_1 \pm (t_{\alpha/2}) \left(\frac{s}{\sqrt{\sum x_i^2}} \right)$$

$100(1 - \alpha)\%$ *confidence interval for $E(y)$:*

$$\hat{y} \pm (t_{\alpha/2}) s_{\hat{y}} = \hat{y} \pm (t_{\alpha/2}) s \left(\frac{x_p}{\sqrt{\sum x_i^2}} \right)$$

$100(1 - \alpha)\%$ *prediction interval for y:*

$$\hat{y} \pm (t_{\alpha/2}) s_{(y - \hat{y})} = \hat{y} \pm (t_{\alpha/2}) s \sqrt{1 + \frac{x_p^2}{\sum x_i^2}}$$

where the distribution of t is based on $(n - 1)$ df

TABLE 3.9 **Sample Data for Example 3.6**

NUMBER OF SIMULTANEOUS USERS	TERMINAL RESPONSE TIME
x	y, *seconds*
10	1.0
15	1.2
20	2.0
20	2.1
25	2.2
30	2.0
30	1.9

central processing unit at the same time the command was sent). For this computer system, response time will be 0 when the number of simultaneous users is 0. Table 3.9 gives the response times and number of simultaneous users for a sample of $n = 7$ terminal requests. Use the data to fit a straight-line regression model through the origin and calculate SSE.

TABLE 3.10   **Preliminary Calculations for Example 3.6**

x_i	y_i	x_i^2	$x_i y_i$	y_i^2
10	1.0	100	10	1.00
15	1.2	225	18	1.44
20	2.0	400	40	4.00
20	2.1	400	42	4.41
25	2.2	625	55	4.84
30	2.0	900	60	4.00
30	1.9	900	57	3.61
Totals		$\sum x_i^2 = 3,550$	$\sum x_i y_i = 282$	$\sum y_i^2 = 23.3$

Solution

The model we want to fit is $y = \beta_1 x + \varepsilon$. Preliminary calculations for estimating β_1 and calculating SSE are given in Table 3.10. The estimate of the slope is

$$\hat{\beta}_1 = \frac{\sum x_i y_i}{\sum x_i^2} = \frac{282}{3,550}$$

$$= .07944$$

and the least squares line is

$$\hat{y} = .07944x$$

The value of SSE for the line is

$$SSE = \sum y_i^2 - \hat{\beta}_1 \sum x_i y_i$$

$$= 23.3 - (.07944)(282)$$

$$= .899$$

Both these values, $\hat{\beta}_1$ and SSE, are highlighted on the SPSS printout of the analysis, Figure 3.28. The graph of the least squares line with the observations is shown in Figure 3.29.   ◆

EXAMPLE   3.7

Refer to Example 3.6. Conduct the appropriate test for model adequacy. If the model is deemed adequate, predict the response time y with a 95% prediction interval, for a terminal request when $x = 23$ users are signed on to the system.

Solution

The appropriate test for model adequacy is

$$H_0 \colon \beta_1 = 0$$

$$H_a \colon \beta_1 > 0$$

FIGURE 3.28

SPSS regression through the origin output

Model Summary

Model	R	R Square[a]	Adjusted R Square	Std. Error of the Estimate
1	.981[b]	.961	.955	.387

▶ a. For regression through the origin (the no-intercept model), R Square measures the proportion of the variability in the dependent variable about the origin explained by regression. This CANNOT be compared to R Square for models which include an intercept.

b. Predictors: USERS

ANOVA[c,d]

Model		Sum of Squares	df	Mean Square	F	Sig.
1	Regression	22.401	1	22.401	149.528	.000[a]
	Residual	.899	6	.150		
	Total	23.300[b]	7			

a. Predictors: USERS

b. This total sum of squares is not corrected for the constant because the constant is zero for regression through the origin.

c. Dependent Variable: RESPTIME

d. Linear Regression through the Origin

Coefficients[a,b]

Model		Unstandardized Coefficients		Standardized Coefficients	t	Sig.
		B	Std. Error	Beta		
1	USERS	7.944E-02	.006	.981	12.228	.000

a. Dependent Variable: RESPTIME

b. Linear Regression through the Origin

FIGURE 3.29

The least squares line $\hat{y} = .07944x$

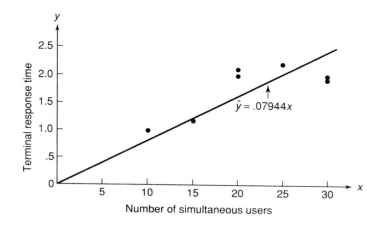

(We choose to do an upper-tailed test since it is reasonable to assume that if a linear relationship exists between number of users x and response time y, it is a positive one.)

To calculate the test statistic, we first compute s, where

$$s = \sqrt{\frac{SSE}{n-1}} = \sqrt{\frac{.899}{6}}$$

$$= .387$$

(This value is highlighted of the top of Figure 3.28.)

Then the test statistic is

$$t = \frac{\hat{\beta}_1}{s/\sqrt{\sum x_i^2}} = \frac{.07944}{.387/\sqrt{3,550}}$$

$$= 12.23$$

Note that this value is also shown (highlighted) on the SPSS printout, Figure 3.28. For $\alpha = .05$, we will reject the null hypothesis if $t > t_{.05}$, where the distribution of t is based on $(n-1) = 6df$. From Table 2 of Appendix C, $t_{.05} = 1.943$. Thus, we will reject H_0 if

$$t > 1.943$$

Since the calculated value of t, 12.23, exceeds the critical value, there is sufficient evidence (at $\alpha = .05$) to conclude that the model is adequate for predicting terminal response time y. [*Note*: The same decision can be reached by comparing $\alpha = .05$ to the p-value of the test, highlighted on the SPSS printout.]

To calculate a 95% prediction interval for the response time y when 20 users are signed on to the system, we first substitute $x = 20$ into the least squares prediction equation $\hat{y} = .07944x$:

$$\hat{y} = .07944(20) = 1.589$$

From Table 2 of Appendix C, $t_{.025} = 2.447$ (based on 6 df). Then, our 95% prediction interval is

$$\hat{y} \pm (t_{.025})s\sqrt{1 + \frac{x_p^2}{\sum x_i^2}}$$

$$= 1.589 \pm 2.447(.387)\sqrt{1 + \frac{(20)^2}{3,550}}$$

$$= 1.589 \pm 2.447(.387)(1.055)$$

$$= 1.589 \pm .999 \quad \text{or } (.589, 2.587)$$

[*Note*: This 95% prediction interval is highlighted on the SPSS spreadsheet printout, Figure 3.30.] We predict with 95% confidence that the terminal response time will range between .589 seconds and 2.587 seconds when 20 users are signed on to the system.   ◆

FIGURE 3.30

SPSS spreadsheet with 95% prediction limits

↓	users	resptime	predict	low95pi	upp95pi	
1	10	1	.79437	-.16597	1.75470	
2	15	1	1.19155	.21491	2.16819	
3	20	2	1.58873	.58971	2.58776	
4	20	2	1.58873	.58971	2.58776	
5	25	2	1.98592	.95883	3.01300	
6	30	2	2.38310	1.32273	3.44347	
7	30	2	2.38310	1.32273	3.44347	

Warning: There are several situations where it is dangerous to fit the model $E(y) = \beta_1 x$. If you are not certain that the regression line passes through the origin, it is a safe practice to fit the more general model $E(y) = \beta_0 + \beta_1 x$. If the line of means does, in fact, pass through the origin, the estimate of β_0 will differ from the true value $\beta_0 = 0$ by only a small amount. For all practical purposes, the least squares prediction equations will be the same.

On the other hand, you may know that the regression passes through the origin (see Example 3.6), but are uncertain about whether the true relationship between y and x is linear or curvilinear. In fact, most theoretical relationships are *curvilinear*. Yet, we often fit a linear model to the data in such situations because we believe that a straight line will make a good approximation to the mean response $E(y)$ over the region of interest. The problem is that this straight line is not likely to pass through the origin (see Figure 3.31). By forcing the regression line through the origin, we may not obtain a very good approximation to $E(y)$. For these reasons, regression through the origin should be used with extreme caution.

FIGURE 3.31

Using a straight line to approximate a curvilinear relationship when the true relationship passes through the origin

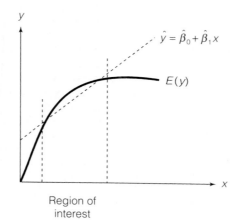

EXERCISES

3.56. Consider the eight data points shown in the table.

⊙ EX3_56

x	-4	-2	0	2	4	6	8	10
y	-12	-7	0	6	14	21	24	31

a. Fit a straight-line model through the origin; i.e., fit $E(y) = \beta_1 x$.
b. Calculate SSE, s^2, and s.
c. Do the data provide sufficient evidence to indicate that x and y are positively linearly related?
d. Construct a 95% confidence interval for β_1.
e. Construct a 95% confidence interval for $E(y)$ when $x = 7$.
f. Construct a 95% prediction interval for y when $x = 7$.

3.57. Consider the five data points shown in the table.

⊙ EX3_57

x	0	1	2	3	4
y	0	-8	-20	-30	-35

a. Fit a straight-line model through the origin; i.e., fit $E(y) = \beta_1 x$.
b. Calculate SSE, s^2, and s.
c. Do the data provide sufficient evidence to indicate that x and y are negatively linearly related?
d. Construct a 95% confidence interval for β_1.
e. Construct a 95% confidence interval for $E(y)$ when $x = 1$.
f. Construct a 95% prediction interval for y when $x = 1$.

3.58. Consider the 10 data points shown in the table.

⊙ EX3_58

x	30	50	70	90	100	120	140	160	180	200
y	4	10	15	21	21	22	29	34	39	41

a. Fit a straight-line model through the origin; i.e., fit $E(y) = \beta_1 x$.
b. Calculate SSE, s^2, and s.
c. Do the data provide sufficient evidence to indicate that x and y are positively linearly related?
d. Construct a 95% confidence interval for β_1.
e. Construct a 95% confidence interval for $E(y)$ when $x = 125$.
f. Construct a 95% prediction interval for y when $x = 125$.

3.59. A pharmaceutical company has developed a new drug designed to reduce a smoker's reliance on tobacco. Since certain dosages of the drug may reduce one's pulse rate to dangerously low levels, the product-testing division of the pharmaceutical company wants to model the relationship between decrease in pulse rate, y (beats/minute), and dosage, x (cubic centimeters). Different dosages of the drug were administered to eight randomly selected patients, and 30 minutes later the decrease in each patient's pulse rate was recorded. The results are given in the accompanying table. Initially, the company considered the model $y = \beta_1 x + \varepsilon$ since, in theory, a patient who receives a dosage of $x = 0$ should show no decrease in pulse rate $(y = 0)$.

⊙ PULSE

PATIENT	DOSAGE x, cubic centimeters	DECREASE IN PULSE RATE y, beats/minute
1	2.0	12
2	4.0	20
3	1.5	6
4	1.0	3
5	3.0	16
6	3.5	20
7	2.5	13
8	3.0	18

a. Fit a straight-line model that passes through the origin.
b. Is there evidence of a linear relationship between drug dosage and decrease in pulse rate? Test at $\alpha = .10$.
c. Find a 99% prediction interval for the decrease in pulse rate corresponding to a dosage of 3.5 cubic centimeters.

3.60. Consider the relationship between the total weight of a shipment of 50-pound bags of flour, y, and the number of bags in the shipment, x. Since a shipment containing $x = 0$ bags (i.e., no shipment at all) has a total weight of $y = 0$, a straight-line model of the relationship between x and y should pass through the point $x = 0$, $y = 0$. Hence, the appropriate model might be

$$y = \beta_1 x + \varepsilon$$

From the records of past flour shipments, 15 shipments were randomly chosen and the data in the following table recorded.

FLOUR

WEIGHT OF SHIPMENT	NUMBER OF 50-POUND BAGS IN SHIPMENT
5,050	100
10,249	205
20,000	450
7,420	150
24,685	500
10,206	200
7,325	150
4,958	100
7,162	150
24,000	500
4,900	100
14,501	300
28,000	600
17,002	400
16,100	400

a. Find the least squares line for the given data under the assumption that $\beta_0 = 0$. Plot the least squares line on a scatterplot of the data.

b. Find the least squares line for the given data using the model

$$y = \beta_0 + \beta_1 x + \varepsilon$$

(i.e., do not restrict β_0 to equal 0). Plot this line on the scatterplot you constructed in part **a**.

c. Refer to part **b**. Why might $\hat{\beta}_0$ be different from 0 even though the true value of β_0 is known to be 0?

d. The estimated standard error of $\hat{\beta}_0$ is equal to

$$s\sqrt{\frac{1}{n} + \frac{\bar{x}^2}{SS_{xx}}}$$

Use the t statistic,

$$t = \frac{\hat{\beta}_0 - 0}{s\sqrt{\dfrac{1}{n} + \dfrac{\bar{x}^2}{SS_{xx}}}}$$

to test the null hypothesis $H_0 : \beta_0 = 0$ against the alternative $H_a : \beta_0 \neq 0$. Use $\alpha = .10$. Should you include β_0 in your model?

3.61. To satisfy the Public Service Commission's energy conservation requirements, an electric utility company must develop a reliable model for projecting the number of residential electricity customers in its service area. The first step is to study the effect of changing population on the number of electricity customers. The information shown in the table was obtained for the service area from 1993 to 2002. Since a service area with 0 population obviously would have 0 residential electricity customers, one could argue that regression through the origin is appropriate.

a. Fit the model $y = \beta_1 x + \varepsilon$ to the data.

b. Is there evidence that x contributes information for the prediction of y? Test using $\alpha = .01$.

c. Now fit the more general model $y = \beta_0 + \beta_1 x + \varepsilon$ to the data. Is there evidence (at $\alpha = .01$) that x contributes information for the prediction of y?

d. Which model would you recommend?

PSC

YEAR	POPULATION IN SERVICE AREA x, hundreds	RESIDENTIAL ELECTRICITY CUSTOMERS IN SERVICE AREA y
1993	262	14,041
1994	319	16,953
1995	361	18,984
1996	381	19,870
1997	405	20,953
1998	439	22,538
1999	472	23,985
2000	508	25,641
2001	547	27,365
2002	592	29,967

3.12
A Summary of the Steps to Follow in a Simple Linear Regression Analysis

We have introduced an extremely useful tool in this chapter—**the method of least squares** for fitting a prediction equation to a set of data. This procedure, along with associated statistical tests and estimations, is called a **regression analysis**. In six steps, we showed how to use sample data to build a model relating a dependent variable y to a single independent variable x.

Steps to Follow in a Simple Linear Regression Analysis

1. The first step is to hypothesize a **probabilistic model**. In this chapter, we confined our attention to the **first-order (straight-line) model**

$$y = \beta_0 + \beta_1 x + \varepsilon$$

2. The second step is to collect the (x, y) pairs for each experimental unit in the sample.

3. The third step is to use the method of least squares to estimate the unknown parameters in the **deterministic component**, $\beta_0 + \beta_1 x$. The least squares estimates yield a model $\hat{y} = \hat{\beta}_0 \, \hat{\beta}_1 x$ with a **sum of squared errors (SSE)** that is smaller than the SSE for any other straight-line model.

4. The fourth step is to specify the probability distribution of the **random error component ε**.

5. The fifth step is to assess the utility of the hypothesized model. Included here are making inferences about the **slope β_1**, interpreting the **coefficient of correlation r**, and interpreting the **coefficient of determination r^2**.

6. Finally, if we are satisfied with the model, we are prepared to use it. We can use the model to **estimate the mean y value**, $E(y)$, for a given x value and to **predict an individual y value** for a specific value of x.

The concepts introduced in this chapter will be developed more fully in Chapter 4.

SUPPLEMENTARY EXERCISES

3.62. Any medical item used in the care of hospital patients is called a *factor*. For example, factors can be intravenous tubing, intravenous fluid, needles, shave kits, bedpans, diapers, dressings, medications, and even code carts. The coronary care unit at Bayonet Point Hospital (St. Petersburg, Florida) investigated the relationship between the number of factors per patient, x, and the patient's length of stay (in days), y. The data for a random sample of 50 coronary care patients are given in the following table, while a SAS printout of the simple linear regression analysis is shown on page 156.

FACTORS

NUMBER OF FACTORS x	LENGTH OF STAY y, days	NUMBER OF FACTORS x	LENGTH OF STAY y, days
231	9	354	11
323	7	142	7
113	8	286	9
208	5	341	10
162	4	201	5

(Continued)

FACTORS

NUMBER OF FACTORS x	LENGTH OF STAY y, days	NUMBER OF FACTORS x	LENGTH OF STAY y, days
117	4	158	11
159	6	243	6
169	9	156	6
55	6	184	7
77	3	115	4
103	4	202	6
147	6	206	5
230	6	360	6
78	3	84	3
525	9	331	9
121	7	302	7
248	5	60	2
233	8	110	2
260	4	131	5
224	7	364	4
472	12	180	7
220	8	134	6
383	6	401	15
301	9	155	4
262	7	338	8

Source: Bayonet Point Hospital, Coronary Care Unit.

SAS Output for Exercise 3.62

```
                              Dependent Variable: LOS

                                  Analysis of Variance

                                       Sum of          Mean
       Source              DF         Squares        Square    F Value    Pr > F

       Model                1       126.58393     126.58393      28.68    <.0001
       Error               48       211.83607       4.41325
       Corrected Total     49       338.42000


                Root MSE               2.10077    R-Square     0.3740
                Dependent Mean         6.54000    Adj R-Sq     0.3610
                Coeff Var             32.12193


                                  Parameter Estimates

                      Parameter      Standard
  Variable     DF      Estimate         Error    t Value   Pr > |t|    95% Confidence Limits

  Intercept     1       3.30603       0.67297       4.91    <.0001     1.95293      4.65914
  FACTORS       1       0.01475       0.00276       5.36    <.0001     0.00922      0.02029

                                       Predictions

                                                    Lower         Upper
                                       Predicted    prediction    prediction
           Obs     FACTORS     LOS        LOS        limit of LOS  limit of LOS

            1        231        9       6.7144       6.11348       7.3153
            2        354       11       8.5292       7.57292       9.4856
            3        323        7       8.0718       7.24266       8.9010
            4        142        7       5.4012       4.66664       6.1358
            5        113        8       4.9733       4.13502       5.8116
```

a. Construct a scatterplot of the data.

b. Find the least squares line for the data and plot it on your scatterplot.

c. Define β_1 in the context of this problem.

d. Test the hypothesis that the number of factors per patient (x) contributes no information for the prediction of the patient's length of stay (y) when a linear model is used (use $\alpha = .05$). Draw the appropriate conclusions.

e. Find a 95% confidence interval for β_1. Interpret your results.

f. Find the coefficient of correlation for the data. Interpret your results.

g. Find the coefficient of determination for the linear model you constructed in part b. Interpret your result.

h. Find a 95% prediction interval for the length of stay of a coronary care patient who is administered a total of $x = 231$ factors.

i. Explain why the prediction interval obtained in part **h** is so wide. How could you reduce the width of the interval?

3.63. Civil engineers often use the straight-line equation $E(y) = \hat{\beta}_0 + \hat{\beta}_1 x$ to model the relationship between the mean shear strength $E(y)$ of masonry joints and precompression stress x. To test this theory, a series of stress tests was performed on solid bricks arranged in triplets and joined with mortar (*Proceedings of the Institute of Civil Engineers*, Mar. 1990). The precompression stress was varied for each triplet, and the ultimate shear load just before failure (called the shear strength) was recorded. The stress results for seven triplets (measured in newtons per square millimeter) is shown in the table.

◎ STRESS

TRIPLET TEST	1	2	3	4	5	6	7
Shear strength, y	1.00	2.18	2.24	2.41	2.59	2.82	3.06
Precompression stress, x	0	.60	1.20	1.33	1.43	1.75	1.75

Source: Riddington, J. R., and Ghazali, M. Z. "Hypothesis for shear failure in masonry joints." *Proceedings of the Institute of Civil Engineers, Part 2*, Mar. 1990. Vol. 89, p. 96 (Fig. 7).

a. Graph the seven data points in a scatterplot. Does the relationship between shear strength and precompression stress appear to be linear?

b. Use the method of least squares to estimate the parameters of the linear model.

c. Interpret the values of $\hat{\beta}_0$ and $\hat{\beta}_1$.

3.64. Refer to the *Chemosphere* (Vol. 20, 1990) study of Vietnam veterans exposed to Agent Orange (and the dioxin 2,3,7,8-TCDD), Exercise 1.87 (p. 76). The table below gives the amounts of 2,3,7,8-TCDD (measured in parts per trillion) in both blood plasma and fat tissue drawn from each of the 20 veterans studied. One goal of the researchers is to determine the degree of linear association between the level of dioxin found in blood plasma and fat tissue. If a linear association between the two variables can be established, the researchers want to build models to predict (1) the blood plasma level of 2,3,7,8-TCDD from the observed level of 2,3,7,8-TCDD in fat tissue and (2) the fat tissue level from the observed blood plasma level.

a. Find the prediction equations for the researchers. Interpret the results.

b. Test the hypothesis that fat tissue level (x) is a useful linear predictor of blood plasma level (y). Use $\alpha = .05$.

c. Test the hypothesis that blood plasma level (x) is a useful linear predictor of fat tissue level (y). Use $\alpha = .05$.

d. Intuitively, why must the results of the tests, parts **b** and **c**, agree?

◎ AGORANGE

VETERAN	TCDD LEVELS IN PLASMA	TCDD LEVELS IN FAT TISSUE
1	2.5	4.9
2	3.1	5.9
3	2.1	4.4
4	3.5	6.9
5	3.1	7.0
6	1.8	4.2
7	6.0	10.0
8	3.0	5.5
9	36.0	41.0
10	4.7	4.4
11	6.9	7.0
12	3.3	2.9
13	4.6	4.6
14	1.6	1.4
15	7.2	7.7
16	1.8	1.1
17	20.0	11.0
18	2.0	2.5
19	2.5	2.3
20	4.1	2.5

Source: Schecter, A., et al. "Partitioning of 2,3,7,8-chlorinated dibenzo-*p*-dioxins and dibenzofurans between adipose tissue and plasma-lipid of 20 Massachusetts Vietnam veterans." *Chemosphere*, Vol. 20, Nos. 7–9, 1990, pp. 954–955 (Tables I and II).

Results for Exercise 3.65

CONGENER	$y =$ FAT TISSUE LEVEL $x =$ BLOOD PLASMA LEVEL	$y =$ BLOOD PLASMA LEVEL $x =$ FAT TISSUE LEVEL	t VALUE FOR TESTING β_1
2,3,4,7,8-P_n CDF	$\hat{y} = .8109 + .9713x$	$\hat{y} = .9855 + .7605x$	7.13
H_x CDD	$\hat{y} = 18.1565 + .7377x$	$\hat{y} = 5.2009 + .9018x$	5.98
OCDD	$\hat{y} = 118.6057 + .3679x$	$\hat{y} = 167.723 + 1.5752x$	4.98

Source: Schecter, A., et al. "Partitioning of 2,3,7,8-chlorinated dibenzo-*p*-dioxins and dibenzofurans between adipose tissue and plasma lipid of 20 Massachusetts Vietnam veterans." *Chemosphere*, Vol. 20, Nos. 7–9, 1990, pp. 954–955 (Table III).

3.65. Refer to Exercise 3.64. The blood plasma and fat tissue levels of several other types of dioxin (called congeners) were also measured for each of the 20 Vietnam veterans. For each congener, a simple linear regression analysis was conducted to predict (1) fat tissue level from blood plasma level and (2) blood plasma level from fat tissue level. The results for three of these congeners are shown in the table above.

a. For the congener 2,3,4,7,8,-P_n CDF, are the two regression models statistically adequate for predicting y? Test both using $\alpha = .05$.

b. Repeat part **a** for the congener H_x CDD.

c. Repeat part **a** for the congener OCDD.

d. Use the regression results to predict the level of 2,3,4,7,8,-P_n CDF in the blood plasma for a veteran with a fat tissue level of 8.0 ppt.

e. Use the regression results to predict the level of H_x CDD in fat tissue for a veteran with a blood plasma level of 24.0 ppt.

f. Use the regression results to predict the level of OCDD in blood plasma for a veteran with a fat tissue level of 776 ppt.

3.66. Investors in real estate investment trust (REIT) stock generally prefer REITs that are internally managed and that reward management both for share performance and for performance relative to industry competitors. To better understand management compensation patterns in the REIT industry, the *Real Estate Review* (Summer 1993) studied a sample of 16 internally managed REITs. The main purpose was to establish a relationship between CEO compensation and REIT performance. Data collected on the 16 REITs from the National Association of Real Estate Investment Trusts revealed a correlation coefficient of $r = .328$ between CEO cash compensation and the 1-year annualized total return

for the REIT. Is there sufficient evidence to establish a positive linear relationship between CEO cash compensation and REIT performance? Test using $\alpha = .05$.

3.67. Researchers at the University of North Carolina–Greensboro investigated a model for the rate of seed germination (*Journal of Experimental Botany*, Jan. 1993). In one experiment, alfalfa seeds were placed in a specially constructed germination chamber. Eleven hours later, the seeds were examined and the change in free energy (a measure of germination rate) recorded. The results for seeds germinated at seven different temperatures are given in the table. The data were used to fit a simple linear regression model, with $y =$ change in free energy and $x =$ temperature.

⬤ SEEDGERM

CHANGE IN FREE ENERGY, kj/mol	TEMPERATURE, °K
7	295
6.2	297.5
9	291
9.5	2895
8.5	3015
7.8	2935
11.2	286.5

Source: Hageseth, G. T., and Cody, A. L. "Energy-level model for isothermal seed germination." *Journal of Experimental Botany*, Vol. 44, No. 258, Jan. 1993, p. 123 (Figure 9).

a. Graph the points in a scatterplot.

b. Find the least squares prediction equation.

c. Plot the least squares line, part **b**, on the scatterplot of part **a**.

d. Conduct a test of model adequacy. Use $\alpha = .01$.

e. Use the plot, part **c**, to locate any unusual data points (outliers).

f. Eliminate the outlier, part **e**, from the data set, and repeat parts **a–d**.

3.68. Passive exposure to environmental tobacco smoke has been associated with growth suppression and an increased frequency of respiratory tract infections in normal children. Is this association more pronounced in children with cystic fibrosis? To answer this question, 43 children (18 girls and 25 boys) attending a 2-week summer camp for cystic fibrosis patients were studied (*New England Journal of Medicine*, Sept. 20, 1990). Among several variables measured were the child's weight percentile (y) and the number of cigarettes smoked per day in the child's home (x).

a. For the 18 girls, the coefficient of correlation between y and x was reported as $r = -.50$. Interpret this result.

b. Refer to part a. The p-value for testing $H_0: \rho = 0$ against $H_a: \rho \neq 0$ was reported as $p = .03$. Interpret this result.

c. For the 25 boys, the coefficient of correlation between y and x was reported as $r = -.12$. Interpret this result.

d. Refer to part c. The p-value for testing $H_0: \rho = 0$ against $H_a: \rho \neq 0$ was reported as $p = .57$. Interpret this result.

3.69. The Environmental Protection Agency (EPA) evaluates state pollution control policies through the use of an emissions-to-job (E/J) ratio. The E/J ratio is obtained by dividing the amount (in pounds) of annual toxic emissions of an industry in a state by the number of jobs the state provides in that industry. *Environmental Technology* (Oct. 1993) investigated the relationship between the E/J ratio and spending on pollution control in the chemical industry. Data collected for $n = 19$ large chemical-producing states were used to conduct a simple linear regression analysis, where x = a state's pollution abatement capital expenditures (PACE), in millions of dollars, and y = a state's chemical industry E/J ratio, in pounds per job. [*Note*: Positive x represents overspending on pollution control, whereas negative x represents underspending.] The analysis yielded a least squares line with a negative slope and $r^2 = .587$.

a. Interpret the value of r^2.

b. Calculate the correlation coefficient r and interpret the result.

c. In theory, underspending on pollution control will result in higher emissions and fewer jobs (i.e., a higher E/J ratio). Test the theory (at

$\alpha = .01$) using the results of the straight-line regression.

3.70. The *American Scientist* (July–Aug. 1998) reported on a study of the relationship between self-avoiding and unrooted walks. A self-avoiding walk is one where you never retrace or cross your own path; an unrooted walk is a path in which the starting and ending points are impossible to distinguish. The possible number of walks of each type of various lengths are reproduced in the table. Consider the straight-line model $y = \beta_0 + \beta_1 x + \varepsilon$, where x is walk length (number of steps).

⊛ WALKS

WALK LENGTH (number of steps)	UNROOTED WALKS	SELF-AVOIDING WALKS
1	1	4
2	2	12
3	4	36
4	9	100
5	22	284
6	56	780
7	147	2,172
8	388	5,916

Source: Hayes, B. "How to avoid yourself." *American Scientist*, Vol. 86, No. 4, July–Aug. 1988, p. 317 (Figure 5).

a. Use the method of least squares to fit the model to the data if y is the possible number of unrooted walks.

b. Interpret $\hat{\beta}_0$ and $\hat{\beta}_1$ in the estimated model, part **a**.

c. Repeat parts **a** and **b** if y is the possible number of self-avoiding walks.

d. Find a 99% confidence interval for the number of unrooted walks possible when walk length is four steps.

e. Would you recommend using simple linear regression to predict the number of walks possible when walk length is 15 steps? Explain.

3.71. Common maize rust is a serious disease of sweet corn. Researchers in New York state have developed an action threshold for initiation of fungicide applications based on a regression equation relating maize rust incidence to severity of the disease (*Phytopathology*, Vol. 80, 1990). In one particular field, data were collected on more than 100 plants of the sweet corn hybrid Jubilee. For each plant, incidence was measured as the percentage of leaves infected (x) and severity was calculated as the log (base 10) of the average number of infections per leaf (y). A simple linear regression analysis of the data produced the following results:

$$\hat{y} = -.939 + .020x$$

$$r^2 = .816$$

$$s = .288$$

a. Interpret the value of $\hat{\beta}_1$.
b. Interpret the value of r^2.
c. Interpret the value of s.
d. Calculate the value of r and interpret it.
e. Use the result, part **d**, to test the utility of the model. Use $\alpha = .05$. (Assume $n = 100$.)
f. Predict the severity of the disease when the incidence of maize rust for a plant is 80%. [*Note:* Take the antilog (base 10) of $\hat{y}$ to obtain the predicted average number of infections per leaf.

3.72. At temperatures approaching absolute zero ($-273°C$), helium exhibits traits that seem to defy many laws of Newtonian physics. An experiment has been conducted with helium in solid form at various temperatures near absolute zero. The solid helium is placed in a dilution refrigerator along with a solid impure substance, and the fraction (in weight) of the impurity passing through the solid helium is recorded. (This phenomenon of solids passing directly through solids is known as *quantum tunneling*.) The data are given in the table.

💿 HELIUM

TEMPERATURE, $x(°C)$	PROPORTION OF IMPURITY, y
−262.0	.315
−265.0	.202
−256.0	.204
−267.0	.620
−270.0	.715
−272.0	.935
−272.4	.957
−272.7	.906
−272.8	.985
−272.9	.987

a. Find the least squares estimates of the intercept and slope. Interpret them.
b. Use a 95% confidence interval to estimate the slope β_1. Interpret the interval in terms of this application. Does the interval support the hypothesis that temperature contributes information about the proportion of impurity passing through helium?
c. Interpret the coefficient of determination for this model.

d. Find a 95% prediction interval for the percentage of impurity passing through solid helium at $-273°C$. Interpret the result.
e. Note that the value of x in part **d** is outside the experimental region. Why might this lead to an unreliable prediction?

3.73. Neurologists have found that the hippocampus, a structure found in the brain, plays an important role in short-term memory. The *American Journal of Psychiatry* (July 1995) published a study of the relationship between hippocampal volume and short-term verbal memory of 21 Vietnam vets with combat related post-traumatic stress disorder (PTSD). Magnetic resonance imaging was used to measure the volume, x, of the right hippocampus (in cubic millimeters) of each subject while verbal memory retention, y, of each subject was measured by the percent retention subscale of the Wechsler Memory Scale. The scatterplot for the data is reproduced here.

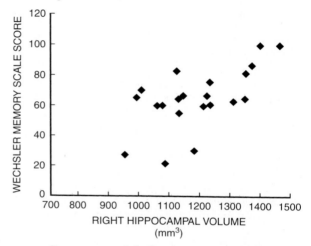

a. Propose a straight-line model relating verbal memory y with right hippocampal volume x to support the theory that smaller hippocampal volume would be associated with deficits in short-term verbal memory in patients with PTSD.
b. Based on the theory, would you expect the slope of the line to be positive or negative? Explain.
c. The coefficient of correlation between right hippocampal volume and verbal retention was $r = .64$. Interpret this value.
d. A statistical test of $H_0: \beta_1 = 0$ versus $H_a: \beta_1 > 0$ resulted in a p-value smaller than .05. Interpret this result.

3.74. Is there a link between the loneliness of parents and their offspring? Psychologists J. Lobdell and D. Perlman examined this question in an article published in the *Journal of Marriage and the Family* (Aug. 1986). The participants in the study were 130 female college undergraduates and their parents. Each triad of daughter, mother, and father completed the UCLA Loneliness Scale, a 20-item questionnaire designed to assess loneliness and several variables theoretically related to loneliness, such as social accessibility to others, difficulty in making friends, and depression. Pearson product moment correlations relating a daughter's loneliness score to her parents' loneliness scores as well as the other variables were calculated. The results are summarized below.

| | CORRELATION (r) BETWEEN DAUGHTER'S LONELINESS AND PARENTAL VARIABLES | |
VARIABLE	MOTHER	FATHER
Loneliness	.26	.19
Depression	.11	.06
Self-esteem	−.14	−.06
Assertiveness	−.05	.01
Number of friends	−.21	−.10
Quality of friendships	−.17	.01

Source: Lobdell, J., and Perlman, D. "The intergenerational transmission of loneliness: A study of college females and their parents." *Journal of Marriage and the Family*, Vol. 48, No. 8, Aug. 1986, p. 592. Copyright 1986 by the National Council on Family Relations, 3989 Central Ave., N.E., Suite #550, Minneapolis, MN 55421.

a. Lobdell and Perlman conclude that "mother and daughter loneliness scores were (positively) significantly correlated at $\alpha = .01$." Do you agree?

b. Determine which, if any, of the other sample correlations are large enough to indicate (at $\alpha = .01$) that linear correlation exists between the daughter's loneliness score and the variable measured.

c. Explain why it would be dangerous to conclude that a causal relationship exists between a mother's loneliness and her daughter's loneliness.

d. Explain why it would be dangerous to conclude that the variables with nonsignificant correlations in the table are unrelated.

3.75. At major colleges and universities, administrators (e.g., deans, chairpersons, provosts, vice presidents, and presidents) are among the highest-paid state employees. Is there a relationship between the raises administrators receive and their performance on the job? This was the question of interest to a group of faculty union members at the University of South Florida called the United Faculty of Florida (UFF). The UFF compared

the April 1990 ratings of 15 University of South Florida administrators (as determined by faculty in a survey) to their subsequent raises in August 1990. The data for the analysis is listed in the accompanying table. [*Note*: Ratings are measured on a 5-point scale, where $1 =$ very poor and $5 =$ very good.] According to the UFF, the "relationship is inverse; i.e., the lower the rating by the faculty, the greater the raise. Apparently, bad administrators are more valuable than good administrators."* (With tongue in cheek, the UFF refers to this phenomenon as "the SOB effect.") The UFF based its conclusions on a simple linear regression analysis of the data in the table, where $y =$ administrator's raise and $x =$ average rating of administrator.

 UFFSAL

ADMINISTRATOR	PRAISE[a]	AVERAGE RATING (5-pt scale)[b]
1	$18,000	2.76
2	16,700	1.52
3	15,787	4.40
4	10,608	3.10
5	10,268	3.83
6	9,795	2.84
7	9,513	2.10
8	8,459	2.38
9	6,099	3.59
10	4,557	4.11
11	3,751	3.14
12	3,718	3.64
13	3,652	3.36
14	3,227	2.92
15	2,808	3.00

Source: [a]Faculty and A&P Salary Report, University of South Florida, Resource Analysis and Planning, 1990. [b]Administrative Compensation Survey, *Chronicle of Higher Education*, Jan. 1991.

a. Initially, the UFF conducted the analysis using all 15 data points in the table. Fit a straight-line model to the data. Is there evidence to support the UFF's claim of an inverse relationship between raise and rating?

b. A second simple linear regression was performed using only 14 of the data points in the table. The data for administrator #3 was eliminated because he was promoted to dean in the middle of the 1989–1990 academic year. (No other reason was given for removing this data point from the analysis.) Perform the simple linear regression analysis using the remaining 14 data points in the table. Is

*UFF Faculty Forum, University of South Florida Chapter, Vol. 3, No. 5, May 1991.

there evidence to support the UFF's claim of an inverse relationship between raise and rating?

c. Based on the results of the regression, part **b**, the UFF computed estimated raises for selected faculty ratings of administrators. These are shown in the following table. What problems do you perceive with using this table to estimate administrators' raises at the University of South Florida?

d. The ratings of administrators listed in this table were determined by surveying the faculty at the University of South Florida. All faculty are mailed the survey each year, but the response rate is typically low (approximately 10–20%). The danger with such a survey is that only disgruntled faculty, who are more apt to give a low rating to an administrator, will respond. Many of the faculty also believe that they are underpaid and that the administrators

are overpaid. Comment on how such a survey could bias the results shown here.

RATINGS		RAISE
Very Poor	1.00	$15,939
	1.50	13,960
Poor	2.00	11,980
	2.50	10,001
Average	3.00	8,021
	3.50	6,042
Good	4.00	4,062
	4.50	2,083
Very Good	5.00	103

e. Based on your answers to the previous questions, would you support the UFF's claim?

REFERENCES

DRAPER, N., and SMITH, H. *Applied Regression Analysis*, 3rd ed. New York: Wiley, 1987.

MONTGOMERY, D. C., PECK, E. A., and VINING, G. G. *Introduction to Linear Regression Analysis*. 3rd ed. New York: Wiley, 2001.

CHATTERJEE, S., and PRICE, B. *Regression Analysis by Example*, 2nd ed. New York: Wiley, 1991.

MENDENHALL, W. *Introduction to Linear Models and the Design and Analysis of Experiments*. Belmont, Calif: Wadsworth, 1968.

MOSTELLER, F., and TUKEY, J. W. *Data Analysis and Regression: A Second Course in Statistics*. Reading, Mass.: Addison-Wesley, 1977.

NETER, J., KUTNER, M., NACHTSHEIM, C., and WASSERMAN, W. *Applied Linear Statistical Models*, 4th ed. Homewood, Ill.: Richard Irwin, 1996.

ROUSSEEUW, P. J., and LEROY, A. M. *Robust Regression and Outlier Detection*. New York: Wiley, 1987.

CHAPTER

4

Multiple Regression Models

CONTENTS

OBJECTIVE

To extend the methods of Chapter 3; to develop a procedure for predicting a response y based on the values of two or more independent variables; to illustrate the types of practical inferences that can be drawn from this type of analysis

---

4.1 General Form of a Multiple Regression Model

Most practical applications of regression analysis utilize models that are more complex than the first-order (straight-line) model. For example, a realistic probabilistic model for monthly sales revenue would include more than just the advertising expenditure discussed in Chapter 3 to provide a good predictive model for sales. Factors such as season, inventory on hand, sales force, and productivity are a few of the many variables that might influence sales. Thus, we would want to incorporate these and other potentially important independent variables into the model if we need to make accurate predictions.

Probabilistic models that include more than one independent variable are called **multiple regression models**. The general form of these models is shown in the box.

The dependent variable y is now written as a function of k independent variables, $x_1, x_2, \ldots, x_k$. The random error term is added to make the model probabilistic rather than deterministic. The value of the coefficient β_i determines the contribution of the independent variable x_i, given that the other $(k-1)$ independent variables are held constant, and β_0 is the y-intercept. The coefficients $\beta_0, \beta_1, \ldots, \beta_k$ will usually be unknown, since they represent population parameters.

General Form of the Multiple Regression Model

$$y = \beta_0 + \beta_1 x_1 + \beta_2 x_2 + \cdots + \beta_k x_k + \varepsilon$$

where y is the dependent variable

$x_1, x_2, \ldots, x_k$ are the independent variables

$E(y) = \beta_0 + \beta_1 x_1 + \beta_2 x_2 + \cdots + \beta_k x_k$ is the deterministic portion of the model

β_i determines the contribution of the independent variable x_i

Note: The symbols $x_1, x_2, \ldots, x_k$ may represent higher-order terms. For example, x_1 might represent the current interest rate, x_2 might represent x_1^2, and so forth.

At first glance it might appear that the regression model shown here would not allow for anything other than straight-line relationships between y and the independent variables, but this is not true. Actually, $x_1, x_2, \ldots, x_k$ can be functions of variables as long as the functions do not contain unknown parameters. For example, the carbon monoxide content y of smoke emitted from a cigarette could be a function of the independent variables

$$x_1 = \text{Tar content}$$

$$x_2 = (\text{Tar content})^2 = x_1^2$$

$$x_3 = 1 \text{ if a filter cigarette, } 0 \text{ if a nonfiltered cigarette}$$

The x_2 term is called a **higher-order term** because it is the value of a quantitative variable (x_1) squared (i.e., raised to the second power). The x_3 term is a **coded variable** representing a qualitative variable (filter type). The multiple regression model is quite versatile and can be made to model many different types of response variables.

The steps we followed in developing a straight-line model are applicable to the multiple regression model.

Analyzing a Multiple Regression Model

STEP 1   Collect the sample data, i.e., the values of $y, x_1, x_2, \ldots, x_k$, for each experimental unit in the sample.

STEP 2   Hypothesize the form of the model, i.e., the deterministic component, $E(y)$. This involves choosing which independent variables to include in the model.

STEP 3   Use the method of least squares to estimate the unknown parameters $\beta_0, \beta_1, \ldots, \beta_k$.

STEP 4   Specify the probability distribution of the random error component ε and estimate its variance σ^2.

STEP 5   Statistically evaluate the utility of the model.

STEP 6   Check that the assumptions on σ are satisfied and make model modifi-cations, if necessary.

STEP 7   Finally, if the model is deemed adequate, use the fitted model to estimate the mean value of y or to predict a particular value of y for given values of the independent variables, and to make other inferences.

Hypothesizing the form of the model (step 2) is the subject of Chapter 5. In this chapter, we will assume that the form of the model is known, and we will discuss steps 3–6 for a given model.

4.2 Model Assumptions

We noted in Section 4.1 that the multiple regression model is of the form

$$y = \beta_0 + \beta_1 x_1 + \beta_2 x_2 + \cdots + \beta_k x_k + \varepsilon$$

where y is the response variable that you want to predict; $\beta_0, \beta_1, \ldots, \beta_k$ are parameters with unknown values; $x_1, x_2, \ldots, x_k$ are independent information-contributing variables that are measured without error; and ε is a random error component. Since $\beta_0, \beta_1, \ldots, \beta_k$ and $x_1, x_2, \ldots, x_k$ are nonrandom, the quantity

$$\beta_0 + \beta_1 x_1 + \beta_2 x_2 + \cdots + \beta_k x_k$$

represents the deterministic portion of the model. Therefore, y is made up of two com-ponents—one fixed and one random—and, consequently, y is a random variable.

$$y = \overbrace{\beta_0 + \beta_1 x_1 + \beta_2 x_2 + \cdots + \beta_k x_k}^{\substack{\text{Deterministic} \\ \text{portion of model}}} + \overbrace{\varepsilon}^{\substack{\text{Random} \\ \text{error}}}$$

We will assume (as in Chapter 3) that the random error can be positive or negative and that for any setting of the x values, $x_1, x_2, \ldots, x_k, \varepsilon$ has a normal probability distribution with mean equal to 0 and variance equal to σ^2. Further, we assume that the random errors associated with any (and every) pair of y values are probabilistically independent. That is, the error ε associated with any one y value is independent of the error associated with any other y value. These assumptions are summarized in the accompanying box.

Assumptions About the Random Error ε

1. For any given set of values of $x_1, x_2, \ldots, x_k, \varepsilon$ has a normal probability distribution with mean equal to 0 [i.e., $E(\varepsilon) = 0$] and variance equal to σ^2 [i.e., $\text{Var}(\varepsilon) = \sigma^2$].

2. The random errors are independent (in a probabilistic sense).

The assumptions that we have described for a multiple regression model imply that the mean value $E(y)$ for a given set of values of $x_1, x_2, \cdots, x_k$ is equal to

$$E(y) = \beta_0 + \beta_1 x_1 + \beta_2 x_2 + \cdots + \beta_k x_k$$

Models of this type are called **linear statistical models** because $E(y)$ is a *linear function* of the unknown parameters $\beta_0, \beta_1, \ldots, \beta_k$.

All the estimation and statistical test procedures described in this chapter depend on the data satisfying the assumptions described in this section. Since we will rarely, if ever, know for certain whether the assumptions are actually satisfied in practice, we will want to know how well a regression analysis works, and how much faith we can place in our inferences when certain assumptions are not satisfied. We will have more to say on this topic in Chapters 7 and 8. First, we need to discuss the methods of a regression analysis more thoroughly and show how they are used in a practical situation.

4.3
A First-Order Model with Quantitative Predictors

A model that includes only terms for *quantitative* independent variables, called a **first-order model**, is described in the box. Note that the first-order model does not include any higher-order terms (such as x_1^2). The term *first-order* is derived from the fact that each x in the model is raised to the first power.

A First-Order Model in Five Quantitative Independent Variables

$$E(y) = \beta_0 + \beta_1 x_1 + \beta_2 x_2 + \beta_3 x_3 + \beta_4 x_4 + \beta_5 x_5$$

where $x_1, x_2, \ldots, x_5$ are all quantitative variables that *are not* functions of other independent variables.

Note: β_i represents the slope of the line relating y to x_i when all the other x's are held fixed.

Recall that in the straight-line model (Chapter 3)

$$y = \beta_0 + \beta_1 x + \varepsilon$$

β_0 represents the y-intercept of the line and β_1 represents the slope of the line. From our discussion in Chapter 3, β_1 has a practical interpretation—it represents the mean change in y for every 1-unit increase in x. When the independent variables are quantitative, the β parameters in the first-order model specified in the box have similar interpretations. The difference is that when we interpret the β that multiplies one of the variables (e.g., x_1), we must be certain to hold the values of the remaining independent variables (e.g., x_2, x_3) fixed.

To see this, suppose that the mean $E(y)$ of a response y is related to two quantitative independent variables, x_1 and x_2, by the first-order model

$$E(y) = 1 + 2x_1 + x_2$$

In other words, $\beta_0 = 1$, $\beta_1 = 2$, and $\beta_2 = 1$.

Now, when $x_2 = 0$, the relationship between $E(y)$ and x_1 is given by

$$E(y) = 1 + 2x_1 + (0) = 1 + 2x_1$$

A graph of this relationship (a straight line) is shown in Figure 4.1. Similar graphs of the relationship between $E(y)$ and x_1 for $x_2 = 1$,

$$E(y) = 1 + 2x_1 + (1) = 2 + 2x_1$$

and for $x_2 = 2$,

$$E(y) = 1 + 2x_1 + (2) = 3 + 2x_1$$

also are shown in Figure 4.1. Note that the slopes of the three lines are all equal to $\beta_1 = 2$, the coefficient that multiplies x_1.

FIGURE 4.1

Graphs of $E(y) = 1 + 2x_1 + x_2$ for $x_2 = 0, 1, 2$

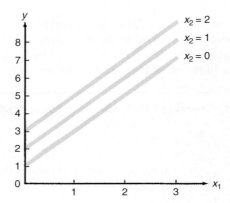

Figure 4.1 exhibits a characteristic of all first-order models: If you graph $E(y)$ versus any one variable—say, x_1—for fixed values of the other variables, the result will always be a *straight line* with slope equal to β_1. If you repeat the process for other values of the fixed independent variables, you will obtain a set of *parallel* straight lines. This indicates that the effect of the independent variable x_i on $E(y)$ is independent of all the other independent variables in the model, and this effect is measured by the slope β_i (as stated in the box).

The first-order model is the most basic multiple regression model encountered in practice. In the next several sections, we present an analysis of this model.

4.4
Fitting the Model: The Method of Least Squares

The method of fitting multiple regression models is identical to that of fitting the straight-line model of Chapter 3—namely, the method of least squares. That is, we choose the estimated model

$$\hat{y} = \hat{\beta}_0 + \hat{\beta}_1 x_1 + \cdots + \hat{\beta}_k x_k$$

that minimizes

$$\text{SSE} = \sum (y_i - \hat{y}_i)^2$$

As in the case of the straight-line model, the sample estimates $\hat{\beta}_0, \hat{\beta}_1, \ldots, \hat{\beta}_k$ will be obtained as solutions to a set of simultaneous linear equations.*

The primary difference between fitting the simple and multiple regression models is computational difficulty. The $(k+1)$ simultaneous linear equations that must be solved to find the $(k+1)$ estimated coefficients $\hat{\beta}_0, \hat{\beta}_1, \ldots, \hat{\beta}_k$ are often difficult (tedious and time-consuming) to solve with a calculator. Consequently, we resort to the use of statistical computer software and present output from, SAS, SPSS, and MINITAB in examples and exercises.

EXAMPLE   4.1

Suppose a property appraiser wants to model the relationship between the sale price of a residential property in a mid-size city and the following three independent variables: (1) appraised land value of the property, (2) appraised value of improvements (i.e., home value) on the property, and (3) area of living space on the property (i.e., home size). Consider the first-order model

$$y = \beta_0 + \beta_1 x_1 + \beta_2 x_2 + \beta_3 x_3 + \varepsilon$$

where y = Sale price (dollars)
   x_1 = Appraised land value (dollars)
   x_2 = Appraised improvements (dollars)
   x_3 = Area (square feet)

To fit the model, the appraiser selected a random sample of $n = 20$ properties from the thousands of properties that were sold in a particular year. The resulting data are given in Table 4.1 (p. 169).

a. Use scatterplots to graph the sample data. Interpret the plots.

b. Use the method of least squares to estimate the unknown parameters β_0, β_1, β_2, and β_3 in the model.

c. Find the value of SSE that is minimized by the least squares method.

Solution

a. MINITAB scatterplots for examining the bivariate relationships between y and x_1, y and x_2, and y and x_3 are shown in Figure 4.2. Of the three variables, appraised improvements (x_2) appears to have the strongest linear relationship with sale price (y).

b. The model hypothesized above is fit to the data of Table 4.1 using SAS. A portion of the SAS printout is reproduced in Figure 4.3. The least squares estimates of the β parameters (highlighted on the printout) are $\hat{\beta}_0 = 1,470.27592$, $\hat{\beta}_1 = .81449$, $\hat{\beta}_2 = .82044$, and $\hat{\beta}_3 = 13.52865$. Therefore, the equation that minimizes SSE for this data set (i.e., the **least squares prediction equation**) is

$$\hat{y} = 1,470.28 + .8145x_1 + .8204x_2 + 13.53x_3$$

c. The minimum value of the sum of squared errors, highlighted in Figure 4.3, is SSE = 1,003,491,259.   ◆

---

*Students who are familiar with calculus should note that $\hat{\beta}_0, \hat{\beta}_1, \ldots, \hat{\beta}_k$ are the solutions to the set of equations $\partial \text{SSE}/\partial \beta_0 = 0$, $\partial \text{SSE}/\partial \beta_1 = 0, \ldots, \partial \text{SSE}/\partial \beta_k = 0$. The solution, given in matrix notation, is presented in Appendix A.

FIGURE 4.2
MINITAB scatterplots for the
data of Table 4.1

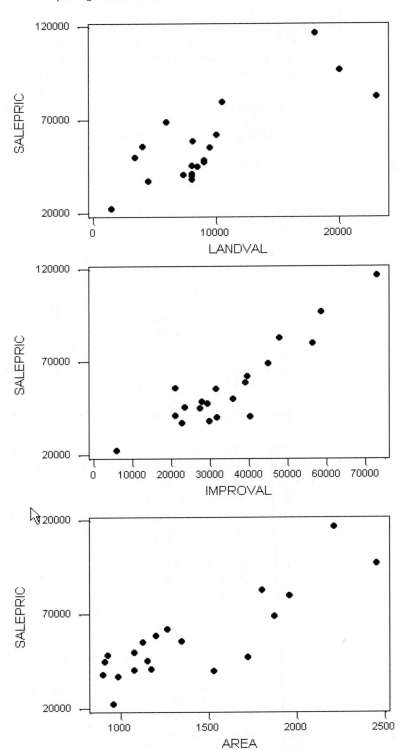

REALESTATE

TABLE 4.1   Real Estate Appraisal Data for 20 Properties 4.2

PROPERTY # (OBS.)	SALE PRICE y	LAND VALUE x_1	IMPROVEMENTS VALUE x_2	AREA x_3
1	68,900	5,960	44,967	1,873
2	48,500	9,000	27,860	928
3	55,500	9,500	31,439	1,126
4	62,000	10,000	39,592	1,265
5	116,500	18,000	72,827	2,214
6	45,000	8,500	27,317	912
7	38,000	8,000	29,856	899
8	83,000	23,000	47,752	1,803
9	59,000	8,100	39,117	1,204
10	47,500	9,000	29,349	1,725
11	40,500	7,300	40,166	1,080
12	40,000	8,000	31,679	1,529
13	97,000	20,000	58,510	2,455
14	45,500	8,000	23,454	1,151
15	40,900	8,000	20,897	1,173
16	80,000	10,500	56,248	1,960
17	56,000	4,000	20,859	1,344
18	37,000	4,500	22,610	988
19	50,000	3,400	35,948	1,076
20	22,400	1,500	5,779	962

Source: Alachua County (Florida) Property Appraisers Office.

FIGURE 4.3

SAS regression printout for the sale price model,
Example 4.1

```
                    Dependent Variable: SALEPRIC

                        Analysis of Variance

                                Sum of          Mean
Source                  DF      Squares         Square      F Value    Pr > F

Model                    3    8779676741      2926558914     46.66    <.0001
Error                   16    1003491259        62718204
Corrected Total         19    9783168000


           Root MSE              7919.48254    R-Square    0.8974
           Dependent Mean            56660    Adj R-Sq    0.8782
           Coeff Var              13.97720


                        Parameter Estimates

                    Parameter      Standard
Variable     DF     Estimate         Error      t Value    Pr > |t|

Intercept     1    1470.27592     5746.32458      0.26      0.8013
LANDVAL       1       0.81449        0.51222      1.59      0.1314
IMPROVAL      1       0.82044        0.21118      3.88      0.0013
AREA          1      13.52865        6.58568      2.05      0.0567
```

EXAMPLE   4.2

Refer to the first-order model for sale price y considered in Example 4.1. Interpret the estimates of the β parameters in the model.

Solution

The least squares prediction equation, as given in Example 4.1, is $\hat{y} = 1{,}470.28 + .8145x_1 + .8204x_2 + 13.53x_3$. We know that with first-order models β_1 represents the slope of the y versus x_1 line for fixed x_2 and x_3. That is, β_1 measures the change in $E(y)$ for every 1-unit increase in x_1 when all other independent variables in the model are held fixed. Similar statements can be made about β_2 and β_3; e.g., β_2 measures the change in $E(y)$ for every 1-unit increase in x_2 when all other x's in the model are held fixed. Consequently, we obtain the following interpretations:

$\hat{\beta}_1 = .8145$: We estimate the mean sale price of a property, $E(y)$, to increase .8145 dollar for every \$1 increase in appraised land value (x_1) when both appraised improvements (x_2) and area (x_3) are held fixed.

$\hat{\beta}_2 = .8204$: We estimate the mean sale price of a property, $E(y)$, to increase .8204 dollar for every \$1 increase in appraised improvements (x_2) when both appraised land value (x_1) and area (x_3) are held fixed.

$\hat{\beta}_3 = 13.53$: We estimate the mean sale price of a property, $E(y)$, to increase \$13.53 for each additional square foot of living area (x_3) when both appraised land value (x_1) and appraised improvements (x_2) are held fixed.

The value $\hat{\beta}_0 = 1{,}470.28$ does not have a meaningful interpretation in this example. To see this, note that $\hat{y} = \hat{\beta}_0$ when $x_1 = x_2 = x_3 = 0$. Thus, $\hat{\beta}_0 = 1{,}470.28$ represents the estimated mean sale price when the values of all the independent variables are set equal to 0. Since a residential property with these characteristics—appraised land value of \$0, appraised improvements of \$0, and 0 square feet of living area—is not practical, the value of $\hat{\beta}_0$ has no meaningful interpretation. In general, $\hat{\beta}_0$ will not have a practical interpretation unless it makes sense to set the values of the x's simultaneously equal to 0.   ◆

4.5
Estimation of σ^2, the Variance of ε

Recall that σ^2 is the variance of the random error ε. As such, σ^2 is an important measure of model utility. If $\sigma^2 = 0$, all the random errors will equal 0 and the prediction equation $\hat{y}$ will be identical to $E(y)$; i.e., $E(y)$ will be estimated without error. In contrast, a large value of σ^2 implies large (absolute) values of ε and larger deviations between the prediction equation $\hat{y}$ and the mean value $E(y)$. Consequently, the larger the value of σ^2, the greater will be the error in estimating the model parameters $\beta_0, \beta_1, \ldots, \beta_k$ and the error in predicting a value of y for a specific set of values of $x_1, x_2, \ldots, x_k$. Thus, σ^2 plays a major role in making inferences about $\beta_0, \beta_1, \ldots, \beta_k$, in estimating $E(y)$, and in predicting y for specific values of $x_1, x_2, \ldots, x_k$.

Since the variance σ^2 of the random error ε will rarely be known, we must use the results of the regression analysis to estimate its value. Recall that σ^2 is the variance of the probability distribution of the random error ε for a given set of values for $x_1, x_2, \ldots, x_k$; hence, it is the mean value of the squares of the deviations of the y values (for given values of $x_1, x_2, \ldots, x_k$) about the mean value $E(y)$.*Since the predicted value $\hat{y}$ estimates $E(y)$ for each of the data points, it seems natural to use

$$\text{SSE} = \sum (y_i - \hat{y}_i)^2$$

to construct an estimator of σ^2.

Estimator of σ^2 for Multiple Regression Model with k Independent Variables

$$s^2 = \text{MSE} = \frac{\text{SSE}}{n - \text{Number of estimated } \beta \text{ parameters}}$$

$$= \frac{\text{SSE}}{n - (k + 1)}$$

For example, in the first-order model of Example 4.1, we found that SSE = 1,003,491,259. We now want to use this quantity to estimate the variance of ε. Recall that the estimator for the straight-line model is $s^2 = \text{SSE}/(n - 2)$ and note that the denominator is $(n - \text{Number of estimated } \beta \text{ parameters})$, which is $(n - 2)$ in the straight-line model. Since we must estimate four parameters, $\beta_0, \beta_1, \beta_2,$ and β_3 for the first-order model, the estimator of σ^2 is

$$s^2 = \frac{\text{SSE}}{n - 4}$$

The numerical estimate for this example is

$$s^2 = \frac{\text{SSE}}{20 - 4} = \frac{1,003,491,259}{16} = 62,718,204$$

In many computer printouts and textbooks, s^2 is called the **mean square for error (MSE)**. This estimate of σ^2 is highlighted in the SAS printout in Figure 4.3.

The units of the estimated variance are squared units of the dependent variable y. Since the dependent variable y in this example is sale price in dollars, the units of s^2 are (dollars)^2. This makes meaningful interpretation of s^2 difficult, so we use the standard deviation s to provide a more meaningful measure of variability. In this example,

$$s = \sqrt{62,718,204} = 7,919.5$$

which is highlighted on the SAS printout in Figure 4.3 (next to **Root MSE**). One useful interpretation of the estimated standard deviation s is that the interval $\pm 2s$ will provide a rough approximation to the accuracy with which the model will predict future values of y for given values of x. Thus, in Example 4.1, we expect the model

---

*Because $y = E(y) + \varepsilon$, then ε is equal to the deviation $y - E(y)$. Also, by definition, the variance of a random variable is the expected value of the square of the deviation of the random variable from its mean. According to our model, $E(\varepsilon) = 0$. Therefore, $\sigma^2 = E(\varepsilon^2)$.

to provide predictions of sale price to within about $\pm 2s = \pm 2(7,919.5) = \pm 15,839$ dollars.[†]

For the general multiple regression model

$$y = \beta_0 + \beta_1 x_1 + \beta_2 x_2 + \cdots + \beta_k x_k + \varepsilon$$

we must estimate the $(k + 1)$ parameters $\beta_0, \beta_1, \beta_2, \ldots, \beta_k$. Thus, the estimator of σ^2 is SSE divided by the quantity $(n - \text{Number of estimated } \beta \text{ parameters})$.

We will use MSE, the estimator of σ^2, both to check the utility of the model (Sections 4.6 and sections 4.8) and to provide a measure of the reliability of predictions and estimates when the model is used for those purposes (Section 4.11). Thus, you can see that the estimation of σ^2 plays an important part in the development of a regression model.

4.6
Inferences About the β Parameters

Inferences about the individual β parameters in a model are obtained using either a confidence interval or a test of hypothesis, as outlined in the following two boxes.[*]

Test of an Individual Parameter Coefficient in the Multiple Regression Model

ONE-TAILED TEST   TWO-TAILED TEST

H_0: $\beta_i = 0$ H_0: $\beta_i = 0$

H_a: $\beta_i < 0$ [or H_a: $\beta_i > 0$] H_a: $\beta_i \neq 0$

$$\text{Test statistic}: t = \frac{\hat{\beta}_i}{s_{\hat{\beta}_i}}$$

Rejection region: $t < -t_\alpha$ Rejection region: $|t| > t_{\alpha/2}$

[or $t > t_\alpha$ when H_a: $\beta_i > 0$]

where t_α and $t_{\alpha/2}$ are based on $n - (k + 1)$ degrees of freedom and

$n = \text{Number of observations}$
$k + 1 = \text{Number of } \beta \text{ parameters in the model}$

Assumptions: See Section 4.2 for assumptions about the probability distribution for the random error component ε.

[†]The $\pm 2s$ approximation will improve as the sample size is increased. We will provide more precise methodology for the construction of prediction intervals in Section 4.11.

[*]The formulas for computing $\hat{\beta}_i$ and its standard error are so complex, the only reasonable way to present them is by using matrix algebra. We do not assume a prerequisite of matrix algebra for this text and, in any case, we think the formulas can be omitted in an introductory course without serious loss. They are programmed into all statistical software packages with multiple regression routines and are presented in some of the texts listed in the references.

A 100 $(1 - \alpha)$% Confidence Interval for a β Parameter

$$\hat{\beta}_i \pm (t_{\alpha/2})s_{\hat{\beta}_i}$$

where $t_{\alpha/2}$ is based on $n - (k + 1)$ degrees of freedom and

n = Number of observations

$k + 1$ = Number of β parameters in the model

We illustrate these methods with another example.

EXAMPLE 4.3

A collector of antique grandfather clocks knows that the price received for the clocks increases linearly with the age of the clocks. Moreover, the collector hypothesizes that the auction price of the clocks will increase linearly as the number of bidders increases. Thus, the following first-order model is hypothesized:

$$y = \beta_0 + \beta_1 x_1 + \beta_2 x_2 + \varepsilon$$

where y = Auction price (dollars)

x_1 = Age of clock (years)

x_2 = Number of bidders

A sample of 32 auction prices of grandfather clocks, along with their age and the number of bidders, is given in Table 4.2. The model $y = \beta_0 + \beta_1 x_1 + \beta_2 x_2 + \varepsilon$ is fit to the data, and a portion of the MINITAB printout is shown in Figure 4.4.

 GFCLOCKS

TABLE 4.2 **Auction price Data**

AGE x_1	NUMBER OF BIDDERS x_2	AUCTION PRICE y	AGE x_1	NUMBER OF BIDDERS x_2	AUCTION PRICE y
127	13	$1,235	170	14	$2,131
115	12	1,080	182	8	1,550
127	7	845	162	11	1,884
150	9	1,522	184	10	2,041
156	6	1,047	143	6	845
182	11	1,979	159	9	1,483
156	12	1,822	108	14	1,055
132	10	1,253	175	8	1,545
137	9	1,297	108	6	729
113	9	946	179	9	1,792
137	15	1,713	111	15	1,175
117	11	1,024	187	8	1,593
137	8	1,147	111	7	785
153	6	1,092	115	7	744
117	13	1,152	194	5	1,356
126	10	1,336	168	7	1,262

FIGURE 4.4

MINITAB regression printout for
grandfather clock model,
Example 4.3

```
The regression equation is
PRICE = - 1339 + 12.7 AGE + 86.0 BIDDERS

Predictor        Coef      SE Coef          T         P
Constant      -1339.0        173.8      -7.70     0.000
AGE           12.7406       0.9047      14.08     0.000
BIDDERS        85.953        8.729       9.85     0.000

S = 133.5        R-Sq = 89.2%      R-Sq(adj) = 88.5%

Analysis of Variance

Source           DF           SS          MS          F         P
Regression        2      4283063     2141531     120.19     0.000
Residual Error   29       516727       17818
Total            31      4799790
```

a. Test the hypothesis that the mean auction price of a clock increases as the number of bidders increases when age is held constant, that is, $\beta_2 > 0$. Use $\alpha = .05$.

b. Form a 90% confidence interval for β_1 and interpret the result.

Solution

a. The hypotheses of interest concern the parameter β_2. Specifically,

$$H_0: \beta_2 = 0$$

$$H_a: \beta_2 > 0$$

The test statistic is a t statistic formed by dividing the sample estimate $\hat{\beta}_2$ of the parameter β_2 by the estimated standard error of $\hat{\beta}_2$ (denoted $s_{\hat{\beta}_2}$). These estimates, $\hat{\beta}_2 = 85.953$ and $s_{\hat{\beta}_2} = 8.729$, as well as the calculated t value, are highlighted on the MINITAB printout, Figure 4.4

$$\textit{Test statistic}: \ t = \frac{\hat{\beta}_2}{s_{\hat{\beta}_2}} = \frac{85.953}{8.729} = 9.85$$

The rejection region for the test is found in exactly the same way as the rejection regions for the t-tests in previous chapters. That is, we consult Table 2 in Appendix A to obtain an upper-tail value of t. This is a value t_α such that $P(t > t_\alpha) = \alpha$. We can then use this value to construct rejection regions for either one-tailed or two-tailed tests.

For $\alpha = .05$ and $n - (k + 1) = 32 - (2 + 1) = 29$ df, the critical t value obtained from Table 2 is $t_{.05} = 1.699$. Therefore,

$$\textit{Rejection region}: \ t > 1.699 \quad \text{(see Figure 4.5)}$$

Since the test statistic value, $t = 9.85$, falls in the rejection region, we have sufficient evidence to reject H_0. Thus, the collector can conclude that the mean auction price of a clock increases as the number of bidders increases, when age is held constant. Note that the observed significance level of the test is also highlighted on the printout. Since p-value = 0, any nonzero α will lead us to reject H_0.

FIGURE 4.5

Rejection region for
$H_0: \beta_2 = 0$ vs $H_a: \beta_2 > 0$

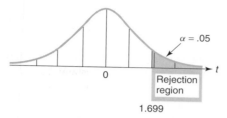

b. A 90% confidence interval for β_1 is (from the box):

$$\hat{\beta}_1 \pm (t_{\alpha/2})s_{\hat{\beta}_1} = \hat{\beta}_1 \pm (t_{.05})s_{\hat{\beta}_1}$$

Substituting $\hat{\beta}_1 = 12.74$, $s_{\hat{\beta}_i} = .905$ (both obtained from the MINITAB printout, Figure 4.4) and $t_{.05} = 1.699$ (from part **a**) into the equation, we obtain

$$12.74 \pm (1.699)(.905) = 12.74 \pm 1.54$$

or $(11.20, 14.28)$. Thus, we are 90% confident that β_1 falls between 11.20 and 14.28. Since β_1 is the slope of the line relating auction price (y) to age of the clock (x_1), we conclude that price increases between \$11.20 and \$14.28 for every 1-year increase in age, holding number of bidders (x_2) constant. ◆

In Example 4.3, be careful not to try to give a practical interpretation for the estimated intercept $\hat{\beta}_0 = -1,339.0$. You might think that this implies a negative price for clocks 0 years of age with 0 bidders. However, these zeros are meaningless numbers in this example, since the ages range from 108 to 194 and the number of bidders ranges from 5 to 15. Keep in mind that we are modeling y within the range of values observed for the predictor variables and that interpretations of the models for values of the independent variables outside their sampled ranges can be very misleading.

We conclude this section with a caution about conducting t-tests on individual β parameters in a model.

Caution

Extreme care should be exercised when conducting t-tests on the individual β parameters in a *first-order linear model* for the purpose of determining which independent variables are useful for predicting y and which are not. If you fail to reject $H_0: \beta_i = 0$, several conclusions are possible:

1. There is no relationship between y and x_i.
2. A straight-line relationship between y and x exists (holding the other x's in the model fixed), but a Type II error occurred.
3. A relationship between y and x_i (holding the other x's in the model fixed) exists, but is more complex than a straight-line relationship (e.g., a curvilinear relationship may be appropriate). The most you can say about a β parameter test is that there is either sufficient (if you reject $H_0: \beta_i = 0$) or insufficient (if you do not reject $H_0: \beta_i = 0$) evidence of a *linear (straight-line)* relationship between y and x_i.

EXERCISES

4.1. How is the number of degrees of freedom available for estimating σ^2, the variance of ε, related to the number of independent variables in a regression model?

4.2. Perfectionists are persons who set themselves standards and goals that cannot be reasonably met or accomplished. One theory suggests that those individuals who are depressed have a tendency toward perfectionism. To study this phenomenon, 76 members of an introductory psychology class completed questionnaires that measure four different scales: (1) the ASO scale, designed to measure self-acceptance, (2) the Burns scale, designed to measure perfectionism, (3) the Zung scale, designed to measure depression, and (4) the Rotter scale, designed to measure perceptions between actions and reinforcement (*The Journal of Adlerian Theory, Research, and Practice*, Mar. 1986).

   a. Write a first-order model relating depression (Zung scale) to self-acceptance (ASO scale), perfectionism (Burns scale), and reinforcement (Rotter scale).

   b. A t test for the perfectionism (Burns scale) variable resulted in a (two-tailed) p-value of .87. Interpret this value.

4.3. In *Chance* (Fall 2000), statistician Scott Berry built a multiple regression model for predicting total number of runs scored by a Major League Baseball team during a season. Using data on all teams from 1990–1998 (a sample of $n = 234$), the results in the next table were obtained.

   a. Write the least squares prediction equation for $y =$ total number of runs scored by a team in a season.

   b. Conduct a test of $H_0: \beta_7 = 0$ against $H_a: \beta_7 < 0$ at $\alpha = .05$. Interpret the results.

   c. Form a 95% confidence interval for β_5. Interpret the results.

INDEPENDENT VARIABLE	β ESTIMATE	STANDARD ERROR
Intercept	3.70	15.00
Walks (x_1)	.34	.02
Singles (x_2)	.49	.03
Doubles (x_3)	.72	.05
Triples (x_4)	1.14	.19
Home Runs (x_5)	1.51	.05
Stolen Bases (x_6)	.26	.05
Caught Stealing (x_7)	−.14	.14
Strikeouts (x_8)	−.10	.01
Outs (x_9)	−.10	.01

Source: Berry, S.M. "A statistician reads the sports pages: Modeling offensive ability in baseball." *Chance*, Vol. 13, No. 4, Fall 2000 (Table 2).

   d. Predict the number of runs scored by your favorite Major League Baseball team last year. How close is the predicted value to the actual number of runs scored by your team? (*Note*: You can find data on your favorite team on the Internet at www.majorleaguebaseball.com.)

4.4. Detailed interviews were conducted with over 1,000 street vendors in the city of Puebla, Mexico, in order to study the factors influencing vendors' incomes (*World Development*, Feb. 1998). Vendors were defined as individuals working in the street, and included vendors with carts and stands on wheels and excluded beggars, drug dealers, and prostitutes. The researchers collected data on gender, age, hours worked per day, annual earnings, and education level. A subset of these data appears in the table.

   a. Write a first-order model for mean annual earnings, $E(y)$, as a function of age (x_1) and hours worked (x_2).

⬤ STREETVEN

VENDOR NUMBER	ANNUAL EARNINGS y	AGE x_1	HOURS WORKED PER DAY x_2
21	$2841	29	12
53	1876	21	8
60	2934	62	10
184	1552	18	10
263	3065	40	11
281	3670	50	11
354	2005	65	5
401	3215	44	8
515	1930	17	8
633	2010	70	6
677	3111	20	9
710	2882	29	9
800	1683	15	5
914	1817	14	7
997	4066	33	12

Source: Adapted from Smith, P.A., and Metzger, M.R. "The return to education: Street vendors in Mexico." *World Development*, Vol. 26, No. 2, Feb. 1998, pp. 289–296.

   b. The model was fit to the data using SAS. Find the least squares prediction equation on the printout shown on p. 177.

   c. Interpret the estimated β coefficients in your model.

   d. Is age (x_1) a statistically useful predictor of annual earnings? Test using $\alpha = .01$.

   e. Find a 95% confidence interval for β_2. Interpret the interval in the words of the problem.

SAS output for Exercise 4.4

Dependent Variable: EARNINGS

Analysis of Variance

Source	DF	Sum of Squares	Mean Square	F Value	Pr > F
Model	2	5018232	2509116	8.36	0.0053
Error	12	3600196	300016		
Corrected Total	14	8618428			

Root MSE	547.73748	R-Square	0.5823	
Dependent Mean	2577.13333	Adj R-Sq	0.5126	
Coeff Var	21.25375			

Parameter Estimates

Variable	DF	Parameter Estimate	Standard Error	t Value	Pr > |t|	95% Confidence Limits	
Intercept	1	-20.35201	652.74532	-0.03	0.9756	-1442.56189	1401.85787
AGE	1	13.35045	7.67168	1.74	0.1074	-3.36470	30.06559
HOURS	1	243.71446	63.51174	3.84	0.0024	105.33428	382.09465

4.5. Empirical research was conducted to investigate the variables that impact the size distribution of manufacturing firms in international markets (*World Development*, Vol. 20, 1992). Data collected on $n = 54$ countries were used to model the country's size distribution y, measured as the share of manufacturing firms in the country with 100 or more workers. The model studied was $E(y) = \beta_0 + \beta_1 x_1 + \beta_2 x_2 + \beta_3 x_3 + \beta_4 x_4 + \beta_5 x_5$, where

$x_1 = $ Natural logarithm of Gross National Product (LGNP)
$x_2 = $ Geographic area per capita (in thousands of square meters) (AREAC)
$x_3 = $ Share of heavy industry in manufacturing value added (SVA)
$x_4 = $ Ratio of credit claims on the private sector to Gross Domestic Product (CREDIT)
$x_5 = $ Ratio of stock equity shares to Gross Domestic Product (STOCK)

a. The researchers hypothesized that the higher the credit ratio of a country, the smaller the size distribution of manufacturing firms. Explain how to test this hypothesis.
b. The researchers hypothesized that the higher the stock ratio of a country, the larger the size distribution of manufactoring firms. Explain how to test this hypothesis.

4.6. Residential property appraisers make extensive use of multiple regression in their evaluation of property. Typically, the sale price (y) of a property is modeled as a function of several home-related conditions (e.g., gross living area, location, number of bedrooms). However, appraisers are not interested in the predicted price, $\hat{y}$. Rather, they use the regression model as a tool for making value adjustments to the property. These adjustments are derived from the parameter estimates of the model. The *Real Estate Appraiser* (April 1992) reported the results of a multiple regression on the price (y) of $n = 157$ residential properties recently sold in a northern Virginia subdivision. The SAS printout of the analysis is reproduced on p. 178. Note that there are 27 independent variables in the model.

a. One of the independent variables in the model is gross living area (GLA), measured in square feet. A 95% confidence interval for the β coefficient associated with GLA is shown on the printout. Interpret this interval.
b. Interpret the t test for testing the variable lot size (LOTSIZE).
c. Interpret the t test for testing the variable bay window (BAYWIND).
d. Demonstrate that the inferences derived in parts **b** and **c** can also be obtained by examining the corresponding 95% confidence interval for the β.

SAS output for Exercise 4.6

Dependent Variable: SALEPRIC

Analysis of Variance

Source	DF	Sum of Squares	Mean Square	F Value	Pr > F
Model	27	24184211898	895711552	20.91	<.0001
Error	129	5524834283	42828173		
Corrected Total	156	29709046181			

Root MSE	6544.324	R-Square	0.8140	
Dependent Mean	173157.5	Adj R-Sq	0.7751	
Coeff Var	3.779404			

Parameter Estimates

Variable	DF	Parameter Estimate	Standard Error	t Value	Pr > |t|	95% Confidence Limits	
Intercept	1	96603	12530	7.71	<.0001	71794	121412
TIME	1	150	123	1.22	0.2248	-94	394
LOTSIZE	1	.80	.30	2.02	0.0452	0.01	1.19
AGE	1	381	502	0.76	0.4501	-613	1375
GLA	1	22.40	3.67	6.10	<.0001	15.13	29.67
BEDROOMS	1	2263	1609	1.41	0.1619	-923	5499
HALFBATH	1	5962	2934	2.03	0.0442	153	11771
CORNRLOT	1	-1481	1692	-0.88	0.3829	-4831	1869
CULDESAC	1	-56	2557	-0.02	0.9825	-5119	5007
BACKWOOD	1	4086	2044	1.99	0.0477	39	8133
DECK	1	2408	2167	1.11	0.2686	-1883	6699
FENCE	1	2896	1271	2.23	0.0243	379	5413
SHED	1	70	1343	0.05	0.9588	-2589	2729
PATIO	1	2377	1671	1.42	0.1572	-932	5686
PORTICO	1	-906	2963	-0.31	0.7603	-6773	4961
SCRPORCH	1	5021	2038	2.46	0.0151	986	9056
INGRPOOL	1	7570	3028	2.50	0.0137	1575	13565
GARAGE	1	2989	1446	2.07	0.0407	126	5852
DRIVEWAY	1	-1844	3222	-0.57	0.5681	-8224	4536
FIREPLAC	1	1290	1277	1.01	0.3144	-1238	3818
BRICKFAC	1	-2140	2369	-0.90	0.3680	-6381	2551
UPDKITCH	1	4171	1470	2.84	0.0053	1260	7082
REMKITCH	1	6091	2367	2.57	0.0112	1404	10778
INTERCOM	1	1933	2146	0.90	0.3693	-2316	6182
CENVACUM	1	-4636	2166	-2.14	0.0342	-8925	-347
SKYLITES	1	7744	2622	-2.95	0.0037	2552	12936
AIRFILT	1	874	2506	0.35	0.7280	-4088	5836
BAYWIND	1	-3174	2086	-1.52	0.1305	-7304	956

Source: Gilson, S.J. "A case study—Comparing the results: Multiple regression analysis vs. matched pairs in residential subdivision." *The Real Estate Appraiser*. Apr. 1992, p. 37 (Table 4).

4.7. *Artificial Intelligence (AI) Applications* (Jan. 1993) discussed the use of computer-based technologies in building explanation systems for regression models. As an example, the authors presented a model for predicting the scenic beauty (y) of southeastern pine stands (measured on a numeric scale) as a function of age (x_1) of the dominant stand, stems per acre (x_2) in trees, and basal area (x_3) per acre in hardwoods. A user of the AI system simply inputs the values of x_1, x_2, and x_3, and the system uses the least squares equation to predict the scenic beauty (y) value.

The AI system generates information on how each independent variable can be manipulated to effect changes in the dependent variable. For example, "if all else were held constant in the stand, allowing the age (x_1) of the dominant trees in the stand to mature by 1 year will *increase* scenic beauty (y)." From what portion of the regression analysis would the AI system extract this type of information?

4.8. Refer to the *Journal of Applied Ecology* (Vol. 32, 1995) study of the feeding habits of baby snow geese, Exercise 3.42 (p. 128). The data on gosling weight change, digestion efficiency, acid-detergent fibre (all measured as percentages) and diet (plants or duck chow) for 42 feeding trials are reproduced in the next table. The botanists were interested in predicting weight change (y) as a function of the other variables. The first-order model $E(y) = \beta_0 + \beta_1 x_1 + \beta_2 x_2$, where x_1 is digestion efficiency and x_2 is acid-detergent fibre, was fit to the data. The MINITAB printout is given on p. 180.

a. Find the least squares prediction equation for weight change, y.
b. Interpret the β-estimates in the equation, part a.
c. Conduct a test to determine if digestion efficiency, x_1, is a useful linear predictor of weight change. Use $\alpha = .01$.
d. Form a 99% confidence interval for β_2. Interpret the result.

SNOWGEESE

FEEDING TRIAL	DIET	WEIGHT CHANGE(%)	DIGESTION EFFICIENCY (%)	ACID-DETERGENT FIBRE (%)
1	Plants	−6	0	28.5
2	Plants	−5	2.5	27.5
3	Plants	−4.5	5	27.5
4	Plants	0	0	32.5
5	Plants	2	0	32
6	Plants	3.5	1	30
7	Plants	−2	2.5	34
8	Plants	−2.5	10	36.5
9	Plants	−3.5	20	28.5
10	Plants	−2.5	12.5	29
11	Plants	−3	28	28
12	Plants	−8.5	30	28
13	Plants	−3.5	18	30
14	Plants	−3	15	31
15	Plants	−2.5	17.5	30
16	Plants	−.5	18	22
17	Plants	0	23	22.5
18	Plants	1	20	24
19	Plants	2	15	23
20	Plants	6	31	21
21	Plants	2	15	24
22	Plants	2	21	23
23	Plants	2.5	30	22.5
24	Plants	2.5	33	23
25	Plants	0	27.5	30.5
26	Plants	.5	29	31
27	Plants	−1	32.5	30
28	Plants	−3	42	24
29	Plants	−2.5	39	25
30	Plants	−2	35.5	25
31	Plants	.5	39	20
32	Plants	5.5	39	18.5
33	Plants	7.5	50	15
34	Duck Chow	0	62.5	8
35	Duck Chow	0	63	8
36	Duck Chow	2	69	7
37	Duck Chow	8	42.5	7.5
38	Duck Chow	9	59	8.5
39	Duck Chow	12	52.5	8
40	Duck Chow	8.5	75	6

(continued overleaf)

(*Continued*)

FEEDING TRIAL	DIET	WEIGHT CHANGE(%)	DIGESTION EFFICIENCY (%)	ACID-DETERGENT FIBRE (%)
41	Duck Chow	10.5	72.5	6.5
42	Duck Chow	14	69	7

Source: Gadallah, F.L., and Jefferies, R.L. "Forage quality in brood rearing areas of the lesser snow goose and the growth of captive goslings." *Journal of Applied Ecology*, Vol. 32, No. 2, 1995, pp. 281–282 (adapted from Figures 2 and 3).

MINITAB output for Exercise 4.8

```
The regression equation is
WTCHANGE = 12.2 - 0.0265 DIGEST - 0.458 ADFIBRE

Predictor      Coef    SE Coef        T      P
Constant     12.180      4.402     2.77  0.009
DIGEST      -0.02654    0.05349    -0.50  0.623
ADFIBRE     -0.4578      0.1283    -3.57  0.001


S = 3.519     R-Sq = 52.9%     R-Sq(adj) = 50.5%

Analysis of Variance

Source           DF        SS        MS        F      P
Regression        2    542.03    271.02    21.88  0.000
Residual Error   39    483.08     12.39
Total            41   1025.12
```

4.9. A quasar is a distant celestial object (at least four billion light-years away) that provides a powerful source of radio energy. The *Astronomical Journal* (July 1995) reported on a study of 90 quasars detected by a deep space survey. The survey enabled astronomers to measure several different quantitative characteristics of each quasar, including redshift range, line flux (erg/cm^2·s), line luminosity (erg/s), AB$_{1450}$ magnitude, absolute magnitude, and rest frame equivalent width. The data for a sample of 25 large (redshift) quasars is listed in the table below.

a. Hypothesize a first-order model for equivalent width, y, as a function of the first four variables in the table.

b. The first-order model is fit to the data using SPSS. The printout is provided here. Give the least squares prediction equation.

c. Interpret the β estimates in the model.

d. Test to determine whether redshift (x_1) is a useful linear predictor of equivalent width (y), using $\alpha = .05$.

SPSS output for Exercise 4.9

Coefficients[a]

Model		Unstandardized Coefficients B	Std. Error	Standardized Coefficients Beta	t	Sig.
1	(Constant)	21087.951	18553.161		1.137	.269
	REDSHIFT	108.451	88.740	1.102	1.222	.236
	LINEFLUX	557.910	315.990	2.786	1.766	.093
	LINELUM	-340.166	320.763	-1.412	-1.060	.302
	AB1450	85.681	6.273	1.230	13.658	.000

a. Dependent Variable: RFEWIDTH

💿 QUASAR

QUASAR	REDSHIFT (x_1)	LINE FLUX (x_2)	LINE LUMINOSITY (x_3)	AB$_{1450}$ (x_4)	ABSOLUTE MAGNITUDE (x_5)	REST FRAME EQUIVALENT WIDTH y
1	2.81	−13.48	45.29	19.50	−26.27	117
2	3.07	−13.73	45.13	19.65	−26.26	82
3	3.45	−13.87	45.11	18.93	−27.17	33

(Continued)

QUASAR	REDSHIFT (x_1)	LINE FLUX (x_2)	LINE LUMINOSITY (x_3)	AB_{1450} (x_4)	ABSOLUTE MAGNITUDE (x_5)	REST FRAME EQUIVALENT WIDTH y
4	3.19	−13.27	45.63	18.59	−27.39	92
5	3.07	−13.56	45.30	19.59	−26.32	114
6	4.15	−13.95	45.20	19.42	−26.97	50
7	3.26	−13.83	45.08	19.18	−26.83	43
8	2.81	−13.50	45.27	20.41	−25.36	259
9	3.83	−13.66	45.41	18.93	−27.34	58
10	3.32	−13.71	45.23	20.00	−26.04	126
11	2.81	−13.50	45.27	18.45	−27.32	42
12	4.40	−13.96	45.25	20.55	−25.94	146
13	3.45	−13.91	45.07	20.45	−25.65	124
14	3.70	−13.85	45.19	19.70	−26.51	75
15	3.07	−13.67	45.19	19.54	−26.37	85
16	4.34	−13.93	45.27	20.17	−26.29	109
17	3.00	−13.75	45.08	19.30	−26.58	55
18	3.88	−14.17	44.92	20.68	−25.61	91
19	3.07	−13.92	44.94	20.51	−25.41	116
20	4.08	−14.28	44.86	20.70	−25.67	75
21	3.62	−13.82	45.20	19.45	−26.73	63
22	3.07	−14.08	44.78	19.90	−26.02	46
23	2.94	−13.82	44.99	19.49	−26.35	55
24	3.20	−14.15	44.75	20.89	−25.09	99
25	3.24	−13.74	45.17	19.17	−26.83	53

Source: Schmidt, M., Schneider, D.P., and Gunn, J.E. "Spectroscopic CCD surveys for quasars at large redshift." *The Astronomical Journal*, Vol. 110, No. 1, July 1995, p. 70 (Table 1).

4.10. *Sociology of Sport Journal* (Spring 1992) published research on Proposition 48, the NCAA regulation that requires a minimum of 700 on the Scholastic Assessment Test (SAT) for eligibility for college athletics. Critics have claimed that the SAT is biased against black student athletes and is not a valid predictor of academic success. To investigate the validity of the SAT as a predictor of academic success, high school GPA and study time were also considered as potential predictors of college GPA. The following variables are defined:

> y = College GPA (the best measure of academic success)
> x_1 = High school GPA
> x_2 = Hours of studying in a season x_3 = SAT score

The study conducted separate regression analyses for black and white student athletes. Some of the results are shown in the accompanying table.

a. Write the model relating the mean value of the college GPA as a linear function of the three identified explanatory variables. Explain the meaning of each of the β parameters in your model.

b. Note that the estimate of the β parameter for SAT score for both models is .001. Interpret this estimate.

c. The reported *p*-value for testing $H_0: \beta_1 = 0$ is .734 for the black athletes' model and .000 for the white athletes' model. Interpret these values.

	PARAMETER ESTIMATE	STANDARD ERROR
Blacks		
Intercept	2.245	.09
High School GPA	−.040	.01
Study Hours	.14	.10
SAT Score	.001	.0025
Whites		
Intercept	1.970	.16
High School GPA	−.089	.01
Study Hours	.20	.23
SAT Score	.001	.0002

4.11. In the oil industry, water that mixes with crude oil during production and transportation must be removed. Chemists have found that the oil can be extracted from the water/oil mix electrically. Researchers at the University of Bergen (Norway) conducted a series of experiments to study the factors that influence the voltage (y) required to separate the water from the oil (*Journal of Colloid and Interface Science*, Aug. 1995). The seven independent variables investigated in the study are listed in the table. (Each variable was measured at two levels—a "low" level and a "high"

level.) Sixteen water/oil mixtures were prepared using different combinations of the independent variables; then each emulsion was exposed to a high electric field. In addition, three mixtures were tested when all independent variables were set to 0. The data for all 19 experiments are also given in the table below.

a. Propose a first-order model for y as a function of all seven independent variables.

b. Use a statistical software package to fit the model to the data in the table.

c. Fully interpret the β estimates.

⬤ WATEROIL

EXPERIMENT NUMBER	VOLTAGE y (kw/cm)	DISPERSE PHASE VOLUME x_1 (%)	SALINITY x_2 (%)	TEMPERATURE x_3 (°C)	TIME DELAY x_4 (hours)	SURFACTANT CONCENTRATION x_5 (%)	SPAN:TRITON x_6	SOLID PARTICLES x_7 (%)
1	.64	40	1	4	.25	2	.25	.5
2	.80	80	1	4	.25	4	.25	2
3	3.20	40	4	4	.25	4	.75	.5
4	.48	80	4	4	.25	2	.75	2
5	1.72	40	1	23	.25	4	.75	2
6	.32	80	1	23	.25	2	.75	.5
7	.64	40	4	23	.25	2	.25	2
8	.68	80	4	23	.25	4	.25	.5
9	.12	40	1	4	24	2	.75	2
10	.88	80	1	4	24	4	.75	.5
11	2.32	40	4	4	24	4	.25	2
12	.40	80	4	4	24	2	.25	.5
13	1.04	40	1	23	24	4	.25	.5
14	.12	80	1	23	24	2	.25	2
15	1.28	40	4	23	24	2	.75	.5
16	.72	80	4	23	24	4	.75	2
17	1.08	0	0	0	0	0	0	0
18	1.08	0	0	0	0	0	0	0
19	1.04	0	0	0	0	0	0	0

Source: Førdedal, H., *et al.* "A multivariate analysis of W/O emulsions in high external electric fields as studied by means of dielectric time domain spectroscopy." *Journal of Colloid and Interface Science*, Vol. 173, No. 2, Aug. 1995, p. 398 (Table 2).

4.7
The Multiple Coefficient of Determination, R^2

Recall from Chapter 3 that the coefficient of determination, r^2, is a measure of how well a straight-line model fits a data set. To measure how well a multiple regression model fits a set of data, we compute the multiple regression equivalent of r^2, called the **multiple coefficient of determination** and denoted by the symbol $\boldsymbol{R^2}$.

> **Definition 4.1**
>
> The **multiple coefficient of determination**, R^2, is defined as
>
> $$R^2 = 1 - \frac{\text{SSE}}{\text{SS}_{yy}} \quad 0 \le R^2 \le 1$$
>
> where $\text{SSE} = \sum(y_i - \hat{y}_i)^2$, $\text{SS}_{yy} = \sum(y_i - \bar{y})^2$, and $\hat{y}_i$ is the predicted value of y_i for the multiple regression model.

Just as for the simple linear model, R^2 represents the fraction of the sample variation of the y values (measured by SS_{yy}) that is explained by the least squares regression model. Thus, $R^2 = 0$ implies a complete lack of fit of the model to the data, and $R^2 = 1$ implies a perfect fit, with the model passing through every data point. In general, the closer the value of R^2 is to 1, the better the model fits the data.

To illustrate, the value $R^2 = .8974$ for the sale price model of Example 4.1 is highlighted in Figure 4.6. This high value of R^2 implies that using the independent variables land value, appraised improvements, and home size in a first-order model explains 89.7% of the total *sample variation* (measured by SS_{yy}) of sale price y. Thus, R^2 is a sample statistic that tells how well the model fits the data and thereby represents a measure of the usefulness of the entire model.

FIGURE 4.6

SAS regression printout for the sale price model

Dependent Variable: SALEPRIC

Analysis of Variance

Source	DF	Sum of Squares	Mean Square	F Value	Pr > F
Model	3	8779676741	2926558914	46.66	<.0001
Error	16	1003491259	62718204		
Corrected Total	19	9783168000			

Root MSE	7919.48254	R-Square	0.8974
Dependent Mean	56660	Adj R-Sq	0.8782
Coeff Var	13.97720		

Parameter Estimates

Variable	DF	Parameter Estimate	Standard Error	t Value	Pr > |t|
Intercept	1	1470.27592	5746.32458	0.26	0.8013
LANDVAL	1	0.81449	0.51222	1.59	0.1314
IMPROVAL	1	0.82044	0.21118	3.88	0.0013
AREA	1	13.52865	6.58568	2.05	0.0567

A large value of R^2 computed from the *sample* data does not necessarily mean that the model provides a good fit to all of the data points in the *population*. For example, a first-order linear model that contains three parameters will provide a perfect fit to a sample of three data points and R^2 will equal 1. Likewise, you will always obtain a perfect fit ($R^2 = 1$) to a set of n data points if the model contains exactly n parameters. Consequently, if you want to use the value of R^2 as a measure of how useful the model will be for predicting y, it should be based on a sample that contains substantially more data points than the number of parameters in the model.

Caution

In a multiple regression analysis, use the value of R^2 as a measure of how useful a linear model will be for predicting y only if the sample contains substantially more data points than the number of β parameters in the model.

As an alternative to using R^2 as a measure of model adequacy, the **adjusted multiple coefficient of determination**, denoted R_a^2, is often reported. The formula for R_a^2 is shown in the box.

Definition 4.2

The **adjusted multiple coefficient of determination** is given by

$$R_a^2 = 1 - \left[\frac{(n-1)}{n-(k+1)} \right] \left(\frac{\text{SSE}}{\text{SS}_{yy}} \right)$$

$$= 1 - \left[\frac{(n-1)}{n-(k+1)} \right] (1 - R^2)$$

Note: $R_a^2 \leq R^2$

R^2 and R_a^2 have similar interpretations. However, unlike R^2, R_a^2 takes into account ("adjusts" for) both the sample size n and the number of β parameters in the model. R_a^2 will always be smaller than R^2, and more importantly, cannot be "forced" to 1 by simply adding more and more independent variables to the model. Consequently, analysts prefer the more conservative R_a^2 when choosing a measure of model adequacy. The value of R_a^2 is also highlighted in Figure 4.5. Note that $R_a^2 = .8782$, a value only slightly smaller than R^2.

Despite their utility, R^2 and R_a^2 are only sample statistics. Consequently, it is dangerous to judge the usefulness of the model based solely on these values. We discuss a more formal method of checking the predictive ability of a general linear model—a statistical test of hypothesis—in the following section.

4.8
Testing the Utility of a Model: The Analysis of Variance F Test

The objective of step 5 in a multiple regression analysis is to conduct a test of the utility of the model—that is, a test to determine whether the model is adequate for predicting y. Conducting t tests on each β parameter in a model (Section 4.6) is generally **not** a good way to determine whether the overall model is contributing information for the prediction of y. If we were to conduct a series of t tests to determine whether the independent variables are contributing to the predictive relationship, we would be very likely to make one or more errors in deciding which terms to retain in the model and which to exclude.

Suppose you fit a first-order model with 10 quantitative independent variables, $x_1, x_2, \ldots, x_{10}$, and decide to conduct t tests on all 10 individual β's in the model, each at $\alpha = .05$. Even if all the β parameters (except β_0) in the model are equal to 0, approximately 40% of the time you will incorrectly reject the null hypothesis at least once and conclude that some β parameter is nonzero.* In other words, the overall Type I error is about .40, not .05!

Thus, in multiple regression models for which a large number of independent variables are being considered, conducting a series of t tests may cause the experimenter to include a large number of insignificant variables and exclude some useful ones. If we want to test the utility of a multiple regression model, we will need a **global test** (one that encompasses all the β parameters).

In particular, for the sale price model (Example 4.1) with three independent variables, we would test

$$H_0: \beta_1 = \beta_2 = \beta_3 = 0$$

$$H_a: \text{At least one of the coefficients is nonzero}$$

The test statistic used to test this hypothesis is an F statistic, and several equivalent versions of the formula can be used (although we will usually rely on the computer to calculate the F statistic):

$$Test\ statistic\colon\ F = \frac{(\text{SS}_{yy} - \text{SSE})/k}{\text{SSE}/[n - (k + 1)]} = \frac{R^2/k}{(1 - R^2)/[n - (k + 1)]}$$

Both these formulas indicate that the F statistic is the ratio of the *explained* variability divided by the model degrees of freedom to the *unexplained* variability divided by the error degrees of freedom. Thus, the larger the proportion of the total variability accounted for by the model, the larger the F statistic.

To determine when the ratio becomes large enough that we can confidently reject the null hypothesis and conclude that the model is more useful than no model at all for predicting y, we compare the calculated F statistic to a tabulated F value with k df in the numerator and $[n - (k + 1)]$ df in the denominator. Recall that tabulations of the F-distribution for various values of α are given in Tables 3, 4, 5, and 6 of Appendix C.

Rejection region: $F > F_\alpha$, where F is based on k numerator and $n - (k + 1)$ denominator degrees of freedom.

For the sale price example [$n = 20, k = 3, n - (k + 1) = 16$, and $\alpha = .05$], we will reject $H_0: \beta_1 = \beta_2 = \beta_3 = 0$ if

$$F > F_{.05} = 3.24 \quad \text{(see Figure 4.7)}$$

---

*The proof of this result proceeds as follows:

$$P(\text{Reject } H_0 \text{ at least once } |\beta_1 = \beta_2 = \cdots = \beta_{10} = 0)$$

$$= 1 - P(\text{Reject } H_0 \text{ no times } |\beta_1 = \beta_2 = \cdots = \beta_{10} = 0)$$

$$\leq 1 - [P(\text{Accept } H_0\colon \beta_1 = 0|\beta_1 = 0) \times P(\text{Accept } H_0\colon \beta_2 = 0|\beta_2 = 0) \cdots$$

$$\times P(\text{Accept } H_0\colon \beta_{10} = 0|\beta_{10} = 0)]$$

$$= 1 - [(1 - \alpha)^{10}] = 1 - (.95)^{10} = .401$$

FIGURE 4.7

Rejection region for the F statistic with $\nu_1 = 3$, $\nu_2 = 16$, and $\alpha = .05$

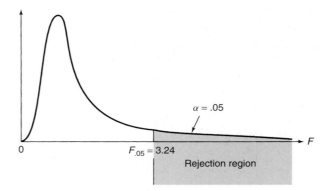

The computed F value, highlighted in Figure 4.6, is $F = 46.66$. Since this value greatly exceeds the tabulated value of 3.24, we conclude that at least one of the model coefficients β_1, β_2, and β_3 is nonzero. Therefore, this **global F-test** indicates that the first-order model $y = \beta_0 + \beta_1 x_1 + \beta_2 x_2 + \beta_3 x_3 + \varepsilon$ is useful for predicting sale price.

Like SAS, most other regression packages give the F value in a portion of the printout called the "Analysis of Variance." This is an appropriate descriptive term, since the F statistic relates the explained and unexplained portions of the total variance of y. For example, the elements of the SAS printout in Figure 4.6 that lead to the calculation of the F value are:

$$F \text{ value} = \frac{\text{Sum of squares (Model)/df(Model)}}{\text{Sum of squares (Error)/df(Error)}} = \frac{\text{Mean square (Model)}}{\text{Mean square (Error)}}$$

$$= \frac{8{,}779{,}676{,}741/3}{1{,}003{,}491{,}259/16} = \frac{2{,}926{,}558{,}914}{62{,}718{,}204} = 46.66$$

Note, too, that the observed significance level for the F statistic (highlighted in Figure 4.6) is .0001, which means that we would reject the null hypothesis $H_0 \colon \beta_1 = \beta_2 = \beta_3 = 0$ at any α value greater than .0001.

The analysis of variance F-test for testing the usefulness of the model is summarized in the next box.

Testing Global Usefulness of the Model: The Analysis of Variance F-Test

$H_0 \colon \beta_1 = \beta_2 = \cdots = \beta_k = 0$ (All model terms are unimportant for predicting y)
$H_a \colon$ At least one $\beta_i \neq 0$ (At least one model term is useful for predicting y)

Test statistic: $F = \dfrac{(\text{SS}_{yy} - \text{SSE})/k}{\text{SSE}/[n - (k + 1)]} = \dfrac{R^2/k}{(1 - R^2)/[n - (k + 1)]}$

$= \dfrac{\text{Mean square (Model)}}{\text{Mean square (Error)}}$

where n is the sample size and k is the number of terms in the model.

Rejection region: $F > F_\alpha$, with k numerator degrees of freedom and $[n - (k + 1)]$ denominator degrees of freedom.

Assumptions: The standard regression assumptions about the random error component (Section 4.2).

Caution

A rejection of the null hypothesis H_0: $\beta_1 = \beta_2 = \cdots = \beta_k$ in the global F-test leads to the conclusion [with $100(1 - \alpha)\%$ confidence] that the model is statistically useful. However, statistically "useful" does not necessarily mean "best." Another model may prove even more useful in terms of providing more reliable estimates and predictions. This global F-test is usually regarded as a test that the model *must* pass to merit further consideration.

EXAMPLE 4.4

Refer to Example 4.3, in which an antique collector modeled the auction price y of grandfather clocks as a function of the age of the clock, x_1, and the number of bidders, x_2. The hypothesized first-order model is

$$y = \beta_0 + \beta_1 x_1 + \beta_2 x_2 + \varepsilon$$

A sample of 32 observations is obtained, with the results summarized in the MINITAB printout repeated in Figure 4.8.

FIGURE 4.8

MINITAB regression printout for grandfather clock model

```
The regression equation is
PRICE = - 1339 + 12.7 AGE + 86.0 BIDDERS

Predictor        Coef      SE Coef         T          P
Constant      -1339.0        173.8      -7.70      0.000
AGE           12.7406       0.9047      14.08      0.000
BIDDERS        85.953        8.729       9.85      0.000

S = 133.5      R-Sq = 89.2%     R-Sq(adj) = 88.5%

Analysis of Variance

Source          DF          SS          MS          F          P
Regression       2     4283063     2141531     120.19      0.000
Residual Error  29      516727       17818
Total           31     4799790
```

a. Find and interpret the adjusted coefficient of determination R_a^2 for this example.
b. Conduct the global F-test of model usefulness at the $\alpha = .05$ level of significance.

Solution

a. The R_a^2 value (highlighted in Figure 4.8) is .885. This implies that the least squares model has explained about 88.5% of the total sample variation in y values (auction prices), after adjusting for sample size and number of independent variables in the model.

b. The elements of the global test of the model follow:

$$H_0: \beta_1 = \beta_2 = 0 \qquad [\text{Note}: k = 2]$$

H_a: At least one of the two model coefficients is nonzero

Test statistic: $F = 120.19$ (see Figure 4.8)

p-value $= .000$

Conclusion: Since $\alpha = .05$ exceeds the observed significance level, $p = .000$, the data provide strong evidence that at least one of the model coefficients is nonzero. The overall model appears to be statistically useful for predicting auction prices. ◆

Can we be sure that the best prediction model has been found if the global F-test indicates that a model is useful? Unfortunately, we cannot. The addition of other independent variables may improve the usefulness of the model. (See the box.) We consider more complex multiple regression models in Sections 4.9, 4.10, and 4.12.

To summarize the discussion in this and the previous section, both R^2 and R_a^2 are indicators of how well the prediction equation fits the data. Intuitive evaluations of the contribution of the model based on R^2 must be examined with care. Unlike R_a^2, the value of R^2 increases as more and more variables are added to the model. Consequently, you could force R^2 to take a value very close to 1 even though the model contributes no information for the prediction of y. In fact, R^2 equals 1 when the number of terms in the model (including β_0) equals the number of data points. Therefore, you should not rely solely on the value of R^2 (or even R_a^2) to tell you whether the model is useful for predicting y. Use the F-test for testing the global utility of the model.

After we have determined that the overall model is useful for predicting y using the F-test, we may elect to conduct one or more t-tests on the individual β parameters (see Section 4.6). However, the test (or tests) to be conducted should be decided *a priori*, that is, prior to fitting the model. Also, we should limit the number of t-tests conducted to avoid the potential problem of making too many Type I errors. Generally, the regression analyst will conduct t-tests only on the "most important" β's. We provide insight in identifying the most important β's in a linear model in the next several sections.

Recommendation for Checking the Utility of a Multiple Regression Model

1. First, conduct a test of overall model adequacy using the F-test, that is, test
   $$H_0: \beta_1 = \beta_2 = \cdots = \beta_k = 0$$
   If the model is deemed adequate (that is, if you reject H_0), then proceed to step 2. Otherwise, you should hypothesize and fit another model. The new model may include more independent variables or higher-order terms.

2. Conduct t-tests on those β parameters in which you are particularly interested (that is, the "most important" β's). These usually involve only the β's associated with higher-order terms (x^2, $x_1 x_2$, etc.). However, it is a safe practice to limit the number of β's that are tested. Conducting a series of t-tests leads to a high overall Type I error rate α.

EXERCISES

4.12. Refer to the *Journal of Adlerian Theory, Research, and Practice* study of depression, Exercise 4.2. Recall that the researchers fitted the model

$$E(y) = \beta_0 + \beta_1 x_1 + \beta_2 x_2 + \beta_3 x_3$$

where y = Zung scale of depression
$\quad\quad\quad\quad x_1$ = ASO scale of self-acceptance
$\quad\quad\quad\quad x_2$ = Burns scale of perfectionism
$\quad\quad\quad\quad x_3$ = Rotter scale of reinforcement

a. The model was fitted to the $n = 76$ points and resulted in a coefficient of determination of $R^2 = .70$. Interpret this value.
b. Is there sufficient evidence to indicate that the model is useful for predicting depression (Zung scale) score? Test using $\alpha = .05$.

4.13. Because the coefficient of determination R^2 always increases when a new independent variable is added to the model, it is tempting to include many variables in a model to force R^2 to be near 1. However, doing so reduces the degrees of freedom available for estimating σ^2, which adversely affects our ability to make reliable inferences. As an example, suppose you want to use the responses to a survey consisting of 18 demographic, social, and economic questions to model a college student's intelligence quotient (IQ). You fit the model

$$y = \beta_0 + \beta_1 x_1 + \beta_2 x_2 + \cdots + \beta_{17} x_{17} + \beta_{18} x_{18} + \varepsilon$$

where y = IQ and $x_1, x_2, \ldots, x_{18}$ are the 18 independent variables. Data for only 20 students ($n = 20$) are used to fit the model, and you obtain $R^2 = .95$.
a. Test to see whether this impressive-looking R^2 is large enough for you to infer that this model is useful, i.e., that at least one term in the model is important for predicting IQ. Use $\alpha = .05$.
b. Calculate R_a^2 and interpret its value.

4.14. Refer to the *World Development* (Feb. 1998) study of street vendors in the city of Puebla, Mexico, Exercise 4.4 (p. 176). Recall that the vendors' mean annual earnings, $E(y)$, was modeled as a first-order function of age (x_1) and hours worked (x_2). Refer to the SAS printout on p. 177 and answer the following:
a. Interpret the value of R^2.
b. Interpret the value of R_a^2. Explain the relationship between R^2 and R_a^2.
c. Conduct a test of the global utility of the model (at $\alpha = .01$). Interpret the result.

4.15. *Professional Geographer* (Feb. 2000) published a study of urban and rural counties in the western United States. University of Nevada (Reno) researchers asked a sample of 256 county commissioners to rate their "home" county on a scale of 1 (most rural) to 10 (most urban). The urban/rural rating (y) was used as the dependent variable in a first-order multiple regression model with six independent variables: total county population (x_1), population density (x_2), population concentration (x_3), population growth (x_4), proportion of county land in farms (x_5), and 5-year change in agricultural land base (x_6). Some of the regression results are shown in the table.

INDEPENDENT VARIABLE	β ESTIMATE	p-VALUE
x_1: Total population	0.110	0.045
x_2: Population density	0.065	0.230
x_3: Population concentration	0.540	0.000
x_4: Population growth	−0.009	0.860
x_5: Farm land	−0.150	0.003
x_6: Agricultural change	−0.027	0.580

Overall model: $R^2 = .44$ $R_a^2 = .43$
$F = 32.47$ p-value $< .001$

Source: Berry, K.A., *et al.* "Interpreting what is rural and urban for western U.S. counties." *Professional Geographer*, Vol. 52, No. 1, Feb. 2000 (Table 2).

a. Write the least squares prediction equation for y.
b. Give the null hypothesis for testing overall model adequacy.
c. Conduct the test, part **b**, at $\alpha = .01$ and give the appropriate conclusion.
d. Interpret the values of R^2 and R_a^2.
e. Give the null hypothesis for testing the contribution of population growth (x_4) to the model.
f. Conduct the test, part **e**, at $\alpha = .01$ and give the appropriate conclusion.

4.16. Refer to *The Real Estate Appraiser* multiple regression model of sale price of a property, Exercise 4.6 (p. 177) and the SAS printout (p. 178).
a. Interpret the values of **F Value, Root MSE, R-Square**, and **Adj R-Sq** shown on the printout.
b. Identify the independent variables with β coefficients that are significantly different from 0 (at $\alpha = .05$). Would you advise the property appraiser to ignore any value adjustments based on nonsignificant independent variables? Explain.

4.17. Refer to the *Journal of Applied Ecology* study of the feeding habits of baby snow geese, Exercise 4.8 (p. 178). Recall that weight change (y) was related to digestion-efficiency (x_1) and acid-detergent fibre (x_2) using a first-order model.

   a. Locate R^2 and R_a^2 on the MINITAB printout (p. 180). Interpret these values. Which statistic is the preferred measure of model fit? Explain.

   b. Locate the global F value for testing the overall model on the MINITAB printout. Use the statistic to test the null hypothesis $H_0: \beta_1 = \beta_2 = 0$.

4.18. Refer to the *Astronomical Journal* study of quasars, Exercise 4.9 (p. 180). A portion of the SPSS printout for the multiple regression model relating equivalent width (y) to four independent variables is reproduced below.

   a. Locate R^2 and R_a^2 on the SPSS printout. Interpret these values. Which statistic is the preferred measure of model fit? Explain.

   b. Locate the global F-value for testing the overall model on the SPSS printout. Use the statistic to test the null hypothesis $H_0: \beta_1 = \beta_2 = \cdots = \beta_5 = 0$.

Model Summary

Model	R	R Square	Adjusted R Square	Std. Error of the Estimate
1	.955ª	.912	.894	15.416

a. Predictors: (Constant), AB1450, REDSHIFT, LINELUM, LINEFLUX

ANOVAᵇ

Model		Sum of Squares	df	Mean Square	F	Sig.
1	Regression	49162.671	4	12290.668	51.720	.000ª
	Residual	4752.769	20	237.638		
	Total	53915.440	24			

a. Predictors: (Constant), AB1450, REDSHIFT, LINELUM, LINEFLUX

b. Dependent Variable: RFEWIDTH

4.19. The *Journal of Quantitative Criminology* (Vol. 8, 1992) published a paper on the determinants of area property crime levels in the United Kingdom. Several multiple regression models for property crime prevalence, y, measured as the percentage of residents in a geographical area who were victims of at least one property crime, were examined. The results for one of the models, based on a sample of $n = 313$ responses collected for the British Crime Survey, are shown in the next table. [*Note:* All variables except Density are expressed as a percentage of the base area.]

   a. Test the hypothesis that the density (x_1) of a region is positively linearly related to crime prevalence (y), holding the other independent variables constant.

   b. Do you advise conducting t-tests on each of the 18 independent variables in the model to determine which variables are important predictors of crime prevalence? Explain.

   c. The model yielded $R^2 = .411$. Use this information to conduct a test of the global utility of the model. Use $\alpha = .05$.

Results for Exercise 4.19

VARIABLE	$\hat{\beta}$	t	p-VALUE
x_1 = Density (population per hectare)	.331	3.88	$p < .01$
x_2 = Unemployed male population	−.121	−1.17	$p > .10$
x_3 = Professional population	−.187	−1.90	$.01 < p < .10$
x_4 = Population aged less than 5	−.151	−1.51	$p > .10$
x_5 = Population aged between 5 and 15	.353	3.42	$p < .01$
x_6 = Female population	.095	1.31	$p > .10$
x_7 = 10-year change in population	.130	1.40	$p > .10$
x_8 = Minority population	−.122	−1.51	$p > .10$
x_9 = Young adult population	.163	5.62	$p < .01$
x_{10} = 1 if North region, 0 if not	.369	1.72	$.01 < p < .10$
x_{11} = 1 if Yorkshire region, 0 if not	−.210	−1.39	$p > .10$
x_{12} = 1 if East Midlands region, 0 if not	−.192	−0.78	$p > .10$
x_{13} = 1 if East Anglia region, 0 if not	−.548	−2.22	$.01 < p < .10$
x_{14} = 1 if South East region, 0 if not	.152	1.37	$p > .10$
x_{15} = 1 if South West region, 0 if not	−.151	−0.88	$p > .10$
x_{16} = 1 if West Midlands region, 0 if not	−.308	−1.93	$.01 < p < .10$
x_{17} = 1 if North West region, 0 if not	.311	2.13	$.01 < p < .10$
x_{18} = 1 if Wales region, 0 if not	−.019	−0.08	$p > .10$

Source: Osborn, D.R., Tickett, Al., and Elder, R. "Area characteristics and regional variates as determinants of area property crime." *Journal of Quantitative Criminology*, Vol. 8, No. 3, 1992, Plenum Publishing Corp.

4.20. An important goal in occupational safety is "active caring." Employees demonstrate active caring (AC) about the safety of their co-workers when they identify environmental hazards and unsafe work practices and then implement appropriate corrective actions for these unsafe conditions or behaviors. Three factors hypothesized to increase the propensity for an employee to actively care for safety are (1) high self-esteem, (2) optimism, and (3) group cohesiveness. *Applied & Preventive Psychology* (Winter 1995) attempted to establish empirical support for the AC hypothesis by fitting the model $E(y) = \beta_0 + \beta_1 x_1 + \beta_2 x_2 + \beta_3 x_3$, where

$\qquad y =$ AC score (measuring active caring on a 15-point scale)
$\qquad x_1 =$ Self-esteem score
$\qquad x_2 =$ Optimism score
$\qquad x_3 =$ Group cohesion score

The regression analysis, based on data collected for $n = 31$ hourly workers at a large fiber-manufacturing plant, yielded a multiple coefficient of determination of $R^2 = .362$.

a. Interpret the value of R^2.

b. Use the R^2 value to test the global utility of the model. Use $\alpha = .05$.

4.21. According to the 1990 census, the number of homeless people in the United States is more than a quarter of a million. Yet little is known about what causes homelessness. Economists at the City University of New York used multiple regression to assist in determining the factors that cause homelessness in

American intercities (*American Economic Review*, Mar. 1993). Data on the number y of homeless per 100,000 population in $n = 50$ metropolitan areas were obtained from the Department of Housing and Urban Development. In addition, the 16 independent variables listed in the table below were measured for each city and a multiple regression analysis performed by fitting the first-order model: $E(y) = \beta_0 + \beta_1 x_1 + \beta_2 x_2 + \cdots + \beta_{16} x_{16}$

a. Interpret the β estimate for the independent variable, rental price.

b. Test the hypothesis that the incidence of homelessness decreases as employment growth increases. Use $\alpha = .05$.

c. Test (at $\alpha = .05$) each of the 16 independent variables to determine which are significantly related to homelessness.

d. What is the danger in performing the t tests, part c?

e. For this model, $R_a^2 = .83$. Interpret this result.

INDEPENDENT VARIABLE	β ESTIMATE	t VALUE
Intercept	307.54	—
Rental price (10% percentile)	2.87	3.93
Vacancy rate (10% percentile)	−872.90	−1.58
Rent-control law (yes or no)	−15.50	−.23
Employment growth	−859.09	−2.71
Share of employment in service industries	−347.69	−1.33
Size of low-skill labor market	−1,003.87	−.38
Households (per 100,000) below poverty level	.013	1.22
Public welfare expenditures	.11	.59
AFDC benefits	−.95	−2.58
SSI benefits	1.07	2.14
Percent reduction in AFDC (nonpoor percents)	146.62	1.49
AFDC accuracy rate	98.15	.13
Mental health in-patients (per 100,000)	−.83	−1.50
Fraction of births to teenage mothers	−1,173.00	−1.39
Blacks (per 100,000)	.004	1.78
1984 population (100,000s)	1.22	1.44

Source: Honig, M., and Filer, R.K. "Causes of intercity variation in homelessness." *American Economic Review*, Vol. 83, No. 1, Mar. 1993, p. 251 (Table 2).

4.9

An Interaction Model with Quantitative Predictors

In Section 4.3, we demonstrated the relationship between $E(y)$ and the independent variables in a first-order model. When $E(y)$ is graphed against any one variable (say, x_1) for fixed values of the other variables, the result is a set of *parallel* straight lines (see Figure 4.1). When this situation occurs (as it always does for a first-order model), we say that the relationship between $E(y)$ and any one independent variable *does not depend* on the values of the other variables in the model.

However, if the relationship between $E(y)$ and x_1 does, in fact, depend on the values of the remaining x's held fixed, then the first-order model is not appropriate

FIGURE 4.9

Graphs of $1 + 2x_1 - x_2 + x_1x_2$ for $x_2 = 0, 1, 2$

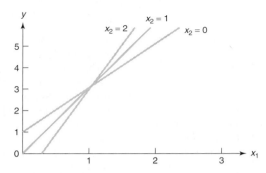

for predicting y. In this case, we need another model that will take into account this dependence. Such a model includes the *cross products* of two or more x's.

For example, suppose that the mean value $E(y)$ of a response y is related to two quantitative independent variables, x_1 and x_2, by the model

$$E(y) = 1 + 2x_1 - x_2 + x_1x_2$$

A graph of the relationship between $E(y)$ and x_1 for $x_2 = 0, 1,$ and 2 is displayed in Figure 4.9.

Note that the graph shows three nonparallel straight lines. You can verify that the slopes of the lines differ by substituting each of the values $x_2 = 0, 1,$ and 2 into the equation. For $x_2 = 0$:

$$E(y) = 1 + 2x_1 - (0) + x_1(0) = 1 + 2x_1 \quad (\text{slope} = 2)$$

For $x_2 = 1$:
$$E(y) = 1 + 2x_1 - (1) + x_1(1) = 3x_1 \quad (\text{slope} = 3)$$

For $x_2 = 2$:
$$E(y) = 1 + 2x_1 - (2) + x_1(2) = -1 + 4x_1 \quad (\text{slope} = 4)$$

Note that the slope of each line is represented by $\beta_1 + \beta_3x_2 = 2 + x_2$. Thus, the effect on $E(y)$ of a change in x_1 (i.e., the slope) now *depends* on the value of x_2. When this situation occurs, we say that x_1 and x_2 **interact**. The cross-product term, x_1x_2, is called an **interaction term**, and the model $E(y) = \beta_0 + \beta_1x_1 + \beta_2x_2 + \beta_3x_1x_2$ is called an **interaction model** with two quantitative variables.

An Interaction Model Relating $E(y)$ to Two Quantitative Independent Variables

$$E(y) = \beta_0 + \beta_1x_1 + \beta_2x_2 + \beta_3x_1x_2$$

where

$(\beta_1 + \beta_3x_2)$ represents the change in $E(y)$ for every 1-unit increase in x_1, holding x_2 fixed

$(\beta_2 + \beta_3x_1)$ represents the change in $E(y)$ for every 1-unit increase in x_2, holding x_1 fixed

---

FIGURE 4.10

Examples of no-interaction and interaction models

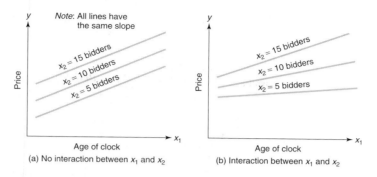

EXAMPLE 4.5

Refer to Examples 4.3 and 4.4. Suppose the collector of grandfather clocks, having observed many auctions, believes that the *rate of increase* of the auction price with age will be driven upward by a large number of bidders. Thus, instead of a relationship like that shown in Figure 4.10a, in which the rate of increase in price with age is the same for any number of bidders, the collector believes the relationship is like that shown in Figure 4.10b. Note that as the number of bidders increases from 5 to 15, the slope of the price versus age line increases.

Consequently, the interaction model is proposed:

$$y = \beta_0 + \beta_1 x_1 + \beta_2 x_2 + \beta_3 x_1 x_2 + \varepsilon$$

The 32 data points listed in Table 4.2 were used to fit the model with interaction. A portion of the MINITAB printout is shown in Figure 4.11.

a. Test the overall utility of the model using the global F-test at $\alpha = .05$.

b. Test the hypothesis (at $\alpha = .05$) that the price-age slope increases as the number of bidders increases—that is, that age and number of bidders, x_2, interact positively.

c. Estimate the change in auction price of a 150-year-old grandfather clock, y, for each additional bidder.

FIGURE 4.11

MINITAB regression printout for grandfather clock model with interaction

```
The regression equation is
PRICE = 320 + 0.88 AGE - 93.3 BIDDERS + 1.30 AGEBID

Predictor      Coef      SE Coef        T         P
Constant      320.5        295.1     1.09     0.287
AGE           0.878        2.032     0.43     0.669
BIDDERS      -93.26        29.89    -3.12     0.004
AGEBID       1.2978       0.2123     6.11     0.000

S = 88.91      R-Sq = 95.4%      R-Sq(adj) = 94.9%

Analysis of Variance

Source          DF        SS         MS        F        P
Regression       3   4578427    1526142   193.04    0.000
Residual Error  28    221362       7906
Total           31   4799790
```

Solution

a. The global F-test is used to test the null hypothesis

$$H_0: \beta_1 = \beta_2 = \beta_3 = 0$$

The test statistic and p-value of the test (highlighted on the MINITAB printout) are $F = 193.04$ and $p = 0$, respectively. Since $\alpha = .05$ exceeds the p-value, there is sufficient evidence to conclude that the model fit is a statistically useful predictor of auction price, y.

b. The hypotheses of interest to the collector concern the interaction parameter β_3. Specifically,

$$H_0: \beta_3 = 0$$

$$H_a: \beta_3 > 0$$

Since we are testing an individual β parameter, a t-test is required. The test statistic and two-tailed p-value (highlighted on the printout) are $t = 6.11$ and $p = 0$, respectively. The upper-tailed p-value, obtained by dividing the two-tailed p-value in half, is $0/2 = 0$. Since $\alpha = .05$ exceeds the p-value, the collector can reject H_0 and conclude that the rate of change of the mean price of the clocks with age increases as the number of bidders increases; that is, x_1 and x_2 interact positively. Thus, it appears that the interaction term should be included in the model.

c. To estimate the change in auction price, y, for every 1-unit increase in number of bidders, x_2, we need to estimate the slope of the line relating y to x_2 when the age of the clock, x_1, is 150 years old. An analyst who is not careful may estimate this slope as $\hat{\beta}_2 = -93.26$. Although the coefficient of x_2 is negative, this does *not* imply that auction price decreases as the number of bidders increases. Since interaction is present, the rate of change (slope) of mean auction price with the number of bidders *depends* on x_1, the age of the clock. Thus, the estimated rate of change of y for a unit increase in x_2 (one new bidder) for a 150-year-old clock is

$$\text{Estimated } x_2 \text{ slope} = \hat{\beta}_2 + \hat{\beta}_3 x_1 = -93.26 + 1.30(150) = 101.74$$

In other words, we estimate that the auction price of a 150-year-old clock will *increase* by about \$101.74 for every additional bidder. Although the rate of increase will vary as x_1 is changed, it will remain positive for the range of values of x_1 included in the sample. Extreme care is needed in interpreting the signs and sizes of coefficients in a multiple regression model.   ◆

Example 4.5 illustrates an important point about conducting t-tests on the β parameters in the interaction model. The "most important" β parameter in this model is the interaction β, β_3. [Note that this β is also the one associated with the highest-order term in the model, $x_1 x_2$.*] Consequently, we will want to test $H_0: \beta_3 = 0$ after we have determined that the overall model is useful for predicting y. Once interaction is detected (as in Example 4.5), however, tests on the first-order

---

*The order of a term is equal to the sum of the exponents of the quantitative variables included in the term. Thus, when x_1 and x_2 are both quantitative variables, the cross product, $x_1 x_2$, is a second-order term.

terms x_1 and x_2 should *not* be conducted since they are meaningless tests; the presence of interaction implies that both x's are important.

Caution

Once interaction has been deemed important in the model $E(y) = \beta_0 + \beta_1 x_1 + \beta_2 x_2 + \beta_3 x_1 x_2$, do not conduct t-tests on the β coefficients of the first-order terms x_1 and x_2. These terms should be kept in the model regardless of the magnitude of their associated p-values shown on the printout.

EXERCISES

4.22. Refer to the *World Development* (Feb. 1998) study of street vendors in the city of Puebla, Mexico, Exercise 4.4 (p. 176). Recall that the vendors' mean annual earnings, $E(y)$, was modeled as a first-order function of age (x_1) and hours worked (x_2). Now, consider the interaction model $E(y) = \beta_0 + \beta_1 x_1 + \beta_2 x_2 + \beta_3 x_1 x_2$. The SAS printout for the model is displayed below.

The REG Procedure
Dependent Variable: EARNINGS

Analysis of Variance

Source	DF	Sum of Squares	Mean Square	F Value	Pr > F
Model	3	5287427	1762476	5.82	0.0124
Error	11	3331000	302818		
Corrected Total	14	8618428			

Root MSE	550.28921	R-Square	0.6135	
Dependent Mean	2577.13333	Adj R-Sq	0.5081	
Coeff Var	21.35276			

Parameter Estimates

Variable	DF	Parameter Estimate	Standard Error	t Value	Pr > |t|
Intercept	1	1041.89440	1303.59326	0.80	0.4411
AGE	1	-13.23762	29.23395	-0.45	0.9595
HOURS	1	103.30564	162.01356	0.64	0.5368
AGEHRS	1	3.62096	3.84044	0.94	0.3660

a. Give the least squares prediction equation.
b. What is the estimated slope relating annual earnings (y) to age (x_1) when number of hours worked (x_2) is 10? Interpret the result.
c. What is the estimated slope relating annual earnings (y) to hours worked (x_2) when age (x_1) is 40? Interpret the result.
d. Give the null hypothesis for testing whether age (x_1) and hours worked (x_2) interact.
e. Find the p-value of the test, part d.

f. Refer to part e. Give the appropriate conclusion in the words of the problem.

4.23. To what degree do the attitudes of your peers influence your behavior? A study presented in *Social Psychology Quarterly* (Vol. 50, 1987) included a sample of $n = 143$ adult drinkers in an urban setting characterized by high physical availability of alcoholic beverages. The goal of the study was to build a model relating frequency of drinking alcoholic beverages, y, to attitude toward drinking (x_1) and social support (x_2). Consider the interaction model

$$E(y) = \beta_0 + \beta_1 x_1 + \beta_2 x_2 + \beta_3 x_1 x_2$$

a. Interpret the phrase "x_1 and x_2 interact" in terms of the problem.
b. Write the null and alternative hypotheses for determining whether attitude (x_1) and social support (x_2) interact.
c. The reported p-value for the test, part b, was $p < .001$. Interpret this result.

4.24. Licensed therapists are mandated, by law, to report child abuse by their clients. This requires the therapist to breach confidentiality and possibly lose the client's trust. A national survey of licensed psychotherapists was conducted to investigate clients' reactions to legally mandated child-abuse reports (*American Journal of Orthopsychiatry*, Jan. 1997). The sample consisted of 303 therapists who had filed a child-abuse report against one of their clients. The researchers were interested in finding the best predictors of a client's reaction (y) to the report, where y is measured on a 30-point scale. (The higher the value, the more favorable the client's response to the report.) The independent variables found to have the most predictive power are listed here.

x_1: Therapist's age (years)

x_2: Therapist's gender (1 if male, 0 if female)

x_3: Degree of therapist's role strain (25-point scale)

x_4: Strength of client-therapist relationship (40-point scale)

x_5: Type of case (1 if family, 0 if not)

x_1x_2: Age × Gender interaction

a. Hypothesize a first-order model relating y to each of the five independent variables.

b. Give the null hypothesis for testing the contribution of x_4, strength of client-therapist relationship, to the model.

c. The test statistic for the test, part **b**, was $t = 4.408$ with an associated p-value of .001. Interpret this result.

d. The estimated β coefficient for the x_1x_2 interaction term was positive and highly significant ($p < .001$). According to the researchers, "this interaction suggests that ... as the age of the therapist increased, ... male therapists were less likely to get negative client reactions than were female therapists." Do you agree?

e. For this model, $R^2 = .2946$. Interpret this value.

4.25. *Trichuristrichiura*, a parasitic worm, affects millions of school-age children each year, especially children from developing countries. A study was conducted to determine the effects of treatment of the parasite on school achievement in 407 school-age Jamaican children infected with the disease (*Journal of Nutrition*, July 1995). About half the children in the sample received the treatment, while the others received a placebo. Multiple regression was used to model spelling test score y, measured as number correct, as a function of the following independent variables:

$$\text{Treatment (T): } x_1 = \begin{cases} 1 & \text{if treatment} \\ 0 & \text{if placebo} \end{cases}$$

$$\text{Disease intensity (I): } x_2 = \begin{cases} 1 & \text{if more than 7,000 eggs} \\ & \text{per gram of stool} \\ 0 & \text{if not} \end{cases}$$

a. Propose a model for $E(y)$ that includes interaction between treatment and disease intensity.

b. The estimates of the β's in the model, part **a**, and the respective p-values for t-tests on the β's are given in the table. Is there sufficient evidence to indicate that the effect of the treatment on spelling score depends on disease intensity? Test using $\alpha = .05$.

VARIABLE	β ESTIMATE	p-VALUE
Treatment (x_1)	−.1	.62
Intensity (x_2)	−.3	.57
T × I(x_1x_2)	1.6	.02

c. Based on the result, part **b**, explain why the analyst should avoid conducting t-tests for the treatment (x_1) and intensity (x_2) β's or interpreting these β's individually.

4.26. Does extensive media coverage of a military crisis influence public opinion on how to respond to the crisis? Political scientists at UCLA researched this question and reported their results in *Communication Research* (June 1993). The military crisis of interest was the 1990 Persian Gulf War, precipitated by Iraqi leader Saddam Hussein's invasion of Kuwait. The researchers used multiple regression analysis to model the level y of support Americans had for a military (rather than a diplomatic) response to the crisis. Values of y ranged from 0 (preference for a diplomatic response) to 4 (preference for a military response). The following independent variables were used in the model:

x_1 = Level of TV news exposure in a selected week (number of days)

x_2 = Knowledge of seven political figures (1 point for each correct answer)

x_3 = Gender (1 if male , 0 if female)

x_4 = Race (1 if nonwhite , 0 if white)

x_5 = Partisanship (0–6 scale, where 0 = strong Democrat and 6 = strong Republican)

x_6 = Defense spending attitude (1–7 scale, where 1 = greatly decrease spending and 7 = greatly increase spending)

x_7 = Education level (1–7 scale, where 1 = less than eight grades and 7 = college)

Data from a survey of 1,763 U.S. citizens were used to fit the model

$$E(y) = \beta_0 + \beta_1 x_1 + \beta_2 x_2 + \beta_3 x_3 + \beta_4 x_4$$
$$+ \beta_5 x_5 + \beta_6 x_6 + \beta_7 x_7 + \beta_8 x_2 x_3 + \beta_9 x_2 x_4$$

The regression results are shown in the table no p. 197.

a. Interpret the β estimate for the variable x_1, TV news exposure.

b. Conduct a test to determine whether an increase in TV news exposure is associated with an increase in support for a military resolution of the crisis. Use $\alpha = .05$.

c. Is there sufficient evidence to indicate that the relationship between support for a military resolution

(y) and gender (x_1) depends on political knowledge (x_2)? Test using $\alpha = .05$.

d. Is there sufficient evidence to indicate that the relationship between support for a military resolution (y) and race (x_4) depends on political knowledge (x_2)? Test using $\alpha = .05$.

VARIABLE	β EST.	STD. ERROR	2-TAILED p-VALUE
TV news exposure (x_1)	.02	.01	.03
Political knowledge (x_2)	.07	.03	.03
Gender (x_3)	.67	.11	<.001
Race (x_4)	−.76	.13	<.001
Partisanship (x_5)	.07	.01	<.001
Defense spending (x_6)	.20	.02	<.001
Education (x_7)	.07	.02	<.001
Knowledge × Gender ($x_2 x_3$)	−.09	.04	.02
Knowledge × Race ($x_2 x_4$)	.10	.06	.08

Source: Iyengar, S., and Simon, A. "News coverage of the Gulf crisis and public opinion." *Communication Research*, Vol. 20, No. 3, June 1993, p. 380 (Table 2).

e. The coefficient of determination for the model was $R^2 = .194$. Interpret this value.

f. Use the value of R^2, part **e**, to conduct a global test for model utility. Use $\alpha = .05$.

WATEROIL

4.27. Refer to the *Journal of Colloid and Interface Science* study of water/oil mixtures, Exercise 4.11 (p. 181). Recall that three of the seven variables used to predict voltage (y) were volume (x_1), salinity (x_2), and surfactant concentration (x_5). The model the researchers fit is

$$E(y) = \beta_0 + \beta_1 x_1 + \beta_2 x_2 + \beta_3 x_5 + \beta_4 x_1 x_2 + \beta_5 x_1 x_5.$$

a. Note that the model includes interaction between disperse phase volume (x_1) and salinity (x_2) as well as interaction between disperse phase volume (x_1) and surfactant concentration (x_5). Discuss how these interaction terms affect the hypothetical relationship between y and x_1. Draw a sketch to support your answer.

b. Fit the interaction model to the WATEROIL data. Does this model appear to fit the data better than the first-order model in Exercise 4.11? Explain.

c. Interpret the β estimates of the interaction model.

4.10 A Quadratic (Second-Order) Model with a Quantitative Predictor

All of the models discussed in the previous sections proposed straight-line relationships between $E(y)$ and each of the independent variables in the model. In this section, we consider models that allow for curvature in the relationships. Each of these models is a **second-order model** because it will include an x^2 term.

First, we consider a model that includes only one independent variable x. The form of this model, called the **quadratic model**, is

$$y = \beta_0 + \beta_1 x + \beta_2 x^2 + \varepsilon$$

The term involving x^2, called a **quadratic term** (or **second-order term**), enables us to hypothesize curvature in the graph of the response model relating y to x. Graphs of the quadratic model for two different values of β_2 are shown in Figure 4.12. When the curve opens upward, the sign of β_2 is positive (see Figure 4.12a); when the curve opens downward, the sign of β_2 is negative (see Figure 4.12b).

FIGURE 4.12

Graphs for two quadratic models

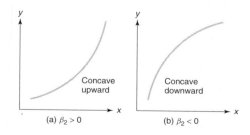

Concave upward

(a) $\beta_2 > 0$

Concave downward

(b) $\beta_2 < 0$

A Quadratic (Second-Order) Model in a Single Quantitative Independent Variable

$$E(y) = \beta_0 + \beta_1 x + \beta_2 x^2$$

where β_0 is the y-intercept of the curve
          β_1 is a shift parameter
          β_2 is the rate of curvature

EXAMPLE   4.6

A physiologist wants to investigate the impact of exercise on the human immune system. The physiologist theorizes that the amount of immunoglobulin y in blood (called IgG, an indicator of long-term immunity) is related to the maximal oxygen uptake x (a measure of aerobic fitness level) of a person by the model

$$y = \beta_0 + \beta_1 x + \beta_2 x^2 + \varepsilon$$

To fit the model, values of y and x were measured for each of 30 human subjects. The data are shown in Table 4.3.

a. Construct a scatterplot for the data. Is there evidence to support the use of a quadratic model?

b. Use the method of least squares to estimate the unknown parameters β_0, β_1, and β_2 in the quadratic model.

c. Graph the prediction equation and assess how well the model fits the data, both visually and numerically.

d. Interpret the β estimates.

e. Is the overall model useful (at $\alpha = .01$) for predicting IgG y?

 AEROBIC

TABLE 4.3   **Data on Immunity and Fitness Level of 30 Subjects**

SUBJECT	IGG *y, milligrams*	MAXIMAL OXYGEN UPTAKE *x, milliliters per kilogram*	SUBJECT	IGG *y, milligrams*	MAXIMAL OXYGEN UPTAKE *x, milliliters per kilogram*
1	881	34.6	16	1,660	52.5
2	1,290	45.0	17	2,121	69.9
3	2,147	62.3	18	1,382	38.8
4	1,909	58.9	19	1,714	50.6
5	1,282	42.5	20	1,959	69.4
6	1,530	44.3	21	1,158	37.4
7	2,067	67.9	22	965	35.1
8	1,982	58.5	23	1,456	43.0
9	1,019	35.6	24	1,273	44.1
10	1,651	49.6	25	1,418	49.8
11	752	33.0	26	1,743	54.4
12	1,687	52.0	27	1,997	68.5
13	1,782	61.4	28	2,177	69.5
14	1,529	50.2	29	1,965	63.0
15	969	34.1	30	1,264	43.2

f. Is there sufficient evidence of concave downward curvature in the immunity-fitness level? Test using $\alpha = .01$.

Solution

a. A scatterplot for the data of Table 4.3, produced using SPSS, is shown in Figure 4.13. The figure illustrates that the appears to increase in a curvilinear manner with the. This provides some support for the inclusion of the quadratic term x^2 in the model.

FIGURE 4.13

SPS scatterplot for data of example 4.6

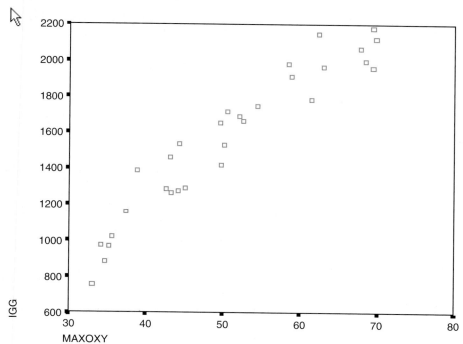

b. We also used SPSS to fit the model to the data in Table 4.3. Part of the SPSS regression output is displayed in Figure 4.14. The least squares estimates of the β parameters (highlighted at the bottom of the printout) are $\hat{\beta}_0 = -1{,}464.404$, $\hat{\beta}_1 = 88.307$, and $\hat{\beta}_2 = -.536$. Therefore, the equation that minimizes the SSE for the data is

$$\hat{y} = -1{,}464.4 + 88.307x - .536x^2$$

c. Figure 4.15 is a graph of the least squares prediction equation. Note that the graph provides a good fit to the data of Table 4.3. A numerical measure of fit is obtained with the adjusted coefficient of determination, R_a^2. From the SPSS printout, $R_a^2 = .933$. This implies that about 93% of the sample variation in IgG (y) can be explained by the quadratic model (after adjusting for sample size and degrees of freedom).

FIGURE 4.14

SPSS output for quadratic model
of example 4.6

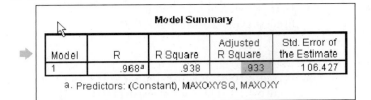

Model Summary

Model	R	R Square	Adjusted R Square	Std. Error of the Estimate
1	.968[a]	.938	.933	106.427

a. Predictors: (Constant), MAXOXYSQ, MAXOXY

ANOVA[b]

Model		Sum of Squares	df	Mean Square	F	Sig.
1	Regression	4602211	2	2301105.316	203.159	.000[a]
	Residual	305818.3	27	11326.605		
	Total	4908029	29			

a. Predictors: (Constant), MAXOXYSQ, MAXOXY

b. Dependent Variable: IGG

Coefficients[a]

Model		Unstandardized Coefficients		Standardized Coefficients	t	Sig.
		B	Std. Error	Beta		
1	(Constant)	-1464.404	411.401		-3.560	.001
	MAXOXY	88.307	16.474	2.574	5.361	.000
	MAXOXYSQ	-.536	.158	-1.628	-3.390	.002

a. Dependent Variable: IGG

FIGURE 4.15

Least squares fit of the quadratic
model

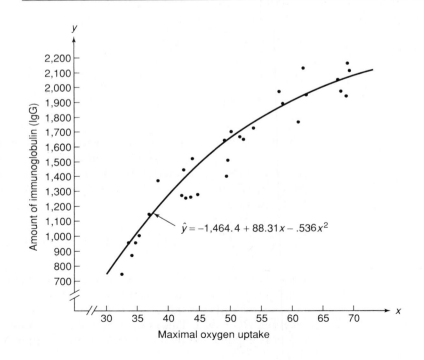

$$\hat{y} = -1,464.4 + 88.31x - .536x^2$$

d. The interpretation of the estimated coefficients in a quadratic model must be undertaken cautiously. First, the estimated y-intercept, $\hat{\beta}_0$, can be meaningfully interpreted only if the range of the independent variable includes zero—that is, if $x = 0$ is included in the sampled range of x. Although $\hat{\beta}_0 = -1,464.4$ seems to imply that the estimated immunity level is negative when $x = 0$, this zero point is not in the range of the sample (the lowest value of maximal oxygen uptake x is 33 milliliters per kilogram), and the value is nonsensical (a person with 0 aerobic fitness level); thus, the interpretation of $\hat{\beta}_0$ is not meaningful.

The estimated coefficient of x is $\hat{\beta}_1 = 88.31$, but it no longer represents a slope in the presence of the quadratic term x^2.* The estimated coefficient of the first-order term x will not, in general, have a meaningful interpretation in the quadratic model.

The sign of the coefficient, $\hat{\beta}_2 = -.536$, of the quadratic term, x^2, is the indicator of whether the curve is concave downward (mound-shaped) or concave upward (bowl-shaped). A negative $\hat{\beta}_2$ implies downward concavity, as in this example (Figure 4.15), and a positive $\hat{\beta}_2$ implies upward concavity. Rather than interpreting the numerical value of $\hat{\beta}_2$ itself, we utilize a graphical representation of the model, as in Figure 4.15, to describe the model.

Note that Figure 4.15 implies that the estimated immunity level (IgG) is leveling off as the aerobic fitness levels increase beyond 70 milliliters per kilogram. In fact, the concavity of the model would lead to decreasing usage estimates if we were to display the model out to $x = 120$ and beyond (see Figure 4.16). However, model interpretations are not meaningful outside the range of the independent variable, which has a maximum value of 69.9 in this example. Thus, although the model appears to support the hypothesis that the *rate of increase* of IgG with maximal oxygen uptake *decreases* for subjects with aerobic fitness levels near the high end of the sampled values, the conclusion that IgG will actually begin to decrease for very large aerobic fitness levels would be a *misuse* of the model, since no subjects with x-values of 70 or more were included in the sample.

e. To test whether the quadratic model is statistically useful, we conduct the global F-test:

$$H_0: \beta_1 = \beta_2 = 0$$

$$H_a: \text{At least one of the above coefficients is nonzero}$$

From the SPSS printout, Figure 4.14, the test statistic is $F = 203.159$ with an associated p-value of 0. For any reasonable α, we reject H_0 and conclude that the overall model is a useful predictor of immunity level, y.

f. Figure 4.15 shows concave downward curvature in the relationship between immunity level and aerobic fitness level in the sample of 30 data points. To

---

*For students with knowledge of calculus, note that the slope of the quadratic model is the first derivative $\partial y / \partial x = \beta_1 + 2\beta_2 x$. Thus, the slope varies as a function of x, rather than the constant slope associated with the straight-line model.

FIGURE 4.16
Potential misuse of quadratic
model

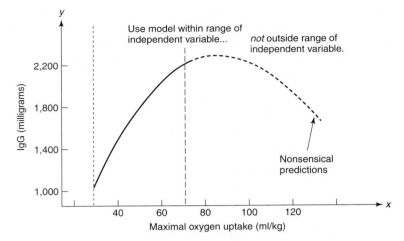

determine if this type of curvature exists in the population, we want to test

$$H_0: \beta_2 = 0 \text{ (no curvature in the response curve)}$$

$$H_a: \beta_2 < 0 \text{ (downward concavity exists in the response curve)}$$

The test statistic for testing β_2, highlighted on the SPSS printout (Figure 4.14), is $t = -3.39$ and the associated two-tailed p-value is .002. Since this is a one-tailed test, the appropriate p-value is $.002/2 = .001$. Now $\alpha = .01$ exceeds this p-value. Thus, there is very strong evidence of downward curvature in the population, that is, immunity level (IgG) increases more slowly per unit increase in maximal oxygen uptake for subjects with high aerobic fitness than for those with low fitness levels.

Note that the SPSS printout in Figure 4.14 also provides the t-test statistic and corresponding two-tailed p-values for the tests of $H_0: \beta_0 = 0$ and $H_0: \beta_1 = 0$. Since the interpretation of these parameters is not meaningful for this model, the tests are not of interest.   ◆

EXERCISES

4.28. A quadratic model was applied to motor vehicle toxic emissions data collected between 1984 and 1999 in Mexico City (*Environmental Science & Engineering*, Sept. 1, 2000). The following equation was used to predict the percentage (y) of motor vehicles without catalytic converters in the Mexico City fleet for a given year (x): $\hat{y} = 325{,}790 - 321.67x + 0.794x^2$.

  a. Explain why the value $\hat{\beta}_0 = 325{,}790$ has no practical interpretation.

  b. Explain why the value $\hat{\beta}_1 = -321.67$ should not be interpreted as a slope.

  c. Examine the value of $\hat{\beta}_2$ to determine the nature of the curvature (upward or downward) in the sample data.

  d. The researchers used the model to estimate "that just after the year 2021 the fleet of cars with catalytic converters will completely disappear." Comment on the danger of using the model to predict y in the year 2021.

4.29. *Fisheries Science* (Feb. 1995) reported on a study of the variables that affect endogenous nitrogen excretion (ENE) in carp raised in Japan. Carp were divided into groups of 2 to 15 fish each according to body weight and each group placed in a separate tank. The carp were then fed a protein-free diet three times daily for a period of 20 days. One day after terminating the feeding experiment, the amount of ENE in each tank was measured. The table gives the mean body weight (in grams) and ENE amount (in milligrams per 100 grams of body weight per day) for each carp group.

🔘 CARP

TANK	BODY WEIGHT x	ENE y
1	11.7	15.3
2	25.3	9.3
3	90.2	6.5
4	213.0	6.0
5	10.2	15.7
6	17.6	10.0
7	32.6	8.6
8	81.3	6.4
9	141.5	5.6
10	285.7	6.0

Source: Watanabe, T., and Ohta, M. "Endogenous nitrogen excretion and non-fecal energy losses in carp and rainbow trout." *Fisheries Science*, Vol. 61, No. 1, Feb. 1995, p. 56 (Table 5).

a. Graph the data in a scatterplot. Do you detect a pattern?

b. The quadratic model $E(y) = \beta_0 + \beta_1 x + \beta_2 x^2$ was fit to the data using MINITAB. The MINITAB printout is displayed below. Conduct the test $H_0: \beta_2 = 0$ against $H_a: \beta_2 \neq 0$ using $\alpha = .10$. Give the conclusion in the words of the problem.

4.30. Engineers at the University of Massachusetts studied the feasibility of using semiconductor lasers for solar lighting in spaceborne applications (*Journal of Applied Physics*, Sept. 1993). A series of $n = 8$ experiments with quantum-well lasers yielded the data on solar pumping threshold current (y) and waveguide A1 mole fraction (x) shown below.

a. The researchers theorize that the relationship between threshold current (y) and waveguide A1 composition (x) will be represented by a U-shaped curve. Hypothesize a model that corresponds to this theory.

b. Graph the data points in a scatterplot. Comment on the researchers' theory, part **a**.

c. Use the SPSS printout on p. 204 to test the theory, part **a**.

🔘 LASERS

THRESHOLD CURRENT y, A/cm^2	WAVEGUIDE A1 MOLE FRACTION x
273	.15
175	.20
146	.25
166	.30
162	.35
165	.40
245	.50
314	.60

Source: Unnikrishnan, S., and Anderson, N.G. "Quantum-well lasers for direct solar photopumping." *Journal of Applied Physics*, Vol. 74, No. 6, Sept. 15, 1993, p. 4226 (adapted from Figure 2).

MINITAB output for Exercise 4.29

```
The regression equation is
ENE = 13.7 - 0.102 BODYWT +0.000273 BODYWTSQ

Predictor        Coef       SE Coef          T         P
Constant       13.713         1.306      10.50     0.000
BODYWT        -0.10184       0.02881      -3.53     0.010
BODYWTSQ     0.0002735     0.0001016       2.69     0.031

S = 2.194      R-Sq = 73.7%      R-Sq(adj) = 66.2%

Analysis of Variance

Source          DF         SS         MS         F        P
Regression       2     94.659     47.329      9.83    0.009
Residual Error   7     33.705      4.815
Total            9    128.364
```

SPSS output for Exercise 4.30

Model Summary

Model	R	R Square	Adjusted R Square	Std. Error of the Estimate
1	.938[a]	.880	.832	25.647

a. Predictors: (Constant), XSQ, X

ANOVA[b]

Model		Sum of Squares	df	Mean Square	F	Sig.
1	Regression	24162.596	2	12081.298	18.367	.005[a]
	Residual	3288.904	5	657.781		
	Total	27451.500	7			

a. Predictors: (Constant), XSQ, X
b. Dependent Variable: Y

Coefficients[a]

Model		Unstandardized Coefficients		Standardized Coefficients	t	Sig.
		B	Std. Error	Beta		
1	(Constant)	438.310	60.537		7.240	.001
	X	−16.843	3.573	−4.094	−4.714	.005
	XSQ	.250	0.47	4.618	5.318	.003

a. Dependent Variable: Y

4.31. Newpaper cartoons, although designed to be funny, often evoke hostility, pain, and/or aggression in readers, especially those cartoons that are violent. A study was undertaken to determine how violence in cartoons is related to aggression or pain (*Motivation and Emotion*, Vol. 10, 1986). A group of volunteers (psychology students) rated each of 32 violent newspaper cartoons (16 "Herman" and 16 "Far Side" cartoons) on three dimensions:

$y =$ Funniness (0 = not funny , ... , 9 = very funny)

$x_1 =$ Pain (0 = none, ... , 9 = a very great deal)

$x_2 =$ Aggression/hostility (0 = none , ... , 9 = a very great deal)

The ratings of the students on each dimension were averaged and the resulting $n = 32$ observations were subjected to a multiple regression analysis. Based on the underlying theory (called the *inverted-U theory*) that the funniness of a joke will increase at low levels of aggression or pain, level off, and then decrease at high levels of aggressiveness or pain, the following quadratic models were proposed:

Model 1: $E(y) = \beta_0 + \beta_1 x_1 + \beta_2 x_1^2$,
$R^2 = .099$, $F = 1.60$

Model 2: $E(y) = \beta_0 + \beta_1 x_2 + \beta_2 x_2^2$,
$R^2 = .100$, $F = 1.61$

a. According to the theory, what is the expected sign of β_2 in either model?

b. Is there sufficient evidence to indicate that the quadratic model relating pain to funniness rating is useful? Test at $\alpha = .05$.

c. Is there sufficient evidence to indicate that the quadratic model relating aggression/hostility to funniness rating is useful? Test at $\alpha = .05$.

4.32. *the movie times* periodically lists the all-time gross revenues for major motion pictures produced in North America. The table on p. 205 gives both the U.S. and worldwide gross revenues for the top 25 movies of all time.

a. Write a first-order model for worldwide gross revenues, y, as a function of movie budget, x.

b. Write a second-order model for worldwide gross revenues, y, as a function of movie budget, x.

c. Fit the model of part c to the data and investigate its usefulness. Is there evidence of a curvilinear relationship between worldwide gross revenues and movie budget? Test using $\alpha = .05$.

d. Based on your analysis in part **c**, which of the two models better explains the variation in worldwide gross revenues?

e. Repeat parts **a–d**, but use U.S. gross revenues as the dependent variable, y.

⊙ MOVIEGROSS

RANK MOVIE		YEAR RELEASED	GROSS IN U.S. DOLLARS SINCE OPENING (millions)	MPAA RATING USA	EST. BUDGET (millions)	EST WORLDWIDE GROSS (millions)
1	Titanic	1997	$600.8	PG-13	$200	$1,835
2	Star Wars	1977	$460.9	PG	$11	$784
3	Star Wars: The Phantom Menace	1999	$431.1	PG	$110	$922
4	E.T.	1982	$399.8	PG	NA	$705
5	Jurassic Park	1993	$356.8	PG-13	$63	$920
6	Forrest Gump	1994	$329.7	PG-13	$55	$680
7	The Lion King	1994	$312.9	G	NA	$767
8	Return of the Jedi	1983	$309.1	PG	$33	$573
9	Independence Day	1996	$306.1	PG-13	$75	$811
10	The Sixth Sense	1999	$293.5	PG-13	$40	$660
11	The Empire Strikes Back	1980	$290.2	PG	$18	$513
12	Home Alone	1990	$285.0	PG	NA	$533
13	How The Grinch Stole Christmas	2000	$260.0	PG	$123	$340
14	Jaws	1975	$260.0	PG	$12	$471
15	Batman	1989	$251.2	PG-13	NA	$413
16	Men in Black	1997	$250.1	PG-13	$90	$586
17	Toy Story 2	1999	$245.8	G	$90	$485
18	Raiders of the Lost Ark	1981	$242.4	PG	$20	$384
19	Twister	1996	$241.7	PG-13	$92	$495
20	Shrek	2001	$240.6	PG	$60	$289
21	Beverly Hills Cop	1984	$234.8	R	NA	$316
22	Cast Away	2000	$233.6	PG-13	$90	$408
23	The Lost World	1997	$229.1	PG-13	$73	$614
24	Ghostbusters	1984	$220.9	PG	$30	$274
25	Mrs. Doubtfire	1993	$219.2	PG-13	NA	$423

Source: the movie times, July 11, 2001 (http://www.the-movie-times.com).

4.33. Spinocerebellar ataxia type 1 (SCA1) is an inherited neurodegenerative disorder characterized by dysfunction of the brain. From a DNA analysis of SCA1 chromosomes, researchers discovered the presence of repeat gene sequences (*Cell Biology*, Feb. 1995). In general, the more repeat sequences observed, the earlier the onset of the disease (in years of age). The scatterplot below shows this relationship for data collected on 113 individuals diagnosed with SCA1.

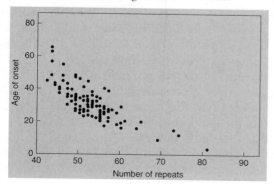

a. Suppose you want to model the age y of onset of the disease as a function of number x of repeat gene sequences in SCA1 chromosomes. Propose a quadratic model for y.

b. Will the sign of β_2 in the model, part **a**, be positive or negative? Base your decision on the results shown in the scatterplot.

c. The researchers reported a correlation of $r = -.815$ between age and number of repeats. Since $r^2 = (-.815)^2 = .664$, they concluded that about "66% of the variability in the age of onset can be accounted for by the number of repeats." Does this statement apply to the quadratic model $E(y) = \beta_0 + \beta_1 x + \beta_2 x^2$? If not, give the equation of the model for which it does apply.

4.34. Refer to the *American Scientist* (July–Aug. 1998) study of the relationship between self-avoiding and unrooted walks, Exercise 3.70 (p. 158). Recall that in a self-avoiding walk you never retrace or cross your own path, while an unrooted walk is a path in which the starting and ending points are impossible to distinguish.

The possible number of walks of each type of various lengths are reproduced in the table. In Exercise 3.70 you analyzed the straight-line model relating number of unrooted walks (y) to walk length (x). Now consider the quadratic model $E(y) = \beta_0 + \beta_1 x + \beta_2 x^2$. A SAS printout for the regression analysis is provided below. Is there sufficient evidence of an upward concave curvilinear relationship between y and x? Test at $\alpha = .10$.

⬤ WALKS

WALK LENGTH (number of steps)	UNROOTED WALKS	SELF-AVOIDING WALKS
1	1	4
2	2	12
3	4	36
4	9	100
5	22	284
6	56	780
7	147	2,172
8	388	5,916

Source: Hayes, B. "How to avoid yourself." *American Scientist*, Vol. 86, No. 4, July–Aug. 1998, p. 317 (Figure 5).

SAS output for Exercise 4.34

Dependent Variable: UNROOT

Analysis of Variance

Source	DF	Sum of Squares	Mean Square	F Value	Pr > F
Model	2	115583	57792	26.66	0.0022
Error	5	10837	2167.33452		
Corrected Total	7	126420			

Root MSE	46.55464	R-Square	0.9143	
Dependent Mean	78.62500	Adj R-Sq	0.8800	
Coeff Var	59.21099			

Parameter Estimates

Variable	DF	Parameter Estimate	Standard Error	t Value	Pr > |t|
Intercept	1	112.12500	64.95046	1.73	0.1449
LENGTH	1	-93.01786	33.11447	-2.81	0.0376
LENGTHSQ	1	15.10119	3.59177	4.20	0.0085

4.11
Using the Model for Estimation and Prediction

In Section 3.9, we discussed the use of the least squares line for estimating the mean value of y, $E(y)$, for some value of x, say, $x = x_p$. We also showed how to use the same fitted model to predict, when $x = x_p$, some value of y to be observed in the future. Recall that the least squares line yielded the same value for both the estimate of $E(y)$ and the prediction of some future value of y. That is, both are the result obtained by substituting x_p into the prediction equation $\hat{y} = \hat{\beta}_0 + \hat{\beta}_1 x$ and calculating $\hat{y}$. There the equivalence ends. The confidence interval for the mean $E(y)$ was narrower than the prediction interval for y, because of the additional uncertainty attributable to the random error ε when predicting some future value of y.

These same concepts carry over to the multiple regression model. As a follow-up to Example 4.6, suppose we want to estimate the mean IgG in blood for a given value of maximal oxygen uptake, say, $x_p = 40$ milliliters per kilogram. Assuming the quadratic model fit in Example 4.6 represents the true relationship between IgG and maximal oxygen uptake, we want to estimate

$$E(y) = \beta_0 + \beta_1 x_p + \beta_2 x_p^2 = \beta_0 + \beta_1(40) + \beta_2(40)^2$$

Substituting into the least squares prediction equation yields the estimate of $E(y)$:

$$\hat{y} = \hat{\beta}_0 + \hat{\beta}_1(40) + \hat{\beta}_2(40)^2$$
$$= -1,464.4 - 88.31(40) - .536(40)^2 = 1,209.9$$

To form a confidence interval for the mean, we must know the standard deviation of the sampling distribution for the estimator $\hat{y}$. For multiple regression models, the form of this standard deviation is rather complex. However, most statistical software packages allow us to obtain the confidence intervals for mean values of y at any given setting of the independent variables. The MINITAB output for the IgG example is shown in Figure 4.17. The value $\hat{y} = 1,209.9$ when $x_p = 40$, is highlighted on the printout. The corresponding 95% confidence interval for the true mean of y, highlighted under **95.0% CI**, is (1,156.2, 1,263.6). Thus, we are 95% confident that the mean IgG level for all people with a maximal oxygen uptake of 40 ml/kg falls between 1,156.2 and 1,263.6 milligrams.

FIGURE 4.17

MINITAB output for quadratic model with 95% confidence and prediction intervals

```
The regression equation is
IgG = - 1464 + 88.3 MaxOxy - 0.536 MaxOxySQ


Predictor        Coef       SE Coef          T          P
Constant       -1464.4         411.4      -3.56      0.001
MaxOxy           88.31         16.47       5.36      0.000
MaxOxySQ       -0.5362        0.1582      -3.39      0.002


S = 106.4        R-Sq = 93.8%       R-Sq(adj) = 93.3%

Analysis of Variance

Source           DF          SS          MS          F          P
Regression        2     4602211     2301105     203.16      0.000
Residual Error   27      305818       11327
Total            29     4908029



Predicted Values for New Observations

New Obs      Fit      SE Fit          95.0% CI               95.0% PI
1         1209.9        26.2    ( 1156.2,   1263.6)   (   985.0,   1434.8)

Values of Predictors for New Observations

New Obs    MaxOxy   MaxOxySQ
1            40.0       1600
```

If we were interested in predicting IgG in the blood of a person with a maximal oxygen uptake of 40 milliliters per kilogram, $\hat{y} = 1,209.9$ would also be used as the predicted value. However, the prediction interval for a particular value of y will be wider than the confidence interval for the mean value. This interval, highlighted on Figure 4.17 under **95.0% PI**, is (985.0, 1,434.8). We can say, with 95% confidence, that the IgG level of a particular person with a maximal oxygen uptake of 40 ml/kg will fall between 985 and 1,434.8 milligrams.

As in simple linear regression, the confidence interval for $E(y)$ will always be narrower than the corresponding prediction interval for y. The relationship between the two intervals for the IgG example is shown graphically in Figure 4.18.

FIGURE 4.18

Confidence interval for mean IgG and prediction interval for IgG, when $x_p = 40$

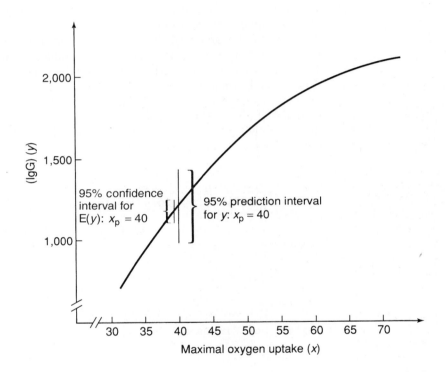

EXERCISES

4.35. Refer to the *World Development* (Feb. 1998) study of street vendors' earnings (y), Exercise 4.4 (p. 176). The SAS printout (p. 209) shows both a 95% prediction interval for y and a 95% confidence interval for $E(y)$ for a 45-year-old vendor who works 10 hours a day (i.e., for $x_1 = 45$ and $x_2 = 10$).

a. Interpret the 95% prediction interval for y in the words of the problem.
b. Interpret the 95% confidence interval for $E(y)$ in the words of the problem.
c. Note that the interval of part **a** is wider than the interval of part **b**. Will this always be true? Explain.

SAS output for Exercise 4.35

Dependent Variable: EARNINGS

Output Statistics

Obs	AGE	HOURS	Dep Var EARNINGS	Predicted Value	Std Error Mean Predict	95% CL Predict	
16	45	10	.	3018	182.3519	1760	4275

Obs	AGE	HOURS	Dep Var EARNINGS	Predicted Value	Std Error Mean Predict	95% CL Mean	
16	45	10	.	3018	182.3519	2620	3415

4.36. Refer to the *Journal of Applied Ecology* study of the feeding habits of baby snow geese, Exercise 4.8 (p. 178). Recall that a first-order model was used to relate gosling weight change (y) to digestion efficiency (x_1) and acid-detergent fibre (x_2). The MINITAB printout below shows both a confidence interval for $E(y)$ and a prediction interval for y when $x_1 = 40\%$ and $x_2 = 15\%$.
a. Interpret the confidence interval for $E(y)$.
b. Interpret the prediction interval for y.

4.37. Refer to *The Astronomical Journal* study of quasars, Exercise 4.9 (p. 180). Recall that a first-order model was used to relate a quasar's equivalent width (y) to redshift (x_1), line flux (x_2), line luminosity (x_3), and AB_{1450} (x_4). A portion of the SPSS spreadsheet showing 95% prediction intervals for y for the first five observations in the data set is reproduced no p. 210. Interpret the interval corresponding to the fifth observation.

MINITAB output for Exercise 4.36

```
The regression equation is
WTCHANGE = 12.2 - 0.0265 DIGEST - 0.458 ADFIBRE

Predictor        Coef      SE Coef         T        P
Constant       12.180        4.402      2.77    0.009
DIGEST        -0.02654      0.05349     -0.50    0.623
ADFIBRE       -0.4578        0.1283     -3.57    0.001

S = 3.519      R-Sq = 52.9%      R-Sq(adj) = 50.5%

Analysis of Variance

Source          DF          SS          MS         F        P
Regression       2      542.03      271.02     21.88    0.000
Residual Error  39      483.08       12.39
Total           41     1025.12


Predicted Values for New Observations

New Obs      Fit      SE Fit        95.0% CI              95.0% PI
1          4.251      0.776   (  2.683,   5.820)  ( -3.038,  11.541)

Values of Predictors for New Observations

New Obs     DIGEST     ADFIBRE
1            40.0        15.0
```

SPSS output for Exercise 4.37

	quasar	redshift	lineflux	linelum	ab1450	rfewidth	lcl_95	ucl_95
1	1	2.81	-13.48	45.29	19.50	117	101.29	172.22
2	2	3.07	-13.73	45.13	19.65	82	59.62	125.89
3	3	3.45	-13.87	45.11	18.93	33	-35.09	37.04
4	4	3.19	-13.27	45.63	18.59	92	63.76	139.25
5	5	3.07	-13.56	45.30	19.59	114	90.69	158.57

4.38. Refer to the *Artificial Intelligence (AI) Applications* study, Exercise 4.7 (p. 178). Recall that the authors use AI to build a regression model relating scenic beauty (y) of southwestern pine stands to age (x_1) of the dominant stand, stems per acre (x_2) in trees, and basal area (x_3) per acre in hardwoods. The AI system is designed to check the values of the input variables (x_1, x_2, and x_3) with the sample data ranges. If the input data value is "out-of-range," a warning is issued about the potential inaccuracy of the predicted y value. Explain the reasoning behind this warning.

WATEROIL

4.39. Refer to Exercise 4.11 (p. 181). The researchers concluded that "in order to break a water-oil mixture with the lowest possible voltage, the volume fraction of the disperse phase (x_1) should be high, while the salinity (x_2) and the amount of surfactant (x_5) should be low." Use this information and the first-order model of Exercise 4.11 to find a 95% prediction interval for this "low" voltage y. Interpret the interval.

4.40. In a production facility, an accurate estimate of hours needed to complete a task is crucial to management in making such decisions as the proper number of workers to hire, an accurate deadline to quote a client, or cost-analysis decisions regarding budgets. A manufacturer of boiler drums wants to use regression to predict the number of hours needed to erect the drums in future projects. To accomplish this, data for 35 boilers were collected. In addition to hours (y), the variables measured were boiler capacity ($x_1 = $ lb/hr), boiler design pressure ($x_2 = $ pounds per square inch or psi), boiler type ($x_3 = 1$ if industry field erected, 0 if utility field erected), and drum type ($x_4 = 1$ if steam, 0 if mud). The data are provided in the table below. A MINITAB printout for the model $E(y) = \beta_0 + \beta_1 x_1 + \beta_2 x_2 + \beta_3 x_3 + \beta_4 x_4$ is given on p. 211.

a. Conduct a test for the global utility of the model. Use $\alpha = .01$.

b. Both a 95% confidence interval for $E(y)$ and a 95% prediction interval for y when $x_1 = 150,000$, $x_2 = 500$, $x_3 = 1$, and $x_4 = 0$ are shown at the bottom of the MINITAB printout. Interpret both of these intervals.

c. Which of the intervals, part **b**, would you use if you want to estimate the average hours required to erect all industrial, mud boilers with a capacity of 150,000 lb/hr and a design pressure of 500 psi?

BOILERS

HOURS y	BOILER CAPACITY x_1	DESIGN PRESSURE x_2	BOILER TYPE x_3	DRUM TYPE x_4
3,137	120,000	375	1	1
3,590	65,000	750	1	1
4,526	150,000	500	1	1
10,825	1,073,877	2,170	0	1
4,023	150,000	325	1	1
7,606	610,000	1,500	0	1
3,748	88,200	399	1	1
2,972	88,200	399	1	1
3,163	88,200	399	1	1
4,065	90,000	1,140	1	1
2,048	30,000	325	1	1

(*Continued*)

HOURS y	BOILER CAPACITY x_1	DESIGN PRESSURE x_2	BOILER TYPE x_3	DRUM TYPE x_4
6,500	441,000	410	1	1
5,651	441,000	410	1	1
6,565	441,000	410	1	1
6,387	441,000	410	1	1
6,454	627,000	1,525	0	1
6,928	610,000	1,500	0	1
4,268	150,000	500	1	1
14,791	1,089,490	2,170	0	1
2,680	125,000	750	1	1
2,974	120,000	375	1	0
1,965	65,000	750	1	0
2,566	150,000	500	1	0
1,515	150,000	250	1	0
2,000	150,000	500	1	0
2,735	150,000	325	1	0
3,698	610,000	1,500	0	0
2,635	90,000	1,140	1	0
1,206	30,000	325	1	0
3,775	441,000	410	1	0
3,120	441,000	410	1	0
4,206	441,000	410	1	0
4,006	441,000	410	1	0
3,728	627,000	1,525	0	0
3,211	610,000	1,500	0	0
1,200	30,000	325	1	0

Source: Dr. Kelly Uscategui, University of Connecticut.

MINITAB output for Exercise 4.40

```
The regression equation is
MANHRS = - 3783 + 0.00875 CAPACITY + 1.93 PRESSURE + 3444 BOILER + 2093 DRUM

Predictor        Coef      SE Coef        T        P
Constant         -3783        1205     -3.14    0.004
CAPACITY     0.0087490    0.0009035      9.68    0.000
PRESSURE        1.9265       0.6489      2.97    0.006
BOILER          3444.3        911.7      3.78    0.001
DRUM            2093.4        305.6      6.85    0.000

S = 894.6      R-Sq = 90.3%    R-Sq(adj) = 89.0%

Analysis of Variance

Source           DF          SS          MS        F        P
Regression        4   230854854    57713714    72.11    0.000
Residual Error   31    24809761      800315
Total            35   255664615

Predicted Values for New Observations

New Obs     Fit     SE Fit        95.0% CI              95.0% PI
1          1936        239    (   1449,    2424)  (    48,    3825)

Values of Predictors for New Observations

New Obs  CAPACITY  PRESSURE    BOILER      DRUM
1          150000       500      1.00  0.000000
```

4.12
More Complex Multiple Regression Models (Optional)

In the preceding sections, we have demonstrated the methods of multiple regression analysis by fitting several basic models, including a first-order model, a quadratic model, and an interaction model. In this section we introduce more advanced models that are useful for relating a response variable y to a set of data.*

MODELS WITH QUANTITATIVE X'S   We begin with a discussion of models using **quantitative** independent variables. We have already encountered several basic models of this type in previous sections. These models are summarized in the following boxes.

A First-Order Model Relating $E(y)$ to Five Quantitative x's

$$E(y) = \beta_0 + \beta_1 x_1 + \beta_2 x_2 + \cdots + \beta_5 x_5$$

A Quadratic (Second-Order) Model Relating $E(y)$ to One Quantitative x

$$E(y) = \beta_0 + \beta_1 x + \beta_2 x^2$$

An Interaction Model Relating $E(y)$ to Two Quantitative x's

$$E(y) = \beta_0 + \beta_1 x_1 + \beta_2 x_2 + \beta_3 x_1 x_2$$

Now, we consider a model for $E(y)$ that incorporates both interaction and curvature. Suppose $E(y)$ is related to two quantitative x's, x_1 and x_2, by the equation:

$$E(y) = 1 + 7x_1 - 10x_2 + 5x_1 x_2 - x_1^2 + 3x_2^2$$

Note that this model contains all of the terms in the interaction model, plus the second-order terms, x_1^2 and x_2^2. Figure 4.19 shows a graph of the relationship between $E(y)$ and x_1 for $x_2 = 0$, 1, and 2. You can see that there are three curvilinear relationships—one for each value of x_2 held fixed—and the curves have different shapes. The model $E(y) = 1 + 7x_1 - 10x_2 + 5x_1 x_2 - x_1^2 + 3x_2^2$ is an example of a **complete second-order model** in two quantitative independent variables. A complete second-order model contains all of the terms in a first-order model and, in addition, the second-order terms involving cross-products (interaction terms) and squares of the independent variables. (Note that an interaction model is a special case of a second-order model, where the β coefficients of x_1^2 and x_2^2 are both equal to 0.)

---

*A more complete discussion of general linear models and their role in model building is provided in Chapter 5.

FIGURE 4.19

Graph of $E(y) =$
$1 + 7x_1 - 10x_2 + 5x_1x_2 - x_1^2 + 3x_2^2$

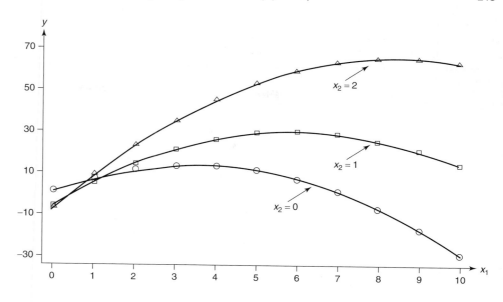

> **A Complete Second-Order Model with Two Quantitative x's**
>
> $$E(y) = \beta_0 + \beta_1 x_1 + \beta_2 x_2 + \beta_3 x_1 x_2 + \beta_4 x_1^2 + \beta_5 x_2^2$$

How can you choose an appropriate model to fit a set of quantitative data? Since most relationships in the real world are curvilinear (at least to some extent), a good first choice would be a second-order model. If you are fairly certain that the relationships between $E(y)$ and the individual quantitative independent variables are approximately first-order and that the independent variables do not interact, you could select a first-order model for the data. If you have prior information that suggests there is moderate or very little curvature over the region in which the independent variables are measured, you could use the interaction model described previously. However, keep in mind that for all multiple regression models, the number of data points must exceed the number of parameters in the model. Thus, you may be forced to use a first-order model rather than a second-order model simply because you do not have sufficient data to estimate all of the parameters in the second-order model.

A practical example of choosing and fitting a linear model with two quantitative independent variables follows.

EXAMPLE 4.7

Although a regional express delivery service bases the charge for shipping a package on the package weight and distance shipped, its profit per package depends on the package size (volume of space that it occupies) and the size and nature of the load on the delivery truck. The company recently conducted a study to investigate the relationship between the cost, y, of shipment (in dollars) and the variables that control the shipping charge—package weight, x_1 (in pounds), and distance shipped, x_2 (in miles). Twenty packages were randomly selected

EXPRESS

TABLE 4.4   Cost of Shipment Data for Example 4.7

PACKAGE	WEIGHT x_1 (lbs)	DISTANCE x_2 (miles)	COST y (dollars)	PACKAGE	WEIGHT x_1 (lbs)	DISTANCE x_2 (miles)	COST y (dollars)
1	5.9	47	2.60	11	5.1	240	11.00
2	3.2	145	3.90	12	2.4	209	5.00
3	4.4	202	8.00	13	.3	160	2.00
4	6.6	160	9.20	14	6.2	115	6.00
5	.75	280	4.40	15	2.7	45	1.10
6	.7	80	1.50	16	3.5	250	8.00
7	6.5	240	14.50	17	4.1	95	3.30
8	4.5	53	1.90	18	8.1	160	12.10
9	.60	100	1.00	19	7.0	260	15.50
10	7.5	190	14.00	20	1.1	90	1.70

from among the large number received for shipment and a detailed analysis of the cost of shipment was made for each package, with the results shown in Table 4.4.

a. Give an appropriate linear model for the data.

b. Fit the model to the data and give the prediction equation.

c. Find the value of s and interpret it.

d. Find the value of R_a^2 and interpret it.

e. Is the model statistically useful for the prediction of shipping cost y? Find the value of the F statistic on the printout and give the observed significance level (p-value) for the test.

f. Find a 95% prediction interval for the cost of shipping a 5-pound package a distance of 100 miles.

Solution

a. Since we have no reason to expect that the relationship between y and x_1 and x_2 would be first-order, we will allow for curvature in the response surface and fit the complete second-order model

$$y = \beta_0 + \beta_1 x_1 + \beta_2 x_2 + \beta_3 x_1 x_2 + \beta_4 x_1^2 + \beta_5 x_2^2 + \varepsilon$$

The mean value of the random error term ε is assumed to equal 0. Therefore, the mean value of y is

$$E(y) = \beta_0 + \beta_1 x_1 + \beta_2 x_2 + \beta_3 x_1 x_2 + \beta_4 x_1^2 + \beta_5 x_2^2$$

b. The SAS printout for fitting the model to the $n = 20$ data points is shown in Figure 4.20. The parameter estimates (highlighted on the printout) are:

$$\hat{\beta}_0 = .82702 \quad \hat{\beta}_1 = -.60914 \quad \hat{\beta}_2 = .00402$$

$$\hat{\beta}_3 = .00733 \quad \hat{\beta}_4 = .08975 \quad \hat{\beta}_5 = .00001507$$

FIGURE 4.20

SAS multiple regression output for Example 4.7

Dependent Variable: Y

Analysis of Variance

Source	DF	Sum of Squares	Mean Square	F Value	Pr > F
Model	5	449.34076	89.86815	458.39	<.0001
Error	14	2.74474	0.19605		
Corrected Total	19	452.08550			

Root MSE	0.44278	R-Square	0.9939	
Dependent Mean	6.33500	Adj R-Sq	0.9918	
Coeff Var	6.98940			

Parameter Estimates

Variable	DF	Parameter Estimate	Standard Error	t Value	Pr > |t|
Intercept	1	0.82702	0.70229	1.18	0.2586
X1	1	-0.60914	0.17990	-3.39	0.0044
X2	1	0.00402	0.00800	0.50	0.6230
X1X2	1	0.00733	0.00063743	11.49	<.0001
X1SQ	1	0.08975	0.02021	4.44	0.0006
X2SQ	1	0.00001507	0.00002243	0.67	0.5127

Output Statistics

Obs	X1	X2	Dep Var Value Y	Predicted Value	Std Error Mean Predict	95% CL Predict	
1	5.9	47	2.6000	2.6114	0.2990	1.4655	3.7573
2	3.2	145	3.9000	4.0964	0.2111	3.0443	5.1486
3	4.4	202	8.0000	7.8238	0.1994	6.7822	8.8654
4	6.6	160	9.2000	9.4828	0.1818	8.4562	10.5093
5	0.75	280	4.4000	4.2666	0.3992	2.9880	5.5452
6	0.7	80	1.5000	1.2730	0.2453	0.1874	2.3587
7	6.5	240	14.5000	13.9229	0.2127	12.8693	14.9764
8	4.5	53	1.9000	1.9063	0.2284	0.8378	2.9748 -
9	0.6	100	1.0000	1.4862	0.2259	0.4201	2.5523
10	7.5	190	14.0000	13.0560	0.2155	11.9998	14.1122
11	5.1	240	11.0000	10.8562	0.1962	9.8175	11.8949
12	2.4	209	5.0000	5.0559	0.1999	4.0140	6.0979
13	0.3	160	2.0000	2.0332	0.2602	0.9316	3.1347
14	6.2	115	6.0000	6.3863	0.1954	5.3482	7.4243
15	2.7	45	1.1000	0.9383	0.2572	-0.1600	2.0366
16	3.5	250	8.0000	8.1527	0.2117	7.1001	9.2054
17	4.1	95	3.3000	3.2101	0.1803	2.1847	4.2356
18	8.1	160	12.1000	12.3066	0.3015	11.1577	13.4555
19	7	260	15.5000	16.3603	0.3109	15.1999	17.5207
20	1.1	90	1.7000	1.4749	0.1993	0.4335	2.5163
21	5	100	.	4.2414	0.1837	3.2133	5.2695

Therefore, the prediction equation that relates the predicted shipping cost, $\hat{y}$, to weight of package, x_1, and distance shipped, x_2, is

$$\hat{y} = .82702 - .60914x_1 + .00402x_2 + .00733x_1x_2 + .08975x_1^2$$
$$+ .00001507x_2^2$$

c. The value of s (shaded on the printout) is .44278. Since s estimates the standard deviation σ of the random error term, our interpretation is that approximately 95% of the sampled shipping cost values fall within $2s = .886$, or about 89¢, of their respective predicted values.

d. The value of R_a^2 (highlighted) is .9918. This means that after adjusting for sample size and the number of model parameters, about 99% of the total sample variation in shipping cost (y) is explained by the model; the remainder is explained by random error.

e. The test statistic for testing whether the model is useful for predicting shipping cost is

$$F = \frac{\text{Mean square for model}}{\text{Mean square for error}} = \frac{\text{SS (Model)}/k}{\text{SSE}/[n - (k + 1)]}$$

where $n = 20$ is the number of data points and $k = 5$ is the number of parameters (excluding β_0) contained in the model. This value of F, highlighted on the printout, is $F = 458.39$. The observed significance level (p-value) for the test also highlighted is less than .0001. This means that if the model contributed no information for the prediction of y, the probability of observing a value of the F statistic as large as 458.39 would be only .0001. Thus, we would reject the null hypothesis for all values of α larger than .0001 and conclude that the model is useful for predicting shipping cost y.

f. The predicted value of y for $x_1 = 5.0$ pounds and $x_2 = 100$ miles is obtained by computing

$$\hat{y} = .82702 - .60914(5.0) + .00402(100) + .00733(5.0)(100)$$
$$+ .08975(5.0)^2 + .00001507(100)^2$$

This quantity is shown (shaded) on the printout as $\hat{y} = 4.2414$. The corresponding 95% prediction interval (shaded) is given as 3.2133 to 5.2695. Therefore, if we were to select a 5-pound package and ship it 100 miles, we are 95% confident that the actual cost will fall between \$3.21 and \$5.27.   ◆

MODELS WITH QUALITATIVE X'S   Multiple regression models can also include **qualitative** (or **categorical**) independent variables. (You have encountered some of these models in the exercises.) Qualitative variables, unlike quantitative variables, cannot be measured on a numerical scale. Therefore, we need to code the values of the qualitative variable (called **levels**) as numbers before we can fit the model. These coded qualitative variables are called **dummy variables** since the numbers assigned to the various levels are arbitrarily selected.

For example, consider a salary discrimination case where there exists a claim of sex discrimination—specifically, the claim that male executives at a large company receive higher average salaries than female executives with the same credentials and qualifications.

To test this claim, we might propose a multiple regression model for executive salaries using the gender of an executive as one of the independent variables. The dummy variable used to describe gender may be coded as follows:

$$x_3 = \begin{cases} 1 & \text{if male} \\ 0 & \text{if female} \end{cases}$$

The advantage of using a 0–1 coding scheme is that the β coefficients associated with the dummy variables are easily interpreted. To illustrate, consider the following model for executive salary y:

$$E(y) = \beta_0 + \beta_1 x$$

where

$$x = \begin{cases} 1 & \text{if male} \\ 0 & \text{if female} \end{cases}$$

This model allows us to compare the mean executive salary $E(y)$ for males with the corresponding mean for females:

$Males(x = 1):$ $E(y) = \beta_0 + \beta_1(1) = \beta_0 + \beta_1$
$Females(x = 0):$ $E(y) = \beta_0 + \beta_1(0) = \beta_0$

First note that β_0 represents the mean salary for females (say, μ_F). When using a 0–1 coding convention, β_0 will always represent the mean response associated with the level of the qualitative variable assigned the value 0 (called the **base level**). The difference between the mean salary for males and the mean salary for females, $\mu_M - \mu_F$, is represented by β_1—that is,

$$\mu_M - \mu_F = (\beta_0 + \beta_1) - (\beta_0) = \beta_1$$

Therefore, with the 0–1 coding convention, β_1 will always represent the difference between the mean response for the level assigned the value 1 and the mean for the base level. Thus, for the executive salary model we have

$$\beta_0 = \mu_F$$
$$\beta_1 = \mu_M - \mu_F$$

If β_1 exceeds 0, then $\mu_M > \mu_F$ and evidence of sex discrimination at the company exists.

The model relating a mean response $E(y)$ to a qualitative independent variable at two levels is shown in the box.

A Model Relating $E(y)$ to a Qualitative Independent Variable with Two Levels

$$E(y) = \beta_0 + \beta_1 x$$

where

$$x = \begin{cases} 1 & \text{if level A} \\ 0 & \text{if level B} \end{cases}$$

Interpretation of β's:

$$\beta_0 = \mu_B \ (\text{Mean for base level})$$
$$\beta_1 = \mu_A - \mu_B$$

For models that involve qualitative independent variables at more than two levels, additional dummy variables must be created. In general, the number of dummy variables used to describe a qualitative variable will be one less than the number of levels of the qualitative variable. The next box presents a model that includes a qualitative independent variable at three levels.

A Model Relating $E(y)$ to a Qualitative Independent Variable with Three Levels

$$E(y) = \beta_0 + \beta_1 x_1 + \beta_2 x_2$$

where

$$x_1 = \begin{cases} 1 & \text{if level A} \\ 0 & \text{if not} \end{cases} \qquad x_2 = \begin{cases} 1 & \text{if level B} \\ 0 & \text{if not} \end{cases} \qquad \text{Base level} = \text{Level C}$$

Interpretation of β's:

$$\beta_0 = \mu_C \text{ (Mean for base level)}$$
$$\beta_1 = \mu_A - \mu_C$$
$$\beta_2 = \mu_B - \mu_C$$

EXAMPLE 4.8

Refer to the problem of modeling the shipment cost, y, of a regional express delivery service, described in Example 4.7. Suppose we want to model $E(y)$ as a function of cargo type, where cargo type has three levels—fragile, semifragile, and durable. Costs for 15 packages of approximately the same weight and same distance shipped, but of different cargo types, are listed in Table 4.5.

a. Write a model relating $E(y)$ to cargo type.

⊙ CARGO

TABLE 4.5   **Data for Example 4.8**

PACKAGE	COST, y	CARGO TYPE	x_1	x_2
1	$17.20	Fragile	1	0
2	11.10	Fragile	1	0
3	12.00	Fragile	1	0
4	10.90	Fragile	1	0
5	13.80	Fragile	1	0
6	6.50	Semifragile	0	1
7	10.00	Semifragile	0	1
8	11.50	Semifragile	0	1
9	7.00	Semifragile	0	1
10	8.50	Semifragile	0	1
11	2.10	Durable	0	0
12	1.30	Durable	0	0
13	3.40	Durable	0	0
14	7.50	Durable	0	0
15	2.00	Durable	0	0

b. Interpret the estimated β coefficients in the model.

c. A MINITAB printout for the model, part **a**, is shown in Figure 4.21. Conduct the F test for overall model utility using $\alpha = .05$. Explain the practical significance of the result.

FIGURE 4.21

MINITAB multiple regression output for Example 4.8

```
The regression equation is
Y = 3.26 + 9.74 X1 + 5.44 X2

Predictor        Coef        SE Coef          T          P
Constant        3.260          1.075       3.03      0.010
X1              9.740          1.521       6.41      0.000
X2              5.440          1.521       3.58      0.004

S = 2.404        R-Sq = 77.4%      R-Sq(adj) = 73.7%

Analysis of Variance

Source          DF           SS          MS          F          P
Regression       2       238.25      119.13      20.61      0.000
Residual Error  12        69.37        5.78
Total           14       307.62
```

Solution

a. Since the qualitative variable of interest, cargo type, has three levels, we must create $(3 - 1) = 2$ dummy variables. First, select (arbitrarily) one of the levels to be the base level—say, durable cargo. Then each of the remaining levels is assigned the value 1 in one of the two dummy variables as follows:

$$x_1 = \begin{cases} 1 & \text{if fragile} \\ 0 & \text{if not} \end{cases} \qquad x_2 = \begin{cases} 1 & \text{if semifragile} \\ 0 & \text{if not} \end{cases}$$

(Note that for the base level, durable cargo, $x_1 = x_2 = 0$.) The values of x_1 and x_2 for each package are given in Table 4.5. Then, the appropriate model is

$$E(y) = \beta_0 + \beta_1 x_1 + \beta_2 x_2$$

b. To interpret the β's, first write the mean shipment cost $E(y)$ for each of the three cargo types as a function of the β's:

Fragile $(x_1 = 1, x_2 = 0)$:

$$E(y) = \beta_0 + \beta_1(1) + \beta_2(0) = \beta_0 + \beta_1 = \mu_F$$

Semifragile $(x_1 = 0, x_2 = 1)$:

$$E(y) = \beta_0 + \beta_1(0) + \beta_2(1) = \beta_0 + \beta_2 = \mu_S$$

Durable ($x_1 = 0$, $x_2 = 0$):

$$E(y) = \beta_0 + \beta_1(0) + \beta_2(0) = \beta_0 = \mu_D$$

Then we have

$$\beta_0 = \mu_D \text{ (Mean of the base level)}$$

$$\beta_1 = \mu_F - \mu_D$$

$$\beta_2 = \mu_S - \mu_D$$

Note that the β's associated with the non–base levels of cargo type (fragile and semifragile) represent differences between a pair of means. As always, β_0 represents a single mean—the mean response for the base level (durable). Now, the estimated β's (highlighted on the MINITAB printout, Figure 4.21) are:

$$\hat{\beta}_0 = 3.26, \quad \hat{\beta}_1 = 9.74, \quad \hat{\beta}_2 = 5.44$$

Consequently, the estimated mean shipping cost for durable cargo ($\hat{\beta}_0$) is \$3.26; the difference between the estimated mean costs for fragile and durable cargo ($\hat{\beta}_1$) is \$9.74; and the difference between the estimated mean costs for semifragile and durable cargo ($\hat{\beta}_2$) is \$5.44.

c. The F test for overall model utility tests the null hypothesis

$$H_0 = \beta_1 = \beta_2 = 0$$

Note that $\beta_1 = 0$ implies that $\mu_F = \mu_D$ and $\beta_2 = 0$ implies that $\mu_S = \mu_D$. Therefore, $\beta_1 = \beta_2 = 0$ implies that $\mu_F = \mu_S = \mu_D$. Thus, a test for model utility is equivalent to a test for equality of means, i.e.,

$$H_0: \mu_F = \mu_S = \mu_D$$

From the MINITAB printout, Figure 4.21, $F = 20.61$. Since the p-value of the test (.000) is less than $\alpha = .05$, the null hypothesis is rejected. Thus, there is evidence of a difference between any two of the three mean shipment costs; i.e., cargo type is a useful predictor of shipment cost y.   ◆

MULTIPLICATIVE MODELS   In all the models presented so far, the random error component has been assumed to be *additive*. An additive error is one for which the response is equal to the mean $E(y)$ plus random error,

$$y = E(y) + \varepsilon$$

Another useful type of model for business, economic, and scientific data is the **multiplicative model**. In this model, the response is written as a *product* of its mean and the random error component, i.e.,

$$y = [E(y)] \cdot \varepsilon$$

Researchers have found multiplicative models to be useful when the change in the response y for every 1-unit change in an independent variable x is better represented by a percentage increase (or decrease) rather than a constant amount increase (or decrease).[†] For example, economists often want to predict a percentage change in the price of a commodity or a percentage increase in the salary of a worker. Consequently, a multiplicative model is used rather than an additive model.

A multiplicative model in two independent variables can be specified as

$$y = (e^{\beta_0})(e^{\beta_1 x_1})(e^{\beta_2 x_2})(e^{\varepsilon})$$

where β_0, β_1, and β_2 are population parameters that must be estimated from the sample data and e^x is a notation for the antilogarithm of x. Note, however, that the multiplicative model is not a linear statistical model as defined in Section 4.1. To use the method of least squares to fit the model to the data, we must transform the model into the form of a linear model. Taking the natural logarithm (denoted ln) of both sides of the equation, we obtain

$$\ln(y) = \beta_0 + \beta_1 x_1 + \beta_2 x_2 + \varepsilon$$

which is now in the form of a linear (additive) model.

When the dependent variable is $\ln(y)$, rather than y, the β parameters and other key regression quantities have slightly different interpretations, as the next example illustrates.

EXAMPLE 4.9   Towers, Perrin, Forster & Crosby (TPF&C), an international management consulting firm, has developed a unique and interesting application of multiple regression analysis. Many firms are interested in evaluating their management salary structure, and TPF&C uses multiple regression models to accomplish this salary evaluation. The Compensation Management Service, as TPF&C calls it, measures both the internal and external consistency of a company's pay policies to determine whether they reflect the management's intent.

The dependent variable y used to measure executive compensation is annual salary. The independent variables used to explain salary structure include the variables listed in Table 4.6. The management at TPF&C has found that executive compensation models that use the natural logarithm of salary as the dependent variable are better predictors than models that use salary as the dependent variable. This is probably because salaries tend to be incremented in *percentages* rather than dollar values. Thus, the multiplicative model we propose (in its linear form) is

$$\ln(y) = \beta_0 + \beta_1 x_1 + \beta_2 x_2 + \beta_3 x_3 + \beta_4 x_4 + \beta_5 x_5 + \beta_6 x_1^2 + \beta_7 x_3 x_4 + \varepsilon$$

We have included a second-order term, x_1^2, to account for a possible curvilinear relationship between log(salary) and years of experience, x_1. Also, the interaction term $x_3 x_4$ is included to account for the fact that the relationship between the

---

[†]Multiplicative models are also found to be useful when the standard regression assumption of equal variances is violated. We discuss this application of multiplicative models in Chapter 8.

TABLE 4.6   **List of Independent Variables for Executive Compensation Example**

INDEPENDENT VARIABLE	DESCRIPTION
x_1	Years of experience
x_2	Years of education
x_3	1 if male; 0 if female
x_4	Number of employees supervised
x_5	Corporate assets (millions of dollars)
x_6	x_1^2
x_7	$x_3 x_4$

number of employees supervised, x_4, and corporate salary may depend on gender, x_3. For example, as the number of supervised employees increases, a male's salary (with all other factors being equal) might rise more rapidly than a female's. (If this is found to be true, the firm will take steps to remove the apparent discrimination against female executives.)

A sample of 100 executives is selected and the variables y and $x_1, x_2, \ldots, x_5$ are recorded. (The data are saved in the CD file named EXECSAL.) The multiplicative model is fit to the data using MINITAB, with the results shown in the MINITAB printout, Figure 4.22.

FIGURE 4.22

MINITAB multiple regression output for Example 4.9

```
The regression equation is
LNSAL = 9.86 + 0.0436 EXP + 0.0309 EDUC + 0.117 SEX +0.000326 NUMSUP
             + 0.00239 ASSETS -0.000635 EXPSQ +0.000302 SEX_SUP

Predictor        Coef      SE Coef         T        P
Constant      9.86182      0.09703    101.63    0.000
EXP          0.043642     0.003761     11.60    0.000
EDUC         0.030937     0.002950     10.49    0.000
SEX           0.11661      0.03696      3.16    0.002
NUMSUP     0.00032593   0.00007850      4.15    0.000
ASSETS      0.0023912    0.0004440      5.39    0.000
EXPSQ       -0.0006347    0.0001384     -4.59    0.000
SEX_SUP    0.00030198   0.00009239      3.27    0.002

S = 0.06596     R-Sq = 94.0%     R-Sq(adj) = 93.6%

Analysis of Variance

Source          DF          SS         MS        F        P
Regression       7     6.28238    0.89748   206.27    0.000
Residual Error  92     0.40029    0.00435
Total           99     6.68267


Predicted Values for New Observations

New Obs     Fit     SE Fit        95.0% CI            95.0% PI
1       11.3021     0.0140   ( 11.2742, 11.3300)  ( 11.1681, 11.4360)

Values of Predictors for New Observations

New Obs      EXP       EDUC       SEX    NUMSUP    ASSETS    EXPSQ    SEX_SUP
1           12.0       16.0  0.000000       400       160      144   0.000000
```

a. Find the least squares prediction equation, and interpret the estimate of β_2.

b. Locate the estimate of s and interpret its value.

c. Locate R_a^2 and interpret its value.

d. Conduct a test of overall model utility using $\alpha = .05$.

e. Test for evidence of sex discrimination at the firm.

f. Use the model to predict the salary of an executive with the characteristics shown in Table 4.7.

TABLE 4.7   Values of Independent Variables for a Particular Executive

---

$x_1 = 12$ years of experience
$x_2 = 16$ years of education
$x_3 = 0$ (female)
$x_4 = 400$ employees supervised
$x_5 = \$160 \times$ million (the firm's asset value)
$x_1^2 = 144$
$x_3x_4 = 0$

Solution

a. The least squares model (highlighted on the MINITAB printout) is

$$\widehat{\ln(y)} = 9.86 + .0436x_1 + .0309x_2 + .117x_3 + .000326x_4 + .00239x_5$$
$$- .000635x_6 + .000302x_7$$

Because we are using the logarithm of salary as the dependent variable, the β estimates have different interpretations than previously discussed. In general, a parameter β in a multiplicative (log) model represents the percentage increase (or decrease) in the dependent variable for a 1-unit increase in the corresponding independent variable. The percentage change is calculated by taking the antilogarithm of the β estimate and subtracting 1, i.e., $e^{\hat{\beta}} - 1$.[‡] For example, the percentage change in executive compensation associated with a 1-unit (i.e., 1-year) increase in years of education x_2 is $(e^{\hat{\beta}_2} - 1) = (e^{.0309} - 1) = .031$. Thus, when all other independent variables are held constant, we estimate executive salary to increase 3.1% for each additional year of education.

[‡]The result is derived by expressing the percentage change in salary y, as $(y_1 - y_0)/y_0$, where $y_1 =$ the value of y when, say, $x = 1$, and $y_0 =$ the value of y when $x = 0$. Now let $y^* = \ln(y)$ and assume the log model is $y^* = \beta_0 + \beta_1 x$. Then

$$y = e^{y^*} = e^{\beta_0}e^{\beta_1 x} = \begin{cases} e^{\beta_0} & \text{when } x = 0 \\ e^{\beta_0}e^{\beta_1} & \text{when } x = 1 \end{cases}$$

Substituting, we have

$$\frac{y_1 - y_0}{y_0} = \frac{e^{\beta_0}e^{\beta_1} - e^{\beta_0}}{e^{\beta_0}} = e^{\beta_1} - 1$$

A Multiplicative (Log) Model Relating y to Several Independent Variables

$$\ln(y) = \beta_0 + \beta_1 x_1 + \beta_2 x_2 + \cdots + \beta_k x_k + \varepsilon$$

where $\ln(y)$ = natural logarithm of y

Interpretation of β's

$$(e^{\beta_i} - 1) \times 100\% = \text{Percentage change in } y \text{ for every}$$

$$1 \text{ unit increase in } x_i, \text{ holding all other } x's \text{ fixed}$$

---

b. The estimate of the standard deviation σ (shaded on the printout) is $s = .066$. Our interpretation is that most of the observed $\ln(y)$ values (logarithms of salaries) lie within $2s = 2(.066) = .132$ of their least squares predicted values. A more practical interpretation (in terms of salaries) is obtained, however, if we take the antilog of this value and subtract 1, similar to the manipulation in part **a**. That is, we expect most of the observed executive salaries to lie within $e^{2s} - 1 = e^{.132} - 1 = .141$, or 14.1% of their respective least squares predicted values.

c. The adjusted R^2 value (highlighted on the printout) is $R_a^2 = .936$. This implies that, after taking into account sample size and the number of independent variables, almost 94% of the variation in the logarithm of salaries for these 100 sampled executives is accounted for by the model.

d. The test for overall model utility is conducted as follows:

H_0: $\beta_1 = \beta_2 = \cdots = \beta_7 = 0$
H_a: At least one of the model coefficients is nonzero.

Test statistic: $F = \dfrac{\text{Mean square for model}}{\text{MSE}} = 206.27$ (shaded on Figure 4.22)

p-value $= .000$

Since $\alpha = .05$ exceeds the p-value of the test, we conclude that the model does contribute information for predicting executive salaries. It appears that at least one of the β parameters in the model differs from 0.

e. If the firm is (knowingly or unknowingly) discriminating against female executives, then the mean salary for females (denoted μ_F) will be less than the mean salary for males (denoted μ_M) with the same qualifications (e.g., years of experience, years of education, etc.) From our previous discussion of dummy variables, this difference will be represented by β_3, the β coefficient multiplied by x_3 if we set number of employees supervised, x_4, equal to 0. Since $x_3 = 1$ if male, 0 if female, then $\beta_3 = (\mu_M - \mu_F)$ for fixed values of x_1, x_2 and x_5, and $x_4 = 0$. Consequently, a test of

H_0: $\beta_3 = 0$ versus H_a: $\beta_3 > 0$

is one way to test the discrimination hypothesis.[§] The p-value for this one-tailed test is one-half the p-value shown on the MINITAB printout, i.e., $.002/2 = .001$. With such a small p-value, there is strong evidence to reject H_0 and claim that some form of sex discrimination exists at the firm.

f. The least squares model can be used to obtain a predicted value for the logarithm of salary. Substituting the values of the x's shown in Table 4.7, we obtain

$$\widehat{\ln(y)} = \hat{\beta}_0 + \hat{\beta}_1(12) + \hat{\beta}_2(16) + \hat{\beta}_3(0) + \hat{\beta}_4(400) + \hat{\beta}_5(160x)$$
$$+\hat{\beta}_6(144) + \hat{\beta}_7(0)$$

This predicted value is given at the bottom of the MINITAB printout, Figure 4.22, $\widehat{\ln(y)} = 11.3021$. The 95% prediction interval, from 11.1681 to 11.4360, is also highlighted on the printout. To predict the salary of an executive with these characteristics, we take the antilog of these values. That is, the predicted salary is $e^{11.3021} = \$80,992$ (rounded to the nearest hundred) and the 95% prediction interval is from $e^{11.1681}$ to $e^{11.4360}$ (or from \$70,834 to \$92,596). Thus, an executive with the characteristics in Table 4.7 should be paid between \$70,834 and \$92,596 to be consistent with the sample data.   ◆

Warning: To decide whether a log transformation on the dependent variable is necessary, naive researchers sometimes compare the R^2 values for the two models

$$y = \beta_0 + \beta_1 x_1 + \cdots + \beta_k x_k + \varepsilon$$

and

$$\ln(y) = \beta_0 + \beta_1 x_1 + \cdots + \beta_k x_k + \varepsilon$$

and choose the model with the larger R^2. But these R^2 values *are not comparable* since the dependent variables are not the same! One way to generate comparable R^2 values is to calculate the predicted values, $\widehat{\ln(y)}$, for the log model and then compute the corresponding $\hat{y}$ values using the inverse transformation $\hat{y} = e^{\widehat{\ln(y)}}$. A pseudo-$R^2$ for the log model can then be calculated in the usual way:

$$R^2_{\ln(y)} = 1 - \frac{\Sigma(y_i - \hat{y}_i)^2}{\Sigma(y_i - \overline{y}_i)^2}$$

$R^2_{\ln(y)}$ is now comparable to the R^2 for the untransformed model. See Maddala (1988) for a discussion of more formal methods for comparing the two models.

[§] A test for discrimination could also include testing the interaction term, $\beta_7 x_3 x_4$. If, as number of employees supervised (x_4) increases, the rate of increase in salary for males exceeds the rate for females, then $\beta_7 > 0$. Thus, rejecting $H_0: \beta_7 = 0$ in favor of $H_a: \beta_7 > 0$ would also suggest discrimination against female executives.

EXERCISES

4.41. Write a first-order linear model relating the mean value of y, $E(y)$, to
  a. two quantitative independent variables
  b. four quantitative independent variables

4.42. Write a complete second-order linear model relating the mean value of y, $E(y)$, to
  a. two quantitative independent variables
  b. three quantitative independent variables

4.43. Write a model relating $E(y)$ to a qualitative independent variable with
  a. two levels, A and B
  b. four levels, A, B, C, and D
  Interpret the β parameters in each case.

4.44. Consider the first-order equation

$$y = 1 + 2x_1 + x_2$$

  a. Graph the relationship between y and x_1 for $x_2 = 0$, 1, and 2.
  b. Are the graphed curves in part a first-order or second-order?
  c. How do the graphed curves in part a relate to each other?
  d. If a linear model is first-order in two independent variables, what type of geometric relationship will you obtain when $E(y)$ is graphed as a function of one of the independent variables for various values of the other independent variable?

4.45. Consider the first-order equation

$$y = 1 + 2x_1 + x_2 - 3x_3$$

  a. Graph the relationship between y and x_1 for $x_2 = 1$ and $x_3 = 3$.
  b. Repeat part **a** for $x_2 = -1$ and $x_3 = 1$.
  c. If a linear model is first-order in three independent variables, what type of geometric relationship will you obtain when $E(y)$ is graphed as a function of one of the independent variables for various values of the other independent variables?

4.46. Consider the second-order model

$$y = 1 + x_1 - x_2 + 2x_1^2 + x_2^2$$

  a. Graph the relationship between y and x_1 for $x_2 = 0$, 1, and 2.
  b. Are the graphed curves in part **a** first-order or second-order?

c. How do the graphed curves in part **a** relate to each other?

d. Do the independent variables x_1 and x_2 interact? Explain.

4.47. Consider the second-order model

$$y = 1 + x_1 - x_2 + x_1x_2 + 2x_1^2 + x_2^2$$

  a. Graph the relationship between y and x_1 for $x_2 = 0$, 1, and 2.
  b. Are the graphed curves in part **a** first-order or second-order?
  c. How do the graphed curves in part **a** relate to each other?
  d. Do the independent variables x_1 and x_2 interact? Explain.
  e. Note that the model used in this exercise is identical to the noninteraction model of Exercise 4.46, except that it contains the term involving x_1x_2. What does the term x_1x_2 introduce into the model?

4.48. *Chance* (Winter 2001) published a study of students who paid a private tutor (or coach) to help them improve their Scholastic Assessment Test (SAT) scores. Multiple regression was used to estimate the effect of coaching on SAT-Mathematics scores. Data on 3,492 students (573 of whom were coached) were used to fit the model $E(y) = \beta_0 + \beta_1x_1 + \beta_2x_2$, where y = SAT-Math score, x_1 = score on PSAT, and $x_2 = \{1$ if student was coached, 0 if not$\}$.
  a. The fitted model had an adjusted R^2 value of .76. Interpret this result.
  b. The estimate of β_2 in the model was 19, with a standard error of 3. Use this information to form a 95% confidence interval for β_2. Interpret the interval.
  c. Based on the interval, part **b**, what can you say about the effect of coaching on SAT-Math scores?

4.49. *New Scientist* (Apr. 3, 1993) published an article on strategies for foiling assassination attempts on politicians. The strategies are based on the findings of researchers at Middlesex University (United Kingdom), who used a multiple regression model for predicting the level y of assassination risk. One of the variables used in the model was political status of a country (communist, democratic, or dictatorship).
  a. Propose a model for $E(y)$ as a function of political status.
  b. Interpret the β's in the model, part **a**.

4.50. One of the most promising methods for extracting crude oil employs a carbon dioxide (CO_2) flooding technique. When flooded into oil pockets, CO_2 enhances oil recovery by displacing the crude oil. In a microscopic investigation of the CO_2 flooding process, flow tubes were dipped into sample oil pockets containing a known amount of oil. The oil pockets were flooded with CO_2 and the percentage of oil displaced was recorded. The experiment was conducted at three different flow pressures and three different dipping angles. The displacement test data are recorded in the table.

◉ CRUDEOIL

PRESSURE x_1, pounds per square inch	DIPPING ANGLE x_2, degrees	OIL RECOVERY y, percentage
1,000	0	60.58
1,000	15	72.72
1,000	30	79.99
1,500	0	66.83
1,500	15	80.78
1,500	30	89.78
2,000	0	69.18
2,000	15	80.31
2,000	30	91.99

Source: Wang, G.C. "Microscopic investigation of CO_2 flooding process." *Journal of Petroleum Technology*, Vol. 34, No. 8, Aug. 1982, pp. 1789–1797. Copyright © 1982, Society of Petroleum Engineers, American Institute of Mining. First published in the *JPT* Aug. 1982.

a. Write the complete second-order model relating percentage oil recovery y to pressure x_1 and dipping angle x_2.

b. Graph the sample data on a scatterplot, with percentage oil recovery y on the vertical axis and pressure x_1 on the horizontal axis. Connect the points corresponding to the same value of dipping angle x_2. Based on the scatterplot, do you believe a complete second-order model is appropriate?

c. The SAS printout for the interaction model

$$y = \beta_0 + \beta_1 x_1 + \beta_2 x_2 + \beta_3 x_1 x_2 + \varepsilon$$

is provided below. Give the prediction equation for this model.

d. Construct a plot similar to the scatterplot of part **b**, but use the predicted values from the interaction model on the vertical axis. Compare the two plots. Do you believe the interaction model will provide an adequate fit?

e. Check model adequacy using a statistical test with $\alpha = .05$.

f. Is there evidence of interaction between pressure x_1 and dipping angle x_2? Test using $\alpha = .05$.

4.51. The *Journal of Hazardous Materials* (July 1995) presented a literature review of models designed to predict oil spill evaporation. One model discussed in the article used boiling point (x_1) and API specific gravity (x_2) to predict the molecular weight (y) of the oil that is spilled. A complete second-order model for y was proposed.

a. Write the equation of the model.

b. Identify the terms in the model that allow for curvilinear relationships.

SAS output for Exercise 4.50

Dependent Variable: OILRECOV

Analysis of Variance

Source	DF	Sum of Squares	Mean Square	F Value	Pr > F
Model	3	843.19083	281.06361	44.67	0.0005
Error	5	31.45997	6.29199		
Corrected Total	8	874.65080			

Root MSE	2.50838	R-Square	0.9640	
Dependent Mean	76.90667	Adj R-Sq	0.9425	
Coeff Var	3.26160			

Parameter Estimates

Variable	DF	Parameter Estimate	Standard Error	t Value	Pr > |t|
Intercept	1	54.50000	5.03416	10.83	0.0001
PRESSURE	1	0.00770	0.00324	2.38	0.0634
DIPANGLE	1	0.55411	0.25996	2.13	0.0862
PRESS_ANG	1	0.00011333	0.00016723	0.68	0.5280

⬤ REPELLENT

INSECT REPELLENT	TYPE	COST/USE	MAXIMUM PROTECTION
Amway Hourguard 12	Lotion/Cream	$2.08	13.5 hours
Avon Skin-So-Soft	Aerosol/Spray	0.67	0.5
Avon Bug Guard Plus	Lotion/Cream	1.00	2.0
Ben's Backyard Formula	Lotion/Cream	0.75	7.0
Bite Blocker	Lotion/Cream	0.46	3.0
BugOut	Aerosol/Spray	0.11	6.0
Cutter Skinsations	Aerosol/Spray	0.22	3.0
Cutter Unscented	Aerosol/Spray	0.19	5.5
Muskol Ultra 6 Hours	Aerosol/Spray	0.24	6.5
Natrapel	Aerosol/Spray	0.27	1.0
Off! Deep Woods	Aerosol/Spray	1.77	14.0
Off! Skintastic	Lotion/Cream	0.67	3.0
Sawyer Deet Formula	Lotion/Cream	0.36	7.0
Repel Permanone	Aerosol/Spray	2.75	24.0

Source: "Buzz off." *Consumer Reports*, June 2000.

4.52. Which insect repellents protect best against mosquitoes? *Consumer Reports* (June 2000) tested 14 products that all claim to be an effective mosquito repellent. Each product was classified as either lotion/cream or aerosol/spray. The cost of the product (in dollars) was divided by the amount of the repellent needed to cover exposed areas of the skin (about 1/3 ounce) to obtain a cost-per-use value. Effectiveness was measured as the maximum number of hours of protection (in half-hour increments) provided when human testers exposed their arms to 200 mosquitoes. The data from the report are listed in the table above.

a. Suppose you want to use repellent type to model the cost per use (y). Create the appropriate number of dummy variables for repellent type and write the model.

b. Fit the model, part **a**, to the data.

c. Give the null hypothesis for testing whether repellent type is a useful predictor of cost per use (y).

d. Conduct the test, part **c**, and give the appropriate conclusion. Use $\alpha = .10$.

e. Repeat parts **a–d** if the dependent variable is maximum number of hours of protection (y).

4.53. The vineyards in the Bordeaux region of France are known for producing excellent red wines. However, the uncertainty of the weather during the growing season, the phenomenon that wine tastes better with age, and the fact that some Bordeaux vineyards produce better wines than others, encourages speculation concerning the value of a case of wine produced by a certain vineyard during a certain year (or vintage). As a result, many wine experts attempt to predict the auction price of a case of Bordeaux wine. The publishers of a newsletter titled *Liquid Assets: The International Guide to Fine Wine* discussed a multiple regression approach to predicting the London auction price of red Bordeaux wine in *Chance* (Fall 1995). The natural logarithm of the price y (in dollars) of a case containing a dozen bottles of red wine was modeled as a function of weather during growing season and age of vintage using data collected for the vintages of 1952–1980. Three models were fit to the data. The results of the regressions are summarized in the table on p. 229.

a. For each model, conduct a t-test for each of the β parameters in the model. Interpret the results.

b. Interpret the β estimates of each model.

c. The three models for auction price (y) have R^2 and s values as shown in the table. Based on this information, which of the three models would you use to predict red Bordeaux wine prices? Explain.

Results for Exercise 4.53

INDEPENDENT VARIABLES	BETA ESTIMATES (STANDARD ERRORS)		
	MODEL 1	MODEL 2	MODEL 3
x_1 = Vintage year	.0354 (.0137)	.0238 (.00717)	.0240 (.00747)
x_2 = Average growing season temperature (°C)	(not included)	.616 (.0952)	.608 (.116)
x_3 = Sept./Aug. rainfall (cm)	(not included)	−.00386 (.00081)	−.00380 (.00095)
x_4 = Rainfall in months preceding vintage (cm)	(not included)	.0001173 (.000482)	.00115 (.000505)
x_5 = Average Sept. temperature (°C)	(not included)	(not included)	.00765 (.565)
	$R^2 = .212$	$R^2 = .828$	$R^2 = .828$
	$s = .575$	$s = .287$	$s = .293$

Source: Ashenfelter, O., Ashmore, D., and LaLonde, R. "Bordeaux wine vintage quality and weather." *Chance*, Vol. 8, No. 4, Fall 1995, p. 116 (Table 2).

4.54. External auditors are hired to review and analyze the financial and other records of an organization and to attest to the integrity of the organization's financial statements. In recent years, the fees charged by auditors have come under increasing scrutiny. S. Butterworth and K. A. Houghton, two University of Melbourne (Australia) researchers, investigated the effects of several variables on the fee charged by auditors. These variables are described here.

y = Logarithm of audit fee charged to auditee (FEE)

$$x_1 = \begin{cases} 1 & \text{if auditee changed auditors after} \\ & \text{one year (CHANGE)} \\ 0 & \text{if not} \end{cases}$$

x_2 = Logarithm of auditee's total assets (SIZE)
x_3 = Number of subsidiaries of auditee (COMPLEX)

$$x_4 = \begin{cases} 1 & \text{if auditee receives an audit qualification} \\ & \text{(RISK)} \\ 0 & \text{if not} \end{cases}$$

$$x_5 = \begin{cases} 1 & \text{if auditee in mining industry} \\ & \text{(INDUSTRY)} \\ 0 & \text{if not} \end{cases}$$

$$x_6 = \begin{cases} 1 & \text{if auditor is a member of a "Big 8" firm} \\ & \text{(BIG8)} \\ 0 & \text{if not} \end{cases}$$

x_7 = Logarithm of dollar-value of non-audit services provided by auditor (NAS)

The multiple regression model $E(y) = \beta_0 + \beta_1 x_1 + \beta_2 x_2 + \beta_3 x_3 + \cdots + \beta_7 x_7$ was fit to data collected for $n = 268$ companies. The results are summarized in the table below.

INDEPENDENT VARIABLE	EXPECTED SIGN OF β	β ESTIMATE	t VALUE	LEVEL OF SIGNIFICANCE (p-VALUE)
Constant	−	−4.30	−3.45	.001 (two-tailed)
CHANGE	+	−.002	−0.049	.961 (one-tailed)
SIZE	+	.336	9.94	.000 (one-tailed)
COMPLEX	+	.384	7.63	.000 (one-tailed)
RISK	+	.067	1.76	.079 (one-tailed)
INDUSTRY	−	−.143	−4.05	.000 (one-tailed)
BIG8	+	.081	2.18	.030 (one-tailed)
NAS	+/−	.134	4.54	.000 (two-tailed)
$R^2 = .712$		$F = 111.1$		

Source: Butterworth, S., and Houghton, K.A. "Auditor switching: The pricing of audit services." *Journal of Business Finance and Accounting*, Vol. 22, No. 3, April 1995, p. 334 (Table 4).

a. Write the least squares prediction equation.

b. Assess the overall fit of the model.

c. Interpret the estimate of β_3.

d. The researchers hypothesized the direction of the effect of each independent variable on audit fees. These hypotheses are given in the "Expected Sign of β" column in the table. (For example, if the expected sign is negative, the alternative hypothesis is $H_a: \beta_i < 0$.) Interpret the results of the hypothesis test for β_4. Use $\alpha = .05$.

e. The main objective of the analysis was to determine whether new auditors charge less than incumbent auditors in a given year. If this hypothesis is true, then the true value of β_1 is negative. Is there evidence to support this hypothesis? Explain.

4.55. As a result of the U.S. surgeon general's warnings about the health hazards of smoking, Congress banned television and radio advertising of cigarettes in January 1971. The banning of prosmoking messages, however, also led to the virtual elimination of antismoking messages. In theory, if the pre-1971 antismoking commercials had been more effective than prosmoking commercials, the net effect of the Congressional ban should have been an increase in the consumption of cigarettes, thus benefiting the tobacco industry. To test this hypothesis, researchers at the University of Houston built a cigarette demand model based on data collected for 46 states over the 18-year period from 1963 to 1980 (*The Review of Economics and Statistics*, Feb. 1986). For each state–year, the following independent variables were recorded:

x_1 = Natural log of price of a carton of cigarettes

x_2 = Natural log of minimum price of a carton of cigarettes in any neighboring state (This variable was included to measure the effect of "bootlegging" cigarettes in nearby states with lower tax rates.)

x_3 = Natural log of real disposable income per capita

x_4 = Per capita index of expenditures for cigarette advertising on television and radio (This value is 0 for the years 1971–1980, when the ban was in effect.)

The dependent variable of interest is y, the natural log of per capita consumption of cigarettes by persons of smoking age (14 years and older). The multiple regression model

$$E(y) = \beta_0 + \beta_1 x_1 + \beta_2 x_2 + \beta_3 x_3 + \beta_4 x_4$$

was estimated using $n = 828$ observations (48 states $\times$ 18 years) with the following results:

$$R^2 = .95$$

$$s = .047$$

a. Test the hypothesis that the model is useful for predicting y. Use $\alpha = .05$.

b. Interpret the value of s.

c. Give the null and alternative hypotheses appropriate for testing whether a decrease in per capita cigarette advertising expenditures is accompanied by an increase in per capita consumption of cigarettes over the period 1963–1980.

d. The value of $\hat{\beta}_4$ was determined to be .033. Interpret this value.

e. Does the value $\hat{\beta}_4 = .033$ support the alternative hypothesis in part **c**? Explain.

4.56. Where do you look when you are listening to someone speak? Researchers have discovered that listeners tend to gaze at the eyes or mouth of the speaker. In a study published in *Perception & Psychophysics* (Aug. 1998), subjects watched a videotape of a speaker giving a series of short monologues at a social gathering (e.g., a party). The level of background noise (multilingual voices and music) was varied during the listening sessions. Each subject wore a pair of clear plastic goggles on which an infrared corneal detection system was mounted, enabling the researchers to monitor the subject's eye movements. One response variable of interest was the proportion y of times the subject's eyes fixated on the speaker's mouth.

a. The researchers wanted to estimate $E(y)$ for four different noise levels: none, low, medium, and high. Hypothesize a model that will allow the researchers to obtain these estimates.

b. Interpret the β's in the model, part **a**.

c. Explain how to test the hypothesis of no differences in the mean proportions of mouth fixations for the four background noise levels.

4.57. *The Archives of Disease in Childhood* (Apr. 2000) published a study of whether height influences a child's progression through elementary school. Australian school children were divided into equal thirds (tertiles) based on age (youngest third, middle third, and oldest third). The average heights of the three groups (where all height measurements were standardized using z-scores), by gender, are shown in the table below.

	SAMPLE SIZE	YOUNGEST TERTILE MEAN HEIGHT	MIDDLE TERTILE MEAN HEIGHT	OLDEST TERTILE MEAN HEIGHT
Boys	1439	0.33	0.33	0.16
Girls	1409	0.27	0.18	0.21

Source: Wake, M., Coghlan, D., and Hesketh, K. "Does height influence progression through primary school grades?" *The Archives of Disease in Childhood*, Vol. 82, Apr. 2000 (Table 2).

a. Propose a regression model that will enable you to compare the average heights of the three age groups for boys.

b. Find the estimates of the β's in the model, part **a**.

c. Repeat parts **a** and **b** for girls.

4.58. A study reported in *Human Factors* (Apr. 1990) investigated the effects of recognizer accuracy and vocabulary size on the performance of a computerized speech recognition device. Accuracy (x_1) of the device, measured as the percentage of correctly recognized spoken utterances, was set at three levels: 90%, 95%, and 99%. Vocabulary size (x_2), measured as the percentage of words needed for the task, was also set at three levels: 75%, 87.5%, and 100%.

The dependent variable of primary interest was task completion time (y, in minutes), measured from when a user of the recognition device spoke the first input until the recognizer displayed the last spoken word of the task. Data collected for $n = 162$ trials were used to fit a complete second-order model for task completion time (y), as a function of the quantitative independent variables accuracy (x_1) and vocabulary (x_2). The coefficient of determination for the model was $R^2 = .75$.

a. Write the complete second-order model for $E(y)$.

b. Interpret the value of R^2.

c. Conduct a test of overall model adequacy. Use $\alpha = .05$.

4.13
A Test for Comparing Nested Models

In regression analysis, we often want to determine (with a high degree of confidence) which one among a set of candidate models best fits the data. In this section, we present such a method for **nested models**.

> **Definition 4.3**
> Two models are **nested** if one model contains all the terms of the second model and at least one additional term. The more complex of the two models is called the **complete** (or **full**) model. The simpler of the two models is called the **reduced** (or **restricted**) model.

To illustrate, suppose you have collected data on a response, y, and two quantitative independent variables, x_1 and x_2, and you are considering the use of either a straight-line, interaction model or a curvilinear model to relate $E(y)$ to x_1 and x_2. Will the curvilinear model provide better predictions of y than the straight-line model? To answer this question, examine the two models, and note that the curvilinear model contains all the terms in the straight-line, interaction model plus two additional terms—those involving β_4, and β_5:

Straight-line, interaction model: $\quad E(y) = \beta_0 + \beta_1 x_1 + \beta_2 x_2 + \beta_3 x_1 x_2$

$$\text{Quadratic terms}$$

Curvilinear model: $\quad E(y) = \beta_0 + \beta_1 x_1 + \beta_2 x_2 + \beta_3 x_1 x_2 + \overbrace{\beta_4 x_1^2 + \beta_5 x_2^2}$

Consequently, these are nested models. Since the straight-line model is the simpler of the two, we say that the *straight-line model is nested within the more complex curvilinear model*. Also, the straight-line model is called the **reduced** model while the curvilinear model is called the **complete** (or **full**) **model**.

Asking whether the curvilinear (or *complete*) model contributes more information for the prediction of y than the straight-line (or *reduced*) model is equivalent to asking whether at least one of the parameters, β_4, or β_5, differs from 0, i.e., whether the terms involving β_4, and β_5 should be retained in the model. Therefore, to test whether the quadratic terms should be included in the model, we test the null hypothesis

$$H_0: \beta_4 = \beta_5 = 0$$

(i.e., the quadratic terms do not contribute information for the prediction of y) against the alternative hypothesis

   H_a: At least one of the parameters, β_4, or β_5, differs from 0.

(i.e., at least one of the quadratic terms contributes information for the prediction of y).

The procedure for conducting this test is intuitive: First, we use the method of least squares to fit the reduced model and calculate the corresponding sum of squares for error, SSE_R (the sum of squares of the deviations between observed and predicted y values). Next, we fit the complete model and calculate its sum of squares for error, SSE_C. Then, we compare SSE_R to SSE_C by calculating the difference $SSE_R - SSE_C$. If the quadratic terms contribute to the model, then SSE_C should be much smaller than SSE_R, and the difference $SSE_R - SSE_C$ will be large. The larger the difference, the greater the weight of evidence that the complete model provides better predictions of y than does the reduced model.

The sum of squares for error will always decrease when new terms are added to the model since the total sum of squares, $SS_{yy} = \Sigma(y - \bar{y})^2$, remains the same. The question is whether this decrease is large enough to conclude that it is due to more than just an increase in the number of model terms and to chance. To test the null hypothesis that the curvature coefficients β_4 and β_5 simultaneously equal 0, we use an F statistic. For our example, this F statistic is:

$$F = \frac{\text{Drop in SSE/Number of } \beta \text{ parameters being tested}}{s^2 \text{ for larger model}}$$

$$= \frac{(SSE_R - SSE_C)/2}{SSE_C/[n - (5 + 1)]}$$

When the assumptions listed in Sections 3.4 and 4.2 about the error term ε are satisfied and the β parameters for curvature are all 0 (H_0 is true), this F statistic has an F distribution with $\nu_1 = 2$ and $\nu_2 = n - 6$ df. Note that ν_1 is the number of β parameters being tested and ν_2 is the number of degrees of freedom associated with s^2 in the larger, second-order model.

If the quadratic terms *do* contribute to the model (H_a is true), we expect the F statistic to be large. Thus, we use a one-tailed test and reject H_0 if F exceeds some critical value, F_α, as shown in Figure 4.23.

FIGURE 4.23

Rejection region for the F test
H_0: $\beta_4 = \beta_5 = 0$

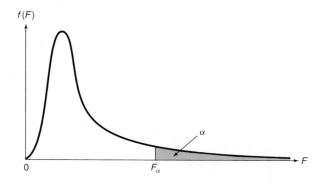

F **Test for Comparing Nested Models**

Reduced model: $E(y) = \beta_0 + \beta_1 x_1 + \cdots + \beta_g x_g$

Complete model: $E(y) = \beta_0 + \beta_1 x_1 + \cdots + \beta_g x_g$
$$+ \beta_{g+1} x_{g+1} + \cdots + \beta_k x_k$$

H_0: $\beta_{g+1} = \beta_{g+2} = \cdots = \beta_k = 0$

H_a: At least one of the β parameters being tested in nonzero.

Test statistic: $F = \dfrac{(\text{SSE}_R - \text{SSE}_C)/(k - g)}{\text{SSE}_C/[n - (k+1)]}$

$$= \dfrac{(\text{SSE}_R - \text{SSE}_C)/\text{Number of } \beta\text{'s tested}}{\text{MSE}_C}$$

where $\text{SSE}_R =$ Sum of squared errors for the reduced model
$\text{SSE}_C =$ Sum of squared errors for the complete model
$\text{MSE}_C =$ Mean square error for the complete model
$k - g =$ Number of β parameters specified in H_0
i.e., number of β's tested)
$k + 1 =$ Number of β parameters in the complete model
(including β_0)

$n =$ Total sample size

Rejection region: $F > F_\alpha$ where

$$\nu_1 = k - g = \text{Degrees of freedom for the numerator}$$

$$\nu_2 = n - (k + 1) = \text{Degrees of freedom for the denominator}$$

EXAMPLE   4.10

In Example 4.7, we fit the complete second-order model for a set of $n = 20$ data points relating shipment cost to package weight and distance shipped. The SAS printout for this model, called the *complete* model, is reproduced in Figure 4.24. Figure 4.25 shows the SAS printout for the straight-line interaction model (the *reduced* model) fit to the same $n = 20$ data points. Referring to the printouts, we find the following:

Straight-line interaction (reduced) model:

$$\text{SSE}_R = 6.63331 \text{ (shaded on Figure 4.25)}$$

Second-order (complete) model:

$$\text{SSE}_C = 2.74474 \text{ (shaded on Figure 4.24)}$$

Test the hypothesis that the quadratic terms, $\beta_4 x_1^2$ and $\beta_5 x_2^2$, do not contribute information for the prediction of y.

FIGURE 4.24

SAS output for complete model, Example 4.10

Dependent Variable: Y

Analysis of Variance

Source	DF	Sum of Squares	Mean Square	F Value	Pr > F
Model	5	449.34076	89.86815	458.39	<.0001
Error	14	2.74474	0.19605		
Corrected Total	19	452.08550			

Root MSE	0.44278	R-Square	0.9939
Dependent Mean	6.33500	Adj R-Sq	0.9918
Coeff Var	6.98940		

Parameter Estimates

Variable	DF	Parameter Estimate	Standard Error	t Value	Pr > |t|
Intercept	1	0.82702	0.70229	1.18	0.2586
X1	1	-0.60914	0.17990	-3.39	0.0044
X2	1	0.00402	0.00800	0.50	0.6230
X1X2	1	0.00733	0.00063743	11.49	<.0001
X1SQ	1	0.08975	0.02021	4.44	0.0006
X2SQ	1	0.00001507	0.00002243	0.67	0.5127

Test CURVE Results for Dependent Variable Y

Source	DF	Mean Square	F Value	Pr > F
Numerator	2	1.94428	9.92	0.0021
Denominator	14	0.19605		

FIGURE 4.25

SAS output for reduced model, Example 4.10

Dependent Variable: Y

Analysis of Variance

Source	DF	Sum of Squares	Mean Square	F Value	Pr > F
Model	3	445.45219	148.48406	358.15	<.0001
Error	16	6.63331	0.41458		
Corrected Total	19	452.08550			

Root MSE	0.64388	R-Square	0.9853
Dependent Mean	6.33500	Adj R-Sq	0.9826
Coeff Var	10.16385		

Parameter Estimates

Variable	DF	Parameter Estimate	Standard Error	t Value	Pr > |t|
Intercept	1	-0.14050	0.64810	-0.22	0.8311
X1	1	0.01909	0.15821	0.12	0.9055
X2	1	0.00772	0.00391	1.98	0.0656
X1X2	1	0.00780	0.00089766	8.68	<.0001

Solution

The null and alternative hypotheses for our test are:

H_0: $\beta_4 = \beta_5 = 0$

H_a: Either β_4 or β_5 (or both) are non zero.

The test statistic is

$$F = \frac{(\text{SSE}_R - \text{SSE}_C)/2}{\text{SSE}_C/(20 - 6)}$$

$$= \frac{(6.63331 - 2.74474)/2}{2.74474/14} = \frac{1.94428}{.19605} = 9.92$$

The critical value of F for $\alpha = .05$, $\nu_1 = 2$, and $\nu_2 = 14$ is found in Table 4 (Appendix C) to be

$$F_{.05} = 3.74$$

Since the calculated $F = 9.92$ exceeds 3.74, we reject H_0 and conclude that the quadratic terms contribute to the prediction of y, shipment cost per package. The curvature terms should be retained in the model. [*Note*: The test statistic and p-value for this nested model F test can be obtained in SAS with the proper SAS commands. Both the F-value and p-value are highlighted at the bottom of the SAS printout, Figure 4.24. Since p-value $= .0021$ is less than $\sigma = .05$, we arrive at the same conclusion: reject H_0.]   ◆

Suppose the F test in Example 4.10 yielded a test statistic that did not fall in the rejection region. That is, suppose there was insufficient evidence (at $\alpha = .05$) to say that the curvature terms contribute information for the prediction of product quality. As with any statistical test of hypothesis, we must be cautious about accepting H_0 since the probability of a Type II error is unknown. Nevertheless, most practitioners of regression analysis adopt the principle of **parsimony**. That is, in situations where two competing models are found to have essentially the same predictive power, the model with the lower number of β's (i.e., the more **parsimonious model**) is selected. The principle of parsimony would lead us to choose the simpler (reduced) model over the more complex complete model when we fail to reject H_0 in the F test for nested models.

Definition 4.4

A **parsimonious model** is a model with a small number of β parameters. In situations where two competing models have essentially the same predictive power (as determined by an F test), choose the more parsimonious of the two.

When the candidate models in model building are nested models, the F test developed in this section is the appropriate procedure to apply to compare the models. However, if the models are not nested, this F test is not applicable. In this situation, the analyst must base the choice of the best model on statistics such as R_a^2 and s. It is important to remember that decisions based on these and other numerical descriptive measures of model adequacy cannot be supported with a measure of reliability and are often very subjective in nature.

EXERCISES

4.59. Determine which pairs of the following models are "nested" models. For each pair of nested models, identify the complete and reduced model.
   a. $E(y) = \beta_0 + \beta_1 x_1 + \beta_2 x_2$
   b. $E(y) = \beta_0 + \beta_1 x_1$
   c. $E(y) = \beta_0 + \beta_1 x_1 + \beta_2 x_1^2$
   d. $E(y) = \beta_0 + \beta_1 x_1 + \beta_2 x_2 + \beta_3 x_1 x_2$
   e. $E(y) = \beta_0 + \beta_1 x_1 + \beta_2 x_2 + \beta_3 x_1 x_2 + \beta_4 x_1^2 + \beta_5 x_2^2$

4.60. Consider the second-order model relating $E(y)$ to three quantitative independent variables, x_1, x_2, and x_3:

$$E(y) = \beta_0 + \beta_1 x_1 + \beta_2 x_2 + \beta_3 x_3 + \beta_4 x_1 x_2 + \beta_5 x_1 x_3$$
$$+ \beta_6 x_2 x_3 + \beta_7 x_1^2 + \beta_8 x_2^2 + \beta_9 x_3^2$$

   a. Specify the parameters involved in a test of the hypothesis that no curvature exists in the response surface.
   b. State the hypothesis of part **a** in terms of the model parameters.
   c. What hypothesis would you test to determine whether x_3 is useful for the prediction of $E(y)$?

4.61. The *American Educational Research Journal* (Fall 1998) published a study of students' perceptions of their science ability in hands-on classrooms. A first-order, main effects model that was used to predict ability perception (y) included the following independent variables:

Control Variables
   $x_1 =$ Prior science attitude score
   $x_2 =$ Science ability test score
   $x_3 = 1$ if boy, 0 if girl
   $x_4 = 1$ if classroom 1 student, 0 if not
   $x_5 = 1$ if classroom 3 student, 0 if not
   $x_6 = 1$ if classroom 4 student, 0 if not
   $x_7 = 1$ if classroom 5 student, 0 if not
   $x_8 = 1$ if classroom 6 student, 0 if not

Performance Behaviors
   $x_9 =$ Active-leading behavior score
   $x_{10} =$ Passive-assisting behavior score
   $x_{11} =$ Active-manipulating behavior score

   a. Hypothesize the equation of the first-order, main effects model for $E(y)$.
   b. The researchers also considered a model that included all possible interactions between the control variables and the performance behavior variables. Write the equation for this model for $E(y)$.
   c. The researchers determined that the interaction terms in the model, part **b**, were not significant,

and therefore used the model, part **a**, to make inferences. Explain the best way to conduct this test for interaction. Give the null hypothesis of the test.

4.62. An article in the *Community Mental Health Journal* (Aug. 2000) used multiple regression analysis to model the level of community adjustment of clients of the Department of Mental Health and Addiction Services in Connecticut. The dependent variable, community adjustment (y), was measured quantitatively based on staff ratings of the clients. (Lower scores indicate better adjustment.) The complete model was a first-order model with 21 independent variables. The independent variables were categorized as Demographic (four variables), Diagnostic (seven variables), Treatment (four variables), and Community (six variables).
   a. Write the equation of $E(y)$ for the complete model.
   b. Give the null hypothesis for testing whether the seven Diagnostic variables contribute information for the prediction of y.
   c. Give the equation of the reduced model appropriate for the test, part **b**.
   d. The test, part **b**, resulted in a test statistic of $F = 59.3$ and p-value $< .0001$. Interpret this result in the words of the problem.

🔘 BEANIE

4.63. Refer to Exercise 3.55 (p. 144) and the data on the values of 50 beanie babies collector's items, published in *Beanie World Magazine*. Suppose we want to predict the market value of a beanie baby using age (in months since Sept. 1998) and whether the beanie baby has been retired or is current (i.e., still in production).
   a. Write a complete second-order model for market value as a function of age and current/retired status.
   b. Specify the null hypothesis for testing whether the quadratic terms in the model, part **a**, are important for predicting market value.
   c. Specify the null hypothesis for testing whether the interaction terms in the model, part **a**, are important for predicting market value.
   d. Three models were fit to the data using SPSS. The relevant SPSS printouts are shown on p. 237. Use this information to conduct the tests specified in parts **b** and **c**. Interpret the results.

SPSS output for Exercise 4.63

ANOVA[b]

Model		Sum of Squares	df	Mean Square	F	Sig.
1	Regression	1186549	5	237309.713	2.885	.024[a]
	Residual	3618994	44	82249.862		
	Total	4805543	49			

a. Predictors: (Constant), AGESQ_ST, RCSTATUS, AGE, AGESQ, AGE_STAT

b. Dependent Variable: VALUE

ANOVA[b]

Model		Sum of Squares	df	Mean Square	F	Sig.
1	Regression	1116017	3	372005.590	4.638	.006[a]
	Residual	3689526	46	80207.081		
	Total	4805543	49			

a. Predictors: (Constant), AGE_STAT, AGE, RCSTATUS

b. Dependent Variable: VALUE

ANOVA[b]

Model		Sum of Squares	df	Mean Square	F	Sig.
1	Regression	1082210	3	360736.761	4.457	.008[a]
	Residual	3723332	46	80942.005		
	Total	4805543	49			

a. Predictors: (Constant), RCSTATUS, AGESQ, AGE

b. Dependent Variable: VALUE

4.64. Refer to the *Motivation and Emotion* study on the relationship between cartoon funniness ratings (y) and pain (x_1) or aggression (x_2), Exercise 4.31 (p. 204). Since no evidence of an inverted-U relationship was found, the researchers fitted the following two first-order models to the $n = 32$ data points:

Model 1: $E(y) = \beta_0 + \beta_1 x_1$, SSE = 26.01
Model 2: $E(y) = \beta_0 + \beta_1 x_1 + \beta_2 x_2$, SSE = 25.44

a. What hypothesis would you test to determine whether the addition of the aggression (x_2) rating improves the predictive ability of the model?

b. Use the F test discussed in this section to test the hypothesis, part **a**. [Note that this test is equivalent to a t test on β_2.]

4.65. Refer to the *Chance* (Winter 2001) study of students who paid a private tutor (or coach) to help them improve their Scholastic Assessment Test (SAT) scores, Exercise 4.48 (p. 226). Recall that the baseline model, $E(y) = \beta_0 + \beta_1 x_1 + \beta_2 x_2$, where y = SAT-Math score, x_1 = score on PSAT, and x_2 = {1 if student was coached, 0 if not}, had the following results: $R_a^2 = .76$, $\hat{\beta}_2 = 19$, and $s_{\hat{\beta}_2} = 3$. As an alternative model, the researcher added several "control" variables, including dummy variables for student ethnicity (x_3, x_4, and x_5), a socioeconomic status index variable (x_6), two variables that measured high school performance (x_7 and x_8), the number of math courses taken in high school (x_9), and the overall GPA for the math courses (x_{10}).

a. Write the hypothesized equation for $E(y)$ for the alternative model.

b. Give the null hypothesis for a nested model F-test comparing the initial and alternative models.

c. The nested model F-test, part **b**, was statistically significant at $\alpha = .05$. Practically interpret this result.

d. The alternative model, part **a**, resulted in $R_a^2 = .79$, $\hat{\beta}_2 = 14$, and $s_{\hat{\beta}_2} = 3$. Interpret the value of R_a^2.

e. Refer to part **d**. Find and interpret a 95% confidence interval for β_2.

f. The researcher concluded that "the estimated effect of SAT coaching decreases from the baseline model when control variables are added to the model." Do you agree? Justify your answer.

g. As a modification to the model of part **a**, the researcher added all possible interactions between the coaching variable (x_2) and the other independent variables in the model. Write the equation for $E(y)$ for this modified model.

h. Give the null hypothesis for comparing the models, parts **a** and **g**. How would you perform this test?

4.66. Since 1978, when the U.S. airline industry was deregulated, researchers have questioned whether the deregulation has ensured a truly competitive environment. If so, the profitability of any major airline would be related only to overall industry conditions (e.g., disposable income and market share) but not to any unchanging feature of that airline. This profitability hypothesis was tested in *Transportation Journal*

(Winter 1990) using multiple regression. Data for $n = 234$ carrier-years were used to fit the model

$$E(y) = \beta_0 + \beta_1 x_1 + \beta_2 x_2 + \beta_3 x_3 + \ldots + \beta_{30} x_{30}$$

where

$$y = \text{Profit rate}$$
$$x_1 = \text{Real personal disposable income}$$
$$x_2 = \text{Industry market share}$$
$$x_3 \text{ through } x_{30} = \text{Dummy variables (coded 0–1) for the 29 air carriers investigated in the study}$$

The results of the regression are summarized in the table. Interpret the results. Is the profitability hypothesis supported?

VARIABLE	β ESTIMATE	t VALUE	p-VALUE
Intercept	1.2642	.09	.9266
x_1	−.0022	−.99	.8392
x_2	4.8405	3.57	.0003
$x_3 - x_{30}$	(not given)	—	—

$R^2 = .3402$
$F(\text{Model}) = 3.49, \quad p\text{-value} = .0001$
$F(\text{Carrier dummies}) = 3.59, \quad p\text{-value} = .0001$

Source: Leigh, L.E. "Contestability in deregulated airline markets: Some empirical tests." *Transportation Journal*, Winter 1990, p. 55 (Table 4).

4.67. The *Journal of Human Stress* (Summer 1987) reported on a study of "psychological response of firefighters to chemical fire." It is thought that the following complete second-order model will be adequate to describe the relationship between emotional distress and years of experience for two groups of firefighters—those exposed to a chemical fire and those unexposed.

$$E(y) = \beta_0 + \beta_1 x_1 + \beta_2 x_1^2 + \beta_3 x_2 + \beta_4 x_1 x_2 + \beta_5 x_1^2 x_2$$

where $\quad y = \text{Emotional distress}$
$\qquad x_1 = \text{Experience (years)}$
$\qquad x_2 = 1 \text{ if exposed to chemical fire, 0 if not}$

a. What hypothesis would you test to determine whether the *rate* of increase of emotional distress with experience is different for the two groups of firefighters?

b. What hypothesis would you test to determine whether there are differences in mean emotional distress levels that are attributable to exposure group?

c. Data for a sample of 200 firefighters were used to fit the complete model as well as the reduced model, $E(y) = \beta_0 + \beta_1 x_1 + \beta_2 x_1^2$. The resulting SAS printouts are shown on below and on p. 239. Is there sufficient evidence to support the claim that the mean emotional distress levels differ for the two groups of firefighters? Use $\alpha = .05$.

SAS output for complete model, Exercise 4.67

The REG Procedure
Dependent Variable: DISTRESS

Analysis of Variance

Source	DF	Sum of Squares	Mean Square	F Value	Pr > F
Model	5	2351.70	470.34	116.42	<.0001
Error	194	783.90	4.04		
Corrected Total	199	3135.60			

Root MSE	2.0102	R-Square 0.7500
Dependent Mean	24.221	Adj R-Sq 0.7436
Coeff Var	8.229	

SAS output for reduced model, Exercise 4.67

The REG Procedure
Dependent Variable: DISTRESS

Analysis of Variance

Source	DF	Sum of Squares	Mean Square	F Value	Pr > F
Model	2	2340.37	1170.18	289.87	<.0001
Error	197	795.23	4.04		
Corrected Total	199	3135.60			

Root MSE	2.0092	R-Square	0.7464
Dependent Mean	24.221	Adj R-Sq	0.7438
Coeff Var	8.295		

4.14
A Complete Example

The basic elements of multiple regression analysis have been presented in Sections 4.1–4.13. Now we assemble these elements by applying them to a practical problem.

In the United States, commercial contractors bid for the right to construct state highways and roads. A state government agency, usually the Department of Transportation (DOT), notifies various contractors of the state's intent to build a highway. Sealed bids are submitted by the contractors, and the contractor with the lowest bid (building cost) is awarded the road construction contract. The bidding process works extremely well in competitive markets, but has the potential to increase construction costs if the markets are noncompetitive or if collusive practices are present. The latter occurred in the 1970s and 1980s in Florida. Numerous contractors either admitted or were found guilty of price-fixing, i.e., setting the cost of construction above the fair, or competitive, cost through bid-rigging or other means.

In this section, we apply multiple regression to a data set obtained from the office of the Florida Attorney General. Our objective is to build and test the adequacy of a model designed to predict the cost y of a road construction contract awarded using the sealed-bid system in Florida.

STEP 1 Based on the opinions of several experts in road construction and bid-rigging, two important predictors of contract cost (y) are the DOT engineer's estimate of the cost (x_1) and the fixed-or-competitive status of the bid contract (x_2). Since x_2 is a qualitative variable, we create the dummy variable

$$x_2 = \begin{cases} 1 & \text{if fixed} \\ 0 & \text{if competitive} \end{cases}$$

Data collected on these two predictors and contract cost for a sample of $n = 235$ contracts are saved in the file named FLAG. (A description of the full data set, which includes several other potential predictors, is provided in Appendix G.) Contract cost y and DOT estimate x_1 one both recorded in thousands of dollars.

STEP 2   In Chapter 5, we will learn that a good initial choice is the complete second-order model. For one quantitative variable (x_1) and one qualitative variable (x_2), the model has the following form:

$$E(y) = \beta_0 + \beta_1 x_1 + \beta_2 x_1^2 + \beta_3 x_2 + \beta_4 x_1 x_2 + \beta_5 x_1^2 x_2$$

The SAS printout for the complete second-order model is shown in Figure 4.26. The β estimates, shaded on the printout, yield the following least squares prediction equation:

$$\hat{y} = -2.975 + .9155 x_1$$
$$+ .00000072 x_1^2 - 36.724 x_2$$
$$+ .324 x_1 x_2 - .0000358 x_1^2 x_2$$

FIGURE 4.26

SAS output for complete second-order model for road cost

Dependent Variable: COST

Analysis of Variance

Source	DF	Sum of Squares	Mean Square	F Value	Pr > F
Model	5	866723202	173344640	1969.85	<.0001
Error	229	20151771	87999		
Corrected Total	234	886874973			

Root MSE	296.64625	R-Square	0.9773	
Dependent Mean	1268.70221	Adj R-Sq	0.9768	
Coeff Var	23.38187			

Parameter Estimates

Variable	DF	Parameter Estimate	Standard Error	t Value	Pr > |t|
Intercept	1	-2.97527	30.89144	-0.10	0.9234
DOTEST	1	0.91553	0.02917	31.39	<.0001
DOTEST2	1	7.186888E-7	0.00000340	0.21	0.8330
STATUS	1	-36.72420	74.77310	-0.49	0.6238
STA_DOT	1	0.32421	0.11917	2.72	0.0070
STA_DOT2	1	-0.00003576	0.00002478	-1.44	0.1504

Test CURV Results for Dependent Variable COST

Source	DF	Mean Square	F Value	Pr > F
Numerator	2	9.15987	1.04	0.3548
Denominator	229	8.79990		

STEP 3   Before we can make inferences about model adequacy, we should be sure that the standard regression assumptions about the random error ε are satisfied. For given values of x_1 and x_2, the random errors ε have a normal distribution with mean 0, constant variance σ^2, and are independent. We learn how to check these assumptions in Chapter 8. For now, we are satisfied with estimating σ and interpreting its value.

The value of s, shaded on Figure 4.26, is $s = 296.65$. Our interpretation is that the complete second-order model can predict contract costs to within $2s = 593.3$ thousand dollars of its true value.

STEP 4   To check the adequacy of the complete second-order model, we conduct the analysis of variance F test. The elements of the test are as follows:

$H_0:$ $\beta_1 = \beta_2 = \beta_3 = \beta_4 = \beta_5 = 0$
$H_a:$ At least one $\beta \neq 0$

Test statistic: $F = 1,969.85$ (shaded in Figure 4.26)

p-value: p $= .0001$ (shaded in Figure 4.26)

Conclusion: The extremely small p-value indicates that the model is statistically adequate (at $\alpha = .01$) for predicting contract cost, y.

Are all the terms in the model statistically significant predictors? For example, is it necessary to include the curvilinear terms, $\beta_2 x_1^2$ and $\beta_5 x_1^2 x_2$, in the model? If not, the model can be simplified by dropping these curvature terms. The hypothesis we want to test is

$H_0:$ $\beta_2 = \beta_5 = 0$
$H_a:$ At least one of the curvature β's is nonzero.

To test this subset of β's, we compare the complete second-order model to a model without the curvilinear terms. The reduced model takes the form

$$E(y) = \beta_0 + \beta_1 x_1 + \beta_3 x_2 + \beta_4 x_1 x_2$$

The results of this nested model (or partial) F-test are shown at the bottom of the SAS printout, Figure 4.26. The test statistic and p-value (shaded) are $F = 1.04$ and p-value $= .3548$. Since the p-value exceeds, say, $\alpha = .01$, we fail to reject H_0. That is, there is insufficient evidence (at $\alpha = .01$) to indicate that the curvature terms are useful predictors of construction cost, y.

The results of the partial F test lead us to select the reduced model as the better predictor of cost. The SAS printout for the reduced model is shown in Figure 4.27. The least squares prediction equation, highlighted on the printout, is

$$\hat{y} = -6.429 + .921 x_1 + 28.671 x_2 + .163 x_1 x_2$$

Note that we cannot simplify the model any further. The t test for the interaction term $\beta_3 x_1 x_2$ is highly significant (p-value $< .0001$, shaded on Figure 4.27). Thus, our best model for construction cost proposes interaction between the DOT estimate (x_1) and status (x_5) of the contract, but only a linear relationship between cost and DOT estimate.

To demonstrate the impact of the interaction term, we find the least squares line for both fixed and competitive contracts.

Competitive $(x_2 = 0):$ $\hat{y} = -6.429 + .921 x_1 + 28.671(0) + .163 x_1(0)$
$= -6.429 + .921 x_1$
Fixed $(x_2 = 1):$ $\hat{y} = -6.429 + .921 x_1 + 28.671(1) + .163 x_1(1)$
$= 22.242 + 1.084 x_1$

FIGURE 4.27

SAS output for reduced model for road coast

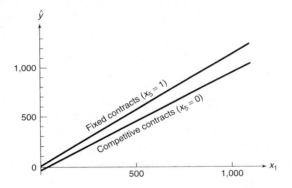

```
                        Dependent Variable: COST

                           Analysis of Variance

                              Sum of          Mean
Source              DF       Squares        Square      F Value    Pr > F

Model                3     866540004      288846668     3281.22    <.0001
Error              231      20334968         88030
Corrected Total    234     886874973


        Root MSE              296.69878    R-Square      0.9771
        Dependent Mean       1268.70221    Adj R-Sq      0.9768
        Coeff Var             23.38601

                          Parameter Estimates

                         Parameter      Standard
Variable      DF          Estimate         Error     t Value    Pr > |t|

Intercept      1          -6.42905      26.20856       -0.25      0.8064
DOTEST         1           0.92134       0.00972       94.75      <.0001
STATUS         1          28.67148      58.66234        0.49      0.6255
STA_DOT        1           0.16328       0.04043        4.04      <.0001
```

FIGURE 4.28

Plot of the least squares lines for the reduced model

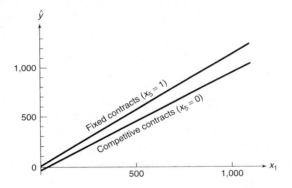

A plot of the least squares lines for the reduced model is shown in Figure 4.28. You can see that the model proposes two straight lines (one for fixed contracts and one for competitive contracts) with different slopes. The estimated slopes of the $y - x_1$ lines are computed and interpreted as follows:

Competitive contracts$(x_2 = 0)$:   Estimated slope $= \hat{\beta}_1 = .921$
For every \$1,000 increase in DOT estimate, we estimate contract cost to increase \$921.

Fixed contracts$(x_2 = 1)$:   Estimated slope $= \hat{\beta}_1 + \hat{\beta}_4 = .921 + .163 = 1.084$
For every \$1,000 increase in DOT estimate, we estimate contract cost to increase \$1,084.

Before deciding to use the interaction model for estimation and/or prediction (step 5), we should check R^2_{adj} and s for the model. $R^2_{adj} = .9768$ (shaded on Figure 4.27) indicates that nearly adj 98% of the variation in the sample of construction costs can be "explained" by the model. The value of s (also shaded) implies that we can predict construction cost to within about $2s = 2(296.7) = 593.40$ thousand dollars of its true value using the model. Although the adjusted

R^2 value is high, the large $2s$ value suggests that the predictive ability of the model might be improved by additional independent variables.

STEP 5   A portion of the SAS printout for the interaction (reduced) model not shown earlier is presented in Figure 4.29. The printout gives predicted values and 95% prediction intervals for the first 10 contracts in the sample. The shaded portion gives the 95% prediction interval for contract cost when the DOT estimate is $1,386,290 ($x_1 = 1,386.29$) and the contract is fixed ($x_2 = 1$). For a contract with these characteristics, we predict the cost to fall between 933.72 thousand dollars and 2,118.0 thousand dollars, with 95% confidence.

FIGURE 4.29

SAS output showing prediction intervals for reduced model

```
                         Dependent Variable: COST

                             Output Statistics

                     Dep Var  Predicted    Std Error
Obs  DOTEST   STATUS     COST  Value  Mean Predict      95% CL Predict
 1   1386.29     1       1379   1526     47.7923    933.7200      2118
 2     85.71     1    134.0300  115.2050  50.5334  -477.7949   708.2050
 3    248.89     0    202.3300  222.8822  24.9485  -363.7625   809.5269
 4    467.49     0    397.1200  424.2862  23.9891  -162.2031      1011
 5    117.72     1    158.5400  149.9236  49.8442  -442.8499   742.6971
 6   1008.91     1       1128   1117     42.7285    525.9115      1707
 7    472.98     1    400.3300  535.2449  43.9161   -55.7058      1126
 8    785.39     0    581.6400  717.1788  22.8762    130.8621      1303
 9    370.02     0    353.9600  334.4836  24.3987  -252.0713   921.0385
10    174.25     0    138.7100  154.1137  25.3087  -432.5909   740.8183
```

4.15
A Summary of the Steps to Follow in a Multiple Regression Analysis

We have discussed some of the methodology of **multiple regression analysis,** a technique for modeling a dependent variable y as a function of several independent variables $x_1, x_2, \ldots, x_k$. The steps we follow in constructing and using multiple regression models, which are much the same as those for the simple straight-line models, are listed here:

1. **The set of independent variables to be included in the model is identified.** (In Chapter 6 we will learn that if the number of independent variables is large, you may want to use a variable selection technique such as **stepwise regression** to screen out those that do not seem important for the prediction of y.)

2. **The form of the probabilistic model is hypothesized.** The model may include **second-order** terms for quantitative variables, **interaction** terms, and **dummy variables** for qualitative variables. Remember that a model with no interaction terms implies that each of the independent variables affects the response y independently of the other independent variables. **Quadratic** (or **second-order**) terms add curvature to the response curve when $E(y)$ is plotted as a function of the quantitative independent variable. Dummy variables allow the mean response to differ for the different levels of the qualitative variable.

3. **The model coefficients are estimated using the method of least squares.**

4. **The probability distribution of ε is specified and σ^2 is estimated.**

5. **The utility of the model is checked using the analysis of variance F test and the multiple coefficient of determination R^2.** The F test for testing a partial set of β

parameters and t tests on individual β parameters aid in deciding the final form of the model.

6. **If the model is deemed useful, it may be used to make estimates of $E(y)$ and to predict future values of y.**

Subsequent chapters extend the methods of this chapter to special applications and problems encountered during a regression analysis. One of the most common problems faced by regression analysts is the problem of multicollinearity, i.e., intercorrelations among the independent variables. Methods for detecting and overcoming multicollinearity, as well as other problems, are discussed in Chapter 7. Another aspect of regression analysis is the analysis of the residuals, i.e., the deviations between the observed and the predicted values of y. An analysis of residuals (Chapter 8) may indicate that the data do not comply with the assumptions of Section 4.2 and may suggest appropriate procedures for modifying the data analysis.

SUPPLEMENTARY EXERCISES

4.68. After a regression model is fit to a set of data, a confidence interval for the mean value of y at a given setting of the independent variables will *always* be narrower than the corresponding prediction interval for a particular value of y at the same setting of the independent variables. Why?

4.69. A disabled person's acceptance of a disability is critical to the rehabilitation process. The *Journal of Rehabilitation* (Sept. 1989) published a study that investigated the relationship between assertive behavior level and acceptance of disability in 160 disabled adults. The dependent variable, assertiveness (y), was measured using the Adult Self Expression Scale (ASES). Scores on the ASES range from 0 (no assertiveness) to 192 (extreme assertiveness). The model analyzed was $E(y) = \beta_0 + \beta_1 x_1 + \beta_2 x_2 + \beta_3 x_3$, where

x_1 = Acceptance of disability (AD) score
x_2 = Age (years)
x_3 = Length of disability (years)

The regression results are shown in the table.

INDEPENDENT VARIABLE	t	TWO-TAILED p-VALUE
AD score (x_1)	5.96	.0001
Age (x_2)	0.01	.9620
Length (x_3)	1.91	.0576

a. Is there sufficient evidence to indicate that AD score is positively linearly related to assertiveness level,

once age and length of disability are accounted for? Test using $\alpha = .05$.

b. Test the hypothesis $H_0: \beta_2 = 0$ against $H_a: \beta_2 \neq 0$. Use $\alpha = .05$. Give the conclusion in the words of the problem.

c. Test the hypothesis $H_0: \beta_3 = 0$ against $H_a: \beta_3 > 0$. Use $\alpha = .05$. Give the conclusion in the words of the problem.

4.70. An Educational Testing Service (ETS) research scientist used multiple regression analysis to model y, the final grade point average (GPA) of business and management doctoral students (*Journal of Educational Statistics*, Spring 1993). A list of the potential independent variables measured for each doctoral student in the study is given below.

1. Quantitative Graduate Management Aptitude Test (GMAT) score
2. Verbal GMAT score
3. Undergraduate GPA
4. First-year graduate GPA
5. Student cohort (i.e., year in which student entered doctoral program: 1988, 1990, or 1992)

a. Identify the variables as quantitative or qualitative.

b. For each quantitative variable, give your opinion on whether the variable is positively or negatively related to final GPA.

c. For each of the qualitative variables, set up the appropriate dummy variable.

d. Write a first-order, main-effects model relating final GPA, y, to the five independent variables.

e. Interpret the β's in the model, part **d**.

f. Write a first-order model for final GPA, y, that allows for a different slope for each student cohort.

g. For each quantitative independent variable in the model, part **f**, give the slope of the line (in terms of the β's) for the 1988 cohort.

4.71. The *Journal of Consulting and Clinical Psychology* (June 1995) reported on a study of emergency service (EMS) rescue workers who responded to the I-880 freeway collapse during the 1989 San Francisco earthquake. The goal of the study was to identify the predictors of symptomatic distress in the EMS workers. One of the distress variables studied was the Global Symptom Index (GSI). Several models for GSI, y, were considered based on the following independent variables:

x_1 = Critical Incident Exposure scale (CIE)
x_2 = Hogan Personality Inventory-Adjustment scale (HPI-A)
x_3 = Years of experience (EXP)
x_4 = Locus of Control scale (LOC)
x_5 = Social Support scale (SS)
x_6 = Dissociative Experiences scale (DES)
x_7 = Peritraumatic Dissociation Experiences Questionnaire, self-report (PDEQ-SR)

a. Write a first-order model for $E(y)$ as a function of the first five independent variables, $x_1 - x_5$.

b. The model of part **a**, fitted to data collected for $n = 147$ EMS workers, yielded the following results: $R^2 = .469$, $F = 34.47$, p-value $< .001$. Interpret these results.

c. Write a first-order model for $E(y)$ as a function of all seven independent variables, $x_1 - x_7$.

d. The model, part **c**, yielded $R^2 = .603$. Interpret this result.

e. The t-tests for testing the DES and PDEQ-SR variables both yielded a p-value of .001. Interpret these results.

4.72. The *Journal of Applied Phycology* (Dec. 1994) published research on the seasonal growth activity of algae in an indoor environment. The daily growth rate y was regressed against temperature x using the quadratic model $E(y) = \beta_0 + \beta_1 x + \beta_2 x^2$. A particular algal strain was grown in a glass dish indoors at temperatures ranging from $10°$ to $32°$ Celsius. The data for $n = 33$ such experiments were used to fit the quadratic model, with the following results:

$$\hat{y} = -2.51 + .55x - .01x^2, \quad R^2 = .67$$

a. Sketch the least squares prediction equation. Interpret the graph.

b. Interpret the value of R^2.

c. Is the model useful for predicting algae growth rate, y? Test using $\alpha = .05$. $F = 30.45$

4.73. *Zoning* is defined as the distribution of vacant land to residential and nonresidential uses via policy set by local governments. Although the negative effects of zoning have been studied (e.g., distorting urban property markets, creating barriers to residential mobility, and impeding economic and social integration), little empirical evidence exists that identifies the factors that encourage restrictive zoning practices. A study reported in the *Journal of Urban Economics* (Vol. 21, 1987) developed a series of multiple regression models that hypothesize several determinants of zoning. One of the models studied took the form

$$E(y) = \beta_0 + \beta_1 x_1 + \beta_2 x_1^2 + \beta_3 x_2$$

where

y = Percentage of vacant land zoned for residential use
x_1 = Proportion of existing land in nonresidential use
x_2 = Proportion of total tax base derived from nonresidential property

The model was fit to data collected for $n = 185$ municipal communities in northeastern New Jersey, with the following results:

INDEPENDENT VARIABLE	PARAMETER ESTIMATE	STANDARD ERROR OF ESTIMATE	t VALUE	p-VALUE
Intercept	92.26	3.07	30.05	$p < .01$
x_1	−96.35	46.59	−2.07	$p < .05$
$(x_1)^2$	166.80	120.88	1.38	$p > .10$
x_2	−75.51	13.35	−5.66	$p < .01$

Adjusted $R^2 = .25$ $F = 21.86$ $(p < .01)$

Source: Rolleston, B.S. "Determinants of restrictive suburban zoning: An empirical analysis." *Journal of Urban Economics*, Vol. 21, 1987, p. 15, Table 4.

a. Construct a 95% confidence interval for β_3. Interpret the result.

b. Test the hypothesis that a curvilinear relationship exists between percentage (y) of land zoned for residential use and proportion (x_1) of existing land in nonresidential use.

c. Interpret the adjusted R^2 value.

d. Is the overall model statistically useful for predicting y?

4.74. The amount of fructose in an athlete's bloodstream is critical to performance. A researcher ran an experiment to determine whether diet had any effect on the level of fructose in the blood after 10 minutes of running on a treadmill. Twenty-one subjects were selected, and each subject's fructose level was measured after 10 minutes on a treadmill. Each subject was randomly assigned to one of three diets. One diet was high in protein, one high in carbohydrates, and one high in fruits. After one month on the diet, each subject was again asked to run on the treadmill for 10 minutes, and the fructose level in the bloodstream was measured. The dependent variable was the difference in the level of fructose in the blood between the second and the first runs on the treadmill.

a. Identify the independent variables in the experiment.

b. Write an appropriate regression model relating mean difference in fructose in the blood, $E(y)$, to the independent variables. Identify and code all dummy variables.

4.75. Location is one of the most important decisions for hotel chains and lodging firms. Researchers S. E. Kimes (Cornell University) and J. A. Fitzsimmons (University of Texas) studied the site selection process of La Quinta Motor Inns, a moderately priced hotel chain (*Interfaces*, Mar.–Apr. 1990). Using data collected on 57 mature inns owned by La Quinta, the researchers built a regression model designed to predict the profitability for sites under construction. The least squares model is:

$$\hat{y} = 39.05 - 5.41x_1 + 5.86x_2 + 3.09x_3 + 1.75x_4$$

where

$y = $ operating margin (measured as a percentage)

$x_1 = $ state population (in thousands) divided by the total number of inns in the state

$x_2 = $ room rate ($) for the inn

$x_3 = $ square root of the median income of the area (in $ thousands)

$x_4 = $ number of college students within four miles of the inn

All variables were "standardized" to have a mean of 0 and a standard deviation of 1.

a. Interpret the β estimates of the model. Comment on the effect of each independent variable on operating margin, y. [*Note:* A profitable inn is defined as one with an operating margin of over 50%.]

b. The model yielded $R^2 = .51$. Give a descriptive measure of model adequacy.

c. Make an inference about model adequacy by conducting the appropriate test. Use $\alpha = .05$.

4.76. Research on the relationship between job performance and job turnover has yielded conflicting results. Some early studies found a negative relationship (i.e., the lower the performance, the greater the likelihood of turnover) among all types of workers, whereas others detected a positive relationship (i.e., the higher the performance, the greater the likelihood of turnover) among those employed in white-collar positions. These early studies, however, focused on the linear (first-order) relationship between these variables. The possibility of a curvilinear (second-order) relationship between job performance and turnover was subsequently investigated both for white-collar workers (accountants) and for blue-collar workers (truck drivers). For each sample of workers the quadratic model $E(y) = \beta_0 + \beta_1 x + \beta_2 x^2$ was fitted, where

$x = $ Performance rating (1 = poor, ... , 4 = outstanding)

$y = $ Probability of turnover (i.e., likelihood of worker leaving his or her job within 1 year)

The results are shown in the table below.

ACCOUNTANTS ($n = 169$)	TRUCK DRIVERS ($n = 107$)
$\hat{\beta}_1 = -1.40 \ (t = -3.88)$	$\hat{\beta}_1 = 1.50 \ (t = -3.83)$
$\hat{\beta}_2 = \ \ \ 1.13 \ (t = 3.23)$	$\hat{\beta}_2 = 1.22 \ (t = 4.70)$
$R^2 = \ \ \ .114$	$R^2 = .298$

Source: Jackofsky, E.F., Ferris, K.R., and Breckenridge, B.G. "Evidence for a curvilinear relationship between job performance and turnover." *Journal of Management*, Vol. 12, No. 1, 1986, pp. 105–111.

a. Conduct a test of model adequacy for each of the two groups of workers. Use $\alpha = .05$.

b. Interpret the β estimates for each of the two groups of workers. Which of the $\hat{\beta}$'s have practical interpretations?

c. Is there evidence of upward curvature in the relationship between turnover and performance for accountants? Use $\alpha = .05$. What is the practical implication of this result?

d. Repeat part **c** for truck drivers.

4.77. Unions in the United States officially opposed a free trade agreement with Mexico, fearing a decrease in wages. However, in an article in the *Journal of Labor Research* (Spring 1993), one researcher used regression analysis to predict that a Mexican free trade agreement should have little influence on union wages and should increase nonunion wages. The model for union wages (y) included the independent variable years of completed education (x_1) among others.

a. The researcher hypothesized a curvilinear relationship between union wages (y) and education (x_1). Write a model for $E(y)$ as a function of x_1 that incorporates this hypothesis.

b. Refer to the model, part **a**. The researcher also hypothesized that education (x_1) will have a positive net impact on union wages (y). Specifically, wages (y) will increase with education (x_1), but at a decreasing rate. Explain how to test this hypothesis.

4.78. Most academic theorists advocate group decision making as a way to resolve conflicts among a manager's subordinates. Many managers reject this proposition in practice, however, believing that airing conflict in groups is counterproductive. A study was conducted to examine this contradiction between accepted normative theory and current practice in Australia (*Organizational Behavior & Human Decision Processes*, Vol. 39, 1987). For one part of the study, multiple regression analysis was used to test "the proposition that the effective use of group discussion methods to resolve conflict depends on the manager's ability and willingness to encourage subordinates to confront conflict." A sample of 89 upper-level managers were asked to complete a questionnaire that measured the following (on a 7-point Likert scale):

y = Average performance of manager's subordinates (i.e., subordinate performance)

x_1 = Manager's preferred level of subordinate participation in decision making when conflict is present (i.e., group decision method)

x_2 = Average of subordinates' perceptions of manager's inclination to legitimize conflict (i.e., conflict legitimization)

The interaction model $E(y) = \beta_0 + \beta_1 x_1 + \beta_2 x_2 + \beta_3 x_1 x_2$ was fit to the 89 data points, with the following results (t values in parentheses):

$$\hat{y} = 7.09 - .44x_1 - .01x_2 + .06x_1x_2 \quad R^2 = .22$$

$$(-1.86) \quad (-.01) \quad (1.85)$$

a. Conduct a test to determine whether the model is adequate for predicting subordinate performance y. Use $\alpha = .10$.

b. Use the least squares prediction equation to graph the estimated relationships between subordinate performance (y) and group decision method (x_1) for low conflict legitimization ($x_2 = 1$) and high conflict legitimization ($x_2 = 7$). Interpret the graphs.

c. Conduct a test to determine whether the relationship between subordinate performance (y) and a manager's use of a group decision method (x_1)

depends on the manager's legitimization of conflict (x_2). Use $\alpha = .10$.

d. Based on the result of part **c**, would you recommend that the researchers conduct t tests on β_1 and β_2? Explain.

4.79. Much research—and much litigation—has been conducted on the disparity between the salary levels of men and women. Research reported in *Work and Occupations* (Nov. 1992) analyzes the salaries for a sample of 191 Illinois managers using a regression analysis with the following independent variables:

$$x_1 = \text{Gender of manager} = \begin{cases} 1 & \text{if male} \\ 0 & \text{if not} \end{cases}$$

$$x_2 = \text{Race of manager} = \begin{cases} 1 & \text{if white} \\ 0 & \text{if not} \end{cases}$$

x_3 = Education level (in years)

x_4 = Tenure with firm (in years)

x_5 = Number of hours worked per week

The regression results are shown in the table as they were reported in the article.

VARIABLE	$\hat{\beta}$	p-VALUE
x_1	12.774	<.05
x_2	.713	>.10
x_3	1.519	<.05
x_4	.320	<.05
x_5	.205	<.05
Constant	15.491	—
$R^2 = .240 \quad n = 191$		

a. Write the hypothesized model that was used, and interpret each of the β parameters in the model.

b. Write the least squares equation that estimates the model in part **a**, and interpret each of the β estimates.

c. Interpret the value of R^2. Test to determine whether the model is useful for predicting annual salary. Test using $\alpha = .05$.

d. Test to determine whether the gender variable indicates that male managers are paid more than female managers, even after adjusting for and holding constant the other four factors in the model. Test using $\alpha = .05$. [*Note:* The p-values given in the table are two-tailed.]

e. Why would one want to adjust for these other factors before conducting a test for salary discrimination?

4.80. To determine whether extra personnel are needed for the day, the owners of a water adventure park would like to find a model that would allow them to predict the day's attendance each morning before opening based on the day of the week and weather conditions. The model is of the form

$$E(y) = \beta_0 + \beta_1 x_1 + \beta_2 x_2 + \beta_3 x_3$$

where

y = Daily admissions

$$x_1 = \begin{cases} 1 & \text{if weekend} \\ 0 & \text{otherwise} \end{cases}$$

$$x_2 = \begin{cases} 1 & \text{if sunny} \\ 0 & \text{if overcast} \end{cases}$$

x_3 = Predicted daily high temperature($°F$)

After collecting 30 days of data, the following least squares model is obtained:

$$\hat{y} = -105 + 25x_1 + 100x_2 + 10x_3$$

with $s_{\hat{\beta}_1} = 10$, $s_{\hat{\beta}_2} = 30$, and $s_{\hat{\beta}_3} = 4$. Also, $R^2 = .65$.

a. Interpret the model coefficients.

b. Is there sufficient evidence to conclude that this model is useful in the prediction of daily attendance? Use $\alpha = .05$.

c. Is there sufficient evidence to conclude that mean attendance increases on weekends? Use $\alpha = .10$.

d. Use the model to predict the attendance on a sunny weekday with a predicted high temperature of 95°F.

e. Suppose the 90% prediction interval for part d is (645, 1,245). Interpret this interval.

4.81. Refer to Exercise 4.80. The owners of the water adventure park are advised that the prediction model could probably be improved if interaction terms were added. In particular, it is thought that the *rate* of increase in mean attendance with increases in predicted high temperature will be greater on weekends than on weekdays. The following model is therefore proposed:

$$E(y) = \beta_0 + \beta_1 x_1 + \beta_2 x_2 + \beta_3 x_3 + \beta_4 x_1 x_3$$

The same 30 days of data used in Exercise 4.58 are again used to obtain the least squares model

$$\hat{y} = 250 - 700x_1 + 100x_2 + 5x_3 + 15x_1 x_3$$

with $s_{\hat{\beta}_4} = 3.0$ and $R^2 = .96$.

a. Graph the predicted day's attendance, $\hat{y}$, against the day's predicted high temperature, x_3, for a sunny weekday and for a sunny weekend day. Plot both on the same graph for x_3 between 70°F and 100°F. Note the increase in slope for the weekend day.

b. Do the data indicate that the interaction term is a useful addition to the model? Use $\alpha = .05$.

c. Use this model to predict the attendance on a sunny weekday with a predicted high temperature of 95°F.

d. Suppose the 90% prediction interval for part c is (800, 850). Compare this with the prediction interval for the model without interaction in Exercise 4.58, part e. Do the relative widths of the confidence intervals support or refute your conclusion about the utility of the interaction term (part b)?

e. The owners, noting that the coefficient $\hat{\beta}_1 = -700$, conclude that the model is ridiculous because it seems to imply that the mean attendance will be 700 less on weekends than on weekdays. Refute their argument.

4.82. Refer to Exercise 4.81. Suppose the second-order model

$$E(y) = \beta_0 + \beta_1 x_1 + \beta_2 x_2 + \beta_3 x_3 + \beta_4 x_1 x_3 + \beta_5 x_3^2$$
$$+ \beta_6 x_1 x_3^2$$

is fit to the $n = 30$ observations on daily admissions.

a. What hypothesis would you test to determine whether the quadratic terms for predicted daily high temperature x_3 are important?

b. Use the SSEs for the interaction model (Exercise 4.81) and the second-order model given here to test the hypothesis of part a. Use $\alpha = .05$.

Interaction model: $SSE_1 = 585,000$
Second-order model: $SSE_2 = 530,000$

4.83. *Environmental Science & Technology* (Oct. 1993) published an article that investigated the variables that affect the sorption of organic vapors on clay minerals. The independent variables and levels considered in the study are listed here. Identify the type (quantitative or qualitative) of each.

a. Temperature (50°, 60°, 75°, 90°)

b. Relative humidity (30%, 50%, 70%)

c. Organic compound (benzene, toluene, chloroform, methanol, anisole)

d. Refer to part c. Write a model for $E(y)$ as a function of organic compound at five levels.

e. Interpret the β parameters in the model, part d.

f. Explain how to test for differences among the mean retention coefficients of the five organic compounds.

4.84. A naval base is considering modifying or adding to its fleet of 48 standard aircraft. The final decision regarding the type and number of aircraft to be added depends on a comparison of cost versus effectiveness of the modified fleet. Consequently, the naval base would like to model the projected percentage increase y in fleet effectiveness by the end of the decade as a function of the cost x of modifying the fleet. A first proposal is the quadratic model

$$E(y) = \beta_0 + \beta_1 x + \beta_2 x^2$$

The data provided in the table were collected on 10 naval bases of similar size that recently expanded their fleets.

NAVALBASE

PERCENTAGE IMPROVEMENT AT END OF DECADE y	COST OF MODIFYING FLEET x, millions of dollars
18	125
32	160
9	80
37	162
6	110
3	90
30	140
10	85
25	150
2	50

a. Fit the quadratic model to the data.
b. Interpret the value of R_a^2 on the printout.
c. Find the value of s and interpret it.
d. Perform a test of overall model adequacy. Use $\alpha = .05$.
e. Is there sufficient evidence to conclude that the percentage improvement y increases more quickly for more costly fleet modifications than for less costly fleet modifications? Test with $\alpha = .05$.
f. Now consider the complete second-order model

$$E(y) = \beta_0 + \beta_1 x_1 + \beta_2 x_1^2 + \beta_3 x_2 + \beta_4 x_1 x_2 + \beta_5 x_1^2 x_2$$

where
$x_1 = $ Cost of modifying the fleet
$x_2 = \begin{cases} 1 & \text{if U.S. base} \\ 0 & \text{if foreign base} \end{cases}$
Fit the complete model to the data. Is there sufficient evidence to indicate that type of base (U.S. or foreign) is a useful predictor of percentage improvement y? Test using $\alpha = .05$.

4.85. The *Journal of Personal Selling & Sales Management* (Summer 1990) published a study of gender differences in the industrial sales force. A sample of 244 male sales managers and a sample of 153 female sales managers participated in the survey. One objective of the research was to assess how supervisory behavior affects intrinsic job satisfaction. Initially, the researchers fitted the following reduced model to the data on each gender group:

$$E(y) = \beta_0 + \beta_1 x_1 + \beta_2 x_2 + \beta_3 x_3 + \beta_4 x_4$$

where
$y = $ Intrinsic job satisfaction(measured on a scale of 0 to 40)
$x_1 = $ Age (years)
$x_2 = $ Education level (years)
$x_3 = $ Firm experience (months)
$x_4 = $ Sales experience (months)

To determine the effects of supervisory behavior, four variables (all measured on a scale of 0 to 50) were added to the model: $x_5 = $ contingent reward behavior, $x_6 = $ noncontingent reward behavior, $x_7 = $ contingent punishment behavior, and $x_8 = $ noncontingent punishment behavior. Thus, the complete model is

$$E(y) = \beta_0 + \beta_1 x_1 + \beta_2 x_2 + \beta_3 x_3 + \beta_4 x_4 + \beta_5 x_5 + \beta_6 x_6 + \beta_7 x_7 + \beta_8 x_8$$

a. For each gender, specify the null hypothesis and rejection region ($\alpha = .05$) for testing whether any of the four supervisory behavior variables affect intrinsic job satisfaction.
b. The R^2 values for the four models (reduced and complete model for both samples) are given in the accompanying table. Interpret the results. For each gender, does it appear that the supervisory behavior variables have an impact on intrinsic job satisfaction? Explain.

	R^2	
MODEL	MALES	FEMALES
Reduced	.218	.268
Complete	.408	.496

Source: Schul, *et al.* "Assessing gender differences in relationships between supervisory behaviors and job related outcomes in industrial sales force." *Journal of Personal Selling & Sales Management*, Vol. X, Summer 1990, p. 9 (Table 4).

c. The F statistics for comparing the two models are $F_{males} = 13.00$ and $F_{females} = 9.05$. Conduct the tests, part **a**, and interpret the results.

4.86. Refer to Exercise 4.85. One way to test for gender differences in the industrial sales force is to incorporate a dummy variable for gender into the model for intrinsic job satisfaction, y, and then fit the model to the data for the combined sample of males and females.

a. Write a model for y as a function of the independent variables, x_1 through x_8, and the gender dummy variable. Include interactions between gender and each of the other independent variables in the model.

b. Based on the model, part **a**, what is the null hypothesis for testing whether gender has an effect on job satisfaction?

c. Explain how to conduct the test, part **b**.

REFERENCES

DRAPER, N., and SMITH, H. *Applied Regression Analysis*, 3rd ed. New York: Wiley, 1998.

MADDALA, G.S., *Introduction to Econometrics*. New York: Macmillan, 1988.

MONTGOMERY, D., PECK, E., and VINING, G. *Introduction to Linear Regression Analysis*, 3rd ed. New York: Wiley, 2001.

CHATTERJEE, S., and PRICE, B. *Regression Analysis by Example*, 2nd ed. New York: Wiley, 1991.

GRAYBILL, F. *Theory and Application of the Linear Model.* North Scituate, Mass.: Duxbury, 1976.

MENDENHALL, W. *Introduction to Linear Models and the Design and Analysis of Experiments*. Belmont, Ca.: Wadsworth, 1968.

MOSTELLER, F., and TUKEY, J.W. *Data Analysis and Regression: A Second Course in Statistics.* Reading, Mass.: Addison-Wesley, 1977.

NETER, J., KUTNER, M., NACHTSHEIM, C., and WASSERMAN, W. *Applied Linear Statistical Models*, 4th ed. Homewood, Ill.: Richard Irwin, 1996.

WEISBERG, S. *Applied Linear Regression*, 2nd ed. New York Wiley, 1985.

CHAPTER 5

Model Building

CONTENTS

OBJECTIVE

To show you why the choice of the deterministic portion of a linear model is crucial to the acquisition of a good prediction equation; to present some basic concepts and procedures for constructing good linear models

5.1 Introduction: Why Model Building Is Important

We have emphasized in both Chapters 3 and 4 that one of the first steps in the construction of a regression model is to hypothesize the form of the deterministic portion of the probabilistic model. This *model-building*, or model-construction, stage is the key to the success (or failure) of the regression analysis. If the hypothesized model does not reflect, at least approximately, the true nature of the relationship between the mean response $E(y)$ and the independent variables $x_1, x_2, \ldots, x_k$, the modeling effort will usually be unrewarded.

By **model building**, we mean writing a model that will provide a good fit to a set of data and that will give good estimates of the mean value of y and good predictions of future values of y for given values of the independent variables. To illustrate, several years ago, a nationally recognized educational research group issued a report concerning the variables related to academic achievement for a certain type of college student. The researchers selected a random sample of students and recorded a measure of academic achievement, y, at the end of the senior year together with data on an extensive list of independent variables, $x_1, x_2, \ldots, x_k$, that they thought were related to y. Among these independent variables were the student's IQ, scores on mathematics and verbal achievement examinations, rank in class, etc. They fit the model

$$E(y) = \beta_0 + \beta_1 x_1 + \beta_2 x_2 + \cdots + \beta_k x_k$$

to the data, analyzed the results, and reached the conclusion that none of the independent variables was "significantly related" to y. The **goodness of fit** of the model, measured by the coefficient of determination R^2, was not particularly good, and t tests on individual parameters did not lead to rejection of the null hypotheses that these parameters equaled 0.

How could the researchers have reached the conclusion that there is no significant relationship, when it is evident, just as a matter of experience, that some of the independent variables studied are related to academic achievement? For example, achievement on a college mathematics placement test should be related to achievement in college mathematics. Certainly, many other variables will affect achievement—motivation, environmental conditions, and so forth—but generally speaking, there will be a positive correlation between entrance achievement test scores and college academic achievement. So, what went wrong with the educational researchers' study?

Although you can never discard the possibility of computing error as a reason for erroneous answers, most likely the difficulties in the results of the educational study were caused by the use of an improperly constructed model. For example, the model

$$E(y) = \beta_0 + \beta_1 x_1 + \beta_2 x_2 + \cdots + \beta_k x_k$$

assumes that the independent variables $x_1, x_2, \ldots, x_k$ affect mean achievement $E(y)$ independently of each other.* Thus, if you hold all the other independent variables constant and vary only x_1, $E(y)$ will increase by the amount β_1 for every unit increase in x_1. A 1-unit change in any of the other independent variables will increase $E(y)$ by the value of the corresponding β parameter for that variable.

Do the assumptions implied by the model agree with your knowledge about academic achievement? First, is it reasonable to assume that the effect of time spent on study is independent of native intellectual ability? We think not. No matter how much effort some students invest in a particular subject, their rate of achievement is low. For others, it may be high. Therefore, assuming that these two variables—effort and native intellectual ability—affect $E(y)$ independently of each other is likely to be an erroneous assumption. Second, suppose that x_5 is the amount of time a student devotes to study. Is it reasonable to expect that a 1-unit increase in x_5 will always produce the same change β_5 in $E(y)$? The changes in $E(y)$ for a 1-unit increase in x_5 might depend on the value of x_5 (for example, the law of diminishing returns). Consequently, it is quite likely that the assumption of a constant rate of change in $E(y)$ for 1-unit increases in the independent variables will not be satisfied.

---

*Keep in mind that we are discussing the deterministic portion of the model and that the word *independent* is used in a mathematical rather than a probabilistic sense.

Clearly, the model

$$E(y) = \beta_0 + \beta_1 x_1 + \beta_2 x_2 + \cdots + \beta_k x_k$$

was a poor choice in view of the researchers' prior knowledge of some of the variables involved. Terms have to be added to the model to account for interrelationships among the independent variables and for curvature in the response function. Failure to include needed terms causes inflated values of SSE, nonsignificance in statistical tests, and, often, erroneous practical conclusions.

In this chapter, we discuss the most difficult part of a multiple regression analysis—the formulation of a good model for $E(y)$. Although many of the models presented in this chapter have already been introduced in optional Section 4.12, we assume the reader has little or no background in model building. This chapter serves as a basic reference guide to model building for teachers, students, and practitioners of multiple regression analysis.

5.2
The Two Types of Independent Variables: Quantitative and Qualitative

The independent variables that appear in a linear model can be one of two types. Recall from Chapter 1 that a *quantitative* variable is one that assumes numerical values corresponding to the points on a line. (Definition 1.4). An independent variable that is not quantitative i.e., one that is categorical in nature, is called *qualitative* (Definition 1.5).

The nicotine content of a cigarette, prime interest rate, number of defects in a product, and IQ of a student are all examples of quantitative independent variables. On the other hand, suppose three different styles of packaging, A, B, and C, are used by a manufacturer. This independent variable, style of packaging, is qualitative, since it is not measured on a numerical scale. Certainly, style of packaging is an independent variable that may affect sales of a product, and we would want to include it in a model describing the product's sales, y.

Definition 5.1

The different values of an independent variable used in regression are called its **levels**.

For a quantitative variable, the levels correspond to the numerical values it assumes. For example, if the number of defects in a product ranges from 0 to 3, the independent variable assumes four levels: 0, 1, 2, and 3.

The levels of a qualitative variable are not numerical. They can be defined only by describing them. For example, the independent variable style of packaging was observed at three levels: A, B, and C.

EXAMPLE 5.1

In Chapter 4, we considered the problem of predicting executive salary as a function of several independent variables. Consider the following four independent variables that may affect executive salaries:

a. Years of experience
b. Gender of the employee

c. Firm's net asset value

d. Rank of the employee

For each of these independent variables, give its type and describe the nature of the levels you would expect to observe.

Solution

a. The independent variable for the number of years of experience is quantitative, since its values are numerical. We would expect to observe levels ranging from 0 to 40 (approximately) years.

b. The independent variable for gender is qualitative, since its levels can only be described by the nonnumerical labels "female" and "male."

c. The independent variable for the firm's net asset value is quantitative, with a very large number of possible levels corresponding to the range of dollar values representing various firms' net asset values.

d. Suppose the independent variable for the rank of the employee is observed at three levels: supervisor, assistant vice president, and vice president. Since we cannot assign a realistic measure of relative importance to each position, rank is a qualitative independent variable.   ◆

Quantitative and qualitative independent variables are treated differently in regression modeling. In the next section, we will see how quantitative variables are entered into a regression model.

EXERCISES

5.1. *Chance* (Winter, 2000) published a study of men's and women's winning times in the Boston Marathon. The researchers built a model for predicting winning time (y) of the marathon as a function of year (x_1) in which the race is run and gender (x_2) of the winning runner. Classify each variable in the model as quantitative or qualitative.

5.2. Writing in the *Journal of Applied Ecology* (Vol. 32, 1995), botanists used multiple regression to model the weight change (y) of a baby snow goose subjected to a feeding trial. Three independent variables measured are listed below. Classify each as quantitative or qualitative.

a. Diet type (plant or duck chow)

b. Digestion efficiency (percentage)

c. Amount of acid-fiber detergent added to diet (percentage)

5.3. *Business Horizons* (Jan.–Feb. 1993) conducted a comprehensive study of 800 chief executive officers who run the country's largest global corporations. The purpose of the study was to build a profile of the CEOs based on characteristics of each CEO's social

background. Several of the variables measured for each CEO are listed here. Classify each variable as quantitative or qualitative.

a. State of birth   **b.** Age

c. Education level   **d.** Tenure with firm

e. Total compensation   **f.** Area of expertise

5.4. The *Journal of Human Stress* (Summer 1987) reported on a study of "psychological response of firefighters to chemical fire." The researchers used multiple regression to predict emotional distress as a function of the following independent variables. Identify each independent variable as quantitative or qualitative. For qualitative variables, suggest several levels that might be observed. For quantitative variables, give a range of values (levels) for which the variable might be observed.

a. Number of preincident psychological symptoms

b. Years of experience

c. Cigarette smoking behavior

d. Level of social support

e. Marital status

f. Age

g. Ethnic status
h. Exposure to a chemical fire
i. Education level
j. Distance lived from site of incident
k. Gender

5.5. Which of the assumptions about ε (Section 4.2) prohibit the use of a qualitative variable as a dependent variable? (We present a technique for modeling a qualitative dependent variable in Chapter 9.)

5.3
Models with a Single Quantitative Independent Variable

To write a prediction equation that provides a good model for a response (one that will eventually yield good predictions), we have to know how the response might vary as the levels of an independent variable change. Then we have to know how to write a mathematical equation to model it. To illustrate (with a simple example), suppose we want to model a student's score on a statistics exam, y, as a function of the single independent variable x, the amount of study time invested. It may be that exam score, y, increases in a straight line as the amount of study time, x, varies from 1 hour to 6 hours, as shown in Figure 5.1a. If this were the entire range of x values for which you wanted to predict y, the model

$$E(y) = \beta_0 + \beta_1 x$$

would be appropriate.

FIGURE 5.1

Modeling exam score, y, as a function of study time, x

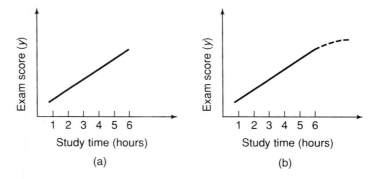

Now, suppose you want to expand the range of values of x to $x = 8$ or $x = 10$ hours of studying. Will the straight-line model

$$E(y) = \beta_0 + \beta_1 x$$

be satisfactory? Perhaps, but making this assumption could be risky. As the amount of studying, x, is increased, sooner or later the point of diminishing returns will be reached. That is, the increase in exam score for a unit increase in study time will decrease, as shown by the dashed line in Figure 5.1b. To produce this type of curvature, you must know the relationship between models and graphs, and how types of terms will change the shape of the curve.

A response that is a function of a single quantitative independent variable can often be modeled by the first few terms of a polynomial algebraic function. The equation relating the mean value of y to a polynomial of order p in one independent variable x is shown in the box.

A *p*th-Order Polynomial with One Independent Variable

$$E(y) = \beta_0 + \beta_1 x + \beta_2 x^2 + \beta_3 x^3 + \cdots + \beta_p x^p$$

where p is an integer and $\beta_0, \beta_1, \ldots, \beta_p$ are unknown parameters that must be estimated.

As we mentioned in Chapters 3 and 4, a **first-order polynomial** in x (i.e., $p = 1$),

$$E(y) = \beta_0 + \beta_1 x$$

graphs as a straight line. The β interpretations of this model are provided in the next box.

First-Order (Straight-Line) Model with One Independent Variable

$$E(y) = \beta_0 + \beta_1 x$$

Interpretation of model parameters

β_0: y-intercept; the value of $E(y)$ when $x = 0$
β_1: Slope of the line; the change in $E(y)$ for a 1-unit increase in x

---

In Chapter 4 we also covered a **second-order polynomial** model ($p = 2$), called a **quadratic**. For convenience, the model is repeated in the following box.

A Second-Order (Quadratic) Model with One Independent Variable

$$E(y) = \beta_0 + \beta_1 x + \beta_2 x^2$$

where β_0, β_1, and β_2 are unknown parameters that must be estimated.

Interpretation of model parameters

β_0: y-intercept; the value of $E(y)$ when $x = 0$
β_1: Shift parameter; changing the value of β_1 shifts the parabola to the right or left (increasing the value of β_1 causes the parabola to shift to the right)
β_2: Rate of curvature

Graphs of two quadratic models are shown in Figure 5.2. As we learned in Chapter 4, the quadratic model is the equation of a **parabola** that opens either upward, as in Figure 5.2a, or downward, as in Figure 5.2b. If the coefficient of x^2 is positive, it opens upward; if it is negative, it opens downward. The parabola may be

FIGURE 5.2

Graphs for two second-order polynomial models

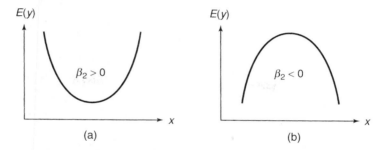

(a)　　　　　　　　　(b)

shifted upward or downward, left or right. The least squares procedure uses only the portion of the parabola that is needed to model the data. For example, if you fit a parabola to the data points shown in Figure 5.3, the portion shown as a solid curve passes through the data points. The outline of the unused portion of the parabola is indicated by a dashed curve.

FIGURE 5.3

Example of the use of a quadratic model

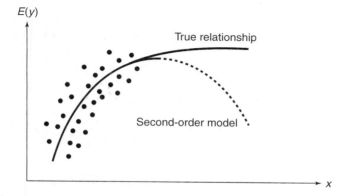

Figure 5.3 illustrates an important limitation on the use of prediction equations: The model is valid only over the range of x values that were used to fit the model. For example, the response might rise, as shown in the figure, until it reaches a plateau. The second-order model might fit the data very well over the range of x values shown in Figure 5.3, but would provide a very poor fit if data were collected in the region where the parabola turns downward.

How do you decide the order of the polynomial you should use to model a response if you have no prior information about the relationship between $E(y)$ and x? If you have data, construct a scatterplot of the data points, and see whether you can deduce the nature of a good approximating function. A pth-order polynomial, when graphed, will exhibit $(p-1)$ peaks, troughs, or reversals in direction. Note that the graphs of the second-order model shown in Figure 5.2 each have $(p-1) = 1$ peak (or trough). Likewise, a third-order model (shown in the box) will have $(p-1) = 2$ peaks or troughs, as illustrated in Figure 5.4.

The graphs of most responses as a function of an independent variable x are, in general, curvilinear. Nevertheless, if the rate of curvature of the response curve is very small over the range of x that is of interest to you, a straight line might provide an excellent fit to the response data and serve as a very useful prediction equation. If the curvature is expected to be pronounced, you should try a second-order model. Third- or higher-order models would be used only where you expect more than one

FIGURE 5.4

Graphs of two third-order
polynomial models

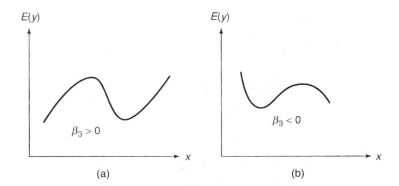

reversal in the direction of the curve. These situations are rare, except where the response is a function of time. Models for forecasting over time are presented in Chapter 10.

Third-Order Model with One Independent Variable

$$E(y) = \beta_0 + \beta_1 x + \beta_2 x^2 + \beta_3 x^3$$

Interpretation of model parameters

β_0 : y-intercept; the value of $E(y)$ when $x = 0$

β_1 :   Shift parameter (shifts the polynomial right or left on the x-axis)

β_2 :   Rate of curvature

β_3 :   The magnitude of β_3 controls the rate of reversal of curvature for the polynomial

---

EXAMPLE   5.2

To operate efficiently, power companies must be able to predict the peak power load at their various stations. Peak power load is the maximum amount of power that must be generated each day to meet demand. A power company wants to use daily high temperature, x, to model daily peak power load, y, during the summer months when demand is greatest. Although the company expects peak load to increase as the temperature increases, the *rate* of increase in $E(y)$ might not remain constant as x increases. For example, a 1-unit increase in high temperature from 100°F to 101°F might result in a larger increase in power demand than would a 1-unit increase from 80°F to 81°F. Therefore, the company postulates that the model for $E(y)$ will include a second-order (quadratic) term and, possibly, a third-order (cubic) term.

A random sample of 25 summer days is selected and both the peak load (measured in megawatts) and high temperature (in degrees) recorded for each day. The data are listed in Table 5.1.

a. Construct a scatterplot for the data. What type of model is suggested by the plot?

⊙ POWERLOADS

TABLE 5.1   **Power Load Data**

TEMPERATURE °F	PEAK LOAD megawatts	TEMPERATURE °F	PEAK LOAD megawatts	TEMPERATURE °F	PEAK LOAD megawatts
94	136.0	106	178.2	76	100.9
96	131.7	67	101.6	68	96.3
95	140.7	71	92.5	92	135.1
108	189.3	100	151.9	100	143.6
67	96.5	79	106.2	85	111.4
88	116.4	97	153.2	89	116.5
89	118.5	98	150.1	74	103.9
84	113.4	87	114.7	86	105.1
90	132.0				

b. Fit the third-order model, $E(y) = \beta_0 + \beta_1 x + \beta_2 x^2 + \beta_3 x^3$, to the data. Is there evidence that the cubic term, $\beta_3 x^3$, contributes information for the prediction of peak power load? Test at $\alpha = .05$.

c. Fit the second-order model, $E(y) = \beta_0 + \beta_1 x + \beta_2 x^2$, to the data. Test the hypothesis that the power load increases at an increasing rate with temperature. Use $\alpha = .05$.

d. Give the prediction equation for the second-order model, part **c**. Are you satisfied with using this model to predict peak power loads?

Solution

a. The scatterplot of the data, produced using MINITAB, is shown in Figure 5.5. The nonlinear, upward curving trend indicates that a second-order model would likely fit the data well.

FIGURE 5.5

MINITAB scatterplot for power load data

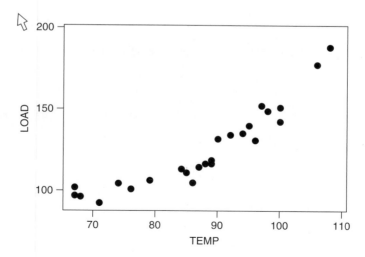

b. The third-order model is fit to the data using MINITAB and the resulting printout is shown in Figure 5.6. The p-value for testing

$$H_o: \beta_3 = 0$$
$$H_a: \beta_3 \neq 0$$

FIGURE 5.6

MINITAB output for third-order
model of power load

```
The regression equation is
LOAD = 331 - 6.4 TEMP + 0.038 TEMP2 +0.000084 TEMP3

Predictor       Coef      SE Coef         T         P
Constant       331.3        477.1      0.69     0.495
TEMP           -6.39        16.79     -0.38     0.707
TEMP2         0.0378       0.1945      0.19     0.848
TEMP3      0.0000843    0.0007426      0.11     0.911

S = 5.501      R-Sq = 95.9%      R-Sq(adj) = 95.4%

Analysis of Variance

Source            DF          SS         MS         F         P
Regression         3     15012.2     5004.1    165.36     0.000
Residual Error    21       635.5       30.3
Total             24     15647.7
```

highlighted on the printout, is .911. Since this value exceeds $\alpha = .05$, there is insufficient evidence of a third-order relationship between peak load and high temperature. Consequently, we will drop the cubic term, $\beta_3 x^3$, from the model.

c. The second-order model is fit to the data using MINITAB and the resulting printout is shown in Figure 5.7. For this quadratic model, if β_2 is positive, then the peak power load y increases at an increasing rate with temperature x. Consequently, we test

$H_o: \beta_2 = 0$
$H_a: \beta_2 > 0$

FIGURE 5.7

MINITAB output for
second-order model of power load

```
The regression equation is
LOAD = 385 - 8.29 TEMP + 0.0598 TEMP2

Predictor       Coef      SE Coef         T         P
Constant      385.05        55.17      6.98     0.000
TEMP          -8.293        1.299     -6.38     0.000
TEMP2       0.059823     0.007549      7.93     0.000

S = 5.376      R-Sq = 95.9%      R-Sq(adj) = 95.6%

Analysis of Variance

Source            DF          SS         MS         F         P
Regression         2     15011.8     7505.9    259.69     0.000
Residual Error    22       635.9       28.9
Total             24     15647.7
```

The test statistic, $t = 7.93$, and two-tailed p-value, are both highlighted on Figure 5.7. Since the one-tailed p-value, $p = 0/2 = 0$, is less than $\alpha = .05$, we reject H_o and conclude that peak power load increases at an increasing rate with temperature.

d. The prediction equation for the quadratic model, highlighted on Figure 5.7, is $\hat{y} = 385 - 8.29x + .0598x^2$. The adjusted-$R^2$ and standard deviation for the model, also highlighted, are $R^2_{adj} = .956$ and $s = 5.376$. These values imply that (1) more than 95% of the sample variation in peak power loads can be explained by the second-order model, and (2) the model can predict peak load to within about $2s = 10.75$ megawatts of its true value. Based on this high value of R^2_{adj} and reasonably small value of $2s$, we recommend using this equation to predict peak power loads for the company. ◆

EXERCISES

5.6. The accompanying graphs depict pth-order polynomials for one independent variable.

i. $E(y)$

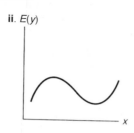

ii. $E(y)$

iii. $E(y)$

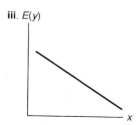

iv. $E(y)$

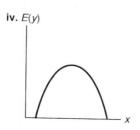

a. For each graph, identify the order of the polynomial.
b. Using the parameters $\beta_0, \beta_1, \beta_2$, etc., write an appropriate model relating $E(y)$ to x for each graph.
c. The signs (+ or −) of many of the parameters in the models of part **b** can be determined by examining the graphs. Give the signs of those parameters that can be determined.

5.7. The optomotor responses of tree frogs were studied in the *Journal of Experimental Zoology* (Sept. 1993). Microspectrophotometry was used to measure the threshold quantal flux (the light intensity at which the optomotor response was first observed) of tree frogs tested at different spectral wavelengths. The data revealed the following relationship between the log of quantal flux (y) and wavelength (x). Hypothesize a model for $E(y)$ that corresponds to the following graph.

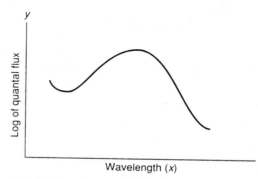

5.8. Underinflated or overinflated tires can increase tire wear and decrease gas mileage. A new tire was tested for wear at different pressures with the results shown in the table.

◉ TIRES2

PRESSURE	MILEAGE
x, pounds per square inch	y, thousands
30	29
31	32
32	36
33	38
34	37
35	33
36	26

a. Graph the data in a scatterplot.
b. If you were given the information for $x = 30, 31, 32, 33$ only, what kind of model would you suggest? For $x = 33, 34, 35, 36$? For all the data?

5.9. A company is considering having the employees on its assembly line work 4 10-hour days instead of 5 8-hour days. Management is concerned that the effect of fatigue as a result of longer afternoons of work might increase assembly times to an unsatisfactory level. An experiment with the 4-day week is planned in which time studies will be conducted on some of the workers during the afternoons. It is believed that an adequate model of the relationship between assembly time, y, and time since lunch, x, should allow for the average assembly time to decrease for a while after lunch before it starts to increase as the workers become tired. Write a model relating $E(y)$ and x that would reflect the management's belief, and sketch the hypothesized shape of the model.

5.4
First-Order Models with Two or More Quantitative Independent Variables

Like models for a single independent variable, models with two or more independent variables are classified as first-order, second-order, and so forth, but it is difficult (most often impossible) to graph the response because the plot is in a multidimensional space. For example, with one quantitative independent variable, x, the response y traces a curve. But for two quantitative independent variables, x_1 and x_2, the plot of y traces a surface over the x_1, x_2-plane (see Figure 5.8). For three or more quantitative independent variables, the response traces a surface in a four- or higher-dimensional space. For these, we can construct two-dimensional graphs of y versus one independent variable for fixed levels of the other independent variables, or three-dimensional plots of y versus two independent variables for fixed levels of the remaining independent variables, but this is the best we can do in providing a graphical description of a response.

FIGURE 5.8

Response surface for first-order model with two quantitative independent variables

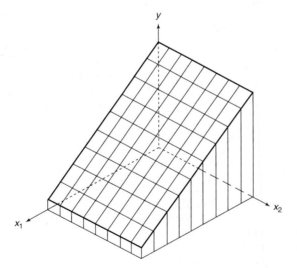

A **first-order model** in k quantitative variables is a first-order polynomial in k independent variables. For $k = 1$, the graph is a straight line. For $k = 2$, the response surface is a plane (usually tilted) over the x_1, x_2-plane.

If we use a first-order polynomial to model a response, we are assuming that there is no curvature in the response surface and that the variables affect the response independently of each other. For example, suppose the true relationship between the mean response and the independent variables x_1 and x_2 is given by the equation

$$E(y) = 1 + 2x_1 + x_2$$

First-Order Model in k Quantitative Independent Variables

$$E(y) = \beta_0 + \beta_1 x_1 + \beta_2 x_2 + \cdots + \beta_k x_k$$

where $\beta_0, \beta_1, \ldots, \beta_k$ are unknown parameters that must be estimated.

Interpretation of model parameters

β_0: y-intercept of $(k+1)$-dimensional surface; the value of $E(y)$ when $x_1 = x_2 = \cdots = x_k = 0$

β_1: Change in $E(y)$ for a 1-unit increase in x_1, when $x_2, x_3, \ldots, x_k$ are held fixed

β_2: Change in $E(y)$ for a 1-unit increase in x_2, when $x_1, x_3, \ldots, x_k$ are held fixed

$\vdots$

β_k: Change in $E(y)$ for a 1-unit increase in x_k, when $x_1, x_2, \ldots, x_{k-1}$ are held fixed

---

In Section 4.3, we graphed this expression for $x_2 = 1$, 2, and 3. The graphs, reproduced in Figure 5.9, are called **contour lines**. You can see from Figure 5.9 that regardless of the value of x_2, $E(y)$ graphs as a straight line with a slope of 2. Changing x_2 changes only the y-intercept (the constant in the equation). Consequently, assuming that a first-order model will adequately model a response is equivalent to assuming that a 1-unit change in one independent variable will have the same effect on the mean value of y regardless of the levels of the other independent variables. That is, *the contour lines will be parallel.* In Chapter 4, we stated that independent variables that have this property *do not interact.*

FIGURE 5.9

Contour lines of $E(y)$ for $x_2 = 1$, 2, 3 (first-order model)

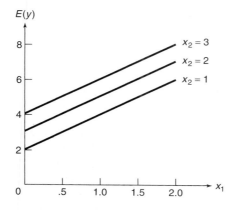

Except in cases where the ranges of levels for all independent variables are very small, the implication of no curvature in the response surface and the independence of variable effects on the response restrict the applicability of first-order models.

5.5
Second-Order Models with Two or More Quantitative Independent Variables

Second-order models with two or more independent variables permit curvature in the response surface. One important type of second-order term accounts for **interaction** between two variables.* Consider the two-variable model

$$E(y) = \beta_0 + \beta_1 x_1 + \beta_2 x_2 + \beta_3 x_1 x_2$$

This interaction model traces a ruled surface (twisted plane) in a three-dimensional space (see Figure 5.10). The second-order term $\beta_3 x_1 x_2$ is called the **interaction term**, and it permits the contour lines to be *nonparallel.*

In Section 4.9, we demonstrated the effect of interaction on the model:

$$E(y) = 1 + 2x_1 - x_2 + x_1 x_2$$

The slope of the line relating x_1 to $E(y)$ is $\beta_1 + \beta_3 x_2 = 2 + x_2$, while the y-intercept is $\beta_0 + \beta_2 x_2 = 1 - x_2$.

FIGURE 5.10

Response surface for an interaction model (second-order)

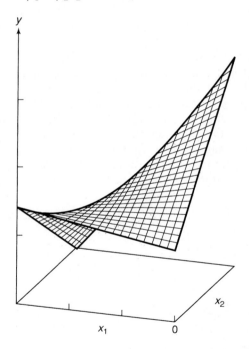

Thus, when interaction is present in the model, both the y-intercept and the slope change as x_2 changes. Consequently, *the contour lines are not parallel. The presence of an interaction term implies that the effect of a 1-unit change in one independent variable will depend on the level of the other independent variable.* The contour lines for $x_2 = 1$, 2, and 3 are reproduced in Figure 5.11. You can see that when $x_2 = 1$, the line has slope $2 + 1 = 3$ and y-intercept $1 - 1 = 0$. But when $x_2 = 3$, the slope is $2 + 3 = 5$ and the y-intercept is $1 - 3 = -2$.

*The order of a term involving two or more *quantitative* independent variables is equal to the sum of their exponents. Thus, $\beta_3 x_1 x_2$ is a second-order term, as is $\beta_4 x_1^2$. A term of the form $\beta_i x_1 x_2 x_3$ is a third-order term.

FIGURE 5.11

Contour lines of $E(y)$ for $x_2 = 1$, 2, 3 (first-order model plus interaction)

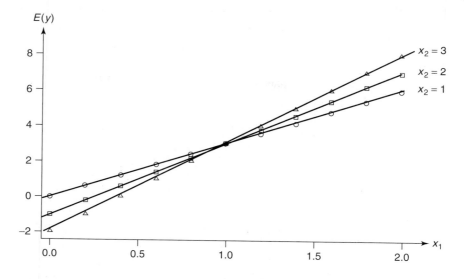

Interaction (Second-Order) Model with Two Independent Variables

$$E(y) = \beta_0 + \beta_1 x_1 + \beta_2 x_2 + \beta_3 x_1 x_2$$

Interpretation of Model Parameters

β_0: y-intercept; the value of $E(y)$ when $x_1 = x_2 = 0$

β_1 and β_2: Changing β_1 and β_2 causes the surface to shift along the x_1 and x_2 axes

β_3: Controls the rate of twist in the ruled surface (see Figure 5.10)

When one independent variable is held fixed, the model produces straight lines with the following slopes:

$\beta_1 + \beta_3 x_2$: Change in $E(y)$ for a 1-unit increase in x_1, when x_2 is held fixed

$\beta_2 + \beta_3 x_1$: Change in $E(y)$ for a 1-unit increase in x_2, when x_1 is held fixed

---

Definition 5.2

Two variables x_1 and x_2 are said to **interact** if the change in $E(y)$ for a 1-unit change in x_1 (when x_2 is held fixed) is dependent on the value of x_2.

We can introduce even more flexibility into a model by the addition of quadratic terms. The complete second-order model includes the constant β_0, all linear (first-order) terms, all two-variable interactions, and all quadratic terms. This complete second-order model for two quantitative independent variables is shown in the box.

Complete Second-Order Model with Two Independent Variables

$$E(y) = \beta_0 + \beta_1 x_1 + \beta_2 x_2 + \beta_3 x_1 x_2 + \beta_4 x_1^2 + \beta_5 x_2^2$$

Interpretation of Model Parameters

β_0: y-intercept; the value of $E(y)$ when $x_1 = x_2 = 0$

β_1 and β_2:   Changing β_1 and β_2 causes the surface to shift along the x_1 and x_2 axes

β_3:   The value of β_3 controls the rotation of the surface

β_4 and β_5:   Signs and values of these parameters control the type of surface and the rates of curvature

Three types of surfaces may be produced by a second-order model.*
    A paraboloid that opens upward (Figure 5.12a)
    A paraboloid that opens downward (Figure 5.12b)
    A saddle-shaped surface (Figure 5.12c)

The quadratic terms $\beta_4 x_1^2$ and $\beta_5 x_2^2$ in the second-order model imply that the response surface for $E(y)$ will possess curvature (see Figure 5.12). The interaction term $\beta_3 x_1 x_2$ allows the contours depicting $E(y)$ as a function of x_1 to have different shapes for various values of x_2. For example, suppose the complete second-order model relating $E(y)$ to x_1 and x_2 is

$$E(y) = 1 + 2x_1 + x_2 - 10x_1 x_2 + x_1^2 - 2x_2^2$$

Then the contours of $E(y)$ for $x_2 = -1, 0$, and 1 are shown in Figure 5.13. When we substitute $x_2 = -1$ into the model, we get

$$\begin{aligned} E(y) &= 1 + 2x_1 + x_2 - 10x_1 x_2 + x_1^2 - 2x_2^2 \\ &= 1 + 2x_1 - 1 - 10x_1(-1) + x_1^2 - 2(-1)^2 \\ &= -2 + 12x_1 + x_1^2 \end{aligned}$$

FIGURE 5.12

Graphs of three second-order surfaces

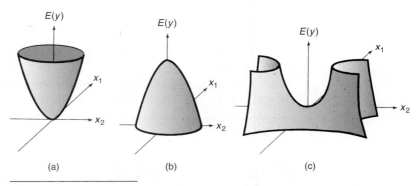

(a)   (b)   (c)

---

*The saddle-shaped surface (Figure 5.12c) is produced when $\beta_3^2 > 4\beta_4\beta_5$. For $\beta_3^2 < 4\beta_4\beta_5$, the paraboloid opens upward (Figure 5.12a) when $\beta_4 + \beta_5 > 0$ and opens downward (Figure 5.12b) when $\beta_4 + \beta_5 < 0$.

For $x_2 = 0$,

$$E(y) = 1 + 2x_1 + (0) - 10x_1(0) + x_1^2 - 2(0)^2$$
$$= 1 + 2x_1 + x_1^2$$

Similarly, for $x_2 = 1$,

$$E(y) = -8x_1 + x_1^2$$

Note how the shapes of the three contour curves in Figure 5.13 differ, indicating that the β parameter associated with the x_1x_2 (interaction) term differs from 0.

FIGURE 5.13

Contours of $E(y)$ for $x_2 = -1, 0, 1$ (complete second-order model)

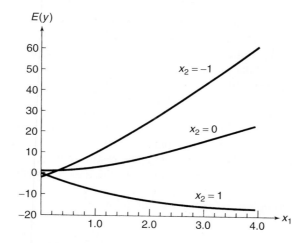

The complete second-order model for three independent variables is shown in the next box.

Complete Second-Order Model with Three Quantitative Independent Variables

$$E(y) = \beta_0 + \beta_1 x_1 + \beta_2 x_2 + \beta_3 x_3 + \beta_4 x_1 x_2 + \beta_5 x_1 x_3 + \beta_6 x_2 x_3$$
$$+ \beta_7 x_1^2 + \beta_8 x_2^2 + \beta_9 x_3^2$$

where $\beta_0, \beta_1, \beta_2, \ldots, \beta_9$ are unknown parameters that must be estimated.

This second-order model in three independent variables demonstrates how you would write a second-order model for any number of independent variables. Always include the constant β_0 and then all first-order terms corresponding to $x_1, x_2, \ldots$. Then add the interaction terms for all pairs of independent variables $x_1 x_2, x_1 x_3, x_2 x_3, \ldots$. Finally, include the second-order terms $x_1^2, x_2^2, \ldots$.

For any number, say, p, of quantitative independent variables, the response traces a surface in a $(p + 1)$-dimensional space, which is impossible to visualize. In spite of this handicap, the prediction equation can still tell us much about the phenomenon being studied.

⬤ PRODQUAL

TABLE 5.2   Temperature, Pressure, and Quality of the Finished Product

x_1, °F	x_2, psi	y	x_1, °F	x_2, psi	y	x_1, °F	x_2, psi	y
80	50	50.8	90	50	63.4	100	50	46.6
80	50	50.7	90	50	61.6	100	50	49.1
80	50	49.4	90	50	63.4	100	50	46.4
80	55	93.7	90	55	93.8	100	55	69.8
80	55	90.9	90	55	92.1	100	55	72.5
80	55	90.9	90	55	97.4	100	55	73.2
80	60	74.5	90	60	70.9	100	60	38.7
80	60	73.0	90	60	68.8	100	60	42.5
80	60	71.2	90	60	71.3	100	60	41.4

EXAMPLE   5.3

Many companies manufacture products that are at least partially produced using chemicals (for example, steel, paint, gasoline). In many instances, the quality of the finished product is a function of the temperature and pressure at which the chemical reactions take place.

Suppose you wanted to model the quality, y, of a product as a function of the temperature, x_1, and the pressure, x_2, at which it is produced. Four inspectors independently assign a quality score between 0 and 100 to each product, and then the quality, y, is calculated by averaging the four scores. An experiment is conducted by varying temperature between 80° and 100°F and pressure between 50 and 60 pounds per square inch (psi). The resulting data ($n = 27$) are given in Table 5.2. Fit a complete second-order model to the data and sketch the response surface.

Solution

The complete second-order model is

$$E(y) = \beta_0 + \beta_1 x_1 + \beta_2 x_2 + \beta_3 x_1 x_2 + \beta_4 x_1^2 + \beta_5 x_2^2$$

The data in Table 5.2 were used to fit this model. A portion of the SAS output is shown in Figure 5.14.

FIGURE 5.14

SAS output for complete second-order model of quality

Dependent Variable: QUALITY

Analysis of Variance

Source	DF	Sum of Squares	Mean Square	F Value	Pr > F
Model	5	8402.26454	1680.45291	596.32	<.0001
Error	21	59.17843	2.81802		
Corrected Total	26	8461.44296			

Root MSE	1.67870	R-Square	0.9930
Dependent Mean	66.96296	Adj R-Sq	0.9913
Coeff Var	2.50690		

Parameter Estimates

Variable	DF	Parameter Estimate	Standard Error	t Value	Pr > |t|
Intercept	1	-5127.89907	110.29601	-46.49	<.0001
TEMP	1	31.09639	1.34441	23.13	<.0001
PRESSURE	1	139.74722	3.14005	44.50	<.0001
TEMPRESS	1	-0.14550	0.00969	-15.01	<.0001
TEMPSQ	1	-0.13339	0.00685	-19.46	<.0001
PRESSQ	1	-1.14422	0.02741	-41.74	<.0001

The least squares model is

$$\hat{y} = -5{,}127.90 + 31.10x_1 + 139.75x_2 - .146x_1x_2 - .133x_1^2 - 1.14x_2^2$$

A three-dimensional graph of this prediction model is shown in Figure 5.15. The mean quality seems to be greatest for temperatures of about $85° - 90°F$ and for pressures of about 55–57 pounds per square inch.[†] Further experimentation in these ranges might lead to a more precise determination of the optimal temperature—pressure combination.

A look at the adjusted coefficient of determination, $R^2_{adj} = .991$, the F value for testing the entire model, $F = 596.32$, and the p-value for the test, $p = .0001$ (shaded in Figure 5.14), leaves little doubt that the complete second-order model is useful for explaining mean quality as a function of temperature and pressure. This, of course, will not always be the case. The additional complexity of second-order models is worthwhile only if a better model results. To determine whether the quadratic terms are important, we would test $H_0: \beta_4 = \beta_5 = 0$ using the F test for comparing nested models outlined in Section 4.13.   ◆

FIGURE 5.15

Plot of second-order least squares model for Example 5.3

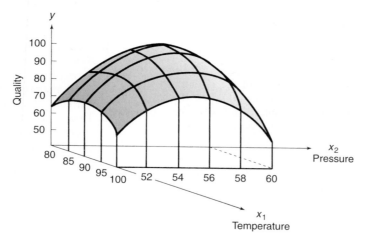

EXERCISES

5.10. An exploration seismologist wants to develop a model to estimate the average signal-to-noise ratio of an earthquake's seismic wave, y, as a function of two independent variables:

$x_1 = $ Frequency (cycles per second)

$x_2 = $ Amplitude of the wavelet

a. Identify the independent variables as quantitative or qualitative.

b. Write the first-order model for $E(y)$.

c. Write a model for $E(y)$ that contains all first-order and interaction terms. Sketch typical response curves showing $E(y)$, the mean signal-to-noise ratio,

---

[†]Students with knowledge of calculus should note that we can determine the exact temperature and pressure that maximize quality in the least squares model by solving $\partial\hat{y}/\partial x_1 = 0$ and $\partial\hat{y}/\partial x_2 = 0$ for x_1 and x_2. These estimated optional values are $x_1 = 86.25°F$ and $x_2 = 55.58$ pounds per square inch. Remember, however, that these are only sample estimates of the coordinates for the optional value.

versus x_2, the amplitude of the wavelet, for different values of x_1 (assume that x_1 and x_2 interact).

d. Write the complete second-order model for $E(y)$.

5.11. Refer to Exercise 5.10. Suppose the model from part **c** is fit, with the following result:

$$\hat{y} = 1 + .05x_1 + x_2 + .05x_1x_2$$

Graph the estimated signal-to-noise ratio $\hat{y}$ as a function of the wavelet amplitude, x_2, over the range $x_2 = 10$ to $x_2 = 50$ for frequencies of $x_1 = 1$, 5, and 10. Do these functions agree (approximately) with the graphs you drew for Exercise 5.9, part **c**?

5.12. Refer to Exercise 5.10. Suppose an additional independent variable is considered, as follows:

x_3 = Time interval (seconds) between seismic waves

a. Write the first-order model plus interaction for $E(y)$ as a function of x_1, x_2, and x_3.

b. Write the complete second-order model plus interaction for $E(y)$ as a function of x_1, x_2, and x_3.

5.13. Researchers at the Upjohn Company utilized multiple regression analysis in the development of a sustained-release tablet. One of the objectives of the research was to develop a model relating the dissolution y of a tablet (i.e., the percentage of the tablet dissolved over a specified period of time) to the following independent variables:

x_1 = Excipient level (i.e., amount of nondrug ingredient in the tablet)

x_2 = Process variable (e.g., machine setting under which tablet is processed)

a. Write the complete second-order model for $E(y)$.

b. Write a model that hypothesizes straight-line relationships between $E(y)$, x_1, and x_2. Assume that x_1 and x_2 do not interact.

c. Repeat part **b**, but add interaction to the model.

d. For the model in part **c**, what is the slope of the $E(y)$, x_1 line for fixed x_2?

e. For the model in part **c**, what is the slope of the $E(y)$, x_2 line for fixed x_1?

5.14. Refer to the *World Development* (Feb. 1998) study of street vendors in the city of Puebla, Mexico, Exercise 4.4 (p. 176). Recall that the vendors mean annual earnings, $E(y)$, was modeled as a first-order function of age (x_1) and hours worked (x_2). The data for the study are reproduced on p. 271.

a. Write a complete second-order model for mean annual earnings, $E(y)$, as a function of age (x_1) and hours worked (x_2).

b. The model was fit to the data using MINITAB. Find the least squares prediction equation on the accompanying printout.

c. Is the model statistically useful for predicting annual earnings? Test using $\alpha = .05$.

d. How would you test the hypothesis that the second-order terms in the model are not necessary for predicting annual earnings?

e. Carry out the test, part **d**. Interpret the results.

MINITAB Output for Exercise 5.14

```
The regression equation is
EARNINGS = 606 + 120 AGE - 140 HOURS + 2.66 AGEHOURS - 1.57 AGESQ + 8.1 HOURSQ

Predictor       Coef      SE Coef        T        P
Constant         606         2331     0.26    0.801
AGE           119.68        64.58     1.85    0.097
HOURS         -139.8        491.6    -0.28    0.783
AGEHOURS       2.662        3.420     0.78    0.456
AGESQ        -1.5710       0.6911    -2.27    0.049
HOURSQ          8.08        26.73     0.30    0.769


S = 482.8      R-Sq = 75.7%      R-Sq(adj) = 62.1%


Analysis of Variance

Source          DF          SS          MS        F        P
Regression       5     6520245     1304049     5.59    0.013
Residual Error   9     2098183      233131
Total           14     8618428
```

STREETVEN

VENDOR NUMBER	ANNUAL EARNINGS, y	AGE, x_1	HOURS WORKED PER DAY, x_2
21	$2841	29	12
53	1876	21	8
60	2934	62	10
184	1552	18	10
263	3065	40	11
281	3670	50	11
354	2005	65	5
401	3215	44	8
515	1930	17	8
633	2010	70	6
677	3111	20	9
710	2882	29	9
800	1683	15	5
914	1817	14	7
997	4066	33	12

Source: Adapted from Smith, Paula A and Michael R. Metzger, "The Return to Education: Street Vendors in Mexico." *World Development*, Vol. 26, No. 2, Feb. 1998, pp. 289–296.

5.15. Refer to *The Astronomical Journal* (July 1995) study of quasars detected by a deep space survey, Exercise 4.9 (p. 180). Recall that several quantitative independent variables were used to model the quasar characteristic, rest frame equivalent width (y). The data for 25 quasars are reproduced in the table.

a. Write a complete second-order model for y as a function of redshift (x_1), lineflux (x_2), and AB_{1450} (x_4).

b. Fit the model, part a, to the data using a statistical software package. Is the overall model statistically useful for predicting y?

c. Conduct a test to determine if any of the curvilinear terms in the model, part **a**, are statistically useful predictors of y.

QUASAR

QUASAR	REDSHIFT (x_1)	LINEFLUX (x_2)	LINE LUMINOSITY (x_3)	AB_{1450} (x_4)	ABSOLUTE MAGNITUDE (x_5)	REST FRAME EQUIVALENT WIDTH (y)
1	2.81	−13.48	45.29	19.50	−26.27	117
2	3.07	−13.73	45.13	19.65	−26.26	82
3	3.45	−13.87	45.11	18.93	−27.17	33
4	3.19	−13.27	45.63	18.59	−27.39	92
5	3.07	−13.56	45.30	19.59	−26.32	114
6	4.15	−13.95	45.20	19.42	−26.97	50
7	3.26	−13.83	45.08	19.18	−26.83	43
8	2.81	−13.50	45.27	20.41	−25.36	259
9	3.83	−13.66	45.41	18.93	−27.34	58
10	3.32	−13.71	45.23	20.00	−26.04	126
11	2.81	−13.50	45.27	18.45	−27.32	42
12	4.40	−13.96	45.25	20.55	−25.94	146
13	3.45	−13.91	45.07	20.45	−25.65	124
14	3.70	−13.85	45.19	19.70	−26.51	75
15	3.07	−13.67	45.19	19.54	−26.37	85
16	4.34	−13.93	45.27	20.17	−26.29	109
17	3.00	−13.75	45.08	19.30	−26.58	55
18	3.88	−14.17	44.92	20.68	−25.61	91
19	3.07	−13.92	44.94	20.51	−25.41	116
20	4.08	−14.28	44.86	20.70	−25.67	75
21	3.62	−13.82	45.20	19.45	−26.73	63
22	3.07	−14.08	44.78	19.90	−26.02	46
23	2.94	−13.82	44.99	19.49	−26.35	55
24	3.20	−14.15	44.75	20.89	−25.09	99
25	3.24	−13.74	45.17	19.17	−26.83	53

Source: Schmidt, M., Schneider, D. P., & Gunn, J. E. "Spectroscopic CCD surveys for quasars at large redshift." *The Astronomical Journal*, Vol. 110, No. 1, July 1995, p. 70 (Table 1).

5.6
Coding Quantitative Independent Variables (Optional)

In fitting higher-order polynomial regression models (e.g., second- or third-order models), it is often a good practice to code the quantitative independent variables. For example, suppose one of the independent variables in a regression analysis is level of competence in performing a task, C, measured on a 20-point scale, and C is observed at three levels: 5, 10, and 15. We can code (or transform) the competence measurements using the formula

$$x = \frac{C - 10}{5}$$

Then the coded levels $x = -1, 0$, and 1 correspond to the original C levels 5, 10, and 15.

In a general sense, *coding* means transforming a set of independent variables (qualitative or quantitative) into a new set of independent variables. For example, if we observe two independent variables,

C = Competence level

S = Satisfaction level

then we can transform C and S into two new coded variables, x_1 and x_2, where x_1 and x_2 are related to C and S by two functional equations:

$$x_1 = f_1(C, S) \qquad x_2 = f_2(C, S)$$

The functions f_1 and f_2, which are frequently expressed as equations, establish a one-to-one correspondence between combinations of levels of C and S with combinations of the coded values of x_1 and x_2.

Since qualitative independent variables are not numerical, it is necessary to code their values to fit the regression model. We demonstrate the coding scheme in Section 5.7. However, you might ask why we would bother to code the quantitative independent variables. There are two related reasons for coding quantitative variables. At first glance, it would appear that a computer would be oblivious to the values assumed by the independent variables in a regression analysis, but this is not the case. To calculate the estimates of the model parameters using the method of least squares, the computer must invert a matrix of numbers, called the **coefficient** (or **information**) **matrix** (see Appendix A). Considerable rounding error may occur during the inversion process if the numbers in the coefficient matrix vary greatly in absolute value. This can produce sizable errors in the computed values of the least squares estimates, $\hat{\beta}_0, \hat{\beta}_1, \hat{\beta}_2, \ldots$. Coding makes it computationally easier for the computer to invert the matrix, thus leading to more accurate estimates.

A second reason for coding quantitative variables pertains to a problem we will discuss in detail in Chapter 7: the problem of independent variables (x's) being intercorrelated (called **multicollinearity**). When polynomial regression models (e.g., second-order models) are fit, the problem of multicollinearity is unavoidable, especially when higher-order terms are fit. For example, consider the quadratic model

$$E(y) = \beta_0 + \beta_1 x + \beta_2 x^2$$

If the range of the values of x is narrow, then the two variables, $x_1 = x$ and $x_2 = x^2$, will generally be highly correlated. As we point out in Chapter 7, the likelihood of rounding errors in the regression coefficients is increased in the presence of these highly correlated independent variables.

The following procedure is the best way to cope with the problem of rounding error:

1. Code the quantitative variable so that the new coded origin is in the center of the coded values. For example, by coding competence level, C, as

$$x = \frac{C - 10}{5}$$

we obtain coded values $-1, 0, 1$. This places the coded origin, 0, in the middle of the range of coded values (-1 to 1).

2. Code the quantitative variable so that the range of the coded values is approximately the same for all coded variables. You need not hold exactly to this requirement. The range of values for one independent variable could be double or triple the range of another without causing any difficulty, but it is not desirable to have a sizable disparity in the ranges, say, a ratio of 100 to 1.

When the data are observational (the values assumed by the independent variables are uncontrolled), the coding procedure described in the box satisfies, reasonably well, these two requirements. The coded variable u is similar to the standardized normal z statistic of Section 1.6. Thus, the u value is the deviation (the distance) between an x value and the mean of the x values, $\bar{x}$, expressed in units of s_x.* Since we know that most (approximately 95%) measurements in a set will lie within 2 standard deviations of their mean, it follows that most of the coded u values will lie in the interval -2 to $+2$.

Coding Procedure for Observational Data

Let

$x = $ Uncoded quantitative independent variable
$u = $ Coded quantitative independent variable

Then if x takes values $x_1, x_2, \ldots, x_n$ for the n data points in the regression analysis, let

$$u_i = \frac{x_i - \bar{x}}{s_x}$$

where s_x is the standard deviation of the x values, i.e.,

$$s_x = \sqrt{\frac{\sum_{i=1}^{n}(x_i - \bar{x})^2}{n - 1}}$$

---

*The divisor of the deviation, $x - \bar{x}$, need not equal s_x exactly. Any number approximately equal to s_x would suffice. Other candidate denominators are the range (R), $R/2$, and the interquartile range (IQR).

If you apply this coding to each quantitative variable, the range of values for each will be approximately -2 to $+2$. The variation in the absolute values of the elements of the coefficient matrix will be moderate, and rounding errors generated in finding the inverse of the matrix will be reduced. Additionally, the correlation between x and x^2 will be reduced.[†]

EXAMPLE 5.4

Carbon dioxide–baited traps are typically used by entomologists to monitor mosquito populations. An article in the *Journal of the American Mosquito Control Association* (Mar. 1995) investigated whether temperature influences the number of mosquitoes caught in a trap. Six mosquito samples were collected on each of nine consecutive days. For each day, two variables were measured: x = average temperature (in degrees Centigrade) and y = mosquito catch ratio (the number of mosquitoes caught in each sample divided by the largest sample caught). The data are reported in Table 5.3.

⊚ MOSQUITO

TABLE 5.3   **Data for Example 5.4**

DATE	AVERAGE TEMPERATURE, x	CATCH RATIO, y
July 24	16.8	.66
25	15.0	.30
26	16.5	.46
27	17.7	.44
28	20.6	.67
29	22.6	.99
30	23.3	.75
31	18.2	.24
Aug. 1	18.6	.51

Source: Petric, D., *et al.* "Dependence of CO_2-baited suction trap captures on temperature variations." *Journal of the American Mosquito Control Association*, Vol. 11, No. 1, Mar. 1995, p. 8.

The researchers are interested in relating catch ratio y to average temperature x. Suppose we consider using a quadratic model.

a. Give the equation relating the coded variable u to the temperature x using the coding system for observational data.

b. Calculate the coded values, u, for the eight x values.

c. Find the sum of the $n = 9$ values for u.

Solution

a. We first find $\bar{x}$ and s_x. From the MINITAB printout, Figure 5.16, which provides summary statistics for temperature, x, we obtain

$$\bar{x} = 18.811 \quad \text{and} \quad s_x = 2.812$$

[†] Another by-product of coding is that the β coefficients of the model have slightly different interpretations. For example, in the model $E(y) = \beta_0 + \beta_1 u$, where $u = (x - 10)/5$, the change in y for every 1-unit increase in x is not β_1, but $\beta_1/5$. In general, for first-order models with coded independent quantitative variables, the slope associated with x_i is represented by β_i/s_{x_i}, where s_{x_i} is the divisor of the coded x_i.

FIGURE 5.16

MINITAB descriptive statistics for temperature, x

Variable	N	Mean	Median	TrMean	StDev	SE Mean
X	9	18.811	18.200	18.811	2.812	0.937

Variable	Minimum	Maximum	Q1	Q3
X	15.000	23.300	16.650	21.600

Then the equation relating u and x is

$$u = \frac{x - 18.8}{2.8}$$

b. When temperature $x = 16.8$

$$u = \frac{x - 18.8}{2.8} = \frac{16.8 - 18.8}{2.8} = -.71$$

Similarly, when $x = 15.0$

$$u = \frac{x - 18.8}{2.8} = \frac{15.0 - 18.8}{2.8} = -1.36$$

Table 5.4 gives the coded values for all $n = 9$ observations. [*Note*: You can see that all the $n = 9$ values for u lie in the interval from -2 to $+2$.]

TABLE 5.4 **Coded Values of** x,
Example 5.4

TEMPERATURE, x	CODED VALUES, u
16.8	−.71
15.0	−1.36
16.5	−.82
17.7	−.39
20.6	.64
22.6	1.36
23.3	1.61
18.2	−.21
18.6	−.07

c. If you ignore rounding error, the sum of the $n = 8$ values for u will equal 0. This is because the sum of the deviations of a set of measurements about their mean is always equal to 0. ◆

To illustrate the advantage of coding, consider fitting the second-order model

$$E(y) = \beta_0 + \beta_1 x + \beta_2 x^2$$

to the data of Example 5.4. The coefficient of correlation between the two variables, x and x^2, shown at the top of the MINITAB printout displayed in Figure 5.17, is $r = .998$. However, the coefficient of correlation between the corresponding coded values, u and u^2, shown at the bottom of Figure 5.17, is only $r = .448$. Thus, we can avoid potential rounding error caused by highly correlated x values by fitting, instead, the model

$$E(y) = \beta_0^* + \beta_1^* u + \beta_2^* u^2$$

FIGURE 5.17

MINITAB correlations for temperature, x, coded temperature, u

---

Correlations: X, XSQ

Pearson correlation of X and XSQ = 0.998
P-Value = 0.000

Correlations: U, USQ

Pearson correlation of U and USQ = 0.448
P-Value = 0.227

Other methods of coding have been developed to reduce rounding errors and multicollinearity. One of the more complex coding systems involves fitting **orthogonal polynomials**. An orthogonal system of coding guarantees that the coded independent variables will be uncorrelated. For a discussion of orthogonal polynomials, consult the references given at the end of this chapter.

EXERCISES

5.16. Suppose you want to use the coding system for observational data to fit a second-order model to the tire pressure–automobile mileage data of Exercise 5.8 (p. 261), which are repeated in the table.

PRESSURE	MILEAGE
x, pounds per square inch	y, thousands
30	29
31	32
32	36
33	38
34	37
35	33
36	26

a. Give the equation relating the coded variable u to pressure, x, using the coding system for observational data.

b. Calculate the coded values, u.

c. Calculate the coefficient of correlation r between the variables x and x^2.

d. Calculate the coefficient of correlation r between the variables u and u^2. Compare this value to the value computed in part **c**.

e. Fit the model

$$E(y) = \beta_0 + \beta_1 u + \beta_2 u^2$$

using available statistical software. Interpret the results.

5.17. As part of the first-year evaluation for new salespeople, a large food-processing firm projects the second-year sales for each salesperson based on his or her sales for the first year. Data for $n = 8$ salespeople are shown in the table at the top of page 277.

a. Give the equation relating the coded variable u to first-year sales, x, using the coding system for observational data.

b. Calculate the coded values, u.

c. Calculate the coefficient of correlation r between the variables x and x^2.

d. Calculate the coefficient of correlation r between the variables u and u^2. Compare this value to the value computed in part **c**.

SALES

FIRST-YEAR SALES	SECOND-YEAR SALES
x, thousands of dollars	y, thousands of dollars
75.2	99.3
91.7	125.7
100.3	136.1
64.2	108.6
81.8	102.0
110.2	153.7
77.3	108.8
80.1	105.4

 LASERS

THRESHOLD CURRENT	WAVEGUIDE Al MOLE FRACTION
y, A/cm^2	x
273	.15
175	.20
146	.25
166	.30
162	.35
165	.40
245	.50
314	.60

Source: Unnikrishnan, S., and Anderson, N. G. "Quantum-well lasers for direct solar photopumping." *Journal of Applied Physics*, Vol. 74, No. 6, Sept. 15, 1993, p. 4226 (data adapted from Figure 2).

5.18. Refer to the *Journal of Applied Physics* (Sept. 1993) study of solar lighting with semiconductor lasers, Exercise 4.30 (p. 203). The data for the analysis are repeated here.

a. Give the equation relating the coded variable u to waveguide, x, using the coding system for observational data.

b. Calculate the coded values, u.

c. Calculate the coefficient of correlation r between the variables x and x^2.

d. Calculate the coefficient of correlation r between the variables u and u^2. Compare this value to the value computed in part **c**.

e. Fit the model

$$E(y) = \beta_0 + \beta_1 u + \beta_2 u^2$$

using available statistical software. Interpret the results.

5.7
Models with One Qualitative Independent Variable

Suppose we want to write a model for the mean performance, $E(y)$, of a diesel engine as a function of type of fuel. (For the purpose of explanation, we will ignore other independent variables that might affect the response.) Further suppose there are three fuel types available: a petroleum-based fuel (P), a coal-based fuel (C), and a blended fuel (B). The fuel type is a single qualitative variable with three levels corresponding to fuels P, C, and B. Note that with a qualitative independent variable, we cannot attach a quantitative meaning to a given level. All we can do is describe it.

To simplify our notation, let μ_P be the mean performance for fuel P, and let μ_C and μ_B be the corresponding mean performances for fuels C and B. Our objective is to write a single prediction equation that will give the mean value of y for the three fuel types. A coding scheme that yields useful β-interpretations is the following:

$$E(y) = \beta_0 + \beta_1 x_1 + \beta_2 x_2$$

where

$$x_1 = \begin{cases} 1 & \text{if fuel P is used} \\ 0 & \text{if not} \end{cases}$$

$$x_2 = \begin{cases} 1 & \text{if fuel C is used} \\ 0 & \text{if not} \end{cases}$$

The values of x_1 and x_2 for each of the three fuel types are shown in Table 5.5.

TABLE 5.5   **Mean Response for the Model with Three Diesel Fuel Types**

FUEL TYPE	x_1	x_2	MEAN RESPONSE, $E(y)$
Blended (B)	0	0	$\beta_0 = \mu_B$
Petroleum (P)	1	0	$\beta_0 + \beta_1 = \mu_P$
Coal (C)	0	1	$\beta_0 + \beta_2 = \mu_C$

The variables x_1 and x_2 are not meaningful independent variables as in the case of the models containing quantitative independent variables. Instead, they are **dummy (indicator) variables** that make the model work. To see how they work, let $x_1 = 0$ and $x_2 = 0$. This condition will apply when we are seeking the mean response for fuel B (neither fuel P nor C is used; hence, it must be B). Then the mean value of y when fuel B is used is

$$\mu_B = E(y) = \beta_0 + \beta_1(0) + \beta_2(0)$$
$$= \beta_0$$

This tells us that the mean performance level for fuel B is β_0. Or, it means that

$$\beta_0 = \mu_B.$$

Now suppose we want to represent the mean response, $E(y)$, when fuel P is used. Checking the dummy variable definitions, we see that we should let $x_1 = 1$ and $x_2 = 0$:

$$\mu_P = E(y) = \beta_0 + \beta_1(1) + \beta_2(0)$$
$$= \beta_0 + \beta_1$$

or, since $\beta_0 = \mu_B$,

$$\mu_P = \mu_B + \beta_1$$

Then it follows that the interpretation of β_1 is

$$\beta_1 = \mu_P - \mu_B$$

which is the difference in the mean performance levels for fuels P and B.

Finally, if we want the mean value of y when fuel C is used, we let $x_1 = 0$ and $x_2 = 1$:

$$\mu_C = E(y) = \beta_0 + \beta_1(0) + \beta_2(1)$$
$$= \beta_0 + \beta_2$$

or, since $\beta_0 = \mu_B$,

$$\mu_C = \mu_B + \beta_2$$

Then it follows that the interpretation of β_2 is

$$\beta_2 = \mu_C - \mu_B$$

Note that we were able to describe *three levels* of the qualitative variable with only *two dummy variables*, because the mean of the base level (fuel B, in this case) is accounted for by the intercept β_0.

Now, carefully examine the model for a single qualitative independent variable with three levels, because we will use exactly the same pattern for any number of levels. Arbitrarily select one level to be the base level, then set up 1-0 dummy variables for the remaining levels. This setup always leads to the interpretation of the parameters given in the box.

Procedure for Writing a Model with One Qualitative Independent Variable at k Levels (A, B, C, D, ...)

$$E(y) = \beta_0 + \beta_1 x_1 + \beta_2 x_2 + \cdots + \beta_{k-1} x_{k-1}$$

where

$$x_i = \begin{cases} 1 & \text{if qualitative variable at level } i + 1 \\ 0 & \text{otherwise} \end{cases}$$

The number of dummy variables for a single qualitative variable is always 1 less than the number of levels for the variable. Then, assuming the base level is A, the mean for each level is

$\mu_A = \beta_0$
$\mu_B = \beta_0 + \beta_1$
$\mu_C = \beta_0 + \beta_2$
$\mu_D = \beta_0 + \beta_3$
$\vdots$

β Interpretations:

$\beta_0 = \mu_A$
$\beta_1 = \mu_B - \mu_A$
$\beta_2 = \mu_C - \mu_A$
$\beta_3 = \mu_D - \mu_A$
$\vdots$

EXAMPLE 5.5

A large consulting firm markets a computerized system for monitoring road construction bids to various state departments of transportation. Since the high cost of maintaining the system is partially absorbed by the firm, the firm wants to compare the mean annual maintenance costs accrued by system users in three different states: Kansas, Kentucky, and Texas. A sample of 10 users is selected from each state installation and the maintenance cost accrued by each is recorded, as shown in Table 5.6.

BIDMAINT

TABLE 5.6   Annual Maintenance Costs

	STATE INSTALLATION		
	KANSAS	KENTUCKY	TEXAS
	$ 198	$ 563	$ 385
	126	314	693
	443	483	266
	570	144	586
	286	585	178
	184	377	773
	105	264	308
	216	185	430
	465	330	644
	203	354	515
Totals	$2,796	$3,599	$4,778

a. Do the data provide sufficient evidence (at $\alpha = .05$) to indicate that the mean annual maintenance costs accrued by system users differ for the three state installations?

b. Find and interpret a 95% confidence interval for the difference between the mean cost in Texas and the mean cost in Kentucky.

Solution

a. The model relating $E(y)$ to the single qualitative variable, state installation, is

$$E(y) = \beta_0 + \beta_1 x_1 + \beta_2 x_2$$

where

$$x_1 = \begin{cases} 1 & \text{if Kentucky} \\ 0 & \text{if not} \end{cases} \qquad x_2 = \begin{cases} 1 & \text{if Texas} \\ 0 & \text{if not} \end{cases}$$

and

$$\beta_1 = \mu_2 - \mu_1$$
$$\beta_2 = \mu_3 - \mu_1$$

where μ_1, μ_2, and μ_3 are the mean responses for Kansas, Kentucky, and Texas, respectively. Testing the null hypothesis that the means for three states are equal, i.e., $\mu_1 = \mu_2 = \mu_3$, is equivalent to testing

$H_0: \beta_1 = \beta_2 = 0$

because if $\beta_1 = \mu_2 - \mu_1 = 0$ and $\beta_2 = \mu_3 - \mu_1 = 0$, then μ_1, μ_2, and μ_3 must be equal. The alternative hypothesis is

$H_a:$ At least one of the parameters, β_1 or β_2, differs from 0.

There are two ways to conduct this test. We can fit the complete model shown previously and the reduced model (discarding the terms involving β_1 and β_2),

$$E(y) = \beta_0$$

and conduct the nested model F test described in Section 4.13 (we leave this as an exercise for you). Or, we can use the global F test of the complete model (Section 4.8), which tests the null hypothesis that all parameters in the model, with the exception of β_0, equal 0. Either way you conduct the test, you will obtain the same computed value of F. The SPSS printout for fitting the complete model,

$$E(y) = \beta_0 + \beta_1 x_1 + \beta_2 x_2$$

is shown in Figure 5.18. The value of the F statistic for testing the complete model (shaded on Figure 5.18) is $F = 3.482$, the p-value for the test (also shaded) is $p = .045$. Since our choice of α, $\alpha = .05$, exceeds the p-value, we reject H_0 and conclude that at least one of the parameters, β_1 or β_2, differs from 0. Or equivalently, we conclude that the data provide sufficient evidence to indicate that the mean user maintenance cost does vary among the three state installations.

FIGURE 5.18

SPSS printout for dummy variable model, Example 5.5

Model Summary

Model	R	R Square	Adjusted R Square	Std. Error of the Estimate
1	.453[a]	.205	.146	168.948

a. Predictors: (Constant), X2, X1

ANOVA[b]

Model		Sum of Squares	df	Mean Square	F	Sig.
1	Regression	198772.5	2	99386.233	3.482	.045[a]
	Residual	770670.9	27	28543.367		
	Total	969443.4	29			

a. Predictors: (Constant), X2, X1
b. Dependent Variable: COST

Coefficients[a]

Model		Unstandardized Coefficients		Standardized Coefficients	t	Sig.	95% Confidence Interval for B	
		B	Std. Error	Beta			Lower Bound	Upper Bound
1	(Constant)	279.600	53.426		5.233	.000	169.979	389.221
	X1	80.300	75.556	.211	1.063	.297	-74.728	235.328
	X2	198.200	75.556	.520	2.623	.014	43.172	353.228

a. Dependent Variable: COST

b. Since $\beta_3 = \mu_3 - \mu_1 =$ the difference between the mean costs of Texas and Kentucky, we want a 95% confidence interval for β_3. The interval, highlighted on Figure 5.18, is (43.172, 353.228). Consequently, we are 95% confident that the difference, $\mu_3 - \mu_1$, falls in our interval. This implies that the mean cost of users in Texas is anywhere from \$43.17 to \$353.23 higher than the mean cost of Kentucky users. ◆

We will demonstrate how to write a model with two qualitative independent variables and then, in Section 5.9, we will explain how to use this technique to write models with any number of qualitative independent variables.

Let us return to the example used in Section 5.7, where we wrote a model for the mean performance, $E(y)$, of a diesel engine as a function of one qualitative independent variable, fuel type. Now suppose the performance is also a function of engine brand and we want to compare the top two brands. Therefore, this second qualitative independent variable, brand, will be observed at two levels. To simplify our notation, we will change the symbols for the three fuel types from B, D, C, to F_1, F_2, F_3, and we will let B_1 and B_2 represent the two brands. The six population means of performance measurements (measurements of y) are symbolically represented by the six cells in the two-way table shown in Table 5.7. Each μ subscript corresponds to one of the six fuel type–brand combinations.

TABLE 5.7   **The Six Combinations of Fuel Type and Diesel Engine Brand**

		BRAND	
		B_1	B_2
	F_1	μ_{11}	μ_{12}
FUEL TYPE	F_2	μ_{21}	μ_{22}
	F_3	μ_{31}	μ_{32}

First we will write a model in its simplest form—where the two qualitative variables affect the response independently of each other. To write the model for mean performance, $E(y)$, we start with a constant β_0 and then add *two* dummy variables for the three levels of fuel type in the manner explained in Section 5.7. These terms, which are called the **main effect terms** for fuel type, F, account for the effect of F on $E(y)$ when fuel type, F, and brand, B, affect $E(y)$ independently. Then,

$$E(y) = \beta_0 + \overbrace{\beta_1 x_1 + \beta_2 x_2}^{\substack{\text{Main effect} \\ \text{terms for } F}}$$

where

$$x_1 = \begin{cases} 1 & \text{if fuel type } F_2 \text{ was used} \\ 0 & \text{if not} \end{cases}$$

$$x_2 = \begin{cases} 1 & \text{if fuel type } F_3 \text{ was used} \\ 0 & \text{if not} \end{cases}$$

Now let level B_1 be the base level of the brand variable. Since there are two levels of this variable, we will need only one dummy variable to include the brand in the model:

$$E(y) = \beta_0 + \overbrace{\beta_1 x_1 + \beta_2 x_2}^{\substack{\text{Main effect} \\ \text{terms for } F}} + \overbrace{\beta_3 x_3}^{\substack{\text{Main effect} \\ \text{term for } B}}$$

where the dummy variables x_1 and x_2 are as defined previously and

$$x_3 = \begin{cases} 1 & \text{if engine brand } B_2 \text{ used} \\ 0 & \text{if engine brand } B_1 \text{ used} \end{cases}$$

If you check the model, you will see that by assigning specific values to x_1, x_2, and x_3, you create a model for the mean value of y corresponding to one of the cells of Table 5.7. We will illustrate with two examples.

EXAMPLE 5.6

Give the values of x_1, x_2, and x_3 and the model for the mean performance, $E(y)$, when using fuel type F_1 in engine brand B_1.

Solution

Checking the coding system, you will see that F_1 and B_1 occur when $x_1 = x_2 = x_3 = 0$. Then,

$$\begin{aligned} E(y) &= \beta_0 + \beta_1 x_1 + \beta_2 x_2 + \beta_3 x_3 \\ &= \beta_0 + \beta_1(0) + \beta_2(0) + \beta_3(0) \\ &= \beta_0 \end{aligned}$$

Therefore, the mean value of y at levels F_1 and B_1, which we represent as μ_{11}, is

$$\mu_{11} = \beta_0 \qquad \blacklozenge$$

EXAMPLE 5.7

Give the values of x_1, x_2, and x_3 and the model for the mean performance, $E(y)$, when using fuel type F_3 in engine brand B_2.

Solution

Checking the coding system, you will see that for levels F_3 and B_2,

$$x_1 = 0 \qquad x_2 = 1 \qquad x_3 = 1$$

Then, the mean performance for fuel F_3 used in engine brand B_2, represented by the symbol μ_{32} (see Table 5.7), is

$$\begin{aligned} \mu_{32} = E(y) &= \beta_0 + \beta_1 x_1 + \beta_2 x_2 + \beta_3 x_3 \\ &= \beta_0 + \beta_1(0) + \beta_2(1) + \beta_3(1) \\ &= \beta_0 + \beta_2 + \beta_3 \end{aligned} \qquad \blacklozenge$$

Note that in the model described previously, we assumed the qualitative independent variables for fuel type and engine brand affect the mean response, $E(y)$, independently of each other. This type of model is called a **main effects model** and is shown in the box. Changing the level of one qualitative variable will have the same effect on $E(y)$ for any level of the second qualitative variable. In other words, the effect of one qualitative variable on $E(y)$ is independent (in a mathematical sense) of the level of the second qualitative variable.

When two independent variables affect the mean response independently of each other, you may obtain the pattern shown in Figure 5.19. Note that the difference in mean performance between any two fuel types (levels of F) is the same, *regardless* of the engine brand used. That is, the main effects model assumes that the relative effect of fuel type on performance is the same in both engine brands.

Main Effects Model with Two Qualitative Independent Variables, One at Three Levels (F_1, F_2, F_3) and the Other at Two Levels (B_1, B_2)

$$E(y) = \beta_0 + \overbrace{\beta_1 x_1 + \beta_2 x_2}^{\substack{\text{Main effect} \\ \text{terms for } F}} + \overbrace{\beta_3 x_3}^{\substack{\text{Main effect} \\ \text{term for } B}}$$

where

$$x_1 = \begin{cases} 1 & \text{if } F_2 \\ 0 & \text{if not} \end{cases} \qquad x_2 = \begin{cases} 1 & \text{if } F_3 \\ 0 & \text{if not} \end{cases} \qquad (F_1 \text{ is base level})$$

$$x_3 = \begin{cases} 1 & \text{if } B_2 \\ 0 & \text{if } B_1 \end{cases} \quad \text{(base level)}$$

Interpretation of Model Parameters

$\beta_0 = \mu_{11}$ (Mean of the combination of base levels)
$\beta_1 = \mu_{2j} - \mu_{1j}$, for any level $B_j (j = 1, 2)$
$\beta_2 = \mu_{3j} - \mu_{1j}$, for any level $B_j (j = 1, 2)$
$\beta_3 = \mu_{i2} - \mu_{i1}$, for any level $F_i (i = 1, 2, 3)$

FIGURE 5.19

Main effects model: Mean response as a function of F and B when F and B affect $E(y)$ independently

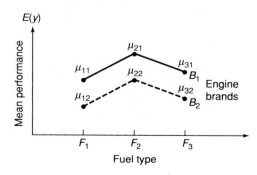

If F and B do not affect $E(y)$ independently of each other, then the response function might appear as shown in Figure 5.20. Note the difference between the

mean response functions for Figures 5.19 and 5.20. When F and B affect the mean response in a dependent manner (Figure 5.20), the response functions differ for each brand. This means that you cannot study the effect of one variable on $E(y)$ without considering the level of the other. When this situation occurs, we say that the qualitative independent variables **interact**. The interaction model is shown in the box. In this example, interaction might be expected if one fuel type tends to perform better in engine B_1, whereas another performs better in engine B_2.

FIGURE 5.20

Interaction model: Mean response as a function of F and B when F and B interact to affect $E(y)$

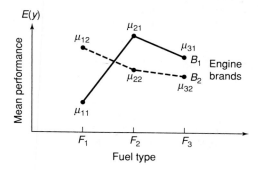

Interaction Model with Two Qualitative Independent Variables, One at Three Levels (F_1, F_2, F_3) and the Other at Two Levels (B_1, B_2)

$$E(y) = \beta_0 + \overbrace{\beta_1 x_1 + \beta_2 x_2}^{\substack{\text{Main effect} \\ \text{terms for } F}} + \overbrace{\beta_3 x_3}^{\substack{\text{Main effect} \\ \text{term for } B}} + \overbrace{\beta_4 x_1 x_3 + \beta_5 x_2 x_3}^{\substack{\text{Interaction} \\ \text{terms}}}$$

where the dummy variables x_1, x_2, and x_3 are defined in the same way as for the main effects model.

Interpretation of Model Parameters

$\beta_0 = \mu_{11}$ (Mean of the combination of base levels)
$\beta_1 = \mu_{21} - \mu_{11}$ (i.e., for base level B_1 only)
$\beta_2 = \mu_{31} - \mu_{11}$ (i.e., for base level B_1 only)
$\beta_3 = \mu_{12} - \mu_{11}$ (i.e., for base level F_1 only)
$\beta_4 = (\mu_{22} - \mu_{12}) - (\mu_{21} - \mu_{11})$
$\beta_5 = (\mu_{32} - \mu_{12}) - (\mu_{31} - \mu_{11})$

When qualitative independent variables interact, the model for $E(y)$ must be constructed so that it is able (if necessary) to give a different mean value, $E(y)$, for every cell in Table 5.7. We do this by adding **interaction terms** to the main effects model. These terms will involve all possible two-way cross-products between each of the two dummy variables for F, x_1, and x_2, and the one dummy variable for B, x_3. The number of interaction terms (for two independent variables) will equal the number of main effect terms for the one variable times the number of main effect terms for the other.

When F and B interact, the model contains six parameters: the two main effect terms for F, one main effect term for B, $(2)(1) = 2$ interaction terms, and β_0. This will make it possible, by assigning the various combinations of values to the dummy variables x_1, x_2, and x_3, to give six different values for $E(y)$ that will correspond to the means of the six cells of Table 5.7.

EXAMPLE 5.8

In Example 5.6, we gave the mean response when fuel F_1 was used in engine B_1, where we assumed that F and B affected $E(y)$ independently (no interaction). Now give the value of $E(y)$ for the model where F and B interact to affect $E(y)$.

Solution

When F and B interact,

$$E(y) = \beta_0 + \beta_1 x_1 + \beta_2 x_2 + \beta_3 x_3 + \beta_4 x_1 x_3 + \beta_5 x_2 x_3$$

For levels F_1 and B_1, we have agreed (according to our system of coding) to let $x_1 = x_2 = x_3 = 0$. Substituting into the equation for $E(y)$, we have

$$E(y) = \beta_0$$

(the same as for the main effects model).   ◆

EXAMPLE 5.9

In Example 5.7, we gave the mean response for fuel type F_3 and brand B_2, when F and B affected $E(y)$ independently. Now assume that F and B interact, and give the value for $E(y)$ when fuel F_3 is used in engine brand B_2.

Solution

When F and B interact,

$$E(y) = \beta_0 + \beta_1 x_1 + \beta_2 x_2 + \beta_3 x_3 + \beta_4 x_1 x_3 + \beta_5 x_2 x_3$$

To find $E(y)$ for F_3 and B_2, we set $x_1 = 0$, $x_2 = 1$, and $x_3 = 1$:

$$E(y) = \beta_0 + \beta_1(0) + \beta_2(1) + \beta_3(1) + \beta_4(0)(1) + \beta_5(1)(1)$$
$$= \beta_0 + \beta_2 + \beta_3 + \beta_5$$

This is the value of μ_{32} in Table 5.7. Note the difference in $E(y)$ for the model assuming independence between F and B versus this model, which assumes interaction between F and B. The difference is β_5.   ◆

EXAMPLE 5.10

The performance, y (measured as mass burning rate per degree of crank angle), for the six combinations of fuel type and engine brand is shown in Table 5.8. The number of test runs per combination varies from one for levels (F_1, B_2) to three for levels (F_1, B_1). A total of twelve test runs are sampled.

a. Assume the interaction between F and B is negligible. Fit the model for $E(y)$ with interaction terms omitted.

b. Fit the complete model for $E(y)$ allowing for the fact that interactions might occur.

🐟 DIESEL

TABLE 5.8 **Performance Data for Combinations of Fuel Type and Diesel Engine Brand**

		BRAND	
		B_1	B_2
	F_1	65	36
		73	
		68	
FUEL TYPE	F_2	78	50
		82	43
	F_3	48	61
		46	62

c. Use the prediction equation for the model, part **a** to estimate the mean engine performance when fuel F_3 is used in brand B_2. Then calculate the sample mean for this cell of Table 5.8. Repeat for the model, part **b**. Explain the discrepancy between the sample mean for levels (F_3, B_2) and the estimate(s) obtained from one or both of the two prediction equations.

Solution

a. A portion of the SAS printout for main effects model

$$E(y) = \beta_0 + \overbrace{\beta_1 x_1 + \beta_2 x_2}^{\substack{\text{Main effect} \\ \text{terms for } F}} + \overbrace{\beta_3 x_3}^{\substack{\text{Main effect} \\ \text{term for } B}}$$

is given in Figure 5.21. The least squares prediction equation is (after rounding):

$$\hat{y} = 64.45 + 6.70x_1 - 2.30x_2 - 15.82x_3$$

b. The SAS printout for the complete model is given in Figure 5.22. Recall that the complete model is

$$E(y) = \beta_0 + \beta_1 x_1 + \beta_2 x_2 + \beta_3 x_3 + \beta_4 x_1 x_3 + \beta_5 x_2 x_3$$

The least squares prediction equation is (after rounding):

$$\hat{y} = 68.67 + 11.33x_1 - 21.67x_2 - 32.67x_3 - .83x_1 x_3 + 47.17x_2 x_3$$

c. To obtain the estimated mean response for cell (F_3, B_2), we let $x_1 = 0, x_2 = 1$, and $x_3 = 1$. Then, for the main effects model, we find

$$\hat{y} = 64.45 + 6.70(0) - 2.30(1) - 15.82(1) = 46.34$$

The 95% confidence interval for the true mean performance (shaded in Figure 5.21) is (27.82, 64.85).

FIGURE 5.21

SAS printout for main effects
model, Example 5.10

```
                              Dependent Variable: PERFORM

                                Analysis of Variance

                                           Mean
                                Sum of     Square    F Value    Pr > F
          Source          DF   Squares

          Model            3   858.25758   286.08586    1.51     0.2838
          Error            8   1512.40909  189.05114
          Corrected Total  11  2370.66667


                    Root MSE          13.74959    R-Square    0.3620
                    Dependent Mean    59.33333    Adj R-Sq    0.1228
                    Coeff Var         23.17346


                              Parameter Estimates

                       Parameter    Standard
          Variable  DF  Estimate     Error     t Value    Pr > |t|

          Intercept  1   64.45455    7.18049     8.98     <.0001
          X1         1    6.70455    9.94093     0.67      0.5190
          X2         1   -2.29545    9.94093    -0.23      0.8232
          X3         1  -15.81818    8.29131    -1.91      0.0928


                              Output Statistics

                    Dep Var   Predicted   Std Error
     Obs  FUELBRND   PERFORM     Value   Mean Predict     95% CL Mean     Residual

      1    F1B1     65.0000    64.4545     7.1805     47.8963   81.0128    0.5455
      2    F1B1     73.0000    64.4545     7.1805     47.8963   81.0128    8.5455
      3    F1B1     68.0000    64.4545     7.1805     47.8963   81.0128    3.5455
      4    F1B2     36.0000    48.6364     9.2700     27.2598   70.0130  -12.6364
      5    F2B1     78.0000    71.1591     8.0280     52.6464   89.6718    6.8409
      6    F2B1     82.0000    71.1591     8.0280     52.6464   89.6718   10.8409
      7    F2B2     50.0000    55.3409     8.0280     36.8282   73.8536   -5.3409
      8    F2B2     43.0000    55.3409     8.0280     36.8282   73.8536  -12.3409
      9    F3B1     48.0000    62.1591     8.0280     43.6464   80.6718  -14.1591
     10    F3B1     46.0000    62.1591     8.0280     43.6464   80.6718  -16.1591
     11    F3B2     61.0000    46.3409     8.0280     27.8282   64.8536   14.6591
     12    F3B2     62.0000    46.3409     8.0280     27.8282   64.8536   15.6591


                    Sum of Residuals                          0
                    Sum of Squared Residuals           1512.40909
                    Predicted Residual SS (PRESS)      3615.37520
```

For the complete model, we find

$$\hat{y} = 68.67 + 11.33(0) - 21.67(1) - 32.67(1) - .83(0)(1) = 47.17(1)(1) = 61.50$$

The 95% confidence interval for true mean performance (shaded in Figure 5.22) is
(55.69, 67.31). The mean for the cell (F_3, B_2) in Table 5.8 is

$$\bar{y}_{32} = \frac{61 + 62}{2} = 61.5$$

which is precisely what is estimated by the complete (interaction) model. However,
the main effects model yields a different estimate, 46.34. The reason for the
discrepancy is that the main effects model assumes the two qualitative independent
variables affect $E(y)$ independently of each other. That is, the change in $E(y)$
produced by a change in levels of one variable is the same regardless of the level
of the other variable. In contrast, the complete model contains six parameters
$(\beta_0, \beta_1, \ldots, \beta_5)$ to describe the six cell populations, so that each population cell
mean will be estimated by its sample mean. Thus, the complete model estimate for
any cell mean is equal to the observed (sample) mean for that cell.   ◆

Example 5.10 demonstrates an important point. If we were to ignore the least
squares analysis and calculate the six sample means of Table 5.8 directly, we would

FIGURE 5.22

SAS printout for interaction model, Example 5.10

```
                              Dependent Variable: PERFORM
                                 Analysis of Variance

                                        Sum of           Mean
     Source                  DF         Squares          Square    F Value   Pr > F

     Model                    5      2303.00000        460.60000    40.84    0.0001
     Error                    6        67.66667         11.27778
     Corrected Total         11      2370.66667

                    Root MSE                 3.35824    R-Square    0.9715
                    Dependent Mean          59.33333    Adj R-Sq    0.9477
                    Coeff Var                5.65996

                              Parameter Estimates

                              Parameter        Standard
     Variable        DF        Estimate           Error    t Value   Pr > |t|

     Intercept        1        68.66667         1.93888     35.42     <.0001
     X1               1        11.33333         3.06564      3.70     0.0101
     X2               1       -21.66667         3.06564     -7.07     0.0004
     X3               1       -32.66667         3.87776     -8.42     0.0002
     X1X3             1        -0.83333         5.12980     -0.16     0.8763
     X2X3             1        47.16667         5.12980      9.19     <.0001

                                Output Statistics

                  Dep Var    Predicted    Std Error
 Obs   FUELBRND    PERFORM        Value  Mean Predict      95% CL Mean        Residual

  1    F1B1       65.0000      68.6667       1.9389   63.9224   73.4109      -3.6667
  2    F1B1       73.0000      68.6667       1.9389   63.9224   73.4109       4.3333
  3    F1B1       68.0000      68.6667       1.9389   63.9224   73.4109      -0.6667
  4    F1B2       36.0000      36.0000       3.3582   27.7827   44.2173      -7.11E-15
  5    F2B1       78.0000      80.0000       2.3746   74.1895   85.8105      -2.0000
  6    F2B1       82.0000      80.0000       2.3746   74.1895   85.8105       2.0000
  7    F2B2       50.0000      46.5000       2.3746   40.6895   52.3105       3.5000
  8    F2B2       43.0000      46.5000       2.3746   40.6895   52.3105      -3.5000
  9    F3B1       48.0000      47.0000       2.3746   41.1895   52.8105       1.0000
 10    F3B1       46.0000      47.0000       2.3746   41.1895   52.8105      -1.0000
 11    F3B2       61.0000      61.5000       2.3746   55.6895   67.3105      -0.5000
 12    F3B2       62.0000      61.5000       2.3746   55.6895   67.3105       0.5000

                    Sum of Residuals                              0
                    Sum of Squared Residuals                67.66667
                    Predicted Residual SS (PRESS)          213.50000
```

obtain estimates of $E(y)$ exactly the same as those obtained by a least squares analysis for the case where the interaction between F and B is assumed to exist. We would not obtain the same estimates if the model assumes that interaction does not exist.

Also, the estimates of means raise important questions. Do the data provide sufficient evidence to indicate that F and B interact? For our example, does the effect of fuel type on diesel engine performance depend on which engine brand is used? The plot of all six sample means, shown in Figure 5.23, seems to indicate interaction, since fuel types F_1 and F_2 appear to operate more effectively in engine brand B_1, whereas the mean performance of F_3 is higher in brand B_2. Can these sample facts be reliably generalized to conclusions about the populations?

To answer this question, we will want to perform a test for interaction between the two qualitative independent variables, fuel type and engine brand. Since allowance for interaction between fuel type and brand in the complete model was provided by the addition of the terms $\beta_4 x_1 x_3$ and $\beta_5 x_2 x_3$, it follows that the null hypothesis that the independent variables fuel type and brand do not interact is equivalent to the hypothesis that the terms $\beta_4 x_1 x_3$ and $\beta_5 x_2 x_3$ are not needed in the model for $E(y)$—or equivalently, that $\beta_4 = \beta_5 = 0$. Conversely, the alternative hypothesis that fuel type and brand do interact is equivalent to stating that at least one of the two parameters, β_4 or β_5, differs from 0.

FIGURE 5.23

Graph of sample means for engine
performance example

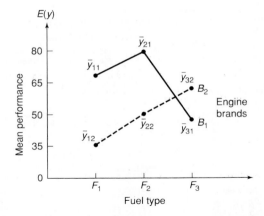

The appropriate procedure for testing a portion of the model parameters, a nested model F test, was discussed in Section 4.13. The F test is carried out as follows:

$H_0: \beta_4 = \beta_5 = 0$

$H_a:$ At least one of β_4 and β_5 differs from 0

Test statistic: $F = \dfrac{(\text{SSE}_R - \text{SSE}_C)/g}{\text{SSE}_C/[n - (k + 1)]}$

where

$$\text{SSE}_R = \text{SSE for reduced model (main effects model)}$$
$$\text{SSE}_C = \text{SSE for complete model (interaction model)}$$
$$g = \text{Number of } \beta\text{'s tested}$$
$$= \text{Numerator df for the } F \text{ statistic}$$
$$n - (k + 1) = \text{df for error for complete model}$$
$$= \text{Denominator df for the } F \text{ statistic}$$

For this example, we have

$$\text{SSE}_R = 1{,}512.41 \text{ (highlighted on Figure 5.21)}$$
$$\text{SSE}_C = 67.67 \text{ (highlighted on Figure 5.22)}$$
$$g = 2 \quad \text{and} \quad n - (k + 1) = 6$$

Then

$$F = \frac{(1{,}512.41 - 67.67)/2}{67.67/6}$$
$$= \frac{722.37}{11.28}$$
$$= 64.05$$

The test statistic, $F = 64.05$, is highlighted on the SAS printout of the analysis, Figure 5.24. The p-value of the test, also highlighted, is less than .0001. Thus, we

FIGURE 5.24

SAS printout for nested model F test of interaction

Test INTERACT Results for Dependent Variable PERFORM

Source	DF	Mean Square	F Value	Pr > F
Numerator	2	722.37121	64.05	<.0001
Denominator	6	11.27778		

are confident (at $\alpha = .05$) in concluding that the interaction terms contribute to the prediction of y, engine performance. Equivalently, there is sufficient evidence to conclude that factors F and B do interact.

EXERCISES

5.19. Refer to the *Journal of Human Stress* study of fire-fighters, Exercise 5.4 (p. 254). Consider using the qualitative variable, level of social support, as a predictor of emotional stress y. Suppose that four social support levels were studied: none, low, moderate, and high.
  a. Write a model for $E(y)$ as a function of social support at four levels.
  b. Interpret the β parameters in the model.
  c. Explain how to test for differences among the emotional stress means for the four social support levels.

5.20. Because of the hot, humid weather conditions in Florida, the growth rates of beef cattle and the milk production of dairy cows typically decline during the summer. However, agricultural and environmental engineers have found that a well-designed shade structure can significantly increase the milk production of dairy cows. In one experiment, 30 cows were selected and divided into three groups of 10 cows each. Group 1 cows were provided with a man-made shade structure, group 2 cows with tree shade, and group 3 cows with no shade. Of interest was the mean milk production (in gallons) of the cows in each group.
  a. Identify the independent variables in the experiment.
  b. Write a model relating the mean milk production, $E(y)$, to the independent variables. Identify and code all dummy variables.
  c. Interpret the β parameters of the model.

5.21. Each semester, the University of Florida's Career Resource Center collects information on the job status and starting salary of graduating seniors. Data recently collected over a two-year period included over 900 seniors who had found employment at the time of graduation. This information was used to model starting salary y as a function of two qualitative independent variables: college at five levels (Business Administration, Engineering, Liberal Arts & Sciences, Journalism, and Nursing) and gender at two levels

(male and female). A main effects model relating starting salary, y, to both college and gender is

$$E(y) = \beta_0 + \beta_1 x_1 + \beta_2 x_2 + \beta_3 x_3 + \beta_4 x_4 + \beta_5 x_5$$

where

$$x_1 = \begin{cases} 1 & \text{if Business Administration} \\ 0 & \text{if not} \end{cases}$$

$$x_2 = \begin{cases} 1 & \text{if Engineering} \\ 0 & \text{if not} \end{cases}$$

$$x_3 = \begin{cases} 1 & \text{if Liberal Arts \& Sciences} \\ 0 & \text{if not} \end{cases}$$

$$x_4 = \begin{cases} 1 & \text{if Journalism} \\ 0 & \text{if not} \end{cases}$$

$$x_5 = \begin{cases} 1 & \text{if female} \\ 0 & \text{if male} \end{cases}$$

  a. Write the equation relating mean starting salary, $E(y)$, to college, for male graduates only.
  b. Interpret β_1 in the model, part **a**.
  c. Interpret β_2 in the model, part **a**.
  d. Interpret β_3 in the model, part **a**.
  e. Interpret β_4 in the model, part **a**.
  f. Write the equation relating mean starting salary, $E(y)$, to college, for female graduates only.
  g. Interpret β_1 in the model, part **f**. Compare to your answer, part **b**.
  h. Interpret β_2 in the model, part **f**. Compare to your answer, part **c**.
  i. Interpret β_3 in the model, part **f**. Compare to your answer, part **d**.
  j. Interpret β_4 in the model, part **f**. Compare to your answer, part **e**.

k. For a given college, interpret the value of β_5 in the model.

l. A multiple regression analysis revealed the following statistics for the β_5 term in the model: $\hat{\beta}_5 = -1,142.17$, $s_{\hat{\beta}_5} = 419.58$, t (for $H_0: \beta_5 = 0) = -2.72$, p-value $= .0066$. Make a statement about whether gender has an effect on average starting salary.

5.22. Refer to Exercise 5.21.

a. Write an interaction model relating starting salary, y, to both college and gender. Use the dummy variables assignments made in Exercise 5.21.

b. Interpret β_1 in the model, part **a**.

c. Interpret β_2 in the model, part **a**.

d. Interpret β_3 in the model, part **a**.

e. Interpret β_4 in the model, part **a**.

f. Interpret β_5 in the model, part **a**.

g. Explain how to test to determine whether the difference between the mean starting salaries of male and female graduates depends on college.

5.23. Refer to *The Archives of Disease in Childhood* (Apr., 2000) study of whether height influences a child's progression through elementary school, Exercise 4.57 (p. 230). Recall that Australian school children were divided into equal thirds (tertiles) based on age (youngest third, middle third, and oldest third). The average heights of the three groups (where all heights measurements were standardized using z-scores), by gender, are repeated in the table.

a. Write a main effects model for the mean standardized height, $E(y)$, as a function of age tertile and gender.

	YOUNGEST TERTILE MEAN HEIGHT	MIDDLE TERTILE MEAN HEIGHT	OLDEST TERTILE MEAN HEIGHT
BOYS	0.33	0.33	0.16
GIRLS	0.27	0.18	0.21

Source: Wake, M., Coghlan, D., & Hesketh, K. "Does height influence progression through primary school grades?" *The Archives of Disease in Childhood*, Vol. 82, Apr. 2000 (Table 2).

b. Interpret the β's in the main effects model, part **a**.

c. Write a model for the mean standardized height, $E(y)$, that includes interaction between age tertile and gender.

d. Use the information in the table to find estimates of the β's in the interaction model, part **c**.

e. How would you test the hypothesis that the difference between the mean standardized heights of boys and girls is the same across all three age tertiles?

5.24. The administration of a midwestern university commissioned a salary equity study to help establish benchmarks for faculty salaries. The administration utilized the following regression model for annual salary, y: $E(y) = \beta_0 + \beta_1 x$, where $x = 0$ if lecturer, 1 if assistant professor, 2 if associate professor, and 3 if full professor. The administration wanted to use the model to compare the mean salaries of professors in the different ranks. Explain the flaw in the model. Propose an alternative model that will achieve the administration's objective.

5.9
Models with Three or More Qualitative Independent Variables

We construct models with three or more qualitative independent variables in the same way that we construct models for two qualitative independent variables, except that we must add three-way interaction terms if we have three qualitative independent variables, three-way and four-way interaction terms for four independent variables, and so on. In this section, we will explain what we mean by three-way, four-way, etc., interactions, and we will demonstrate the procedure for writing the model for any number, say, k, of qualitative independent variables. The pattern used to write the model is shown in the box.

Recall that a two-way interaction term is formed by multiplying the dummy variable associated with one of the main effect terms of one (call it the first) independent variable by the dummy variable from a main effect term of another (the second) independent variable. Three-way interaction terms are formed in a similar way, by forming the product of three dummy variables, one from a main effect term from each of the three independent variables. Similarly, four-way interaction terms are formed by taking the product of four dummy variables, one from a main effect term from each of four independent variables. We will illustrate with three examples.

> **Pattern of the Model Relating $E(y)$ to k Qualitative Independent Variables**
>
> $E(y) = \beta_0 +$ Main effect terms for all independent variables
>
> $+$ All two-way interaction terms between pairs of independent variables
>
> $+$ All three-way interaction terms between different groups of three independent variables
>
> $+$
>
> $\vdots$
>
> $+$ All k-way interaction terms for the k independent variables

EXAMPLE 5.11

Suppose you have three qualitative independent variables, the first at three levels, A_1, A_2, and A_3, the second at three levels, B_1, B_2, and B_3, and the third at two levels, C_1 and C_2. Write a model for $E(y)$ that includes all main effect and interaction terms for the independent variables.

Solution

First write a model containing the main effect terms for the three variables:

$$E(y) = \beta_0 + \overbrace{\beta_1 x_1 + \beta_2 x_2}^{\substack{\text{Main effect}\\\text{terms for } A}} + \overbrace{\beta_3 x_3 + \beta_4 x_4}^{\substack{\text{Main effect}\\\text{terms for } B}} + \overbrace{\beta_5 x_5}^{\substack{\text{Main effect}\\\text{term for } C}}$$

where

$$x_1 = \begin{cases} 1 & \text{if level } A_2 \\ 0 & \text{if not} \end{cases} \qquad x_3 = \begin{cases} 1 & \text{if level } B_2 \\ 0 & \text{if not} \end{cases} \qquad x_5 = \begin{cases} 1 & \text{if level } C_2 \\ 0 & \text{if level } C_1 \end{cases}$$

$$x_2 = \begin{cases} 1 & \text{if level } A_3 \\ 0 & \text{if not} \end{cases} \qquad x_4 = \begin{cases} 1 & \text{if level } B_3 \\ 0 & \text{if not} \end{cases}$$

The next step is to add two-way interaction terms. These will be of three types—those for the interaction between A and B, between A and C, and between B and C. Thus,

$$E(y) = \beta_0 + \overbrace{\beta_1 x_1 + \beta_2 x_2}^{\substack{\text{Main effect}\\A}} + \overbrace{\beta_3 x_3 + \beta_4 x_4}^{\substack{\text{Main effect}\\B}} + \overbrace{\beta_5 x_5}^{\substack{\text{Main effect}\\C}}$$

$$+ \overbrace{\beta_6 x_1 x_3 + \beta_7 x_1 x_4 + \beta_8 x_2 x_3 + \beta_9 x_2 x_4}^{AB \text{ interaction terms}}$$

$$\overbrace{\phantom{AC \text{ interaction terms}}}^{AC \text{ interaction terms}}$$

$$+ \quad \beta_{10}x_1x_5 + \beta_{11}x_2x_5$$

$$\overbrace{\phantom{BC \text{ interaction terms}}}^{BC \text{ interaction terms}}$$

$$+ \quad \beta_{12}x_3x_5 + \beta_{13}x_4x_5$$

Finally, since there are three independent variables, we must include terms for the interaction of A, B, and C. These terms are formed as the products of dummy variables, one from each of the A, B, and C main effect terms. The complete model for $E(y)$ is

$$E(y) = \beta_0 + \overbrace{\beta_1 x_1 + \beta_2 x_2}^{\substack{\text{Main effect} \\ A}} + \overbrace{\beta_3 x_3 + \beta_4 x_4}^{\substack{\text{Main effect} \\ B}} + \overbrace{\beta_5 x_5}^{\substack{\text{Main effect} \\ C}}$$

$$+ \overbrace{\beta_6 x_1 x_3 + \beta_7 x_1 x_4 + \beta_8 x_2 x_3 + \beta_9 x_2 x_4}^{AB \text{ interaction terms}}$$

$$+ \overbrace{\beta_{10} x_1 x_5 + \beta_{11} x_2 x_5}^{AC \text{ interaction terms}}$$

$$+ \overbrace{\beta_{12} x_3 x_5 + \beta_{13} x_4 x_5}^{BC \text{ interaction terms}}$$

$$+ \overbrace{\beta_{14} x_1 x_3 x_5 + \beta_{15} x_1 x_4 x_5 + \beta_{16} x_2 x_3 x_5 + \beta_{17} x_2 x_4 x_5}^{\text{Three-way } ABC \text{ interaction terms}}$$

Note that the complete model in Example 5.11 contains 18 parameters, one for each of the $3 \times 3 \times 2$ combinations of levels for A, B, and C. There are 18 linearly independent linear combinations of these parameters, *one corresponding to each of the means of the $3 \times 3 \times 2$ combinations of levels of A, B, and C.* We illustrate with an example. ◆

EXAMPLE   5.12

Refer to Example 5.11 and give the expression for the mean value of y for observations taken at the second level of A, the first level of B, and the second level of C, i.e., at (A_2, B_1, C_2).

Solution

Check the coding for the dummy variables (given in Example 5.11) and you will see that they assume the following values:

For level A_2: $x_1 = 1$, $x_2 = 0$
For level B_1: $x_3 = 0$, $x_4 = 0$
For level C_2: $x_5 = 1$

Substituting these values into the expression for $E(y)$, we obtain

$$E(y) = \beta_0 + \beta_1(1) + \beta_2(0) + \beta_3(0) + \beta_4(0) + \beta_5(1)$$

$$+ \beta_6(1)(0) + \beta_7(1)(0) + \beta_8(0)(0) + \beta_9(0)(0)$$

$$+ \beta_{10}(1)(1) + \beta_{11}(0)(1) + \beta_{12}(0)(1) + \beta_{13}(0)(1)$$
$$+ \beta_{14}(1)(0)(1) + \beta_{15}(1)(0)(1) + \beta_{16}(0)(0)(1) + \beta_{17}(0)(0)(1)$$
$$= \beta_0 + \beta_1 + \beta_5 + \beta_{10}$$

Thus, the mean value of y observed at levels A_2, B_1, and C_2 is $\beta_0 + \beta_1 + \beta_5 + \beta_{10}$. You could find the mean values of y for the other 17 combinations of levels of A, B, and C by substituting the appropriate values of the dummy variables into the expression for $E(y)$ in the same manner. Each of the 18 means is a unique linear combination of the 18 β parameters in the model. ◆

EXAMPLE 5.13

Suppose you want to test the hypothesis that the three qualitative independent variables discussed in Example 5.11 do not interact, i.e., the hypothesis that the effect of any one of the variables on $E(y)$ is independent of the level settings of the other two variables. Formulate the appropriate test of hypothesis about the model parameters.

Solution

No interaction among the three qualitative independent variables implies that the main effects model,

$$E(y) = \beta_0 + \overbrace{\beta_1 x_1 + \beta_2 x_2}^{\substack{\text{Main effects} \\ A}} + \overbrace{\beta_3 x_3 + \beta_4 x_4}^{\substack{\text{Main effects} \\ B}} + \overbrace{\beta_5 x_5}^{\substack{\text{Main effect} \\ C}}$$

is appropriate for modeling $E(y)$ or, equivalently, that all interaction terms should be excluded from the model. This situation will occur if

$$\beta_6 = \beta_7 = \cdots = \beta_{17} = 0$$

Consequently, we will test the null hypothesis

$$H_0: \beta_6 = \beta_7 = \cdots = \beta_{17} = 0$$

against the alternative hypothesis that at least one of these β parameters differs from 0, or equivalently, that some interaction among the independent variables exists. This statistical test was described in Section 4.13. ◆

Of what value is this section? If you are modeling a response, say, profit of a corporation, and you believe that several qualitative independent variables affect the response, then you must know how to enter these variables into your model. You must understand the implication of the interaction (or lack of it) among a subset of independent variables and how to write the appropriate terms in the model to account for it. Failure to write a good model for your response will usually lead to inflated values of the SSE and s^2 (with a consequent loss of information), and it also can lead to biased estimates of $E(y)$ and biased predictions of y.

5.10
Models with Both
Quantitative
and Qualitative
Independent
Variables

Perhaps the most interesting data analysis problems are those that involve both quantitative and qualitative independent variables. For example, suppose mean performance of a diesel engine is a function of one qualitative independent variable, fuel type at levels F_1, F_2, and F_3 and one quantitative independent variable, engine speed in revolutions per minute (rpm). We will proceed to build a model in stages, showing graphically the interpretation that we would give to the model at each stage. This will help you see the contribution of various terms in the model.

At first we assume that the qualitative independent variable has no effect on the response (i.e., the mean contribution to the response is the same for all three fuel types), but the mean performance, $E(y)$, is related to engine speed. In this case, one response curve, which might appear as shown in Figure 5.25, would be sufficient to characterize $E(y)$ for all three fuel types. The following second-order model would likely provide a good approximation to $E(y)$:

$$E(y) = \beta_0 + \beta_1 x_1 + \beta_2 x_1^2$$

where x_1 is speed in rpm. This model has some distinct disadvantages. If differences in mean performance exist for the three fuel types, they cannot be detected (because the model does not contain any parameters representing differences among fuel types). Also, the differences would inflate the SSE associated with the fitted model and consequently would increase errors of estimation and prediction.

FIGURE 5.25

Model for $E(y)$ as a function of engine speed

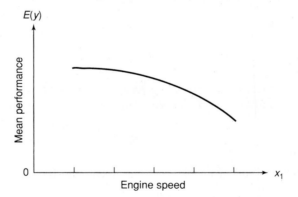

The next stage in developing a model for $E(y)$ is to assume that the qualitative independent variable, fuel type, does affect mean performance, but the effect on $E(y)$ is independent of speed. In other words, the assumption is that the two independent variables do not interact. This model is obtained by adding main effect terms for fuel type to the second-order model we used in the first stage. Therefore, using the methods of Sections 5.7 and 5.8, we choose F_1 as the base level and add two terms to the model corresponding to levels F_2 and F_3:

$$E(y) = \beta_0 + \beta_1 x_1 + \beta_2 x_1^2 + \beta_3 x_2 + \beta_4 x_3$$

where

$$x_1 = \text{Engine speed} \quad x_2 = \begin{cases} 1 & \text{if } F_2 \\ 0 & \text{if not} \end{cases} \quad x_3 = \begin{cases} 1 & \text{if } F_3 \\ 0 & \text{if not} \end{cases}$$

What effect do these terms have on the graph for the response curve(s)? Suppose we want to model $E(y)$ for level F_1. Then we let $x_2 = 0$ and $x_3 = 0$. Substituting into the model equation, we have

$$E(y) = \beta_0 + \beta_1 x_1 + \beta_2 x_1^2 + \beta_3(0) + \beta_4(0)$$
$$= \beta_0 + \beta_1 x_1 + \beta_2 x_1^2$$

which would graph as a second-order curve similar to the one shown in Figure 5.25.

Now suppose that we use one of the other two fuel types, for example, F_2. Then $x_2 = 1$, $x_3 = 0$, and

$$E(y) = \beta_0 + \beta_1 x_1 + \beta_2 x_1^2 + \beta_3(1) + \beta_4(0)$$
$$= (\beta_0 + \beta_3) + \beta_1 x_1 + \beta_2 x_1^2$$

This is the equation of exactly the same parabola that we obtained for fuel type F_1 except that the y-intercept has changed from β_0 to $(\beta_0 + \beta_3)$. Similarly, the response curve for F_3 is

$$E(y) = (\beta_0 + \beta_4) + \beta_1 x_1 + \beta_2 x_1^2$$

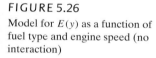

FIGURE 5.26

Model for $E(y)$ as a function of fuel type and engine speed (no interaction)

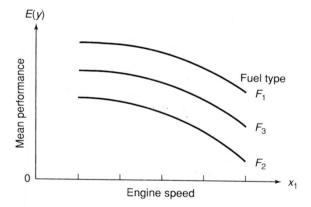

Therefore, the three response curves for levels F_1, F_2, and F_3 (shown in Figure 5.26) are identical except that they are shifted vertically upward or downward in relation to each other. The curves depict the situation when the two independent variables do not interact, i.e., the effect of speed on mean performance is the same regardless of the fuel type used, and the effect of fuel type on mean performance is the same for all speeds (the relative distances between the curves is constant).

This noninteractive second-stage model has drawbacks similar to those of the simple first-stage model. It is highly unlikely that the response curves for the three fuel types would be identical except for differing y-intercepts. Because the model does not contain parameters that measure interaction between engine speed and fuel type, we cannot test to see whether a relationship exists. Also, if interaction does exist, it will cause the SSE for the fitted model to be inflated and will consequently increase the errors of estimating model parameters $E(y)$.

This leads us to the final stage of the model-building process—adding interaction terms to allow the three response curves to differ in shape:

$$E(y) = \beta_0 + \overbrace{\beta_1 x_1 + \beta_2 x_1^2}^{\substack{\text{Main effect} \\ \text{terms for} \\ \text{engine speed}}} + \overbrace{\beta_3 x_2 + \beta_4 x_3}^{\substack{\text{Main effect} \\ \text{terms for} \\ \text{fuel type}}}$$

$$\underbrace{+ \beta_5 x_1 x_2 + \beta_6 x_1 x_3 + \beta_7 x_1^2 x_2 + \beta_8 x_1^2 x_3}_{\text{Interaction terms}}$$

where

$$x_1 = \text{Engine speed} \quad x_2 = \begin{cases} 1 & \text{if } F_2 \\ 0 & \text{if not} \end{cases} \quad x_3 = \begin{cases} 1 & \text{if } F_3 \\ 0 & \text{if not} \end{cases}$$

Notice that this model graphs as three different second-order curves.* If fuel type F_1 is used, we substitute $x_2 = x_3 = 0$ into the formula for $E(y)$, and all but the first three terms equal 0. The result is

$$E(y) = \beta_0 + \beta_1 x_1 + \beta_2 x_1^2$$

If F_2 is used, $x_2 = 1$, $x_3 = 0$, and

$$E(y) = \beta_0 + \beta_1 x_1 + \beta_2 x_1^2 + \beta_3(1) + \beta_4(0)$$
$$+ \beta_5 x_1(1) + \beta_6 x_1(0) + \beta_7 x_1^2(1) + \beta_8 x_1^2(0)$$
$$= (\beta_0 + \beta_3) + (\beta_1 + \beta_5)x_1 + (\beta_2 + \beta_7)x_1^2$$

The y-intercept, the coefficient of x_1, and the coefficient of x_1^2 differ from the corresponding coefficients in $E(y)$ at level F_1. Finally, when F_3 is used, $x_2 = 0$, $x_3 = 1$, and the result is

$$E(y) = (\beta_0 + \beta_4) + (\beta_1 + \beta_6)x_1 + (\beta_2 + \beta_8)x_1^2$$

A graph of the model for $E(y)$ might appear as shown in Figure 5.27. Compare this figure with Figure 5.25, where we assumed the response curves were identical for all three fuel types, and with Figure 5.26, where we assumed no interaction between the independent variables. Note in Figure 5.27 that the second-order curves may be completely different.

Now that you know how to write a model for two independent variables—one qualitative and one quantitative—we ask a question. Why do it? Why not write a separate second-order model for each level of fuel type where $E(y)$ is a function of engine speed only? *One reason we write the single model representing all three*

---

*Note that the model remains a second-order model for the quantitative independent variable x_1. The terms involving $x_1^2 x_2$ and $x_1^2 x_3$ appear to be third-order terms, but they are not because x_2 and x_3 are dummy variables.

FIGURE 5.27

Graph of $E(y)$ as a function of fuel type and engine speed (interaction)

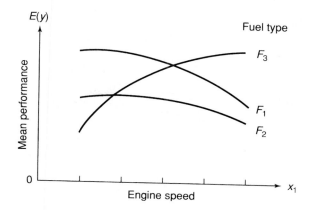

response curves is so that we can test to determine whether the curves are different. For example, we might want to know whether the effect of fuel type depends on engine speed. Thus, one fuel type might be especially efficient at low engine speeds, but less so at high speeds. The reverse might be true for one of the other two fuel types. The hypothesis that the independent variables, fuel type and engine speed, affect the response independently of one another (a case of no interaction) is equivalent to testing the hypothesis that $\beta_5 = \beta_6 = \beta_7 = \beta_8 = 0$ [i.e., that the model in Figure 5.26 adequately characterizes $E(y)$] using the partial F test discussed in Section 4.13. *A second reason for writing a single model is that we obtain a pooled estimate of σ^2, the variance of the random error component ε.* If the variance of ε is truly the same for each fuel type, the pooled estimate is superior to calculating three estimates by fitting a separate model for each fuel type.

In conclusion, suppose you want to write a model relating $E(y)$ to several quantitative and qualitative independent variables. Proceed in exactly the same manner as for two independent variables, one qualitative and one quantitative. First, write the model (using the methods of Sections 5.4 and 5.5) that you want to use to describe the quantitative independent variables. Then introduce the main effect and interaction terms for the qualitative independent variables. This gives a model that represents a set of identically shaped response surfaces, one corresponding to each combination of levels of the qualitative independent variables. If you could visualize surfaces in multidimensional space, their appearance would be analogous to the response curves of Figure 5.26. To complete the model, add terms for the interaction between the quantitative and qualitative variables. This is done by interacting *each* qualitative variable term with *every* quantitative variable term. We will demonstrate with an example.

EXAMPLE 5.14

A marine biologist wished to investigate the effects of three factors on the level of the contaminant DDT found in fish inhabiting a polluted lake. The factors were

1. Species of fish (two levels)
2. Location of capture (two levels)
3. Fish length (centimeters)

Write a model for the DDT level, y, found in contaminated fish.

FIGURE 5.28

A graphical portrayal of three
factors—two qualitative and one
quantitative—on DDT level

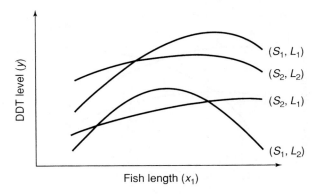

Solution

The response y is affected by two qualitative factors (species and location), each at two levels, and one quantitative factor (length). Fish of each of the two species, S_1 and S_2, could be captured at each of the two locations, L_1 and L_2, giving $2 \times 2 = 4$ possible combinations—call them (S_1, L_1), (S_1, L_2), (S_2, L_1), (S_2, L_2). For each of these combinations, you would obtain a curve that graphs DDT level as a function of the quantitative factor x_1, fish length (see Figure 5.28). The stages in writing the model for the response y shown in Figure 5.28 are listed here.

STAGE 1   *Write a model relating y to the quantitative factor(s)*. It is likely that an increase in the value of the single quantitative factor x_1, length, will yield an increase in DDT level. However, this increase is likely to be slower for larger fish, and will eventually level off once a fish reaches a certain length, thus producing the curvature shown in Figure 5.28. Consequently, we will model the mean DDT level, $E(y)$, with the second-order model.

$$E(y) = \beta_0 + \beta_1 x_1 + \beta_2 x_1^2$$

This is the model we would use if we were certain that the DDT curves were identical for all species–location combinations (S_i, L_j). The model would appear as shown in Figure 5.29a.

FIGURE 5.29

DDT curves for stages 1 and 2

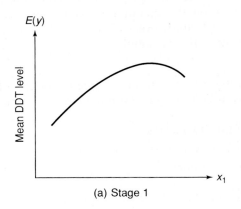

(a) Stage 1

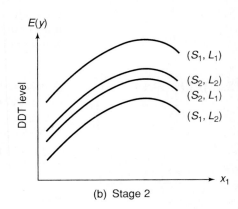

(b) Stage 2

STAGE 2 *Add the terms, both main effect and interaction, for the qualitative factors.*

$$E(y) = \beta_0 + \overbrace{\beta_1 x_1 + \beta_2 x_1^2}^{\text{Terms for quantitative factor}}$$

$$+ \overbrace{\beta_3 x_2}^{\substack{\text{Main effect} \\ S}} + \overbrace{\beta_4 x_3}^{\substack{\text{Main effect} \\ L}} + \overbrace{\beta_5 x_2 x_3}^{SL \text{ interaction}}$$

where

$$x_2 = \begin{cases} 1 & \text{if species } S_2 \\ 0 & \text{if species } S_1 \end{cases} \qquad x_3 = \begin{cases} 1 & \text{if location } L_2 \\ 0 & \text{if location } L_1 \end{cases}$$

This model implies that the DDT curves are identically shaped for each of the (S_i, L_j) combinations but that they possess different y-intercepts, as shown in Figure 5.29b.

STAGE 3 *Add terms to allow for interaction between the quantitative and qualitative factors.* This is done by interacting every pair of terms—one quantitative and one qualitative. Thus, the complete model, which graphs as four different-shaped second-order curves (see Figure 5.28), is

$$E(y) = \overbrace{\beta_0 + \beta_1 x_1 + \beta_2 x_1^2}^{\text{First-stage terms}}$$

$$+ \overbrace{\beta_3 x_2 + \beta_4 x_3 + \beta_5 x_2 x_3}^{\text{Second-stage terms}}$$

$$+ \overbrace{\beta_6 x_1 x_2 + \beta_7 x_1 x_3 + \beta_8 x_1 x_2 x_3 + \beta_9 x_1^2 x_2 + \beta_{10} x_1^2 x_3 + \beta_{11} x_1^2 x_2 x_3}^{\text{Third-stage terms}} \qquad \blacklozenge$$

EXAMPLE 5.15

Use the model of Example 5.14 to find the equation relating $E(y)$ to x_1 for species S_1 and location L_2.

Solution

Checking the coding for the model, we see (noted at the second stage) that when DDT level y is measured on a fish of species S_1 captured at location L_2, we set $x_2 = 0$ and $x_3 = 1$. Substituting these values into the complete model, we obtain

$$E(y) = \beta_0 + \beta_1 x_1 + \beta_2 x_1^2$$

$$+ \beta_3 x_2 + \beta_4 x_3 + \beta_5 x_2 x_3$$

$$+ \beta_6 x_1 x_2 + \beta_7 x_1 x_3 + \beta_8 x_1 x_2 x_3 + \beta_9 x_1^2 x_2 + \beta_{10} x_1^2 x_3 + \beta_{11} x_1^2 x_2 x_3$$

$$= \beta_0 + \beta_1 x_1 + \beta_2 x_1^2$$

$$+ \beta_3(0) + \beta_4(1) + \beta_5(0)(1)$$

$$+ \beta_6 x_1(0) + \beta_7 x_1(1) + \beta_8 x_1(0)(1)$$
$$+ \beta_9 x_1^2(0) + \beta_{10} x_1^2(1) + \beta_{11} x_1^2(0)(1)$$
$$= (\beta_0 + \beta_4) + (\beta_1 + \beta_7) x_1 + (\beta_2 + \beta_{10}) x_1^2$$

Note that this equation graphs as a portion of a parabola with y-intercept equal to $(\beta_0 + \beta_4)$. The coefficient of x_1 is $(\beta_1 + \beta_7)$, and the curvature coefficient (the coefficient of x_1^2) is $(\beta_2 + \beta_{10})$.   ◆

EXAMPLE   5.16

Suppose you have two qualitative independent variables, A and B, and A is at two levels and B is at three levels. You also have two quantitative independent variables, C and D, each at three levels. Further suppose you plan to fit a second-order response surface as a function of the quantitative independent variables C and D, and that you want your model for $E(y)$ to allow for different shapes of the second-order surfaces for the six (2×3) combinations of levels corresponding to the qualitative independent variables A and B. Write a model for $E(y)$.

Solution

STAGE 1   *Write the second-order model corresponding to the two quantitative independent variables.* If we let

$x_1 =$ Level for independent variable C
$x_2 =$ Level for independent variable D

then

$$E(y) = \beta_0 + \beta_1 x_1 + \beta_2 x_2 + \beta_3 x_1 x_2 + \beta_4 x_1^2 + \beta_5 x_2^2$$

This is the model you would use if you believed that the six response surfaces, corresponding to the six combinations of levels of A and B, were identical.

STAGE 2   *Add the main effect and interaction terms for the qualitative independent variables.* These are

$$+ \underbrace{\beta_6 x_6}_{\substack{\text{Main effect} \\ \text{term for } A}} + \underbrace{\beta_7 x_7 + \beta_8 x_8}_{\substack{\text{Main effect} \\ \text{terms for } B}} + \underbrace{\beta_9 x_6 x_7 + \beta_{10} x_6 x_8}_{\substack{AB \text{ interaction} \\ \text{terms}}}$$

where $x_6 = \begin{cases} 1 & \text{if at level } A_2 \\ 0 & \text{if not} \end{cases}$ $x_7 = \begin{cases} 1 & \text{if at level } B_2 \\ 0 & \text{if not} \end{cases}$ $x_8 = \begin{cases} 1 & \text{if at level } B_3 \\ 0 & \text{if not} \end{cases}$

The addition of these terms to the model produces six identically shaped second-order surfaces, one corresponding to each of the six combinations of levels of A and B. They differ only in their y-intercepts.

STAGE 3   *Add terms that allow for interaction between the quantitative and qualitative independent variables.* This is done by interacting each of the five qualitative independent variable terms (both main effect and interaction) with each term (except β_0) of the quantitative first-stage model. Thus,

$$E(y) = \beta_0 + \beta_1 x_1 + \beta_2 x_2 + \beta_3 x_1 x_2 + \beta_4 x_1^2 + \beta_5 x_2^2 \qquad \left.\begin{array}{l} \text{First-stage} \\ \text{model} \end{array}\right\}$$

$$
+ \underbrace{\beta_6 x_6}_{\substack{\text{Main effect} \\ A}} + \underbrace{\beta_7 x_7 + \beta_8 x_8}_{\substack{\text{Main effect} \\ B}} + \underbrace{\beta_9 x_6 x_7 + \beta_{10} x_6 x_8}_{AB \text{ interaction}}
\qquad \left.\begin{array}{l} \text{Portion} \\ \text{added to} \\ \text{form second-} \\ \text{stage model} \end{array}\right\}
$$

$$
+ \beta_{11} x_6 x_1 + \beta_{12} x_6 x_2 + \beta_{13} x_6 x_1 x_2 + \beta_{14} x_6 x_1^2 + \beta_{15} x_6 x_2^2
\qquad \left.\begin{array}{l} \text{Interacting } x_6 \\ \text{with the} \\ \text{quantitative} \\ \text{terms} \end{array}\right\}
$$

$$
+ \beta_{16} x_7 x_1 + \beta_{17} x_7 x_2 + \beta_{18} x_7 x_1 x_2 + \beta_{19} x_7 x_1^2 + \beta_{20} x_7 x_2^2
\qquad \left.\begin{array}{l} \text{Interacting } x_7 \\ \text{with the} \\ \text{quantitative} \\ \text{terms} \end{array}\right\}
$$

$$+ \cdots \qquad\qquad\qquad\qquad\qquad\qquad\qquad\qquad\qquad \cdots$$

$$
\begin{aligned}
&+ \beta_{31} x_6 x_8 x_1 + \beta_{32} x_6 x_8 x_2 + \beta_{33} x_6 x_8 x_1 x_2 \\
&+ \beta_{34} x_6 x_8 x_1^2 + \beta_{35} x_6 x_8 x_2^2
\end{aligned}
\qquad \left.\begin{array}{l} \text{Interacting} \\ x_6 x_8 \text{ with the} \\ \text{quantitative} \\ \text{terms} \end{array}\right\}
$$

Note that the complete model contains 36 terms, one for β_0, five needed to complete the second-order model in the two quantitative variables, five for the two qualitative variables, and $5 \times 5 = 25$ terms for the interactions between the quantitative and qualitative variables. ◆

To see how the model gives different second-order surfaces—one for each combination of the levels of variables A and B—consider the next example.

EXAMPLE   5.17

Refer to Example 5.16. Find the response surface that portrays $E(y)$ as a function of the two quantitative independent variables C and D for the (A_1, B_2) combination of levels of the qualitative independent variables.

Solution

Checking the coding, we see that when y is observed at the first level of A (level A_1) and the second level of B (level B_2), the dummy variables take the following values: $x_6 = 0$, $x_7 = 1$, $x_8 = 0$. Substituting these values into the formula for the complete model (and deleting the terms that equal 0), we obtain

$$E(y) = \beta_0 + \beta_1 x_1 + \beta_2 x_2 + \beta_3 x_1 x_2 + \beta_4 x_1^2 + \beta_5 x_2^2 + \beta_7 + \beta_{16} x_1 + \beta_{17} x_2$$
$$+ \beta_{18} x_1 x_2 + \beta_{19} x_1^2 + \beta_{20} x_2^2$$
$$= (\beta_0 + \beta_7) + (\beta_1 + \beta_{16}) x_1 + (\beta_2 + \beta_{17}) x_2 + (\beta_3 + \beta_{18}) x_1 x_2$$
$$+ (\beta_4 + \beta_{19}) x_1^2 + (\beta_5 + \beta_{20}) x_2^2$$

Note that this is the equation of a second-order model for $E(y)$. It graphs the response surface for $E(y)$ when the qualitative independent variables A and B are at levels A_1 and B_2. ◆

EXERCISES

5.25. Refer to the *Chance* (Winter 2000) study of winning Boston Marathon times, Exercise 5.1 (p. 254). The independent variables used to model winning time y are:

x_1 = year in which race is run (expressed as number of years since 1880)

x_2 = {1 if winning runner is male, 0 if female}

a. Write a first-order, main effects model for $E(y)$ as a function of year and gender.

b. Interpret the β parameters of the model, part a.

c. Write a model for $E(y)$ as a function of year and gender that hypothesizes different winning time-year slopes for male and female runners. Sketch the model.

5.26. Refer to Exercise 5.25. Consider adding the following independent quantitative variable to the model of winning time y:

x_3 = number of marathons run prior to the Boston Marathon during the year

a. Write a complete second-order model that relates $E(y)$ to x_1 and x_3.

b. Add the main effect term for gender to the model of part **a**.

c. Add terms to the model of part **b** to allow for interaction between quantitative and qualitative terms.

d. Under what circumstances will the response curves of the model of part **c** possess the same shape but have different y-intercepts when x_1 is held constant?

e. Under what circumstances will the response curves of the model of part **c** be parallel lines when x_1 is held constant?

f. Under what circumstances will the response curves of the model of part **c** be identical when x_1 is held constant?

5.27. An experiment was conducted to evaluate the performances of a diesel engine run on synthetic (coal-derived) and petroleum-derived fuel oil (*Journal of Energy Resources Technology*, Mar. 1990). The petroleum-derived fuel used was a number 2 diesel fuel (DF-2) obtained from Phillips Chemical Company. Two synthetic fuels were used: a blended fuel (50% coal-derived and 50% DF-2) and a blended fuel with advanced timing. The brake power (kilowatts) and fuel type were varied in test runs, and engine performance was measured. The following table gives the experimental results for the performance measure, mass burning rate per degree of crank angle.

⬭ SYNFUELS

BRAKE POWER, x_1	FUEL TYPE	MASS BURNING RATE, y
4	DF-2	13.2
4	Blended	17.5
4	Advanced Timing	17.5
6	DF-2	26.1
6	Blended	32.7
6	Advanced Timing	43.5
8	DF-2	25.9
8	Blended	46.3
8	Advanced Timing	45.6
10	DF-2	30.7
10	Blended	50.8
10	Advanced Timing	68.9
12	DF-2	32.3
12	Blended	57.1

Source: Litzinger, T. A., and Buzza, T. G. "Performance and emissions of a diesel engine using a coal-derived fuel." *Journal of Energy Resources Technology*, Vol. 112, Mar. 1990, p. 32, Table 3.

SAS Output for Exercise 5.27

Dependent Variable: BURNRATE

Analysis of Variance

Source	DF	Sum of Squares	Mean Square	F Value	Pr > F
Model	5	3253.97929	650.79586	25.65	<.0001
Error	8	203.01000	25.37625		
Corrected Total	13	3456.98929			

Root MSE		5.03748	R-Square	0.9413	
Dependent Mean		36.29286	Adj R-Sq	0.9046	
Coeff Var		13.88010			

Parameter Estimates

Variable	DF	Parameter Estimate	Standard Error	t Value	Pr > |t|
Intercept	1	-10.83000	8.27743	-1.31	0.2271
POWER	1	7.81500	1.12642	6.94	0.0001
X2	1	19.35000	10.68612	1.81	0.1078
X3	1	12.79000	10.68612	1.20	0.2656
POWERX2	1	-5.67500	1.37957	-4.11	0.0034
POWERX3	1	-2.95000	1.37957	-2.14	0.0649

Test INTERACT Results for Dependent Variable BURNRATE

Source	DF	Mean Square	F Value	Pr > F
Numerator	2	223.03750	8.79	0.0096
Denominator	8	25.37625		

a. The researchers fit the interaction model

$$E(y) = \beta_0 + \beta_1 x_1 + \beta_2 x_2 + \beta_3 x_3 + \beta_4 x_1 x_2 + \beta_5 x_1 x_3$$

where

$$y = \text{Mass burning rate}$$

$$x_1 = \text{Brake power (kW)}$$

$$x_2 = \begin{cases} 1 & \text{if DF-2 fuel} \\ 0 & \text{if not} \end{cases}$$

$$x_3 = \begin{cases} 1 & \text{if blended fuel} \\ 0 & \text{if not} \end{cases}$$

The results shown in the SAS printout above.

a. Conduct a test to determine whether brake power and fuel type interact. Test using $\alpha = .01$.

b. Refer to the model, part **a**. Give the estimates of the slope of the y–x_1 line for each of the three fuel types.

5.28. Refer to the *Chance* (Winter 2001) study of students who paid a private tutor (or coach) to help them improve their Standardized Admission Test (SAT)

scores, Exercise 4.48 (p. 226). Recall that multiple regression was used to estimate the effect of coaching on SAT-Mathematics scores, where

$$y = \text{SAT-Math score}$$

$$x_1 = \text{score on PSAT}$$

$$x_2 = \{1 \text{ if student was coached, 0 if not}\}.$$

a. Write a complete second-order model for $E(y)$ as a function of x_1 and x_2.

b. Give the equation of the curve relating $E(y)$ to x_1 for noncoached students. Identify the y-intercept, shift parameter, and rate of curvature in the equation.

c. Repeat part **b** for students who have been coached on the SAT.

d. How would you test to determine if coaching has an effect on SAT-Math scores?

5.29. A study of the atmospheric pollution on the slopes of the Blue Ridge Mountains (Tennessee) was conducted. The file LEADMOSS contains the levels of lead found in 70 fern moss specimens (in micrograms of lead per gram of moss tissue) collected from the mountain slopes, as well as the elevation of the moss specimen (in feet) and the direction (1 if east, 0 if west) of the slope face. The first five and last five observations of the data set are listed in the table.

🌊 LEADMOSS

SPECIMEN	LEAD LEVEL	ELEVATION	SLOPE FACE
1	3.475	2000	0
2	3.359	2000	0
3	3.877	2000	0
4	4.000	2500	0
5	3.618	2500	0
⋮	⋮	⋮	⋮
66	5.413	2500	1
67	7.181	2500	1
68	6.589	2500	1
69	6.182	2000	1
70	3.706	2000	1

Source: Schilling, J. "Bioindication of atmospheric heavy metal deposition in the Blue Ridge using the moss, *Thuidium delicatulum*." Master of Science Thesis, Spring 2000.

a. Write the equation of a first-order model relating mean lead level, $E(y)$, to elevation (x_1) and slope face (x_2). Include interaction between elevation and slope face in the model.

b. Graph the relationship between mean lead level and elevation for the different slope faces that is hypothesized by the model, part **a**.

c. In terms of the β's of the model, part **a**, give the change in lead level for every one foot increase in elevation for moss specimens on the east slope.

d. Fit the model, part **a**, to the data using an available statistical software package. Is the overall model statistically useful for predicting lead level? Test using $\alpha = .10$.

e. Write the equation of the complete second order model relating mean lead level, $E(y)$, to elevation (x_1) and slope face (x_2).

5.30. Since glass is not subject to radiation damage, encapsulation of waste in glass is considered to be one of the most promising solutions to the problem of low-level nuclear waste in the environment. However, glass undergoes chemical changes when exposed to extreme environmental conditions, and certain of its constituents can leach into the surroundings. In addition, these chemical reactions may weaken the glass. These concerns led to a study undertaken jointly by the Department of Materials Science and Engineering at the University of Florida and the U.S. Department of Energy to assess the utility of glass as a waste encapsulant material.[†] Corrosive chemical solutions (called corrosion baths) were prepared and applied directly to glass samples containing one of three types of waste (TDS-3A, FE, and AL); the chemical reactions were observed over time. A few of the key variables measured were

$y =$ Amount of silicon (in parts per million) found in solution at end of experiment. (This is both a measure of the degree of breakdown in the glass and a proxy for the amount of radioactive species released into the environment.)

$x_1 =$ Temperature (°C) of the corrosion bath

$$x_2 = \begin{cases} 1 & \text{if waste} \\ & \text{type TDS-3A} \\ 0 & \text{if not} \end{cases} \quad x_3 = \begin{cases} 1 & \text{if waste type FE} \\ 0 & \text{if not} \end{cases}$$

Waste type AL is the base level. Suppose we want to model amount y of silicon as a function of temperature (x_1) and type of waste (x_2, x_3).

a. Write a model that proposes parallel straight-line relationships between amount of silicon and temperature, one line for each of the three waste types.

b. Add terms for the interaction between temperature and waste type to the model of part **a**.

c. Refer to the model of part **b**. For each waste type, give the slope of the line relating amount of silicon to temperature.

d. Explain how you could test for the presence of temperature-waste type interaction.

5.11	Regression analysis is one of the most widely used statistical tools for estima-
External Model Validation (Optional)	tion and prediction. All too frequently, however, a regression model deemed to be an adequate predictor of some response y performs poorly when applied

---

[†]The background information for this exercise was provided by Dr. David Clark, Department of Materials Science and Engineering, University of Florida.

in practice. For example, a model developed for forecasting new housing starts, although found to be statistically useful based on a test for overall model adequacy, may fail to take into account any extreme changes in future home mortgage rates generated by new government policy. This points out an important problem. *Models that fit the sample data well may not be successful predictors of y when applied to new data.* For this reason, it is important to assess the **validity** of the regression model in addition to its **adequacy** before using it in practice.

In Chapter 4, we presented several techniques for checking *model adequacy* (for example, tests of overall model adequacy, partial F tests, R^2_{adj} and s). In short, checking model adequacy involves determining whether the regression model adequately fits the *sample data*. **Model validation**, however, involves an assessment of how the fitted regression model will perform in practice—that is, how successful it will be when applied to new or future data. A number of different model validation techniques have been proposed, several of which are briefly discussed in this section. You will need to consult the references for more details on how to apply these techniques.

1. *Examining the predicted values*: Sometimes, the predicted values $\hat{y}$ of the fitted regression model can help to identify an invalid model. Nonsensical or unreasonable predicted values may indicate that the form of the model is incorrect or that the β coefficients are poorly estimated. For example, a model for a binary response y, where y is 0 or 1, may yield predicted probabilities that are negative or greater than 1. In this case, the user may want to consider a model that produces predicted values between 0 and 1 in practice (One such model, called the *logistic regression model*, is covered in Chapter 9.) On the other hand, if the predicted values of the fitted model all seem reasonable, the user should refrain from using the model in practice until further checks of model validity are carried out.

2. *Examining the estimated model parameters*: Typically, the user of a regression model has some knowledge of the relative size and sign (positive or negative) of the model parameters. This information should be used as a check on the estimated β coefficients. Coefficients with signs opposite to what is expected or with unusually small or large values, or unstable coefficients (i.e., coefficients with large standard errors) forewarn that the final model may perform poorly when applied to new or different data.

3. *Collecting new data for prediction*: One of the most effective ways of validating a regression model is to use the model to predict y for a new sample. By directly comparing the predicted values to the observed values of the new data, we can determine the accuracy of the predictions and use this information to assess how well the model performs in practice.

   Several measures of model validity have been proposed for this purpose. One simple technique is to calculate the percentage of variability in the new data explained by the model, denoted $R^2_{prediction}$, and compare it to the coefficient of determination R^2 for the least squares fit of the final model. Let $y_1, y_2, \ldots, y_n$ represent the n observations used to build and fit the final regression model and $y_{n+1}, y_{n+2}, \ldots, y_{n+m}$ represent the m observations in the new data set. Then

$$R^2_{prediction} = 1 - \left\{ \frac{\sum\limits_{i=n+1}^{n+m} (y_i - \hat{y}_i)^2}{\sum\limits_{i=n+1}^{n+m} (y_i - \overline{y})^2} \right\}$$

where $\hat{y}_i$ is the predicted value for the ith observation using the β estimates from the fitted model and $\overline{y}$ is the sample mean of the original data.* If $R^2_{prediction}$ compares favorably to R^2 from the least squares fit, we will have increased confidence in the usefulness of the model. However, if a significant drop in R^2 is observed, we should be cautious about using the model for prediction in practice.

A similar type of comparison can be made between the mean square error, MSE, for the least squares fit and the mean squared prediction error

$$MSE_{prediction} = \frac{\sum\limits_{i=n+1}^{n+m} (y_i - \hat{y}_i)^2}{m - (k+1)}$$

where k is the number of β coefficients (excluding β_0) in the model. Whichever measure of model validity you decide to use, the number of observations in the new data set should be large enough to reliably assess the model's prediction performance. Montgomery, Peck, and Vining (2001), for example, recommend 15–20 new observations, *at minimum.*

4. *Data-splitting (cross-validation)*: For those applications where it is impossible or impractical to collect new data, the original sample data can be split into two parts, with one part used to estimate the model parameters and the other part used to assess the fitted model's predictive ability. **Data-splitting** (or **cross-validation**, as it is sometimes known) can be accomplished in a variety of ways. A common technique is to randomly assign half the observations to the estimation data set and the other half to the prediction data set.† Measures of model validity, such as $R^2_{prediction}$ or $MSE_{prediction}$ can then be calculated. Of course, a sufficient number of observations must be available for data-splitting to be effective. For the estimation and prediction data sets of equal size, it has been recommended that the entire sample consist of *at least* $n = 2k + 25$ observations, where k is the number of β parameters in the model [see Snee (1977)].

5. *Jackknifing*: In situations where the sample data set is too small to apply data-splitting, a method called the **jackknife** can be applied. Let $y_{(i)}$ denote the predicted value for the ith observation obtained when the regression model is fit with the data point for y_i omitted (or deleted) from the sample. The jackknife

---

*Alternatively, the sample mean of the new data may be used.
†Random splits are usually applied in cases where there is no logical basis for dividing the data. Consult the references for other, more formal. data-splitting techniques.

method involves leaving each observation out of the data set, one at a time, and calculating the difference, $y_i - \hat{y}_{(i)}$, for all n observations in the data set. Measures of model validity, such as R^2 and MSE, are then calculated:

$$R^2_{\text{jackknife}} = \frac{\Sigma(y_i - \hat{y}_{(i)})^2}{\Sigma(y_i - \overline{y})^2}$$

$$\text{MSE}_{\text{jackknife}} = \frac{\Sigma(y_i - \hat{y}_{(i)})^2}{n - (k + 1)}$$

The numerator of both $R^2_{\text{jackknife}}$ and $\text{MSE}_{\text{jackknife}}$ is called the **prediction sum of squares**, or **PRESS**. In general, PRESS will be larger than the SSE of the fitted model. Consequently, $R^2_{\text{jackknife}}$ will be smaller than the R^2 of the fitted model and $\text{MSE}_{\text{jackknife}}$ will be larger than the MSE of the fitted model. These jackknife measures, then, give a more conservative (and more realistic) assessment of the ability of the model to predict future observations than the usual measures of model adequacy.

The appropriate model validation technique(s) will vary from application to application. Keep in mind that a favorable result is still no guarantee that the model will always perform successfully in practice. However we have much greater confidence in a validated model than in one that simply fits the sample data well.

5.12
Model Building: An Example

We illustrate the modeling techniques outlined in this chapter with an example from an actual trucking deregulation study. Consider the problem of modeling the price charged for motor transport service (such as trucking) in a particular state. In the early 1980s, several states removed regulatory constraints on the rate charged for intrastate trucking services. (Florida was the first state to embark on a deregulation policy, on July 1, 1980.) One of the goals of the regression analysis is to assess the impact of state deregulation on the supply price charged per ton-mile.

Data for 134 shipments made by a particular Florida carrier are saved in the TRUCKING file. The dependent variable of interest is y, the natural logarithm of the price charged per ton-mile. A careful variable-screening process, including stepwise regression, was conducted to determine which independent variables are the "best" predictor's of supply price. The SAS stepwise regression printout is shown in Figure 5.30. This analysis leads us to select the following variables to begin the model building process:

1. Distance shipped (hundreds of miles)
2. Weight of product shipped (thousands of pounds)

FIGURE 5.30

Portion of SAS stepwise regression output

Summary of Stepwise Selection

Step	Variable Entered	Variable Removed	Number Vars In	Partial R-Square	Model R-Square	C(p)	F Value	Pr > F
1	DISTANCE		1	0.2969	0.2969	417.090	55.74	<.0001
2	DEREG		2	0.3127	0.6096	175.795	104.91	<.0001
3	WEIGHT		3	0.1897	0.7993	30.1997	122.84	<.0001
4	ORIGIN		4	0.0362	0.8355	4.0122	28.40	<.0001

3. Deregulation in effect (yes or no)

4. Origin of shipment (Miami or Jacksonville)

Distance shipped and weight of product are quantitative variables since they each assume numerical values (miles and pounds, respectively) corresponding to the points on a line. Deregulation and origin are qualitative, or categorical, variables that we must describe with dummy (or coded) variables. The variable assignments are given as follows:

$$x_1 = \text{Distance shipped}$$

$$x_2 = \text{Weight of product}$$

$$x_3 = \begin{cases} 1 & \text{if deregulation in effect} \\ 0 & \text{if not} \end{cases}$$

$$x_4 = \begin{cases} 1 & \text{if originate in Miami} \\ 0 & \text{if originate in Jacksonville} \end{cases}$$

Note that in defining the dummy variables, we have arbitrarily chosen "no" and "Jacksonville" to be the base levels of deregulation and origin, respectively.

We begin the model-building process by specifying four models. These models, named Model 1–4, are shown in Table 5.9. Notice that the Model 1 is the complete second-order model. Recall from Section 5.10 that the complete second-order model contains quadratic (curvature) terms for quantitative variables and interactions among the quantitative and qualitative terms. For the trucking data, Model 1 traces a parabolic surface for mean natural log of price, $E(y)$, as a function of distance (x_1) and weight (x_2), and the response surfaces differ for the $2 \times 2 = 4$ combinations of the levels of deregulation (x_3) and origin (x_4). Generally, the complete second-order model is a good place to start the model-building process since most real-world relationships are curvilinear. (Keep in mind, however, that you must have a sufficient number of data points to find estimates of all the parameters in the model.) Model 1 is fit to the data for the 134 shipments in the TRUCKING file using SAS. The results are shown in Figure 5.31. Note that the p-value for the global model F-test is less than .0001, indicating that the complete second-order model is statistically useful for predicting trucking price.

Model 2 contains all the terms of Model 1, except that the quadratic terms (i.e., terms involving x_1^2 and x_2^2) are dropped. This model also proposes four different response surfaces for the combinations of levels of deregulation and origin, but the surfaces are twisted planes (see Figure 5.10) rather than paraboloids. A direct comparison of Models 1 and 2 will allow us to test for the importance of the curvature terms.

Model 3 contains all the terms of Model 1, except that the quantitative-qualitative interaction terms are omitted. This model proposes four curvilinear paraboloids corresponding to the four deregulation-origin combinations, that differ only with respect to the y-intercept. By directly comparing Models 1 and 3, we can test for the importance of all the quantitative-qualitative interaction terms.

Model 4 is identical to Model 1, except that it does not include any interactions between the quadratic terms and the two qualitative variables, deregulation (x_3) and origin (x_4). Although curvature is included in this model, the rates of curvature

---

TABLE 5.9 Hypothesized Models for Natural log of Trucking Price

---

Model 1: $E(y) = \beta_0 + \beta_1 x_1 + \beta_2 x_2 + \beta_3 x_1 x_2 + \beta_4 x_1^2 + \beta_5 x_2^2$
$\qquad + \beta_6 x_3 + \beta_7 x_4 + \beta_8 x_3 x_4$
$\qquad + \beta_9 x_1 x_3 + \beta_{10} x_1 x_4 + \beta_{11} x_1 x_3 x_4$
$\qquad + \beta_{12} x_2 x_3 + \beta_{13} x_2 x_4 + \beta_{14} x_2 x_3 x_4$
$\qquad + \beta_{15} x_1 x_2 x_3 + \beta_{16} x_1 x_2 x_4 + \beta_{17} x_1 x_2 x_3 x_4$
$\qquad + \beta_{18} x_1^2 x_3 + \beta_{19} x_1^2 x_4 + \beta_{20} x_1^2 x_3 x_4$
$\qquad + \beta_{21} x_2^2 x_3 + \beta_{22} x_2^2 x_4 + \beta_{23} x_2^2 x_3 x_4$

Model 2: $E(y) = \beta_0 + \beta_1 x_1 + \beta_2 x_2 + \beta_3 x_1 x_2$
$\qquad + \beta_6 x_3 + \beta_7 x_4 + \beta_8 x_3 x_4$
$\qquad + \beta_9 x_1 x_3 + \beta_{10} x_1 x_4 + \beta_{11} x_1 x_3 x_4$
$\qquad + \beta_{12} x_2 x_3 + \beta_{13} x_2 x_4 + \beta_{14} x_2 x_3 x_4$
$\qquad + \beta_{15} x_1 x_2 x_3 + \beta_{16} x_1 x_2 x_4 + \beta_{17} x_1 x_2 x_3 x_4$

Model 3: $E(y) = \beta_0 + \beta_1 x_1 + \beta_2 x_2 + \beta_3 x_1 x_2 + \beta_4 x_1^2 + \beta_5 x_2^2$
$\qquad + \beta_6 x_3 + \beta_7 x_4 + \beta_8 x_3 x_4$

Model 4: $E(y) = \beta_0 + \beta_1 x_1 + \beta_2 x_2 + \beta_3 x_1 x_2 + \beta_4 x_1^2 + \beta_5 x_2^2$
$\qquad + \beta_6 x_3 + \beta_7 x_4 + \beta_8 x_3 x_4$
$\qquad + \beta_9 x_1 x_3 + \beta_{10} x_1 x_4 + \beta_{11} x_1 x_3 x_4$
$\qquad + \beta_{12} x_2 x_3 + \beta_{13} x_2 x_4 + \beta_{14} x_2 x_3 x_4$
$\qquad + \beta_{15} x_1 x_2 x_3 + \beta_{16} x_1 x_2 x_4 + \beta_{17} x_1 x_2 x_3 x_4$

Model 5: $E(y) = \beta_0 + \beta_1 x_1 + \beta_2 x_2 + \beta_3 x_1 x_2 + \beta_4 x_1^2 + \beta_5 x_2^2$
$\qquad + \beta_6 x_3$
$\qquad + \beta_9 x_1 x_3 + \beta_{12} x_2 x_3 + \beta_{15} x_1 x_2 x_3$

Model 6: $E(y) = \beta_0 + \beta_1 x_1 + \beta_2 x_2 + \beta_3 x_1 x_2 + \beta_4 x_1^2 + \beta_5 x_2^2$
$\qquad + \beta_7 x_4$
$\qquad + \beta_{10} x_1 x_4 + \beta_{13} x_2 x_4 + \beta_{16} x_1 x_2 x_4$

Model 7: $E(y) = \beta_0 + \beta_1 x_1 + \beta_2 x_2 + \beta_3 x_1 x_2 + \beta_4 x_1^2 + \beta_5 x_2^2$
$\qquad + \beta_6 x_3 + \beta_7 x_4$
$\qquad + \beta_9 x_1 x_3 + \beta_{10} x_1 x_4$
$\qquad + \beta_{12} x_2 x_3 + \beta_{13} x_2 x_4$
$\qquad + \beta_{15} x_1 x_2 x_3 + \beta_{16} x_1 x_2 x_4$

---

for both distance (x_1) and weight (x_2) are the same for all levels of deregulation and origin.

Figure 5.32 shows the results of the nested model F-tests described in the above paragraphs. Each of these tests is summarized as follows:

FIGURE 5.31

SAS regression printout for
Model 1

Dependent Variable: LNPRICE

Analysis of Variance

Source	DF	Sum of Squares	Mean Square	F Value	Pr > F
Model	23	83.90934	3.64823	65.59	<.0001
Error	110	6.11860	0.05562		
Corrected Total	133	90.02794			

Root MSE	0.23585	R-Square	0.9320	
Dependent Mean	10.57621	Adj R-Sq	0.9178	
Coeff Var	2.22997			

Parameter Estimates

Variable	DF	Parameter Estimate	Standard Error	t Value	Pr > |t|
Intercept	1	12.51593	0.95441	13.11	<.0001
X1	1	-0.89923	0.73410	-1.22	0.2232
X2	1	0.02421	0.02886	0.84	0.4034
X1X2	1	-0.02071	0.00673	-3.08	0.0026
X1SQ	1	0.15145	0.13455	1.13	0.2628
X2SQ	1	-0.00010196	0.00076963	-0.13	0.8948
X3	1	-1.12650	1.49104	-0.76	0.4516
X4	1	0.27615	0.96332	0.29	0.7749
X3X4	1	0.49697	1.50290	0.33	0.7415
X1X3	1	0.48205	1.15882	0.42	0.6782
X1X4	1	0.06958	0.73882	0.09	0.9251
X1X3X4	1	-0.54037	1.16440	-0.46	0.6435
X2X3	1	-0.09486	0.04477	-2.12	0.0363
X2X4	1	-0.05261	0.03528	-1.49	0.1387
X2X3X4	1	0.06826	0.05220	1.31	0.1937
X1X2X3	1	0.02207	0.01078	2.05	0.0429
X1X2X4	1	0.02355	0.00709	3.32	0.0012
X1X2X3X4	1	-0.02694	0.01127	-2.39	0.0185
X1SQX3	1	-0.11674	0.21918	-0.53	0.5954
X1SQX4	1	-0.07276	0.13510	-0.54	0.5913
X1SQX3X4	1	0.13424	0.21984	0.61	0.5427
X2SQX3	1	0.00043756	0.00119	0.37	0.7127
X2SQX4	1	0.00011095	0.00107	0.10	0.9174
X2SQX3X4	1	-0.00027597	0.00157	-0.18	0.8609

TEST FOR SIGNIFICANCE OF ALL QUADRATIC TERMS (MODEL 1 VS. MODEL 2)

H_0: $\beta_4 = \beta_5 = \beta_{18} = \beta_{19} = \beta_{20} = \beta_{21} = \beta_{22} = \beta_{23} = 0$
H_a: At least one of the quadratic β's in Model 1 differs from 0

$F = 13.61$, p-value $< .0001$ (shaded at the top of Figure 5.32)

Conclusion: There is sufficient evidence (at $\alpha = .01$) of curvature in the relationships between $E(y)$ and distance (x_1) and weight (x_2). Model 1 is a statistically better predictor of trucking price than Model 2.

TEST FOR SIGNIFICANCE OF ALL QUANTITATIVE-QUALITATIVE INTERACTION TERMS (MODEL 1 VS. MODEL 3)

H_0: $\beta_9 = \beta_{10} = \beta_{11} = \beta_{12} = \beta_{13} = \beta_{14} = \beta_{15} = \beta_{16} = \beta_{17} = \beta_{18} = \beta_{19} = \beta_{20} = \beta_{21} = \beta_{22} = \beta_{23} = 0$
H_a: At least one of the QN×QL interaction β's in Model 1 differs from 0

$F = 4.60$, p-value $< .0001$ (shaded in the middle of Figure 5.32)

FIGURE 5.32

SAS nested model F-tests for terms in Model 1

```
         Test QUADRATIC Results for Dependent Variable LNPRICE

                                     Mean
       Source              DF       Square      F Value    Pr > F

       Numerator            8      0.75727       13.61     <.0001
       Denominator        110      0.05562
```

```
            Test QN_QL_INTERACT Results for
                Dependent Variable LNPRICE

                                     Mean
       Source              DF       Square      F Value    Pr > F

       Numerator           15      0.25574        4.60     <.0001
       Denominator        110      0.05562
```

```
          Test QL_QUAD_INTERACT Results
            for Dependent Variable LNPRICE

                                     Mean
       Source              DF       Square      F Value    Pr > F

       Numerator            6      0.01407        0.25     0.9572
       Denominator        110      0.05562
```

Conclusion: There is sufficient evidence (at $\alpha = .01$) of interaction between the quantitative variables, distance (x_1) and weight (x_2), and the qualitative variables, deregulation (x_3) and origin (x_4). Model 1 is a statistically better predictor of trucking price than Model 3.

TEST FOR SIGNIFICANCE OF QUALITATIVE-QUADRATIC INTERACTION (MODEL 1 VS. MODEL 4)

$H_0: \beta_{18} = \beta_{19} = \beta_{20} = \beta_{21} = \beta_{22} = \beta_{23} = 0$

$H_a:$ At least one of the qualitative-quadratic interaction β's in Model 1 differs from 0

$F = .25$, p-value $= .9572$ (shaded at the bottom of Figure 5.32)

Conclusion: There is insufficient evidence (at $\alpha = .01$) of interaction between the quadratic terms for distance (x_1) and weight (x_2), and the qualitative variables, deregulation (x_3) and origin (x_4). Since these terms are not statistically useful, we will drop these terms from Model 1 and conclude that Model 4 is a statistically better predictor of trucking price.*

Based on the three nested-model F-tests, we found Model 4 to be the "best" of the first four models. The SAS printout for Model 4 is shown in Figure 5.33. Looking at the results of the global F-test (p-value less than .0001), you can see that the

---

*There is always danger in dropping terms from the model. Essentially, we are accepting $H_0: \beta_{18} = \beta_{19} = \beta_{20} = \cdots = \beta_{23} = 0$ when $P(\text{Type II error}) = P(\text{Accepting } H_0 \text{ when } H_0 \text{ is false}) = \beta$ is unknown. In practice, however, many researchers are willing to risk making a Type II error rather than use a more complex model for $E(y)$ when simpler models that are nearly as good as predictors (and easier to apply and interpret) are available. Note that we used a relatively large amount of data ($n = 134$) in fitting our models and that R^2_{adj} for Model 4 is actually larger than R^2_{adj} for Model 1. If the quadratic interaction terms are, in fact, important (i.e., we have made a Type II error), there is little lost in terms of explained variability in using Model 4.

overall model is statistically useful for predicting trucking price. Also, $R^2_{\text{adj}} = .9210$ implies that about 92% of the sample variation in the natural log of trucking price can be explained by the model. Although these model statistics are impressive, we may be able to find a simpler model that fits the data just as well.

FIGURE 5.33

SAS regression printout for Model 4

Dependent Variable: LNPRICE

Analysis of Variance

Source	DF	Sum of Squares	Mean Square	F Value	Pr > F
Model	17	83.82495	4.93088	92.21	<.0001
Error	116	6.20299	0.05347		
Corrected Total	133	90.02794			

Root MSE	0.23124	R-Square	0.9311
Dependent Mean	10.57621	Adj R-Sq	0.9210
Coeff Var	2.18646		

Parameter Estimates

Variable	DF	Parameter Estimate	Standard Error	t Value	Pr > |t|
Intercept	1	12.08482	0.25871	46.71	<.0001
X1	1	-0.55296	0.09648	-5.73	<.0001
X2	1	0.01889	0.02120	0.89	0.3748
X1X2	1	-0.02041	0.00649	-3.15	0.0021
X1SQ	1	0.08738	0.00827	10.56	<.0001
X2SQ	1	0.00008202	0.00037354	0.22	0.8266
X3	1	-0.38504	0.40009	-0.96	0.3379
X4	1	0.76041	0.27135	2.80	0.0059
X3X4	1	-0.35311	0.42661	-0.83	0.4095
X1X3	1	-0.13163	0.14172	-0.93	0.3549
X1X4	1	-0.33374	0.08995	-3.71	0.0003
X1X3X4	1	0.18215	0.14746	1.24	0.2192
X2X3	1	-0.08259	0.02956	-2.79	0.0061
X2X4	1	-0.04830	0.02049	-2.36	0.0201
X2X3X4	1	0.05937	0.03217	1.85	0.0675
X1X2X3	1	0.02136	0.01043	2.05	0.0428
X1X2X4	1	0.02320	0.00685	3.39	0.0010
X1X2X3X4	1	-0.02601	0.01090	-2.39	0.0186

Table 5.9 gives three additional models. Model 5 is identical to Model 4, but all terms for the qualitative variable origin(x_4) have been dropped. A comparison of Model 4 to Model 5 will allow us to determine whether origin really has an impact on trucking price. Similarly, Model 6 is identical to Model 4, but now all terms for the qualitative variable deregulation (x_3) have been dropped. By comparing Model 4 to Model 6, we can determine whether deregulation has an impact on trucking price. Finally, we propose Model 7, which is obtained by dropping all the qualitative-qualitative interaction terms. A comparison of Model 4 to Model 7 will allow us to see whether deregulation and origin interact to effect the natural log of trucking price.

Figure 5.34 shows the results of the nested model F-tests described above. A summary of each of these tests follows:

TEST FOR SIGNIFICANCE OF ALL ORIGIN TERMS (MODEL 4 VS. MODEL 5)

H_0: $\beta_7 = \beta_8 = \beta_{10} = \beta_{11} = \beta_{13} = \beta_{14} = \beta_{16} = \beta_{17} = 0$

H_a: At least one of the origin β's in Model 4 differs from 0

FIGURE 5.34

SAS nested model F-tests for terms in Model 4

Test ORIGIN Results for Dependent Variable LNPRICE

Source	DF	Mean Square	F Value	Pr > F
Numerator	8	0.18987	3.55	0.0010
Denominator	116	0.05347		

Test DEREG Results for Dependent Variable LNPRICE

Source	DF	Mean Square	F Value	Pr > F
Numerator	8	4.03417	75.44	<.0001
Denominator	116	0.05347		

Test ORG_DEREG_INTERACTION Results
for Dependent Variable LNPRICE

Source	DF	Mean Square	F Value	Pr > F
Numerator	4	0.11367	2.13	0.0820
Denominator	116	0.05347		

$$F = 3.55, \ p\text{-value} = .001 \text{ (shaded at the top of Figure 5.34)}$$

Conclusion: There is sufficient evidence (at $\alpha = .01$) to indicate that origin (x_4) has an impact on trucking price. Model 4 is a statistically better predictor of trucking price than Model 5.

TEST FOR SIGNIFICANCE OF ALL DEREGULATION TERMS (MODEL 4 VS. MODEL 6)

$H_0: \beta_6 = \beta_8 = \beta_9 = \beta_{11} = \beta_{12} = \beta_{14} = \beta_{15} = \beta_{17} = 0$
$H_a:$ At least one of the deregulation β's in Model 4 differs from 0

$$F = 75.44, \ p\text{-value} < .0001 \text{ (shaded in the middle of Figure 5.34)}$$

Conclusion: There is sufficient evidence (at $\alpha = .01$) to indicate that deregulation (x_3) has an impact on trucking price. Model 4 is a statistically better predictor of trucking price than Model 6.

TEST FOR SIGNIFICANCE OF ALL DEREGULATION-ORIGIN INTERACTION TERMS (MODEL 4 VS. MODEL 7)

$H_0: \beta_8 = \beta_{11} = \beta_{14} = \beta_{17} = 0$
$H_a:$ At least one of the QL × QL interaction β's in Model 4 differs from 0

$$F = 2.13, \ p\text{-value} = .0820 \text{ (shaded at the bottom of Figure 5.34)}$$

Conclusion: There is insufficient evidence (at $\alpha = .01$) to indicate that deregulation (x_3) and origin (x_4) interact. Thus, we will drop these interaction terms from Model 4 and conclude that Model 7 is a statistically better predictor of trucking price.

In summary, the nested model F-tests suggest that Model 7 is the best for modeling the natural log of trucking price. The SAS printout for Model 7 is shown in Figure 5.35. The β-estimates used for making predictions of trucking price are highlighted on the printout.

FIGURE 5.35

SAS regression printout for Model 7

```
                        Dependent Variable: LNPRICE

                              Analysis of Variance

                                  Sum of           Mean
      Source            DF       Squares         Square      F Value    Pr > F

      Model             13      83.37026        6.41310       115.59    <.0001
      Error            120       6.65767        0.05548
      Corrected Total  133      90.02794


              Root MSE              0.23554     R-Square     0.9260
              Dependent Mean      10.57621     Adj R-Sq     0.9180
              Coeff Var            2.22710


                            Parameter Estimates

                             Parameter        Standard
      Variable      DF        Estimate           Error    t Value    Pr > |t|

      Intercept      1        12.19150         0.21583      56.49    <.0001
      X1             1        -0.59800         0.08425      -7.10    <.0001
      X2             1        -0.00598         0.01857      -0.32     0.7480
      X1X2           1        -0.01078         0.00530      -2.03     0.0442
      X1SQ           1         0.08575         0.00834      10.28    <.0001
      X2SQ           1         0.00014207    0.00037728       0.38     0.7072
      X3             1        -0.78192         0.12900      -6.06    <.0001
      X4             1         0.67679         0.21035       3.22     0.0017
      X1X3           1         0.03991         0.03999       1.00     0.3203
      X1X4           1        -0.27464         0.07267      -3.78     0.0002
      X2X3           1        -0.02094         0.01045      -2.00     0.0473
      X2X4           1        -0.02619         0.01610      -1.63     0.1063
      X1X2X3         1        -0.00332         0.00303      -1.10     0.2757
      X1X2X4         1         0.01298         0.00544       2.39     0.0186
```

A note of caution: Just as with t tests on individual β parameters, you should avoid conducting too many partial F tests. Regardless of the type of test (t test or F test), the more tests that are performed, the higher the overall Type I error rate will be. In practice, you should limit the number of models that you propose for $E(y)$ so that the overall Type I error rate α for conducting partial F tests remains reasonably small.[†]

Summary

Although this chapter on **model building** covered many topics, only experience can make you competent in this fascinating area of statistics. Successful model building requires a delicate blend of knowledge of the process being modeled, geometry, and formal statistical testing.

1. Identify the **response variable y** and the **set of independent variables**.
2. Classify each independent variable as either **quantitative** or **qualitative**.
3. Define **dummy variables** to represent the qualitative independent variables.

[†]A technique suggested by Bonferroni is often applied to maintain control of the overall Type I error rate α. If c tests are to be performed, then conduct each individual test at significance level α/c. This will guarantee an overall Type I error rate less than or equal to α. For example, conducting each of $c = 5$ tests at the $.05/5 = .01$ level of significance guarantees an overall $\alpha \leq .05$.

When the number of independent variables is manageable, we are ready to consider what level of complexity is appropriate.

4. Consider **second-order models**—those containing **two-way interactions** and **quadratic terms** in the quantitative variables. Remember that a model with no interaction terms implies that each of the independent variables affects the response independently of the other independent variables.

5. Consider **quadratic terms that add curvature** to the contour curves when $E(y)$ is plotted as a function of the independent variable.

6. Consider **coding the quantitative independent variables** in higher-order models, thereby reducing rounding error and the built-in multicollinearity problem.

Many problems can arise in regression modeling, and the intermediate steps are often tedious and frustrating. However, the end result of a careful and determined modeling effort is very rewarding—you will have a better understanding of the process and will have a predictive model for the dependent variable y.

SUPPLEMENTARY EXERCISES

[Exercises from the optional sections are identified by an asterisk (*).]

5.31. Psychiatrists keep personnel files that contain important information on each client's background. The data in these files could be used to predict the probability that therapy will be successful. Identify the independent variables listed here as qualitative or quantitative.

a. Age
b. Years in therapy
c. Highest educational degree
d. Job classification
e. Religious preference
f. Marital status
g. IQ
h. Gender

5.32. *Multinational* is the term given to an industry with foreign investors. A study of 216 manufacturing industries in Mexico found that multinational presence in a firm has a positive influence on market concentration (*World Development*, Vol. 14, 1986). The result was revealed in a multiple regression analysis on the dependent variable y, market concentration index, using the following quantitative independent variables:

x_1 = Market size

x_2 = Market rate of growth

x_3 = Gross production in largest plants (expressed as a percentage of total gross production)

x_4 = Capital intensity (ratio of total assets to total number of employees)

x_5 = Advertising intensity (ratio of advertising to value added)

x_6 = Foreign share (i.e., gross output produced by foreign subsidiaries)

a. Write a first-order model for $E(y)$ as a function of $x_1 - x_6$.

b. Interpret β_6 in the model in part **a**.

c. Based on the results of the study, is β_6 positive or negative?

d. Write a second-order model for $E(y)$ that proposes interaction between the independent variables but with no curvature.

e. Using the model in part **d**, how would you test the hypothesis that effect of a multinational presence on market concentration is independent of the other independent variables in the model?

5.33. As a result of the dramatic decline in the cost of computer hardware, it is becoming economically feasible to build computers with thousands of processors. However, the scheduling of computer jobs on these advanced computers can be a difficult task. Parallel scheduling algorithms have been designed to solve this problem. A *parallel algorithm* is a set of scheduling instructions designed to minimize the number of tardy jobs in the system and to minimize the mean finish time of the entire job stream. Suppose three different scheduling algorithms (A, B, and C) have been proposed for minimizing the mean finish time of n jobs in a system with a large number of processors.

a. Write a main effects model with interaction to relate the mean finish time, $E(y)$, the number of jobs (x_1) and scheduling algorithm (A, B, or C).

SAS output for Exercise 5.33

Analysis of Variance

Source	DF	Sum of Squares	Mean Square	F Value	Pr > F
Model	5	87.473	17.495	40.90	0.0394
Error	6	21.443	3.574		
Corrected Total	11	108.916			

Root MSE	1.8905	R-Square	0.8031
Dependent Mean	12.604	Adj R-Sq	0.6390
Coeff Var	14.999		

b. The model of part a was fitted to data collected on 12 simulated systems (four systems for each of the three algorithms) with the results shown in the SAS printout above. Test whether the model is useful in predicting mean finish time. Use $\alpha = .05$.

c. Write the main effects model (with no interaction) relating mean finish time to number of jobs and scheduling algorithm.

d. The main effects (reduced) model was fit to the data and produced SSE = 38.289. Does this provide sufficient evidence at the $\alpha = .05$ level of significance to indicate that the interaction terms should be kept in the model?

5.34. Refer to the *Journal of Educational Statistics* (Spring, 1993) study of the variables related to the final grade point average of doctoral students, Exercise 4.70 (p. 244). Consider a model for mean GPA, $E(y)$, as a function of: quantitative GMAT score, verbal GMAT score, and student cohort (1988, 1990, or 1992).

a. Write the complete second-order model for $E(y)$.

b. Write a second-order model for $E(y)$ that proposes three parallel paraboloids, one for each of the student cohorts.

c. Write a model for $E(y)$ that proposes linear relationships between GPA and the two GMAT scores, such that the slopes of the lines depend on student cohort but not on the other GMAT score.

5.35. A company wants to model the total weekly sales, y, of its product as a function of the variables packaging and location. Two types of packaging, P_1 and P_2, are used in each of four locations, L_1, L_2, L_3, and L_4.

a. Write a main effects model to relate $E(y)$ to packaging and location. What implicit assumption are we making about the interrelationships between sales, packaging, and location when we use this model?

b. Now write a model for $E(y)$ that includes interaction between packaging and location. How many parameters are in this model (remember to include β_0)? Compare this number to the number of packaging—location combinations being modeled.

c. Suppose the main effects and interaction models are fit for 40 observations on weekly sales. The values of SSE are

SSE for main effects model = 422.36
SSE for interaction model = 346.65

Determine whether the data indicate that the interaction between location and packaging is important in estimating mean weekly sales. Use $\alpha = .05$. What implications does your conclusion have for the company's marketing strategy?

5.36. To make a product more appealing to the consumer, an automobile manufacturer is experimenting with a new type of paint that is supposed to help the car maintain its new-car look. The durability of this paint depends on the length of time the car body is in the oven after it has been painted. In the initial experiment, three groups of 10 car bodies each were baked for three different lengths of time—12, 24, and 36 hours—at the standard temperature setting. Then, the paint finish of each of the 30 cars was analyzed to determine a durability rating, y.

a. Write a quadratic model relating the mean durability, $E(y)$, to the length of baking.

b. Could a cubic model be fit to the data? Explain.

c. Suppose the research and development department develops three new types of paint to be tested. Thus, 90 cars are to be tested—30 for each type of paint. Write the complete second-order model for $E(y)$ as a function of the type of paint and bake time.

5.37. Economic research has established evidence of a positive correlation between earnings and educational attainment (*Economic Inquiry*, Jan. 1984). However, it is unclear whether higher wage rates for better educated workers reflect an individual's added value or merely the employer's use of higher education as a screening device in the recruiting process. One version of this "sheepskin screening" hypothesis supported by many economists is that wages will rise faster with extra years of education when the extra years culminate in a certificate (e.g., master's or Ph.D. degree, CPA certificate, or actuarial degree).

a. Write a first-order, main effects model for mean wage rate $E(y)$ of an employer as a function of employee's years of education and whether or not the employee is certified.

b. Write a first-order model for $E(y)$ that corresponds to the "sheepskin screening" hypothesis.

c. Write the complete second-order model for $E(y)$ as a function of the two independent variables.

*5.38 Use the coding system for observational data to fit a second-order model to the data on demand y and price p given in the following table. Show that the inherent multicollinearity problem with fitting a polynomial model is reduced when the coded values of p are used.

🔷 DEMAND

DEMAND								
y, pounds	1,120	999	932	884	807	760	701	688
PRICE								
p, dollars	3.00	3.10	3.20	3.30	3.40	3.50	3.60	3.70

5.39. One factor that must be considered in developing a shipping system that is beneficial to both the customer and the seller is time of delivery. A manufacturer of farm equipment can ship its products by either rail or truck. Quadratic models are thought to be adequate in relating time of delivery to distance to be shipped for both modes of transportation. Consequently, it has been suggested that the following model be fit to begin the model-building process:

$$E(y) = \beta_0 + \beta_1 x_1 + \beta_2 x_1^2 + \beta_3 x_2 + \beta_4 x_1 x_2 + \beta_5 x_1^2 x_2$$

where

$$y = \text{Time of delivery}$$

$$x_1 = \text{Distance to be shipped}$$

$$x_2 = \begin{cases} 1 & \text{if rail} \\ 0 & \text{if truck} \end{cases}$$

a. Sketch the proposed relationships between delivery time y and distance x_1 for both modes of transportation.

b. What hypothesis would you test to determine whether the data indicate that the quadratic distance terms are useful in the model, i.e., whether curvature is present in the relationship between mean delivery time and distance?

c. What hypothesis would you test to determine whether there is a difference in mean delivery time by rail and by truck?

5.40. Eli Lilly and Company has developed three methods (G, R_1, and R_2) for estimating the shelf life of its drug products based on potency. One way to compare the three methods is to build a regression model for the dependent variable, estimated shelf life y (as a percentage of true shelf life), with potency of the drug (x_1) as a quantitative predictor and method as a qualitative predictor.

a. Write a first-order, main effects model for $E(y)$ as a function of potency (x_1) and method.

b. Interpret the β coefficients of the model, part **a**.

c. Write a first-order model for $E(y)$ that will allow the slopes to differ for the three methods.

d. Refer to part **c**. For each method, write the slope of the $y-x_1$ line in terms of the β's.

5.41. The performance of an industry is often measured by the level of excess (or unutilized) capacity within the industry. Researchers examined the relationship between excess capacity y and several market variables in 273 U.S. manufacturing industries (*Quarterly Journal of Business and Economics*, Summer 1986). Two qualitative independent variables considered in the study were Market concentration (low, moderate, and high) and Industry type (producer or consumer).

a. Write the main effects model for $E(y)$ as a function of the two qualitative variables.

b. Interpret the β coefficients in the main effects model.

c. Write the model for $E(y)$ that includes interaction between market concentration and industry type.

d. Interpret the β coefficients in the interaction model.

e. How would you test the hypothesis that the difference between the mean excess capacity levels of producer and consumer industry types is the same across all three market concentrations?

*5.42 Use the coding system for observational data to fit a complete second-order model to the data of Example 5.3. which are repeated on p. 320.

a. Give the coded values u_1 and u_2 for x_1 and x_2, respectively.

b. Compare the coefficient of correlation between x_1 and x_1^2 with the coefficient of correlation between u_1 and u_1^2.

PRODQUAL

x_1	x_2	y	x_1	x_2	y	x_1	x_2	y
80	50	50.8	90	50	63.4	100	50	46.6
80	50	50.7	90	50	61.6	100	50	49.1
80	50	49.4	90	50	63.4	100	50	46.4
80	55	93.7	90	55	93.8	100	55	69.8
80	55	90.9	90	55	92.1	100	55	72.5
80	55	90.9	90	55	97.4	100	55	73.2
80	60	74.5	90	60	70.9	100	60	38.7
80	60	73.0	90	60	68.8	100	60	42.5
80	60	71.2	90	60	71.3	100	60	41.4

c. Compare the coefficient of correlation between x_2 and x_2^2 with the coefficient of correlation between u_2 and u_2^2.

d. Give the prediction equation.

5.43. Research conducted at Ohio State University focused on the factors that influence the allocation of black and white men in labor market positions (*American Sociological Review*, June 1986). Data collected for each of 837 labor market positions were used to build a regression model for y, defined as the natural logarithm of the ratio of the proportion of blacks employed in a labor market position to the corresponding proportion of whites employed (called the *black-white log odds ratio*). Positive values of y indicate that blacks have a greater likelihood of employment than whites. Several independent variables were considered, including the following:

$x_1 =$ Market power (a quantitative measure of the size and visibility of firms in the labor market)

$x_2 =$ Percentage of workers in the labor market who are union members

$$x_3 = \begin{cases} 1 & \text{if labor market position} \\ & \text{includes craft occupations} \\ 0 & \text{if not} \end{cases}$$

a. Write the first-order main effects model for E(y) as a function of x_1, x_2, and x_3.

b. One theory hypothesized by the researchers is that the mean log odds ratio E(y) is smaller for craft occupations than for noncraft occupations. (That is, the likelihood of black employment is less for craft occupations.) Explain how to test this hypothesis using the model in part **a**.

c. Write the complete second-order model for E(y) as a function of x_1, x_2, and x_3.

d. Using the model in part **c**, explain how to test the hypothesis that level of market power x_1 has no effect on black—white log odds ratio y.

e. Holding x_2 fixed, sketch the contour lines relating y to x_1 for the following model:

$$E(y) = \beta_0 + \beta_1 x_1 + \beta_2 x_2 + \beta_3 x_3 + \beta_4 x_1 x_3 + \beta_5 x_2 x_3$$

REFERENCES

DRAPER, N., and SMITH, H. *Applied Regression Analysis*, 3rd ed. New York: Wiley, 1998.

GRAYBILL, F. A. *Theory and Application of the Linear Model*. North Scituate, Mass.: Duxbury, 1976.

GEISSER, S. "The predictive sample reuse method with applications," *Journal of the American Statistical Association*, Vol. 70, 1975.

MENDENHALL, W. *Introduction to Linear Models and the Design and Analysis of Experiments*. Belmont, Calif.: Wadsworth, 1968.

MONTGOMERY, D., PECK, E., and VINING, G. *Introduction to Linear Regression Analysis*, 3rd ed. New York: Wiley, 2001.

NETER, J., KUTNER, M., NACHTSHEIM, C., and WASSERMAN, W. *Applied Linear Statistical Models*, 4th ed. Homewood, Ill.: Richard D. Irwin, 1996.

SNEE, R., "Validation of regression models: Methods and examples." *Technometrics*, Vol. 19, 1977.

Variable Screening Methods

CONTENTS

OBJECTIVE

To introduce methods designed to select the most important independent variables for modeling the mean response, $E(y)$; to learn when these methods are appropriate to apply.

6.1
Introduction: Why Use a Variable Screening Method?

Researches often will collect a data set with a large number of independent variables, each of which is a potential predictor of some dependent variable, y. The problem of deciding which x's in a large set of independent variables to include in a multiple regression model for $E(y)$ is common, for instance, when the dependent variable is profit of a firm, a college student's grade point average, or an economic variable reflecting the state of the economy (e.g., inflation rate).

Consider the problem of predicting the annual salary y of an executive. In Example 4.9 (p. 221) we examined a model with several predictors of y. Suppose we have collected data for 10 potential predictors of an executive's salary. Assume that the list includes seven quantitative x's and three qualitative x's (each of the qualitative x's at two levels). Now consider a complete second-order model for $E(y)$. From our discussion of model building in Chapter 5, we can write the model as follows:

$$E(y) = \beta_0 + \underbrace{\beta_1 x_1 + \beta_2 x_2 + \beta_3 x_3 + \beta_4 x_4 + \beta_5 x_5 + \beta_6 x_6 + \beta_7 x_7}_{\text{(first-order terms for quantitative variables)}}$$

$$+ \beta_8 x_1 x_2 + \beta_9 x_1 x_3 + \beta_{10} x_1 x_4 + \beta_{11} x_1 x_5 + \beta_{12} x_1 x_6 + \beta_{13} x_1 x_7$$

$$+ \underbrace{\beta_{14}x_2x_3 + \beta_{15}x_2x_4 + \beta_{16}x_2x_5 + \beta_{17}x_2x_6 + \beta_{18}x_2x_7 + \cdots + \beta_{28}x_6x_7}$$

(two-way interaction terms for quantitative variables)

$$+ \quad \underbrace{\beta_{29}x_1^2 + \beta_{30}x_2^2 + \beta_{31}x_3^2 + \beta_{32}x_4^2 + \beta_{33}x_5^2 + \beta_{34}x_6^2 + \beta_{35}x_7^2}$$

(quadratic [second-order] terms for quantitative variables)

$$+ \beta_{36}x_8 + \beta_{37}x_9 + \beta_{38}x_{10} \quad \text{(dummy variables for qualitative variables)}$$

$$+ \underbrace{\beta_{39}x_8x_9 + \beta_{40}x_8x_{10} + \beta_{41}x_9x_{10} + \beta_{42}x_8x_9x_{10}}$$

(interaction terms for qualitative variables)

$$+ \beta_{43}x_1x_8 + \beta_{44}x_2x_8 + \beta_{45}x_3x_8 + \beta_{46}x_4x_8 + \beta_{47}x_5x_8 + \beta_{48}x_6x_8 + \beta_{49}x_7x_8$$

$$+ \beta_{50}x_1x_2x_8 + \beta_{51}x_1x_3x_8 + \beta_{52}x_1x_4x_8 + \cdots + \beta_{70}x_6x_7x_8$$

$$+ \quad\quad\quad\quad\quad \underbrace{\beta_{71}x_1^2x_8 + \beta_{72}x_2^2x_8 + \cdots + \beta_{77}x_7^2x_8}$$

(interactions between quantitative terms and qualitative variable x_8)

$$+ \quad\quad\quad\quad \underbrace{\beta_{78}x_1x_9 + \beta_{79}x_2x_9 + \beta_{80}x_3x_9 + \cdots + \beta_{112}x_7^2x_9}$$

(interactions between quantitative terms and qualitative variable x_9)

$$+ \quad\quad\quad \underbrace{\beta_{113}x_1x_{10} + \beta_{114}x_2x_{10} + \beta_{115}x_3x_{10} + \cdots + \beta_{147}x_7^2x_{10}}$$

(interactions between quantitative terms and qualitative variable x_{10})

$$+ \quad\quad \underbrace{\beta_{148}x_1x_8x_9 + \beta_{149}x_2x_8x_9 + \beta_{150}x_3x_8x_9 + \cdots + \beta_{182}x_7^2x_8x_9}$$

(interactions between quantitative terms and qualitative term x_8x_9)

$$+ \quad\quad \underbrace{\beta_{183}x_1x_8x_{10} + \beta_{184}x_2x_8x_{10} + \beta_{185}x_3x_8x_{10} + \cdots + \beta_{217}x_7^2x_8x_{10}}$$

(interactions between quantitative terms and qualitative term x_8x_{10})

$$+ \quad\quad \underbrace{\beta_{218}x_1x_9x_{10} + \beta_{219}x_2x_9x_{10} + \beta_{220}x_3x_9x_{10} + \cdots + \beta_{252}x_7^2x_9x_{10}}$$

(interactions between quantitative terms and qualitative term x_9x_{10})

$$+ \underbrace{\beta_{253}x_1x_8x_9x_{10} + \beta_{254}x_2x_8x_9x_{10} + \beta_{255}x_3x_8x_9x_{10} + \cdots + \beta_{287}x_7^2x_8x_9x_{10}}$$

(interactions between quantitative terms and qualitative term $x_8x_9x_{10}$)

To fit this model, we would need to collect data for, at minimum, 289 executives! Otherwise, we will have 0 degrees of freedom for estimating σ^2, the variance of the random error term. Even if we could obtain a data set this large, the task of interpreting the β parameters in the model is a daunting one. This model, with its numerous multivariable interactions and squared terms, is way too complex to be of use in practice.

In this chapter, we consider two systematic methods designed to reduce a large list of potential predictors to a more manageable one. These techniques, known as **variable screening procedures**, objectively determine which independent variables in the list are the most important predictors of y and which are the least important predictors. The most widely used method, *stepwise regression*, is discussed in Section 6.2, while another popular method, the *all-possible-regressions-selection* procedure, is the topic of Section 6.3. In Section 6.4, several caveats of these methods are identified.

6.2
Stepwise Regression

One of the most widely used variable screening methods is known as **stepwise regression**. To run a stepwise regression, the user first identifies the dependent variable (response) y, and the set of potentially important independent variables, $x_1, x_2, \ldots, x_k$, where k is generally large. [*Note*: This set of variables could include both first-order and higher-order terms; as well as interactions.] The data are entered into the computer software, and the stepwise procedure begins.

STEP 1   The software program fits all possible one-variable models of the form

$$E(y) = \beta_0 + \beta_1 x_i$$

to the data, where x_i is the ith independent variable, $i = 1, 2, \ldots, k$. For each model, the test of the null hypothesis

$$H_0: \beta_1 = 0$$

against the alternative hypothesis

$$H_a : \beta_1 \neq 0$$

is conducted using the t-test (or the equivalent F-test) for a single β parameter. The independent variable that produces the largest (absolute) t value is declared the best one-variable predictor of y.*Call this independent variable x_1.

STEP 2   The stepwise program now begins to search through the remaining $(k-1)$ independent variables for the best two-variable model of the form

$$E(y) = \beta_0 + \beta_1 x_1 + \beta_2 x_i$$

This is done by fitting all two-variable models containing x_1 (the variable selected in the first step) and each of the other $(k - 1)$ options for the second variable x_i. The t values for the test $H_0 : \beta_2 = 0$ are computed for each of the $(k - 1)$ models (corresponding to the remaining independent variables, $x_i, i = 2, 3, \ldots, k$), and the variable having the largest t is retained. Call this variable x_2.

Before proceeding to Step 3, the stepwise routine will go back and check the t value of $\hat{\beta}_1$ after $\hat{\beta}_2 x_2$ has been added to the model. If the t value has become nonsignificant at some specified α level (say $\alpha = .05$), the variable x_1 is removed and a search is made for the independent variable with a β parameter that will yield the most significant t value in the presence of $\hat{\beta}_2 x_2$.

The reason the t value for x_1 may change from step 1 to step 2 is that the meaning of the coefficient $\hat{\beta}_1$ changes. In step 2, we are approximating a complex response surface in two variables with a plane. The best-fitting plane may yield a

---

*Note that the variable with the largest t value is also the one with the largest (absolute) Pearson product moment correlation, r (Section 3.7), with y.

different value for $\hat{\beta}_1$ than that obtained in step 1. Thus, both the value of $\hat{\beta}_1$ and its significance usually changes from step 1 to step 2. For this reason, stepwise procedures that recheck the t values at each step are preferred.

STEP 3   The stepwise regression procedure now checks for a third independent variable to include in the model with x_1 and x_2. That is, we seek the best model of the form

$$E(y) = \beta_0 + \beta_1 x_1 + \beta_2 x_2 + \beta_3 x_i$$

To do this, the computer fits all the $(k - 2)$ models using x_1, x_2, and each of the $(k - 2)$ remaining variables, x_i, as a possible x_3. The criterion is again to include the independent variable with the largest t value. Call this best third variable x_3. The better programs now recheck the t values corresponding to the x_1 and x_2 coefficients, replacing the variables that yield nonsignificant t values. This procedure is continued until no further independent variables can be found that yield significant t values (at the specified α level) in the presence of the variables already in the model.

The result of the stepwise procedure is a model containing only those terms with t values that are significant at the specified α level. Thus, in most practical situations only several of the large number of independent variables remain. However, it is very important *not* to jump to the conclusion that all the independent variables important for predicting y have been identified or that the unimportant independent variables have been eliminated. Remember, the stepwise procedure is using only *sample estimates* of the true model coefficients (β's) to select the important variables. An extremely large number of single β parameter t-tests have been conducted, and the probability is very high that one or more errors have been made in including or excluding variables. That is, we have very probably included some unimportant independent variables in the model (Type I errors) and eliminated some important ones (Type II errors).

There is a second reason why we might not have arrived at a good model. When we choose the variables to be included in the stepwise regression, we may often omit high-order terms (to keep the number of variables manageable). Consequently, we may have initially omitted several important terms from the model. Thus, we should recognize stepwise regression for what it is: an objective *variable screening* procedure.

Successful model builders will now consider second-order terms (for quantitative variables) and other interactions among variables screened by the stepwise procedure. It would be best to develop this response surface model with a second set of data independent of that used for the screening, so the results of the stepwise procedure can be partially verified with new data. This is not always possible, however, because in many modeling situations only a small amount of data is available.

Do not be deceived by the impressive-looking t values that result from the stepwise procedure—it has retained only the independent variables with the largest t values. Also, be certain to consider second-order terms in systematically developing the prediction model. Finally, if you have used a first-order model for your stepwise procedure, remember that it may be greatly improved by the addition of higher-order terms.

> **Caution**
> Be wary of using the results of stepwise regression to make inferences about the relationship between $E(y)$ and the independent variables in the resulting first-order model. First, an extremely large number of t-tests have been conducted, leading to a high probability of making one or more Type I or Type II errors. Second it is typical to enter only first-order and main effect terms as candidate variables in the stepwise model. Consequently, the final stepwise model will not include any higher-order or interaction terms. Stepwise regression should be used only when necessary, that is, when you want to determine which of a large number of potentially important independent variables should be used in the model-building process.

EXAMPLE 6.1

Refer to Example 4.9 (p. 221) and the multiple regression model for executive salary. A preliminary step in the construction of this model is the determination of the most important independent variables. For one firm, 10 potential independent variables (seven quantitative and three qualitative) were measured in a sample of 100 executives. The data, described in Table 6.1, are saved in the EXECSAL2 file. Since it would be very difficult to construct a complete second-order model with all of the 10 independent variables, use stepwise regression to decide which of the 10 variables should be included in the building of the final model for the natural log of executive salaries.

 EXECSAL2

TABLE 6.1   Independent Variables in the Executive Salary Example

INDEPENDENT VARIABLE	DESCRIPTION
x_1	Experience (years)—quantitative
x_2	Education (years)—quantitative
x_3	Gender (1 if male, 0 if female)—quantitative
x_4	Number of employees supervised—quantitative
x_5	Corporate assets (millions of dollars)—quantitative
x_6	Board member (1 if yes, 0 if no)—qualitative
x_7	Age (years)—quantitative
x_8	Company profits (past 12 months, millions of dollars)—quantitative
x_9	Has international responsibility (1 if yes, 0 if no)—qualitative
x_{10}	Company's total sales (past 12 months, millions of dollars)—quantitative

Solution

We will use stepwise regression with the main effects of the 10 independent variables to identify the most important variables. The dependent variable y is the natural logarithm of the executive salaries. The MINITAB stepwise regression printout is shown in Figure 6.1. MINITAB automatically enters the constant term (β_0) into the model in the first step. The remaining steps follow the procedure outlined earlier in this section.

In Step 1, MINITAB fits all possible one-variable models of the form,

$$E(y) = \beta_0 + \beta_1 x_i.$$

FIGURE 6.1

MINITAB stepwise regression
results for executive salaries

Stepwise Regression: Y versus X1, X2, X3, X4, X5, X6, X7, X8, X9, X10

Alpha-to-Enter: 0.15   Alpha-to-Remove: 0.15

Response is   Y   on 10 predictors, with N =   100

Step	1	2	3	4	5
Constant	11.091	10.968	10.783	10.278	9.962
X1	0.0278	0.0273	0.0273	0.0273	0.0273
T-Value	12.62	15.13	18.80	24.68	26.50
P-Value	0.000	0.000	0.000	0.000	0.000
X3		0.197	0.233	0.232	0.225
T-Value		7.10	10.17	13.30	13.74
P-Value		0.000	0.000	0.000	0.000
X4			0.00048	0.00055	0.00052
T-Value			7.32	10.92	11.06
P-Value			0.000	0.000	0.000
X2				0.0300	0.0291
T-Value				8.38	8.72
P-Value				0.000	0.000
X5					0.00196
T-Value					3.95
P-Value					0.000
S	0.161	0.131	0.106	0.0807	0.0751
R-Sq	61.90	74.92	83.91	90.75	92.06
R-Sq(adj)	61.51	74.40	83.41	90.36	91.64
C-p	343.9	195.5	93.8	16.8	3.6
PRESS	2.66387	1.78796	1.17124	0.695637	0.610197
R-Sq(pred)	60.14	73.24	82.47	89.59	90.87

You can see from Figure 6.1 that the first variable selected is x_1, years of experience. Thus, x_1 has the largest (absolute) t-value associated with a test of H_0: $\beta_1 = 0$. This value, $t = 12.62$, is highlighted on the printout.

Next (step 2), MINITAB fits all possible two-variable models of the form,

$$E(y) = \beta_0 + \beta_1 x_1 + \beta_2 x_i.$$

(Note that the variable selected in the first step, x_1, is automatically included in the model.) The variable with the largest (absolute) t-value associated with a test of H_0: $\beta_2 = 0$ is the dummy variable for gender, x_3. This t-value, $t = 7.10$, is also highlighted on the printout.

In Step 3, all possible three-variable models of the form

$$E(y) = \beta_0 + \beta_1 x_1 + \beta_2 x_3 + \beta_3 x_i$$

are fit. (Note that x_1 and x_3 are included in the model.) MINITAB selects x_4, number of employees supervised, based on the value $t = 7.32$ (highlighted on the printout) associated with a test of $H_0: \beta_3 = 0$.

In Steps 4 and 5, the variables x_2 (years of education) and x_5 (corporate assets), respectively, are selected for inclusion into the model. The t-values for the tests of the appropriate β's are highlighted on Figure 6.1. MINITAB stopped after five steps because none of the other independent variables met the criterion for admission to the model. As a default, MINITAB (and most other statistical software packages) uses $\alpha = .15$ in the tests conducted. In other words, if the p-value associated with a test of a β-coefficient is greater than $\alpha = .15$, then the corresponding variable is not included in the model.

The results of the stepwise regression suggest that we should concentrate on the five variables, x_1, x_2, x_3, x_4, and x_5 in our final modeling effort. Models with curvilinear terms as well as interactions should be proposed and evaluated (as demonstrated in Chapter 5) to determine the best model for predicting executive salaries. ◆

There are several other stepwise regression techniques designed to select the most important independent variables. One of these, called **forward selection**, is nearly identical to the stepwise procedure previously outlined. The only difference is that the forward selection technique provides no option for rechecking the t values corresponding to the x's that have entered the model in an earlier step. Thus, stepwise regression is preferred to forward selection in practice.

Another technique, called **backward elimination**, initially fits a model containing terms for all potential independent variables. That is, for k independent variables, the model $E(y) = \beta_0 + \beta_1 x_1 + \beta_2 x_2 + \cdots + \beta_k x_k$ is fit in step 1. The variable with the smallest t (or F) statistic for testing $H_0: \beta_i = 0$ is identified and dropped from the model if the t value is less than some specified critical value. The model with the remaining $(k - 1)$ independent variables is fit in step 2, and again, the variable associated with the smallest nonsignificant t value is dropped. This process is repeated until no further nonsignificant independent variables can be found. This approach can be an advantage when at least one of the candidate independent variables is a qualitative variable at three or more levels (requiring at least two dummy variables), since the backward procedure tests the contribution of each dummy variable after the others have been entered into the model. The real disadvantage of using the backward elimination technique is that you need a sufficiently large number of data points to fit the initial model in step 1.

6.3 **All-Possible-** **Regressions** **Selection Procedure**	In Section 6.2, we presented stepwise regression as an objective screening procedure for selecting the most important predictors of y. Other, more subjective, variable selection techniques have been developed in the literature for the purpose of identifying important independent variables. The most popular of these procedures are those that consider all possible regression models given the set of potentially important predictors. Such a procedure is commonly known as an **all-possible-regressions selection procedure**. The techniques differ with respect to the criteria for selecting the "best" subset of variables. In this section, we describe four criteria widely used in practice, then give an example illustrating the four techniques.

R^2 **CRITERION:**   Consider the set of potentially important variables, $x_1, x_2, x_3, \ldots,$ x_k. We learned in Section 4.7 that the multiple coefficient of determination

$$R^2 = 1 - \frac{\text{SSE}}{\text{SS(Total)}}$$

will increase when independent variables are added to the model. Therefore, the model that includes all k independent variables

$$E(y) = \beta_0 + \beta_1 x_1 + \beta_2 x_2 + \cdots + \beta_k x_k$$

will yield the largest R^2. Yet, we have seen examples (Chapter 5) where adding terms to the model does not yield a significantly better prediction equation. The objective of the R^2 criterion is to find a subset model (i.e., a model containing a subset of the k independent variables) so that adding more variables to the model will yield only small increases in R^2. In practice, the best model found by the R^2 criterion will rarely be the model with the largest R^2. Generally, you are looking for a simple model that is as good as, or nearly as good as, the model with all k independent variables. But unlike that in stepwise regression, the decision about when to stop adding variables to the model is a subjective one.

ADJUSTED R^2 **OR MSE CRITERION:**   One drawback to using the R^2 criterion, you will recall, is that the value of R^2 does not account for the number of β parameters in the model. If enough variables are added to the model so that the sample size n equals the total number of β's in the model, you will force R^2 to equal 1. Alternatively, we can use the adjusted R^2. It is easy to show that R_a^2 is related to MSE as follows:

$$R_a^2 = 1 - (n-1) \left[\frac{\text{MSE}}{\text{SS(Total)}} \right]$$

Note that R_a^2 increases only if MSE decreases [since SS(Total) remains constant for all models]. Thus, an equivalent procedure is to search for the model with the minimum, or near minimum, MSE.

C_p **CRITERION:**   A third option is based on a quantity called the **total mean square error** (**TMSE**) for the fitted regression model:

$$\text{TMSE} = E \left\{ \sum_{i=1}^{n} [\hat{y}_i - E(y_i)]^2 \right\} = \sum_{i=1}^{n} [E(\hat{y}_i) - E(y_i)]^2 + \sum_{i=1}^{n} \text{Var}(\hat{y}_i)$$

where $E(\hat{y}_i)$ is the mean response for the subset (fitted) regression model and $E(y_i)$ is the mean response for the true model. The objective is to compare the TMSE for the subset regression model with σ^2, the variance of the random error for the true model, using the ratio

$$\Gamma = \frac{\text{TMSE}}{\sigma^2}$$

Small values of Γ imply that the subset regression model has a small total mean square error relative to σ^2. Unfortunately, both TMSE and σ^2 are unknown, and we must rely on sample estimates of these quantities. It can be shown (proof omitted) that a good estimator of the ratio Γ is given by

$$C_p = \frac{\text{SSE}_p}{\text{MSE}_k} + 2(p+1) - n$$

where n is the sample size, p is the number of independent variables in the subset model, k is the total number of potential independent variables, SSE_p is the SSE for the subset model, and MSE_k is the MSE for the model containing all k independent variables. The statistical software packages discussed in this text have routines that calculate the C_p statistic. In fact, the C_p value is automatically printed at each step in the SAS and MINITAB stepwise regression printouts (see Figure 6.1).

The C_p criterion selects as the best model the subset model with (1) a small value of C_p (i.e., a small total mean square error) and (2) a value of C_p near $p + 1$, a property that indicates that slight or no bias exists in the subset regression model.*

Thus, the C_p criterion focuses on minimizing total mean square error and the regression bias. If you are mainly concerned with minimizing total mean square error, you will want to choose the model with the smallest C_p value, as long as the bias is not large. On the other hand, you may prefer a model that yields a C_p value slightly larger than the minimum but which has slight (or no) bias.

PRESS CRITERION: A fourth criterion used to select the best subset regression model is the PRESS statistic, introduced in Section 5.11. Recall that the PRESS (or, prediction sum of squares) statistic for a model is calculated as follows:

$$\text{PRESS} = \sum_{i=1}^{n} [y_i - \hat{y}_{(i)}]^2$$

where $\hat{y}_{(i)}$ denotes the predicted value for the ith observation obtained when the regression model is fit with the data point for the ith observation omitted (or deleted) from the sample.† Thus, the candidate model is fit to the sample data n times, each time omitting one of the data points and obtaining the predicted value of y for that data point. Since small differences $y_i - \hat{y}_{(i)}$ indicate that the model is predicting well, we desire a model with a small PRESS.

Computing the PRESS statistic may seem like a tiresome chore, since repeated regression runs (a total of n runs) must be made for each candidate model. However, most statistical software packages have options for computing PRESS automatically.‡

---

*A model is said to be *unbiased* if $E(\hat{y}) = E(y)$. We state (without proof) that for an unbiased regression model, $E(C_p) \approx p + 1$. In general, subset models will be biased since $k - p$ independent variables are omitted from the fitted model. However, when C_p is near $p + 1$, the bias is small and can essentially be ignored.

†The quantity $y_i - \hat{y}_{(i)}$ is called the "deleted" residual for the ith observation. We discuss deleted residuals in more detail in Chapter 8.

‡PRESS can also be calculated using the results from a regression run on all n data points. The formula is

$$\text{PRESS} = \sum_{i=1}^{n} \left(\frac{y_i - \hat{y}_i}{1 - h_{ii}} \right)^2$$

Plots aid in the selection of the best subset regression model using the all-possible-regressions procedure. The criterion measure, either R^2, MSE, C_p, or PRESS, is plotted on the vertical axis against p, the number of independent variables in the subset model, on the horizontal axis. We illustrate all three variable selection techniques in an example.

EXAMPLE   6.2

Refer to Example 6.1 and the data on executive salaries. Recall that we want to identify the most important independent variables for predicting the natural log of salary from the list of 10 variables given in Table 6.1. Apply the all-possible-regressions selection procedure to find the most important independent variables.

Solution

We entered the executive salary data into MINITAB and used MINITAB's all-possible-regressions selection routine to obtain the printout shown in Figure 6.2. For $p = 10$ independent variables, there exists 1,023 possible subset first-order models. Although MINITAB fits all of these models, the output in Figure 6.2 shows only the results for the "best" model for each value of p. From the printout, you can see that the best one-variable model includes x_1 (years of experience); the best two-variable model includes x_1 and x_3 (gender); the best three-variable model includes x_1, x_3 and x_4 (number supervised); and so on.

FIGURE 6.2

MINITAB all-possible-regressions selection results for executive salaries

Best Subsets Regression: Y versus X1, X2,

Response is Y

					X X X X X X X X X X 1
Vars	R-Sq	R-Sq(adj)	C-p	S	1 2 3 4 5 6 7 8 9 0
1	61.9	61.5	343.9	0.16119	X
2	74.9	74.4	195.5	0.13145	X X
3	83.9	83.4	93.8	0.10583	X X X
4	90.7	90.4	16.8	0.080676	X X X X
5	92.1	91.6	3.6	0.075118	X X X X X
6	92.2	91.7	4.0	0.074857	X X X X X X
7	92.3	91.7	5.4	0.075022	X X X X X X X
8	92.3	91.6	7.2	0.075326	X X X X X X X X
9	92.3	91.5	9.1	0.075707	X X X X X X X X X
10	92.3	91.4	11.0	0.076084	X X X X X X X X X X

These "best subset" models are summarized in Table 6.2. In addition to the variables included in each model, the table gives the values of R^2, adjusted-R^2, MSE, C_p, and PRESS. To determine which subset model to select, we plot these quantities against the number of variables, p. The MINITAB graphs for R^2, adjusted-R^2, C_p, and PRESS are shown in Figures 6.3a–d, respectively.

In Figure 6.3a, we see that the R^2 values tend to increase in very small amounts for models with more than $p = 5$ predictors. A similar pattern is shown in Figure 6.3b

---

where h_{ii} is a function of the independent variables in the model. In Chapter 8, we show how h_{ii} (called *leverage*) can be used to detected influential observations.

TABLE 6.2   **Results for Best Subset Models**

NUMBER OF PREDICTORS p	VARIABLES IN THE MODEL	R^2	ADJ-R^2	MSE	C_p	PRESS
1	x_1	.619	.615	.0260	343.9	2.664
2	x_1, x_3	.749	.744	.0173	195.5	1.788
3	x_1, x_3, x_4	.839	.834	.0112	93.8	1.171
4	x_1, x_2, x_3, x_4	.907	.904	.0065	16.8	.696
5	x_1, x_2, x_3, x_4, x_5	.921	.916	.0056	3.6	.610
6	$x_1, x_2, x_3, x_4, x_5, x_9$	.922	.917	.0056	4.0	.610
7	$x_1, x_2, x_3, x_4, x_5, x_6, x_9$	.923	.917	.0056	5.4	.620
8	$x_1, x_2, x_3, x_4, x_5, x_6, x_8, x_9$	.923	.916	.0057	7.2	.629
9	$x_1, x_2, x_3, x_4, x_5, x_6, x_7, x_8, x_9$	.923	.915	.0057	9.1	.643
10	$x_1, x_2, x_3, x_4, x_5, x_6, x_7, x_8, x_9, x_{10}$	.923	.914	.0058	11.0	.654

FIGURE 6.3

MINITAB plots of all-possible-regressions selection criteria for Example 6.2

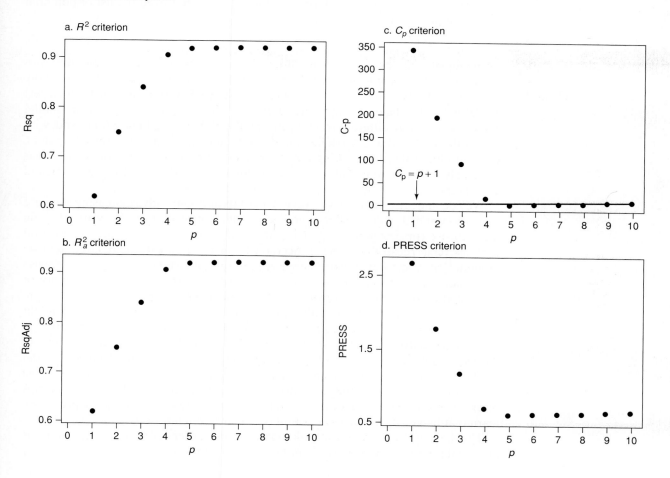

a. R^2 criterion

b. R_a^2 criterion

c. C_p criterion

d. PRESS criterion

for R^2_{adj}. Thus, both the R^2 and R^2_{adj} criteria suggest that the model containing the five predictors $x_1, x_2, x_3, x_4,$ and x_5 is a good candidate for the best subset regression model.

Figure 6.3c shows the plotted C_p values and the line $C_p = p + 1$. Notice that the subset models with $p \geq 5$ independent variables all have relatively small C_p values and vary tightly about the line $C_p = p+1$. This implies that these models have a small total mean square error and a negligible bias. The model corresponding to $p = 4$, although certainly outperforming the models $p \leq 3$, appears to fall short of the larger models according to the C_p criterion. From Figure 6.3d you can see that the PRESS is smallest for the five-variable model with $x_1, x_2, x_3, x_4,$ and x_5 (PRESS $= .610$).

According to all four criteria, the variables $x_1, x_2, x_3, x_4,$ and x_5 should be included in the group of the most important predictors.   ◆

In summary, variable selection procedures based on the all-possible-regressions selection criterion will assist you in identifying the most important independent variables for predicting y. Keep in mind, however, that these techniques lack the objectivity of a stepwise regression procedure. Furthermore, you should be wary of concluding that the best model for predicting y has been found, since, in practice, interactions and higher-order terms are typically omitted from the list of potential important predictors.

6.4
Caveats

Both stepwise regression and the all-possible-regressions selection procedure are useful *variable screening methods*. Many regression analysts, however, tend to apply these procedures as *model building methods*. Why? The stepwise (or best subset) model will often have a high value of R^2 and all the β coefficients in the model will be significantly different from 0 with small p-values. (see Figure 6.1). And, with very little work (other than collecting the data and entering it into the computer), you can obtain the model using a statistical software package. Consequently, it is extremely tempting to use the stepwise model as the *final* model for predicting and making inferences about the dependent variable, y.

We conclude this chapter with several caveats and some advice on using stepwise regression and the all-possible-regressions selection procedure. Be wary of using the stepwise (or best subset) model as the final model for predicting y for several reasons. First, recall that either procedure tends to fit an extremely large number of models and perform an extremely large number of tests (objectively, in stepwise regression, and subjectively, in best subsets regression). Thus, the probability of making at least one Type I error or at least one Type II error is often quite high. That is, you are very likely to either include at least one unimportant independent variable or leave out at least one important independent variable in the final model!

Second, analysts typically do not include higher-order terms or interactions in the list of potential predictors for stepwise regression. Therefore, if no real model building is performed, the final model will be a first-order, main effects model. Most real-world relationships between variables are not linear, and these relationships often are moderated by another variable (i.e., interaction exists). In Chapter 8, we learn that higher-order terms are often revealed through residual plotting.

Third, even if the analyst includes some higher-order terms and interactions as potential predictors, the stepwise and best subsets procedures will more than likely select a nonsensical model. For example, consider the stepwise model

$$E(y) = \beta_0 + \beta_1 x_1 + \beta_2 x_2 x_5 + \beta_3 x_3^2.$$

The model includes an interaction for x_2 and x_5, but omits the main effects for these terms, and it includes a quadratic term for x_3 but omits the first-order (shift parameter) term. Also, this strategy requires the analyst to "guess" or have an intuitive feel for which interactions may be the most important. If all possible interactions and squared terms are included in the list of potential predictors, the problem discussed in Section 6.1 (lacking sufficient data to estimate all the model parameters) will arise, especially in the all-possible-regressions selection method.

All of these problems can be avoided if we use stepwise or all-possible-regressions as they were originally intended—as objective methods of screening independent variables from a long list of potential predictors of y. Once the "most important" variables have been selected, *begin* the model-building phase of the analysis.

Summary

When the number of potential independent variables in a regression is large, fitting a complete second-order model for y is problematic because of the extremely high number of terms in the model. Either we have an insufficient sample size to estimate all the β's in the model, or the model is too complex to be of practical use. The problem of deciding which x's in a large set of independent variables are "most important" is known as **variable screening**.

Two widely used variable screening methods are **stepwise regression** and the **all-possible-regressions selection procedure**. In both methods, all possible 1-variable, 2-variable, 3-variable, etc., models are fit to the data. Stepwise regression determines the independent variable to add at each step by objectively testing the β-value associated with the x-value under consideration. The x with the largest absolute t-value is entered into the model. With all-possible-regressions, the analyst subjectively decides, at each step which variable(s) should be included in the model based on criteria such as R^2, adjusted-R^2, MSE, the C_p **statistic**, and **PRESS**.

You should avoid using these variable-screening methods alone to determine the "final" or "best" model for y. Numerous tests have been performed, highly inflating the overall probability of a Type I error, and the models that result from stepwise regression and the all-possible-regressions selection method typically do not include important interactions and higher-order terms. After screening the variables, use substantive theory and the model building methods of Chapter 5 to determine the "best" model for y.

SUPPLEMENTARY EXERCISES

6.1. There are six independent variables, $x_1, x_2, x_3, x_4, x_5,$ and x_6, that might be useful in predicting a response y. A total of $n = 50$ observations are available, and it is decided to employ stepwise regression to help in selecting the independent variables that appear to be useful. The computer fits all possible one-variable models of the form

$$E(y) = \beta_0 + \beta_1 x_i$$

where x_i is the ith independent variable, $i = 1, 2, \ldots, 6$. The information in the table is provided from the computer printout.

INDEPENDENT VARIABLE	$\hat{\beta}_i$	$s_{\hat{\beta}_i}$
x_1	1.6	.42
x_2	−.9	.01
x_3	3.4	1.14
x_4	2.5	2.06
x_5	−4.4	.73
x_6	.3	.35

a. Which independent variable is declared the best one-variable predictor of y? Explain.

b. Would this variable be included in the model at this stage? Explain.

c. Describe the next phase that a stepwise procedure would execute.

6.2. *Benefits Quarterly* (First Quarter, 1995) published a study of entry level job preferences. A number of independent variables were used to model the job preferences (measured on a 10-point scale) of 164 business school graduates. Suppose stepwise regression is used to build a model for job preference score (y) as a function of the following independent variables:

$$x_1 = \begin{cases} 1 & \text{if flextime position} \\ 0 & \text{if not} \end{cases}$$

$$x_2 = \begin{cases} 1 & \text{if day care support required} \\ 0 & \text{if not} \end{cases}$$

$$x_3 = \begin{cases} 1 & \text{if spousal transfer support required} \\ 0 & \text{if not} \end{cases}$$

x_4 = Number of sick days allowed

$$x_5 = \begin{cases} 1 & \text{if applicant married} \\ 0 & \text{if not} \end{cases}$$

x_6 = Number of children of applicant

$$x_7 = \begin{cases} 1 & \text{if male applicant} \\ 0 & \text{if female applicant} \end{cases}$$

a. How many models are fit to the data in step 1? Give the general form of these models.

b. How many models are fit to the data in step 2? Give the general form of these models.

c. How many models are fit to the data in step 3? Give the general form of these models.

d. Explain how the procedure determines when to stop adding independent variables to the model.

e. Describe two major drawbacks to using the final stepwise model as the "best" model for job preference score (y).

6.3. A marine biologist was hired by the EPA to determine whether the hot-water runoff from a particular power plant located near a large gulf is having an adverse effect on the marine life in the area. The biologist's goal is to acquire a prediction equation for the number of marine animals located at certain designated areas, or stations, in the gulf. Based on past experience, the EPA considered the following environmental factors as predictors for the number of animals at a particular station:

x_1 = Temperature of water (TEMP)

x_2 = Salinity of water (SAL)

x_3 = Dissolved oxygen content of water (DO)

x_4 = Turbidity index, a measure of the turbidity of the water (TI)

x_5 = Depth of the water at the station (ST_DEPTH)

x_6 = Total weight of sea grasses in sampled area (TGRSWT)

As a preliminary step in the construction of this model, the biologist used a stepwise regression procedure to identify the most important of these six variables. A total of 716 samples were taken at different stations in the gulf, producing the SAS printout shown on pages 335–337. (The response measured was y, the logarithm of the number of marine animals found in the sampled area.)

a. According to the SAS printout, which of the six independent variables should be used in the model? [*Note:* SAS uses the $F = t^2$ statistic for testing terms.]

b. Are we able to assume that the marine biologist has identified all the important independent variables for the prediction of y? Why?

c. Using the variables identified in part **a**, write the first-order model with interaction that may be used to predict y.

d. How would the marine biologist determine whether the model specified in part **c** is better than the first-order model?

e. Note the small value of R^2. What action might the biologist take to improve the model?

SAS output for Exercise 6.3

The REG Procedure
Dependent Variable: NUMBER

Stepwise Selection: Step 1

Variable ST_DEPTH Entered: R-Square = 0.1223 and C(p) = 51.57

Analysis of Variance

Source	DF	Sum of Squares	Mean Square	F Value	Pr > F
Model	1	57.44	57.44	99.47	<.0001
Error	714	412.33	0.58		
Corrected Total	715	469.77			

Variable	Parameter Estimate	Standard Error	Type II SS	F Value	Pr > F
Intercept	8.38559	5.60286	1.299	2.24	0.1356
ST_DEPTH	-0.43678	0.05763	57.44	99.03	<.0001

Stepwise Selection: Step 2

Variable TGRSWT Entered: R-Square = 0.1821 and C(p) = 1.52

Analysis of Variance

Source	DF	Sum of Squares	Mean Square	F Value	Pr > F
Model	2	85.55	42.78	79.38	<.0001
Error	713	384.22	0.54		
Corrected Total	715	469.77			

Variable	Parameter Estimate	Standard Error	Type II SS	F Value	Pr > F
Intercept	8.07682	7.41024	1.188	2.20	0.1411
ST_DEPTH	-0.35355	0.05966	35.11	65.02	<.0001
TGRSWT	0.00271	0.00051	28.17	52.16	<.0001

6.4. In any production process in which one or more workers are engaged in a variety of tasks, the total time spent in production varies as a function of the size of the work pool and the level of output of the various activities. For example, in a large metropolitan department store, the number of hours worked (y) per day by the clerical staff may depend on the following variables:

x_1 = Number of pieces of mail processed (open, sort, etc.)

x_2 = Number of money orders and gift certificates sold

x_3 = Number of window payments (customer charge accounts) transacted

x_4 = Number of change order transactions processed

x_5 = Number of checks cashed

x_6 = Number of pieces of miscellaneous mail processed on an "as available" basis

x_7 = Number of bus tickets sold

The table of observations at the bottom of p. 337 gives the output counts for these activities on each of 52 working days.

a. Conduct a stepwise regression analysis of the data using an available statistical software package.

b. Interpret the β estimates in the resulting stepwise model.

c. What are the dangers associated with drawing inferences from the stepwise model?

<u>SAS output for Exercise 6.3 (*continued*)</u>

Stepwise Selection: Step 3

Variable TI Entered: R-Square = 0.1870 and C(p) = 3.51

Analysis of Variance

Source	DF	Sum of Squares	Mean Square	F Value	Pr > F
Model	3	87.85	29.28	54.59	<.0001
Error	712	381.92	0.54		
Corrected Total	715	469.77			

Variable	Parameter Estimate	Standard Error	Type II SS	F Value	Pr > F
Intercept	7.38864	5.03901	1.161	2.15	0.1660
ST_DEPTH	-0.31451	0.04764	80.70	43.58	<.0001
TGRSWT	0.00262	0.00038	88.39	47.73	<.0001
TI	0.65774	0.31793	7.93	4.28	0.0389

Stepwise Selection: Step 4

Variable DO Entered: R-Square = 0.1889 and C(p) = 1.03

Analysis of Variance

Source	DF	Sum of Squares	Mean Square	F Value	Pr > F
Model	4	88.76	22.19	41.40	<.0001
Error	711	381.01	0.54		
Corrected Total	715	469.77			

Variable	Parameter Estimate	Standard Error	Type II SS	F Value	Pr > F
Intercept	7.33576	5.35015	1.015	1.88	0.1782
ST_DEPTH	-0.30417	0.04828	21.43	39.69	<.0001
TGRSWT	0.00267	0.00038	26.58	49.23	<.0001
TI	0.67347	0.31783	2.42	4.49	0.0345
DO	0.01769	0.01361	0.91	1.69	0.1946

6.5. Road construction contracts in the state of Florida are awarded on the basis of competitive, sealed bids; the contractor who submits the lowest bid price wins the contract. During the 1980s, the Office of the Florida Attorney General (FLAG) suspected numerous contractors of practicing bid collusion, i.e., setting the winning bid price above the fair, or competitive, price in order to increase profit margin. By comparing the bid prices (and other important bid variables) of the fixed (or rigged) contracts to the competitively bid contracts, FLAG was able to establish invaluable benchmarks for detecting future bid-rigging. FLAG collected data for 279 road construction contracts. For each contract, the following variables shown in the table on p. 337 were measured. (The data are saved in the file named FLAG.)

a. Consider building a model for the low-bid price (y). Apply stepwise regression to the data to find the independent variables most suitable for modeling y.

b. Interpret the β estimates in the resulting stepwise regression model.

c. What are the dangers associated with drawing inferences from the stepwise model?

SAS output for Exercise 6.3 (*continued*)

Stepwise Selection: Step 5

Variable DO Removed: R-Square = 0.1870 and C(p) = 3.51

Analysis of Variance

Source	DF	Sum of Squares	Mean Square	F Value	Pr > F
Model	3	87.85	29.28	54.59	<.0001
Error	712	381.92	0.54		
Corrected Total	715	469.77			

Variable	Parameter Estimate	Standard Error	Type II SS	F Value	Pr > F
Intercept	7.38864	5.03901	1.161	2.15	0.1660
ST_DEPTH	-0.31451	0.04764	80.70	43.58	<.0001
TGRSWT	0.00262	0.00038	88.39	47.73	<.0001
TI	0.65774	0.31793	7.93	4.28	0.0389

All variables left in the model are significant at the 0.0500 level.

No other variable met the 0.0500 significance level for entry into the model.

FLAG

1. Price of contract ($) bid by lowest bidder
2. Department of Transportation (DOT) engineer's estimate of fair contract price ($)
3. Ratio of low (winning) bid price to DOT engineer's estimate of fair price
4. Status of contract (1 if fixed, 0 if competitive)
5. District (1, 2, 3, 4, or 5) in which construction project is located
6. Number of bidders on contract
7. Estimated number of days to complete work
8. Length of road project (miles)
9. Percentage of costs allocated to liquid asphalt
10. Percentage of costs allocated to base material
11. Percentage of costs allocated to excavation
12. Percentage of costs allocated to mobilization
13. Percentage of costs allocated to structures
14. Percentage of costs allocated to traffic control
15. Subcontractor utilization (1 if yes, 0 if no)

6.6. Refer to the data on units of production and time worked for a department store clerical staff in Exercise 6.4. For this exercise, consider only the independent variables x_1, x_2, x_3, and x_4 in an all-possible-regressions select procedure.

a. How many models for $E(y)$ are possible, if the model includes (i) one variable, (ii) two variables, (iii) three variables, and (iv) four variables?

b. For each case in part **a**, use a statistical software package to find the maximum R^2, minimum MSE, minimum C_p, and minimum PRESS.

c. Plot each of the quantities R^2, MSE, C_p, and PRESS in part **b** against p, the number of predictors in the subset model.

d. Based on the plots in part **c**, which variables would you select for predicting total hours worked, y?

6.7. Apply the all-possible-regressions selection method to the FLAG data of Exercise 6.5. Are the variables in the "best subset" model the same as those selected by stepwise regression?

Data for Exercise 6.4
CLERICAL

OBS.	DAY OF WEEK	y	x_1	x_2	x_3	x_4	x_5	x_6	x_7
1	M	128.5	7781	100	886	235	644	56	737
2	T	113.6	7004	110	962	388	589	57	1029
3	W	146.6	7267	61	1342	398	1081	59	830
4	Th	124.3	2129	102	1153	457	891	57	1468

(*continued overleaf*)

(continued)

5	F	100.4	4878	45	803	577	537	49	335
6	S	119.2	3999	144	1127	345	563	64	918
7	M	109.5	11777	123	627	326	402	60	335
8	T	128.5	5764	78	748	161	495	57	962
9	W	131.2	7392	172	876	219	823	62	665
10	Th	112.2	8100	126	685	287	555	86	577
11	F	95.4	4736	115	436	235	456	38	214
12	S	124.6	4337	110	899	127	573	73	484
13	M	103.7	3079	96	570	180	428	59	456
14	T	103.6	7273	51	826	118	463	53	907
15	W	133.2	4091	116	1060	206	961	67	951
16	Th	111.4	3390	70	957	284	745	77	1446
17	F	97.7	6319	58	559	220	539	41	440
18	S	132.1	7447	83	1050	174	553	63	1133
19	M	135.9	7100	80	568	124	428	55	456
20	T	131.3	8035	115	709	174	498	78	968
21	W	150.4	5579	83	568	223	683	79	660
22	Th	124.9	4338	78	900	115	556	84	555
23	F	97.0	6895	18	442	118	479	41	203
24	S	114.1	3629	133	644	155	505	57	781
25	M	88.3	5149	92	389	124	405	59	236
26	T	117.6	5241	110	612	222	477	55	616
27	W	128.2	2917	69	1057	378	970	80	1210
28	Th	138.8	4390	70	974	195	1027	81	1452
29	F	109.5	4957	24	783	358	893	51	616
30	S	118.9	7099	130	1419	374	609	62	957
31	M	122.2	7337	128	1137	238	461	51	968
32	T	142.8	8301	115	946	191	771	74	719
33	W	133.9	4889	86	750	214	513	69	489
34	Th	100.2	6308	81	461	132	430	49	341
35	F	116.8	6908	145	864	164	549	57	902
36	S	97.3	5345	116	604	127	360	48	126
37	M	98.0	6994	59	714	107	473	53	726
38	T	136.5	6781	78	917	171	805	74	1100
39	W	111.7	3142	106	809	335	702	70	1721
40	Th	98.6	5738	27	546	126	455	52	502
41	F	116.2	4931	174	891	129	481	71	737
42	S	108.9	6501	69	643	129	334	47	473
43	M	120.6	5678	94	828	107	384	52	1083
44	T	131.8	4619	100	777	164	834	67	841
45	W	112.4	1832	124	626	158	571	71	627
46	Th	92.5	5445	52	432	121	458	42	313
47	F	120.0	4123	84	432	153	544	42	654
48	S	112.2	5884	89	1061	100	391	31	280
49	M	113.0	5505	45	562	84	444	36	814
50	T	138.7	2882	94	601	139	799	44	907
51	W	122.1	2395	89	637	201	747	30	1666
52	Th	86.6	6847	14	810	230	547	40	614

Source: Adapted from *Work Measurement*, by G. L. Smith, Grid Publishing Co., Columbus, Ohio, 1978 (Table 3.1).

REFERENCES

NETER, J., KUTNER, M. NACHTSHEIM, C., and WASSERMAN, W.
Applied Linear Statistical Models, 4th ed. Homewood, Ill.:
Richard D. Irwin, 1996.

CHAPTER

7

Some Regression Pitfalls

CONTENTS

OBJECTIVE

To identify several potential problems you may encounter when constructing a model for a response y; to help you recognize when these problems exist so that you can avoid some of the pitfalls of multiple regression analysis

---

7.1 Introduction

Multiple regression analysis is recognized by practitioners as a powerful tool for modeling a response y and is therefore widely used. But it is also one of the most abused statistical techniques. The ease with which a multiple regression analysis can be run with statistical computer software has opened the door to many data analysts who have but a limited knowledge of multiple regression and statistics. In practice, building a model for some response y is rarely a simple, straightforward process. There are a number of pitfalls that trap the unwary analyst. In this chapter, we discuss several problems that you should be aware of when constructing a multiple regression model.

7.2 Observational Data Versus Designed Experiments

One problem encountered in using a regression analysis is caused by the type of data that the analyst is often forced to collect. Recall, from Section 2.4, that the data for regression can be either *observational* (where the values of the independent variables are uncontrolled) or *experimental* (where the x's are controlled via a designed experiment). Whether data are observational or experimental is important for the following reasons. First, as you will subsequently learn in Chapter 11, the quantity of information in an experiment is controlled not only by the *amount of data*, but also by the *values of the predictor variables* $x_1, x_2, \ldots, x_k$. Consequently, if you can design the experiment (sometimes this is physically impossible), you may be able to increase greatly the amount of information in the data at no additional cost.

Second, the use of observational data creates a problem involving **randomization**. When an experiment has been designed and we have decided on the various settings of the independent variables to be used, the experimental units are then randomly assigned in such a way that each combination of the independent variables has an equal chance of receiving experimental units with unusually high (or low) readings. (We will illustrate this method of randomization in Chapter 12). This procedure tends to average out any variation within the experimental units. The result is that if the difference between two sample means is statistically significant, then you can infer (with probability of Type I error equal to α) that the population means differ. But more important, you can infer that this difference was due to the settings of the predictor variables, which is what you did to make the two populations different. Thus, you can infer a cause-and-effect relationship.

If the data are observational, a statistically significant relationship between a response y and a predictor variable x does not imply a cause-and-effect relationship. It simply means that x contributes information for the prediction of y, and nothing more. This point is aptly illustrated in the following example.

EXAMPLE 7.1

USA Today (Apr. 16, 2002) published an article titled "Cocaine Use During Pregnancy Linked to Development Problems." The article reports on the results of a study in which researchers gave IQ tests to two groups of infants recently born in Ohio—218 whose mothers admitted using cocaine during pregnancy and 197 who were not exposed to cocaine. The mothers (and their infants) who participated in the study were volunteers. About 80% of the mothers in each group were minorities, and both groups of women used a variety of legal and illegal substances, from alcohol to marijuana. The researchers found that "babies whose mothers use cocaine during pregnancy score lower on early intelligence tests... than those not exposed to the drug." The two variables measured in the study—IQ score at age 2, y, and cocaine use during pregnancy, x (where $x = 1$ if mother admits cocaine use during pregnancy and $x = 0$ if not) —were found to be negatively correlated. Although the quotation does not use the word *cause*, it certainly implies to the casual reader that a low IQ results from cocaine use during pregnancy.

a. Are the data collected in the study observational or experimental?

b. Identify any weaknesses in the study.

Solution

a. The mothers (and their infants) represent the experimental units in the study. Since no attempt was made to control the value of x, cocaine use during pregnancy, the data are observational.

b. The pitfalls provided by the study are apparent. First, the response y (IQ of infant at age 2) is related to only a single variable, x (cocaine use during pregnancy). Second, since the mothers who participated in the study were *not* randomly assigned to one of the two groups —an obvious impossible task —a real possibility exists that mothers with lower IQ and/or lower socioeconomic status tended to fall in the cocaine-use group. In other words, perhaps the study is simply showing that mothers from lower socioeconomic groups, who may not provide the daily care and nurturing that more fortunate children receive, are more likely to have babies with lower IQ scores. Also, many of the babies in the study were premature —a factor known to hinder a baby's development. Which variable, cocaine use during pregnancy, socioeconomic status, or premature status, is the cause of an infant's low IQ is impossible to determine based on the observational data collected in the study. This demonstrates the primary weakness of observational experiments. ◆

The point of the previous example is twofold. If you can control the values of the independent variables in an experiment, it pays to do so. If you cannot control them, you can still learn much from a regression analysis about the relationship between a response y and a set of predictors. In particular, a prediction equation that provides a good fit to your data will almost always be useful. But, **you must be careful about deducing cause-and-effect relationships between the response and the predictors in an observational experiment**.

Caution

With observational data, a statistically significant relationship between a response y and a predictor variable x *does not necessarily* imply a cause-and-effect relationship.

Learning about the design of experiments is useful even if most of your applications of regression analysis involve observational data. Learning how to design an experiment and control the information in the data will improve your ability to assess the quality of observational data. We introduce experimental design in Chapter 11 and present methods for analyzing the data in a designed experiment in Chapter 12.

7.3
Deviating from the Assumptions

When we apply a regression analysis to a set of data, we never know for certain that the assumptions about the random error term ε are satisfied. How far can we deviate from the assumptions and still expect a multiple regression analysis to yield results that will have the reliability stated in Chapter 4? How can we detect departures (if they exist) from the assumptions, and what can we do about them? We will provide some partial answers to these questions in this section and direct you to further discussion in succeeding chapters.

Remember (from Section 4.2) that for a given set of values of $x_1, x_2, \ldots, x_k$,

$$y = \beta_0 + \beta_1 x_1 + \beta_2 x_2 + \cdots + \beta_k x_k + \varepsilon$$

where ε is a random error. The first assumption we made was that the mean value of the random error for *any* given set of values of $x_1, x_2, \ldots, x_k$ is $E(\varepsilon) = 0$.

One consequence of this assumption is that the mean $E(y)$ for a specific set of values of $x_1, x_2, \ldots, x_k$ is

$$E(y) = \beta_0 + \beta_1 x_1 + \beta_2 x_2 + \cdots + \beta_k x_k$$

That is,

$$
y = \overbrace{E(y)}^{\substack{\text{Mean value of } y \\ \text{for specific values} \\ x_1, x_2, \ldots, x_k}} + \overbrace{\varepsilon}^{\substack{\text{Random} \\ \text{error}}}
$$

The second consequence of this assumption (which we state without proof) is that the least squares estimators of the model parameters $\beta_0, \beta_1, \beta_2, \ldots, \beta_k$, will be unbiased regardless of the remaining assumptions that we attribute to the random errors and their probability distributions.

The properties of the sampling distributions of the parameter estimators $\hat{\beta}_0$, $\hat{\beta}_1, \ldots, \hat{\beta}_k$ will depend on the remaining assumptions that we specify concerning the probability distributions of the random errors. You will recall that we assumed that for any given set of values of $x_1, x_2, \ldots, x_k$, ε has a normal probability distribution with mean equal to 0 and variance equal to σ^2. Also, we assumed that the random errors are independent (in a probabilistic sense).

It is unlikely that the assumptions stated above are satisfied exactly for many practical situations. If departures from the assumptions are not too great, experience has shown that a least squares regression analysis produces estimates—predictions and statistical test results—that have, for all practical purposes, the properties specified in Chapter 4. On the other hand, if the assumptions are flagrantly violated, any inferences derived from the regression analysis are suspect.

If the observations are likely to be correlated (as in the case of **time series data**—that is, data collected over time), we must check for correlation between the random errors (a topic to be discussed in Chapter 8) and we may have to modify our methodology if correlation exists. The solution to this problem is to construct a time series model; this will be the subject of Chapter 10. If the variance of the random error ε changes from one setting of the independent x variables to another, we can sometimes transform the data so that the standard least squares methodology will be appropriate. Some techniques for detecting non-homogeneous variances of the random errors (a condition called **heteroscedasticity**) and some methods for treating this type of data are discussed in Chapter 8. The normality assumption is the least restrictive of the assumptions when regression analysis is applied in practice. However, nonnormality can reduce the power* of statistical tests and may result in predicted values that deviate greatly from the observed values. A careful analysis of these extreme values, or **outliers**, is an important component of regression analysis. In Chapter 8, we give some methods for detecting outliers and determining their influence on the prediction equation.

Frequently, the data $(y, x_1, x_2, \ldots, x_k)$ are observational. Do these data violate the assumption that $x_1, x_2, \ldots, x_k$ are fixed? For this particular case, if we can assume that $x_1, x_2, \ldots, x_k$ are *measured without error*, the mean value $E(y)$ can be viewed as a conditional mean. That is, it gives the mean value of y, *given* that the x variables assume a specific set of values. With this modification in our thinking, the least squares regression analysis is applicable to observational data. Keep in mind,

---

*The power of a test is defined as: Power $= 1$-P (Type II error) $=$ P (Reject $H_o | H_a$ is true).

however, that inferences about $E(y)$ have the reliability stated in Chapter 4 only for the given set of x values.

To conclude, remember that when you perform a regression analysis, the reliability you can place in your inferences depends on having satisfied the assumptions prescribed in Section 4.2. We will never know for certain that the random errors satisfy these assumptions, but we will examine the residuals [the deviations $(y_i - \hat{y}_i)$ between the observed and the corresponding predicted values of y] in Chapter 8 to see whether we can discover patterns that suggest correlation, heteroscedasticity, nonnormality, or an improper choice for the deterministic portion of the model. An examination of the residuals will also have another beneficial effect: The magnitudes of the residuals will give you an idea of how well the model is predicting. This should convince you that, although the assumptions may not always be satisfied exactly, a multiple regression analysis is a powerful statistical tool.

7.4
Parameter Estimability and Interpretation

Suppose we want to fit the first-order model

$$E(y) = \beta_0 + \beta_1 x$$

to relate a developmentally challenged child's creativity score y to the child's flexibility score x. Now, suppose we collect data for three such challenged children, and each child has a flexibility score of 5. The data are shown in Figure 7.1. You can see the problem: The parameters of the straight-line model cannot be estimated when all the data are concentrated at a single x value. Recall that it takes two points (x values) to fit a straight line. Thus, the parameters are not estimable when only one x value is observed.

FIGURE 7.1

Creativity and flexibility data for three children

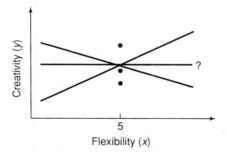

A similar problem would occur if we attempted to fit the second-order model

$$E(y) = \beta_0 + \beta_1 x + \beta_2 x^2$$

to a set of data for which only one *or two* different x values were observed (see Figure 7.2). At least three different x values must be observed before a second-order model can be fitted to a set of data (that is, before all three parameters are estimable). In general, the number of levels of a quantitative independent variable x must be at least one more than the order of the polynomial in x that you want to fit. If two values of x are too close together, you may not be able to estimate a parameter because of rounding error encountered in fitting the model. Remember, also, that the sample size n must be sufficiently large so that the degrees of freedom for estimating σ^2, $df(\text{Error}) = n - (k + 1)$, exceeds 0. In other words, n must exceed the number of β parameters in the model, $k + 1$. The requirements for fitting a pth-order polynomial regression model are shown in the box.

FIGURE 7.2

Only two different x values observed—the second-order model is not estimable

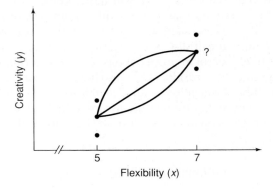

Requirements for Fitting a pth-Order Polynomial Regression Model

$$E(y) = \beta_0 + \beta_1 x + \beta_2 x^2 + \cdots + \beta_p x^p$$

1. The number of levels of x must be greater than or equal to $(p + 1)$.
2. The sample size n must be greater than $(p + 1)$ to allow sufficient degrees of freedom for estimating σ^2.

Most variables are not controlled by the researcher, but the independent variables are usually observed at a sufficient number of levels to permit estimation of the model parameters. However, when the computer program you use is unable to fit a model, the problem is probably inestimable parameters.

Given that the parameters of the model are estimable, it is important to interpret the parameter estimates correctly. A typical misconception is that $\hat{\beta}_i$ always measures the effect of x_i on $E(y)$, *independently* of the other x variables in the model. This may be true for some models, but it is not true in general. We will see in Section 7.5 that when the independent variables are correlated, the values of the estimated β coefficients are often misleading. Even if the independent variables are uncorrelated, the presence of interaction changes the meaning of the parameters. For example, the underlying assumption of the first-order model

$$E(y) = \beta_0 + \beta_1 x_1 + \beta_2 x_2$$

is, in fact, that x_1 and x_2 affect the mean response $E(y)$ independently. Recall from Sections 4.5 and 5.4 that the slope parameter β_1 measures the rate of change of y for a 1-unit increase in x_1, for any given value of x_2. However, if the relationship between $E(y)$ and x_1 depends on x_2 (i.e., if x_1 and x_2 interact), then the interaction model

$$E(y) = \beta_0 + \beta_1 x_1 + \beta_2 x_2 + \beta_3 x_1 x_2$$

is more appropriate. For the interaction model, we showed that the effect of x_1 on $E(y)$, i.e., the slope, is not measured by a single β parameter, but by $\beta_1 + \beta_3 x_2$.

Generally, the interpretation of an individual β parameter becomes increasingly difficult as the model becomes more complex. As we learned in Chapter 5, the individual β's of higher-order models usually have no practical interpretation.

Another misconception about the parameter estimates is that the magnitude of $\hat{\beta}_i$ determines the importance of x_i; that is, the larger (in absolute value) the $\hat{\beta}_i$, the more important the independent variable x_i is as a predictor of y. We learned in Chapter 4, however, that the standard error of the estimate $s_{\hat{\beta}_i}$ is critical in making inferences about the true parameter value. To reliably assess the importance of an individual term in the model, we conduct a test of H_0: $\beta_i = 0$ or construct a confidence interval for β_i using formulas that reflect the magnitude of $s_{\hat{\beta}_i}$.

In addition to the parameter estimates, $\hat{\beta}_i$, some statistical software packages report the **standardized regression coefficients**,

$$\hat{\beta}_i^* = \hat{\beta}_i \left(\frac{s_{x_i}}{s_y} \right)$$

where s_{x_i} and s_y are the standard deviations of the x_i and y values, respectively, in the sample. Unlike $\hat{\beta}_i$, $\hat{\beta}_i^*$ is scaleless. These standardized regression coefficients make it more feasible to compare parameter estimates since the units are the same. However, the problems with interpreting standardized regression coefficients are much the same as those mentioned previously. Therefore, you should be wary of using a standardized regression coefficient as the sole determinant of an x variable's importance. The next example illustrates this point.

EXAMPLE 7.2

Refer to the problem of modeling the auction price y of antique grandfather clocks, Examples 4.1 and 4.2. In Example 4.1, we fit the model

$$E(y) = \beta_0 + \beta_1 x_1 + \beta_2 x_2$$

where x_1 = age of the clock and x_2 = number of bidders. The SPSS printout for the regression analysis is shown in Figure 7.3. Locate the standardized β coefficients on the printout and interpret them.

Solution

The standardized β coefficients are highlighted on the SPSS printout in the column labeled **Beta**. These values, are

$$\hat{\beta}_1^* = .887 \quad \text{and} \quad \hat{\beta}_2^* = .620$$

Compare these values to the unstandardized β coefficients (in the **B** column):

$$\hat{\beta}_1 = 12.741 \quad \text{and} \quad \hat{\beta}_2 = 85.953$$

Based on the fact that $\hat{\beta}_2$ is nearly seven times larger than $\hat{\beta}_1$, we might be tempted to say that number of bidders (x_2) is a more important predictor of auction price than age of the clock (x_1). Once we standardize the β's (i.e., take the units of

FIGURE 7.3
SPSS regression printout for
grandfather clock model

Model Summary

Model	R	R Square	Adjusted R Square	Std. Error of the Estimate
1	.945[a]	.892	.885	133.485

a. Predictors: (Constant), BIDDERS, AGE

ANOVA[b]

Model		Sum of Squares	df	Mean Square	F	Sig.
1	Regression	4283063	2	2141531.480	120.188	.000[a]
	Residual	516726.5	29	17818.157		
	Total	4799790	31			

a. Predictors: (Constant), BIDDERS, AGE

b. Dependent Variable: PRICE

Coefficients[a]

Model		Unstandardized Coefficients		Standardized Coefficients	t	Sig.
		B	Std. Error	Beta		
1	(Constant)	-1338.951	173.809		-7.704	.000
	AGE	12.741	.905	.887	14.082	.000
	BIDDERS	85.953	8.729	.620	9.847	.000

a. Dependent Variable: PRICE

measurement and variation into account), we see that the opposite may, in fact, be true since $\hat{\beta}_1^*$ exceeds $\hat{\beta}_2^*$. Of course, from Example 4.2 we know that the two independent variables, x_1 and x_2, interact to affect y. Consequently, both age and number of bidders are important for predicting auction price and we should resist inferring that one of the variables is more important than the other.   ◆

7.5
Multicollinearity

Often, two or more of the independent variables used in the model for $E(y)$ will contribute redundant information. That is, the independent variables will be correlated with each other. For example, suppose we want to construct a model to predict the gasoline mileage rating, y, of a truck as a function of its load, x_1, and the horsepower, x_2, of its engine. In general, you would expect heavier loads to require greater horsepower and to result in lower mileage ratings. Thus, although both x_1 and x_2 contribute information for the prediction of mileage rating, some of the information is overlapping, because x_1 and x_2 are correlated. When the independent variables are correlated, we say that **multicollinearity** exists. In practice, it is not uncommon to observe correlations among the independent variables. However, a few problems arise when serious multicollinearity is present in the regression analysis.

Definition 7.1

Multicollinearity exists when two or more of the independent variables used in regression are moderately or highly correlated.

First, high correlations among the independent variables (i.e., **extreme** multicollinearity) increase the likelihood of rounding errors in the calculations of the β estimates, standard errors, and so forth.* Second, the regression results may be confusing and misleading.

To illustrate, if the gasoline mileage rating model

$$E(y) = \beta_0 + \beta_1 x_1 + \beta_2 x_2$$

were fitted to a set of data, we might find that the t values for both $\hat{\beta}_1$ and $\hat{\beta}_2$ (the least squares estimates) are nonsignificant. However, the F test for $H_0 : \beta_1 = \beta_2 = 0$ would probably be highly significant. The tests may seem to be contradictory, but really they are not. The t tests indicate that the contribution of one variable, say, $x_1 = $ load , is not significant after the effect of $x_2 = $ horsepower has been discounted (because x_2 is also in the model). The significant F test, on the other hand, tells us that at least one of the two variables is making a contribution to the prediction of y (i.e., β_1, β_2, or both differ from 0). In fact, both are probably contributing, but the contribution of one overlaps with that of the other.

Multicollinearity can also have an effect on the signs of the parameter estimates. More specifically, a value of $\hat{\beta}_i$ may have the opposite sign from what is expected. For example, we expect the signs of both of the parameter estimates for the gasoline mileage rating model to be negative, yet the regression analysis for the model might yield the estimates $\hat{\beta}_1 = .2$ and $\hat{\beta}_2 = -.7$. The positive value of $\hat{\beta}_1$ seems to contradict our expectation that heavy loads will result in lower mileage ratings. We mentioned in the previous section, however, that it is dangerous to interpret a β coefficient when the independent variables are correlated. Because the variables contribute redundant information, the effect of load x_1 on mileage rating is measured only partially by $\hat{\beta}_1$. Also, we warned in Section 6.2 that we cannot establish a cause-and-effect relationship between y and the predictor variables based on observational data. By attempting to interpret the value $\hat{\beta}_1$, we are really trying to establish a cause-and-effect relationship between y and x_1 (by suggesting that a heavy load x_1 will *cause* a lower mileage rating y).

How can you avoid the problems of multicollinearity in regression analysis? One way is to conduct a designed experiment so that the levels of the x variables are uncorrelated (see Section 7.2). Unfortunately, time and cost constraints may prevent you from collecting data in this manner. For these and other reasons, much of the data collected in scientific studies are observational. Since observational data frequently

---

*The result is due to the fact that, in the presence of severe multicollinearity, the computer has difficulty inverting the information matrix $(X'X)$. See Appendix A for a discussion of the $(X'X)$ matrix and the mechanics of a regression analysis.

consist of correlated independent variables, you will need to recognize when multicollinearity is present and, if necessary, make modifications in the analysis.

Several methods are available for detecting multicollinearity in regression. A simple technique is to calculate the coefficient of correlation r between each pair of independent variables in the model. If one or more of the r values is close to 1 or -1, the variables in question are highly correlated and a severe multicollinearity problem may exist.[†] Other indications of the presence of multicollinearity include those mentioned in the beginning of this section—namely, nonsignificant t tests for the individual β parameters when the F test for overall model adequacy is significant, and estimates with opposite signs from what is expected.

A more formal method for detecting multicollinearity involves the calculation of **variance inflation factors** for the individual β parameters. One reason why the t tests on the individual β parameters are nonsignificant is that the standard errors of the estimates, $s_{\hat{\beta}_i}$, are inflated in the presence of multicollinearity. When the dependent and independent variables are appropriately transformed,[‡] it can be shown that

$$s_{\hat{\beta}_i}^2 = s^2 \left(\frac{1}{1 - R_i^2} \right)$$

where s^2 is the estimate of σ^2, the variance of ε, and R_i^2 is the multiple coefficient of determination for the model that regresses the independent variable x_i on the remaining independent variables $x_1, x_2, \ldots, x_{i-1}, x_{i+1}, \ldots, x_k$. The quantity $1/(1 - R_i^2)$ is called the *variance inflation factor* for the parameter β_i, denoted $(\text{VIF})_i$. Note that $(\text{VIF})_i$ will be large when R_i^2 is large—that is, when the independent variable x_i is strongly related to the other independent variables.

Various authors maintain that, in practice, a severe multicollinearity problem exists if the largest of the variance inflation factors for the β's is greater than 10 or, equivalently, if the largest multiple coefficient of determination, R_i^2, is greater than .90.[§] Several of the statistical software packages discussed in this text have options for calculating variance inflation factors.[¶]

The methods for detecting multicollinearity are summarized in the accompanying box. We illustrate the use of these statistics in Example 7.3.

[†]Remember that r measures only the pairwise correlation between x values. Three variables, x_1, x_2, and x_3, may be highly correlated as a group, but may not exhibit large pairwise correlations. Thus, multicollinearity may be present even when all pairwise correlations are not significantly different from 0.

[‡]The transformed variables are obtained as

$$y_i^* = (y_i - \overline{y})/s_y \quad x_{1i}^* = (x_{1i} - \overline{x}_1)/s_1 \quad x_{2i}^* = (x_{2i} - \overline{x}_2)/s_2$$

and so on, where $\overline{y}, \overline{x}_1, \overline{x}_2, \ldots$, and $s_y, s_1, s_2, \ldots$, are the sample means and standard deviations, respectively, of the original variables.

[§]See, for example, Montgomery, Peck, and Vining (2001) or Neter, Kutner, Nachtsheim, and Wasserman, (1996).

[¶]Some software packages calculate an equivalent statistic, called the **tolerance**. The tolerance for a β coefficient is the reciprocal of the variance inflation factor, i.e.,

$$(\text{TOL})_i = \frac{1}{(\text{VIF})_i} = 1 - R_i^2$$

Detecting Multicollinearity in the Regression Model

$$E(y) = \beta_0 + \beta_1 x_1 + \beta_2 x_2 + \cdots + \beta_k x_k$$

The following are indicators of multicollinearity:

1. Significant correlations between pairs of independent variables in the model
2. Nonsignificant t tests for all (or nearly all) the individual β parameters when the F test for overall model adequacy $H_0: \beta_1 = \beta_2 = \cdots = \beta_k = 0$ is significant
3. Opposite signs (from what is expected) in the estimated parameters
4. A variance inflation factor (VIF) for a β parameter greater than 10, where

$$(\text{VIF})_i = \frac{1}{1 - R_i^2}, \quad i = 1, 2, \ldots, k$$

and R_i^2 is the multiple coefficient of determination for the model

$$E(x_i) = \alpha_0 + \alpha_1 x_1 + \alpha_2 x_2 + \cdots + \alpha_{i-1} x_{i-1} + \alpha_{i+1} x_{i+1} + \cdots + \alpha_k x_k$$

EXAMPLE 7.3

The Federal Trade Commission (FTC) annually ranks varieties of domestic cigarettes according to their tar, nicotine, and carbon monoxide contents. The U.S. surgeon general considers each of these three substances hazardous to a smoker's health. Past studies have shown that increases in the tar and nicotine contents of a cigarette are accompanied by an increase in the carbon monoxide emitted from the cigarette smoke. Table 7.1 lists tar, nicotine, and carbon monoxide contents (in milligrams) and weight (in grams) for a sample of 25 (filter) brands tested in a recent year. Suppose we want to model carbon monoxide content, y, as a function of tar content, x_1, nicotine content, x_2, and weight, x_3, using the model

$$E(y) = \beta_0 + \beta_1 x_1 + \beta_2 x_2 + \beta_3 x_3$$

The model is fit to the 25 data points in Table 7.1. A portion of the resulting SAS printout is shown in Figure 7.4. Examine the printout. Do you detect any signs of multicollinearity?

Solution

First, notice that a test of the global utility of the model

$$H_0: \beta_1 = \beta_2 = \beta_3 = 0$$

is highly significant. The F value (shaded on the printout) is very large ($F = 78.98$), and the observed significance level of the test (also shaded) is small ($p < .0001$).

For $R_i^2 > .90$ (the extreme multicollinearity case), $(\text{TOL})_i < .10$. These computer packages allow the user to set tolerance limits, so that any independent variable with a value of $(\text{TOL})_i$ below the tolerance limit will not be allowed to enter into the model.

FTCCIGAR

TABLE 7.1   FTC Cigarette Data for Example 7.3

BRAND	TAR x_1, milligrams	NICOTINE x_2, milligrams	WEIGHT x_3, grams	CARBON MONOXIDE y, milligrams
Alpine	14.1	.86	.9853	13.6
Benson & Hedges	16.0	1.06	1.0938	16.6
Bull Durham	29.8	2.03	1.1650	23.5
Camel Lights	8.0	.67	.9280	10.2
Carlton	4.1	.40	.9462	5.4
Chesterfield	15.0	1.04	.8885	15.0
Golden Lights	8.8	.76	1.0267	9.0
Kent	12.4	.95	.9225	12.3
Kool	16.6	1.12	.9372	16.3
L&M	14.9	1.02	.8858	15.4
Lark Lights	13.7	1.01	.9643	13.0
Marlboro	15.1	.90	.9316	14.4
Merit	7.8	.57	.9705	10.0
Multifilter	11.4	.78	1.1240	10.2
Newport Lights	9.0	.74	.8517	9.5
Now	1.0	.13	.7851	1.5
Old Gold	17.0	1.26	.9186	18.5
Pall Mall Light	12.8	1.08	1.0395	12.6
Raleigh	15.8	.96	.9573	17.5
Salem Ultra	4.5	.42	.9106	4.9
Tareyton	14.5	1.01	1.0070	15.9
True	7.3	.61	.9806	8.5
Viceroy Rich Lights	8.6	.69	.9693	10.6
Virginia Slims	15.2	1.02	.9496	13.9
Winston Lights	12.0	.82	1.1184	14.9

Source: Federal Trade Commission.

Therefore, we can reject H_0 for any α greater than .0001 and conclude that at least one of the parameters, β_1, β_2, and β_3, is nonzero. The t tests for two of the three individual β's, however, are nonsignificant. (The p-values for these tests are shaded on the printout.) Unless tar is the only one of the three variables useful for predicting carbon monoxide content, these results are the first indication of a potential multicollinearity problem.

A second clue to the presence of multicollinearity is the negative value for $\hat{\beta}_2$ and $\hat{\beta}_3$ (shaded on the printout),

$$\hat{\beta}_2 = -2.63 \quad \text{and} \quad \hat{\beta}_3 = -.13$$

From past studies, the FTC expects carbon monoxide content y to increase when either nicotine content x_2 or weight x_3 increases—that is, the FTC expects *positive* relationships between y and x_2, and y and x_3, not negative ones.

A more formal procedure for detecting multicollinearity is to examine the variance inflation factors. Figure 7.4 shows the variance inflation factors (shaded) for each of the three parameters under the column labeled **Variance Inflation**. Note that the variance inflation factors for both the tar and nicotine parameters are greater than 10. The variance inflation factor for the tar parameter, $(\text{VIF})_1 = 21.63$, implies that a model relating tar content x_1 to the remaining two independent variables,

FIGURE 7.4

SAS regression printout for FTC model

```
                           Dependent Variable: CO

                            Analysis of Variance

                                   Sum of          Mean
         Source          DF       Squares         Square     F Value    Pr > F

         Model            3      495.25781      165.08594      78.98    <.0001
         Error           21       43.89259        2.09012
         Corrected Total 24      539.15040


                   Root MSE           1.44573     R-Square    0.9186
                   Dependent Mean    12.52800     Adj R-Sq    0.9070
                   Coeff Var         11.53996


                            Parameter Estimates

                        Parameter      Standard                        Variance
         Variable   DF    Estimate        Error    t Value   Pr > |t|  Inflation

         Intercept   1     3.20219       3.46175      0.93     0.3655          0
         TAR         1     0.96257       0.24224      3.97     0.0007   21.63071
         NICOTINE    1    -2.63166       3.90056     -0.67     0.5072   21.89992
         WEIGHT      1    -0.13048       3.88534     -0.03     0.9735    1.33386
```

nicotine content x_2 and weight x_3, resulted in a coefficient of determination

$$R_1^2 = 1 - \frac{1}{(\text{VIF})_1}$$

$$= 1 - \frac{1}{21.63} = .954$$

All signs indicate that a serious multicollinearity problem exists. To confirm our suspicions, we used SAS to find the coefficient of correlation r for each of the three pairs of independent variables in the model. These values are highlighted on the SAS printout, Figure 7.5. You can see that tar content x_1 and nicotine content x_2 appear to be highly correlated ($r = .977$), whereas weight x_3 appears to be moderately correlated with both tar content ($r = .491$) and nicotine content ($r = .500$). In fact, all three sample correlations are significantly different from 0 based on the small p-values, also shown in Figure 7.5. ◆

FIGURE 7.5

SAS correlation matrix for independent variables in FTC model

```
          Pearson Correlation Coefficients, N = 25
                Prob > |r| under H0: Rho=0

                      TAR        NICOTINE      WEIGHT

         TAR        1.00000       0.97661      0.49077
                                  <.0001       0.0127

         NICOTINE   0.97661       1.00000      0.50018
                    <.0001                     0.0109

         WEIGHT     0.49077       0.50018      1.00000
                    0.0127        0.0109
```

Once you have detected that a multicollinearity problem exists, there are several alternative measures available for solving the problem. The appropriate measure to take depends on the severity of the multicollinearity and the ultimate goal of the regression analysis.

Some researchers, when confronted with highly correlated independent variables, choose to include only one of the correlated variables in the final model. One way of deciding which variable to include is to use *stepwise regression*, a topic discussed in Chapter 6. Generally, only one (or a small number) of a set of multicollinear independent variables will be included in the regression model by the stepwise regression procedure since this procedure tests the parameter associated with each variable in the presence of all the variables already in the model. For example, in fitting the gasoline mileage rating model introduced earlier, if at one step the variable representing truck load is included as a significant variable in the prediction of the mileage rating, the variable representing horsepower will probably never be added in a future step. Thus, if a set of independent variables is thought to be multicollinear, some screening by stepwise regression may be helpful.

If you are interested only in using the model for estimation and prediction, you may decide not to drop any of the independent variables from the model. In the presence of multicollinearity, we have seen that it is dangerous to interpret the individual β's for the purpose of establishing cause and effect. However, confidence intervals for $E(y)$ and prediction intervals for y generally remain unaffected *as long as the values of the independent variables used to predict y follow the same pattern of multicollinearity exhibited in the sample data.* That is, you must take strict care to ensure that the values of the x variables fall within the experimental region. (We discuss this problem in further detail in Section 7.6.) Alternatively, if your goal is to establish a cause-and-effect relationship between y and the independent variables, you will need to conduct a designed experiment to break up the pattern of multicollinearity.

When fitting a polynomial regression model, e.g., the second-order model

$$E(y) = \beta_0 + \beta_1 x + \beta_2 x^2$$

the independent variables $x_1 = x$ and $x_2 = x^2$ will often be correlated. If the correlation is high, the computer solution may result in extreme rounding errors. For this model, the solution is not to drop one of the independent variables but to transform the x variable in such a way that the correlation between the coded x and x^2 values is substantially reduced. Coding the independent quantitative variables as described in optional Section 5.6 is a useful technique for reducing the multicollinearity inherent with polynomial regression models.

Another, more complex, procedure for reducing the rounding errors caused by multicollinearity involves a modification of the least squares method, called **ridge regression**. In ridge regression, the estimates of the β coefficients are biased [that is, $E(\beta_i) \neq \beta_i$] but have significantly smaller standard errors than the unbiased β estimates yielded by the least squares method. Thus, the β estimates for the ridge regression are more stable than the corresponding least squares estimates. Ridge regression is a topic discussed in optional Chapter 9. A summary of the solutions is given in the box on p. 353.

7.6
Extrapolation: Predicting Outside the Experimental Region

By the late 1960s many research economists had developed highly technical models to relate the state of the economy to various economic indices and other independent variables. Many of these models were multiple regression models, where, for example, the dependent variable y might be next year's growth in Gross Domestic Product (GDP) and the independent variables might include this year's rate of inflation, this year's Consumer Price Index, and so forth. In other words, the model might be constructed to predict next year's economy using this year's knowledge.

> **Solutions to Some Problems Created by Multicollinearity**
>
> 1. Drop one or more of the correlated independent variables from the final model. A screening procedure such as stepwise regression is helpful in determining which variables to drop.
> 2. If you decide to keep all the independent variables in the model:
>   (a) Avoid making inferences about the individual β parameters (such as establishing a cause-and-effect relationship between y and the predictor variables).
>   (b) Restrict inferences about $E(y)$ and future y values to values of the independent variables that fall within the experimental region (see Section 7.6).
> 3. If your ultimate objective is to establish a cause-and-effect relationship between y and the predictor variables, use a designed experiment (see Chapters 11 and 12).
> 4. To reduce rounding errors in polynomial regression models, code the independent variables so that first-, second-, and higher-order terms for a particular x variable are not highly correlated (see Section 5.6).
> 5. To reduce rounding errors and stabilize the regression coefficients, use ridge regression to estimate the β parameters (see Section 9.7).

Unfortunately, these models were almost unanimously unsuccessful in predicting the recession in the early 1970s. What went wrong? Well, one of the problems was that regression models were used to predict y for values of the independent variables that were outside the region in which the model was developed. For example, the inflation rate in the late 1960s, when the models were developed, ranged from 6% to 8%. When the double-digit inflation of the early 1970s became a reality, some researchers attempted to use the same models to predict the growth in GDP 1 year hence. As you can see in Figure 7.6, the model may be very accurate for predicting y when x is in the range of experimentation, but the use of the model outside that range is a dangerous (although sometimes unavoidable) practice. A $100(1 - \alpha)\%$ prediction interval for GDP when the inflation rate is, say, 10%, will be less reliable than the stated confidence coefficient $(1 - \alpha)$. How much less is unknown.

FIGURE 7.6

Using a regression model outside the experimental region

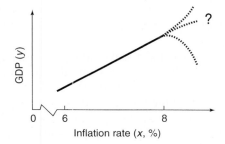

For a single independent variable x, the experimental region is simply the range of the values of x in the sample. Establishing the experimental region for a multiple regression model that includes a number of independent variables may be more difficult. For example, consider a model for GDP (y) using inflation rate (x_1) and prime interest rate (x_2) as predictor variables. Suppose a sample of size $n = 5$ was observed, and the values of x_1 and x_2 corresponding to the five values for GDP were (6, 10), (6.25, 12), (7.25, 10.25), (7.5, 13), and (8, 11.5). Notice that x_1 ranges from 6% to 8% and x_2 ranges from 10% to 13% in the sample data. You may think that the experimental region is defined by the ranges of the individual variables, i.e., $6 \leq x_1 \leq 8$ and $10 \leq x_2 \leq 13$. However, the levels of x_1 and x_2 *jointly* define the region. Figure 7.7 shows the experimental region for our hypothetical data. You can see that an observation with levels $x_1 = 8$ and $x_2 = 10$ clearly falls outside the experimental region, yet is within the ranges of the individual x values. Using the model to predict GDP for this observation may lead to unreliable results.

FIGURE 7.7

Experimental region for modeling GDP (y) as a function of inflation rate (x_1) and prime interest rate (x_2)

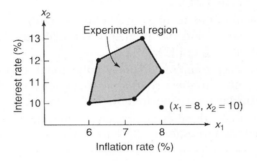

7.7
Data Transformations

The word *transform* means to change the form of some object or thing. Consequently, the phrase *data transformation* means that we have done, or plan to do, something to change the form of the data. For example, if one of the independent variables in a model is the price p of a commodity, we might choose to introduce this variable into the model as $x = 1/p$, $x = \sqrt{p}$, or $x = e^{-p}$. Thus, if we were to let $x = \sqrt{p}$, we would compute the square root of each price value, and these square roots would be the values of x that would be used in the regression analysis.

Data transformations are performed on the y values to make them more nearly satisfy the assumptions of Section 4.2 and, sometimes, to make the deterministic portion of the model a better approximation to the mean value of the transformed response. Transformations of the values of the independent variables are performed solely for the latter reason—that is, to achieve a model that provides a better approximation to $E(y)$. In this section, we discuss transformations on the dependent and independent variables to achieve a good approximation to $E(y)$. (Transformations on the y values for the purpose of satisfying the assumptions will be discussed in Chapter 8.)

Suppose you want to fit a model relating the demand y for a product to its price p. Also, suppose the product is a nonessential item, and you expect the mean demand to decrease as price p increases and then to decrease more slowly as p gets larger (see Figure 7.8). What function of p will provide a good approximation to $E(y)$?

To answer this question, you need to know the graphs of some elementary mathematical functions—there is a one-to-one relationship between mathematical functions and graphs. If we want to model a relationship similar to the one indicated in Figure 7.8, we need to be able to select a mathematical function that will possess a graph similar to the curve shown.

FIGURE 7.8

Hypothetical relation between
demand y and price p

Portions of some curves corresponding to mathematical functions that decrease as p increases are shown in Figure 7.9. Of the seven models shown, the curves in Figure 7.9c, 7.9d, 7.9f, and 7.9g will probably provide the best approximations to $E(y)$. These four graphs all show $E(y)$ decreasing and approaching (but never reaching) 0 as p increases. Figures 7.9c and 7.9d suggest that the independent variable, price, should be transformed using either $x = 1/p$ or $x = e^{-p}$. Then you might try fitting the model

$$E(y) = \beta_0 + \beta_1 x$$

FIGURE 7.9

Graphs of some mathematical
functions relating $E(y)$ to p

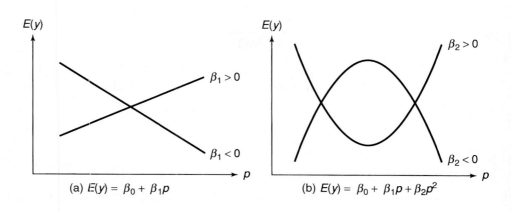

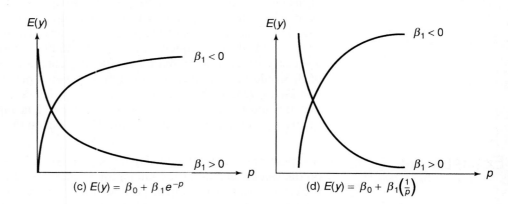

FIGURE 7.9
(*continued*)

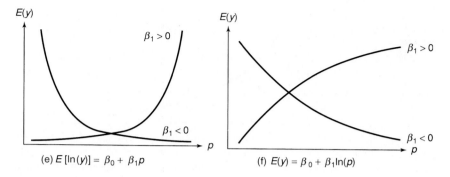

(e) $E\,[\ln(y)] = \beta_0 + \beta_1 p$

(f) $E(y) = \beta_0 + \beta_1 \ln(p)$

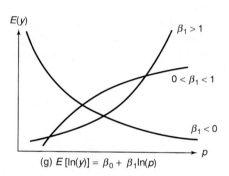

(g) $E\,[\ln(y)] = \beta_0 + \beta_1 \ln(p)$

using the transformed data. Or, as suggested by Figures 7.9f and 7.9g, you might try the transformation $x = \ln(p)$ and fit either of the models

$$E(y) = \beta_0 + \beta_1 x$$

or

$$E\{\ln(y)\} = \beta_0 + \beta_1 x$$

The functions shown in Figure 7.9 produce curves that either rise or fall depending on the sign of the parameter β_1 in parts **a, c, d, e, f**, and **g**, and on β_2 and the portion of the curve used in part **b**. When you choose a model for a regression analysis, you do not have to specify the sign of the parameter(s). The least squares procedure will choose as estimates of the parameters those that minimize the sum of squares of the residuals. Consequently, if you were to fit the model shown in Figure 7.9c to a set of y values that increase in value as p increases, your least squares estimate of β_1 would be negative, and a graph of y would produce a curve similar to curve 2 in Figure 7.9c. If the y values decrease as p increases, your estimate of β_1 will be positive and the curve will be similar to curve 1 in Figure 7.9c. All the curves in Figure 7.9 shift upward or downward depending on the value of β_0.

EXAMPLE   7.4

A supermarket chain conducted an experiment to investigate the effect of price p on the weekly demand (in pounds) for a house brand of coffee. Eight supermarket stores that had nearly equal past records of demand for the product were used in the experiment. Eight prices were randomly assigned to the stores and were

COFFEE

TABLE 7.2   Data for Example 7.4

DEMAND y, pounds	PRICE p, dollars
1,120	3.00
999	3.10
932	3.20
884	3.30
807	3.40
760	3.50
701	3.60
688	3.70

advertised using the same procedures. The number of pounds of coffee sold during the following week was recorded for each of the stores and is shown in Table 7.2.

a. Fit the model

$$E(y) = \beta_0 + \beta_1 x$$

to the data, letting $x = 1/p$.

b. Do the data provide sufficient evidence to indicate that the model contributes information for the prediction of demand?

c. Find a 95% confidence interval for the mean demand when the price is set at $3.20 per pound. Interpret this interval.

Solution

a. The first step is to calculate $x = 1/p$ for each data point. These values are given in Table 7.3. The MINITAB printout* (Figure 7.10) gives

$$\hat{\beta}_0 = -1,180.5 \qquad \hat{\beta}_1 = 6,808.1$$

and

$$\hat{y} = -1,180.5 + 6,808.1x$$

$$= -1,180.5 + 6,808.1 \left(\frac{1}{p}\right)$$

TABLE 7.3   Values of Transformed Price

y	$x = 1/p$
1,120	.3333
999	.3226
932	.3125
884	.3030
807	.2941
760	.2857
701	.2778
688	.2703

(You can verify that the formulas of Section 3.3 give the same answers.) A graph of this prediction equation is shown in Figure 7.11.

b. To determine whether x contributes information for the prediction of y, we test H_0: $\beta_1 = 0$ against the alternative hypothesis H_a: $\beta_1 \neq 0$. The test statistic, shaded in Figure 7.10, is $t = 19.0$. We wish to detect either $\beta_1 > 0$ or $\beta_1 < 0$, thus we will use a two-tailed test. Since the two-tailed p-value shown on the printout, .000, is less than $\alpha = .05$, we reject H_0: $\beta_1 = 0$ and conclude that $x = 1/p$ contributes information for the prediction of demand y.

*MINITAB uses full decimal accuracy for $x = 1/p$. Hence, the results shown in Figure 7.10 differ from results that would be calculated using the four-decimal values for $x = 1/p$ shown in the table.

```
The regression equation is
DEMAND = - 1180 + 6808 X

Predictor      Coef       SE Coef       T          P
Constant     -1180.5       107.7      -10.96     0.000
X             6808.1       358.4       19.00     0.000

S = 20.90       R-Sq = 98.4%      R-Sq(adj) = 98.1%

Analysis of Variance

Source          DF         SS          MS         F         P
Regression       1       157718      157718     360.94     0.000
Residual Error   6         2622         437
Total            7       160340



Predicted Values for New Observations

New Obs      Fit       SE Fit          95.0% CI                95.0% PI
1         947.05        8.66     ( 925.86,  968.24)    ( 891.67, 1002.43)

Values of Predictors for New Observations

New Obs          X
1            0.3125
```

c. For price $p = 3.20, x = 1/p = .3125$. The bottom of the MINITAB printout gives a 95% confidence interval for the mean demand $E(y)$ when price is $p = \$3.20$ (i.e., $x = .3125$). The interval (shaded) is (925.86, 968.24). Thus, we are 95% confident that mean demand will fall between 926 and 968 pounds when the price is set at \$3.20.   ◆

This discussion is intended to emphasize the importance of data transformation and to explain its role in model building. Keep in mind that the symbols, $x_1, x_2, \ldots, x_k$ that appear in the linear models of this text can be transformations

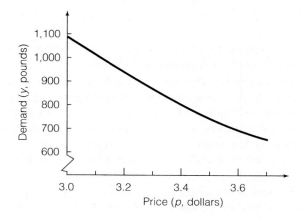

on the independent variables you have observed. These transformations, coupled with the model-building methods of Chapter 5, allow you to use a great variety of mathematical functions to model the mean $E(y)$ for data.

Summary

There are several problems that you should be aware of when constructing a model for a response y. In this chapter we have identified a few of the most important of these problem areas, as follows:

1. **Establishing cause and effect:** When the data used in the regression analysis are observational (i.e., uncontrolled), it is dangerous to deduce a cause-and-effect relationship between y and the independent (predictor) variables. Only when the experiment has been designed properly can you be certain that any changes in y are due solely to the different settings of the predictor variables.

2. **Departures from the assumptions:** In a practical setting, it is unlikely that the standard least squares assumptions about the error term are satisfied exactly. When departures from the assumptions are slight, the model remains a powerful predictor of the response y. On the other hand, the model performs poorly when the assumptions of equal variances and uncorrelated errors are violated.

3. **Parameter estimability and interpretation:** To estimate all the parameters in a pth-order model, the number of levels of x must be greater than or equal to $(p + 1)$. For any model, the sample size n must be sufficiently large to allow degrees of freedom for estimating σ^2 so that a test of model adequacy can be performed. Be wary of interpreting the individual β parameters in the presence of interaction or highly correlated independent variables.

4. **Multicollinearity:** When highly correlated independent variables are present in a regression model, the results may be confusing: The t tests on the individual β's may be nonsignificant even though the F test for overall model adequacy is significant, and the β's may have signs opposite from what is expected. Also, there may be extreme rounding errors in the computation of the β estimates. Variance inflation factors aid in determining whether a serious multicollinearity problem exists. The solution to the problem depends on the severity of the multicollinearity and the ultimate goal of the regression analyst.

5. **Extrapolation:** Predicting y when the x values are outside the range of experimentation is a dangerous practice. The level of reliability associated with any inference derived from the model will be less than the stated level of confidence $(1 - \alpha)$ since the adequacy of the model outside the experimental region is unknown.

6. **Data transformations:** To achieve a model that provides a better approximation to $E(y)$, you may need to transform the values of the independent variables or the value of y. The type of transformation you should make depends on the theoretical relationships between $E(y)$ and the independent variables.

EXERCISES

7.1. Discuss the consequences of fitting multiple regression models when the assumptions of Section 4.2 are violated.

7.2. Why is it dangerous to predict y for values of independent variables that fall outside the experimental region?

7.3. Discuss the problems that result when multicollinearity is present in a regression analysis.

7.4. How can you detect multicollinearity?

7.5. What remedial measures are available when multicollinearity is detected?

7.6. Refer to Example 7.4. Can you think of any other transformations on price that might provide a good fit to the data? Try them and answer the questions of Example 7.4 again.

7.7. Refer to the *Professional Geographer* (Feb. 2000) study of urban and rural counties in the western United States, Exercise 4.15 (p. 189). Recall that six independent variables—total county population (x_1), population density (x_2), population concentration (x_3), population growth (x_4), proportion of county land in farms (x_5), and 5-year change in agricultural land base (x_6)—were used to model the urban/rural rating (y) of a county. Prior to running the multiple regression analysis, the researchers were concerned about possible multicollinearity in the data. Below is a correlation matrix, that is, a table of correlations between all pairs of the independent variables.

INDEPENDENT VARIABLE		x_1	x_2	x_3	x_4	x_5
x_1	Total population					
x_2	Population density	.20				
x_3	Population concentration	.45	.43			
x_4	Population growth	−.05	−.14	−.01		
x_5	Farm land	−.16	−.15	−.07	−.20	
x_6	Agricultural change	−.12	−.12	−.22	−.06	−.06

Source: Berry, K. A., *et al.* "Interpreting what is rural and urban for western U.S. counties." *Professional Geographer*, Vol. 52, No. 1, Feb. 2000 (Table 2).

a. Based on the correlation matrix, is there any evidence of extreme multicollinearity?

b. Refer to the multiple regression results in the table given in Exercise 4.15 (p. 189). Based on the reported tests, is there any evidence of extreme multicollinearity?

7.8. A bioengineer wants to model the amount (y) of carbohydrate solubilized during steam processing of peat as a function of temperature (x_1), exposure time (x_2), and pH value (x_3). Data collected for each of 15 peat samples were used to fit the model

$$E(y) = \beta_0 + \beta_1 x_1 + \beta_2 x_2 + \beta_3 x_3$$

A summary of the regression results follows:

$$\hat{y} = -3,000 + 3.2x_1 - .4x_2 - 1.1x_3 \qquad R^2 = .93$$

$$s_{\hat{\beta}_1} = 2.4 \qquad s_{\hat{\beta}_2} = .6 \qquad s_{\hat{\beta}_3} = .8$$

$$r_{12} = .92 \qquad r_{13} = .87 \qquad r_{23} = .81$$

Based on these results, the bioengineer concludes that none of the three independent variables, x_1, x_2, and x_3, is a useful predictor of carbohydrate amount, y. Do you agree with this statement? Explain.

7.9. The provost of a top research university wants to know what salaries should be paid to the college's top researchers, based on years of experience. An independent consultant has proposed the quadratic model

$$E(y) = \beta_0 + \beta_1 x + \beta_2 x^2$$

where

$$y = \text{Annual salary (thousands of dollars)}$$

$$x = \text{Years of experience}$$

To fit the model, the consultant randomly sampled three researchers at other research universities and recorded the information given in the accompanying table. Give your opinion regarding the adequacy of the proposed model.

	y	x
Researcher 1	60	2
Researcher 2	45	1
Researcher 3	82	5

7.10. A particular meat-processing plant slaughters steers and cuts and wraps the beef for its customers. Suppose a complaint has been filed with the Food and Drug Administration (FDA) against the processing plant. The complaint alleges that the consumer does not get all the beef from the steer he purchases. In particular, one consumer purchased a 300-pound steer but received only 150 pounds of cut and wrapped beef. To settle the complaint, the FDA collected data on the live weights and dressed weights of nine steers processed by a reputable meat-processing plant (not the firm in question). The results are listed in the table.

STEERS

LIVE WEIGHT x, pounds	DRESSED WEIGHT y, pounds
420	280
380	250
480	310
340	210
450	290
460	280
430	270
370	240
390	250

a. Fit the model $E(y) = \beta_0 + \beta_1 x$ to the data.

b. Construct a 95% prediction interval for the dressed weight y of a 300-pound steer.

c. Would you recommend that the FDA use the interval obtained in part **b** to determine whether the dressed weight of 150 pounds is a reasonable amount to receive from a 300-pound steer? Explain.

7.11. Refer to the FTC cigarette data of Example 7.3 (p. 349). The data are saved in the FTCCIGAR file.

a. Fit the model $E(y) = \beta_0 + \beta_1 x_1$ to the data. Is there evidence that tar content x_1 is useful for predicting carbon monoxide content y?

b. Fit the model $E(y) = \beta_0 + \beta_2 x_2$ to the data. Is there evidence that nicotine content x_2 is useful for predicting carbon monoxide content y?

c. Fit the model $E(y) = \beta_0 + \beta_3 x_3$ to the data. Is there evidence that weight x_3 is useful for predicting carbon monoxide content y?

d. Compare the signs of $\hat{\beta}_1$, $\hat{\beta}_2$, and $\hat{\beta}_3$ in the models of parts a, b, and c, respectively, to the signs of the $\hat{\beta}$'s in the multiple regression model fit in Example 7.3. Is the fact that the $\hat{\beta}$'s change dramatically when the independent variables are removed from the model an indication of a serious multicollinearity problem?

7.12. An economist wants to model annual per capita demand, y, for passenger car motor fuel in the United States as a function of the two quantitative independent

variables, average real weekly earnings (x_1) and average price of regular gasoline (x_2). Data on these three variables for the years 1980–2001 are available in the *2002 Statistical Abstract of the United States*. Suppose the economist fits the model $E(y) = \beta_0 + \beta_1 x_1 + \beta_2 x_2$ to the data. Would you recommend that the economist use the least squares prediction equation to predict per capita consumption of motor fuel in 2005? Explain.

7.13. Refer to the Florida Attorney General (FLAG) Office's investigation of bid-rigging in the road construction industry, Exercise 6.5 (p. 336). Recall that FLAG wants to model the price (y) of the contract bid by lowest bidder in hopes of preventing price-fixing in the future.

a. Consider the independent variables selected by the stepwise regression run in Exercise 6.5. Do you detect any multicollinearity in these variables? If so, do you recommend that all of these variables be used to predict low-bid price, y?

b. Using the variables selected in part **a**, fit a full interaction model for $E(y)$ to the data. [*Note*: Do not include squared terms in the model, but be sure to include all possible two-variable interactions.]

7.14. How many levels of x are required to fit the model $E(y) = \beta_0 + \beta_1 x + \beta_2 x^2$? How large a sample size is required to have sufficient degrees of freedom for estimating σ^2?

7.15. How many levels of x_1 and x_2 are required to fit the model $E(y) = \beta_0 + \beta_1 x_1 + \beta_2 x_2 + \beta_3 x_1 x_2$? How large a sample size is required to have sufficient degrees of freedom for estimating σ^2?

7.16. How many levels of x_1 and x_2 are required to fit the model $E(y) = \beta_0 + \beta_1 x_1 + \beta_2 x_2 + \beta_3 x_1 x_2 + \beta_4 x_1^2 + \beta_5 x_2^2$? How large a sample is required to have sufficient degrees of freedom for estimating σ^2?

7.17. A physiologist wanted to investigate the relationship between the physical characteristics of preadolescent boys and their maximal oxygen uptake (measured in milliliters of oxygen per kilogram of body weight). The data shown in the table (p. 362) were collected on a random sample of 10 preadolescent boys. As a first step in the data analysis, the researcher fit the regression model

$$y = \beta_0 + \beta_1 x_1 + \beta_2 x_2 + \beta_3 x_3 + \beta_4 x_4 + \varepsilon$$

to the data. The output for a SAS regression analysis is shown on p. 362.

a. Is the model adequate for predicting maximal oxygen uptake?

YOUNGBOYS

MAXIMAL OXYGEN UPTAKE y	AGE x_1, years	HEIGHT x_2, centimeters	WEIGHT x_3, kilograms	CHEST DEPTH x_4, centimeters
1.54	8.4	132.0	29.1	14.4
1.74	8.7	135.5	29.7	14.5
1.32	8.9	127.7	28.4	14.0
1.50	9.9	131.1	28.8	14.2
1.46	9.0	130.0	25.9	13.6
1.35	7.7	127.6	27.6	13.9
1.53	7.3	129.9	29.0	14.0
1.71	9.9	138.1	33.6	14.6
1.27	9.3	126.6	27.7	13.9
1.50	8.1	131.8	30.8	14.5

b. It seems reasonable to assume that the greater a child's age, the greater should be the maximal oxygen uptake. But note that $\hat{\beta}_1$, the estimated coefficient of age, x_1, is negative. Give an explanation for this result.

c. It would seem that the weight of a child should be positively correlated to lung volume and hence to maximal oxygen uptake. Can you explain the small t value associated with $\hat{\beta}_3$?

d. Calculate the coefficient of correlation r for each pair of independent variables. Does this information confirm your suspicions in parts b and c?

7.18. Consider the data shown in the table on p. 363.

a. Plot the points on a scatterplot. What type of relationship appears to exist between x and y?

b. For each observation calculate $\ln x$ and $\ln y$. Plot the log-transformed data points on a scatterplot. What type of relationship appears to exist between $\ln x$ and $\ln y$?

c. The scatterplot from part **b** suggests that the transformed model

$$\ln y = \beta_0 + \beta_1 \ln x + \varepsilon$$

may be appropriate. Fit the transformed model to the data. Is the model adequate? Test using $\alpha = .05$.

d. Use the transformed model to predict the value of y when $x = 30$. [*Hint*: Use the inverse transformation $y = e^{\ln y}$.]

SAS output for Exercise 7.17

Dependent Variable: MAXOXY

Analysis of Variance

Source	DF	Sum of Squares	Mean Square	F Value	Pr > F
Model	4	0.20604	0.05151	37.20	0.0007
Error	5	0.00692	0.00138		
Corrected Total	9	0.21296			

Root MSE	0.03721	R-Square	0.9675
Dependent Mean	1.49200	Adj R-Sq	0.9415
Coeff Var	2.49391		

Parameter Estimates

Variable	DF	Parameter Estimate	Standard Error	t Value	Pr > |t|	Variance Inflation
Intercept	1	-4.77474	0.86282	-5.53	0.0026	0
AGE	1	-0.03521	0.01539	-2.29	0.0708	1.15862
HEIGHT	1	0.05164	0.00622	8.31	0.0004	3.24542
WEIGHT	1	-0.02342	0.01343	-1.74	0.1416	5.01738
CHEST	1	0.03449	0.08524	0.40	0.7025	5.16380

EX7_18

x	54	42	28	38	25	70	48	41	20	52	65
y	6	16	33	18	41	3	10	14	45	9	5

7.19. D. Hamilton illustrated the multicollinearity problem with an example using the data shown in the accompanying table. The values of x_1, x_2, and y in the table below represent appraised land value, appraised improvements value, and sale price, respectively, of a randomly selected residential property. (All measurements are in thousands of dollars.)

HAMILTON

x_1	x_2	y	x_1	x_2	y
22.3	96.6	123.7	30.4	77.1	128.6
25.7	89.4	126.6	32.6	51.1	108.4
38.7	44.0	120.0	33.9	50.5	112.0
31.0	66.4	119.3	23.5	85.1	115.6
33.9	49.1	110.6	27.6	65.9	108.3
28.3	85.2	130.3	39.0	49.0	126.3
30.2	80.4	131.3	31.6	69.6	124.6
21.4	90.5	114.4			

Source: Hamilton, D. "Sometimes $R^2 > r_{yx_1}^2 + r_{yx_2}^2$: Correlated variables are not always redundant." *The American Statistician*, Vol. 41, No. 2, May 1987, pp. 129–132.

a. Calculate the coefficient of correlation between y and x_1. Is there evidence of a linear relationship between sale price and appraised land value?

b. Calculate the coefficient of correlation between y and x_2. Is there evidence of a linear relationship between sale price and appraised improvements?

c. Based on the results in parts a and b, do you think the model $E(y) = \beta_0 + \beta_1 x_1 + \beta_2 x_2$ will be useful for predicting sale price?

d. Use a statistical computer software package to fit the model in part c, and conduct a test of model adequacy. In particular, note the value of R^2. Does the result agree with your answer to part c?

e. Calculate the coefficient of correlation between x_1 and x_2. What does the result imply?

f. Many researchers avoid the problems of multicollinearity by always omitting all but one of the "redundant" variables from the model. Would you recommend this strategy for this example? Explain. (Hamilton notes that, in this case, such a strategy "can amount to throwing out the baby with the bathwater.")

7.20. *Teaching Sociology* (July 1995) developed a model for the professional socialization of graduate students working toward a Ph.D. in sociology. One of the dependent variables modeled was professional confidence, y, measured on a 5-point scale. The model included over 20 independent variables and was fit to data collected for a sample of 309 sociology graduate students. One concern is whether multicollinearity exists in the data. A matrix of Pearson product moment correlations for ten of the independent variables is shown below. [*Note*: : Each entry in the table is the correlation coefficient r between the variable in the corresponding row and corresponding column.]

a. Examine the correlation matrix and find the independent variables that are moderately or highly correlated.

b. What modeling problems may occur if the variables, part **a**, are left in the model? Explain.

Matrix of correlations for Exercise 7.20

INDEPENDENT VARIABLE	(1)	(2)	(3)	(4)	(5)	(6)	(7)	(8)	(9)	(10)
(1) Father's occupation	1.000	.363	.099	−.110	−.047	−.053	−.111	.178	.078	.049
(2) Mother's education	.363	1.000	.228	−.139	−.216	.084	−.118	.192	.125	.068
(3) Race	.099	.228	1.000	.036	−.515	.014	−.120	.112	.117	.337
(4) Sex	−.110	−.139	.036	1.000	.165	−.256	.173	−.106	−.117	.073
(5) Foreign status	−.047	−.216	−.515	.165	1.000	−.041	.159	−.130	−.165	−.171
(6) Undergraduate GPA	−.053	.084	.014	−.256	−.041	1.000	.032	.028	−.034	.092
(7) Year GRE taken	−.111	−.118	−.120	.173	.159	.032	1.000	−.086	−.602	.016
(8) Verbal GRE score	.178	.192	.112	−.106	−.130	.028	−.086	1.000	.132	.087
(9) Years in graduate program	.078	.125	.117	−.117	−.165	−.034	−.602	.132	1.000	−.071
(10) First-year graduate GPA	.049	.068	.337	.073	−.171	.092	.016	.087	−.071	1.000

Source: Keith, B., and Moore, H. A. "Training sociologists: An assessment of professional socialization and the emergence of career aspirations." *Teaching Sociology*, Vol. 23, No. 3, July 1995, p. 205 (Table 1).

7.21. To model the relationship between y, a dependent variable, and x, an independent variable, a researcher has taken one measurement on y at each of five different x values. Drawing on his mathematical expertise, the researcher realizes that he can fit the fourth-order polynomial model

$$E(y) = \beta_0 + \beta_1 x + \beta_2 x^2 + \beta_3 x^3 + \beta_4 x^4$$

and it will pass exactly through all five points, yielding SSE $= 0$. The researcher, delighted with the "excellent" fit of the model, eagerly sets out to use it to make inferences. What problems will the researcher encounter in attempting to make inferences?

REFERENCES

DRAPER, N., and SMITH, H. *Applied Regression Analysis*, 2nd ed. New York: Wiley, 1981.

MONTGOMERY, D., PECK, E., and VINING, G. *Introduction to Linear Regression Analysis*. 3rd ed. New York: Wiley, 2001.

MOSTELLER, F., and TUKEY, J. W. *Data Analysis and Regression: A Second Course in Statistics*. Reading, Mass.: Addison-Wesley, 1977.

NETER, J., KUTNER, M., NACHTSHEIM, C., and WASSERMAN, W. *Applied Linear Statistical Models*, 4th ed. Homewood, Ill.: Richard D. Irwin, 1996.

CHAPTER

8

Residual Analysis

CONTENTS

OBJECTIVE

To show how residuals can be used to detect departures from the model assumptions and to suggest some procedures for coping with these problems

8.1 Introduction

We have repeatedly stated that the validity of many of the inferences associated with a regression analysis depends on the error term, ε, satisfying certain assumptions. Thus, when we test a hypothesis about a regression coefficient or a set of regression coefficients, or when we form a prediction interval for a future value of y, we must assume that (1) ε is normally distributed with a mean of 0, (2) the variance σ^2 is constant, and (3) all pairs of error terms are uncorrelated.* The objective of this chapter is to provide you with both graphical tools and statistical tests that will aid in checking the validity of these assumptions. In addition, these tools will help you evaluate the utility of the model and, in some cases, may suggest modifications to the model that will allow you to better describe the mean response.

In Section 8.2, we will show how to plot the residuals to reveal model inadequacies. In Section 8.3, we examine the use of these plots and a simple test to detect unequal variances at different levels of the independent variable(s). A graphical analysis

---

*We assumed (Section 4.2) that the random errors associated with the linear model were independent. If two random variables are independent, it follows (proof omitted) that they will be uncorrelated. The reverse is generally untrue, except for normally distributed random variables. If two normally distributed random variables are uncorrelated, it can be shown that they are also independent.

of residuals for checking the normality assumption is the topic of Section 8.4. In Section 8.5, residual plots are used to detect outliers, i.e., observations that are unusually large or small relative to the others; procedures for measuring the influence these outliers may have on the fitted regression model are also presented. Finally, we discuss the use of residuals to test for time series correlation of the error term in Section 8.6.

8.2
Plotting Residuals and Detecting Lack of Fit

The error term in a multiple regression model is, in general, not observable. To see this, consider the model

$$y = \beta_0 + \beta_1 x_1 + \cdots + \beta_k x_k + \varepsilon$$

and solve for the error term:

$$\varepsilon = y - (\beta_0 + \beta_1 x_1 + \cdots + \beta_k x_k)$$

Although you will observe values of the dependent variable and the independent variables $x_1, x_2, \ldots, x_k$, you will not know the true values of the regression coefficients $\beta_0, \beta_1, \ldots, \beta_k$. Therefore, the exact value of ε cannot be calculated.

After we use the data to obtain least squares estimates $\hat{\beta}_0, \hat{\beta}_1, \ldots, \hat{\beta}_k$ of the regression coefficients, we can estimate the value of ε associated with each y value using the corresponding **regression residual**, i.e., the deviation between the observed and the predicted value of y:

$$\hat{\varepsilon}_i = y_i - \hat{y}_i$$

To accomplish this, we must substitute the values of $x_1, x_2, \ldots, x_k$ into the prediction equation for each data point to obtain $\hat{y}$, and then this value must be subtracted from the observed value of y. Remember that you encountered the regression residual in Chapters 3 and 4. In particular, the least squares estimates of $\beta_0, \beta_1, \beta_2, \ldots, \beta_k$ are those that minimize the sum of squares of the residuals,

$$\sum_{i=1}^{n} \hat{\varepsilon}_i^2 = \sum_{i=1}^{n} (y_i - \hat{y}_i)^2$$

Definition 8.1

The **regression residual** is the observed value of the dependent variable minus the predicted value, or

$$\hat{\varepsilon} = y - \hat{y} = y - (\hat{\beta}_0 + \hat{\beta}_1 x_1 + \cdots + \hat{\beta}_k x_k)$$

EXAMPLE 8.1

The data in Table 8.1 represent the level of cholesterol (in milligrams per liter) and average daily intake of saturated fat (in milligrams) for a sample of 10 Olympic

OLYMPIC

TABLE 8.1   Data for Ten Olympic Athletes

ATHLETE	FAT INTAKE x, milligrams	CHOLESTEROL y, milligrams/liter
1	1,290	1,182
2	1,350	1,172
3	1,470	1,264
4	1,600	1,493
5	1,710	1,571
6	1,840	1,711
7	1,980	1,804
8	2,230	1,840
9	2,400	1,956
10	2,930	1,954

athletes. Consider a regression model relating cholesterol level y to fat intake x. Calculate the regression residuals for

a. the straight-line (first-order) model
b. the quadratic (second-order) model

Solution

a. The SAS printout for the regression analysis of the first-order model,

$$y = \beta_0 + \beta_1 x + \varepsilon$$

is shown in Figure 8.1. The least squares model highlighted is

$$\hat{y} = 578.92775 + .54030x$$

Thus, the residual for the first observation, $x = 1,290$ and $y = 1,182$, is obtained by first calculating the predicted value

$$\hat{y} = 578.92775 + .54030(1,290) = 1,275.92$$

and then subtracting from the observed value:

$$\hat{\varepsilon} = y - \hat{y} = 1,182 - 1,275.92$$
$$= -93.92$$

Similar calculations for the other nine observations produce the residuals highlighted at the bottom of Figure 8.1.

b. The SAS printout for the second-order model

$$y = \beta_0 + \beta_1 x + \beta_2 x^2 + \varepsilon$$

FIGURE 8.1

SAS printout for first-order model

Dependent Variable: CHOLES

Analysis of Variance

Source	DF	Sum of Squares	Mean Square	F Value	Pr > F
Model	1	703957	703957	39.54	0.0002
Error	8	142445	17806		
Corrected Total	9	846402			

Root MSE	133.43768	R-Square	0.8317	
Dependent Mean	1594.70000	Adj R-Sq	0.8107	
Coeff Var	8.36757			

Parameter Estimates

Variable	DF	Parameter Estimate	Standard Error	t Value	Pr > |t|
Intercept	1	578.92775	166.96806	3.47	0.0085
FAT	1	0.54030	0.08593	6.29	0.0002

Predictions

Obs	FAT	CHOLES	Predicted CHOLES	Residual of CHOLES
1	1290	1182	1275.92	-93.920
2	1350	1172	1308.34	-136.339
3	1470	1264	1373.18	-109.175
4	1600	1493	1443.41	49.585
5	1710	1571	1502.85	68.152
6	1840	1711	1573.09	137.912
7	1980	1804	1648.73	155.270
8	2230	1840	1783.81	56.193
9	2400	1956	1875.66	80.342
10	2930	1954	2162.02	-208.020

is shown in Figure 8.2. The least squares model is

$$\hat{y} = -1,216.14389 + 2.39893x - .00045004x^2$$

For the first observation, $x = 1,290$ and $y = 1,182$, the predicted cholesterol level is

$$\hat{y} = -1,216.14389 + 2.39893(1,290) - .00045004(1,290)^2$$

$$= 1,129.56^*$$

and the regression residual is

$$\hat{\varepsilon} = y - \hat{y} = 1,182 - 1,129.56$$

$$= 52.44$$

All the regression residuals for the second-order model are highlighted at the bottom of Figure 8.2.* ◆

Graphical displays of regression residuals are useful aids to their interpretation. For example, the regression residual can be plotted on the vertical axis against

---

*The residuals shown in Figure 8.1 and 8.2 have been generated using a computer program. Therefore, the results reported here will differ slightly from hand-calculated residuals because of rounding error.

FIGURE 8.2

SAS printout for second-order model

Dependent Variable: CHOLES

Analysis of Variance

Source	DF	Sum of Squares	Mean Square	F Value	Pr > F
Model	2	831070	415535	189.71	<.0001
Error	7	15333	2190.36480		
Corrected Total	9	846402			

Root MSE	46.80133	R-Square	0.9819	
Dependent Mean	1594.70000	Adj R-Sq	0.9767	
Coeff Var	2.93480			

Parameter Estimates

Variable	DF	Parameter Estimate	Standard Error	t Value	Pr > |t|
Intercept	1	-1216.14389	242.80637	-5.01	0.0016
FAT	1	2.39893	0.24584	9.76	<.0001
FATSQ	1	-0.00045004	0.00005908	-7.62	0.0001

Predictions

Obs	FAT	CHOLES	FATSQ	Predicted CHOLES	Residual of CHOLES
1	1290	1182	1664100	1129.56	52.4359
2	1350	1172	1822500	1202.21	-30.2136
3	1470	1264	2160900	1337.79	-73.7916
4	1600	1493	2560000	1470.04	22.9586
5	1710	1571	2924100	1570.06	0.9359
6	1840	1711	3385600	1674.23	36.7685
7	1980	1804	3920400	1769.40	34.5998
8	2230	1840	4972900	1895.47	-55.4654
9	2400	1956	5760000	1949.06	6.9431
10	2930	1954	8584900	1949.17	4.8287

one of the independent variables on the horizontal axis, or against the predicted value $\hat{y}$ (which is a linear function of the independent variables). If the assumptions concerning the error term ε are satisfied, we expect to see residual plots that have no trends, no dramatic increases or decreases in variability, and only a few residuals (about 5%) more than 2 estimated standard deviations ($2s$) of ε above or below 0. It is a property of the least squares prediction equation that the mean of the regression residuals is always 0. (This property applies only to models that include an intercept term, β_0.) Thus, the least squares prediction equation not only minimizes the SSE, the sum of squared errors (residuals), but also produces residuals whose mean is 0.

Detecting Model Lack of Fit with Residuals

1. Plot the residuals, $\hat{\varepsilon}$, on the vertical axis against each of the independent variables, $x_1, x_2, \ldots, x_k$ on the horizontal axis.

2. Plot the residuals, $\hat{\varepsilon}$, on the vertical axis against the predicted value, $\hat{y}$, on the horizontal axis.

3. In each plot, look for trends, dramatic changes in variability, and/or more than 5% of residuals that lie outside $2s$ of 0. Any of these patterns indicates a problem with model fit.

To illustrate, the residuals $\hat{\varepsilon}$ obtained by fitting the second-order model to the cholesterol data of Table 8.1 are plotted against fat intake, x, in Figure 8.3. The dotted horizontal line in Figure 8.3 locates the mean (0), for the residuals. We detect no distinctive patterns or trends in this plot. Infact, all the residuals lie within $2s$ of the mean (0), and the variability around the mean is consistent for small and large values of x.

FIGURE 8.3

SAS plot of residuals for second-order model

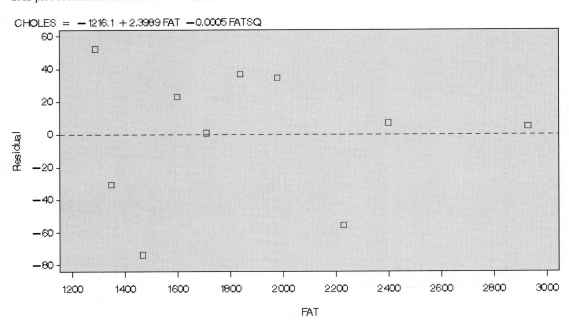

$CHOLES = -1216.1 + 2.3989\,FAT - 0.0005\,FATSQ$

EXAMPLE   8.2

Example 8.1a gives the residuals obtained from fitting a first-order model to the cholesterol data. Plot the residuals obtained in this regression analysis against the fat intake, x (placing x along the horizontal axis). Does the plot suggest model inadequacy or departure from the usual assumptions made about the error term ε?

Solution

The residuals for the first-order model (shown in Figure 8.1) are plotted versus fat intake in Figure 8.4. The distinctive aspect of this plot is the parabolic distribution of the residuals about their mean; i.e., all residuals tend to be positive for athletes with intermediate levels of fat intake x and negative for the athletes with either relatively high or low levels of fat intake. This parabolic appearance of the trend in the residuals suggests that a second-order term may improve the model. In fact, the addition of a second-order term does improve the model: The quadratic (β_2) term in the model is highly significant (see the SAS printout, Figure 8.2), and the residual plot for the second-order model no longer shows an observable pattern (see Figure 8.3). ◆

FIGURE 8.4

SAS plot of residuals for first-order model

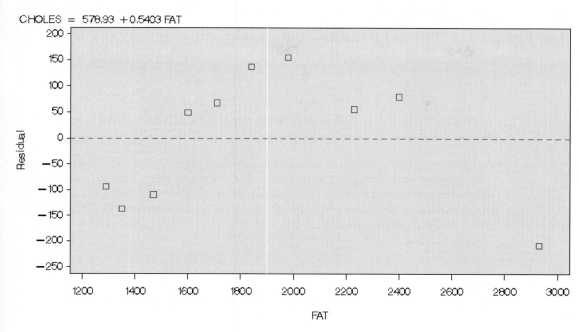

An alternative method of detecting lack of fit in models with more than one independent variable is to construct a partial residual plot. The **partial residuals** for the jth independent variable, x_j, in the model are calculated as follows:

$$\hat{\varepsilon}^* = y - (\hat{\beta}_0 + \hat{\beta}_1 x_1 + \hat{\beta}_2 x_2 + \cdots + \hat{\beta}_{j-1} x_{j-1} + \hat{\beta}_{j+1} x_{j+1} + \cdots + \hat{\beta}_k x_k)$$
$$= \hat{\varepsilon} + \hat{\beta}_j x_j$$

where $\hat{\varepsilon}$ is the usual regression residual.

Partial residuals measure the influence of x_j on the dependent variable y *after the effects of the other independent variables* $(x_1, x_2, \ldots, x_{j-1}, x_{j+1}, \ldots, x_k)$ *have been removed or accounted for.* If the partial residuals $\hat{\varepsilon}^*$ are regressed against x_j in a straight-line model, the resulting least squares slope is equal to $\hat{\beta}_j$—the β estimate obtained from the full model. Therefore, when the partial residuals are plotted against x_j, the points are scattered around a line with slope equal to $\hat{\beta}_j$. Unusual deviations or patterns around this line indicate lack of fit for the variable x_j.

A plot of the partial residuals versus x_j often reveals more information about the relationship between y and x_j than the usual residual plot. In particular, a partial residual plot usually indicates more precisely how to modify the model,[†] as the next example illustrates.

[†]Partial residual plots display the correct functional form of the predictor variables across the relevant range of interest, except in cases where severe multicollinearity exists. See Mansfield and Conerly (1987) for an excellent discussion of the use of residual and partial residual plots.

Definition 8.2

The set of **partial regression residuals** for the jth independent variable x_j is calculated as follows:

$$\hat{\varepsilon}^* = y - (\hat{\beta}_0 + \hat{\beta}_1 x_1 + \hat{\beta}_2 x_2 + \cdots$$
$$+ \hat{\beta}_{j-1} x_{j-1} + \hat{\beta}_{j+1} x_{j+1} + \cdots + \hat{\beta}_k x_k)$$
$$= \hat{\varepsilon} + \hat{\beta}_j x_j$$

where $\hat{\varepsilon} = y - \hat{y}$ is the usual regression residual (see Definition 8.1).

EXAMPLE   8.3

A supermarket chain wants to investigate the effect of price p on the weekly demand y for a house brand of coffee at its stores. Eleven prices were randomly assigned to the stores and were advertised using the same procedures. A few weeks later, the chain conducted the same experiment using no advertisements. The data for the entire study are shown in Table 8.2.

Consider the model

$$E(y) = \beta_0 + \beta_1 p + \beta_2 x_2$$

where

$$x_2 = \begin{cases} 1 & \text{if advertisment used} \\ 0 & \text{if not} \end{cases}$$

 COFFEE2

TABLE 8.2   **Data for Example 8.3**

WEEKLY DEMAND y, pounds	PRICE p, dollars/pound	ADVERTISEMENT x_2
1190	3.0	1
1033	3.2	1
897	3.4	1
789	3.6	1
706	3.8	1
595	4.0	1
512	4.2	1
433	4.4	1
395	4.6	1
304	4.8	1
243	5.0	1
1124	3.0	0
974	3.2	0
830	3.4	0
702	3.6	0
619	3.8	0
529	4.0	0
451	4.2	0
359	4.4	0
296	4.6	0
247	4.8	0
194	5.0	0

a. Fit the model to the data. Is the model adequate for predicting weekly demand y?
b. Plot the residuals versus p. Do you detect any trends?
c. Construct a partial residual plot for the independent variable p. What does the plot reveal?
d. Fit the model $E(y) = \beta_0 + \beta_1 x_1 + \beta_2 x_2$, where $x_1 = 1/p$. Has the predictive ability of the model improved?

Solution

a. The SPSS printout for the regression analysis is shown in Figure 8.5. The F value for testing model adequacy, i.e., $H_0 : \beta_1 = \beta_2 = 0$, is given on the printout (shaded) as $F = 373.71$ with a corresponding p-value (also shaded) of .000. Thus, there is sufficient evidence (at $\alpha = .01$) that the model contributes information for the prediction of weekly demand, y. Also, the coefficient of determination is $R^2 = .975$, meaning that the model explains approximately 97.5% of the sample variation in weekly demand.

FIGURE 8.5

SPSS regression printout for demand model

Model Summary[b]

Model	R	R Square	Adjusted R Square	Std. Error of the Estimate
1	.988[a]	.975	.973	49.876

a. Predictors: (Constant), X2, PRICE
b. Dependent Variable: DEMAND

973

ANOVA[b]

Model		Sum of Squares	df	Mean Square	F	Sig.
1	Regression	1859299	2	929649.475	373.710	.000[a]
	Residual	47264.868	19	2487.625		
	Total	1906564	21			

a. Predictors: (Constant), X2, PRICE
b. Dependent Variable: DEMAND

Coefficients[a]

Model		Unstandardized Coefficients		Standardized Coefficients		
		B	Std. Error	Beta	t	Sig.
1	(Constant)	2400.182	68.914		34.829	.000
	PRICE	-456.295	16.813	-.980	-27.139	.000
	X2	70.182	21.267	.119	3.300	.004

a. Dependent Variable: DEMAND

Recall from Example 7.4, however, that we fit a model with the transformed independent variable $x_1 = 1/p$. That is, we expect the relationship between weekly demand y and price p to be decreasing in a curvilinear fashion and approaching (but never reaching) 0 as p increases. (See Figure 7.9d.) If such a relationship exists, the model (with untransformed price), although statistically useful for predicting demand y, will be inadequate in a practical setting.

b. The regression residuals for the model in part **a** are saved in SPSS and plotted against price p in Figure 8.6. Notice that the plot reveals a clear parabolic trend, implying a lack of fit. Thus, the residual plot supports our hypothesis that the weekly demand–price relationship is curvilinear, not linear. However, the appropriate transformation on price (i.e., $1/p$) is not evident from the plot. In fact, the nature of the curvature in Figure 8.6 may lead you to conclude that the addition of the quadratic term, $\beta_3 p^2$, to the model will solve the problem. In general, a residual plot will detect curvature if it exists, but may not reveal the appropriate transformation.

FIGURE 8.6

SPSS plot of residuals against price for demand model

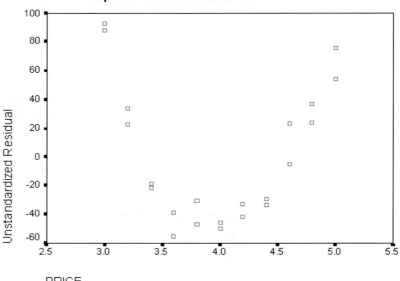

c. Most statistical packages have options for automatic calculation of partial residuals. The partial residual plot for the independent variable p is shown in the SPSS printout, Figure 8.7. SPSS (as does SAS) finds the partial residuals for price, p, by finding (separately) the residuals of the dependent variable y when regressed against advertising x_2, and, the residuals of price, p, when regressed against x_2. A plot of these residuals will look similar to a plot of the partial residuals of Definition 8.2 against price, p. You can see that the partial residual plot of Figure 8.7 also reveals a curvilinear trend but, in addition, displays the correct functional form of weekly demand–price relationship. Notice that the curve is decreasing and approaching (but never reaching) 0 as p increases. This suggests that the appropriate transformation on price is either $1/p$ or e^{-p} (see Figures 7.9c and 7.9d).

d. Using the transformation $x_1 = 1/p$, we refit the model to the data, and the resulting SPSS printout is shown in Figure 8.8. The small p-value (.0001) for testing $H_0: \beta_1 = \beta_2 = 0$ indicates that the model is adequate for predicting y.

FIGURE 8.7
SPSS partial residual plot for price

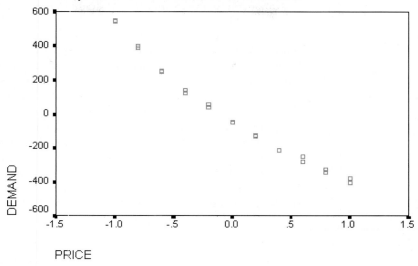

Partial Regression Plot

Dependent Variable: DEMAND

FIGURE 8.8
SPSS regression printout for demand model with transformed price

Model Summary

Model	R	R Square	Adjusted R Square	Std. Error of the Estimate
1	.999a	.999	.999	11.097

a. Predictors: (Constant), X2, X1

ANOVAb

Model		Sum of Squares	df	Mean Square	F	Sig.
1	Regression	1904224	2	952111.957	7731.141	.000a
	Residual	2339.904	19	123.153		
	Total	1906564	21			

a. Predictors: (Constant), X2, X1

b. Dependent Variable: DEMAND

Coefficientsa

Model		Unstandardized Coefficients		Standardized Coefficients	t	Sig.
		B	Std. Error	Beta		
1	(Constant)	-1217.343	14.898		-81.711	.000
	X1	6986.507	56.589	.992	123.460	.000
	X2	70.182	4.732	.119	14.831	.000

a. Dependent Variable. DEMAND

Although the coefficient of determination increased only slightly (from $R^2 = .975$ to $R^2 = .999$), the model standard deviation decreased significantly (from $s = 49.876$ to $s = 11.097$). Thus, whereas the model with untransformed price can predict weekly demand for coffee to within $2s = 2(50) = 100$ pounds, the transformed model can predict demand to within $2(11) = 22$ pounds. ◆

Residual (or partial residual) plots are useful for indicating potential model improvements, but they are no substitute for formal statistical tests of model terms to determine their importance. Thus, a true test of whether the second-order term contributes to the cholesterol model (Example 8.1) is the t test of the null hypothesis $H_0: \beta_2 = 0$. The appropriate test statistic, shown in the printout of Figure 8.2, indicates that the second-order term does contribute information for the prediction of cholesterol level y. We have confidence in this statistical inference because we know the probability α of committing a Type I error (concluding a term is important when, in fact, it is not). In contrast, decisions based on residual plots are subjective, and their reliability cannot be measured. Therefore, we suggest that such plots be used only as indicators of *potential* problems. The final judgment on model adequacy should be based on appropriate statistical tests.[‡]

EXERCISES

8.1. Consider the data on x and y shown in the table.

EX8_1

x	-2	-2	-1	-1	0	0	1	1	2	2	3	3
y	1.1	1.3	2.0	2.1	2.7	2.8	3.4	3.6	4.0	3.9	3.8	3.6

a. Fit the model $E(y) = \beta_0 + \beta_1 x$ to the data.
b. Calculate the residuals for the model.
c. Plot the residuals versus x. Do you detect any trends? If so, what does the pattern suggest about the model?

8.2. Consider the data on x and y shown in the table.

EX8_2

x	2	4	7	10	12	15	18	20	21	25
y	5	10	12	22	25	27	39	50	47	65

a. Fit the model $E(y) = \beta_0 + \beta_1 x$ to the data.
b. Calculate the residuals for the model.
c. Plot the residuals versus x. Do you detect any trends? If so, what does the pattern suggest about the model?

8.3. Refer to Example 3.2 (p. 122). Recall that a manufacturer of a new tire tested the tire for wear at different pressures with the results shown in the table.

TIRES

PRESSURE x, pounds per sq. inch	MILEAGE y, thousands
30	29.5
30	30.2
31	32.1
31	34.5
32	36.3
32	35.0
33	38.2
33	37.6
34	37.7
34	36.1
35	33.6
35	34.2
36	26.8
36	27.4

[‡]A more general procedure for determining whether the straight-line model adequately fits the data tests the null hypothesis $H_0: E(y) = \beta_0 + \beta_1 x$ against the alternative $H_a: E(y) \neq \beta_0 + \beta_1 x$. You can see that this test, called a test for *lack of fit*, does not restrict the alternative hypothesis to second-order models. Lack-of-fit tests are appropriate when the x values are replicated, i.e., when the sample data include two or more observations for several different levels of x. When the data are observational, however, replication rarely occurs. (Note that none of the values of x is repeated in Table 8.1.) For details on how to conduct tests for lack of fit, consult the references given at the end of this chapter.

a. Fit the straight-line model $y = \beta_0 + \beta_1 x + \varepsilon$ to the data.

b. Calculate the residuals for the model.

c. Plot the residuals versus x. Do you detect any trends? If so, what does the pattern suggest about the model?

d. Fit the quadratic model $y = \beta_0 + \beta_1 x + \beta_2 x^2 + \varepsilon$ to the data using an available statistical software package. Has the addition of the quadratic term improved model adequacy?

8.4. Moissanite is a popular abrasive material because of its extreme hardness. Another important property of moissanite is elasticity. The elastic properties of the material were investigated in the *Journal of Applied Physics* (Sept. 1993). A diamond anvil cell was used to compress a mixture of moissanite, sodium chloride, and gold in a ratio of 33:99:1 by volume. The compressed volume, y, of the mixture (relative to the zero-pressure volume) was measured at each of 11 different pressures (GPa). The results are displayed in the table, followed by a MINITAB printout for the straight-line regression model $E(y) = \beta_0 + \beta_1 x$ and a MINITAB residual plot.

a. Calculate the regression residuals.

b. Plot the residuals against x. Do you detect a trend?

MINITAB regression output for Exercise 8.4

```
The regression equation is
VOLUME = 98.6 - 0.256 PRESSURE

Predictor      Coef     SE Coef        T        P
Constant    98.6149      0.4037   244.26    0.000
PRESSURE   -0.255594    0.008646   -29.56    0.000

S = 0.6484     R-Sq = 99.0%     R-Sq(adj) = 98.9%

Analysis of Variance

Source        DF        SS        MS        F        P
Regression     1    367.34    367.34   873.87    0.000
Residual Error 9      3.78      0.42
Total         10    371.12
```

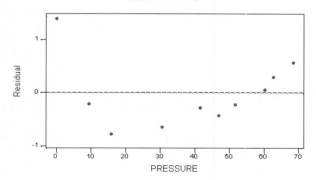

Residuals Versus PRESSURE
(response is VOLUME)

MOISSANITE

COMPRESSED VOLUME y, %	PRESSURE x, GPa
100	0
96	9.4
93.8	15.8
90.2	30.4
87.7	41.6
86.2	46.9
85.2	51.6
83.3	60.1
82.9	62.6
82.9	62.6
81.7	68.4

Source: Bassett, W. A., Weathers, M. S., and Wu, T. C. "Compressibility of SiC up to 68.4 GPa." *Journal of Applied Physics*. Vol. 74. No. 6, Sept. 15, 1993, p. 3825 (Table 1).

c. Propose an alternative model based on the plot, part **b**.

d. Fit and analyze the model, part **c**.

8.5. Refer to the study of man-hours required to erect boiler drams, Exercise 4.40 (p. 210). Recall that the data on 35 boilors were used to fit the model

$$E(y) = \beta_0 + \beta_1 x_1 + \beta_2 x_2 + \beta_3 x_3 + \beta_4 x_4$$

where

$$y = \text{man-hours}$$

$$x_1 = \text{Boiler capacity (lb/hr)}$$

$$x_2 = \text{Design pressure (psi)}$$

$$x_3 = \begin{cases} 1 & \text{if industry erected} \\ 0 & \text{if not} \end{cases}$$

$$x_4 = \begin{cases} 1 & \text{if steam drum} \\ 0 & \text{if mud drum} \end{cases}$$

The data are saved in the BOILERS file.

a. Find the residuals for the model.

b. Plot the residuals versus x_1. Do you detect any trends? If so, what does the pattern suggest about the model?

c. Plot the residuals versus x_2. Do you detect any trends? If so, what does the pattern suggest about the model?

d. Plot the partial residuals for x_1. Interpret the result.

e. Plot the partial residuals for x_2. Interpret the result.

8.6. A certain type of rare gem serves as a status symbol for many of its owners. In theory, then, the demand for the gem would increase as the price increases, decreasing at low prices, leveling off at moderate prices, and increasing at high prices, because obtaining the gem at a high price confers high status on the owner. Although

SAS regression output for Exercise 8.6

Dependent Variable: PRICE

Analysis of Variance

Source	DF	Sum of Squares	Mean Square	F Value	Pr > F
Model	1	46288	46288	0.56	0.4708
Error	10	823912	82391		
Corrected Total	11	870200			

Root MSE	287.03860	R-Square	0.0532	
Dependent Mean	410.00000	Adj R-Sq	-0.0415	
Coeff Var	70.00942			

Parameter Estimates

Variable	DF	Parameter Estimate	Standard Error	t Value	Pr > |t|
Intercept	1	270.62438	203.57433	1.33	0.2133
DEMAND	1	1.17782	1.57139	0.75	0.4708

Predictions

Obs	PRICE	DEMAND	Predicted PRICE	Residual of PRICE
1	100	130	423.741	-323.741
2	700	150	447.298	252.702
3	450	60	341.294	108.706
4	150	120	411.963	-261.963
5	500	50	329.515	170.485
6	800	200	506.189	293.811
7	70	150	447.298	-377.298
8	50	160	459.076	-409.076
9	300	50	329.515	-29.515
10	350	40	317.737	32.263
11	750	180	482.632	267.368
12	700	130	423.741	276.259

a quadratic model would seem to match the theory, the model proposed to explain the demand for the gem by its price is the first-order model

$$y = \beta_0 + \beta_1 x + \varepsilon$$

where y is the demand (in thousands) and x is the retail price per carat (dollars). This model was fit to the 12 data points given in the table, and the results of the analysis are shown in the SAS printout above.

 GEM

x	100	700	450	150	500	800	70	50	300	350	750	700
y	130	150	60	120	50	200	150	160	50	40	180	130

a. Use the least squares prediction equation to verify the values of the regression residuals shown on the printout.

b. Plot the residuals against retail price per carat, x.

c. Can you detect any trends in the residual plot? What does this imply?

8.7. Refer to *The New England Journal of Medicine* study of passive exposure to environmental tobacco smoke in children with cystic fibrosis, Exercise 3.68 (p. 158). Recall that the researchers investigated the relationship between a child's weight percentile (y) and the number of cigarettes smoked per day in the child's home (x). The table on p. 379 lists the data for the 25 boys.

a. Fit the model $E(y) = \beta_0 + \beta_1 x$ to the data.

b. Verify that the sum of the residuals is 0.

c. Plot the residuals against the number of cigarettes smoked per day, x.

d. Do you detect any patterns in the plot, part **b**? What does this imply?

⊙ HOMESMOKE

WEIGHT PERCENTILE y	CIGARETTES SMOKED x	WEIGHT PERCENTILE y	CIGARETTES SMOKED x
6	0	43	0
6	15	49	0
2	40	50	0
8	23	49	22

(*continued*)

11	20	46	30
17	7	54	0
24	3	58	0
25	0	62	0
31	23	83	0
35	10	87	44

Source: Rubin, B. K. "Exposure of children with cystic fibrosis to environmental tobacco smoke." *The New England Journal of Medicine*. Sept. 20, 1990, Vol. 323, No. 12, p. 785 (data extracted from Figure 3).

8.3
Detecting Unequal Variances

Recall that one of the assumptions necessary for the validity of regression inferences is that the error term ε have constant variance σ^2 for all levels of the independent variable(s). Variances that satisfy this property are called **homoscedastic**. Unequal variances for different settings of the independent variable(s) are said to be **heteroscedastic**. Various statistical tests for heteroscedasticity have been developed. However, plots of the residuals will frequently reveal the presence of heteroscedasticity. In this section we will show how residual plots can be used to detect departures from the assumption of equal variances, and then give a simple test for heteroscedasticity. In addition, we will suggest some modifications to the model that may remedy the situation.

When data fail to be homoscedastic, the reason is often that the variance of the response y is a function of its mean $E(y)$. Some examples follow:

1. If the response y is a count that has a Poisson distribution, the variance will be equal to the mean $E(y)$. Poisson data are usually counts per unit volume, area, time, etc. For example, the number of sick days per month for an employee would very likely be a Poisson random variable. If the variance of a response is proportional to $E(y)$, the regression residuals produce a pattern about $\hat{y}$, the least squares estimate of $E(y)$, like that shown in Figure 8.9.

FIGURE 8.9

A plot of residuals for Poisson data

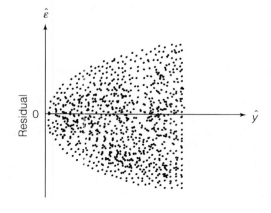

2. Many responses are proportions (or percentages) generated by **binomial exper-iments**. For instance, the proportion of a random sample of 100 convicted felons who are repeat offenders is an example of a binomial response. Binomial propor-tions have variances that are functions of both the true proportion (the mean) and the sample size. In fact, if the observed proportion $y_i = \hat{p}_i$ is generated by a binomial distribution with sample size n_i and true probability p_i, the variance of y_i is

$$\text{Var}(y_i) = \frac{p_i(1 - p_i)}{n_i} = \frac{E(y_i)[1 - E(y_i)]}{n_i}$$

Residuals for binomial data produce a pattern about $\hat{y}$ like that shown in Figure 8.10.

FIGURE 8.10

A plot of residuals for binomial data (proportions or percentages)

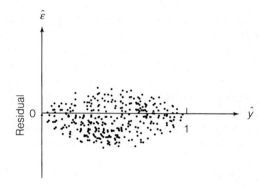

3. The random error component has been assumed to be **additive** in all the models we have constructed. An additive error is one for which the response is equal to the mean $E(y)$ *plus* random error,

$$y = E(y) + \varepsilon$$

Another useful type of model, especially for business and economic data, is the **multiplicative** model. In this model, the response is written as the *product* of its mean and the random error component, i.e.,

$$y = [E(y)]\varepsilon$$

The variance of this response will grow proportionally to the *square* of the mean, i.e.,

$$\text{Var}(y) = [E(y)]^2\sigma^2$$

where σ^2 is the variance of ε. Data subject to multiplicative errors produce a pattern of residuals about $\hat{y}$ like that shown in Figure 8.11.

When the variance of y is a function of its mean, we can often satisfy the least squares assumption of homoscedasticity by transforming the response to some new response that has a constant variance. These are called **variance-stabilizing**

FIGURE 8.11

A plot of residuals for data subject to multiplicative errors

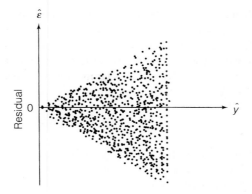

transformations. For example, if the response y is a count that follows a Poisson distribution, the square root transform $\sqrt{y}$ can be shown to have approximately constant variance.* Consequently, if the response is a Poisson random variable, we would let

$$y^* = \sqrt{y}$$

and fit the model

$$y^* = \beta_0 + \beta_1 x_1 + \cdots + \beta_k x_k + \varepsilon$$

This model will satisfy approximately the least squares assumption of homoscedasticity.

Similar transformations that are appropriate for percentages and proportions (binomial data) or for data subject to multiplicative errors are shown in Table 8.3.

TABLE 8.3 **Stabilizing Transformations for Heteroscedastic Responses**

TYPE OF RESPONSE	VARIANCE	STABILIZING TRANSFORMATION
Poisson	$E(y)$	$\sqrt{y}$
Binomial proportion	$\dfrac{E(y)[1 - E(y)]}{n}$	$\sin^{-1}\sqrt{y}$
Multiplicative	$[E(y)]^2 \sigma^2$	$\ln(y)$

*The square root transformation for Poisson responses is derived by finding the integral of $1/\sqrt{E(y)}$. In general, it can be shown (proof omitted) that the appropriate transformation for any response y is

$$y^* = \int \frac{1}{\sqrt{V(y)}} dy$$

where $V(y)$ is an expression for the variance of y.

The transformed responses will satisfy (at least approximately) the assumption of homoscedasticity.

The data in Table 8.4 are the salaries, y, and years of experience, x, for a sample of 50 social workers. If we fit the second-order model $E(y) = \beta_0 + \beta_1 x + \beta_2 x^2$ to the data, we obtain the MINITAB computer printout for the regression analysis shown in Figure 8.12 and the prediction equation

$$\hat{y} = 20{,}242 + 522\,x + 53\,x^2$$

The printout suggests that the second-order model provides an adequate fit to the data. The R^2 value, .816, indicates that the model explains almost 82% of the total variation of the y values about $\bar{y}$. The global F value, $F = 103.99$, is

 SOCWORK

TABLE 8.4   **Salary and Experience Data for 50 Social Workers**

YEARS OF EXPERIENCE x	SALARY y	YEARS OF EXPERIENCE x	SALARY y	YEARS OF EXPERIENCE x	SALARY y
7	$26,075	21	$43,628	28	$99,139
28	79,370	4	16,105	23	52,624
23	65,726	24	65,644	17	50,594
18	41,983	20	63,022	25	53,272
19	62,309	20	47,780	26	65,343
15	41,154	15	38,853	19	46,216
24	53,610	25	66,537	16	54,288
13	33,697	25	67,447	3	20,844
2	22,444	28	64,785	12	32,586
8	32,562	26	61,581	23	71,235
20	43,076	27	70,678	20	36,530
21	56,000	20	51,301	19	52,745
18	58,667	18	39,346	27	67,282
7	22,210	1	24,833	25	80,931
2	20,521	26	65,929	12	32,303
18	49,727	20	41,721	11	38,371
11	33,233	26	82,641		

highly significant ($p = .000$), indicating that the model contributes information for the prediction of y. However, an examination of the salary residuals plotted against the estimated mean salary, $\hat{y}$, as shown in Figure 8.13, reveals a potential problem. Note the "cone" shape of the residual variability; the size of the residuals increases as the estimated mean salary increases. This residual plot indicates that a multiplicative model may be appropriate. We will explore this possibility further in Example 8.4.

FIGURE 8.12

MINITAB regression printout for second-order model of salary

```
The regression equation is
SALARY = 20242 + 522 EXP + 53.0 EXPSQ

Predictor       Coef     SE Coef          T        P
Constant       20242        4423       4.58    0.000
EXP            522.3       616.7       0.85    0.401
EXPSQ          53.01       19.57       2.71    0.009

S = 8123       R-Sq = 81.6%     R-Sq(adj) = 80.8%

Analysis of Variance

Source          DF          SS          MS        F        P
Regression       2  13723582237  6861791118   103.99    0.000
Residual Error  47   3101279310    65984666
Total           49  16824861546
```

FIGURE 8.13

MINITAB residual plot for second-order model of salary

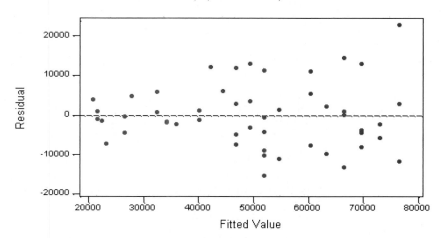

Residuals Versus the Fitted Values

(response is SALARY)

EXAMPLE 8.4

Consider the salary and experience data in Table 8.4. Use the natural log transformation on the dependent variable, and relate $\ln(y)$ to years of experience, x, using the second-order model

$$\ln(y) = \beta_0 + \beta_1 x + \beta_2 x^2 + \varepsilon$$

Evaluate the adequacy of the model.

Solution

The MINITAB printout in Figure 8.14 gives the regression analysis for the $n = 50$ measurements. The prediction equation used in computing the residuals is

$$\widehat{\ln(y)} = 9.84289 + .04969x + .0000094x^2$$

The residual plot in Figure 8.15 indicates that the logarithmic transformation has significantly reduced the heteroscedasticity.[†] Note that the cone shape is gone; there is no apparent tendency of the residual variance to increase as mean salary increases. We therefore are confident that inferences using the logarithmic model are more reliable than those using the untransformed model.

To evaluate model adequacy, we first note that $R^2 = .864$ and that about 86% of the variation in ln(salary) is accounted for by the model. The global F value ($F = 148.67$) and its associated p-value ($p = .000$) indicate that the model significantly improves upon the sample mean as a predictor of ln(salary).

FIGURE 8.14

MINITAB regression printout for second-order model of natural log of salary

```
The regression equation is
LOGY = 9.84 + 0.0497 EXP +0.000009 EXPSQ

Predictor        Coef      SE Coef         T         P
Constant      9.84289      0.08479    116.08     0.000
EXP           0.04969      0.01182      4.20     0.000
EXPSQ       0.0000094    0.0003753      0.03     0.980

S = 0.1557      R-Sq = 86.4%      R-Sq(adj) = 85.8%

Analysis of Variance

Source           DF          SS          MS         F         P
Regression        2      7.2122      3.6061    148.67     0.000
Residual Error   47      1.1400      0.0243
Total            49      8.3522
```

Although the estimate of β_2 is very small, we should check to determine whether the data provide sufficient evidence to indicate that the second-order term contributes information for the prediction of ln(salary). The test of

$H_0: \beta_2 = 0$

$H_a: \beta_2 \neq 0$

is conducted using the t statistic shown in Figure 8.14, $t = .03$. The p-value of the test, .98, is greater than $\alpha = .10$. Consequently, there is insufficient evidence to indicate that the second-order term contributes to the prediction of ln(salary). There is no indication that the second-order model is an improvement over the straight-line model,

$$\ln(y) = \beta_0 + \beta_1 x + \varepsilon$$

for predicting ln(salary).

[†]A printout of the residuals is omitted.

FIGURE 8.15

MINITAB residual plot for second-order model of natural log of salary

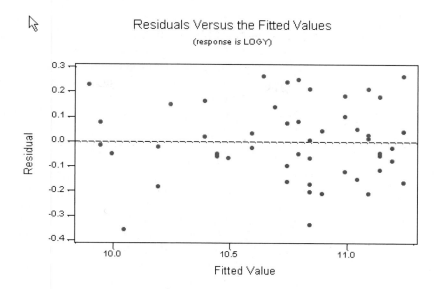

Residuals Versus the Fitted Values

(response is LOGY)

The MINITAB printout for the first-order model (Figure 8.16) shows that the prediction equation for the first-order model is

$$\widehat{\ln y} = 9.84133 + .049978x$$

The value of R^2, 864, is approximately the same as the value of R^2 obtained for the second-order model. The F statistic, computed from the mean squares in Figure 8.16, $F = 303.65(p\text{-value} = .000)$ indicates that the model contributes significantly to the prediction of $\ln(y)$. ◆

FIGURE 8.16

MINITAB regression printout for first-order model of natural log of salary

```
The regression equation is
LOGY = 9.84 + 0.0500 EXP

Predictor        Coef      SE Coef          T          P
Constant      9.84132      0.05636     174.63      0.000
EXP          0.049979     0.002868      17.43      0.000

S = 0.1541      R-Sq = 86.4%      R-Sq(adj) = 86.1%

Analysis of Variance

Source           DF           SS          MS          F          P
Regression        1       7.2121      7.2121     303.65      0.000
Residual Error   48       1.1401      0.0238
Total            49       8.3522
```

When the transformed model of Example 8.4 is used to predict the value of $\ln(y)$, the predicted value of y is the antilog, $\hat{y} = e^{\widehat{\ln y}}$. The endpoints of the prediction interval are similarly transformed back to the original scale, and the interval will retain its meaning. In repeated use, the intervals will contain the observed y value $100(1 - \alpha)\%$ of the time.

Unfortunately, you cannot take antilogs to find the confidence interval for the mean value $E(y)$. The reason for this is that the mean value of $\ln(y)$ is not equal to the natural logarithm of the mean of y. In fact, the antilog of the logarithmic mean of a random variable y is called its **geometric mean**. Thus, the antilogs of the endpoints of the confidence interval for the mean of the transformed response will give a confidence interval for the geometric mean. Similar care must be exercised with other types of transformations. In general, prediction intervals can be transformed back to the original scale without losing their meaning, but confidence intervals for the mean of a transformed response cannot.

The preceding examples illustrate that, in practice, residual plots can be a powerful technique for detecting heteroscedasticity. Furthermore, the pattern of the residuals often suggests the appropriate variance-stabilizing transformation to use. Keep in mind, however, that no measure of reliability can be attached to inferences derived from a graphical technique. For this reason, you may want to rely on a statistical test.

Various tests for heteroscedasticity in regression have been developed. One of the simpler techniques utilizes the F test (discussed in Chapter 1) for comparing population variances. The procedure requires that you divide the sample data in half and fit the regression model to each half. If the regression model fit to one-half the observations yields a significantly smaller or larger MSE than the model fitted to the other half, there is evidence that the assumption of equal variances for all levels of the x variables in the model is being violated. (Recall that MSE, or mean square for error, estimates σ^2, the variance of the random error term.) Where you divide the data depends on where you suspect the differences in variances to be. We illustrate this procedure with an example.

EXAMPLE   8.5

Refer to the data of Table 8.4 and the analysis of the model relating a social worker's salary (y) to years of experience (x). The residual plot for the quadratic model

$$E(y) = \beta_0 + \beta_1 x + \beta_2 x^2$$

indicates that the assumption of equal variances may be violated (see Figure 8.13). Conduct a statistical test of hypothesis to determine whether heteroscedasticity exists. Use $\alpha = .05$.

Solution

The residual plot shown in Figure 8.13 reveals that the residuals associated with larger values of predicted salary tend to be more variable than the residuals associated with smaller values of predicted salary. Therefore, we will divide the sample observations based on the values of $\hat{y}$, or, equivalently, the value of x (since, for the fitted model, $\hat{y}$ increases as x increases). An examination of the data in Table 8.4 reveals that approximately one-half of the 50 observed values of years of experience, x, fall below $x = 20$. Thus, we will divide the data into two subsamples as follows:

SUBSAMPLE 1	SUBSAMPLE 2
$x < 20$	$x \geq 20$
$n_1 = 24$	$n_2 = 26$

Figures 8.17a and 8.17b give the SAS printouts for the quadratic model fit to subsample 1 and subsample 2, respectively. The value of MSE is shaded in each printout.

The null and alternative hypotheses to be tested are

$$H_0: \frac{\sigma_1^2}{\sigma_2^2} = 1 \text{ (Assumption of equal variances satisfied)}$$

$$H_a: \frac{\sigma_1^2}{\sigma_2^2} \neq 1 \text{ (Assumption of equal variances violated)}$$

where
$\sigma_1^2 = $ Variance of the random error term, ε, for subpopulation 1 (i.e., $x < 20$)
$\sigma_2^2 = $ Variance of the random error term, ε, for subpopulation 2 (i.e., $x \geq 20$)

The test statistic for a two-tailed test is given by:

$$F = \frac{\text{Larger } s^2}{\text{Smaller } s^2} = \frac{\text{Larger MSE}}{\text{Smaller MSE}} \quad \text{(see Section 1.11)}$$

where the distribution of F is based on $\nu_1 = $ df(error) associated with the larger MSE and $\nu_2 = $ df(error) associated with the smaller MSE. Recall that for a quadratic model, df(error) $= n - 3$.

From the printouts shown in Figure 8.17a and 8.17b, we have

$$\text{MSE}_1 = 31{,}576{,}998 \quad \text{and MSE}_2 = 94{,}711{,}023$$

FIGURE 8.17a

SAS regression printout for second-order model of salary: Subsample 1 (years of experience < 20)

Dependent Variable: SALARY

Analysis of Variance

Source	DF	Sum of Squares	Mean Square	F Value	Pr > F
Model	2	3231196786	1615598393	51.16	<.0001
Error	21	663116951	31576998		
Corrected Total	23	3894313737			

Root MSE	5619.34139	R-Square	0.8297	
Dependent Mean	37153	Adj R-Sq	0.8135	
Coeff Var	15.12498			

Parameter Estimates

Variable	DF	Parameter Estimate	Standard Error	t Value	Pr > |t|
Intercept	1	20372	3817.87780	5.34	<.0001
EXP	1	263.17043	861.86817	0.31	0.7631
EXPSQ	1	76.76912	40.01402	1.92	0.0687

Therefore, the test statistic is

$$F = \frac{\text{MSE}_2}{\text{MSE}_1} = \frac{94{,}711{,}023}{31{,}576{,}998} = 3.00$$

FIGURE 8.17*b*

SAS Regression printout for second-order model of salary: subsample 2 (years of experience > 20)

Dependent Variable: SALARY

Analysis of Variance

Source	DF	Sum of Squares	Mean Square	F Value	Pr > F
Model	2	2930781596	1465390798	15.47	<.0001
Error	23	2178353538	94711023		
Corrected Total	25	5109135134			

Root MSE	9731.95887	R-Square	0.5736
Dependent Mean	62187	Adj R-Sq	0.5366
Coeff Var	15.64951		

Parameter Estimates

Variable	DF	Parameter Estimate	Standard Error	t Value	Pr > |t|
Intercept	1	-19330	168798	-0.11	0.9098
EXP	1	3004.76433	14415	0.21	0.8367
EXPSQ	1	16.85876	304.22229	0.06	0.9563

Since the MSE for subsample 2 is placed in the numerator of the test statistic, this F value is based on $n_2 - 3 = 26 - 3 = 23$ numerator df and $n_1 - 3 = 24 - 3 = 21$ denominator df. For a two-tailed test at $\alpha = .05$, the critical value for $v_1 = 23$ and $v_2 = 21$ (found in Table 5 of Appendix C) is approximately $F_{.025} = 2.37$.

Since the observed value, $F = 3.00$, exceeds the critical value, there is sufficient evidence (at $\alpha = .05$) to indicate that the error variances differ.[‡] Thus, this test supports the conclusions reached by using the residual plots in the preceding examples. ◆

The test for heteroscedasticity outlined in Example 8.5 is easy to apply when only a single independent variable appears in the model. For a multiple regression model that contains several different independent variables, the choice of the levels of the x variables for dividing the data is more difficult, if not impossible. If you require a statistical test for heteroscedasticity in a multiple regression model, you may need to resort to other, more complex, tests.[§] Consult the references at the end of this chapter for details on how to conduct these tests.

EXERCISES

8.8. Refer to Exercise 8.1 (p. 376). Plot the residuals for the first-order model versus $\hat{y}$. Do you detect any trends? If so, what does the pattern suggest about the model?

8.9. Refer to Exercise 8.2 (p. 376). Plot the residuals for the first-order model versus $\hat{y}$. Do you detect any trends? If so, what does the pattern suggest about the model?

[‡]Most statistical tests require that the observations in the sample be independent. For this F test, the observations are the residuals. Even if the standard least squares assumption of independent errors is satisfied, the regression residuals will be correlated. Fortunately, when n is large compared to the number of β parameters in the model, the correlation among the residuals is reduced and, in most cases, can be ignored.

[§]For example, consider fitting the absolute values of the residuals as a function of the independent variables in the model; i.e., fit the regression model $E\{|\hat{\varepsilon}|\} = \beta_0 + \beta_1 x_1 + \beta_2 x_2 + \cdots + \beta_k x_k$. A nonsignificant global F implies that the assumption of homoscedasticity is satisfied. A significant F, however, indicates that changing the values of the x's will lead to a larger (or smaller) residual variance.

8.10. Chemical engineers at Tokyo Metropolitan University analyzed urban air specimens for the presence of low-molecular-weight dicarboxylic acids (*Environmental Science & Engineering*, Oct. 1993). The dicarboxylic acid (as a percentage of total carbon) and oxidant concentrations (in ppm) for 19 air specimens collected from urban Tokyo are listed in the accompanying table. SAS printouts for the straight-line model relating dicarboxylic acid percentage (y) to oxidant concentration (x) are shown below and on p. 390. Use the information in the SAS printouts to assess the validity of the assumption of equal error variances.

💿 AIRTOKYO

DICARBOXYLIC ACID, %	OXIDANT, ppm	DICARBOXYLIC ACID, %	OXIDANT ppm
.85	78	.50	32
1.45	80	.38	28
1.80	74	.30	25
1.80	78	.70	45
1.60	60	.80	40
1.20	62	.90	45
1.30	57	1.22	41
.20	49	1.00	34
.22	34	1.00	25
.40	36		

Source: Kawamura, K., and Ikushima, K. "Seasonal changes in the distribution of dicarboxylic acids in the urban atmosphere." *Environmental Science & Technology*, Vol. 27, No. 10, Oct. 1993, p. 2,232 (data extracted from Figure 4).

SAS Output for Exercise 8.10

Dependent Variable: DICARBO

Analysis of Variance

Source	DF	Sum of Squares	Mean Square	F Value	Pr > F
Model	1	2.41362	2.41362	17.08	0.0007
Error	17	2.40234	0.14131		
Corrected Total	18	4.81597			

Root MSE	0.37592	R-Square	0.5012	
Dependent Mean	0.92737	Adj R-Sq	0.4718	
Coeff Var	40.53600			

Parameter Estimates

Variable	DF	Parameter Estimate	Standard Error	t Value	Pr > |t|
Intercept	1	-0.02374	0.24577	-0.10	0.9242
OXIDANT	1	0.01958	0.00474	4.13	0.0007

Predictions

Obs	DICARBO	OXIDANT	Predicted DICARBO	Residual of DICARBO
1	0.85	78	1.50339	-0.65339
2	1.45	80	1.54255	-0.09255
3	1.80	74	1.42508	0.37492
4	1.80	78	1.50339	0.29661
5	1.60	60	1.15098	0.44902
6	1.20	62	1.19013	0.00987
7	1.30	57	1.09224	0.20776
8	0.20	49	0.93561	-0.73561
9	0.22	34	0.64193	-0.42193
10	0.40	36	0.68109	-0.28109
11	0.50	32	0.60278	-0.10278
12	0.38	28	0.52446	-0.14446
13	0.30	25	0.46573	-0.16573
14	0.70	45	0.85730	-0.15730
15	0.80	40	0.75941	0.04059
16	0.90	45	0.85730	0.04270
17	1.22	41	0.77898	0.44102
18	1.00	34	0.64193	0.35807
19	1.00	25	0.46573	0.53427

SAS Output for Exercise 8.10 (*continued*)

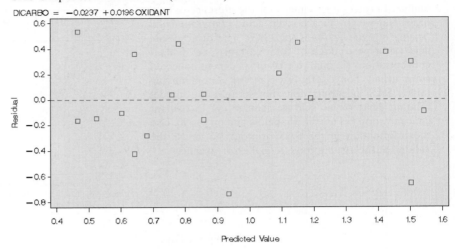

DICARBO = −0.0237 + 0.0196 OXIDANT

8.11. Breakdowns of machines that produce steel cans are very costly. The more breakdowns, the fewer cans produced, and the smaller the company's profits. To help anticipate profit loss, the owners of a can company would like to find a model that will predict the number of breakdowns on the assembly line. The model proposed by the company's statisticians is the following:

$$y = \beta_0 + \beta_1 x_1 + \beta_2 x_2 + \beta_3 x_3 + \beta_4 x_4 + \varepsilon$$

where y is the number of breakdowns per 8-hour shift,

$$x_1 = \begin{cases} 1 & \text{if afternoon shift} \\ 0 & \text{otherwise} \end{cases} \quad x_2 = \begin{cases} 1 & \text{if midnight shift} \\ 0 & \text{otherwise} \end{cases}$$

x_3 is the temperature of the plant (°F), and x_4 is the number of inexperienced personnel working on the assembly line. After the model is fit using the least squares procedure, the residuals are plotted against $\hat{y}$, as shown in the accompanying figure.

$(y - \hat{y})$

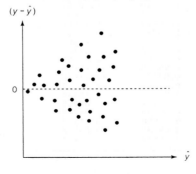

a. Do you detect a pattern in the residual plot? What does this suggest about the least squares assumptions?

b. Given the nature of the response variable y and the pattern detected in part **a**, what model adjustments would you recommend?

8.12. Refer to Exercise 8.11. The regression analysis for the transformed model

$$y^* = \sqrt{y} = \beta_0 + \beta_1 x_1 + \beta_2 x_2 + \beta_3 x_3 + \beta_4 x_4 + \varepsilon$$

produces the prediction equation

$$\hat{y}^* = 1.3 + .008 x_1 - .13 x_2 + .0025 x_3 + .26 x_4$$

a. Use the equation to predict the number of breakdowns during the midnight shift if the temperature of the plant at that time is 87°F and if there is only one inexperienced worker on the assembly line.

b. A 95% prediction interval for y^* when $x_1 = 0$, $x_2 = 0$, $x_3 = 90$°F, and $x_4 = 2$ is $(1.965, 2.125)$. For those same values of the independent variables, find a 95% prediction interval for y, the number of breakdowns per 8-hour shift.

c. A 95% confidence interval for $E(y^*)$ when $x_1 = 0$, $x_2 = 0$, $x_3 = 90$°F, and $x_4 = 2$ is $(1.987, 2.107)$. Using only the information given in this problem, is it possible to find a 95% confidence interval for $E(y)$? Explain.

8.13. The manager of a retail appliance store wants to model the proportion of appliance owners who decide to purchase a service contract for a specific major appliance. Since the manager believes that the proportion y decreases with age x of the appliance (in years), he will fit the first-order model

$$E(y) = \beta_0 + \beta_1 x$$

A sample of 50 purchasers of new appliances are contacted about the possibility of purchasing a service contract. Fifty owners of 1-year-old machines, and 50 owners each of 2-, 3-, and 4-year-old machines are also contacted. One year later, another survey is conducted in a similar manner. The proportion y of owners deciding to purchase the service policy is shown in the table.

⬤ APPLIANCE

Age of Appliance x, years	0	0	1	1	2	2	3	3	4	4
Proportion Buying Service Contract, y	.94	.96	.7	.76	.6	.4	.24	.3	.12	.1

a. Fit the first-order model to the data.
b. Calculate the residuals and construct a residual plot versus $\hat{y}$.
c. What does the plot from part **b** suggest about the variance of y?
d. Explain how you could stabilize the variances.
e. Refit the model using the appropriate variance-stabilizing transformation. Plot the residuals for the transformed model and compare to the plot obtained in part **b**. Does the assumption of homoscedasticity appear to be satisfied?

8.14. Prior to 1980, private homeowners in Hawaii had to lease the land their homes were built on because the law (dating back to the islands' feudal period) required that land be owned only by the big estates. After 1980, however, a new law instituted condemnation proceedings so that citizens could buy their own land. To comply with the 1980 law, one large Hawaiian estate wanted to use regression analysis to estimate the fair market value of its land. Its first proposal was the quadratic model

$$E(y) = \beta_0 + \beta_1 x + \beta_2 x^2$$

where

$y =$ Leased fee value (i.e., sale price of property)
$x =$ Size of property in square feet

Data collected for 20 property sales in a particular neighborhood, given in the accompanying table, were used to fit the model. The least squares prediction equation is

$$\hat{y} = -44.0947 + 11.5339 x - .06378 x^2$$

⬤ HAWAII

PROPERTY	LEASED FEE VALUE y, thousands of dollars	SIZE x, thousands
1	70.7	13.5
2	52.7	9.6
3	87.6	17.6
4	43.2	7.9
5	103.8	11.5
6	45.1	8.2
7	86.8	15.2
8	73.3	12.0
9	144.3	13.8
10	61.3	10.0
11	148.0	14.5
12	85.0	10.2
13	171.2	18.7
14	97.5	13.2
15	158.1	16.3
16	74.2	12.3
17	47.0	7.7
18	54.7	9.9
19	68.0	11.2
20	75.2	12.4

a. Calculate the predicted values and corresponding residuals for the model.
b. Plot the residuals versus $\hat{y}$. Do you detect any trends? If so, what does the pattern suggest about the model?
c. Conduct a test for heteroscedasticity. [*Hint*: Divide the data into two subsamples, $x \leq 12$ and $x > 12$, and fit the model to both subsamples.]
d. Based on your results from parts **b** and **c**, how should the estate proceed?

8.4

Checking the Normality Assumption

Recall from Section 4.2 that all the inferential procedures associated with a regression analysis are based on the assumptions that, for any setting of the independent variables, the random error ε is normally distributed with mean

0 and variance σ^2, and all pairs of errors are independent. Of these assumptions, the normality assumption is the least restrictive when we apply regression analysis in practice. That is, moderate departures from the assumption of normality have very little effect on Type I error rates associated with the statistical tests and on the confidence coefficients associated with the confidence intervals.

Although tests are available to check the normality assumption (see, for example, Stephens, 1974), we discuss only graphical techniques in this section. The simplest way to determine whether the data violate the assumption of normality is to construct a frequency or relative frequency distribution for the residuals using the computer. If this distribution is not badly skewed, you can feel reasonably confident that the measures of reliability associated with your inferences are as stated in Chapter 4. This visual check is not foolproof because we are lumping the residuals together for all settings of the independent variables. It is conceivable (but not likely) that the distribution of residuals might be skewed to the left for some values of the independent variables and skewed to the right for others. Combining these residuals into a single relative frequency distribution could produce a distribution that is relatively symmetric. But, as noted above, we think that this situation is unlikely and that this graphical check is very useful.

To illustrate, consider the $n = 50$ residuals obtained from the model of ln(salary) in Example 8.4.* A MINITAB histogram and stem-and-leaf plot of the residuals are shown in Figures 8.18 and 8.19, respectively. Both graphs show that the distribution is mound-shaped and reasonably symmetric. Consequently, it is unlikely that the normality assumption would be violated using these data.

FIGURE 8.18

MINITAB histogram of residuals from log model of salary

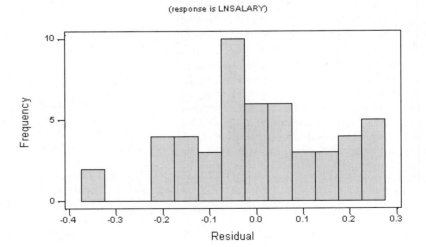

Histogram of the Residuals

(response is LNSALARY)

A third graphical technique for checking the assumption of normality is to construct a **normal probability plot**. In a normal probability plot, the residuals are graphed against the expected values of the residuals under the assumption of

---

*A printout of the residuals is omitted.

FIGURE 8.19

MINITAB stem-and-leaf plot of residuals from log model of salary

```
Stem-and-leaf of RESIDUAL   N   = 50
Leaf Unit = 0.010


    1    -3 5
    2    -3 3
    2    -2
    5    -2 000
   10    -1 86665
   12    -1 11
   18    -0 976655
   (8)   -0 44442221
   24     0 0122344
   17     0 5778
   13     1 044
   10     1 688
```

normality. (These expected values are sometimes called **normal scores**.) When the errors are, in fact, normally distributed, a residual value will approximately equal its expected value. Thus, a linear trend on the normal probability plot suggests that the normality assumption is nearly satisfied, whereas a nonlinear trend indicates that the assumption is most likely violated.

Most statistical software packages have procedures for constructing normal probability plots. Figure 8.20 shows the MINITAB normal probability plot for the residuals of Example 7.4. Notice that the points fall reasonably close to a straight line, indicating that the normality assumption is most likely satisfied. If you do not have access to a statistical software package that contains a normal probability plot option, you can calculate the expected values of the residuals (under normality) using the procedure outlined in the next box.

Nonnormality of the distribution of the random error ε is often accompanied by heteroscedasticity. Both these situations can frequently be rectified by applying

FIGURE 8.20

MINITAB normal probability plot of residuals from log model of salary

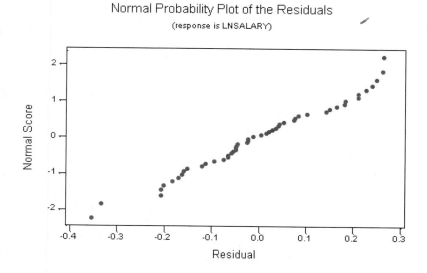

Normal Probability Plot of the Residuals

(response is LNSALARY)

the variance-stabilizing transformations of Section 8.3. For example, if the relative frequency distribution (or stem-and-leaf display) of the residuals is highly skewed to the right (as it would be for Poisson data), the square-root transformation on y will stabilize (approximately) the variance and, at the same time, will reduce skewness in the distribution of the residuals. Thus, for any given setting of the independent variables, the square-root transformation will reduce the larger values of y to a greater extent than the smaller ones. This has the effect of reducing or eliminating the positive skewness.

Constructing a Normal Probability Plot for Regression Residuals

1. List the residuals in ascending order, where $\hat{\varepsilon}_i$ represents the ith ordered residual.

2. For each residual, calculate the corresponding tail area (of the standard normal distribution),

$$A = \frac{i - .375}{n + .25}$$

  where n is the sample size.

3. Calculate the estimated value of $\hat{\varepsilon}_i$ under normality using the following formula:

$$E(\hat{\varepsilon}_i) \approx \sqrt{\text{MSE}}[Z(A)]$$

  where

  MSE = mean square error for the fitted model
  $Z(A)$ = value of the standard normal distribution (z value) that cuts off an area of A in the lower tail of the distribution

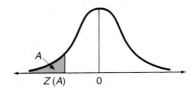

4. Plot the residuals $\hat{\varepsilon}_i$ against the estimated expected residuals, $i = 1, 2, \ldots, n$.

For situations in which the errors are homoscedastic but nonnormal, normalizing transformations are available. This family of transformations on the dependent variable includes $\sqrt{y}$ and $\log(y)$ (Section 8.3), as well as such simple transformations as y^2, $1/\sqrt{y}$, and $1/y$. Box and Cox (1964) have developed a procedure for selecting the appropriate transformation to use. Consult the references to learn details of the Box–Cox approach.

Keep in mind, that regression is *robust* with respect to nonnormal errors. That is, the inferences derived from the regression analysis tend to remain valid even when the assumption of normal errors is not exactly satisfied. However, if the errors come from a "heavy-tailed" distribution, the regression results may be sensitive to a small subset of the data and lead to invalid inferences. Consequently, you may want to search for a normalizing transformation only when the distribution of the regression residuals is highly skewed.

EXERCISES

8.15. Refer to the *Environmental Science & Technology* regression model for dicarboxylic acid percentage, Exercise 8.10 (p. 389). Additional MINITAB printouts of the residual analysis are shown below. Comment on the validity of the assumption of normal errors.

8.16. Refer to Exercise 4.40. (p. 210) and the data saved in the BOILERS file. Use one of the graphical techniques described in this section to check the normality assumption.

8.17. Refer to Exercise 8.14. (p. 391) and the data saved in the HAWAII file. Use one of the graphical techniques described in this section to check the normality assumption.

MINITAB Output for Exercise 8.15

```
Stem-and-leaf of RESIDUAL   N   = 19
Leaf Unit = 0.10


    2     -0 76
    3     -0 4
    4     -0 2
    9     -0 11110
   (3)     0 000
    7      0 2233
    3      0 445
```

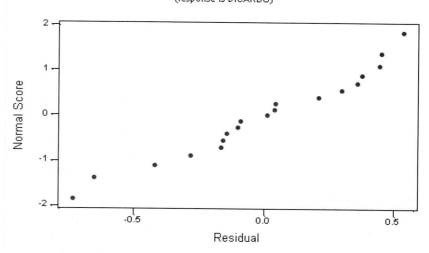

Normal Probability Plot of the Residuals
(response is DICARBO)

SPSS Output for Exercise 8.18

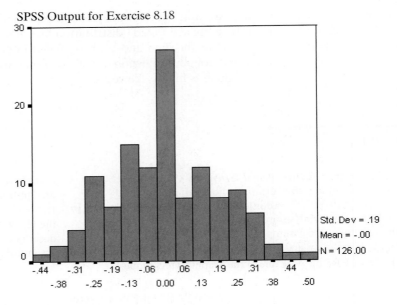

RESIDUAL

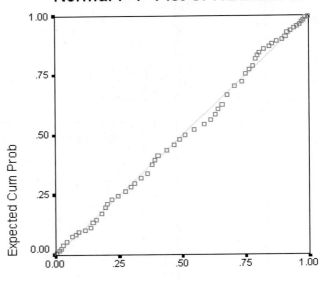

8.18. In the 1980s, L. De Cola conducted an extensive investigation of the geopolitical and socioeconomic processes that shape the urban size distributions of the world's nations. One of the goals of the study was to determine the factors that influence population size in each nation's largest city. Using data collected for a sample of 126 countries, De Cola fit the following log model:

$$E(y) = \beta_0 + \beta_1 x_1 + \beta_2 x_2 + \beta_3 x_3 + \beta_4 x_4$$

$$+ \beta_5 x_5 + \beta_6 x_6 + \beta_7 x_7 + \beta_8 x_8 + \beta_9 x_9 + \beta_{10} x_{10}$$

where

y = Log of population (in thousands) of largest city in country

x_1 = Log of area (in thousands of square kilometers) of country

x_2 = Log of radius (in hundreds of kilometers) of country

x_3 = Log of national population (in thousands)

x_4 = Percentage annual change in national population (1960–1970)

x_5 = Log of energy consumption per capita (in kilograms of coal equivalent)

x_6 = Percentage of nation's population in urban areas

x_7 = Log of population (in thousands) of second largest city in country

$$x_8 = \begin{cases} 1 & \text{if seaport city} \\ 0 & \text{if not} \end{cases}$$

$$x_9 = \begin{cases} 1 & \text{if captial city} \\ 0 & \text{if not} \end{cases}$$

$$x_{10} = \begin{cases} 1 & \text{if city data are for metropolitan area} \\ 0 & \text{if not} \end{cases}$$

[*Note*: All logarithms are to the base 10.]

a. The residuals for five cities selected from the total sample are given in the table. For each of these cities, calculate the estimated expected residuals under the assumption of normality.

CITY	RESIDUAL	RANK
Bangkok	.510	126
Paris	.228	110
London	.033	78
Warsaw	−.132	32
Lagos	−.392	2

Source: De Cola, L. "Statistical determinants of the population of a nation's largest city." *Economic Development and Cultural Change*, Vol. 3. No. 1, Oct. 1984, pp. 71–98.

b. SPSS graphs of all the residuals are shown on p. 396. Does it appear that the assumption of normal errors is satisfied?

8.5 Detecting Outliers and Identifying Influential Observations

We begin this section by defining a **standardized residual** as the value of the residual divided by the model standard deviation s. Since we assume that the residuals have a mean of 0 and a standard deviation estimated by s, you can see from Definition 8.3 that a standardized residual is simply the z-score for a residual (see Section 1.6).

Definition 8.3

The **standardized residual**, denoted z_i, for the ith observation is the residual for the observation divided by s, i.e.,

$$z_i = \hat{\varepsilon}_i/s = (y_i - \hat{y}_i)/s$$

Although we expect almost all the regression residuals to fall within three standard deviations of their mean of 0, sometimes one or several residuals fall outside this interval. Observations with residuals that are extremely large or small (say, more than three standard deviations from 0) are called **outliers**. Consequently, observations with standardized residuals that exceed 3 in absolute value are considered outliers.

Definition 8.4

An observation with a residual that is larger than $3s$ (in absolute value)—or, equivalently, a standardized residual that is larger than 3 (in absolute value)—is considered to be an **outlier**.

Note: As an alternative to standardized residuals, some software packages compute **studentized residuals**, so named because they follow an approximate Student's t-distribution.

Definition 8.5

The **studentized residual**, denoted z_i^*, for the ith observation is

$$z_i^* = \frac{\hat{\varepsilon}_i}{s\sqrt{1 - h_i}} = \frac{(y_i - \hat{y}_i)}{s\sqrt{1 - h_i}}$$

where h_i (called *leverage*) is defined in Definition 8.6

Outliers are usually attributable to one of several causes. The measurement associated with the outlier may be invalid. For example, the experimental procedure used to generate the measurement may have malfunctioned, the experimenter may have misrecorded the measurement, or the data may have been coded incorrectly for entry into the computer. Careful checks of the experimental and coding procedures should reveal this type of problem if it exists, so that we can eliminate erroneous observations from a data set.

For example, Table 8.5 presents the sales, y, in thousands of dollars per week, for fast-food outlets in each of four cities. The objective is to model sales, y, as a function of traffic flow, adjusting for city-to-city variations that might be due to size or other market conditions. We expect a first-order (linear) relationship to exist between mean sales, $E(y)$, and traffic flow. Further, we believe that the level of mean sales will differ from city to city, but that the change in mean sales per unit increase in traffic flow will remain the same for all cities, i.e., that the factors Traffic flow and City do not interact. The model is therefore

$$E(y) = \beta_0 + \beta_1 x_1 + \beta_2 x_2 + \beta_3 x_3 + \beta_4 x_4$$

 FASTFOOD

TABLE 8.5   **Data for Fast-Food Sales**

CITY	TRAFFIC FLOW thousands of cars	WEEKLY SALES y, thousands of dollars	CITY	TRAFFIC FLOW thousands of cars	WEEKLY SALES y, thousands of dollars
1	59.3	6.3	3	75.8	8.2
1	60.3	6.6	3	48.3	5.0
1	82.1	7.6	3	41.4	3.9
1	32.3	3.0	3	52.5	5.4
1	98.0	9.5	3	41.0	4.1
1	54.1	5.9	3	29.6	3.1
1	54.4	6.1	3	49.5	5.4
1	51.3	5.0	4	73.1	8.4
1	36.7	3.6	4	81.3	9.5
2	23.6	2.8	4	72.4	8.7
2	57.6	6.7	4	88.4	10.6
2	44.6	5.2	4	23.2	3.3

$$\text{where } x_1 = \begin{cases} 1 & \text{if city 1} \\ 0 & \text{other} \end{cases} \qquad x_2 = \begin{cases} 1 & \text{if city 2} \\ 0 & \text{other} \end{cases}$$

$$x_3 = \begin{cases} 1 & \text{if city 3} \\ 0 & \text{other} \end{cases} \qquad x_4 = \text{Traffic flow}$$

The MINITAB printout for the regression analysis is shown in Figure 8.21. The regression analysis indicates that the first-order model in traffic flow is inadequate for explaining mean sales. The coefficient of determination, R^2, (highlighted), indicates that only 25.9% of the total sum of squares of deviations of the sales y about their mean $\bar{y}$ is accounted for by the model. The global F value, 1.66 (also shaded) does not indicate that the model is useful for predicting sales. The observed significance level is only .200.

Plots of the studentized residuals against traffic flow and city are shown in Figures 8.22 and 8.23, respectively. The dashed horizontal line locates the mean (0) for the residuals. As you can see, the plots of the residuals are very revealing. Both the plot of residuals against traffic flow in Figure 8.22 and the plot of residuals against city in Figure 8.23 indicate the presence of an outlier. One observation in city 3 (observation 13), with traffic flow of 75.8, has a studentized residual value of 4.36. (This observation is shaded on the MINITAB printout, Figure 8.24). A further check of the observation associated with this residual reveals that the sales value entered into the computer, 82.0, does not agree with the corresponding value of sales, 8.2, that appears in Table 8.5. The decimal point was evidently dropped when the data were entered into MINITAB.

If the correct y value, 8.2, is substituted for the 82.0, we obtain the regression analysis shown in Figure 8.24. Plots of the studentized residuals against traffic flow and city are respectively shown in Figures 8.25 and 8.26. The corrected MINITAB printout indicates the dramatic effect that a single outlier can have on a regression analysis. The value of R^2 is now .979, and the F value that tests the adequacy of the model, 222.17, verifies the strong predictive capability of the model. Further analysis reveals that significant differences exist in the mean sales among cities, and that the estimated mean weekly sales increase by \$104 for every 1,000-car increase in traffic flow ($\hat{\beta}_4 = .104$). The 95% confidence interval for β_4 is

$$\hat{\beta}_4 \pm t_{.025} s_{\hat{\beta}4} = .104 \pm (2.093)(.0041) = .104 \pm .009$$

Thus, a 95% confidence interval for the mean increase in sales per 1,000-car increase in traffic flow is \$95 to \$113.

Outliers cannot always be explained by data entry or recording errors. Extremely large or small residuals may be attributable to skewness (nonnormality) of the probability distribution of the random error, chance, or unassignable causes. Although some analysts advocate elimination of outliers, regardless of whether cause can be assigned, others encourage the correction of only those outliers that can be traced to specific causes. The best philosophy is probably a compromise between these extremes. For example, before deciding the fate of an outlier, you may want to determine how much influence it has on the regression analysis.

FIGURE 8.21
MINITAB regression printout for
model of fast food sales

```
The regression equation is
SALES = - 16.5 + 1.11 X1 + 6.1 X2 + 14.5 X3 + 0.363 TRAFFIC

Predictor        Coef      SE Coef         T        P
Constant       -16.46        13.16     -1.25    0.226
X1              1.106         8.423      0.13    0.897
X2               6.14        11.68       0.53    0.605
X3             14.490         9.288      1.56    0.135
TRAFFIC        0.3629        0.1679      2.16    0.044

S = 14.86     R-Sq = 25.9%     R-Sq(adj) = 10.4%

Analysis of Variance

Source            DF        SS         MS        F        P
Regression         4     1469.8      367.4     1.66    0.200
Residual Error    19     4194.2      220.7
Total             23     5664.0


Obs    TRAFFIC     SALES       Fit     SE Fit    Residual    St Resid
  1      59.3       6.30      6.17      4.95        0.13        0.01
  2      60.3       6.60      6.53      4.96        0.07        0.01
  3      82.1       7.60     14.44      6.32       -6.84       -0.51
  4      32.3       3.00     -3.63      6.65        6.63        0.50
  5      98.0       9.50     20.21      8.25      -10.71       -0.87
  6      54.1       5.90      4.28      5.01        1.62        0.12
  7      54.4       6.10      4.39      5.01        1.71        0.12
  8      51.3       5.00      3.26      5.11        1.74        0.12
  9      36.7       3.60     -2.04      6.18        5.64        0.42
 10      23.6       2.80     -1.75      9.11        4.55        0.39
 11      57.6       6.70     10.59      8.97       -3.89       -0.33
 12      44.6       5.20      5.87      8.59       -0.67       -0.06
 13      75.8      82.00     25.54      7.27       56.46        4.36R
 14      48.3       5.00     15.56      5.62      -10.56       -0.77
 15      41.4       3.90     13.05      5.73       -9.15       -0.67
 16      52.5       5.40     17.08      5.66      -11.68       -0.85
 17      41.0       4.10     12.91      5.75       -8.81       -0.64
 18      29.6       3.10      8.77      6.43       -5.67       -0.42
 19      49.5       5.40     15.99      5.62      -10.59       -0.77
 20      73.1       8.40     10.07      6.71       -1.67       -0.13
 21      81.3       9.50     13.04      7.03       -3.54       -0.27
 22      72.4       8.70      9.81      6.69       -1.11       -0.08
 23      88.4      10.60     15.62      7.50       -5.02       -0.39
 24      23.2       3.30     -8.04     10.00       11.34        1.03

R denotes an observation with a large standardized residual
```

When an accurate outlier (i.e., an outlier that is not due to recording or measurement error) is found to have a dramatic effect on the regression analysis, it may be the model and not the outlier that is suspect. Omission of important independent variables or higher-order terms could be the reason why the model is not predicting well for the outlying observation. Several sophisticated numerical techniques are available for identifying outlying influential observations.

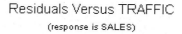

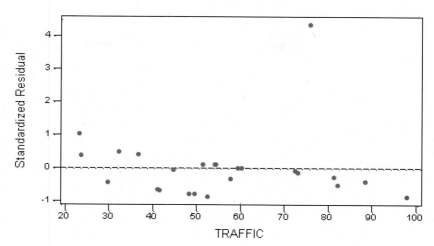

We conclude this section with a brief discussion of some of these methods and an example.

LEVERAGE: This procedure is based on a result (proof omitted) in regression analysis that states that the predicted value for the ith observation, $\hat{y}_i$, can be written as a linear combination of the n observed values $y_1, y_2, \ldots, y_n$:

$$\hat{y}_i = h_1 y_1 + h_2 y_2 + \cdots + h_i y_i + \cdots + h_n y_n, \ i = 1, 2, \ldots, n$$

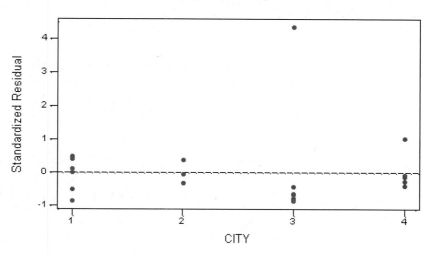

FIGURE 8.24
MINITAB regression printout for
model of fast food sales with
corrected data point

```
The regression equation is
SALES = 1.08 + 0.104 TRAFFIC - 1.22 X1 - 0.531 X2 - 1.08 X3

Predictor        Coef      SE Coef          T        P
Constant       1.0834       0.3210       3.37    0.003
TRAFFIC      0.103673     0.004094      25.32    0.000
X1            -1.2158       0.2054      -5.92    0.000
X2            -0.5308       0.2848      -1.86    0.078
X3            -1.0765       0.2265      -4.75    0.000

S = 0.3623      R-Sq = 97.9%     R-Sq(adj) = 97.5%

Analysis of Variance

Source            DF           SS          MS         F        P
Regression         4      116.656      29.164    222.17    0.000
Residual Error    19        2.494       0.131
Total             23      119.150

Obs    TRAFFIC      SALES         Fit      SE Fit    Residual    St Resid
  1       59.3     6.3000      6.0155      0.1208      0.2845        0.83
  2       60.3     6.6000      6.1191      0.1209      0.4809        1.41
  3       82.1     7.6000      8.3792      0.1541     -0.7792       -2.38R
  4       32.3     3.0000      3.2163      0.1621     -0.2163       -0.67
  5       98.0     9.5000     10.0276      0.2011     -0.5276       -1.75
  6       54.1     5.9000      5.4764      0.1222      0.4236        1.24
  7       54.4     6.1000      5.5075      0.1221      0.5925        1.74
  8       51.3     5.0000      5.1861      0.1245     -0.1861       -0.55
  9       36.7     3.6000      3.6724      0.1507     -0.0724       -0.22
 10       23.6     2.8000      2.9993      0.2222     -0.1993       -0.70
 11       57.6     6.7000      6.5242      0.2188      0.1758        0.61
 12       44.6     5.2000      5.1765      0.2095      0.0235        0.08
 13       75.8     8.2000      7.8653      0.1773      0.3347        1.06
 14       48.3     5.0000      5.0143      0.1369     -0.0143       -0.04
 15       41.4     3.9000      4.2989      0.1398     -0.3989       -1.19
 16       52.5     5.4000      5.4497      0.1380     -0.0497       -0.15
 17       41.0     4.1000      4.2575      0.1402     -0.1575       -0.47
 18       29.6     3.1000      3.0756      0.1569      0.0244        0.07
 19       49.5     5.4000      5.1387      0.1370      0.2613        0.78
 20       73.1     8.4000      8.6619      0.1635     -0.2619       -0.81
 21       81.3     9.5000      9.5120      0.1714     -0.0120       -0.04
 22       72.4     8.7000      8.5893      0.1632      0.1107        0.34
 23       88.4    10.6000     10.2481      0.1829      0.3519        1.13
 24       23.2     3.3000      3.4886      0.2438     -0.1886       -0.70

R denotes an observation with a large standardized residual
```

where the weights $h_1, h_2, \ldots, h_n$ of the observed values are functions of the independent variables. In particular, the coefficient h_i measures the influence of the observed value y_i on its own predicted value $\hat{y}_i$. This value, h_i, is called the **leverage** of the ith observation (with respect to the values of the independent variables). Thus, leverage values can be used to identify influential observations—the larger the leverage value, the more influence the observed y value has on its predicted value.

FIGURE 8.25
MINITAB plot of residuals versus
traffic flow for model with
corrected data point

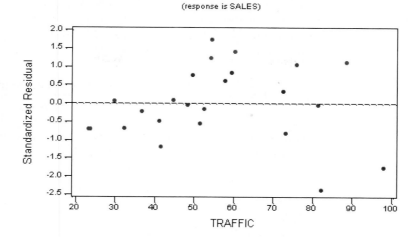

Residuals Versus TRAFFIC
(response is SALES)

FIGURE 8.26
MINITAB plot of residuals versus
city for model with corrected data
point

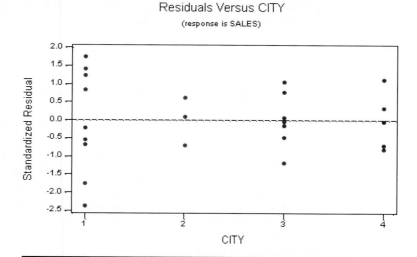

Residuals Versus CITY
(response is SALES)

Definition 8.6

The **leverage** of the ith observation is the weight, h_i, associated with y_i in the equation

$$y_i = h_1 y_1 + h_2 y_2 + h_3 y_3 + \cdots + h_i y_i + \cdots + h_n y_n$$

where $h_1, h_2, h_3 \ldots h_n$ are functions of only the values of the independent variables (x's) in the model.* The leverage, h_i, measures the influence of y_i on its predicted value y_i.

---

*In matrix notation, the leverage values are the diagonals of the H matrix (called the "hat" matrix), where $H = X(X'X)^{-1}X'$. See Appendix A for details on matrix multiplication and definition of the X matrix in regression.

Leverage values for multiple regression models are extremely difficult to calculate without the aid of a computer. Fortunately, most of the statistical software packages discussed in this text have options that give the leverage associated with each observation. The leverage value for an observation is usually compared with the average leverage value of all n observations, $\bar{h}$, where

$$\bar{h} = \frac{k+1}{n} = \frac{\text{Number of } \beta \text{ parameters in the model, including } \beta_0}{n}$$

A good rule of thumb identifies* an observation y_i as influential if its leverage value h_i is more than twice as large as $\bar{h}$, that is, if

$$h_i > \frac{2(k+1)}{n}$$

Rule of Thumb for Detecting Influence with Leverage
    The observed value of y_i is influential if

$$h_i > \frac{2(k+1)}{n}$$

where h_i is the leverage for the ith observation and $k =$ the number of β's in the model (excluding β_0).

THE JACKKNIFE: Another technique for identifying influential observations requires that you delete the observations one at a time, each time refitting the regression model based on only the remaining $n-1$ observations. This method is based on a statistical procedure, called the **jackknife**,[†] that is gaining increasing acceptance among practitioners. The basic principle of the jackknife when applied to regression is to compare the regression results using all n observations to the results with the ith observation deleted to ascertain how much influence a particular observation has on the analysis. Using the jackknife, several alternative influence measures can be calculated.

The **deleted residual**, $d_i = y_i - \hat{y}_{(i)}$, measures the difference between the observed value y_i and the predicted value $\hat{y}_{(i)}$ based on the model with the ith observation deleted. [The notation (i) is generally used to indicate that the observed value y_i was deleted from the regression analysis.] An observation with an unusually large (in absolute value) deleted residual is considered to have large influence on the fitted model.

---

*The proof of this result is beyond the scope of this text. Consult the references given at the end of this chapter. [See Neter, Kutner, Nachtsheim, and Wasserman (1996).]

[†]The procedure derives its name from the Boy Scout jackknife, which serves as a handy tool in a variety of situations when specialized techniques may not be applicable. [See Belsley, Kuh, and Welsch (1980).]

> **Definition 8.7**
>
> A **deleted residual**, denoted d_i, is the difference between the observed response y_i and the predicted value $\hat{y}_{(i)}$ obtained when the data for the ith observation is deleted from the analysis, i.e.,
>
> $$d_i = y_i - \hat{y}_{(i)}$$

A measure closely related to the deleted residual is the difference between the predicted value based on the model fit to all n observations and the predicted value obtained when y_i is deleted, i.e., $\hat{y}_i - \hat{y}_{(i)}$. When the difference $\hat{y}_i - \hat{y}_{(i)}$ is large relative to the predicted value $\hat{y}_i$, the observation y_i is said to influence the regression fit.

A third way to identify an influential observation using the jackknife is to calculate, for each β parameter in the model, the difference between the parameter estimate based on all n observations and that based on only $n - 1$ observations (with the observation in question deleted). Consider, for example, the straightline model $E(y) = \beta_0 + \beta_1 x$. The differences $\hat{\beta}_0 - \hat{\beta}_0^{(i)}$ and $\hat{\beta}_1 - \hat{\beta}_1^{(i)}$ measure how influential the ith observation y_i is on the parameter estimates. [Using the (i) notation defined previously, $\hat{\beta}^{(i)}$ represents the estimate of the β coefficient when the ith observation is omitted from the analysis.] If the parameter estimates change drastically, i.e., if the absolute differences are large, y_i is deemed an influential observation.

Each of the statistical software packages discussed in this text has a jackknife routine that produces one or more of the measures we described.

COOK'S DISTANCE: A measure of the overall influence an outlying observation has on the estimated β coefficients was proposed by R. D. Cook (1979). Cook's distance, D_i, is calculated for the ith observation as follows:

$$D_i = \frac{(y_i - \hat{y}_i)^2}{(k + 1)\text{MSE}} \left[\frac{h_i}{(1 - h_i)^2} \right]$$

Note that D_i depends on both the residual $(y_i - \hat{y}_i)$ and the leverage h_i for the ith observation. Although not obvious from the formula, D_i is a summary measure of the distances between $\hat{\beta}_0$ and $\hat{\beta}_0^{(i)}$, $\hat{\beta}_1$ and $\hat{\beta}_1^{(i)}$, $\hat{\beta}_2$ and $\hat{\beta}_2^{(i)}$, etc. A large value of D_i indicates that the observed y_i value has strong influence on the estimated β coefficients (since the residual, the leverage, or both will be large). Values of D_i can be compared to the values of the F distribution with $v_1 = k + 1$ and $v_2 = n - (k + 1)$ degrees of freedom. Usually, an observation with a value of D_i that falls at or above the 50th percentile of the F distribution is considered to be an influential observation. Like the other numerical measures of influence, options for calculating Cook's distance are available in most statistical software packages.

EXAMPLE 8.6

We now return to the fast-food sales example in which we detected an outlier using residual plots. Recall that the outlier was due to an error in coding the weekly sales value for observation 13 (denoted y_{13}). The SAS regression analysis is run with options for producing influence diagnostics. (An **influence diagnostic** is a number that measures how much influence an observation has on the regression analysis.) The resulting SAS printout is shown in Figure 8.27. Locate and interpret the measures of influence for y_{13} on the printout.

Solution

The influence diagnostics are shown in the bottom portion of the SAS printout in Figure 8.27. Leverage values for each observation are given under the column heading **Hat Diag H**. The leverage value for y_{13} (shaded on the printout) is $h_{13} = .2394$, whereas the average leverage for all $n = 24$ observations is

$$\bar{h} = \frac{k+1}{n}$$

$$= \frac{5}{24}$$

$$= .2083$$

Since the leverage value .2394 does not exceed $2\bar{h} = .4166$, we would not identify y_{13} as an influential observation. At first, this result may seem confusing since we already know the dramatic effect the incorrectly coded value of y_{13} had on the regression analysis. Remember, however, that the leverage values, $h_1, h_2, \ldots, h_{24}$, are functions of the independent variables only. Since we know the values of x_1, x_2, x_3, and x_4 were coded correctly, the relatively small leverage value of .2394 simply indicates that observation 13 is not an outlier with respect to the values of the independent variables.

A better overall measure of the influence of y_{13} on the fitted regression model is Cook's distance, D_{13}. Recall that Cook's distance is a function of both leverage and the magnitude of the residual. This value, $D_{13} = 1.196$ (shaded) is given in the column labeled **Cook's D** located on the right side of the printout. You can see that the value is extremely large relative to the other values of D_i in the printout. [In fact, $D_{13} = 1.196$ falls in the 65th percentile of the F distribution with $v_1 = k + 1 = 5$ and $v_2 = n - (k + 1) = 24 - 5 = 19$ degrees of freedom.] This implies that the observed value y_{13} has substantial influence on the estimates of the model parameters.

A statistic related to the deleted residual of the jackknife procedure is the **Studentized deleted residual** given under the column heading **Rstudent**.

Definition 8.8

The **Studentized deleted residual**, denoted d_i^*, is calculated by dividing the deleted residual d_i by its standard error s_{d_i}:

$$d_i^* = \frac{d_i}{s_{d_i}}$$

Under the assumptions of Section 4.2, the Studentized deleted residual d_i^* has a sampling distribution that is approximated by a Student's t distribution with

FIGURE 8.27

SAS regression analysis with influence diagnostics for fast food sales model

Dependent Variable: SALES

Analysis of Variance

Source	DF	Sum of Squares	Mean Square	F Value	Pr > F
Model	4	1469.76287	367.44072	1.66	0.1996
Error	19	4194.22671	220.74877		
Corrected Total	23	5663.98958			

Root MSE	14.85762	R-Square	0.2595	
Dependent Mean	9.07083	Adj R-Sq	0.1036	
Coeff Var	163.79550			

Parameter Estimates

Variable	DF	Parameter Estimate	Standard Error	t Value	Pr > |t|
Intercept	1	-16.45925	13.16400	-1.25	0.2264
X1	1	1.10609	8.42257	0.13	0.8969
X2	1	6.14277	11.67997	0.53	0.6050
X3	1	14.48962	9.28839	1.56	0.1353
TRAFFIC	1	0.36287	0.16791	2.16	0.0437

Output Statistics

Obs	Dep Var SALES	Predicted Value	Std Error Mean Predict	Residual	Std Error Residual	Student Residual	-2-1 0 1 2	Cook's D
1	6.3000	6.1652	4.9535	0.1348	14.008	0.00962	| | |	0.000
2	6.6000	6.5281	4.9596	0.0719	14.005	0.00513	| | |	0.000
3	7.6000	14.4387	6.3195	-6.8387	13.447	-0.509	| *| |	0.011
4	3.0000	-3.6324	6.6491	6.6324	13.287	0.499	| | |	0.012
5	9.5000	20.2084	8.2476	-10.7084	12.358	-0.866	| *| |	0.067
6	5.9000	4.2783	5.0130	1.6217	13.986	0.116	| | |	0.000
7	6.1000	4.3871	5.0054	1.7129	13.989	0.122	| | |	0.000
8	5.0000	3.2622	5.1069	1.7378	13.952	0.125	| | |	0.000
9	3.6000	-2.0357	6.1807	5.6357	13.511	0.417	| | |	0.007
10	2.8000	-1.7527	9.1137	4.5527	11.734	0.388	| | |	0.018
11	6.7000	10.5850	8.9723	-3.8850	11.843	-0.328	| | |	0.012
12	5.2000	5.8677	8.5897	-0.6677	12.123	-0.0551	| | |	0.000
13	82.0000	25.5362	7.2703	56.4638	12.957	4.358	| |****** |	1.196
14	5.0000	15.5571	5.6157	-10.5571	13.755	-0.767	| *| |	0.020
15	3.9000	13.0533	5.7339	-9.1533	13.707	-0.668	| *| |	0.016
16	5.4000	17.0812	5.6598	-11.6812	13.737	-0.850	| *| |	0.025
17	4.1000	12.9082	5.7479	-8.8082	13.701	-0.643	| *| |	0.015
18	3.1000	8.7714	6.4338	-5.6714	13.392	-0.423	| | |	0.008
19	5.4000	15.9926	5.6193	-10.5926	13.754	-0.770	| *| |	0.020
20	8.4000	10.0668	6.7066	-1.6668	13.258	-0.126	| | |	0.001
21	9.5000	13.0423	7.0271	-3.5423	13.091	-0.271	| | |	0.004
22	8.7000	9.8128	6.6916	-1.1128	13.265	-0.0839	| | |	0.000
23	10.6000	15.6187	7.5002	-5.0187	12.826	-0.391	| | |	0.010
24	3.3000	-8.0406	9.9965	11.3406	10.992	1.032	| |** |	0.176

Obs	RStudent	Hat Diag H	Cov Ratio	DFFITS	Intercept	X1	X2	X3	TRAFFIC
1	0.009366	0.1112	1.4742	0.0033	-0.0001	0.0020	0.0000	0.0000	0.0001
2	0.004998	0.1114	1.4747	0.0018	-0.0001	0.0011	0.0000	0.0000	0.0001
3	-0.4984	0.1809	1.4939	-0.2342	0.1256	-0.1339	-0.0539	-0.0510	-0.1455
4	0.4891	0.2003	1.5339	0.2447	0.1410	0.0780	-0.0604	-0.0572	-0.1633
5	-0.8606	0.3081	1.5482	-0.5743	0.3965	-0.2848	-0.1700	-0.1609	-0.4592
6	0.1129	0.1138	1.4735	0.0405	0.0054	0.0224	-0.0023	-0.0022	-0.0063
7	0.1192	0.1135	1.4724	0.0427	0.0053	0.0237	-0.0023	-0.0022	-0.0062
8	0.1213	0.1181	1.4799	0.0444	0.0094	0.0234	-0.0040	-0.0038	-0.0108
9	0.4079	0.1731	1.5134	0.1866	0.0964	0.0680	-0.0413	-0.0391	-0.1116
10	0.3791	0.3763	2.0190	0.2945	0.0859	-0.0178	0.1667	-0.0348	-0.0995
11	-0.3202	0.3647	2.0048	-0.2426	0.0614	-0.0127	-0.1967	-0.0249	-0.0711
12	-0.0536	0.3342	1.9667	-0.0380	0.0017	-0.0004	-0.0286	-0.0007	-0.0020
13	179.3101	0.2394	0.0000	100.6096	-55.1618	11.4109	23.6508	69.3701	63.8990
14	-0.7589	0.1429	1.3061	-0.3098	-0.0000	-0.0000	0.0000	-0.1873	0.0000
15	-0.6578	0.1489	1.3673	-0.2752	-0.0480	0.0099	0.0206	-0.1435	0.0556
16	-0.8439	0.1451	1.2625	-0.3477	0.0374	-0.0077	-0.0160	-0.2237	-0.0433
17	-0.6327	0.1497	1.3806	-0.2654	-0.0489	0.0101	0.0209	-0.1370	0.0566
18	-0.4141	0.1875	1.5381	-0.1990	-0.0838	0.0173	0.0359	-0.0710	0.0971
19	-0.7616	0.1430	1.3049	-0.3111	0.0096	-0.0020	-0.0041	-0.1919	-0.0112
20	-0.1224	0.2038	1.6389	-0.0619	-0.0237	0.0469	0.0318	0.0409	-0.0084
21	-0.2639	0.2237	1.6557	-0.1417	-0.0278	0.0974	0.0591	0.0797	-0.0461
22	-0.0817	0.2028	1.6408	-0.0412	-0.0164	0.0314	0.0215	0.0276	-0.0049
23	-0.3824	0.2548	1.6888	-0.2236	-0.0104	0.1378	0.0743	0.1054	-0.1037
24	1.0336	0.4527	1.7946	0.9400	0.9216	-0.6183	-0.6154	-0.6930	-0.7023

Sum of Residuals	0
Sum of Squared Residuals	4194.22671
Predicted Residual SS (PRESS)	7303.48386

$(n - 1) - (k + 1)$ df. Note that the Studentized deleted residual for y_{13} (shaded on the printout) is $d^*_{13} = 179.3101$. This extremely large value is another indication that y_{13} is an influential observation.

The **Dffits** column gives the difference between the predicted value when all 24 observations are used and when the ith observation is deleted. The difference, $\hat{y}_i - \hat{y}_{(i)}$, is divided by its standard error so that the differences can be compared more easily. For observation 13, this scaled difference (shaded on the printout) is 100.6096, an extremely large value relative to the other differences in predicted values. Similarly, the changes in the parameter estimates when observation 13 is deleted are given in the **Dfbetas** columns (shaded) immediately to the right of **Dffits** on the printout. (Each difference is also divided by the appropriate standard error.) The large magnitude of these differences provides further evidence that y_{13} is very influential on the regression analysis. ◆

Several techniques designed to limit the influence an outlying observation has on the regression analysis are available. One method produces estimates of the β's that minimize the sum of the absolute deviations, $\sum_{i=1}^{n} |y_i - \hat{y}_i|$.[‡] Because the deviations $(y_i - \hat{y}_i)$ are not squared, this method places less emphasis on outliers than the method of least squares. Regardless of whether you choose to eliminate an outlier or dampen its influence, careful study of residual plots and influence diagnostics are essential for outlier detection.

EXERCISES

8.19. Refer to the data and model of Exercise 8.1 (p. 376). The MSE for the model is .1267. Plot the residuals versus $\hat{y}$. Identify any outliers on the plot.

8.20. Refer to the data and model of Exercise 8.2 (p. 376). The MSE for the model is 17.2557. Plot the residuals versus $\hat{y}$. Identify any outliers on the plot.

8.21. Refer to The *Astronomical Journal* study of quasars detected by a deep space surrey, Exercise 4.9 (p. 180). The following model was fit to data collected on 90 quasars:

$$E(y) = \beta_0 + \beta_1 x_1 + \beta_2 x_2 + \beta_3 x_3 + \beta_4 x_4$$

where y = Rest frame equivalent width
x_1 = redshift
x_2 = line flux
x_3 = line luminosity
x_4 = AB$_{1450}$ magnitude

A portion of the SPSS spreadsheet showing influence diagnostics is shown on p. 409. Do you detect any influential observations?

 GFCLOCKS

8.22. Refer to the grandfather clock example, Example 4.3 (p. 173). The least squares model used to predict auction price, y, from age of the clock, x_1, and number of bidders, x_2, was determined to be

$$\hat{y} = -1,339 + 12.74x_1 + 85.95x_2$$

a. Use this equation to calculate the residuals of each of the prices given in Table 4.2 (p. 173).

b. Calculate the mean and the variance of the residuals. The mean should equal 0, and the variance should be close to the value of MSE given in the MINITAB printout shown in Figure 4.4 (p. 174).

c. Find the proportion of the residuals that fall outside 2 estimated standard deviations ($2s$) of 0 and outside $3s$.

d. Rerun the analysis and request influence diagnostics. Interpret the measures of influence given on the printout.

[‡]The method of absolute deviations requires linear programming techniques that are beyond the scope of this text. Consult the references given at the end of the chapter for details on how to apply this method.

SPSS Output for Exercise 8.21

	quasar	zdelres	cooksd	leverage	zdffits
1	1	-1.49140	.11570	.17631	-.78353
2	2	-.71142	.00685	.02192	-.18278
3	3	2.79116	.40406	.21782	1.64508
4	4	-.77375	.07423	.33791	-.60306
5	5	-.72387	.01380	.07392	-.25955
6	6	.46196	.01327	.18994	.25244
7	7	.58823	.00896	.07133	.20820
8	8	10.35824	2.40433	.37436	8.71278
9	9	.04929	.00011	.14268	.02330
10	10	-.06962	.00022	.13774	-.03237
11	11	-.03333	.00011	.27204	-.02245
12	12	.08399	.00098	.35741	.06821
13	13	.10748	.00068	.17873	.05687
14	14	-.19835	.00073	.04081	-.05881
15	15	-.75670	.00828	.02607	-.20126
16	16	-.50920	.01986	.22942	-.30922
17	17	-.41044	.00528	.09069	-.15914
18	18	-.26351	.00283	.12271	-.11616
19	19	-.62721	.01395	.10675	-.26011
20	20	-.16098	.00152	.17770	-.08492
21	21	.11697	.00031	.05825	.03861
22	22	.58346	.02083	.18833	.31738
23	23	-.52181	.01355	.15343	-.25554
24	24	-.66774	.03060	.21018	-.38571
25	25	.14011	.00038	.04355	.04231

8.23. Refer to the study of the population of the world's largest cities in Exercise 8.18 (p. 396). A multiple regression model for the natural logaritham of the population (in thousands) of each country's largest city was fit to data collected on 126 nations and resulted in s = .19. A SPSS stem-and-leaf plot of the standardized residuals is shown below. Identify any outliers on the plot.

SPSS Output for Exercises 8.23

```
STDRESID Stem-and-Leaf Plot

Frequency      Stem &  Leaf

     2.00        -2 .  04
     5.00        -1 .  55579
    14.00        -1 .  00122222233344
    19.00        -0 .  5555666666677778999
    23.00        -0 .  00011111111222334444444
    24.00         0 .  000000000111111122233444
    14.00         0 .  55555556777789
    15.00         1 .  000001112222234
     7.00         1 .  5566679
     2.00         2 .  03
     1.00         2 .  6

Stem width:      1.00
Each leaf:       1 case(s)
```

8.24. A large manufacturing firm wants to determine whether a relationship exists between y, the number of work-hours an employee misses per year, and x, the employee's annual wages. A sample of 15 employees produced the data in the accompanying table.

   a. Fit the first-order model, $E(y) = \beta_0 + \beta_1 x$, to the data.

   b. Plot the regression residuals. What do you notice?

   c. After searching through its employees' files, the firm has found that employee #13 had been fired but that his name had not been removed from the active employee payroll. This explains the large accumulation of work-hours missed (543) by that employee. In view of this fact, what is your recommendation concerning this outlier?

   d. Measure how influential the observation for employee #13 is on the regression analysis.

   e. Refit the model to the data, excluding the outlier, compare the results to those in part **a**.

8.25. Refer to *The New England Journal of Medicine* regression model relating a child's weight percentile, y, and the number of cigarettes smoked per day in the child's home (x), Exercise 8.7. A SAS regression printout (with residuals) for the straight-line model relating y to x is shown below and on p. 411.

⊙ **MISSWORK**

EMPLOYEE	WORK-HOURS MISSED y	ANNUAL WAGES x, thousands of dollars
1	49	12.8
2	36	14.5
3	127	8.3
4	91	10.2
5	72	10.0
6	34	11.5
7	155	8.8
8	11	17.2
9	191	7.8
10	6	15.8
11	63	10.8
12	79	9.7
13	543	12.1
14	57	21.2
15	82	10.9

   a. Examine the residuals in the printout. Do you detect any outliers?

   b. Influence diagnostics are also given on the SAS printout. Interpret these results.

SAS Output for Exercise 8.25

Dependent Variable: WTPCTILE

Analysis of Variance

Source	DF	Sum of Squares	Mean Square	F Value	Pr > F
Model	1	304.88209	304.88209	0.50	0.4864
Error	23	14011	609.17904		
Corrected Total	24	14316			

Root MSE	24.68155	R-Square	0.0213	
Dependent Mean	37.80000	Adj R-Sq	-0.0213	
Coeff Var	65.29511			

Parameter Estimates

Variable	DF	Parameter Estimate	Standard Error	t Value	Pr > |t|
Intercept	1	41.15266	6.84297	6.01	<.0001
SMOKED	1	-0.26193	0.37024	-0.71	0.4864

SAS Output for Exercise 8.25 (*continued*)

```
                                     Output Statistics

        Dep Var  Predicted     Std Error                 Std Error  Student
Obs    WTPCTILE  Value   Mean Predict   Residual  Residual Residual   -2-1 0 1 2       Cook's
                                                                                          D

 1      6.0000   41.1527      6.8430    -35.1527   23.714  -1.482  |   ** |     |       0.091
 2      6.0000   37.2238      5.0031    -31.2238   24.169  -1.292  |   ** |     |       0.036
 3      2.0000   30.6756     11.2153    -28.6756   21.986  -1.304  |   ** |     |       0.221
 4      8.0000   35.1284      6.2152    -27.1284   23.886  -1.136  |   ** |     |       0.044
 5     11.0000   35.9141      5.6101    -24.9141   24.036  -1.037  |   ** |     |       0.029
 6     17.0000   39.3192      5.3832    -22.3192   24.087  -0.927  |    * |     |       0.021
 7     24.0000   40.3669      6.1264    -16.3669   23.909  -0.685  |    * |     |       0.015
 8     25.0000   41.1527      6.8430    -16.1527   23.714  -0.681  |    * |     |       0.019
 9     17.0000   34.6045      6.6910    -17.6045   23.757  -0.741  |    * |     |       0.022
10     25.0000   35.9141      5.6101    -10.9141   24.036  -0.454  |      |     |       0.006
11     25.0000   37.2238      5.0031    -12.2238   24.169  -0.506  |    * |     |       0.005
12     31.0000   35.1284      6.2152     -4.1284   23.886  -0.173  |      |     |       0.001
13     35.0000   38.5334      5.0440     -3.5334   24.161  -0.146  |      |     |       0.000
14     43.0000   41.1527      6.8430      1.8473   23.714   0.0779 |      |     |       0.000
15     49.0000   41.1527      6.8430      7.8473   23.714   0.331  |      |     |       0.005
16     50.0000   41.1527      6.8430      8.8473   23.714   0.373  |      |     |       0.006
17     49.0000   35.3903      5.9975     13.6097   23.942   0.568  |      | *   |       0.010
18     46.0000   33.2949      8.0573     12.7051   23.329   0.545  |      | *   |       0.018
19     54.0000   41.1527      6.8430     12.8473   23.714   0.542  |      | *   |       0.012
20     58.0000   41.1527      6.8430     16.8473   23.714   0.710  |      | *   |       0.021
21     62.0000   41.1527      6.8430     20.8473   23.714   0.879  |      | *   |       0.032
22     66.0000   41.1527      6.8430     24.8473   23.714   1.048  |      | **  |       0.046
23     66.0000   35.1284      6.2152     30.8716   23.886   1.292  |      | **  |       0.057
24     83.0000   41.1527      6.8430     41.8473   23.714   1.765  |      | *** |       0.130
25     87.0000   29.6279     12.5621     57.3721   21.246   2.700  |      |*****|       1.275
```

```
                  Hat Diag       Cov                 -------DFBETAS-------
Obs    RStudent      H          Ratio      DFFITS     Intercept      SMOKED

 1     -1.5244     0.0769      0.9686     -0.4399     -0.4399       0.3046
 2     -1.3120     0.0411      0.9804     -0.2716     -0.1627      -0.0442
 3     -1.3255     0.2065      1.1812     -0.6762      0.2058      -0.6072
 4     -1.1433     0.0634      1.0398     -0.2975     -0.0453      -0.1808
 5     -1.0383     0.0517      1.0474     -0.2424     -0.0741      -0.1152
 6     -0.9236     0.0476      1.0635     -0.2064     -0.1936       0.0823
 7     -0.6764     0.0616      1.1178     -0.1733     -0.1718       0.1027
 8     -0.6730     0.0769      1.1367     -0.1942     -0.1942       0.1345
 9     -0.7335     0.0735      1.1240     -0.2066     -0.0134      -0.1395
10     -0.4461     0.0517      1.1319     -0.1041     -0.0318      -0.0495
11     -0.4974     0.0411      1.1146     -0.1030     -0.0617      -0.0168
```

```
                  Hat Diag       Cov                 -------DFBETAS-------
Obs    RStudent      H          Ratio      DFFITS     Intercept      SMOKED

12     -0.1691     0.0634      1.1639     -0.0440     -0.0067      -0.0267
13     -0.1431     0.0418      1.1385     -0.0299     -0.0253       0.0061
14      0.0762     0.0769      1.1834      0.0220      0.0220      -0.0152
15      0.3244     0.0769      1.1727      0.0936      0.0936      -0.0648
16      0.3660     0.0769      1.1697      0.1056      0.1056      -0.0731
17      0.5599     0.0590      1.1292      0.1403      0.0281       0.0797
18      0.5361     0.1066      1.1920      0.1852     -0.0195       0.1463
19      0.5333     0.0769      1.1540      0.1539      0.1539      -0.1066
20      0.7026     0.0769      1.1326      0.2027      0.2027      -0.1404
21      0.8746     0.0769      1.1058      0.2524      0.2524      -0.1748
22      1.0501     0.0769      1.0737      0.3030      0.3030      -0.2099
23      1.3126     0.0634      1.0036      0.3415      0.0520       0.2075
24      1.8561     0.0769      0.8851      0.5356      0.5356      -0.3709
25      3.1959     0.2590      0.6880      1.8896     -0.6678       1.7376
```

```
              Sum of Residuals                              0
              Sum of Squared Residuals                  14011
              Predicted Residual SS (PRESS)             18672
```

8.6
Detecting Residual Correlation: The Durbin–Watson Test

Many types of data are observed at regular time intervals. The Consumer Price Index (CPI) is computed and published monthly, the profits of most major corporations are published quarterly, and the *Fortune* 500 list of largest corporations is published annually. Data like these, which are observed over time, are called **time series**. We will often want to construct regression models where the data for the dependent and independent variables are time series.

Regression models of time series may pose a special problem. Because time series tend to follow economic trends and seasonal cycles, the value of a time series at time t is often indicative of its value at time $(t + 1)$. That is, the value of a time series at time t is **correlated** with its value at time $(t + 1)$. If such a series is used as the dependent variable in a regression analysis, the result is that the random errors are correlated. This leads to standard errors of the β-estimates that are seriously underestimated by the formulas given previously. Consequently, we cannot apply the standard least squares inference-making tools and have confidence in their validity. Modifications of the methods, which allow for correlated residuals in time series regression models, will be presented in Chapter 10. In this section, we present a method of testing for the presence of residual correlation.

Consider the time series data in Table 8.6, which gives sales data for the 35-year history of a company. The SAS printout shown in Figure 8.28 gives the regression analysis for the first-order linear model

$$y = \beta_0 + \beta_1 t + \varepsilon$$

This model seems to fit the data very well, since $R^2 = .98$ and the F value (1,615.72) that tests the adequacy of the model is significant. The hypothesis that the coefficient β_1 is positive is accepted at any α level less than .0001 ($t = 40.2$ with 33 df).

 SALES35

TABLE 8.6   **A Firm's Annual Sales Revenue (thousands of dollars)**

YEAR t	SALES y	YEAR t	SALES y	YEAR t	SALES y
1	4.8	13	48.4	25	100.3
2	4.0	14	61.6	26	111.7
3	5.5	15	65.6	27	108.2
4	15.6	16	71.4	28	115.5
5	23.1	17	83.4	29	119.2
6	23.3	18	93.6	30	125.2
7	31.4	19	94.2	31	136.3
8	46.0	20	85.4	32	146.8
9	46.1	21	86.2	33	146.1
10	41.9	22	89.9	34	151.4
11	45.5	23	89.2	35	150.9
12	53.5	24	99.1		

The residuals $\hat{\varepsilon} = y - (\hat{\beta}_0 + \hat{\beta}_1 t)$ are plotted in Figure 8.29. Note that there is a distinct tendency for the residuals to have long positive and negative runs. That is, if the residual for year t is positive, there is a tendency for the residual for year

FIGURE 8.28

SAS regression printout for model of annual sales

```
                    Dependent Variable: SALES

                    Analysis of Variance

                         Sum of          Mean
Source          DF      Squares        Square     F Value    Pr > F

Model            1        65875         65875     1615.72    <.0001
Error           33   1345.45355      40.77132
Corrected Total 34        67221


          Root MSE            6.38524   R-Square    0.9800
          Dependent Mean     77.72286   Adj R-Sq    0.9794
          Coeff Var           8.21540


                    Parameter Estimates

                    Parameter     Standard
Variable    DF       Estimate        Error    t Value    Pr > |t|

Intercept    1        0.40151      2.20571       0.18      0.8567
T            1        4.29563      0.10687      40.20      <.0001


          Durbin-Watson D                    0.821
          Number of Observations               35
          1st Order Autocorrelation         0.590
```

FIGURE 8.29

SAS plot of residuals for model of annual sales

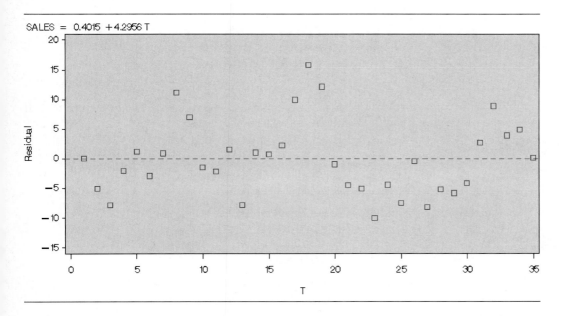

$(t + 1)$ to be positive. These cycles are indicative of possible positive correlation between residuals.

For most economic time series models, we want to test the null hypothesis

H_0: No residual correlation

against the alternative

H_a: Positive residual correlation

since the hypothesis of positive residual correlation is consistent with economic trends and seasonal cycles.

The **Durbin–Watson d statistic** is used to test for the presence of residual correlation. This statistic is given by the formula

$$d = \frac{\sum_{t=2}^{n} (\hat{\varepsilon}_t - \hat{\varepsilon}_{t-1})^2}{\sum_{t=1}^{n} \hat{\varepsilon}_t^2}$$

where n is the number of observations and $(\hat{\varepsilon}_t - \hat{\varepsilon}_{t-1})$ represents the difference between a pair of successive residuals. By expanding the numerator of d, we can also write

$$d = \frac{\sum_{t=2}^{n} \hat{\varepsilon}_t^2}{\sum_{t=1}^{n} \hat{\varepsilon}_t^2} + \frac{\sum_{t=2}^{n} \hat{\varepsilon}_{t-1}^2}{\sum_{t=1}^{n} \hat{\varepsilon}_t^2} - \frac{2 \sum_{t=2}^{n} \hat{\varepsilon}_t \hat{\varepsilon}_{t-1}}{\sum_{t=1}^{n} \hat{\varepsilon}_t^2} \approx 2 - \frac{2 \sum_{t=2}^{n} \hat{\varepsilon}_t \hat{\varepsilon}_{t-1}}{\sum_{t=1}^{n} \hat{\varepsilon}_t^2}$$

If the residuals are uncorrelated,

$$\sum_{t=2}^{n} \hat{\varepsilon}_t \hat{\varepsilon}_{t-1} \approx 0$$

indicating no relationship between $\hat{\varepsilon}_t$ and $\hat{\varepsilon}_{t-1}$, the value of d will be close to 2. If the residuals are highly positively correlated,

$$\sum_{t=2}^{n} \hat{\varepsilon}_t \hat{\varepsilon}_{t-1} \approx \sum_{t=2}^{n} \hat{\varepsilon}_t^2$$

(since $\hat{\varepsilon}_t \approx \hat{\varepsilon}_{t-1}$), and the value of d will be near 0:

$$d \approx 2 - \frac{2 \sum_{t=2}^{n} \hat{\varepsilon}_t \hat{\varepsilon}_{t-1}}{\sum_{t=1}^{n} \hat{\varepsilon}_t^2} \approx 2 - \frac{2 \sum_{t=2}^{n} \hat{\varepsilon}_t^2}{\sum_{t=1}^{n} \hat{\varepsilon}_t^2} \approx 2 - 2 = 0$$

If the residuals are very negatively correlated, then $\hat{\varepsilon}_t \approx -\hat{\varepsilon}_{t-1}$, so that

$$\sum_{t=2}^{n} \hat{\varepsilon}_t \hat{\varepsilon}_{t-1} \approx -\sum_{t=2}^{n} \hat{\varepsilon}_t^2$$

and d will be approximately equal to 4. Thus, d ranges from 0 to 4, with interpretations as summarized in the box.

Definition 8.9

The **Durbin–Watson d statistic** is calculated as follows:

$$d = \frac{\sum_{t=2}^{n} (\hat{\varepsilon}_t - \hat{\varepsilon}_{t-1})^2}{\sum_{t=1}^{n} \hat{\varepsilon}_t^2}$$

The d statistic has the following properties:

1. Range of d: $0 \leq d \leq 4$
2. If residuals are uncorrelated, $d \approx 2$.
3. If residuals are positively correlated, $d < 2$, and if the correlation is very strong, $d \approx 0$.
4. If residuals are negatively correlated, $d > 2$, and if the correlation is very strong, $d \approx 4$.

As an option, we requested SAS to produce the value of d for the annual sales model. The value, $d = .821$, is highlighted at the bottom of Figure 8.28. To determine if this value is close enough to 0 to conclude that positive residual correlation exists in the population, we need to find the rejection region for the test. Durbin and Watson (1951) have given tables for the lower-tail values of the d statistic, which we show in Table 8 ($\alpha = .05$) and Table 9 ($\alpha = .01$) of Appendix C.

Part of Table 8 is reproduced in Table 8.7. For the sales example, we have $k = 1$ independent variable and $n = 35$ observations. Using $\alpha = .05$ for the one-tailed test for positive residual correlation, the table values (shaded) are $d_L = 1.40$ and $d_U = 1.52$. The meaning of these values is illustrated in Figure 8.30. Because of the complexity of the sampling distribution of d, it is not possible to specify a single point that acts as a boundary between the rejection and nonrejection regions, as we did for the z, t, F, and other test statistics. Instead, an upper (d_U) and lower (d_L) bound are specified so that a d value less than d_L definitely *does* provide strong evidence of positive residual correlation at $\alpha = .05$ (recall that small d values indicate positive correlation), a d value greater than d_U *does not* provide evidence of positive correlation at $\alpha = .05$, but a value of d between d_L and d_U *might* be significant at the $\alpha = .05$ level. If $d_L < d < d_U$, more information is needed before we can reach any conclusion about the presence of residual correlation. A summary of the Durbin–Watson d test is given in the box on page 417.

TABLE 8.7   **Reproduction of Part of Table 8 of Appendix C ($\alpha = .05$)**

n	$k = 1$		$k = 2$		$k = 3$		$k = 4$		$k = 5$	
	d_L	d_U	d_L	d_U	d_L	d_U	d_L	d_U	d_L	d_U
31	1.36	1.50	1.30	1.57	1.23	1.65	1.16	1.74	1.09	1.83
32	1.37	1.50	1.31	1.57	1.24	1.65	1.18	1.73	1.11	1.82
33	1.38	1.51	1.32	1.58	1.26	1.65	1.19	1.73	1.13	1.81
34	1.39	1.51	1.33	1.58	1.27	1.65	1.21	1.73	1.15	1.81
35	1.40	1.52	1.34	1.58	1.28	1.65	1.22	1.73	1.16	1.80
36	1.41	1.52	1.35	1.59	1.29	1.65	1.24	1.73	1.18	1.80
37	1.42	1.53	1.36	1.59	1.31	1.66	1.25	1.72	1.19	1.80
38	1.43	1.54	1.37	1.59	1.32	1.66	1.26	1.72	1.21	1.79
39	1.43	1.54	1.38	1.60	1.33	1.66	1.27	1.72	1.22	1.79
40	1.44	1.54	1.39	1.60	1.34	1.66	1.29	1.72	1.23	1.79

FIGURE 8.30

Rejection region for the Durbin–Watson d test: Scale example

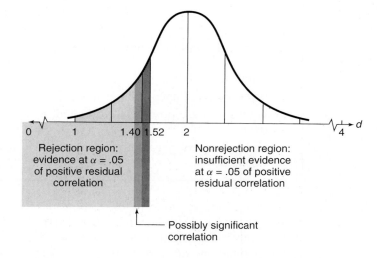

Rejection region: evidence at $\alpha = .05$ of positive residual correlation

Nonrejection region: insufficient evidence at $\alpha = .05$ of positive residual correlation

Possibly significant correlation

As indicated in the printout for the sales example (Figure 8.28), the computed value of d, .821, is less than the tabulated value of d_L, 1.40. Thus, we conclude that the residuals of the straight-line model for sales are positively correlated.

Tests for negative correlation and two-tailed tests can be conducted by making use of the symmetry of the sampling distribution of the d statistic about its mean, 2 (see Figure 8.30). That is, we compare $(4 - d)$ to d_L and d_U and conclude that the residuals are negatively correlated if $(4 - d) < d_L$, that there is insufficient evidence to conclude that the residuals are negatively correlated if $(4 - d) > d_U$, and that the test for negative residual correlation is *possibly* significant if $d_L < (4 - d) < d_U$.

Once strong evidence of residual correlation has been established, as in the case of the sales example, doubt is cast on the least squares results and any inferences drawn from them. In Chapter 10, we present a time series model that accounts for the correlation of the random errors. The residual correlation can be taken into account in a time series model, thereby improving both the fit of the model and the reliability of model inferences.

Durbin–Watson d Test for Residual Correlation

	LOWER-TAILED TEST	TWO-TAILED TEST	UPPER-TAILED TEST

LOWER-TAILED TEST

H_0: No residual correlation
H_a: Positive residual correlation

TWO-TAILED TEST

H_0: No residual correlation
H_a: Positive or negative residual correlation

UPPER-TAILED TEST

H_0: No residual correlation
H_a: Negative residual correlation

Test statistic :

$$d = \frac{\sum_{t=2}^{n}(\hat{\varepsilon}_t - \hat{\varepsilon}_{t-1})^2}{\sum_{t=1}^{n}\hat{\varepsilon}_t^2}$$

Rejection region :
$(4 - d) < d_{L,\alpha}$

Rejection region :
$d < d_{L,\alpha/2}$
or $(4 - d) < d_{L,\alpha/2}$

Rejection region :
$(4 - d) < d_{L,\alpha}$

Nonrejection region :
$(4 - d) > d_{U,\alpha}$

Nonrejection region :
$d > d_{U,\alpha/2}$
or $(4 - d) > d_{U,\alpha/2}$

Nonrejection region :
$(4 - d) > d_{U,\alpha}$

Inconclusive region :
$d_{L,\alpha} \le (4 - d) \le d_{U,\alpha}$

Inconclusive region :
Any other result

Inconclusive region :
$d_{L,\alpha} < (4 - d) < d_{U,\alpha}$

where $d_{L,\alpha}$ and $d_{U,\alpha}$ are the lower and upper tabulated values, respectively, corresponding to k independent variables and n observations.

Assumption: The residuals are normally distributed.

EXERCISES

8.26. Find the values of d_L and d_U from Tables 8 and 9 of Appendix C for each of the following situations:
  a. $n = 30$, $k = 3$, $\alpha = .05$
  b. $n = 40$, $k = 1$, $\alpha = .01$
  c. $n = 35$, $k = 5$, $\alpha = .05$

8.27. Exploratory research published in the *Journal of Professional Services Marketing* (Vol. 5, 1990) examined the relationship between deposit share of a retail bank and several marketing variables. Quarterly deposit share data were collected for five consecutive years for each of nine retail banking

institutions. The model analyzed took the following form:

$$y_t = \beta_0 + \beta_1 P_{t-1} + \beta_2 S_{t-1} + \beta_3 D_{t-1} + \varepsilon_t$$

where

y_t = Deposit share of bank in quarter t, $t = 1, 2, \ldots, 20$

P_{t-1} = Expenditures on promotion-related activities in quarter $t - 1$

S_{t-1} = Expenditures on service-related activities in quarter $t - 1$

D_{t-1} = Expenditures on distribution-related activities in quarter $t - 1$

A separate model was fit for each bank with the results shown in the table.

BANK	R^2	p-VALUE FOR GLOBAL F	DURBIN-WATSON d
1	.914	.000	1.3
2	.721	.004	3.4
3	.926	.000	2.7
4	.827	.000	1.9
5	.270	.155	.85
6	.616	.012	1.8
7	.962	.000	2.5
8	.495	.014	2.3
9	.500	.011	1.1

Note: The values of d shown are approximated based on other information provided in the article.

a. Interpret the value of R^2 for each bank.
b. Test the overall adequacy of the model for each bank.
c. Conduct the Durbin–Watson d test for each bank. Interpret the practical significance of the tests.

8.28. Forecasts of automotive vehicle sales in the United States provide the basis for financial and strategic planning of large automotive corporations. The following forecasting model was developed for y, total monthly passenger car and light truck sales (in thousands):

$$E(y) = \beta_0 + \beta_1 x_1 + \beta_2 x_2 + \beta_3 x_3 + \beta_4 x_4 + \beta_5 x_5$$

where

x_1 = Average monthly retail price of regular gasoline
x_2 = Annual percentage change in GNP per quarter
x_3 = Monthly consumer confidence index
x_4 = Total number of vehicles scrapped (millions) per month
x_5 = Vehicle seasonality

The model was fitted to monthly data collected over a 12-year period (i.e., $n = 144$ months) with the

following results:

$$\hat{y} = -676.42 - 1.93x_1 + 6.54x_2 + 2.02x_3$$
$$+ .08x_4 + 9.82x_5$$
$$R^2 = .856$$

Durbin–Watson $d = 1.01$

a. Is there sufficient evidence to indicate that the model contributes information for the prediction of y? Test using $\alpha = .05$.
b. Is there sufficient evidence to indicate that the regression errors are positively correlated? Test using $\alpha = .05$.
c. Comment on the validity of the inference concerning model adequacy in light of the result of part **b**.

8.29. The consumer purchasing value of the dollar from 1970 to 1997 is illustrated by the data in the accompanying table, where the purchasing power of the dollar (compared to 1982) is listed for each year. The first-order model $E(y_t) = \beta_0 + \beta_1 t$ was fit to the data using SAS. The SAS regression printout and residual plot are shown on p. 419.

BUYPOWER

YEAR	t	VALUE, y	YEAR	t	VALUE, y
1970	1	2.545	1984	15	0.964
1971	2	2.469	1985	16	0.955
1972	3	2.392	1986	17	0.969
1973	4	2.193	1987	18	0.949
1974	5	1.901	1988	19	0.926
1975	6	1.718	1989	20	0.880
1976	7	1.645	1990	21	0.839
1977	8	1.546	1991	22	0.822
1978	9	1.433	1992	23	0.812
1979	10	1.289	1993	24	0.802
1980	11	1.136	1994	25	0.797
1981	12	1.041	1995	26	0.782
1982	13	1.000	1996	27	0.762
1983	14	0.984	1997	28	0.759

Source: *Statistical Abstract of the United States*, 1998.

a. Examine the SAS residual plot. Is there a tendency for the residuals to have long positive and negative runs? How do you account for this?
b. Locate the Durbin–Watson d statistic for this model on the SAS printout. Test the hypothesis that the time series residuals are positively correlated. Test at $\alpha = .05$.

SAS Regression Output for Exercise 8.29

Dependent Variable: VALUE

Analysis of Variance

Source	DF	Sum of Squares	Mean Square	F Value	Pr > F
Model	1	7.00366	7.00366	107.81	<.0001
Error	26	1.68909	0.06496		
Corrected Total	27	8.69275			

Root MSE	0.25488	R-Square	0.8057	
Dependent Mean	1.26107	Adj R-Sq	0.7982	
Coeff Var	20.21156			

Parameter Estimates

Variable	DF	Parameter Estimate	Standard Error	t Value	Pr > |t|
Intercept	1	2.15883	0.09898	21.81	<.0001
T	1	-0.06191	0.00596	-10.38	<.0001

Durbin-Watson D	0.084
Number of Observations	28
1st Order Autocorrelation	0.866

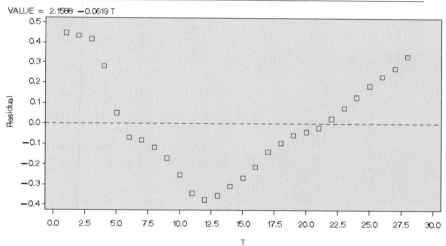

VALUE = 2.1588 −0.0619 T

8.30. The first table on p. 420 gives volume of retail sales (in billions of dollars) of passenger cars by U.S. motor vehicle dealers for the years 1996 and 1997.

 a. Fit the straight-line model $E(y) = \beta_0 + \beta_1 t$ to the data.

 b. Plot the regression residuals against t. Is there a tendency for the residuals to have long positive and negative runs?

 c. Is there evidence at $\alpha = .05$ level of significance that the residuals are autocorrelated?

8.31. A pharmaceutical company based in New Jersey recently introduced a new cold medicine called Coldex. (For proprietary reasons, the actual name of the product is withheld.) It is now sold in drugstores and supermarkets across the United States. Monthly sales for the first two years the product was on the market are reported in the second table on p. 420. Consider the simple linear regression model, $E(y_t) = \beta_0 + \beta_1 t$, where y_t, is the sales in month t.

 a. Fit the simple linear model to the data. Is the model statistically useful for predicting monthly sales?

 b. Construct a plot of the regression residuals against month, t. Does the plot suggest the presence of residual correlation? Explain.

 c. Use the Durbin–Watson test to formally test for correlated errors.

MONCARS

YEAR	MONTH	t	SALES, y
1996	Jan.	1	48.07
	Feb.	2	49.95
	Mar.	3	49.92
	Apr.	4	48.46
	May	5	49.48
	Jun.	6	48.67
	Jul.	7	48.71
	Aug.	8	49.01
	Sep.	9	49.84
	Oct.	10	50.11
	Nov.	11	49.52
	Dec.	12	50.04
1997	Jan.	13	50.84
	Feb.	14	52.69
	Mar.	15	51.86
	Apr.	16	50.45
	May	17	49.93
	Jun.	18	50.75
	Jul.	19	51.89
	Aug.	20	52.75
	Sep.	21	51.93
	Oct.	22	51.32
	Nov.	23	51.85
	Dec.	24	53.03

Source: Standard & Poor's *Current Statistics*, Dec., 1999.

COLDEX

YEAR	MONTH	t	SALES, y_t
1	Jan.	1	3,394
	Feb.	2	4,010
	Mar.	3	924
	Apr.	4	205
	May	5	293
	Jun.	6	1,130
	Jul.	7	1,116
	Aug.	8	4,009
	Sep.	9	5,692
	Oct.	10	3,458
	Nov.	11	2,849
	Dec.	12	3,470
2	Jan.	13	4,568
	Feb.	14	3,710
	Mar.	15	1,675
	Apr.	16	999

(Continued)

	May	17	986
	Jun.	18	1,786
	Jul.	19	2,253
	Aug.	20	5,237
	Sep.	21	6,679
	Oct.	22	4,116
	Nov.	23	4,109
	Dec.	24	5,124

Source: Personal communication from Carol Cowley, Carla Marchesini, and Ginny Wilson, Rutgers University, Graduate School of Management.

8.32. T.C. Chiang considered several time series forecasting models of future foreign exchange rates for U.S. currency (*The Journal of Financial Research*, Summer 1986). One popular theory among financial analysts is that the forward (90-day) exchange rate is a useful predictor of the future spot exchange rate. Using monthly data on exchange rates for the British pound for $n = 81$ months, Chiang fitted the model

$$E(y_t) = \beta_0 + \beta_1 x_{t-1}$$

where

$$y_t = \text{ln (spot rate) in month } t$$
$$x_t = \text{ln (forward rate) in month } t$$

The method of least squares yielded the following results:

$$\hat{y}_t = -.009 + .986 x_{t-1} \quad (t = 47.9)$$

$$s = .0249 \quad R^2 = .957 \quad \text{Durbin–Watson } d = .962$$

a. Is the model useful for predicting future spot exchange rates for the British pound? Test using $\alpha = .05$.

b. Interpret the values of s and R^2.

c. Is there evidence of positive autocorrelation among the residuals? Test using $\alpha = .05$.

d. Based on the results of parts **a–c**, would you recommend using the least squares model to forecast spot exchange rates?

Summary

An analysis of regression **residuals** can play an important role in the modeling process. Plots of the residuals against the independent variables can suggest modifications that will improve the model. These include the **addition of**

quadratic terms to allow for curvature in the response surface and **transforming the dependent variable to stabilize its variance. Histograms, stem-and-leaf displays**, and **normal probability plots** of residuals give visual clues as to whether the normality assumption is satisfied. **Plots of residuals** are also helpful for identification of **outliers**, which can then be traced to determine the cause of an unusually large or small observation. The influence the outlying observation has on the regression analysis can be measured using **leverage, Cook's distance D**, and **deleted residuals**.

When you use residual plots, you should always be aware that conclusions drawn from them are subjective. Therefore, they cannot substitute for formal tests to detect model inadequacies.

The F test (Section 1.9) can be used to detect differences in pairs of variances, and F tests for sets of parameters (Section 4.10) can be used to detect factor interactions and curvature in a response surface. The **Durbin–Watson d test** can be used to test for the presence of **residual correlation**. We will present methods for constructing time series models that allow for correlation of the random errors in Chapter 10.

SUPPLEMENTARY EXERCISES

8.33. Identify the problem(s) in each of the following five residual plots:

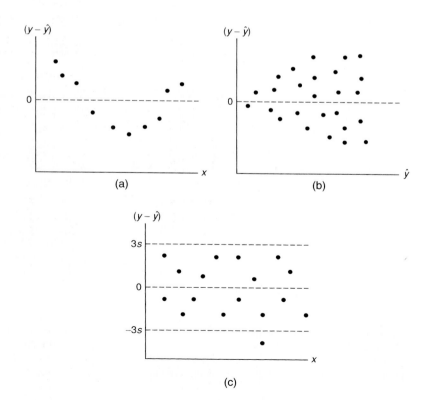

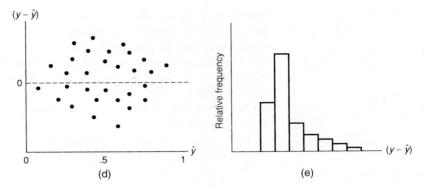

$(y - \hat{y})$

$\hat{y}$

(d)

Relative frequency

$(y - \hat{y})$

(e)

SAS Regression Output for Exercises 8.34

Dependent Variable: IMPROVE

Analysis of Variance

Source	DF	Sum of Squares	Mean Square	F Value	Pr > F
Model	2	1368.77501	684.38750	33.08	0.0003
Error	7	144.82499	20.68928		
Corrected Total	9	1513.60000			

Root MSE	4.54855	R-Square	0.9043	
Dependent Mean	17.20000	Adj R-Sq	0.8770	
Coeff Var	26.44504			

Parameter Estimates

Variable	DF	Parameter Estimate	Standard Error	t Value	Pr > |t|
Intercept	1	10.65904	14.55009	0.73	0.4876
COST	1	-0.28161	0.28088	-1.00	0.3494
COSTSQ	1	0.00267	0.00125	2.13	0.0706

Output Statistics

Obs	Dep Var IMPROVE	Predicted Value	Std Error Mean Predict	Residual	Std Error Residual	Student Residual	-2-1 0 1 2	Cook's D
1	18.0000	17.2073	2.0627	0.7927	4.054	0.196	| |	0.003
2	32.0000	34.0037	2.6525	-2.0037	3.695	-0.542	*|	0.051
3	9.0000	5.2310	2.0601	3.7690	4.055	0.929	|*	0.074
4	37.0000	35.1612	2.8397	1.8388	3.553	0.518	|*	0.057
5	6.0000	12.0128	2.2177	-6.0128	3.971	-1.514	***|	0.238
6	3.0000	6.9572	2.0835	-3.9572	4.043	-0.979	*|	0.085
7	30.0000	23.6042	1.8468	6.3958	4.157	1.539	|***	0.156
8	10.0000	6.0273	2.0492	3.9727	4.061	0.978	|*	0.081
9	25.0000	28.5367	2.0070	-3.5367	4.082	-0.866	*|	0.060
10	2.0000	3.2586	4.1920	-1.2586	1.765	-0.713	*|	0.955

Output Statistics

Obs	RStudent	Hat Diag H	Cov Ratio	DFFITS	DFBETAS Intercept	COST	COSTSQ
1	0.1815	0.2057	1.9665	0.0924	-0.0594	0.0657	-0.0639
2	-0.5129	0.3401	2.1156	-0.3682	-0.1192	0.1505	-0.1871
3	0.9190	0.2051	1.3457	0.4669	0.0286	0.0632	-0.1088
4	0.4886	0.3898	2.3148	0.3904	0.1473	-0.1817	0.2199
5	-1.7093	0.2377	0.6336	-0.9545	0.5920	-0.7014	0.7211
6	-0.9753	0.2098	1.2924	-0.5025	0.1474	-0.2357	0.2724
7	1.7511	0.1649	0.5511	0.7780	-0.3140	0.3143	-0.2585
8	0.9748	0.2030	1.2818	0.4919	-0.0653	0.1582	-0.2007
9	-0.8490	0.1947	1.4030	-0.4174	0.0129	0.0093	-0.0503
10	-0.6854	0.8494	8.4082	-1.6276	-1.4594	1.2838	-1.1539

8.34. A naval base is considering modifying or adding to its fleet of 48 standard aircraft. The final decision regarding the type and number of aircraft to be added depends on a comparison of cost versus effectiveness of the modified fleet. Consequently, the naval base would like to model the projected percentage increase y in fleet effectiveness by the end of the decade as a function of the cost x of modifying the fleet. A first proposal is the quadratic model

$$E(y) = \beta_0 + \beta_1 x + \beta_2 x^2$$

The data provided in the accompanying table were collected on 10 naval bases of a similar size that recently expanded their fleets. The data were used to fit the model, and the SAS printout of the multiple regression analysis is reproduced on p. 422.

⊙ NAVALFLEET

PERCENTAGE IMPROVEMENT AT END OF DECADE y	COST OF MODIFYING FLEET x, millions of dollars
18	125
32	160
9	80
37	162
6	110
3	90
30	140
10	85
25	150
2	50

a. Construct a residual plot versus x. Do you detect any trends? Any outliers?

b. Interpret the influence diagnostics shown on the printout. Are there any observations that have large influence on the analysis?

8.35. B. N. Song compared annual consumption for two lower developed countries (LDCs)—Korea, a poor LDC, and Italy, a rich LDC (*Economic Development and Cultural Change*, Apr. 1981). Using data from the post–Korean War period, Song modeled annual consumption y_t as a function of total labor income x_{1t} and total property income x_{2t}, with the following results (assume data for $n = 40$ years were used in the analysis):

Korea: $\hat{y}_t = 7.81 + .91x_{1t} + .57x_{2t}$
$\quad s = 1.29$
$\quad d = 2.09$
Italy: $\hat{y}_t = 1,043.4 + .85x_{1t} + .40x_{2t}$
$\quad s = 290.5$
$\quad d = 1.07$

a. Is there evidence of positively correlated residuals in the consumption model for Korea? Test using $\alpha = .05$.

b. Is there evidence of positively correlated residuals in the consumption model for Italy? Test using $\alpha = .05$.

8.36. A leading pharmaceutical company that produces a new hypertension pill would like to model annual revenue generated by this product. Company researchers utilized data collected over the previous 15 years to fit the model

$$E(y_t) = \beta_0 + \beta_1 x_t + \beta_2 t$$

where

$\quad y_t$ = Revenue in year t (in millions of dollars)
$\quad x_t$ = Cost per pill in year t
$\quad t$ = Year $(1, 2, \ldots, 15)$

A company statistician suspects that the assumption of independent errors may be violated and that, in fact, the regression residuals are positively correlated. Test this claim using $\alpha = .05$ if the Durbin–Watson d Statistic is $d = .776$.

8.37. Refer to the study of coronary care patients, Exercise 3.62 (p. 154) Recall that simple linear regression was used to model length of stay, y, to number of patient factors, x. Conduct a complete residual analysis for the model. (The data is saved in the FACTORS file.)

8.38. The foreman of a printing shop that has been in business five years is scheduling his work load for next year, and he must estimate the number of employees available for work. He asks the company statistician to forecast next year's absentee rate. Since it is known that quarterly fluctuations exist, the following model is proposed:

$$y = \beta_0 + \beta_1 x_1 + \beta_2 x_2 + \beta_3 x_3 + \varepsilon$$

where

$$y = \text{Absentee rate} = \frac{\text{Total employees absent}}{\text{Total employees}}$$

$$x_1 = \begin{cases} 1 & \text{if quarter 1 (January–March)} \\ 0 & \text{if not} \end{cases}$$

$$x_2 = \begin{cases} 1 & \text{if quarter 2 (April–June)} \\ 0 & \text{if not} \end{cases}$$

$$x_3 = \begin{cases} 1 & \text{if quarter 3 (July–September)} \\ 0 & \text{if not} \end{cases}$$

⊙ PRINTSHOP

YEAR	QUARTER 1	QUARTER 2	QUARTER 3	QUARTER 4
1	.06	.13	.28	.07
2	.12	.09	.19	.09
3	.08	.18	.41	.07
4	.05	.13	.23	.08
5	.06	.07	.30	.05

a. Fit the model to the data given in the table above.

b. Consider the nature of the response variable, y. Do you think that there may be possible violations of the usual assumptions about ε? Explain.

c. Suggest an alternative model that will approximately stabilize the variance of the error term ε.

d. Fit the alternative model. Check R^2 to determine whether model adequacy has improved.

8.39. The stock of Delta Air Lines DAL had the yearly closing prices shown in the table.

⊙ DALPRICES

YEAR	t	PRICE, y_t	YEAR	t	PRICE, y_t
1985	1	17.84	1994	10	27.94
1986	2	18.46	1995	11	27.00
1987	3	24.09	1996	12	34.19
1988	4	19.03	1997	13	39.50
1989	5	25.32	1998	14	57.06
1990	6	29.95	1999	15	54.56
1991	7	34.13	2000	16	46.31
1992	8	33.56			
1993	9	25.94			

Source: Standard & Poor's *NYSE Daily Stock Price Record*, 1985–2000.

a. Fit the model $E(y) = \beta_0 + \beta_1 t$.

b. Calculate the residuals for the model and plot the residuals against t. Do you detect any trends?

c. Test for correlated residuals using $\alpha = .01$.

8.40. The breeding ability of a thoroughbred horse is sometimes a more important consideration to prospective buyers than racing ability. Usually, the longer a horse lives, the greater its value for breeding purposes. Before marketing a group of horses, a breeder would like to be able to predict their life spans. The breeder believes that the gestation period of a thoroughbred horse may be an indicator of its life span. The information in the table below was supplied to the breeder by various stables in the area. (Note that the horse has the greatest variation of gestation period of any species due to seasonal and feed factors.) Consider the first-order model

$$y = \beta_0 + \beta_1 x + \varepsilon,$$

where $y =$ lifespan (in years) and $x =$ gestation period (in months).

⊙ HORSES

HORSE	GESTATION PERIOD x, *days*	LIFE SPAN y, *years*
1	403	30
2	279	22
3	307	7
4	416	31
5	265	21
6	356	27
7	298	25

a. Fit the model to the data.

b. Check model adequacy by interpreting the F and R^2 statistics.

c. Construct a plot of the residuals versus x, gestation period.

d. Check for residuals that lie outside the interval $0 \pm 2s$ or $0 \pm 3s$.

e. The breeder has been informed that the short life span of horse #3 (7 years) was due to a very rare disease. Omit the data for horse #3 and refit the least squares line. Has the omission of this observation improved the model?

8.41. Refer to the *Journal of Colloid and Interface Science* study of the voltage (y) required to separate water from oil, Exercise 4.11 (p. 181). A first-order model for y included seven independent variables. Conduct a complete residual analysis of the model using the data saved in the WATEROIL file.

8.42. The data in the table on p. 425 are the monthly market shares for a product over most of the past year. The least squares line relating market share to television advertising expenditure is found to be

$$\hat{y} = -1.56 + .687x$$

TVSHARE

MONTH	MARKET SHARE y,%	TELEVISION ADVERTISING EXPENDITURE x, thousands of dollars
January	15	23
February	17	27
March	17	25
May	13	21
June	12	20
July	14	24
September	16	26
October	14	23
December	15	25

a. Calculate and plot the regression residuals in the manner outlined in this section.

b. The response variable y, market share, is recorded as a percentage. What does this lead you to believe about the least squares assumption of homoscedasticity? Does the residual plot substantiate this belief?

c. What variance-stabilizing transformation is suggested by the trend in the residual plot? If you have access to a computer package, refit the first-order model using the transformed responses. Calculate and plot these new regression residuals. Is there evidence that the transformation has been successful in stabilizing the variance of the error term, ε?

REFERENCES

BARNETT, V. and LEWIS, T. *Outliers in Statistical Data.* New York: Wiley, 1978.

BELSLEY, D. A., KUH, E., and WELSCH, R. E. *Regression Diagnostics: Identifying Influential Data and Sources of Collinearity.* New York: Wiley, 1980.

BREUSCH, T. S. and PAGAN, A. R. (1979), "A Simple Test for Heteroscedasticity and Random Coefficient Variation," *Econometrica*, Vol. 47, pp. 1287–1294.

BOX, G. E. P. and COX, D. R. "An analysis of transformations." *Journal of the Royal Statistical Society, Series B*, 1964, Vol. 26, pp. 211–243.

COOK, R. D. "Influential observations in linear regression." *Journal of the American Statistical Association*, 1979, Vol. 74, pp. 169–174.

COOK, R. D. and WEISBERG, S. *Residuals and Influence in Regression*, New York: Chapman and Hall, 1982.

DRAPER, N. and SMITH, H. *Applied Regression Analysis*, 2nd ed. New York: Wiley, 1981.

DURBIN, J. and WATSON, G. S. "Testing for serial correlation in least squares regression, I." *Biometrika*, 1950, Vol. 37, pp. 409–428.

DURBIN, J. and WATSON, G. S. "Testing for serial correlation in least squares regression, II." *Biometrika*, 1951, Vol. 38, pp. 159–178.

DURBIN, J. and WATSON, G. S. "Testing for serial correlation in least squares regression, III." *Biometrika*, 1971, Vol. 58, pp. 1–19.

GRANGER, C. W. J., and NEWBOLD, P. *Forecasting Economic Time Series.* New York: Academic Press, 1977.

LARSEN, W. A., and McCLEARY, S. J. "The use of partial residual plots in regression analysis." *Technometrics*, Vol. 14, 1972, pp. 781–790.

MANSFIELD, E. R., and CONERLY, M. D. "Diagnostic value of residual and partial residual plots." *The American Statistician*, Vol. 41, No. 2, May 1987, pp. 107–116.

MENDENHALL, W. *Introduction to Linear Models and the Design and Analysis of Experiments.* Belmont, Calif.: Wadsworth, 1968.

MONTGOMERY, D. C., and PECK, E. A. *Introduction to Linear Regression Analysis.* New York: Wiley, 1982.

NETER, J., KUTNER, M. NACHTSHEIM, C. and WASSERMAN, W. *Applied Linear Statistical Models*, 3rd ed. Homewood, Ill.: Richard D. Irwin, 1996.

STEPHENS, M. A. "EDF statistics for goodness of fit and some comparisons." *Journal of the American Statistical Association*, 1974, Vol. 69, pp. 730–737.

Special Topics in Regression (Optional)

CONTENTS

OBJECTIVE

To introduce a number of special regression techniques for problems that require more advanced methods of analysis

9.1 **Introduction**	The procedures presented in Chapters 3 to 8 provide the tools basic to a regression analysis. An understanding of these techniques will enable you to successfully apply regression analysis to a variety of problems encountered in practice. For some studies, however, you may require more sophisticated techniques. In this chapter, we introduce several special topics in regression for the advanced student.

9.2 **Piecewise Linear Regression**	Occasionally, the linear relationship between a dependent variable y and an independent variable x may differ for different intervals over the range of x. For example, it is known that the compressive strength y of concrete depends on the proportion x of water mixed with the cement. A certain type of concrete, when mixed in batches with varying water/cement ratios (measured as a percentage), may yield compressive strengths (measured in pounds per square inch) that follow the pattern shown in Figure 9.1. Note that the compressive strength decreases at a much faster rate for batches with water/cement ratios greater than 70%. That is, the slope of the relationship between compressive strength (y) and water/cement ratio (x) changes when $x = 70$.

FIGURE 9.1

Relationship between compressive strength (y) and water/cement ratio (x)

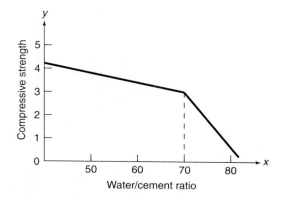

A model that proposes different straight-line relationships for different intervals over the range of x is called a **piecewise linear regression model**. As its name suggests, the linear regression model is fitted in pieces. For the concrete example, the piecewise model would consist of two pieces, $x \le 70$ and $x > 70$. The model can be expressed as follows:

$$y = \beta_0 + \beta_1 x_1 + \beta_2 (x_1 - 70) x_2 + \varepsilon$$

where

$$x_1 = \text{Water/cement ratio}(x)$$

$$x_2 = \begin{cases} 1 & \text{if } x_1 > 70 \\ 0 & \text{if } x_1 \le 70 \end{cases}$$

The value of the dummy variable x_2 controls the values of the slope and y-intercept for each piece. For example, when $x_1 \le 70$, then $x_2 = 0$ and the equation is given by

$$y = \beta_0 + \beta_1 x_1 + \beta_2 (x_1 - 70)(0) + \varepsilon$$

$$= \underbrace{\beta_0}_{y\text{-intercept}} + \underbrace{\beta_1}_{\text{Slope}} x_1 + \varepsilon$$

Conversely, if $x_1 > 70$, then $x_2 = 1$ and we have

$$y = \beta_0 + \beta_1 x_1 + \beta_2 (x_1 - 70)(1) + \varepsilon$$

$$= \beta_0 + \beta_1 x_1 + \beta_2 x_1 - 70\beta_2 + \varepsilon$$

or

$$y = \underbrace{(\beta_0 - 70\beta_2)}_{y\text{-intercept}} + \underbrace{(\beta_1 + \beta_2)}_{\text{Slope}} x_1 + \varepsilon$$

Thus, β_1 and $(\beta_1 + \beta_2)$ represent the slopes of the lines for the two intervals of x, $x \le 70$ and $x > 70$, respectively. Similarly, β_0 and $(\beta_0 - 70\beta_2)$ represent the respective y-intercepts. The slopes and y-intercepts of the two lines are illustrated graphically in Figure 9.2. [*Note*: The value at which the slope changes, 70 in this example, is often referred to as a **knot value**. Usually, the knot values of a piecewise

FIGURE 9.2

Slopes and y-intercepts for piecewise linear regression

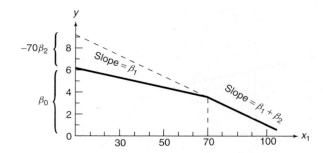

regression are unknown and must be estimated from the sample data. This is often accomplished by visually inspecting the scatterplot for the data and locating the points on the x-axis at which the slope appears to change.]

We can fit piecewise regression models by using the standard multiple regression procedures of most statistical software packages to make the appropriate transformations on the independent variables. For example, consider the data on compressive strength (y) and water/cement ratio (x) for 18 batches of concrete recorded in Table 9.1. (The water/cement ratio is computed by dividing the weight of water used in the mix by the weight of the cement.) To obtain the least squares fit of the piecewise linear regression model for the data of Table 9.1, we specify the model

$$E(y) = \beta_0 + \beta_1 x_1 + \beta_2 x_2^*$$

where

$$x_2^* = (x_1 - 70)x_2 \quad \text{and} \quad x_2 = \begin{cases} 1 & \text{if } x_1 > 70 \\ 0 & \text{if } x_1 \le 70 \end{cases}$$

The SAS printout for the piecewise linear regression is shown in Figure 9.3. The least squares prediction equation (shaded) is:

$$\hat{y} = 7.79198 - .06633x_1 - .10119x_2^*$$

 CEMENT

TABLE 9.1   Data on Compressive Strength and Water/Cement Ratios for 18 Batches of Cement

BATCH	COMPRESSIVE STRENGTH y, pounds per square inch	WATER/CEMENT RATIO x_1, percent	BATCH	COMPRESSIVE STRENGTH y, pounds per square inch	WATER/CEMENT RATIO x_1, percent
1	4.67	47	10	2.21	73
2	3.54	68	11	4.10	60
3	2.25	75	12	1.13	85
4	3.82	65	13	1.67	80
5	4.50	50	14	1.59	75
6	4.07	55	15	3.91	63
7	.76	82	16	3.15	70
8	3.01	72	17	4.37	50
9	4.29	52	18	3.75	57

FIGURE 9.3

SAS piecewise linear regression printout

Dependent Variable: STRENGTH

Analysis of Variance

Source	DF	Sum of Squares	Mean Square	F Value	Pr > F
Model	2	24.71775	12.35888	114.44	<.0001
Error	15	1.61990	0.10799		
Corrected Total	17	26.33765			

Root MSE	0.32862	R-Square	0.9385	
Dependent Mean	3.15500	Adj R-Sq	0.9303	
Coeff Var	10.41595			

Parameter Estimates

Variable	DF	Parameter Estimate	Standard Error	t Value	Pr > |t|
Intercept	1	7.79198	0.67696	11.51	<.0001
X1	1	-0.06633	0.01123	-5.90	<.0001
X2STAR	1	-0.10119	0.02812	-3.60	0.0026

Note that the estimated mean change in compressive strength for a 1% increase in water/cement ratio is $\hat{\beta}_1 = -.06633$ for ratios less than or equal to 70% and $\hat{\beta}_1 + \hat{\beta}_2 = -.06633 + (-.10119) = -.16752$ for ratios greater than 70%.

Piecewise regression is not limited to two pieces, nor is it limited to straight lines. One or more of the pieces may require a quadratic or higher-order fit. Also, piecewise regression models can be proposed to allow for discontinuities or jumps in the regression function. Such models require additional dummy variables to be introduced. Several different piecewise linear regression models relating y to an independent variable x are shown in the following box.

Piecewise Linear Regression Models Relating y to an Independent Variable x_1

TWO STRAIGHT LINES (CONTINUOUS):

$$E(y) = \beta_0 + \beta_1 x_1 + \beta_2(x_1 - k)x_2$$

where

k = Knot value (i.e., the value of the independent variable x_1

at which the slope changes)

$$x_2 = \begin{cases} 1 & \text{if } x_1 > k \\ 0 & \text{if not} \end{cases}$$

	$x_1 \le k$	$x_1 > k$
y-intercept	β_0	$\beta_0 - k\beta_2$
Slope	β_1	$\beta_1 + \beta_2$

(Continued)

THREE STRAIGHT LINES (CONTINUOUS):

$$E(y) = \beta_0 + \beta_1 x_1 + \beta_2(x_1 - k_1)x_2 + \beta_3(x_1 - k_2)x_3$$

where k_1 and k_2 are knot values of the independent variable x_1, $k_1 < k_2$, and

$$x_2 = \begin{cases} 1 & \text{if } x_1 > k_1 \\ 0 & \text{if not} \end{cases} \qquad x_3 = \begin{cases} 1 & \text{if } x_1 > k_2 \\ 0 & \text{if not} \end{cases}$$

	$x_1 \leq k_1$	$k_1 < x_1 \leq k_2$	$x_1 > k_2$
y-intercept	β_0	$\beta_0 - k_1\beta_2$	$\beta_0 - k_1\beta_2 - k_2\beta_3$
Slope	β_1	$\beta_1 + \beta_2$	$\beta_1 + \beta_2 + \beta_3$

TWO STRAIGHT LINES (DISCONTINUOUS)

$$E(y) = \beta_0 + \beta_1 x_1 + \beta_2(x_1 - k)x_2 + \beta_3 x_2$$

where

k = Knot value (i.e., the value of the independent variable x_1 at which the slope changes—also the point of discontinuity)

$$x_2 = \begin{cases} 1 & \text{if } x_1 > k \\ 0 & \text{if not} \end{cases}$$

	$x_1 \leq k$	$x_1 > k$
y-intercept	β_0	$\beta_0 - k\beta_2 + \beta_3$
Slope	β_1	$\beta_1 + \beta_2$

Tests of model adequacy, tests and confidence intervals on individual β parameters, confidence intervals for $E(y)$, and prediction intervals for y for piecewise regression models are conducted in the usual manner.

EXERCISES

9.1. Consider a two-piece linear relationship between y and x with no discontinuity and a slope change at $x = 15$.
   a. Specify the appropriate piecewise linear regression model for y.
   b. In terms of the β coefficients, give the y-intercept and slope for observations with $x \leq 15$; for observations with $x > 15$.

   c. Explain how you could determine whether the two slopes proposed by the model are, in fact, different.

9.2. Consider a three-piece linear relationship between y and x with no discontinuity and slope changes at $x = 1.45$ and $x = 5.20$.
   a. Specify the appropriate piecewise linear regression model for y.

b. In terms of the β coefficients, give the y-intercept and slope for each of the following intervals: $x \leq 1.45$, $1.45 < x \leq 5.20$, and $x > 5.20$.

c. Explain how you could determine whether at least two of the three slopes proposed by the model are, in fact, different.

9.3. Consider a two-piece linear relationship between y and x with discontinuity and slope change at $x = 320$.

a. Specify the appropriate piecewise linear regression model for y.

b. In terms of the β coefficients, give the y-intercept and slope for observations with $x \leq 320$; for observations with $x > 320$.

c. Explain how you could determine whether the two straight lines proposed by the model are, in fact, different.

9.4. In biology, researchers often conduct growth experiments to determine the number of cells that will grow in a petri dish after a certain period of time. The data in the table represent the number of living cells in plant tissue after exposing the specimen to heat for a certain number of hours.

GROWTH

HOURS	CELLS
1	2
5	3
10	4
20	5
30	4
40	6
45	8
50	9
60	10
70	18
80	16
90	18
100	20
110	14
120	12
130	13
140	9

a. Plot number of cells, y, against number of hours, x. What trends are apparent?

b. Propose a piecewise linear model for number of cells. Give the value of x that represents the knot value.

c. Fit the model, part b, to the data.

d. Test the overall adequacy of the model using $\alpha = .05$.

e. What is the estimated rate of growth when heating the tissue for less than 70 hours? More than 70 hours?

f. Conduct a test (at $\alpha = .05$) to determine if the two estimates, part **e**, are statistically different.

9.5. The manager of a packaging plant wants to model the unit cost y of shipping lots of a semifragile product as a linear function of lot size x. Because of economies of scale, the manager believes that the cost per unit will decrease at a faster rate for lot sizes of more than 1,000. Data collected on unit cost and lot size for 15 recent shipments are given in the table.

PACKPLANT

SHIPPING COST y, $ per unit	LOT SIZE x	SHIPPING COST y, $ per unit	LOT SIZE x
1.29	1,150	2.90	520
2.20	840	2.63	670
2.26	900	.55	1,420
2.38	800	2.31	850
1.77	1,070	1.90	1,000
1.25	1,220	2.15	910
1.87	980	1.20	1,230
.71	1,300		

a. Specify the appropriate piecewise linear model for y.

b. Fit the model to the data. Give the least squares prediction equation.

c. Is the model adequate for predicting unit cost y? Test using $\alpha = .10$.

d. Give a 90% confidence interval for the mean increase in shipping cost per unit for every unit increase in lot size for lots with 1,000 or fewer units.

9.3
Inverse Prediction

Often, the goal of regression is to predict the value of one variable when another variable takes on a specified value. For most simple linear regression problems, we are interested in predicting y for a given x. We provided a formula for a prediction interval for y when $x = x_p$ in Section 3.9. In this section, we discuss **inverse prediction**—that is, predicting x for a given value of the dependent variable y.

Inverse prediction has many applications in the engineering and physical sciences, in medical research, and in business. For example, when calibrating a new instrument, scientists often search for approximate measurements y, which are easy and inexpensive to obtain and which are related to the precise, but more expensive and time-consuming measurements x. If a regression analysis reveals that x and y are highly correlated, then the scientist could choose to use the quick and inexpensive approximate measurement value, say, $y = y_p$, to estimate the unknown precise measurement x. (In this context, the problem of inverse prediction is sometimes referred to as a **linear calibration** problem.) Physicians often use inverse prediction to determine the required dosage of a drug. Suppose a regression analysis conducted on patients with high blood pressure showed that a linear relationship exists between decrease in blood pressure y and dosage x of a new drug. Then a physician treating a new patient may want to determine what dosage x to administer to reduce the patient's blood pressure by an amount $y = y_p$. To illustrate inverse prediction in a business setting, consider a firm that sells a particular product. Suppose the firm's monthly market share y is linearly related to its monthly television advertising expenditure x. For a particular month, the firm may want to know how much it must spend on advertising x to attain a specified market share $y = y_p$.

The classical approach to inverse prediction is first to fit the familiar straight-line model

$$y = \beta_0 + \beta_1 x + \varepsilon$$

to a sample of n data points and obtain the least squares prediction equation

$$\hat{y} = \hat{\beta}_0 + \hat{\beta}_1 x$$

Solving the least squares prediction equation for x, we have

$$x = \frac{\hat{y} - \hat{\beta}_0}{\hat{\beta}_1}$$

Now let y_p be an observed value of y in the future with unknown x. Then a point estimate of x is given by

$$\hat{x} = \frac{y_p - \hat{\beta}_0}{\hat{\beta}_1}$$

Although no exact expression for the standard error of $\hat{x}$ (denoted $s_{\hat{x}}$) is known, we can algebraically manipulate the formula for a prediction interval for y given x (see Section 3.9) to form a prediction interval for x given y. It can be shown (proof omitted) that an approximate $(1 - \alpha)$ 100% prediction interval for x when $y = y_p$ is

$$\hat{x} \pm t_{\alpha/2} s_{\hat{x}} \approx \hat{x} \pm t_{\alpha/2} \left(\frac{s}{\hat{\beta}_1} \right) \sqrt{1 + \frac{1}{n} + \frac{(\hat{x} - \bar{x})^2}{\text{SS}_{xx}}}$$

where the distribution of t is based on $(n - 2)$ degrees of freedom, $s = \sqrt{\text{MSE}}$, and

$$\text{SS}_{xx} = \sum x^2 - n(\bar{x})^2$$

This approximation is appropriate as long as the quantity

$$D = \left(\frac{t_{\alpha/2}s}{\hat{\beta}_1}\right)^2 \cdot \frac{1}{SS_{xx}}$$

is small. The procedure for constructing an approximate confidence interval for x in inverse prediction is summarized in the box.

Inverse Prediction: Approximate $(1 - \alpha)$ 100% Prediction Interval for x When $y = y_p$ in Simple Linear Regression

$$\hat{x} \pm t_{\alpha/2} \left(\frac{s}{\hat{\beta}_1}\right) \sqrt{1 + \frac{1}{n} + \frac{(\hat{x} - \overline{x})^2}{SS_{xx}}}$$

where

$$\hat{x} = \frac{y_p - \hat{\beta}_0}{\hat{\beta}_1}$$

$\hat{\beta}_0$ and $\hat{\beta}_1$ are the y-intercept and slope, respectively, of the least squares line

$$n = \text{Sample size}$$

$$\overline{x} = \frac{\sum x}{n}$$

$$SS_{xx} = \sum x^2 - n(\overline{x})^2$$

$$s = \sqrt{MSE}$$

and the distribution of t is based on $(n - 2)$ degrees of freedom.

The approximation is appropriate when the quantity

$$D = \left(\frac{t_{\alpha/2}s}{\hat{\beta}_1}\right)^2 \cdot \frac{1}{SS_{xx}}$$

is small.*

EXAMPLE 9.1

A firm that sells copiers advertises regularly on television. One goal of the firm is to determine the amount it must spend on television advertising in a single month to gain a market share of 10%. For one year, the firm varied its monthly television advertising expenditures (x) and at the end of each month determined its market share (y). The data for the 12 months are recorded in Table 9.2.

a. Fit the straight-line model $y = \beta_0 + \beta_1 x + \varepsilon$ to the data.

b. Is there evidence that television advertising expenditure x is linearly related to market share y? Test using $\alpha = .05$.

c. Use inverse prediction to estimate the amount that must be spent on television advertising in a particular month for the firm to gain a market share of $y =$

---

*Neter, Kutner, Nachtsheim, and Wasserman (1996) and others suggest using the approximation when D is less than .1.

COPIERS

TABLE 9.2   A Firm's Market Share and Television Advertising Expenditure for 12 Months, Example 9.1

MONTH	MARKET SHARE y, percent	TELEVISION ADVERTISING EXPENDITURE x, $ thousands
January	7.5	23
February	8.5	25
March	6.5	21
April	7.0	24
May	8.0	26
June	6.5	22
July	9.5	27
August	10.0	31
September	8.5	28
October	11.0	32
November	10.5	30
December	9.0	29

10%. Construct an approximate 95% prediction interval for monthly television advertising expenditure x.

Solution

a. The MINITAB printout for the simple linear regression is shown in Figure 9.4. The least squares line (shaded) is:

$$\hat{y} = -1.975 + .39685x$$

The least squares line is plotted along with 12 data points in Figure 9.5.

b. To determine whether television advertising expenditure x is linearly related to market share y, we test the hypothesis

$H_0: \beta_1 = 0$
$H_a: \beta_1 \neq 0$

The value of the test statistic, shaded on the printout, is $t = 9.12$, and the associated p-value of the test is $p = .000$ (also shaded). Thus, there is sufficient evidence (at $\alpha = .01$), to indicate that television advertising expenditure x and market share y are linearly related.

   Caution: You should avoid using inverse prediction when there is insufficient evidence to reject the null hypothesis $H_0: \beta_1 = 0$. Inverse predictions made when x and y are *not* linearly related may lead to nonsensical results. Therefore, you should always conduct a test of model adequacy to be sure that x and y are linearly related before you carry out an inverse prediction.

c. Since the model is found to be adequate, we can use the model to predict x from y. For this example, we want to estimate the television advertising expenditure x that yields a market share of $y_p = 10\%$.

FIGURE 9.4

MINITAB printout of straight-line model

```
The regression equation is
MKTSHARE = - 1.97 + 0.397 TVADEXP

Predictor         Coef     SE Coef          T          P
Constant        -1.975       1.163      -1.70      0.120
TVADEXP         0.39685     0.04351       9.12      0.000

S = 0.5204       R-Sq = 89.3%      R-Sq(adj) = 88.2%

Analysis of Variance

Source            DF          SS          MS          F          P
Regression         1      22.521      22.521      83.17      0.000
Residual Error    10       2.708       0.271
Total             11      25.229
```

FIGURE 9.5

Scatterplot of data and least squares line, Example 9.1

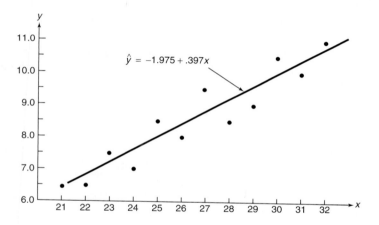

Substituting $y_p = 10$, $\hat{\beta}_0 = -1.975$, and $\hat{\beta}_1 = .397$ into the formula for $\hat{x}$ given in the box on page 433, we have

$$\hat{x} = \frac{y_p - \hat{\beta}_0}{\hat{\beta}_1}$$

$$= \frac{10 - (-1.975)}{.397}$$

$$= 30.16$$

Thus, we estimate that the firm must spend \$30,160 on television advertising in a particular month to gain a market share of 10%.

Before we construct an approximate 95% prediction interval for x, we check to determine whether the approximation is appropriate, that is, whether the quantity

$$D = \left(\frac{t_{\alpha/2}s}{\hat{\beta}_1}\right)^2 \cdot \frac{1}{SS_{xx}}$$

is small. For $\alpha = .05$, $t_{\alpha/2} = t_{.025} = 2.228$ for $n - 2 = 12 - 2 = 10$ degrees of freedom. From the printout (Figure 9.4), $s = .520$ and $\hat{\beta}_1 = .397$. The value of

SS_{xx} is not shown on the printout and must be calculated as follows:

$$SS_{xx} = \sum x^2 - n(\bar{x})^2$$

$$= 8,570 - 12\left(\frac{318}{12}\right)^2$$

$$= 8,570 - 8,427$$

$$= 143$$

Substituting these values into the formula for D, we have

$$D = \left(\frac{t_{\alpha/2} s}{\hat{\beta}_1}\right)^2 \cdot \frac{1}{SS_{xx}}$$

$$= \left[\frac{(2.228)(.520)}{.397}\right]^2 \cdot \frac{1}{143} = .06$$

Since the value of D is small (i.e., less than .1), we may use the formula for the approximate 95% prediction interval given in the box on page 433:

$$\hat{x} \pm t_{\alpha/2}\left(\frac{s}{\hat{\beta}_1}\right)\sqrt{1 + \frac{1}{n} + \frac{(\hat{x} - \bar{x})^2}{SS_{xx}}}$$

$$30.16 \pm (2.228)\frac{(.520)}{.397}\sqrt{1 + \frac{1}{12} + \frac{\left(30.16 - \frac{318}{12}\right)^2}{143}}$$

$$30.16 \pm (2.92)(1.085)$$

$$30.16 \pm 3.17$$

or (26.99, 33.33). Therefore, using the 95% prediction interval, we estimate that the amount of monthly television advertising expenditure required to gain a market share of 10% falls between \$26,999 and \$33,330. ◆

Another approach to the inverse prediction problem is to regress x on y, i.e., fit the model (called the **inverse estimator model**)

$$x = \beta_0 + \beta_1 y + \varepsilon$$

and then use the standard formula for a prediction interval given in Section 3.9. However, in theory this method requires that x be a random variable. In many applications, the value of x is set in advance (i.e., controlled) and therefore is *not* a random variable. (For example, the firm in Example 8.1 selected the amount x spent on advertising *prior* to each month.) Thus, the inverse model above may violate the standard least squares assumptions given in Chapter 4. Some researchers advocate the use of the inverse model despite this caution, whereas others have developed different estimators of x using a modification of the classical approach. Consult the references given at the end of this chapter for details on the various alternative methods of inverse prediction.

EXERCISES

9.6. Refer to the *Brain and Behavior Evolution* (Apr. 2000) study of the feeding behavior of blackbream fish, Exercise 3.12 (p. 103). Recall that the zoologists recorded the number of aggressive strikes of two blackbream fish feeding at the bottom of an aquarium in the 10-minute period following the addition of food. The table listing the weekly number of strikes and age of the fish (in days) is reproduced below. Use inverse prediction to estimate the age (x) of a blackbream fish, which was observed to have $y = 65$ aggressive strikes while feeding in another aquarium. Construct a 99% prediction interval around the estimate and interpret the result.

BLACKBREAM

WEEK	NUMBER OF STRIKES (y)	AGE OF FISH (x, days)
1	85	120
2	63	136
3	34	150
4	39	155
5	58	162
6	35	169
7	57	178
8	12	184
9	15	190

Source: J. Shand, *et al*. "Variability in the location of the retinal ganglion cell area centralis is correlated with ontogenetic changes in feeding behavior in the Blackbream, *Acanthopagrus* 'butcher'," *Brain and Behavior*, Vol. 55, No. 4, April 2000 (Figure H).

9.7. Refer to the study of the relationship between the "sweetness" rating of orange juice and the amount of water soluble pectin in the juice, Exercise 3.13 (p. 103). The data for 24 production runs is reproduced in the table, followed by a MINITAB printout of the least squares regression results for the straight-line model relating sweetness index (y) to pectin amount (x). Use inverse prediction to estimate the amount of pectin required to yield a sweetness index of $y = 5.8$. Construct a 95% prediction interval around the estimate and interpret the result.

OJUICE

RUN	SWEETNESS INDEX	PECTIN (ppm)
1	5.2	220
2	5.5	227
3	6.0	259
4	5.9	210
5	5.8	224
6	6.0	215
7	5.8	231
8	5.6	268
9	5.6	239
10	5.9	212
11	5.4	410
12	5.6	256
13	5.8	306
14	5.5	259
15	5.3	284
16	5.3	383
17	5.7	271
18	5.5	264
19	5.7	227
20	5.3	263
21	5.9	232
22	5.8	220
23	5.8	246
24	5.9	241

MINITAB Output for Exercise 9.7

```
The regression equation is
SWEET = 6.25 - 0.00231 PECTIN

Predictor       Coef      SE Coef          T        P
Constant       6.2521      0.2366      26.42    0.000
PECTIN      -0.0023106   0.0009049      -2.55    0.018

S = 0.2150      R-Sq = 22.9%     R-Sq(adj) = 19.4%

Analysis of Variance

Source          DF         SS          MS       F        P
Regression       1      0.30140     0.30140    6.52    0.018
Residual Error  22      1.01693     0.04622
Total           23      1.31833
```

9.8. The data in Table 3.8 (p. 139). are reproduced below. Use inverse prediction to estimate the distance from nearest fire station, x, for a residential fire that caused $y = \$18,200$ in damages. Construct a 90% prediction interval for x.

◐ FIREDAM

DISTANCE FROM FIRE STATION	FIRE DAMAGE
x, miles	y, thousands of dollars
3.4	26.2
1.8	17.8
4.6	31.3
2.3	23.1
3.1	27.5
5.5	36.0
.7	14.1
3.0	22.3
2.6	19.6
4.3	31.3
2.1	24.0
1.1	17.3
6.1	43.2
4.8	36.4
3.8	26.1

9.9. A pharmaceutical company has developed a new drug designed to reduce a smoker's reliance on tobacco. Since certain dosages of the drug may reduce one's pulse rate to dangerously low levels, the product-testing division of the pharmaceutical company wants to model the relationship between decrease in pulse rate y (beats/minute) and dosage x (cubic centimeters). Different dosages of the drug were administered to eight randomly selected patients, and 30 minutes later the decrease in each patient's pulse rate was recorded, with the results given in the table.

◐ PULSEDRUG

	DOSAGE	DECREASE IN PULSE RATE
PATIENT	x, cubic centimeters	y, beats/minute
1	2.0	12
2	4.0	20
3	1.5	6
4	1.0	3
5	3.0	16
6	3.5	20
7	2.5	13
8	3.0	18

a. Fit the straight-line model $E(y) = \beta_0 + \beta_1 x$ to the data.
b. Conduct a test for model adequacy. Use $\alpha = .05$.
c. Use inverse prediction to estimate the appropriate dosage x to administer to reduce a patient's pulse rate $y = 10$ beats per minute. Construct an approximate 95% prediction interval for x.

9.4
Weighted Least Squares

Consider the general linear model

$$y = \beta_0 + \beta_1 x_1 + \beta_2 x_2 + \cdots + \beta_k x_k + \varepsilon$$

To obtain the least squares estimates of the unknown β parameters, recall (from Section 4.3) that we minimize the quantity

$$SSE = \sum_{i=1}^{n} (y_i - \hat{y}_i)^2 = \sum_{i=1}^{n} [y_i - (\hat{\beta}_0 + \hat{\beta}_1 x_{1i} + \hat{\beta}_2 x_{2i} + \cdots + \hat{\beta}_k x_{ki})]^2$$

with respect to $\hat{\beta}_0, \hat{\beta}_1, \ldots, \hat{\beta}_k$.

The least squares criterion weighs each observation equally in determining the estimates of the β's. Sometimes we will want to weigh some observations more heavily than others. To do this we minimize

$$WSSE = \sum_{i=1}^{n} w_i (y_i - \hat{y}_i)^2$$

$$= \sum_{i=1}^{n} w_i [y_i - (\hat{\beta}_0 + \hat{\beta}_1 x_{1i} + \hat{\beta}_2 x_{2i} + \cdots + \hat{\beta}_k x_{ki})]^2$$

where w_i is the weight assigned to the ith observation. This procedure is known as **weighted least squares** and the resulting parameter estimates are called **weighted least squares estimates**. [Note that the ordinary least squares procedure assigns a weight of $w_i = 1$ to each observation.]

Definition 9.1

Weighted least squares regression is the procedure which obtains estimates of the β's by minimizing WSSE $= \displaystyle\sum_{i=1}^{n} w_i (y_i - \hat{y}_i)^2$, where w_i is the weight assigned to the ith observation. The β-estimates are called **weighted least squares estimates**.

Definition 9.2

The **weighted least squares residuals** are obtained by computing the quantity

$$\sqrt{w_i}(y_i - \hat{y}_i)$$

for each observation, where $\hat{y}_i$ is the predicted value of y obtained using the weight w_i in a weighted least squares regression.

Weighted least squares has applications in the following areas:

1. Stabilizing the variance of ε to satisfy the standard regression assumption of homoscedasticity
2. Limiting the influence of outlying observations on the regression analysis
3. Giving greater weight to more recent observations in time series analysis (the topic of Chapter 10).

Although the applications are related, our discussion of weighted least squares in this section is directed toward the first application.

The regression routines of most statistical software packages have options for conducting a weighted least squares analysis. However, the weights w_i must be specified. When using weighted least squares as a variance-stabilizing technique, the weight for the ith observation should be the reciprocal of the variance of that observation's error term, σ_i^2, i.e.,

$$w_i = \frac{1}{\sigma_i^2}$$

In this manner, observations with larger error variances will receive less weight (and hence have less influence on the analysis) than observations with smaller error variances.

In practice, the actual variances σ_i^2 will usually be unknown. Fortunately, in many applications, the error variance σ_i^2 is proportional to one or more of the levels of the independent variables. This fact will allow us to determine the appropriate weights to use. For example, in a simple linear regression problem, suppose we know that the error variance σ_i^2 increases proportionally with the value of the independent variable x_i, i.e.,

$$\sigma_i^2 = kx_i$$

where k is some unknown constant. Then the appropriate (albeit unknown) weight to use is

$$w_i = \frac{1}{kx_i}$$

Fortunately, it can be shown (proof omitted) that k can be ignored and the weights can be assigned as follows:

$$w_i = \frac{1}{x_i}$$

If the functional relationship between σ_i^2 and x_i is not known prior to conducting the analysis, the weights can be estimated based on the results of an ordinary (unweighted) least squares fit. For example, in simple linear regression, one approach is to divide the regression residuals into several groups of approximately equal size based on the value of the independent variable x and calculate the variance of the observed residuals in each group. An examination of the relationship between the residual variances and several different functions of x (such as x, x^2, and $\sqrt{x}$) may reveal the appropriate weights to use.

EXAMPLE 9.2

A Department of Transportation (DOT) official is investigating the possibility of collusive bidding among the state's road construction contractors. One aspect of the investigation involves a comparison of the winning (lowest) bid price y on a job with the length x of new road construction, a measure of job size. The data listed in Table 9.3 were supplied by the DOT for a sample of 11 new road construction jobs with approximately the same number of bidders.

a. Use the method of least squares to fit the straight-line model

$$E(y) = \beta_0 + \beta_1 x$$

b. Calculate and plot the regression residuals against x. Do you detect any evidence of heteroscedasticity?

 DOT11

TABLE 9.3   **Sample Data for New Road Construction Jobs, Example 9.2**

JOB	LENGTH OF ROAD x, miles	WINNING BID PRICE y, $ thousands	JOB	LENGTH OF ROAD x, miles	WINNING BID PRICE y, $ thousands
1	2.0	10.1	7	7.0	71.1
2	2.4	11.4	8	11.5	132.7
3	3.1	24.2	9	10.9	108.0
4	3.5	26.5	10	12.2	126.2
5	6.4	66.8	11	12.6	140.7
6	6.1	53.8			

c. Use the method described in the preceding paragraph to find the approximate weights necessary to stabilize the error variances with weighted least squares.

d. Carry out the weighted least squares analysis using the weights determined in part **c.**

e. Plot the weighted least squares residuals (see Definition 9.2) against x to determine whether the variances have stabilized.

Solution

a. The simple linear regression analysis was conducted using MINITAB; the resulting printout is given in Figure 9.6. The least squares line (shaded on the printout) is

$$\hat{y} = -15.112 + 12.0687x$$

Note that the model is statistically useful (reject $H_0: \beta_1 = 0$) at $p = .000$.

FIGURE 9.6

MINITAB printout of straight-line model, Example 9.2

```
The regression equation is
BIDPRICE = - 15.1 + 12.1 LENGTH

Predictor         Coef      SE Coef          T         P
Constant       -15.112        3.342      -4.52     0.001
LENGTH         12.0687        0.4138      29.16     0.000

S = 5.374      R-Sq = 99.0%      R-Sq(adj) = 98.8%

Analysis of Variance

Source             DF         SS         MS          F         P
Regression          1      24558      24558     850.45     0.000
Residual Error      9        260         29
Total              10      24818
```

Obs	LENGTH	BIDPRICE	Fit	SE Fit	Residual	St Resid
1	2.0	10.10	9.02	2.65	1.08	0.23
2	2.4	11.40	13.85	2.52	-2.45	-0.52
3	3.1	24.20	22.30	2.31	1.90	0.39
4	3.5	26.50	27.13	2.19	-0.63	-0.13
5	6.4	66.80	62.13	1.64	4.67	0.91
6	6.1	53.80	58.51	1.67	-4.71	-0.92
7	7.0	71.10	69.37	1.62	1.73	0.34
8	11.5	132.70	123.68	2.45	9.02	1.89
9	10.9	108.00	116.44	2.27	-8.44	-1.73
10	12.2	126.20	132.13	2.67	-5.93	-1.27
11	12.6	140.70	136.95	2.81	3.75	0.82

b. The regression residuals are calculated and reported in the bottom portion of the MINITAB printout. A plot of the residuals against the predictor variable x is shown in Figure 9.7. The residual plot clearly shows that the residual variance increases as length of road x increases, strongly suggesting the presence of heteroscedasticity. A procedure such as weighted least squares is needed to stabilize the variances.

FIGURE 9.7

MINITAB plot of residuals
against road length, x

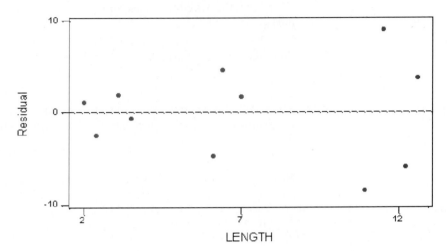

Residuals Versus LENGTH

(response is BIDPRICE)

c. To apply weighted least squares, we must first determine the weights. Since it is not clear what function of x the error variance is proportional to, we will apply the procedure described previously to estimate the weights.

First, we must divide the data into several groups according to the value of the independent variable x. Ideally, we want to form one group of data points for each different value of x. However, unless each value of x is replicated, not all of the group residual variances can be calculated. Therefore, we resort to grouping the data according to "nearest neighbors." One choice would be to use three groups, $2 \leq x \leq 4, 6 \leq x \leq 7$, and $10 \leq x \leq 13$. These groups have approximately the same numbers of observations (namely, 4, 3, and 4 observations, respectively).

Next, we calculate the sample variance s_j^2 of the residuals included in each group. The three residual variances are given in Table 9.4. These variances are compared to three different functions of $\bar{x}$ ($\bar{x}, \bar{x}^2$, and $\sqrt{\bar{x}}$), as shown in Table 9.4, where $\bar{x}_j$ is the mean road length x for group j, $j = 1, 2, 3$.

Note that the ratio $s_j^2/\bar{x}_j^2$ yields a value near .5 for each of the three groups. This result suggests that the residual variance of each group is proportional to $\bar{x}^2$, i.e.,

$$\sigma_j^2 = k\bar{x}_j^2, \quad j = 1, 2, 3$$

TABLE 9.4   Comparison of Residual Variances to Three Functions of $\bar{x}$, Example 9.2

GROUP	RANGE OF x	$\bar{x}_j$	s_j^2	$s_j^2/\bar{x}_j$	$s_j^2/\bar{x}_j^2$	$s_j^2/\sqrt{\bar{x}_j}$
1	$2 \leq x \leq 4$	2.75	3.722	1.353	.492	2.244
2	$6 \leq x \leq 7$	6.5	23.016	3.541	.545	9.028
3	$10 \leq x \leq 13$	11.8	67.031	5.681	.481	19.514

where k is approximately .5. Thus, a reasonable approximation to the weight for each group is

$$w_j = \frac{1}{\overline{x}_j^2}$$

With this weighting scheme, observations associated with large values of length of road x will have less influence on the regression residuals than observations associated with smaller values of x.

d. A weighted least squares analysis was conducted on the data in Table 9.3 using the weights

$$w_{ij} = \frac{1}{\overline{x}_j^2}$$

where w_{ij} is the weight for observation i in group j. The weighted least squares estimates are shown in the MINITAB printout reproduced in Figure 9.8. The prediction equation (shaded) is

$$\hat{y} = -15.274 + 12.1204x$$

Note that the test of model adequacy, $H_0: \beta_1 = 0$, is significant at $p = .000$. Also, the standard error of the model, s, is significantly smaller than the value of s for the unweighted least squares analysis (.669 compared to 5.37). This last result is expected because, in the presence of heteroscedasticity, the unweighted least squares estimates are subject to greater sampling error than the weighted least squares estimates.

FIGURE 9.8

MINITAB printout of weighted least squares fit, Example 9.2

```
Weighted analysis using weights in WEIGHT

The regression equation is
BIDPRICE = - 15.3 + 12.1 LENGTH

Predictor      Coef      SE Coef        T         P
Constant     -15.274       1.601     -9.54     0.000
LENGTH       12.1204      0.3792     31.97     0.000

S = 0.6691      R-Sq = 99.1%      R-Sq(adj) = 99.0%

Analysis of Variance

Source           DF        SS          MS        F         P
Regression        1      457.48      457.48   1021.77    0.000
Residual Error    9        4.03        0.45
Total            10      461.51
```

e. A MINITAB plot of the weighted least squares residuals against x is shown in Figure 9.9. The lack of a discernible pattern in the residual plot suggests that the weighted least squares procedure has corrected the problem of unequal variances. ◆

Before concluding this section, we mention that the "nearest neighbor" technique, illustrated in Example 9.2, will not always be successful in finding the optimal

FIGURE 9.9

MINITAB plot of weighted
residuals against road length, x

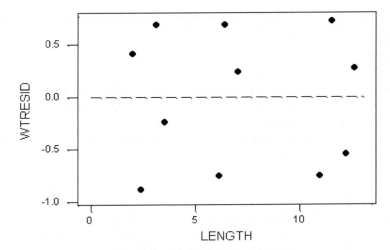

or near-optimal weights in weighted least squares. First, it may not be easy to identify
the appropriate groupings of data points, especially if more than one independent
variable is included in the regression. Second, the relationship between the residual
variance and some preselected function of the independent variables may not reveal
a consistent pattern over groups. In other words, unless the right function (or
approximate function) of x is examined, the weights will be difficult to determine.
More sophisticated techniques for choosing the weights in weighted least squares
are available. Consult the references given at the end of this chapter for details on
how to use these techniques.

EXERCISES

9.10. Consider the straight-line model $y_i = \beta_0 + \beta_1 x_i + \varepsilon_i$.
Give the appropriate weights w_i to use in a weighted
least squares regression if the variance of the random
error ε_i, i.e., σ_i^2, is proportional to

a. x_i^2 **b.** $\sqrt{x_i}$

c. x_i

d. $\frac{1}{n_i}$, where n_i is the number of observations at
level x_i

e. $\frac{1}{x_i}$

9.11. A machine that mass produces rubber gaskets can
be set at one of three different speeds: 100, 150, or
200 gaskets per minute. As part of a quality control
study, the machine was monitored several different
times at each of the three speeds, and the number
of defectives produced per hour was recorded. The
data are provided in the accompanying table. Since
the number of defectives (y) is thought to be linearly
related to speed (x), the following straight-line model
is proposed:

$$y = \beta_0 + \beta_1 x + \varepsilon$$

a. Fit the model using the method of least squares.
Is there evidence that the model is useful for
predicting y? Test using $\alpha = .05$.

b. Plot the residuals from the least squares model
against x. What does the plot reveal about the standard least squares assumption of homoscedasticity?

GASKETS

MACHINE SPEED	NUMBER OF DEFECTIVES	MACHINE SPEED	NUMBER OF DEFECTIVES
x, gaskets per minute	y	x, gaskets per minute	y
100	15	150	35
100	23	150	24
100	11	200	26
100	14	200	48
100	18	200	27
150	19	200	38
150	29	200	39
150	20		

c. Estimate the appropriate weights to use in a weighted least squares regression. [*Hint*: Calculate the variance of the least squares residuals at each level of x.]

d. Refit the model using weighted least squares. Compare the standard deviation of the weighted least squares slope to the standard deviation of the unweighted least squares slope.

e. Plot the weighted residuals against x to determine whether using weighted least squares has corrected the problem of unequal variances.

9.12. Refer to the data on salary (y) and years of experience (x) for 50 social workers, given in Table 8.4. (The data are reproduced here for convenience.) Recall that the least squares fit of the quadratic model $E(y) = \beta_0 + \beta_1 x + \beta_2 x^2$ yielded regression residuals with unequal variances (see Figure 8.13). Apply the method of weighted least squares to correct this problem. [*Hint*: Estimate the weights using the "nearest neighbor" technique outlined in this section.]

 SOCWORK

YEARS OF EXPERIENCE x	SALARY y	YEARS OF EXPERIENCE x	SALARY y	YEARS OF EXPERIENCE x	SALARY y	YEARS OF EXPERIENCE x	SALARY y
7	$26,075	28	$64,785	7	$22,210	26	$65,343
28	79,370	26	61,581	2	20,521	19	46,216
23	65,726	27	70,678	18	49,727	16	54,288
18	41,983	20	51,301	11	33,233	3	20,844
19	62,309	18	39,346	21	43,628	12	32,586
15	41,154	1	24,833	4	16,105	23	71,235
24	53,610	26	65,929	24	65,644	20	36,530
13	33,697	20	41,721	20	63,022	19	52,745
2	22,444	26	82,641	20	47,780	27	67,282
8	32,562	28	99,139	15	38,853	25	80,931
20	43,076	23	52,624	25	66,537	12	32,303
21	56,000	17	50,594	25	67,447	11	38,371
18	58,667	25	53,272				

9.5 Modeling Qualitative Dependent Variables

For all models discussed in the previous sections of this text, the response (dependent) variable y is a *quantitative* variable. In this section, we consider models for which the response y is a **qualitative variable at two levels**, or, as it is sometimes called, a **binary variable**.

For example, a physician may want to relate the success or failure of a new surgical procedure to the characteristics (such as age and severity of the disease) of the patient. The value of the response of interest to the physician is either *yes*, the operation is a success, or *no*, the operation is a failure. Similarly, a state attorney general investigating collusive practices among bidders for road construction contracts may want to determine which contract-related variables (such as number of bidders, bid amount, and cost of materials) are useful indicators of whether a bid is fixed (i.e., whether the bid price is intentionally set higher than the fair market value). Here, the value of the response variable is either *fixed* bid or *competitive* bid.

Just as with qualitative independent variables, we use **dummy** (i.e., coded 0–1) **variables** to represent the qualitative response variable. For example, the response of interest to the entrepreneur is recorded as

$$y = \begin{cases} 1 & \text{if new business a success} \\ 0 & \text{if new business a failure} \end{cases}$$

where the assignment of 0 and 1 to the two levels is arbitrary. The linear statistical model takes the usual form

$$y = \beta_0 + \beta_1 x_1 + \beta_2 x_2 + \cdots + \beta_k x_k + \varepsilon$$

However, when the response is binary, the expected response

$$E(y) = \beta_0 + \beta_1 x_1 + \beta_2 x_2 + \cdots + \beta_k x_k$$

has a special meaning. It can be shown* that $E(y) = \pi$, where π is the probability that $y = 1$ for given values of $x_1, x_2, \ldots, x_k$. Thus, for the entrepreneur, the mean response $E(y)$ represents the probability that a new business with certain owner-related characteristics will be a success.

When the ordinary least squares approach is used to fit models with a binary response, several well-known problems are encountered. A discussion of these problems and their solutions follows.

PROBLEM 1   *Nonnormal errors*: The standard least squares assumption of normal errors is violated since the response y and, hence, the random error ε can take on only two values. To see this, consider the simple model $y = \beta_0 + \beta_1 x + \varepsilon$. Then we can write

$$\varepsilon = y - (\beta_0 + \beta_1 x)$$

Thus, when $y = 1$, $\varepsilon = 1 - (\beta_0 + \beta_1 x)$ and when $y = 0$, $\varepsilon = -\beta_0 - \beta_1 x$.

When the sample size n is large, however, any inferences derived from the least squares prediction equation remain valid in most practical situations even though the errors are nonnormal.[†]

PROBLEM 2   *Unequal variances*: It can be shown[‡] that the variance σ^2 of the random error is a function of π, the probability that the response y equals 1. Specifically,

$$\sigma^2 = V(\varepsilon) = \pi(1 - \pi)$$

Since, for the linear statistical model,

$$\pi = E(y) = \beta_0 + \beta_1 x_1 + \beta_2 x_2 + \cdots + \beta_k x_k$$

---

*The result is a straightforward application of the expectation theorem for a random variable. Let $\pi = P(y = 1)$ and $1 - \pi = P(y = 0)$, $0 \leq \pi \leq 1$. Then, by definition, $E(y) = \Sigma_y y_i \cdot p(y) = (1)P(y = 1) + (0)P(y = 0) = P(y = 1) = \pi$. Students familiar with discrete random variables will recognize y as the **Bernoulli random variable**, i.e., a binomial random variable with $n = 1$.

[†]This property is due to the asymptotic normality of the least squares estimates of the model parameters under very general conditions.

[‡]Using the properties of expected values with the Bernoulli random variable, we obtain $V(y) = E(y^2) - [E(y)]^2 = \Sigma y^2 \cdot p(y) - (\pi)^2 = (1)^2 P(y = 1) + (0)^2 P(y = 0) - \pi^2 = \pi - \pi^2 = \pi(1 - \pi)$. Since in regression, $V(\varepsilon) = V(y)$, the result follows.

this implies that σ^2 is not constant and, in fact, depends on the values of the independent variables; hence, the standard least squares assumption of equal variances is also violated. One solution to this problem is to use weighted least squares (see Section 9.4), where the weights are inversely proportional to σ^2, i.e.,

$$w_i = \frac{1}{\sigma_i^2}$$

$$= \frac{1}{\pi_i(1 - \pi_i)}$$

Unfortunately, the true proportion

$$\pi_i = E(y_i)$$

$$= \beta_0 + \beta_1 x_{1i} + \beta_2 x_{2i} + \cdots + \beta_k x_{ki}$$

is unknown since $\beta_0, \beta_1, \ldots, \beta_k$ are unknown population parameters. However, a technique called **two-stage least squares** can be applied to circumvent this difficulty. Two-stage least squares, as its name implies, involves conducting an analysis in two steps:

STAGE 1 Fit the regression model using the *ordinary least squares* procedure and obtain the predicted values $\hat{y}_i$, $i = 1, 2, \ldots, n$. Recall that $\hat{y}_i$ estimates π_i for the binary model.

STAGE 2 Refit the regression model using *weighted least squares*, where the estimated weights are calculated as follows:

$$w_i = \frac{1}{\hat{y}_i(1 - \hat{y}_i)}$$

Further iterations—revising the weights at each step—can be performed if desired. In most practical problems, however, the estimates of π_i obtained in stage 1 are adequate for use in weighted least squares.

PROBLEM 3 *Restricting the predicted response to be between 0 and 1*: Since the predicted value $\hat{y}$ estimates $E(y) = \pi$, the probability that the response y equals 1, we would like $\hat{y}$ to have the property that $0 \leq \hat{y} \leq 1$. There is no guarantee, however, that the regression analysis will always yield predicted values in this range. Thus, the regression may lead to nonsensical results, i.e., negative estimated probabilities or predicted probabilities greater than 1. To avoid this problem, you may want to fit a model with a mean response function $E(y)$ that automatically falls between 0 and 1. (We consider one such model in the next section.)

In summary, the purpose of this section has been to identify some of the problems resulting from fitting a linear model with a binary response and to suggest ways in which to circumvent these problems. Another approach is to fit a model specially designed for a binary response, called a logistic regression model. Logistic regression models are the subject of Section 9.6.

EXERCISES

9.13. Discuss the problems associated with fitting a model where the response y is recorded as 0 or 1.

9.14. A retailer of hand-held digital organizers conducted a study to relate ownership of the devices with annual income of heads of households. Data collected for a random sample of 20 households were used to fit the straight-line model $E(y) = \beta_0 + \beta_1 x$, where

$$y = \begin{cases} 1 & \text{if own a digital organizer} \\ 0 & \text{if not} \end{cases}$$

$$x = \text{Annual income (in dollars)}$$

The data are shown in the accompanying table. Fit the model using two-stage least squares. Is the model useful for predicting y? Test using $\alpha = .05$.

PALMORG

HOUSEHOLD	y	x	HOUSEHOLD	y	x
1	0	$36,300	11	1	$42,400
2	0	31,200	12	0	30,600
3	0	56,500	13	0	41,400
4	1	41,700	14	0	28,300
5	1	60,200	15	1	47,500
6	0	32,400	16	0	35,700
7	0	35,000	17	0	32,100
8	0	29,200	18	1	79,600
9	1	56,700	19	1	40,200
10	0	82,000	20	0	53,100

9.15. Suppose you are investigating allegations of sex discrimination in the hiring practices of a particular firm. An equal-rights group claims that females are less likely to be hired than males with the same background, experience, and other qualifications. Data (shown in the next table) collected on 28 former applicants will be used to fit the model $E(y) = \beta_0 + \beta_1 x_1 + \beta_2 x_2 + \beta_3 x_3$, where

$$y = \begin{cases} 1 & \text{if hired} \\ 0 & \text{if not} \end{cases}$$

$$x_1 = \text{Years of higher education (4, 6, or 8)}$$

$$x_2 = \text{Years of experience}$$

$$x_3 = \begin{cases} 1 & \text{if male applicant} \\ 0 & \text{if female applicant} \end{cases}$$

a. Interpret each of the β's in the multiple regression model.

b. Fit the multiple regression model using two-stage least squares.

c. Conduct a test of model adequacy. Use $\alpha = .05$.

d. Is there sufficient evidence to indicate that gender is an important predictor of hiring status? Test using $\alpha = .05$.

e. Calculate a 95% confidence interval for the mean response $E(y)$ when $x_1 = 4$, $x_2 = 3$, and $x_3 = 0$. Interpret the interval.

DISCRIM

HIRING STATUS	EDUCATION	EXPERIENCE	GENDER	HIRING STATUS	EDUCATION	EXPERIENCE	GENDER
y	x_1, years	x_2, years	x_3	y	x_1, years	x_2, years	x_3
0	6	2	0	1	4	5	1
0	4	0	1	0	6	4	0
1	6	6	1	0	8	0	1
1	6	3	1	1	6	1	1
0	4	1	0	0	4	7	0
1	8	3	0	0	4	1	1
0	4	2	1	0	4	5	0
0	4	4	0	0	6	0	1
0	6	1	0	1	8	5	1
1	8	10	0	0	4	9	0
0	4	2	1	0	8	1	0
0	8	5	0	0	6	1	1
0	4	2	0	1	4	10	1
0	6	7	0	1	6	12	0

9.6
Logistic Regression

Often, the relationship between a qualitative binary response y and a single predictor variable x is curvilinear. One particular curvilinear pattern frequently encountered in practice is the S-shaped curve shown in Figure 9.10. Points on the curve represent $\pi = P(y = 1)$ for each value of x. A model that accounts for this type of curvature is the **logistic** (or **logit**) **regression model**,

$$E(y) = \frac{\exp(\beta_0 + \beta_1 x)}{1 + \exp(\beta_0 + \beta_1 x)}$$

FIGURE 9.10

Graph of $E(y)$ for the logistic model

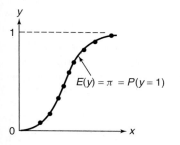

The logistic model was originally developed for use in **survival analysis**, where the response y is typically measured as 0 or 1, depending on whether the experimental unit (for example, a patient) "survives." Note that the curve shown in Figure 9.10 has asymptotes at 0 and 1—that is, the mean response $E(y)$ can never fall below 0 or above 1. Thus, the logistic model ensures that the estimated response $\hat{y}$ (i.e., the estimated probability that $y = 1$) lies between 0 and 1.

In general, the logistic model can be written as shown in the box.

Logistic Regression Model for a Binary Dependent Variable

$$E(y) = \frac{\exp(\beta_0 + \beta_1 x_1 + \beta_2 x_2 + \cdots + \beta_k x_k)}{1 + \exp(\beta_0 + \beta_1 x_1 + \beta_2 x_2 + \cdots + \beta_k x_k)}$$

where

$$y = \begin{cases} 1 & \text{if category A occurs} \\ 0 & \text{if category B occurs} \end{cases}$$

$$E(y) = P(\text{Category A occurs}) = \pi$$

$x_1, x_2, \ldots, x_k$ are quantitative or qualitative independent variables

---

Note that the general logistic model is not a linear function of the β parameters (see Section 4.1). Obtaining the parameter estimates of a **nonlinear regression model**, such as the logistic model, is a numerically tedious process and often requires sophisticated computer programs. In this section we briefly discuss two approaches to the problem, and give an example of a computer printout for the second.

1. *Least squares estimation using a transformation*: One method of fitting the model involves a transformation on the mean response $E(y)$. Recall (from Section 9.5) that for a binary response, $E(y) = \pi$, where π denotes the probability that $y = 1$. Then the logistic model

$$\pi = \frac{\exp(\beta_0 + \beta_1 x_1 + \cdots + \beta_k x_k)}{1 + \exp(\beta_0 + \beta_1 x_1 + \cdots + \beta_k x_k)}$$

implies (proof omitted) that

$$\ln\left(\frac{\pi}{1 - \pi}\right) = \beta_0 + \beta_1 x_1 + \cdots + \beta_k x_k$$

Set

$$\pi^* = \ln \left(\frac{\pi}{1 - \pi} \right)$$

The transformed logistic model

$$\pi^* = \beta_0 + \beta_1 x_1 + \cdots + \beta_k x_k$$

is now linear in the β's and the method of least squares can be applied.

Note: Since $\pi = P(y = 1)$, then $1 - \pi = P(y = 0)$. The ratio

$$\frac{\pi}{1 - \pi} = \frac{P(y = 1)}{P(y = 0)}$$

is known as the **odds** of the event, $y = 1$, occurring. (For example, if $\pi = .8$, then the odds of $y = 1$ occurring are $.8/.2 = 4$, or 4 to 1.) The transformed model, π^*, then, is a model for the natural logarithm of the odds of $y = 1$ occurring and is often called the **log-odds model**.

Definition 9.3

In logistic regression with a binary response y, we define the **odds of the event (y = 1) occurring** as follows:

$$\text{Odds} = \frac{\pi}{1 - \pi} = \frac{P(y = 1)}{P(y = 0)}$$

Although the transformation succeeds in linearizing the response function, two other problems remain. First, since the true probability π is unknown, the values of the log-odds π^*, necessary for input into the regression, are also unknown. To carry out the least squares analysis, we must obtain estimates of π^* for each combination of the independent variables. A good choice is the estimator

$$\pi^* = \ln \left(\frac{\hat{\pi}}{1 - \hat{\pi}} \right)$$

where $\hat{\pi}$ is the sample proportion of 1's for the particular combination of x's. To obtain these estimates, however, *we must have replicated observations of the response y at each combination of the levels of the independent variables.* Thus, the least squares transformation approach is limited to replicated experiments, which occur infrequently in a practical business setting.

The second problem is associated with unequal variances. The transformed logistic model yields error variances that are inversely proportional to $\pi(1 - \pi)$. Since π, or $E(y)$, is a function of the independent variables, the regression errors are heteroscedastic. To stabilize the variances, weighted least squares should be used. This technique also requires that replicated observations be available for each combination of the x's and, in addition, that the number of observations at each combination be relatively large. If the experiment is replicated, with n_j (large) observations at each combination of the levels of the independent variables, then the appropriate weights to use are

$$w_j = n_j \hat{\pi}_j (1 - \hat{\pi}_j)$$

where

$$\hat{\pi}_j = \frac{\text{Number of 1's for combination } j \text{ of the } x\text{'s}}{n_j}$$

2. *Maximum likelihood estimation*: Estimates of the β parameters in the logistic model also can be obtained by applying a common statistical technique, called **maximum likelihood estimation**. Like the least squares estimators, the maximum likelihood estimators have certain desirable properties.* (In fact, when the errors of a linear regression model are normally distributed, the least squares estimates and maximum likelihood estimates are equivalent.) Many of the available statistical computer software packages use maximum likelihood estimation to fit logistic regression models. Therefore, one practical advantage of using the maximum likelihood method (rather than the transformation approach) to fit logistic regression models is that computer programs are readily available. Another advantage is that the data need not be replicated to apply maximum likelihood estimation.

The maximum likelihood estimates of the parameters of a logistic model have distributional properties that are different from the standard F and t distributions of least squares regression. Under certain conditions, the test statistics for testing individual parameters and overall model adequacy have approximate **chi-square** (χ^2) **distributions**. The χ^2 distribution is similar to the F distribution in that it depends on degrees of freedom and is nonnegative, as shown in Figure 9.11. (Critical values of the χ^2 distribution for various values of α and degrees of freedom are given in Table 10, Appendix C.) We illustrate the application of maximum likelihood estimation for logistic regression with an example.

EXAMPLE 9.3

Consider the problem of collusive (i.e., noncompetitive) bidding among road construction contractors. Recall (from Section 9.5) that contractors sometimes scheme to set bid prices higher than the fair market (or competitive) price. Suppose an investigator has obtained information on the bid status (fixed or competitive) for a sample of 31 contracts. In addition, two variables thought to be related to bid status are also recorded for each contract: number of bidders x_1 and the difference between the winning (lowest) bid and the estimated competitive bid (called the engineer's estimate) x_2, measured as a percentage of the estimate. The data appear in Table 9.5, with the response y recorded as follows:

$$y = \begin{cases} 1 & \text{if fixed bid} \\ 0 & \text{if competitive bid} \end{cases}$$

_____

*For details on how to obtain maximum likelihood estimators and what their distributional properties are, consult the references given at the end of the chapter.

FIGURE 9.11

Several chi-square probability
distributions

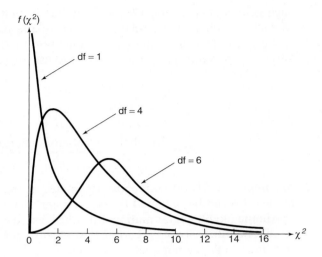

 ROADBIDS

TABLE 9.5   Data for a Sample of 31 Road Construction Bids

CONTRACT	BID STATUS y	NUMBER OF BIDDERS x_1	DIFFERENCE BETWEEN WINNING BID AND ENGINEER'S ESTIMATE x_2, %	CONTRACT	BID STATUS y	NUMBER OF BIDDERS x_1	DIFFERENCE BETWEEN WINNING BID AND ENGINEER'S ESTIMATE x_2, %
1	1	4	19.2	17	0	10	6.6
2	1	2	24.1	18	1	5	−2.5
3	0	4	−7.1	19	0	13	24.2
4	1	3	3.9	20	0	7	2.3
5	0	9	4.5	21	1	3	36.9
6	0	6	10.6	22	0	4	11.7
7	0	2	−3.0	23	1	2	22.1
8	0	11	16.2	24	1	3	10.4
9	1	6	72.8	25	0	2	9.1
10	0	7	28.7	26	0	5	2.0
11	1	3	11.5	27	0	6	12.6
12	1	2	56.3	28	1	5	18.0
13	0	5	−.5	29	0	3	1.5
14	0	3	−1.3	30	1	4	27.3
15	0	3	12.9	31	0	10	−8.4
16	0	8	34.1				

An appropriate model for $E(y)$ is the logistic model

$$E(y) = \frac{\exp(\beta_0 + \beta_1 x_1 + \beta_2 x_2)}{1 + \exp(\beta_0 + \beta_1 x_1 + \beta_2 x_2)}$$

Solution

The model was fitted to the data using the logistic regression option of SAS. The resulting printout is shown in Figure 9.12. Interpret the results.

The maximum likelihood estimates of $\beta_0, \beta_1,$ and β_2 (shaded in the printout) are $\hat{\beta}_0 = 1.4211, \hat{\beta}_1 = -.7553,$ and $\hat{\beta}_2 = .1122$. Therefore, the prediction equation for the probability of a fixed bid [i.e., $\pi = P(y = 1)$] is

$$\hat{y} = \frac{\exp(1.4211 - .7553x_1 + .1122x_2)}{1 + \exp(1.4211 - .7553x_1 + .1122x_2)}$$

FIGURE 9.12

SAS printout of logistic regression on bid status

```
        Probability modeled is STATUS=1.

           Model Convergence Status

   Convergence criterion (GCONV=1E-8) satisfied.

             Model Fit Statistics

                                     Intercept
                         Intercept      and
        Criterion          Only      Covariates

        AIC               43.381       28.843
        SC                44.815       33.145
        -2 Log L          41.381       22.843

      Testing Global Null Hypothesis: BETA=0

   Test                Chi-Square      DF    Pr > ChiSq

   Likelihood Ratio     18.5377        2      <.0001
   Score                13.4661        2      0.0012
   Wald                  6.4289        2      0.0402

        Analysis of Maximum Likelihood Estimates

                              Standard      Wald
   Parameter    DF   Estimate   Error    Chi-Square   Pr > ChiSq

   Intercept     1    1.4211    1.2867     1.2198       0.2694
   NUMBIDS       1   -0.7553    0.3388     4.9703       0.0258
   DOTEST        1    0.1122    0.0514     4.7666       0.0290

              Odds Ratio Estimates

                    Point          95% Wald
        Effect     Estimate     Confidence Limits

        NUMBIDS     0.470      0.242       0.913
        DOTEST      1.119      1.012       1.237

Association of Predicted Probabilities and Observed Responses

   Percent Concordant    90.4    Somers' D    0.807
   Percent Discordant     9.6    Gamma        0.807
   Percent Tied           0.0    Tau-a        0.396
   Pairs                  228    c            0.904

        Wald Confidence Interval for Parameters

   Parameter     Estimate    95% Confidence Limits

   Intercept      1.4211     -1.1008       3.9431
   NUMBIDS       -0.7553     -1.4193      -0.0913
   DOTEST         0.1122      0.0115       0.2129
```

FIGURE 9.12
(*Continued*)

Wald Confidence Interval for Adjusted Odds Ratios

Effect	Unit	Estimate	95% Confidence Limits	
NUMBIDS	1.0000	0.470	0.242	0.913
DOTEST	1.0000	1.119	1.012	1.237

Classification Table

Prob Level	Correct Event	Correct Non-Event	Incorrect Event	Incorrect Non-Event	Percentages Correct	Percentages Sensi-tivity	Percentages Speci-ficity	False POS	False NEG
0.500	9	16	3	3	80.6	75.0	84.2	25.0	15.8

Predictions of STATUS

Obs	STATUS	NUMBIDS	DOTEST	Response Value	Probability	Lower 95% Confidence Limit	Upper 95% Confidence Limit
1	1	4	19.2	1	0.63510	0.32984	0.86023
2	1	2	24.1	1	0.93179	0.53645	0.99384
3	0	4	-7.1	1	0.08342	0.01043	0.44012
4	1	3	3.9	1	0.39958	0.15869	0.70133
5	0	9	4.5	1	0.00760	0.00016	0.26829
6	0	6	10.6	1	0.12771	0.02582	0.44709
7	0	2	-3.0	1	0.39506	0.10273	0.78837
8	0	11	16.2	1	0.00625	0.00007	0.36815
9	1	6	72.8	1	0.99368	0.35196	0.99998
10	0	7	28.7	1	0.34392	0.06138	0.80777
11	1	3	11.5	1	0.60957	0.31578	0.84081
12	1	2	56.3	1	0.99803	0.69691	0.99999
13	0	5	-0.5	1	0.08230	0.01254	0.38784
14	0	3	-1.3	1	0.27078	0.07453	0.63132
15	0	3	12.9	1	0.64625	0.34076	0.86589
16	0	8	34.1	1	0.31103	0.03168	0.86168
17	0	10	6.6	1	0.00453	0.00006	0.26606
18	1	5	-2.5	1	0.06686	0.00852	0.37405
19	0	13	24.2	1	0.00339	0.00001	0.45718
20	0	7	2.3	1	0.02639	0.00166	0.30643
21	1	3	36.9	1	0.96427	0.54748	0.99834
22	0	4	-11.7	1	0.05152	0.00412	0.41605
23	1	2	22.1	1	0.91607	0.51883	0.99103
24	1	3	10.4	1	0.57983	0.29466	0.82010
25	0	2	9.1	1	0.71739	0.33903	0.92627
26	0	5	2.0	1	0.10612	0.02005	0.40786
27	0	6	12.6	1	0.15486	0.03485	0.48182
28	1	5	18.0	1	0.41683	0.17873	0.70127
29	0	3	1.5	1	0.33705	0.11481	0.66587
30	1	4	27.3	1	0.81199	0.40058	0.96541
31	0	10	-8.4	1	0.00085	0.00000	0.15847

In general, the coefficient $\hat{\beta}_i$ in the logistic model estimates the change in the log-odds when x_i is increased by 1 unit, holding all other x's in the model fixed. The antilog of the coefficient, $e^{\hat{\beta}_i}$, then estimates the odds-ratio

$$\frac{\pi_{x+1}/(1 - \pi_{x+1})}{\pi_x/(1 - \pi_x)}$$

where π_x is the value of $P(y = 1)$ for a fixed value x.[†] Typically, analysts compute $(e^{\hat{\beta}_i}) - 1$, which is an estimate of the percentage increase (or decrease) in the odds $\pi = P(y = 1)/P(y = 0)$ for every 1-unit increase in x_i, holding the other x's fixed.

[†]To see this, consider the model $\pi^* = \beta_0 + \beta_1 x$, where $x = 1$ or $x = 0$. When $x = 1$, we have $\pi_1^* = \beta_0 + \beta_1$; when $x = 0, \pi_0^* = \beta_0$. Now replace π_i^* with $\ln[\pi_i/(1 + \pi_i)]$, and take the antilog of each side of the equation. Then we have $\pi_1/(1 - \pi_1) = e^{\beta_0}e^{\beta_1}$ and $\pi_0/(1 - \pi_0) = e^{\beta_0}$. Consequently, the odds-ratio is

$$\frac{\pi_1/(1 - \pi_1)}{\pi_0/(1 - \pi_0)} = e^{\beta_1}$$

This leads to the following interpretations of the β estimates:

$\hat{\beta}_1 = -.7553$; $e^{\hat{\beta}_1} = .47$; $e^{\hat{\beta}_1} - 1 = -.53$: For each additional bidder (x_1), we estimate the odds of a fixed contract to *decrease* by 53%, holding **DOTEST** (x_2) fixed.

$\hat{\beta}_2 = .1122$; $e^{\hat{\beta}_2} = 1.12$; $e^{\hat{\beta}_2} - 1 = .12$: For every 1% increase in **DOTEST** (x_2), we estimate the odds of a fixed contract to *increase* by 12%, holding **NUMBIDS** (x_1) fixed.

Interpretations of β Parameters in the Logistic Model

$$\pi^* = \beta_0 + \beta_1 x_1 + \beta_2 x_2 + \cdots + \beta_k x_k$$

where

$$\pi^* = \ln\left(\frac{\pi}{1-\pi}\right)$$

$$\pi = P(y = 1)$$

β_i = Change in log-odds π^* for every 1-unit increase in x_i, holding all other x's fixed

$e^{\beta_i} - 1$ = Percentage change in odds $\pi/(1-\pi)$ for every 1-unit increase in x_i, holding all other x's fixed

---

The standard errors of the β estimates are given under the column **Standard Error**, and the (squared) ratios of the β estimates to their respective standard errors are given under the column **Wald Chi-Square**. As in regression with a linear model, this ratio provides a test statistic for testing the contribution of each variable to the model (i.e., for testing $H_0: \beta_i = 0$).[‡] The observed significance levels of the tests (i.e., the p-values) are given under the column **Pr > Chi-Square**. Note that both independent variables, **NUMBIDS** (x_1) and **DOTEST** (x_2), have p-values less than .03 (implying that we would reject $H_0: \beta_1 = 0$ and $H_0: \beta_2 = 0$ for $\alpha = .03$).

The test statistic for testing the overall adequacy of the logistic model, i.e., for testing $H_0: \beta_1 = \beta_2 = 0$, is given in the upper portion of the printout (shaded in the **Likelihood Ratio** row) as $\chi^2 = 18.5377$, with observed significance level (shaded) $p < .0001$.[§] Based on the p-value of the test, we can reject H_0 and conclude that at least one of the β coefficients is nonzero. Thus, the model is adequate for predicting bid status y.

[‡]In the logistic regression model, the ratio $(\hat{\beta}_i/s_{\hat{\beta}_i})^2$ has an approximate χ^2 distribution with 1 degree of freedom. Consult the references for more details about the χ^2 distribution and its use in logistic regression.

[§]The test statistic has an approximate χ^2 distribution with $k = 2$ degrees of freedom, where k is the number of β parameters in the model (excluding β_0).

Finally, the bottom portion of the printout gives predicted values and lower and upper 95% prediction limits for each observation used in the analysis. The 95% prediction interval for π for a contract with $x_1 = 3$ bidders and winning bid amount $x_2 = 11.5\%$ above the engineer's estimate is shaded on the printout. We estimate π, the probability of this particular contract being fixed, to fall between .31578 and .84081. Note that all the predicted values and limits lie between 0 and 1, a property of the logistic model.   ◆

In summary, we have presented two approaches to fitting logistic regression models. If the data are replicated, you may want to apply the transformation approach. The maximum likelihood estimation approach can be applied to any data set, but you need access to a statistical software package (such as SAS or SPSS) with logistic regression procedures.

This section should be viewed only as an overview of logistic regression. Many of the details of fitting logistic regression models using either technique have been omitted. Before conducting a logistic regression analysis, we strongly recommend that you consult the references given at the end of this chapter.

EXERCISES

9.16. Refer to Exercise 9.14 (p. 448) and the problem of modeling $y = \{1$ if own a digital organizer , 0 if not$\}$ as a function of annual income, χ.

a. Define π for this problem.
b. Define "odds" for this problem.
c. The data for the random sample of 20 households were used to fit the logit model

$$E(y) = \frac{\exp(\beta_0 + \beta_1 x)}{1 + \exp(\beta_0 + \beta_1 x)}$$

An SPSS printout of the logistic regression is presented below. Interpret the results.

SPSS Output for Exercise 9.16

Omnibus Tests of Model Coefficients

		Chi-square	df	Sig.
Step 1	Step	2.929	1	.087
	Block	2.929	1	.087
	Model	2.929	1	.087

Model Summary

Step	-2 Log likelihood	Cox & Snell R Square	Nagelkerke R Square
1	22.969	.136	.188

Classification Table[a]

			Predicted OWN 0	Predicted OWN 1	Percentage Correct
Step 1	OWN	0	12	1	92.3
		1	5	2	28.6
	Overall Percentage				70.0

a. The cut value is .500

Variables in the Equation

		B	S.E.	Wald	df	Sig.	Exp(B)	95.0% C.I.for EXP(B) Lower	95.0% C.I.for EXP(B) Upper
Step 1	INCOME	.000	.000	2.453	1	.117	1.000	1.000	1.000
	Constant	-3.112	1.675	3.454	1	.063	.044		

a. Variable(s) entered on step 1: INCOME.

9.17. Refer to Exercise 9.15 (p. 448). Use the data in the DISCRIM file to fit the logit model

$$E(y) = \frac{\exp(\beta_0 + \beta_1 x_1 + \beta_2 x_2 + \beta_3 x_3)}{1 + \exp(\beta_0 + \beta_1 x_1 + \beta_2 x_2 + \beta_3 x_3)}$$

where

$$y = \begin{cases} 1 & \text{if hired} \\ 0 & \text{if not} \end{cases}$$

$x_1 =$ Years of higher education (4, 6, or 8)

$x_2 =$ Years of experience

$$x_3 = \begin{cases} 1 & \text{if male applicant} \\ 0 & \text{if female applicant} \end{cases}$$

a. Conduct a test of model adequacy. Use $\alpha = .05$.

b. Is there sufficient evidence to indicate that gender is an important predictor of hiring status? Test using $\alpha = .05$.

c. Find a 95% confidence interval for the mean response $E(y)$ when $x_1 = 4$, $x_2 = 0$, and $x_3 = 1$. Interpret the interval.

9.7
Ridge Regression

When the sample data for regression exhibit multicollinearity, the least squares estimates of the β coefficients may be subject to extreme roundoff error as well as inflated standard errors (see Section 7.5). Since their magnitudes and signs may change considerably from sample to sample, the least squares estimates are said to be *unstable*. A technique developed for stabilizing the regression coefficients in the presence of multicollinearity is **ridge regression**.

Ridge regression is a modification of the method of least squares to allow *biased* estimators of the regression coefficients. At first glance, the idea of biased estimation may not seem very appealing. But consider the sampling distributions of two different estimators of a regression coefficient β, one unbiased and the other biased, shown in Figure 9.13.

FIGURE 9.13

Sampling distributions of two estimators of a regression coefficient β

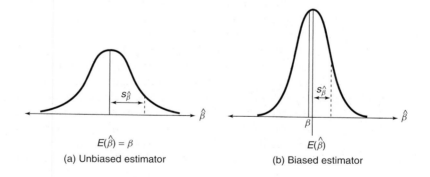

(a) Unbiased estimator $E(\hat{\beta}) = \beta$

(b) Biased estimator $E(\hat{\beta})$

Figure 9.13a shows an unbiased estimator of β with a fairly large variance. In contrast, the estimator shown in Figure 9.13b has a slight bias but is much less variable. In this case, we would prefer the biased estimator over the unbiased estimator since it will lead to more precise estimates of the true β (i.e., narrower confidence intervals for β). One way to measure the "goodness" of an estimator of β is to calculate the **mean square error** of $\hat{\beta}$, denoted by MSE($\hat{\beta}$), where MSE($\hat{\beta}$) is defined as

$$\text{MSE}(\hat{\beta}) = E[(\hat{\beta} - \beta)^2]$$
$$= V(\hat{\beta}) + [E(\hat{\beta}) - \beta]^2$$

The difference $E(\hat{\beta}) - \beta$ is called the **bias** of $\hat{\beta}$. Therefore, MSE$(\hat{\beta})$ is just the sum of the variance of $\hat{\beta}$ and the squared bias:

$$\text{MSE}(\hat{\beta}) = V(\hat{\beta}) + (\text{Bias in} \hat{\beta})^2$$

Let $\hat{\beta}_{\text{LS}}$ denote the least squares estimate of β. Then, since $E(\hat{\beta}_{\text{LS}}) = \beta$, the bias is 0 and

$$\text{MSE}(\hat{\beta}_{\text{LS}}) = V(\hat{\beta}_{\text{LS}})$$

We have previously stated that the variance of the least squares regression coefficient, and hence MSE$(\hat{\beta}_{\text{LS}})$, will be quite large in the presence of multicollinearity. The idea behind ridge regression is to introduce a small amount of bias in the ridge estimator of β, denoted by $\hat{\beta}_{\text{R}}$, so that its mean square error is considerably smaller than the corresponding mean square error for least squares, i.e.,

$$\text{MSE}(\hat{\beta}_{\text{R}}) < \text{MSE}(\hat{\beta}_{\text{LS}})$$

In this manner, ridge regression will lead to narrower confidence intervals for the β coefficients, and hence, more stable estimates.

Although the mechanics of a ridge regression are beyond the scope of this text, we point out that some of the more sophisticated software packages (including SAS) are now capable of conducting this type of analysis. To obtain the ridge regression coefficients, the user must specify the value of a biasing constant c, where $c \geq 0$.* Researchers have shown that as the value of c increases, the bias in the ridge estimates increases while the variance decreases. The idea is to choose c so that the total mean square error for the ridge estimators is smaller than the total mean square error for the least squares estimates. Although such a c exists, the optimal value, unfortunately, is unknown.

Various methods for choosing the value of c have been proposed. One commonly used graphical technique employs a **ridge trace**. Values of the estimated ridge regression coefficients are calculated for different values of c ranging from 0 to 1 and are plotted. The plots for each of the independent variables in the model are overlaid to form the ridge trace. An example of ridge trace for a model with three independent variables is shown in Figure 9.14. Initially, the estimated coefficients may fluctuate dramatically as c is increased from 0 (especially if severe multicollinearity is present). Eventually, however, the ridge estimates will stabilize. After careful examination of the ridge trace, the analyst chooses the smallest value of c for which it appears that all the ridge estimates are stable. The choice of c, therefore, is subjective.

---

*In matrix notation, the ridge estimator $\hat{\beta}_{\text{R}}$ is calculated as follows:

$$\hat{\beta}_{\text{R}} = (\text{X}'\text{X}) + c\text{I})^{-1}\text{X}'\text{Y}$$

When $c = 0$, the least squares estimator

$$\hat{\beta}_{\text{LS}} = (\text{X}'\text{X})^{-1}\text{X}'\text{Y}$$

is obtained. See Appendix A for details on the matrix mechanics of a regression analysis.

FIGURE 9.14

FIGURE 9.14

Ridge trace of β coefficients of a
model with three independent
variables

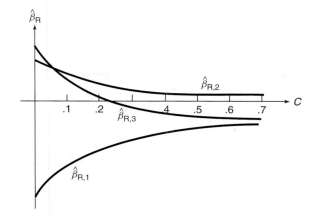

Once the value of c has been determined (using the ridge trace or some other analytical technique), the corresponding ridge estimates may be used in place of the least squares estimates. If the optimal (or near-optimal) value of c has been selected, the new estimates will have reduced variances (which lead to narrower confidence intervals for the β's). Also, some of the other problems associated with multicollinearity (e.g., incorrect signs on the β's) should have been corrected.

In conclusion, we caution that you should not assume that ridge regression is a panacea for multicollinearity or poor data. Although there are probably ridge regression estimates that are better than the least squares estimates when multicollinearity is present, the choice of the biasing constant c is crucial. Unfortunately, much of the controversy in ridge regression centers on how to find the optimal value of c. In addition, the exact distributional properties of the ridge estimators are unknown when c is estimated from the data. For these reasons, some statisticians recommend that ridge regression be used only as an exploratory data analysis tool for identifying unstable regression coefficients, and not for estimating parameters and testing hypotheses in a linear regression model.

9.8
Robust Regression

Consider the problem of fitting the linear regression model

$$y = \beta_0 + \beta_1 x_1 + \beta_2 x_2 + \cdots + \beta_k x_k + \varepsilon$$

by the method of least squares when the errors ε are nonnormal. In practice, moderate departures from the assumption of normality tend to have minimal effect on the validity of the least squares results (see Section 8.4). However, when the distribution of ε is **heavy-tailed** (longer-tailed) compared to the normal distribution, the method of least squares may not be appropriate. For example, the heavy-tailed error distribution shown in Figure 9.15 will most likely produce outliers with strong influence on the regression analysis. Furthermore, since they tend to "pull" the least squares fit too much in their direction, these outliers will have smaller than expected residuals and, consequently, are more difficult to detect.

Robust regression procedures are available for errors that follow a nonnormal distribution. In the context of regression, the term *robust* describes a technique that

FIGURE 9.15
Probability distribution of ε:
Normal versus heavy-tailed

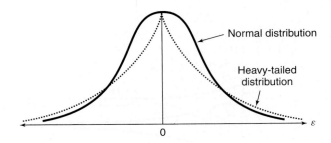

yields estimates for the β's that are nearly as good as the least squares estimates when the assumption of normality is satisfied, and significantly better for a heavy-tailed distribution. Robust regression is designed to dampen the effect of outlying observations that otherwise would exhibit strong influence on the analysis. This has the effect of leaving the residuals of influential observations large so that they may be more easily identified.

A number of different robust regression procedures exist. They fall into one of three general classes: **M estimators, R estimators**, and **L estimators**. Of the three, robust techniques that produce M estimates of the β coefficients receive the most attention in the literature.

The M estimates of the β coefficients are obtained by minimizing the quantity

$$\sum_{i=1}^{n} f(\hat{\varepsilon}_i)$$

where

$$\hat{\varepsilon}_i = y_i - (\hat{\beta}_0 + \hat{\beta}_1 x_{1i} + \hat{\beta}_2 x_{2i} + \cdots + \hat{\beta}_k x_{ki})$$

are the unobservable residuals and $f(\hat{\varepsilon}_i)$ is some function of the residuals. Note that since we are minimizing

$$\sum_{i=1}^{n} f(\hat{\varepsilon}_i) = \sum_{i=1}^{n} \hat{\varepsilon}_i^2$$

$$= \sum_{i=1}^{n} [y_i - (\hat{\beta}_0 + \hat{\beta}_1 x_{1i} + \hat{\beta}_2 x_{2i} + \cdots + \hat{\beta}_k x_{ki})]^2$$

$$= \text{SSE}$$

the function $f(\hat{\varepsilon}_i) = \hat{\varepsilon}_i^2$ yields the ordinary least squares estimates and, therefore, is appropriate when the errors are normal. For errors with heavier-tailed distributions, the analyst chooses some other function $f(\hat{\varepsilon}_i)$ that places less weight on the errors in the tails of the distribution. For example, the function $f(\hat{\varepsilon}_i) = |\hat{\varepsilon}_i|$ is appropriate when the errors follow the heavy-tailed distribution pictured in Figure 9.15. Since we are minimizing

$$\sum_{i=1}^{n} f(\hat{\varepsilon}_i) = \sum_{i=1}^{n} |\hat{\varepsilon}_i|$$

$$= \sum_{i=1}^{n} |y_i - (\hat{\beta}_0 + \hat{\beta}_1 x_{1i} + \hat{\beta}_2 x_{2i} + \cdots + \hat{\beta}_k x_{ki})|$$

the M estimators of robust regression yield the estimates obtained from the **method of absolute deviations** (see Section 8.5).

The other types of robust estimation, R estimation and L estimation, take a different approach. R estimators are obtained by minimizing the quantity

$$\sum_{i=1}^{n} [y_i - (\hat{\beta}_0 + \hat{\beta}_1 x_{1i} + \hat{\beta}_2 x_{2i} + \cdots + \hat{\beta}_k x_{ki})] R_i$$

where R_i is the rank of the ith residual when the residuals are placed in ascending order. L estimation is similar to R estimation because it involves ordering of the data, but it uses measures of location (such as the sample median) to estimate the regression coefficients.

The numerical techniques for obtaining robust estimates (M, R, or L estimates) are quite complex and require sophisticated computer programs. At present, statistical software packages for robust regression are not widely available. However, the growing demand for packaged robust regression programs, especially M estimation procedures, leads us to believe that these programs will be available in the near future.

Much of the current research in the area of robust regression is focused on the distributional properties of the robust estimators of the β coefficients. At present, there is little information available on robust confidence intervals, prediction intervals, and hypothesis testing procedures. For this reason, some researchers recommend that robust regression be used in conjunction with and as a check on the method of least squares. If the results of the two procedures are substantially the same, use the least squares fit since confidence intervals and tests on the regression coefficients can be made. On the other hand, if the two analyses yield quite different results, use the robust fit to identify any influential observations. A careful examination of these data points may reveal the problem with the least squares fit.

9.9
Nonparametric
Regression Models

Recall, in Section 4.1, that the general multiple regression model assumes a "linear" form:

$$E(y) = \beta_0 + \beta_1 x_1 + \beta_2 x_2 + \beta_3 x_3 + \cdots + \beta_k x_k$$

That is, the model is proposed as a linear function of the unknown β's and the method of least squares is used to estimate these β's. In situations where the assumption of linearity may be violated (e.g., with a binary response y), **nonparametric regression models** are available.

In nonparametric regression, the analyst does not necessarily propose a specific functional relationship between y and x_i. Rather, the linear term $\beta_i x_i$ is replaced by a smooth function of x_i that is estimated by visually exploring the data. The general form of a nonparametric regression model is:

$$E(y) = s_0 + s_1(x_1) + s_2(x_2) + s_3(x_3) + \cdots + s_k(x_k),$$

where $s_1(x_1)$ is a smooth function relating y to x_1, $s_2(x_2)$ is a smooth function relating y to x_2, etc. The smooth function $s_i(x_i)$, or **smoother** as it is commonly known, summarizes the trend in the relationship between y and x_i. Figure 9.16 shows a scatterplot for a data set collected on number of exacerbations, y, and age, x, of patients with multiple sclerosis (MS). The smooth function, $s(x)$, that represents a possible trend in the data is shown on the graph.

FIGURE 9.16

Scatterplot of data for MS patients

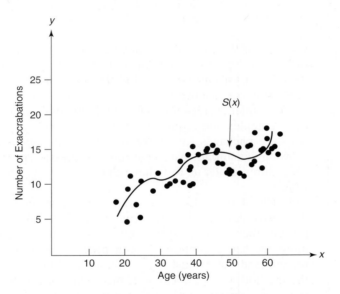

A variety of nonparametric methods have been proposed for estimating smoothers. These include **cubic smoothing splines** and **thin-plate smoothing splines**, where the term *spline* is used to describe a smoothed curve obtained through a complex mathematical optimization technique. Details on these nonparametric methods are beyond the scope of this text. You should consult the references at the end of the chapter if you wish to learn more about them.

As you might expect, as with robust regression, sophisticated computer software is required to fit a nonparametric regression model. Nonparametric regression procedures are available in SAS. (See the SAS Tutorial in the Appendix). Although a nonparametric model can be used as an alternative to the general linear model, the two models really serve two different analytic purposes. The linear statistical model of Chapters 3–8 emphasizes estimation and inferences for the model parameters. In contrast, nonparametric regression models are more suitable for exploring and visualizing relationships among the variables of interest.

Summary

A number of special topics in regression have been introduced in this chapter. **Piecewise linear regression** can be employed when the theoretical relationship between the dependent variable y and a single independent variable x differs for different intervals over the range of x. **Inverse prediction** is a technique used for predicting a value of x for a given value of y. When the response y is **binary** (i.e., takes on only two values, 0 or 1), a **logistic regression model** may be more appropriate than a linear regression model.

Several methods are available for situations when the standard least squares assumptions about the random errors are violated. **Weighted least squares** is a variance-stabilizing technique that can also be used to limit the influence of outlying observations. When the distribution of the errors is nonnormal, **robust regression** yields estimates of the β's with certain optimal properties. **Ridge regression** was developed to stabilize the estimated β coefficients in the presence of multicollinearity.

Finally, **nonparametric regression** can be applied when the assumption of a linear model is questionable or when the analyst is interested in exploring and visualizing relationships in the data.

REFERENCES

AGRESTI, A. *Categorical Data Analysis.* New York: Wiley, 1990.

ANDREWS, D. F. "A robust method for multiple linear regression." *Technometrics*, Vol. 16, 1974, pp. 523–531.

COX, D. R. *The Analysis of Binary Data.* London: Methuen, 1970.

DRAPER, N. R., and VAN NOSTRAND, R. C. "Ridge regression and James–Stein estimation: Review and comments." *Technometrics.* Vol. 21, 1979, p. 451.

GEISSER, S. "The predictive sample reuse method with applications." *Journal of the American Statistical Association*, Vol. 70, 1975, pp. 320–328.

GRAYBILL, F. A. *Theory and Application of the Linear Model.* North Scituate, Mass.: Duxbury Press, 1976.

HALPERIN, M., BLACKWELDER, W. C., and VERTER, J. I. "Estimation of the multivariate logistic risk function: A comparison of the discriminant function and maximum likelihood approaches." *Journal of Chronic Diseases*, Vol. 24, 1971, pp. 125–158.

HASTIE, T. and TIBSHIRANI, R. *Generalized Additive Models*, New York: Chapman and Hall, 1990.

HAUCK, W. W., and DONNER, A. "Wald's test as applied to hypotheses in logit analysis." *Journal of the American Statistical Association*, Vol. 72, 1977, pp. 851–853.

HILL, R. W., and HOLLAND, P. W. "Two robust alternatives to least squares regression." *Journal of the American Statistical Association*, Vol. 72, 1977, pp. 828–833.

HOERL, A. E., and KENNARD, R. W. "Ridge regression: Biased estimation for nonorthogonal problems." *Technometrics*, Vol. 12, 1970, pp. 55–67.

HOERL, A. E., KENNARD, R. W., and BALDWIN, K. F. "Ridge regression: Some simulations." *Communications in Statistics*, Vol. A5, 1976, pp. 77–88.

HOGG, R. V. "Statistical robustness: One view of its use in applications today." *The American Statistician*, Vol. 33, 1979, pp. 108–115.

HOSMER, D. W., and LEMESHOW, S. *Applied Logistic Regression.* (2nd ed.) New York: Wiley, 2000.

MONTGOMERY, D., PECK, E. and VINING, G. *Introduction to Linear Regression Analysis* 3rd ed. New York: Wiley, 2001.

MOSTELLER, F., and TUKEY, J. W. *Data Analysis and Regression: A Second Course in Statistics.* Reading, Mass.: Addison-Wesley, 1977.

NETER, J., KUTNER, M., NACHTSHEIM, C. and WASSERMAN, W., *Applied Linear Statistical Models*, 4th ed. Homewood, Ill.: Richard D. Irwin, 1996.

OBENCHAIN, R. L. "Classical *F*-tests and confidence intervals for ridge regression." *Technometrics*, Vol. 19, 1977, pp. 429–439.

SNEE, R. D. "Validation of regression models: Methods and examples." *Technometrics*, Vol. 19, 1977, pp. 415–428.

STONE, C. J. "Additive regression and other nonparametric models." *Annals of Statistics*, Vol. 13, 1985, p. 689–705.

TSIATIS, A. A. "A note on the goodness-of-fit test for the logistic regression model." *Biometrika*, Vol. 67, 1980, pp. 250–251.

WAHBA, G. *Spline models for observational data*, Philadelphia: Society for Industrial and Applied Mathematics, 1990.

WALKER, S. H., and DUNCAN, D. B. "Estimation of the probability of an event as a function of several independent variables." *Biometrika*, Vol. 54, 1967, pp. 167–179.

Introduction to Time Series Modeling and Forecasting

CONTENTS

OBJECTIVE

To present models that allow for the correlation between observations taken sequentially over time; to show how these models can be used to forecast a future response

---

10.1 What Is a Time Series?

In many business and economic studies, the response variable y is measured sequentially in time. For example, we might record the number y of new housing starts for each month in a particular region. This collection of data is called a **time series**. Other examples of time series are data collected on the quarterly number of highway deaths in the United States, the annual sales for a corporation, and the recorded month-end values of the prime interest rate.

> **Definition 10.1**
> A **time series** is a collection of data obtained by observing a response variable at periodic points in time.

> **Definition 10.2**
> If repeated observations on a variable produce a time series, the variable is called a **time series variable**. We use y_t to denote the value of the variable at time t.

If you were to develop a model relating the number of new housing starts to the prime interest rate over time, the model would be called a **time series model**, because both the dependent variable, new housing starts, and the independent variable, prime interest rate, are measured sequentially over time. Furthermore, time itself would probably play an important role in such a model, because the economic trends and seasonal cycles associated with different points in time would almost certainly affect both time series.

The construction of time series models is an important aspect of business and economic analyses, because many of the variables of most interest to business and economic researchers are time series. This chapter is an introduction to the very complex and voluminous body of material concerned with time series modeling and forecasting future values of a time series.

10.2
Time Series
Components

Researchers often approach the problem of describing the nature of a time series y_t by identifying four kinds of change, or variation, in the time series values. These four components are commonly known as (1) secular trend, (2) cyclical effect, (3) seasonal variation, and (4) residual effect. The components of a time series are most easily identified and explained pictorially.

Figure 10.1a shows a **secular trend** in the time series values. The secular component describes the tendency of the value of the variable to increase or decrease over a long period of time. Thus, this type of change or variation is also known as the **long-term trend**. In Figure 10.1a, the long-term trend is of an increasing nature. However, this does not imply that the time series has always moved upward from month to month and from year to year. You can see that although the series fluctuates, the trend has been an increasing one over that period of time.

FIGURE 10.1

The components of a time series

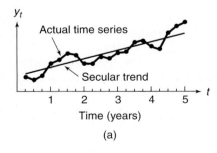
(a)

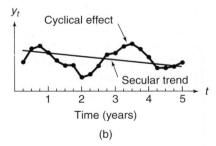
(b)

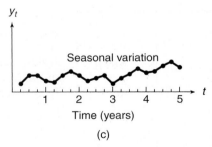
(c)

The **cyclical effect** in a time series, as shown in Figure 10.1b, generally describes the fluctuation about the secular trend that is attributable to business and economic conditions at the time. These fluctuations are sometimes called **business cycles**. During a period of general economic expansion, the business cycle lies above the secular trend, whereas during a recession, when business activity is likely to slump, the cycle lies below the secular trend. You can see that the cyclical variation does not follow any definite trend, but moves rather unpredictably.

The **seasonal variation** in a time series describes the fluctuations that recur during specific portions of each year (e.g., monthly or seasonally). In Figure 10.1c, you can see that the pattern of change in the time series within a year tends to be repeated from year to year, producing a wavelike or oscillating curve.

The final component, the **residual effect**, is what remains after the secular, cyclical, and seasonal components have been removed. This component is not systematic and may be attributed to unpredictable influences such as wars and political unrest, hurricanes and droughts, and the randomness of human actions. Thus, the residual effect represents the random error component of a time series.

Definition 10.3

The **secular trend** (T_t) of a time series is the tendency of the series to increase or decrease over a long period of time. It is also known as the **long-term trend**.

Definition 10.4

The **cyclical fluctuation** (C_t) of a time series is the wavelike or oscillating pattern about the secular trend that is attributable to business and economic conditions at the time. It is also known as a **business cycle**.

Definition 10.5

The **seasonal variation** (S_t) of a time series describes the fluctuations that recur during specific portions of the year (e.g., monthly or seasonally).

Definition 10.6

The **residual effect** (R_t) of a time series is what remains after the secular, cyclical, and seasonal components have been removed.

In many practical applications of time series, the objective is to *forecast* (predict) some *future value or values* of the series. To obtain forecasts, some type of model that can be projected into the future must be used to describe the time series. One of the most widely used models is the *additive model*.*

$$y_t = T_t + C_t + S_t + R_t$$

where $T_t, C_t, S_t,$ and R_t represent the secular trend, cyclical effect, seasonal variation, and residual effect, respectively, of the time series variable y_t. Various methods exist

---

*Another useful model is the **multiplicative model** $y_t = T_t C_t S_t R_t$. Recall (Section 4.12) that this model can be written in the form of an additive model by taking natural logarithms:

$$\ln y_t = \ln T_t + \ln C_t + \ln S_t + \ln R_t$$

for estimating the components of the model and forecasting the time series. These range from simple **descriptive techniques**, which rely on smoothing the pattern of the time series, to complex **inferential models**, which combine regression analysis with specialized time series models. Several descriptive forecasting techniques are presented in Section 10.3, and forecasting using the general linear regression model of Chapter 4 is discussed in Section 10.4. The remainder of the chapter is devoted to the more complex and more powerful time series models.

10.3
Forecasting Using Smoothing Techniques (Optional)

Various descriptive methods are available for identifying and characterizing a time series. Generally, these methods attempt to remove the rapid fluctuations in a time series so that the secular trend can be seen. For this reason, they are sometimes called **smoothing techniques**. Once the secular trend is identified, forecasts for future values of the time series are easily obtained. In this section, we present three of the more popular smoothing techniques.

MOVING AVERAGE METHOD: A widely used smoothing technique is the **moving average method**. A moving average, M_t, at time t is formed by averaging the time series values over adjacent time periods. Moving averages aid in identifying the secular trend of a time series because the averaging modifies the effect of short-term (cyclical or seasonal) variation. That is, a plot of the moving averages yields a "smooth" time series curve that clearly depicts the long-term trend.

For example, consider the 1999–2002 quarterly power loads for a utility company located in a southern part of the United States, given in Table 10.1. A MINITAB graph of the quarterly time series, Figure 10.2, shows the pronounced seasonal

💿 QTRPOWER

TABLE 10.1   **Quarterly Power Loads, 1999–2002**

YEAR	QUARTER	TIME t	POWER LOAD y_t, megawatts
1999	I	1	103.5
	II	2	94.7
	III	3	118.6
	IV	4	109.3
2000	I	5	126.1
	II	6	116.0
	III	7	141.2
	IV	8	131.6
2001	I	9	144.5
	II	10	137.1
	III	11	159.0
	IV	12	149.5
2002	I	13	166.1
	II	14	152.5
	III	15	178.2
	IV	16	169.0

FIGURE 10.2

MINITAB plot of quarterly power loads

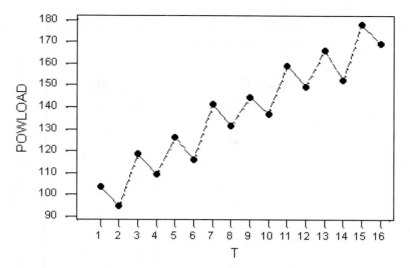

variation, i.e., the fluctuation that recurs from year to year. The quarterly power loads are highest in the summer months (quarter III) with another smaller peak in the winter months (quarter I), and lowest during the spring and fall (quarters II and IV). To clearly identify the long-term trend of the series, we need to average, or "smooth out," these seasonal fluctuations. We will apply the moving average method for this purpose.

The first step in calculating a moving average for quarterly data is to sum the observed time values y_t—in this example, quarterly power loads—for the four quarters during the initial year 1999. Summing the values from Table 10.1, we have

$$y_1 + y_2 + y_3 + y_4 = 103.5 + 94.7 + 118.6 + 109.3$$

$$= 426.1$$

This sum is called a **4-point moving total**, which we denote by the symbol L_t. It is customary to use a subscript t to represent the time period at the midpoint of the four quarters in the total. Since for this sum, the midpoint is between $t = 2$ and $t = 3$, we will use the conventional procedure of "dropping it down one line" to $t = 3$. Thus, our first 4-point moving total is $L_3 = 426.1$.

We find the next moving total by eliminating the first quantity in the sum, $y_1 = 103.5$, and adding the next value in the time series sequence, $y_5 = 126.1$. This enables us to keep four quarters in the total of adjacent time periods. Thus, we have

$$L_4 = y_2 + y_3 + y_4 + y_5 = 94.7 + 118.6 + 109.3 + 126.1 = 448.7$$

Continuing this process of "moving" the 4-point total over the time series until we have included the last value, we find

$$L_5 = y_3 + y_4 + y_5 + y_6 \quad = 118.6 + 109.3 + 126.1 + 116.0 = 470.0$$
$$L_6 = y_4 + y_5 + y_6 + y_7 \quad = 109.3 + 126.1 + 116.0 + 141.2 = 492.6$$
$$\vdots \qquad\qquad \vdots \qquad\qquad\qquad \vdots \quad \vdots$$
$$L_{15} = y_{13} + y_{14} + y_{15} + y_{16} = 166.1 + 152.5 + 178.2 + 169.0 = 665.8$$

TABLE 10.2   **4-Point Moving Average for the Quarterly Power Load Data**

YEAR	QUARTER	TIME t	POWER LOAD y_t	4-POINT MOVING TOTAL L_t	4-POINT MOVING AVERAGE M_t	RATIO y_t/M_t
1999	I	1	103.5	—	—	—
	II	2	94.7	—	—	—
	III	3	118.6	426.1	106.5	1.113
	IV	4	109.3	448.7	112.2	.974
2000	I	5	126.1	470.0	117.5	1.073
	II	6	116.0	492.6	123.2	.942
	III	7	141.2	514.9	128.7	1.097
	IV	8	131.6	533.3	133.3	.987
2001	I	9	144.5	554.4	138.6	1.043
	II	10	137.1	572.2	143.1	.958
	III	11	159.0	590.1	147.5	1.078
	IV	12	149.5	611.7	152.9	.978
2002	I	13	166.1	627.1	156.8	1.059
	II	14	152.5	646.3	161.6	.944
	III	15	178.2	665.8	166.5	1.071
	IV	16	169.0	—	—	—

The complete set of 4-point moving totals is given in the appropriate column of Table 10.2. Notice that three data points will be "lost" in forming the moving totals.

After the 4-point moving totals are calculated, the second step is to determine the **4-point moving average**, denoted by M_t, by dividing each of the moving totals by 4. For example, the first three values of the 4-point moving average for the quarterly power load data are

$$M_3 = \frac{y_1 + y_2 + y_3 + y_4}{4} = \frac{L_3}{4} = \frac{426.1}{4} = 106.5$$

$$M_4 = \frac{y_2 + y_3 + y_4 + y_5}{4} = \frac{L_4}{4} = \frac{448.7}{4} = 112.2$$

$$M_5 = \frac{y_3 + y_4 + y_5 + y_6}{4} = \frac{L_5}{4} = \frac{470.0}{4} = 117.5$$

All of the 4-point moving averages are given in the appropriate column of Table 10.2.

Both the original power load time series and the 4-point moving average are graphed (using MINITAB) in Figure 10.3. Notice that the moving average has smoothed the time series; i.e., the averaging has modified the effects of the short-term or seasonal variation. The plot of the 4-point moving average clearly depicts the secular (long-term) trend component of the time series.

In addition to identifying a long-term trend, moving averages provide us with a measure of the seasonal effects in a time series. The ratio between the observed power load y_t and the 4-point moving average M_t for each quarter measures the seasonal effect (primarily attributable to temperature differences) for that quarter.

FIGURE 10.3

MINITAB plot of quarterly power loads and 4-point moving average

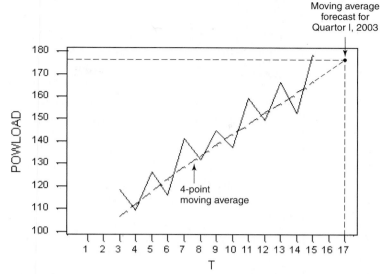

The ratios y_t/M_t are shown in the last column of Table 10.2. Note that the ratio is always greater than 1 in quarters I and III, and always less than 1 in quarters II and IV. The average of the ratios for a particular quarter, multiplied by 100, can be used to form a **seasonal index** for that quarter. For example, the seasonal index for quarter I is

$$100 \left(\frac{1.073 + 1.043 + 1.059}{3} \right) = 105.8$$

implying that the time series value in quarter I is, on the average, 105.8% of the moving average value for that time period.

To forecast a future value of the time series, simply extend the moving average M_t on the graph to the future time period.* For example, a graphical extension of the moving average for the quarterly power loads to quarter I of 2003 ($t = 17$) yields a moving average of approximately $M_{17} = 175$ (see Figure 10.3). Thus, if there were no seasonal variation in the time series, we would expect the power load for quarter I of 2003 to be approximately 175 megawatts. To adjust the forecast for seasonal variation, multiply the future moving average value $M_{17} = 175$ by the seasonal index for quarter I, then divide by 100:

$$F_{17} = M_{17} \left(\frac{\text{Seasonal index for quarter I}}{100} \right)$$

$$= 175 \left(\frac{105.8}{100} \right)$$

$$\approx 185$$

where F_{17} is the forecast of y_{17}. Therefore, the moving average forecast for the power load in quarter I of 2003 is approximately 185 megawatts.

---

*Some statistical software packages (e.g., MINITAB) will use the last moving average in the series as the value of the forecast for any future time period.

Moving averages are not restricted to 4 points. For example, you may wish to calculate a 7-point moving average for daily data, a 12-point moving average for monthly data, or a 5-point moving average for yearly data. Although the choice of the number of points is arbitrary, you should search for the number N that yields a smooth series, but is not so large that many points at the end of the series are "lost." The method of forecasting with a general N-point moving average is outlined in the box.

Forecasting Using an *N*-Point Moving Average

1. Select N, the number of consecutive time series values $y_1, y_2, \ldots, y_N$ that will be averaged. (The time series values must be equally spaced.)

2. Calculate the N-point moving total, L_t, by summing the time series values over N adjacent time periods, where

$$L_t = \begin{cases} y_{t-(N-1)/2} + \cdots + y_t + \cdots + y_{t+(N-1)/2} & \text{if } N \text{ is odd} \\ y_{t-N/2} + \cdots + y_t + \cdots + y_{t+N/2-1} & \text{if } N \text{ is even} \end{cases}$$

3. Compute the N-point moving average, M_t, by dividing the corresponding moving total by N:

$$M_t = \frac{L_t}{N}$$

4. Graph the moving average M_t on the vertical axis with time t on the horizontal axis. (This plot should reveal a smooth curve that identifies the long-term trend of the time series.*) Extend the graph to a future time period to obtain the forecasted value of M_t.

5. For a future time period t, the forecast of y_t is

$$F_t = \begin{cases} M_t & \text{if little or no seasonal variation exists in the time series} \\ M_t \left(\dfrac{\text{Seasonal index}}{100} \right) & \text{otherwise} \end{cases}$$

where the seasonal index for a particular quarter (or month) is the average of past values of the ratios

$$\frac{y_t}{M_t}(100)$$

for that quarter (or month).

EXPONENTIAL SMOOTHING: One problem with using a moving average to forecast future values of a time series is that values at the ends of the series are lost, thereby requiring that we subjectively extend the graph of the moving average into the future. No exact calculation of a forecast is available since the moving average at a future time period t requires that we know one or more future values of the series. **Exponential smoothing** is a technique that leads to forecasts that can be explicitly calculated. Like the moving average method, exponential smoothing deemphasizes

---

*When the number N of points is small, the plot may not yield a very smooth curve. However, the moving average will be smoother (or less variable) than the plot of the original time series values.

(or smooths) most of the residual effects. However, exponential smoothing averages only past and current values of the time series.

To obtain an exponentially smoothed time series, we first need to choose a weight w, between 0 and 1, called the **exponential smoothing constant**. The exponentially smoothed series, denoted E_t, is then calculated as follows:

$$E_1 = y_1$$
$$E_2 = wy_2 + (1 - w)E_1$$
$$E_3 = wy_3 + (1 - w)E_2$$
$$\vdots \quad \vdots$$
$$E_t = wy_t + (1 - w)E_{t-1}$$

You can see that the exponentially smoothed value at time t is simply a weighted average of the current time series value, y_t, and the exponentially smoothed value at the previous time period, E_{t-1}. Smaller values of w give less weight to the current value, y_t, whereas larger values give more weight to y_t.

For example, suppose we want to smooth the quarterly power loads given in Table 10.1 using an exponential smoothing constant of $w = .7$. Then we have

$$E_1 = y_1 = 103.5$$
$$E_2 = .7y_2 + (1 - .7)E_1$$
$$\quad = .7(94.7) + .3(103.5) = 97.34$$
$$E_3 = .7y_3 + (1 - .7)E_2$$
$$\quad = .7(118.6) + .3(97.34) = 112.22$$
$$\vdots$$

The exponentially smoothed values (using $w = .7$) for all the quarterly power loads, obtained using MINITAB, are highlighted on the MINITAB printout, Figure 10.4. Both the actual and the smoothed time series values are graphed on the MINITAB printout, Figure 10.5.

Exponential smoothing forecasts are obtained by using the most recent exponentially smoothed value, E_t. In other words, if n is the last time period in which y_t is observed, then the forecast for a future time period t is given by

$$F_t = E_n$$

As you can see, the right-hand side of the forecast equation does not depend on t; hence, F_t is used to forecast *all* future values of y_t. The MINITAB printout, Figure 10.4, shows that the smoothed value for quarter 4 of 2002 ($t = 16$) is $E_{16} = 169.688$. Therefore, this value represents the forecast for the power load in quarter I of 2003 ($t = 17$), i.e.,

$$F_{17} = E_{16} = 169.688$$

FIGURE 10.4

MINITAB printout listing results of exponential smoothing of quarterly power loads ($w = .7$)

Single Exponential Smoothing

Data	POWLOAD
Length	16.0000
NMissing	0

Smoothing Constant
Alpha: 0.7

Accuracy Measures
MAPE: 7.604
MAD: 10.634
MSD: 177.873

Row	Time	POWLOAD	Smooth	Predict	Error
1	1	103.5	103.500	103.500	0.0000
2	2	94.7	97.340	103.500	-8.8000
3	3	118.6	112.222	97.340	21.2600
4	4	109.3	110.177	112.222	-2.9220
5	5	126.1	121.323	110.177	15.9234
6	6	116.0	117.597	121.323	-5.3230
7	7	141.2	134.119	117.597	23.6031
8	8	131.6	132.356	134.119	-2.5191
9	9	144.5	140.857	132.356	12.1443
10	10	137.1	138.227	140.857	-3.7567
11	11	159.0	152.768	138.227	20.7730
12	12	149.5	150.480	152.768	-3.2681
13	13	166.1	161.414	150.480	15.6196
14	14	152.5	155.174	161.414	-8.9141
15	15	178.2	171.292	155.174	23.0258
16	16	169.0	169.688	171.292	-2.2923

Row	Period	Forecast	Lower	Upper
1	17	169.688	143.634	195.741

This forecast is shown at the bottom of Figure 10.4 and graphically in Figure 10.5. The forecasts for quarter II of 2003 ($t = 18$), quarter III of 2003 ($t = 19$), and all other future time periods will be the same:

$$F_{18} = 169.688$$

$$F_{19} = 169.688$$

$$F_{20} = 169.688$$

$$\vdots$$

FIGURE 10.5

MINITAB plot of exponentially smoothed quarterly power loads ($w = .7$)

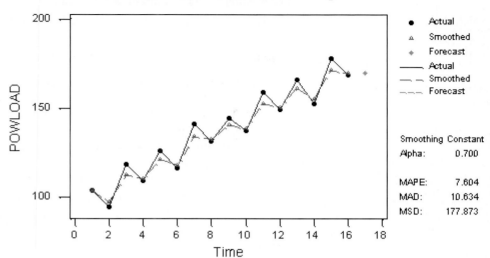

This points out one disadvantage of the exponential smoothing forecasting technique. Since the exponentially smoothed forecast is constant for all future values, any changes in trend and/or seasonality are not taken into account. Therefore, exponentially smoothed forecasts are appropriate only when the trend and seasonal components of the time series are relatively insignificant.

HOLT–WINTERS FORECASTING MODEL: One drawback to the exponential smoothing forecasting method is that the secular trend and seasonal components of a time series are not taken into account. The **Holt–Winters forecasting model** is an extension of the exponential smoothing method that explicitly recognizes the trend and seasonal variation in a time series.

Consider a time series with a trend component, but little or no seasonal variation. Then the Holt–Winters model for y_t is

$$E_t = wy_t + (1 - w)(E_{t-1} + T_{t-1})$$
$$T_t = v(E_t - E_{t-1}) + (1 - v)T_{t-1}$$

where E_t is the exponentially smoothed series, T_t is the trend component, and w and v are smoothing constants between 0 and 1. Note that the trend component T_t is a weighted average of the most recent change in the smoothed value (measured by the difference $E_t - E_{t-1}$) and the trend estimate of the previous time period (T_{t-1}). When seasonal variation is present in the time series, the Holt–Winters model takes the form

$$E_t = w(y_t/S_{t-p}) + (1 - w)(E_{t-1} + T_{t-1})$$
$$T_t = v(E_t - E_{t-1}) + (1 - v)T_{t-1}$$
$$S_t = u(y_t/E_t) + (1 - u)S_{t-P}$$

Forecasting Using Exponential Smoothing

1. The data consist of n equally spaced time series values,

$$y_1, y_2, \ldots, y_n.$$

2. Select a smoothing constant, w, between 0 and 1. (Smaller values of w give less weight to the current value of the series and yield a smoother series. Larger values of w give more weight to the current value of the series and yield a more variable series.)

3. Calculate the exponentially smoothed series, E_t, as follows:*

$$E_1 = y_1$$
$$E_2 = wy_2 + (1-w)E_1$$
$$E_3 = wy_3 + (1-w)E_2$$
$$\vdots$$
$$E_n = wy_n + (1-w)E_{n-1}$$

4. The forecast for any future time period t is:

$$F_t = E_n, \quad t = n+1, n+2, \ldots$$

---

where S_t is the seasonal component, u is a constant between 0 and 1, and P is the number of time periods in a cycle (usually a year). The seasonal component S_t is a weighted average of the ratio y_t/E_t (i.e., the ratio of the actual time series value to the smoothed value) and the seasonal component for the previous cycle. For example, for the quarterly power loads, $P = 4$ (four quarters in a year) and the seasonal component for, say, quarter III of 2000 ($t = 7$) is a weighted average of the ratio y_7/E_7 and the seasonal component for quarter III of 1999 ($t = 3$). That is,

$$S_7 = u(y_7/E_7) + (1-u)S_3$$

Forecasts for future time periods, $t = n+1, n+2, \ldots$, using the Holt–Winters models are obtained by summing the most recent exponentially smoothed component with an estimate of the expected increase (or decrease) attributable to trend. For seasonal models, the forecast is multiplied by the most recent estimate of the seasonal component (similar to the moving average method).

The Holt–Winters forecasting methodology is summarized in the next box. As Example 10.1 illustrates, these calculations can become quite tedious. Most time series analysts will utilize a statistical software package to apply the Holt–Winters forecasting method. Although there are slight variations in their initial computing formulas, SAS, MINITAB, and SPSS all have options for producing Holt–Winters forecasts.

---

*Optionally, the first "smoothed" value, E_1, can be computed as the average value of y_t over the entire series.

Forecasting Using the Holt–Winters Model

Trend Component Only	*Trend and Seasonal Components*

Trend Component Only

1. The data consist of n equally spaced time series values, $y_1, y_2, \ldots, y_n$.
2. Select smoothing constants w and v, where $0 \le w \le 1$ and $0 \le v \le 1$.

3. Calculate the exponentially smoothed component, E_t, and the trend component, T_t, for $t = 2, 3, \ldots, n$ as follows:

$$E_t = \begin{cases} y_2, & t = 2 \\ wy_t + (1-w)(E_{t-1} + T_{t-1}), \\ & t > 2 \end{cases}$$

$$T_t = \begin{cases} y_2 - y_1, & t = 2 \\ v(E_t - E_{t-1}) + (1-v)T_{t-1}, \\ & t > 2 \end{cases}$$

[*Note:* E_1 and T_1 are not defined.]

4. The forecast for a future time period t is given by

$$F_t = \begin{cases} E_n + T_n, & t = n+1 \\ E_n + 2T_n, & t = n+2 \\ \vdots \\ E_n + kT_n, & t = n+k \end{cases}$$

Trend and Seasonal Components

1. The data consist of n equally spaced time series values, $y_1, y_2, \ldots, y_n$.
2. Select smoothing constants w, v, and u, where $0 \le w \le 1, 0 \le v \le 1$, and $0 \le u \le 1$.
3. Determine P, the number of time periods in a cycle. Usually, $P = 4$ for quarterly data and $P = 12$ for monthly data.

4. Calculate the exponentially smoothed component, E_t, the trend component, T_t, and the seasonal component, S_t, for $t = 2, 3, \ldots, n$ as follows:

$$E_t = \begin{cases} y_2, & t = 2 \\ wy_t + (1-w)(E_{t-1} + T_{t-1}), \\ \qquad t = 3, 4, \ldots, P+2 \\ w(y_t/S_{t-P}) + (1-w) \\ \qquad \times (E_{t-1} + T_{t-1}), & t > P+2 \end{cases}$$

$$T_t = \begin{cases} y_2 - y_1, & t = 2 \\ v(E_t - E_{t-1}) + (1-v)T_{t-1}, \\ \qquad t > 2 \end{cases}$$

$$S_t = \begin{cases} y_t/E_t, & t = 2, 3, \ldots, P+2 \\ u(y_t/E_t) + (1-u)S_{t-P}, \\ \qquad t > P+2q \end{cases}$$

[*Note:* E_1, T_1, and S_1 are not defined.]

5. The forecast for a future time period t is given by

$$F_t = \begin{cases} (E_n + T_n)S_{n+1-P}, & t = n+1 \\ (E_n + 2T_n)S_{n+2-P}, & t = n+2 \\ \vdots \\ (E_n + kT_n)S_{n+k-P}, & t = n+k \end{cases}$$

EXAMPLE 10.1

Refer to the 1999–2002 quarterly power loads listed in Table 10.1. Use the Holt–Winters forecasting model with both trend and seasonal components to forecast the utility company's quarterly power loads in 2003. Use the smoothing constants $w = .7$, $v = .5$, and $u = .5$.

Solution

First note that $P = 4$ for the quarterly time series. Following the formulas for E_t, T_t, and S_t given in the box, we calculate

$$E_2 = y_2 = 94.7$$
$$T_2 = y_2 - y_1 = 94.7 - 103.5 = -8.8$$
$$S_2 = y_2/E_2 = 94.7/94.7 = 1$$

$$E_3 = .7y_3 + (1 - .7)(E_2 + T_2)$$
$$= .7(118.6) + .3(94.7 - 8.8) = 108.8$$
$$T_3 = .5(E_3 - E_2) + (1 - .5)T_2$$
$$= .5(108.8 - 94.7) + .5(-8.8) = 2.6$$
$$S_3 = y_3/E_3 = 118.6/108.8 = 1.090$$

$$E_4 = .7y_4 + (1 - .7)(E_3 + T_3)$$
$$= .7(109.3) + .3(108.8 + 2.6) = 109.9$$
$$T_4 = .5(E_4 - E_3) + (1 - .5)T_3$$
$$= .5(109.9 - 108.8) + .5(2.6) = 1.9$$
$$S_4 = y_4/E_4 = 109.3/109.9 = .994$$

$$\vdots$$

The forecast for quarter I of 2003 (i.e., y_{17}) is given by

$$F_{17} = (E_{16} + T_{16})S_{17-4}$$
$$= (E_{16} + T_{16})S_{13} = (168.7 + 4.7)(1.044)$$
$$= 181.0$$

(Remember that beginning with $t = P + 3 = 7$, the formulas for E_t and S_t, shown in the box, are slightly different.) All the values of E_t, T_t, and S_t are given in Table 10.3. Similarly, the forecasts for y_{18}, y_{19}, and y_{20} (quarters II, III, and IV, respectively) are

$$F_{18} = (E_{16} + 2T_{16})S_{18-4}$$
$$= (E_{16} + 2T_{16})S_{14} = [168.7 + 2(4.7)](.959)$$
$$= 170.8$$
$$F_{19} = (E_{16} + 3T_{16})S_{19-4}$$
$$= (E_{16} + 3T_{16})S_{15} = [168.7 + 3(4.7)](1.095)$$
$$= 200.2$$
$$F_{20} = (E_{16} + 4T_{16})S_{20-4}$$
$$= (E_{16} + 4T_{16})S_{16} = [168.7 + 4(4.7)](.999)$$
$$= 187.3$$

◆

TABLE 10.3   Holt–Winters Components for Quarterly Power Load Data

YEAR	QUARTER	TIME t	POWER LOAD y_t	E_t $(w = .7)$	T_t $(v = .5)$	S_t $(u = .5)$
1999	I	1	103.5	—	—	—
	II	2	94.7	94.7	−8.8	1.000
	III	3	118.6	108.8	2.6	1.090
	IV	4	109.3	109.9	1.9	.994
2000	I	5	126.1	121.8	6.9	1.035
	II	6	116.0	119.8	2.5	.968
	III	7	141.2	127.4	5.1	1.100
	IV	8	131.6	132.3	5.0	.995
2001	I	9	144.5	138.9	5.8	1.038
	II	10	137.1	142.6	4.8	.965
	III	11	159.0	145.4	3.8	1.097
	IV	12	149.5	149.9	4.2	.996
2002	I	13	166.1	158.2	6.3	1.044
	II	14	152.5	160.0	4.1	.959
	III	15	178.2	162.9	3.5	1.095
	IV	16	169.0	168.7	4.7	.999

With any of these forecasting methods, forecast errors can be computed once the future values of the time series have been observed. Forecast error is defined as the difference between the actual future value and predicted value at time t, $(y_t - F_t)$. Aggregating the forecast errors into a summary statistic is useful for assessing the overall accuracy of the forecasting method. Formulas for three popular measures of forecast accuracy, the **mean absolute percentage error (MAPE)**, **mean absolute deviation (MAD)**, and **root mean squared error (RMSE)**, are given in the accompanying box. Both MAPE and MAD are summary measures for the "center" of the distribution of forecast errors, while RMSE is a measure of the "variation" in the distribution.

Measures of Overall Forecast Accuracy for m Forecasts

Mean absolute percentage error: $\text{MAPE} = \dfrac{\sum\limits_{t=1}^{m} \left| \dfrac{(y_t - F_t)}{y_t} \right|}{m} \times 100$

Mean absolute deviation: $\text{MAD} = \dfrac{\sum\limits_{t=1}^{m} |y_t - F_t|}{m}$

Root mean square error: $\text{RMSE} = \sqrt{\dfrac{\sum\limits_{t=1}^{m}(y_t - F_t)^2}{m}}$

EXAMPLE 10.2   Refer to the quarterly power load data, Table 10.1. The exponential smoothing and Holt–Winters forecasts for the four quarters of 2003 are listed in Table 10.4, as are the actual quarterly power loads (not previously given) for the year. Compute MAD and RMSE for each of the two forecasting methods. Which method yields more accurate forecasts?

TABLE 10.4   Forecasts and Actual Quarterly Power Loads for 2003

QUARTER	TIME	ACTUAL POWER LOAD	EXPONENTIAL SMOOTHING		HOLT–WINTERS	
	t	y_t	Forecast F_t	Error $(y_t - F_t)$	Forecast F_t	Error $(y_t - F_t)$
I	17	181.5	169.7	11.8	181.0	.5
II	18	175.2	169.7	5.5	170.8	4.4
III	19	195.0	169.7	25.3	200.2	−5.2
IV	20	189.3	169.7	19.6	187.3	2.0

Solution   The first step is to calculate the forecast errors, $y_t - F_t$, for each method. For example, for the exponential smoothing forecast of quarter I ($t = 17$), $y_{17} = 181.5$ and $F_{17} = 169.2$. Thus, the forecast error is $y_{17} - F_{17} = 181.5 - 169.2 = 12.3$. The forecast errors for the remaining exponential smoothing forecasts and the Holt–Winters forecasts are also shown in Table 10.4.

The MAD and RMSE calculations for each method are as follows:

Exponential smoothing:

$$\text{MAPE} = \left\{ \frac{\left|\frac{11.8}{81.5}\right| + \left|\frac{5.5}{175.2}\right| + \left|\frac{25.3}{195.0}\right| + \left|\frac{19.6}{189.3}\right|}{4} \right\} \times 100 = 8.24\%$$

$$\text{MAD} = \frac{|11.8| + |5.5| + |25.3| + |19.6|}{4} = 15.55$$

$$\text{RMSE} = \sqrt{\frac{(11.8)^2 + (5.5)^2 + (25.3)^2 + (19.6)^2}{4}} = 17.27$$

Holt–Winters:

$$\text{MAPE} = \left\{ \frac{\left|\frac{.5}{181.5}\right| + \left|\frac{4.4}{175.2}\right| + \left|\frac{-5.2}{195.0}\right| + \left|\frac{2.0}{189.3}\right|}{4} \right\} \times 100 = 1.63\%$$

$$\text{MAD} = \frac{|.5| + |4.4| + |-5.2| + |2.0|}{4} = 3.03$$

$$\text{RMSE} = \sqrt{\frac{(.5)^2 + (4.4)^2 + (-5.2)^2 + (2.0)^2}{4}} = 3.56$$

The Holt–Winters values of MAPE, MAD, and RMSE are each about one-fifth of the corresponding exponential smoothed values. Overall, the Holt–Winters method clearly leads to more accurate forecasts than exponential smoothing. This, of course, is expected since the Holt–Winters method accounts for both long-term and seasonal variation in the power loads, whereas exponentially smoothing does not. ◆

[*Note*: Most statistical software packages will automatically compute the values of MAPE, MAD, and RMSE (also called the *mean squared deviation, MSD*) for all n observations in the data set. For example, see the highlighted portion at the top of the MINITAB printout, Figure 10.4.]

We conclude this section with a comment. A major disadvantage of forecasting with smoothing techniques (the moving average method, exponential smoothing, or the Holt–Winters models) is that no measure of the forecast error (or reliability) is known *prior* to observing the future value. Although forecast errors can be calculated *after* the future values of the time series have been observed (as in Example 10.2), we prefer to have some measure of the accuracy of the forecast *before* the actual values are observed. One option is to compute forecasts and forecast errors for all n observations in the data set and use these "past" forecast errors to estimate the standard deviation of all forecast errors, i.e., the *standard error of the forecast*. A rough estimate of this standard error is the value of RMSE, and an approximate 95% prediction interval for any future forecast is

$$F_t \pm 2(\text{RMSE})$$

(An interval like this is shown at the bottom of the MINITAB printout, Figure 10.4.) However, because the theoretical distributional properties of the forecast errors with smoothing methods are unknown, many analysts regard smoothing methods as descriptive procedures rather than as inferential ones.

In the preceding chapters, we learned that predictions with inferential regression models are accompanied by well-known measures of reliability. The standard errors of the predicted values allow us to construct 95% prediction intervals. We discuss inferential time series forecasting models in the remaining sections of this chapter.

EXERCISES

10.1. The quarterly numbers of housing starts (in thousands of dwellings) in the United States from 1997 through 2001 are recorded in the accompanying table.

QTRHOUSE

YEAR	QUARTER	HOUSING STARTS
1997	1	297.3
	2	419.0
	3	400.3
	4	357.5
1998	1	324.9
	2	447.8

(continued)

1998	3	445.0
	4	399.3
1999	1	369.5
	2	454.5
	3	453.5
	4	389.0
2000	1	364.5
	2	453.4
	3	405.3
	4	357.5
2001	1	347.8
	2	460.5
	3	429.2
	4	365.8

Source: Standard & Poor's Statistical Service: Current Statistics, Mar. 2002.

a. Plot the quarterly time series. Can you detect a long-term trend? Can you detect any seasonal variation?

b. Calculate the 4-point moving average for the quarterly housing starts.

c. Graph the 4-point moving average on the same set of axes you used for the graph in part **a**. Is the long-term trend more evident? What effects has the moving average method removed or smoothed?

d. Calculate the seasonal index for the number of housing starts in quarter I.

e. Calculate the seasonal index for the number of housing starts in quarter II.

f. Use the moving average method to forecast the number of housing starts in quarters I and II of 2002.

10.2. Refer to the quarterly housing starts data in Exercise 10.1.

a. Calculate the exponentially smoothed series for housing starts using a smoothing constant of $w = .2$.

b. Use the exponentially smoothed series from part **a** to forecast the number of housing starts in the first two quarters of 2002.

c. Use the Holt–Winters forecasting model with both trend and seasonal components to forecast the number of housing starts in the first two quarters of 1994. Use smoothing constants $w = .2$, $v = .5$, and $u = .7$.

10.3. Refer to Exercises 10.1 and 10.2. Suppose the actual numbers of housing starts (in thousands) for quarters I and II of 2002 are 358.2 and 470.3, respectively.

a. Compare the accuracy of the moving average, exponential smoothing, and Holt–Winters forecasts using MAD.

b. Repeat part a using RMSE.

c. Comment on which forecasting method is more accurate.

10.4. The data in the table below are the amounts of crude oil (millions of barrels) imported into the United States from the Organization of Petroleum Exporting Countries (OPEC) for the years 1974–2000.

a. Plot the yearly time series. Can you detect a long-term trend?

b. Calculate and plot a 3-point moving average for annual OPEC oil imports.

c. Calculate and plot the exponentially smoothed series for annual OPEC oil imports using a smoothing constant of $w = .3$.

d. Forecast OPEC oil imports in 2001 using the moving average method.

e. Forecast OPEC oil imports in 2001 using exponential smoothing with $w = .3$.

f. Forecast OPEC oil imports in 2001 using the Holt–Winters forecasting model with trend. Use smoothing constants $w = .3$ and $v = .8$.

g. Actual OPEC crude oil imports in 2001 totaled 1,700 million barrels. Calculate the errors of the forecast, parts **d–f**. Which method yields the most accurate short-term forecast?

10.5. The Consumer Price Index (CPI) measures the increase (or decrease) in the prices of goods and services relative to a base year. The CPI for the years 1987–2000 (using 1984 as a base period) is shown in the table on p. 483.

🔵 OPECOIL

YEAR	t	IMPORTS, y_t	YEAR	t	IMPORTS, y_t	YEAR	t	IMPORTS, y_t
1974	1	926	1984	11	553	1994	21	1,307
1975	2	1,171	1985	12	479	1995	22	1,303
1976	3	1,663	1986	13	771	1996	23	1,258
1977	4	2,058	1987	14	876	1997	24	1,378
1978	5	1,892	1988	15	987	1998	25	1,522
1979	6	1,866	1989	16	1,232	1999	26	1,543
1980	7	1,414	1990	17	1,282	2000	27	1,664
1981	8	1,067	1991	18	1,233			
1982	9	633	1992	19	1,247			
1983	10	540	1993	20	1,339			

Source: Statistical Abstract of the United States, U.S. Bureau of the Census, 2001.

CPI

YEAR	CPI
1987	109.3
1988	113.8
1989	119.4
1990	125.8
1991	129.1
1992	132.8
1993	136.8
1994	147.8
1995	152.4
1996	156.9
1997	160.5
1998	163.0
1999	166.6
2000	171.5

Source: *Survey of Current Business*, U.S. Department of Commerce, Bureau of Economic Analysis.

SP500

YEAR	QUARTER	S&P 500
1995	1	500.7
	2	544.8
	3	584.4
	4	615.9
1996	1	645.5
	2	670.6
	3	687.3
	4	740.7
1997	1	757.1
	2	885.1
	3	947.3
	4	970.4
1998	1	1101.7
	2	1133.8
	3	1017.0
	4	1229.2
1999	1	1286.4
	2	1372.7
	3	1282.7
	4	1469.2
2000	1	1498.6
	2	1454.6
	3	1436.5
	4	1320.3
2001	1	1160.3
	2	1224.4
	3	1040.9
	4	1148.1

Source: *Standard & Poor's Statistical Service: Current Statistics*, Mar. 2002.

a. Graph the time series. Do you detect a long-term trend?

b. Calculate and plot a 5-point moving average for the CPI. Use the moving average to forecast the CPI in 2003.

c. Calculate and plot the exponentially smoothed series for the CPI using a smoothing constant of $w = .4$. Use the exponentially smoothed values to forecast the CPI in 2003.

d. Use the Holt–Winters forecasting model with trend to forecast the CPI in 2003. Use smoothing constants $w = .4$ and $v = .5$.

10.6. Standard & Poor's 500 Composite Stock Index (S&P 500) is a stock market index. Like the Dow Jones Industrial Average, it is an indicator of stock market activity. The next table contains end-of-quarter values of the S&P 500 for the years 1995–2001.

a. Calculate a 4-point moving average for the quarterly S&P 500.

b. Plot the quarterly index and the 4-point moving average on the same graph. Can you identify the long-term trend of the time series? Can you identify any seasonal variations about the secular trend?

c. Use the moving average method to forecast the S&P 500 for the 1st quarter of 2002.

d. Calculate and plot the exponentially smoothed series for the quarterly S&P 500 using a smoothing constant of $w = .3$.

e. Use the exponential smoothing technique with $w = .3$ to forecast the S&P 500 for the 1st quarter of 2002.

f. Use the Holt–Winters forecasting model with trend and seasonal components to forecast the S&P 500 for the 1st quarter of 2002. Use smoothing constants $w = .3$, $v = .8$, and $u = .5$.

10.7. Consider the time series of gold prices recorded in the table on p. 484. (Gold prices are given in dollars per troy ounce.)

GOLD

YEAR	PRICE OF GOLD
1971	41.25
1972	58.61
1973	97.81
1974	159.70
1975	161.40
1976	124.80
1977	148.30
1978	193.50
1979	307.80
1980	606.01
1981	450.63
1982	374.18
1983	449.03
1984	360.29
1985	317.30
1986	367.87
1987	408.91
1988	436.93
1989	381.21
1990	384.07
1991	362.11
1992	343.82
1993	359.77
1994	384.00
1995	383.79
1996	387.81

(*continued*)

1997	331.02
1998	294.24
1999	278.98
2000	279.11
2001	271.04

Source: *Survey of Current Business*, U.S. Department of Commerce, Bureau of Economic Analysis; www.kitco.com

a. Calculate a 3-point moving average for the gold price time series. Plot the gold prices and the 3-point moving average on the same graph. Can you detect the long-term trend and any cyclical patterns in the time series?

b. Use the moving averages to forecast the price of gold in 1999, 2000, and 2001.

c. Calculate and plot the exponentially smoothed gold price series using a smoothing constant of $w = .8$.

d. Use the exponentially smoothed series to forecast the price of gold in 1999, 2000, and 2001.

e. Use the Holt–Winters forecasting model with trend to forecast the price of gold, 1999–2001. Use smoothing constants $w = .8$ and $v = .4$.

f. Use the actual gold prices in 1999–2001 to assess the accuracy of the three forecasting methods, parts **b**, **d**, and **e**.

10.4 Forecasting: The Regression Approach

Many firms use past sales to forecast future sales. Suppose a wholesale distributor of sporting goods is interested in forecasting its sales revenue for each of the next 5 years. Since an inaccurate forecast may have dire consequences to the distributor, some measure of the forecast's reliability is required. To make such forecasts and assess their reliability, an **inferential time series forecasting model** must be constructed. The familiar general linear regression model of Chapter 4 represents one type of inferential model since it allows us to calculate prediction intervals for the forecasts.

To illustrate the technique of forecasting with regression, we'll reconsider the data on annual sales (in thousands of dollars) for a firm (say, the sporting goods distributor) in each of its 35 years of operation. The data, first presented in Table 8.6, is reproduced in Table 10.5. A SAS plot of the data (Figure 10.6) reveals a linearly increasing trend, so the first-order (straight-line) model

$$E(y_t) = \beta_0 + \beta_1 t$$

seems plausible for describing the secular trend. The SAS printout for the model is shown in Figure 10.7. The model apparently provides an excellent fit to the data, with $R^2 = .98$, $F = 1{,}615.724$ (p-value $< .0001$), and $s = 6.38524$. The least squares

SALES 35

TABLE 10.5 A Firm's Yearly Sales Revenue (thousands of dollars)

t	y_t	t	y_t	t	y_t
1	4.8	13	48.4	25	100.3
2	4.0	14	61.6	26	111.7
3	5.5	15	65.6	27	108.2
4	15.6	16	71.4	28	115.5
5	23.1	17	83.4	29	119.2
6	23.3	18	93.6	30	125.2
7	31.4	19	94.2	31	136.3
8	46.0	20	85.4	32	146.8
9	46.1	21	86.2	33	146.1
10	41.9	22	89.9	34	151.4
11	45.5	23	89.2	35	150.9
12	53.5	24	99.1		

prediction equation, whose coefficients are shaded in Figure 10.7, is

$$\hat{y}_t = \hat{\beta}_0 + \hat{\beta}_1 t = .401513 + 4.295630t$$

FIGURE 10.6

SAS scatterplot of sales data

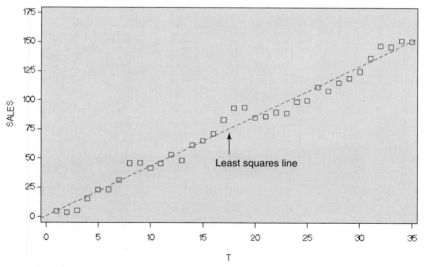

We can obtain sales forecasts and corresponding 95% prediction intervals for years 36–40 by using the formulas given in Section 3.9 or using statistical software These values are given in the bottom portion of the SAS printout shown in Figure 10.7. For example, for $t = 36$, we have $\hat{y}_{36} = 155.0442$ with the 95% prediction interval (141.3001, 168.7883). That is, we predict that sales revenue in year $t = 36$ will fall between $141,300 and $168,788 with 95% confidence.

Note that the prediction intervals for $t = 36, 37, \ldots, 40$ widen as we attempt to forecast farther into the future. Intuitively, we know that the farther into the future we forecast, the less certain we are of the accuracy of the forecast since some unexpected change in business and economic conditions may make the model

FIGURE 10.7

SAS printout for straight-line
model of yearly sales revenue

Dependent Variable: SALES

Analysis of Variance

Source	DF	Sum of Squares	Mean Square	F Value	Pr > F
Model	1	65875	65875	1615.72	<.0001
Error	33	1345.45355	40.77132		
Corrected Total	34	67221			

Root MSE	6.38524	R-Square	0.9800	
Dependent Mean	77.72286	Adj R-Sq	0.9794	
Coeff Var	8.21540			

Parameter Estimates

Variable	DF	Parameter Estimate	Standard Error	t Value	Pr > |t|
Intercept	1	0.40151	2.20571	0.18	0.8567
T	1	4.29563	0.10687	40.20	<.0001

Dependent Variable: SALES

Output Statistics

Obs	T	Dep Var SALES	Predicted Value	Std Error Mean Predict	95% CL Predict		Residual
1	1	4.8000	4.6971	2.1132	-8.9867	18.3809	0.1029
2	2	4.0000	8.9928	2.0220	-4.6339	22.6195	-4.9928
3	3	5.5000	13.2884	1.9325	-0.2844	26.8612	-7.7884
4	4	15.6000	17.5840	1.8448	4.0618	31.1062	-1.9840
5	5	23.1000	21.8797	1.7593	8.4047	35.3546	1.2203
6	6	23.3000	26.1753	1.6761	12.7443	39.6063	-2.8753
7	7	31.4000	30.4709	1.5959	17.0805	43.8614	0.9291
8	8	46.0000	34.7666	1.5189	21.4132	48.1199	11.2334
9	9	46.1000	39.0622	1.4457	25.7425	52.3819	7.0378
10	10	41.9000	43.3578	1.3769	30.0683	56.6473	-1.4578
11	11	45.5000	47.6534	1.3132	34.3907	60.9162	-2.1534
12	12	53.5000	51.9491	1.2554	38.7095	65.1887	1.5509
13	13	48.4000	56.2447	1.2043	43.0248	69.4646	-7.8447
14	14	61.6000	60.5403	1.1609	47.3365	73.7442	1.0597
15	15	65.6000	64.8360	1.1259	51.6447	78.0273	0.7640
16	16	71.4000	69.1316	1.1003	55.9493	82.3139	2.2684
17	17	83.4000	73.4272	1.0846	60.2503	86.6042	9.9728
18	18	93.6000	77.7229	1.0793	64.5477	90.8980	15.8771
19	19	94.2000	82.0185	1.0846	68.8415	95.1954	12.1815
20	20	85.4000	86.3141	1.1003	73.1318	99.4964	-0.9141
21	21	86.2000	90.6097	1.1259	77.4185	103.8010	-4.4097
22	22	89.9000	94.9054	1.1609	81.7016	108.1092	-5.0054
23	23	89.2000	99.2010	1.2043	85.9811	112.4209	-10.0010
24	24	99.1000	103.4966	1.2554	90.2571	116.7362	-4.3966
25	25	100.3000	107.7923	1.3132	94.5295	121.0550	-7.4923
26	26	111.7000	112.0879	1.3769	98.7984	125.3774	-0.3879
27	27	108.2000	116.3835	1.4457	103.0639	129.7032	-8.1835
28	28	115.5000	120.6792	1.5189	107.3258	134.0325	-5.1792
29	29	119.2000	124.9748	1.5959	111.5843	138.3653	-5.7748
30	30	125.2000	129.2704	1.6761	115.8394	142.7014	-4.0704
31	31	136.3000	133.5661	1.7593	120.0911	147.0410	2.7339
32	32	146.8000	137.8617	1.8448	124.3395	151.3839	8.9383
33	33	146.1000	142.1573	1.9325	128.5845	155.7301	3.9427
34	34	151.4000	146.4529	2.0220	132.8263	160.0796	4.9471
35	35	150.9000	150.7486	2.1132	137.0648	164.4324	0.1514
36	36	.	155.0442	2.2057	141.3001	168.7883	.
37	37	.	159.3398	2.2995	145.5322	173.1474	.
38	38	.	163.6355	2.3944	149.7613	177.5097	.
39	39	.	167.9311	2.4903	153.9872	181.8750	.
40	40	.	172.2267	2.5870	158.2101	186.2433	.

inappropriate. Since we have less confidence in the forecast for, say, $t = 40$ than for $t = 36$, it follows that the prediction interval for $t = 40$ must be wider to attain a 95% level of confidence. For this reason, time series forecasting (regardless of the forecasting method) is generally confined to the short term.

Multiple regression models can also be used to forecast future values of a time series with seasonal variation. We illustrate with an example.

EXAMPLE 10.3

Refer to the 1999–2002 quarterly power loads listed in Table 10.1.

a. Propose a model for quarterly power load, y_t, that will account for both the secular trend and seasonal variation present in the series.

b. Fit the model to the data, and use the least squares prediction equation to forecast the utility company's quarterly power loads in 1995. Construct 95% prediction intervals for the forecasts.

Solution

a. A common way to describe seasonal differences in a time series is with dummy variables.* For quarterly data, a model that includes both trend and seasonal components is

$$E(y_t) = \beta_0 + \underbrace{\beta_1 t}_{\substack{\text{Secular} \\ \text{trend}}} + \underbrace{\beta_2 Q_1 + \beta_3 Q_2 + \beta_4 Q_3}_{\text{Seasonal component}}$$

where

$t = $ Time period, ranging from $t = 1$ for quarter I of 1999 to $t = 16$ for quarter IV of 2002

$y_t = $ Power load (megawatts) in time t

$$Q_1 = \begin{cases} 1 & \text{if quarter I} \\ 0 & \text{if not} \end{cases} \qquad Q_2 = \begin{cases} 1 & \text{if quarter II} \\ 0 & \text{if not} \end{cases}$$

$$Q_3 = \begin{cases} 1 & \text{if quarter III} \\ 0 & \text{if not} \end{cases} \qquad \text{Base level} = \text{quarter IV}$$

The β coefficients associated with the seasonal dummy variables determine the mean increase (or decrease) in power load for each quarter, relative to the base level quarter, quarter IV.

b. The model is fit to the data from Table 10.1 using the SAS multiple regression routine. The resulting SAS printout is shown in Figure 10.8. Note that the model appears to fit the data quite well: $R^2 = .9972$, indicating that the model accounts for 99.7% of the sample variation in power loads over the 4-year period; $F = 968.96$ strongly supports the hypothesis that the model has predictive utility (p-value $< .0001$); and the standard deviation, **Root MSE** $= 1.53242$, implies that the model predictions will usually be accurate to within approximately $\pm 2(1.53)$, or about ± 3.06 megawatts.

---

*Another way to account for seasonal variation is with trigonometric (sine and cosine) terms. We discuss seasonal models with trigonometric terms in Section 10.7.

FIGURE 10.8

SAS printout for model of quarterly power loads

Dependent Variable: POWLOAD

Analysis of Variance

Source	DF	Sum of Squares	Mean Square	F Value	Pr > F
Model	4	9101.67800	2275.41950	968.96	<.0001
Error	11	25.83138	2.34831		
Corrected Total	15	9127.50938			

Root MSE	1.53242	R-Square	0.9972	
Dependent Mean	137.30625	Adj R-Sq	0.9961	
Coeff Var	1.11606			

Parameter Estimates

Variable	DF	Parameter Estimate	Standard Error	t Value	Pr > |t|
Intercept	1	90.20625	1.14931	78.49	<.0001
T	1	4.96438	0.08566	57.95	<.0001
Q1	1	10.09313	1.11364	9.06	<.0001
Q2	1	-4.84625	1.09704	-4.42	0.0010
Q3	1	14.36438	1.08696	13.22	<.0001

Output Statistics

Obs	YEAR_QTR	Dep Var POWLOAD	Predicted Value	Std Error Mean Predict	95% CL Predict		Residual
1	1999_1	103.5000	105.2638	0.9226	101.3268	109.2007	-1.7637
2	1999_2	94.7000	95.2887	0.9226	91.3518	99.2257	-0.5887
3	1999_3	118.6000	119.4637	0.9226	115.5268	123.4007	-0.8637
4	1999_4	109.3000	110.0637	0.9226	106.1268	114.0007	-0.7637
5	2000_1	126.1000	125.1213	0.7851	121.3315	128.9110	0.9787
6	2000_2	116.0000	115.1463	0.7851	111.3565	118.9360	0.8538
7	2000_3	141.2000	139.3213	0.7851	135.5315	143.1110	1.8788
8	2000_4	131.6000	129.9212	0.7851	126.1315	133.7110	1.6788
9	2001_1	144.5000	144.9788	0.7851	141.1890	148.7685	-0.4788
10	2001_2	137.1000	135.0038	0.7851	131.2140	138.7935	2.0963
11	2001_3	159.0000	159.1788	0.7851	155.3890	162.9685	-0.1788
12	2001_4	149.5000	149.7787	0.7851	145.9890	153.5685	-0.2787
13	2002_1	166.1000	164.8363	0.9226	160.8993	168.7732	1.2637
14	2002_2	152.5000	154.8613	0.9226	150.9243	158.7982	-2.3613
15	2002_3	178.2000	179.0363	0.9226	175.0993	182.9732	-0.8363
16	2002_4	169.0000	169.6362	0.9226	165.6993	173.5732	-0.6362
17	2003_1	.	184.6938	1.1493	180.4777	188.9098	.
18	2003_2	.	174.7188	1.1493	170.5027	178.9348	.
19	2003_3	.	198.8938	1.1493	194.6777	203.1098	.
20	2003_4	.	189.4937	1.1493	185.2777	193.7098	.

Forecasts and corresponding 95% prediction intervals for the 2003 power loads are reported in the bottom portion of the printout in Figure 10.8. For example, the forecast for power load in quarter I of 2003 is 184.7 megawatts with the 95% prediction interval (180.5, 188.9). Therefore, using a 95% prediction interval, we expect the power load in quarter I of 2003 to fall between 180.5 and 188.9 megawatts. Recall from Table 10.4 in Example 10.2 that the actual 2003 quarterly power loads are 181.5, 175.2, 195.0, and 189.3, respectively. Note that each of these falls within its respective 95% prediction interval shown in Figure 10.8. ◆

Many descriptive forecasting techniques have proved their merit by providing good forecasts for particular applications. Nevertheless, the advantage of forecasting using the regression approach is clear: Regression analysis provides us with a measure of reliability for each forecast through prediction intervals. However, there are two problems associated with forecasting time series using a multiple regression model.

PROBLEM 1: We are using the least squares prediction equation to forecast values outside the region of observation of the independent variable, t. For example, in Example 10.3, we are forecasting for values of t between 17 and 20 (the four quarters of 2003), even though the observed power loads are for t values between 1 and 16. As noted in Chapter 7, it is risky to use a least squares regression model for prediction outside the range of the observed data because some unusual change—economic, political, etc.—may make the model inappropriate for predicting future events. Because forecasting always involves predictions about future values of a time series, this problem obviously cannot be avoided. However, it is important that the forecaster recognize the dangers of this type of prediction.

PROBLEM 2: Recall the standard assumptions made about the random error component of a multiple regression model (Section 4.2). We assume that the errors have mean 0, constant variance, normal probability distributions, and are *independent*. The latter assumption is often violated in time series that exhibit short-term trends. As an illustration, refer to the plot of the sales revenue data shown in Figure 10.6. Notice that the observed sales tend to deviate about the least squares line in positive and negative runs. That is, if the difference between the observed sales and predicted sales in year t is positive (or negative), the difference in year $t + 1$ tends to be positive (or negative). Since the variation in the yearly sales is systematic, the implication is that the errors are correlated. In fact, the Durbin–Watson test for correlated errors (see Section 8.6.) supports this inference. Violation of this standard regression assumption could lead to unreliable forecasts.

Time series models have been developed specifically for the purpose of making forecasts when the errors are known to be correlated. These models include an **autoregressive term** for the correlated errors that result from cyclical, seasonal, or other short-term effects. Time series autoregressive models are the subject of Sections 10.5–10.11.

EXERCISES

10.8. The accompanying table records the acreage of wheat (in thousands) harvested in the United States for the period 1981–2000. A farmers' marketing cooperative is interested in detecting the long-term trend of the wheat harvest.

◎ WHEAT

YEAR	t	WHEAT HARVESTED(1,000 ACRES)
1981	1	88
1982	2	86
1983	3	76
1984	4	79
1985	5	75
1986	6	72
1987	7	66
1988	8	66
1989	9	77
1990	10	77
1991	11	70
1992	12	72
1993	13	72
1994	14	70
1995	15	69
1996	16	75
1997	17	70
1998	18	66

Source: United States Crop Production, 1998 Report, USDA, National Agricultural Statistical Service.

a. Graph the wheat harvest time series.

b. Propose a model for the long-term linear trend of the time series.

c. Fit the model, using the method of least squares. Plot the least squares line on the graph of part **a**. Can you identify the long-term trend?

d. How well does the linear model describe the long-term trend? [*Hint:* Check the value of R^2.]

e. Use the least squares model to forecast the volume of wheat harvested in 1999. Construct a 95% prediction interval for the forecast.

10.9. A realtor working in a large city wants to identify the secular trend in the weekly number of single-family houses sold by her firm. For the past 15 weeks she has collected data on her firm's home sales, as shown in the table at the bottom of the page.

a. Plot the time series. Is there visual evidence of a quadratic trend?

b. The realtor hypothesizes the model $E(y_t) = \beta_0 + \beta_1 t + \beta_2 t^2$ for the secular trend of the weekly time series. Fit the model to the data, using the method of least squares.

c. Plot the least squares model on the graph of part **a**. How well does the quadratic model describe the secular trend?

d. Use the model to forecast home sales in week 16 with a 95% prediction interval.

10.10. Refer to the quarterly S&P 500 values given in Exercise 10.6 (p. 483).

a. Hypothesize a time series model to account for trend and seasonal variation.

b. Fit the model in part **a** to the data.

c. Use the least squares model from part **b** to forecast the S&P 500 for all four quarters of 2002. Obtain 95% prediction intervals for the forecasts.

10.11. Information on intercity passenger traffic (excluding travel by private automobiles) from 1940 to 2000 is given in the table on p. 491. The data are recorded as percentages of total passenger-miles traveled.

◎ HOMESALES

WEEK t	HOMES SOLD y_t	WEEK t	HOMES SOLD y_t	WEEK t	HOMES SOLD y_t
1	59	6	137	11	88
2	73	7	106	12	75
3	70	8	122	13	62
4	82	9	93	14	44
5	115	10	86	15	45

⊙ INTERCITY

YEAR	TIME	RAILROADS	BUSES	AIR CARRIERS
1940	1	67.1	26.5	2.8
1945	2	74.3	21.4	2.7
1950	3	46.3	37.7	14.3
1955	4	36.5	32.4	28.9
1960	5	28.6	25.7	42.1
1965	6	17.9	24.2	54.7
1970	7	7.3	16.9	73.1
1975	8	5.8	14.2	77.7
1980	9	4.7	11.4	83.9
1985	10	3.6	7.9	88.4
1990	11	3.3	5.8	90.9
1995	12	3.1	6.1	90.8
2000	13	2.3	5.7	92.0

Source: *Statistical Abstract of the United States*, 2000. Interstate Commerce Commission, Civil Aeronautics Board.

a. Let y_t be the percentage of total passenger-miles at time t for a particular mode of transportation. Consider the linear model $E(y_t) = \beta_0 + \beta_1 t$. Which modes of transportation do you think have a secular trend adequately represented by this model?

b. Fit the model in part **a** to the data for each mode of transportation, using the method of least squares.

c. Plot the data and the least squares line for each mode of transportation. Which models adequately describe the secular trend of percentage of total passenger-miles traveled? Does this agree with your answer to part **a**?

d. Refer to your answer for part **c**. Use the least squares prediction equations to forecast the percentage of total passenger-miles to be traveled for the respective modes of transportation in 2005. Obtain 95% prediction intervals. What are the risks associated with this forecasting procedure?

10.12. The average annual price (in dollars per metric ton) of steel from 1971 to 1998 is shown in the table.

⊙ STEEL

YEAR	t	STEEL PRICE
1971	1	168.70
1972	2	185.80
1973	3	207.20
1974	4	244.70
1975	5	288.80
1976	6	313.30
1977	7	343.40
1978	8	395.90

(continued)

1979	9	441.10
1980	10	477.40
1981	11	534.00
1982	12	557.10
1983	13	577.40
1984	14	602.10
1985	15	608.50
1986	16	546.70
1987	17	560.00
1988	18	560.00
1989	19	555.60
1990	20	577.50
1991	21	600.20
1992	22	531.20
1993	23	539.80
1994	24	570.30
1995	25	592.30
1996	26	633.90
1997	27	655.90
1998	28	625.20

Source: U.S. Geological Survey, *Minerals Information*, 1999.

a. Plot the time series. Is there visual evidence of a linear trend? A quadratic trend? Propose a regression model that is likely to fit the data well.

b. Fit the model, part **a**, to the data, using the method of least squares.

c. Plot the least squares prediction equation on the graph of part **a**. How well does the model describe the time series?

d. Use the fitted least squares model to forecast the price of steel for the years 1999–2002. Obtain 95% prediction intervals for the forecasts and verify that the width of the interval increases the farther you forecast into the future.

10.13. An analysis of seasonality in returns of stock traded on the London Stock Exchange was published in the *Journal of Business* (Vol. 60, 1987). One of the objectives was to determine whether the introduction of a capital gains tax in 1965 affected rates of return. The following model was fitted to data collected over the years 1956–1980:

$$y_t = \beta_0 + \beta_1 D_1 + \varepsilon_t$$

where y_t is the difference between the April rates of return of the two stocks on the exchange with the largest and smallest returns in year t, and D_t is a dummy variable that takes on the value 1 in the posttax period (1966–1980) and the value 0 in the pretax period (1956–1965).

a. Interpret the value of β_1.

b. Interpret the value of β_0.

YEAR ONE	ROOMS OCCUPIED (%)		YEAR TWO	ROOMS OCCUPIED (%)	
MONTH	ATLANTA	PHOENIX	MONTH	ATLANTA	PHOENIX
January	59	67	January	64	72
February	63	85	February	69	91
March	68	83	March	73	87
April	70	69	April	67	75
May	63	63	May	68	70
June	59	52	June	71	61
July	68	49	July	67	46
August	64	49	August	71	44
September	62	56	September	65	63
October	73	69	October	72	73
November	62	63	November	63	71
December	47	48	December	47	51

Source: Trends in the Hotel Industry.

c. The least squares prediction equation was found to be $\hat{y}_t = -.55 + 3.08D_t$. Use the equation to estimate the mean difference in April rates of returns of the two stocks during the pretax period.

d. Repeat part c for the posttax period.

e. Obtain a forecast of the difference in April rates of return of the two stocks in 1995.

10.14. A traditional indicator of the economic health of the accommodations (hotel–motel) industry is the trend in room occupancy. Average monthly occupancies for 2 recent years are given in the above table for hotels and motels in the cities of Atlanta, Georgia, and Phoenix, Arizona. Let y_t = occupancy rate for Phoenix in month t.

a. Propose a model for $E(y_t)$ that accounts for possible seasonal variation in the monthly series. [*Hint:* Consider a model with dummy variables for the 12 months, January, February, etc.]

b. Fit the model of part a to the data.

c. Test the hypothesis that the monthly dummy variables are useful predictors of occupancy rate. [*Hint:* Conduct a partial F test.]

d. Use the fitted least squares model from part b to forecast the Phoenix occupancy rate in January of year 3 with a 95% prediction interval.

e. Repeat parts a–d for the Atlanta monthly occupancy rates.

10.15. The Employee Retirement Income Security Act (ERISA) of 1974 was originally established to enhance retirement security income. J. Ledolter (University of

Iowa) and M. L. Power (Iowa State University) investigated the effects of ERISA on the growth in the number of private retirement plans (*Journal of Risk and Insurance*, Dec. 1983). Using quarterly data from 1956 through the third quarter of 1982 ($n = 107$ quarters), Ledolter and Power fitted quarterly time series models for the number of pension qualifications and the number of profit-sharing plan qualifications. One of several models investigated was the quadratic model $E(y_t) = \beta_0 + \beta_1 t + \beta_2 t^2$, where y_t is the logarithm of the dependent variable (number of pension or number of profit-sharing qualifications) in quarter t. The results (modified for the purpose of this exercise) are summarized here:

Pension plan qualifications:

$$\hat{y}_t = 6.19 + .039t - .00024t^2$$

$$t(\text{for } H_0: \beta_2 = 0) = -1.39$$

Profit-sharing plan qualifications:

$$\hat{y}_t = 6.22 + .035t - .00021t^2$$

$$t(\text{for } H_0: \beta_2 = 0) = -1.61$$

a. Is there evidence that the quarterly number of pension plan qualifications increases at a decreasing rate over time? Test using $\alpha = .05$. [*Hint:* Test $H_0: \beta_2 = 0$ against $H_a: \beta_2 < 0$.]

b. Forecast the number of pension plan qualifications for the fourth quarter of 1982 (i.e., $t = 108$). [*Hint:* Since y_t is the logarithm of the number of pension plan qualifications, to obtain the forecast you must take the antilogarithm of $\hat{y}_{108}$, i.e., $e^{\hat{y}_{108}}$.]

c. Is there evidence that the quarterly number of profit-sharing plan qualifications increases at a decreasing rate over time? Test using $\alpha = .05$. [*Hint:* Test $H_0: \beta_2 = 0$ against $H_a: \beta_2 < 0$.]

d. Forecast the number of profit-sharing plan qualifications for the fourth quarter of 1982 (i.e., $t = 108$). [*Hint:* Since y_t is the logarithm of the number of profit-sharing plan qualifications, to obtain the forecast you must take the antilogarithm of $\hat{y}_{108}$, i.e., $e^{\hat{y}_{108}}$.]

10.5 Autocorrelation and Autoregressive Error Models

In Chapter 8, we presented the Durbin–Watson test for detecting correlated residuals in a regression analysis. Correlated residuals are quite common when the response is a *time series* variable. Correlation of residuals for a regression model with a time series response is called **autocorrelation**, because the correlation is between residuals from the *same* time series model at different points in time.

A special case of autocorrelation that has many applications to business and economic phenomena is the case in which neighboring residuals one time period apart (say, at times t and $t + 1$) are correlated. This type of correlation is called **first-order autocorrelation**. In general, correlation between time series residuals m time periods apart is mth-order autocorrelation.

Definition 10.7

Autocorrelation is the correlation between time series residuals at different points in time. The special case in which neighboring residuals one time period apart (at times t and $t + 1$) are correlated is called **first-order autocorrelation**.

To see how autocorrelated residuals affect the regression model, we will assume a model similar to the linear statistical model of Chapter 4,

$$y_t = E(y_t) + R_t$$

where $E(y_t)$ is the regression model

$$E(y_t) = \beta_0 + \beta_1 x_1 + \cdots + \beta_k x_k$$

and R_t represents the random residual. We assume that the residual R_t has mean 0 and constant variance σ^2, but that it is autocorrelated. The effect of autocorrelation on the general linear model depends on the pattern of the autocorrelation. One of the most common patterns is that the autocorrelation between residuals at consecutive time points is positive. Thus, when the residual at time t, R_t, indicates that the observed value y_t is more than the mean value $E(y_t)$, then the residual at time $(t + 1)$ will have a tendency (probability greater than .5) to be positive. This would occur, for example, if you were to model a monthly economic index (e.g., the Consumer Price Index) with a straight-line model. In times of recession, the observed values of the index will tend to be less than the predictions of a straight line for most or all of the months during the period. Similarly, in extremely inflationary periods, the residuals are likely to be positive because the observed value of the index will lie above the straight-line model. In either case, the fact that residuals at consecutive time points tend to have the same sign implies that they are **positively correlated**.

A second property commonly observed for autocorrelated residuals is that the size of the autocorrelation between values of the residual R at two different points in time diminishes rapidly as the distance between the time points increases. Thus, the autocorrelation between R_t and R_{t+m} becomes smaller (i.e., weaker) as the distance m between the time points becomes larger.

A residual model that possesses this property—positive autocorrelation diminishing rapidly as distance between time points increases—is the **first-order autoregressive error model**:

$$R_t = \phi R_{t-1} + \varepsilon_t, \quad -1 < \phi < 1$$

where ε_t, a residual called **white noise**, is uncorrelated with any and all other residual components. Thus, the value of the residual R_t is equal to a constant multiple, ϕ (Greek letter "phi"), of the previous residual, R_{t-1}, plus random error. In general, the constant ϕ is between -1 and $+1$, and the numerical value of ϕ determines the sign (positive or negative) and strength of the autocorrelation. In fact, it can be shown (proof omitted) that the autocorrelation (abbreviated AC) between two residuals that are m time units apart, R_t, and R_{t+m}, is

$$AC(R_t, R_{t+m}) = \phi^m$$

Since the absolute value of ϕ will be less than 1, the autocorrelation between R_t and R_{t+m}, ϕ^m, will decrease as m increases. This means that neighboring values of R_t, i.e., $m = 1$, will have the highest correlation, and the correlation diminishes rapidly as the distance m between time points is increased. This points to an interesting property of the autoregressive time series model. The autocorrelation function depends only on the distance m between R values, and not on the time t. Time series models that possess this property are said to be **stationary**.

Definition 10.8

A **stationary time series model** for regression residuals is one that has mean 0, constant variance, and autocorrelations that depend only on the distance between time points.

The autocorrelation function of first-order autoregressive models is shown for several values of ϕ in Figure 10.9. Note that positive values of ϕ yield positive autocorrelation for all residuals, whereas negative values of ϕ imply negative correlation for neighboring residuals, positive correlation between residuals two time points apart, negative correlation for residuals three time points apart, and so forth. The appropriate pattern will, of course, depend on the particular application, but the occurrence of a positive autocorrelation pattern is more common.

Although the first-order autoregressive error model provides a good representation for many autocorrelation patterns, more complex patterns can be described by

FIGURE 10.9

Autocorrelation functions for several first-order autoregressive error models: $R_t = \phi_1 R_{t-1} + \varepsilon_t$

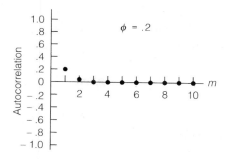

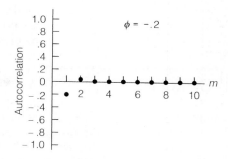

(a) Weak autocorrelation

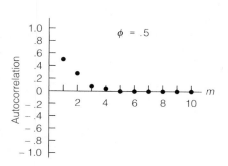

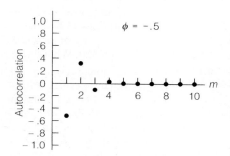

(b) Moderate autocorrelation

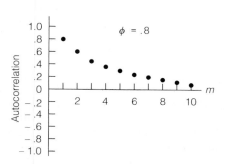

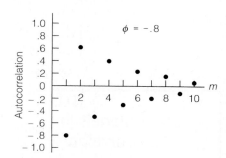

(c) Strong autocorrelation

higher-order autoregressive models. The general form of a pth-order autoregressive error model is

$$R_t = \phi_1 R_{t-1} + \phi_2 R_{t-2} + \cdots + \phi_p R_{t-p} + \varepsilon_t$$

The inclusion of p parameters, $\phi_1, \phi_2, \dots, \phi_p$, permits more flexibility in the pattern of autocorrelations exhibited by a residual time series. When an autoregressive model is used to describe residual autocorrelations, the observed autocorrelations are used to estimate these parameters. Methods for estimating these parameters will be presented in Section 10.8.

EXERCISES

10.16. Suppose that the random component of a time series model follows the first-order autoregressive model $R_t = \phi R_{t-1} + \varepsilon_t$, where ε_t is a white-noise process. Consider four versions of this model: $\phi = .9$, $\phi = -.9$, $\phi = .2$, and $\phi = -.2$.

  a. Calculate the first 10 autocorrelations, $\text{AC}(R_t, R_{t+m})$, $m = 1, 2, 3, \ldots, 10$, for each of the four models.

  b. Plot the autocorrelations against the distance in time separating the R values (m) for each case.

  c. Examine the rate at which the correlation diminishes in each plot. What does this imply?

10.17. When using time series to analyze quarterly data (data in which seasonal effects are present), it is highly possible that the random component of the model R_t also exhibits the same seasonal variation as the dependent variable. In these cases, the following non–first-order autoregressive model is sometimes postulated for the correlated error term, R_t:

$$R_t = \phi R_{t-4} + \varepsilon_t$$

where $|\phi| < 1$ and ε_t is a white-noise process. The autocorrelation function for this model is given by

$$\text{AC}(R_t, R_{t+m}) = \begin{cases} \phi^{m/4} & \text{if } m = 4, 8, 12, 16, 20, \ldots \\ 0 & \text{if otherwise} \end{cases}$$

  a. Calculate the first 20 autocorrelations ($m = 1, 2, \ldots, 20$) for the model with constant coefficient $\phi = .5$.

  b. Plot the autocorrelations against m, the distance in time separating the R values. Compare the rate at which the correlation diminishes with the first-order model $R_t = .5R_{t-1} + \varepsilon_t$.

10.18. Consider the autocorrelation pattern shown in the figure. Write a first-order autoregressive model that exhibits this pattern.

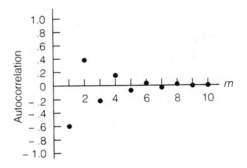

10.19. Write the general form for a fourth-order autoregressive model.

<table>
<tr><td>

10.6

Other Models for Autocorrelated Errors (Optional)

</td><td>

There are many models for autocorrelated residuals in addition to the autoregressive model, but the autoregressive model provides a good approximation for the autocorrelation pattern in many applications. Recall that the autocorrelations for autoregressive models diminish rapidly as the time distance m between the residuals increases. Occasionally, residual autocorrelations appear to change abruptly from nonzero for small values of m to 0 for larger values of m. For example, neighboring residuals ($m = 1$) may be correlated, whereas residuals that are farther apart ($m > 1$) are uncorrelated. This pattern can be described by the **first-order moving average model**

</td></tr>
</table>

$$R_t = \varepsilon_t + \theta \varepsilon_{t-1}$$

Note that the residual R_t is a linear combination of the current and previous *uncorrelated* (white-noise) residuals. It can be shown that the autocorrelations for this model are

$$AC(R_t, R_{t+m}) = \begin{cases} \frac{\theta}{1+\theta^2} & \text{if } m = 1 \\ 0 & \text{if } m > 1 \end{cases}$$

This pattern is shown in Figure 10.10

FIGURE 10.10

Autocorrelations for the first-order moving average model: $R_t = \varepsilon_t + \theta\varepsilon_{t-1}$

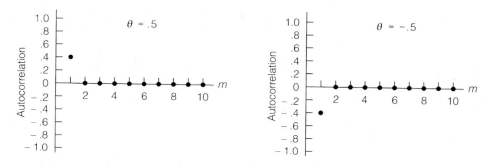

More generally, a qth-order moving average model is given by

$$R_t = \varepsilon_t + \theta_1\varepsilon_{t-1} + \theta_2\varepsilon_{t-2} + \cdots + \theta_q\varepsilon_{t-q}$$

Residuals within q time points are correlated, whereas those farther than q time points apart are uncorrelated. For example, a regression model for the quarterly earnings per share for a company may have residuals that are autocorrelated when within 1 year ($m = 4$ quarters) of one another, but uncorrelated when farther apart. An example of this pattern is shown in Figure 10.11.

Some autocorrelation patterns require even more complex residual models. A more general model is a combination of the **autoregressive—moving average (ARMA) models**,

$$R_t = \phi_1 R_{t-1} + \cdots + \phi_p R_{t-p} + \varepsilon_t + \theta_1\varepsilon_{t-1} + \cdots + \theta_q\varepsilon_{t-q}$$

Like the autoregressive model, the ARMA model has autocorrelations that diminish as the distance m between residuals increases. However, the patterns that can be described by ARMA models are more general than those of either autoregressive or moving average models.

FIGURE 10.11

Autocorrelations for a fourth-order moving average model

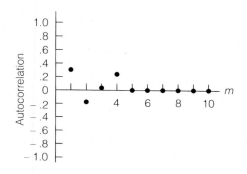

In Section 10.8, we present a method for estimating the parameters of an autoregressive residual model. The method for fitting time series models when the residual is either moving average or ARMA is more complicated, however. Consult the references at the end of the chapter for details of these methods.

10.7
Constructing Time Series Models

Recall that the general form of the times series model is

$$y_t = E(y_t) + R_t$$

We are assuming that the expected value of y_t is

$$E(y_t) = \beta_0 + \beta_1 x_1 + \beta_2 x_2 + \cdots + \beta_k x_k$$

where $x_1, x_2, \ldots, x_k$ are independent variables, which themselves may be time series, and the residual component, R_t, accounts for the pattern of autocorrelation in the residuals. Thus, a time series model consists of a pair of models: one model for the deterministic component $E(y_t)$ and one model for the autocorrelated residuals R_t.

CHOOSING THE DETERMINISTIC COMPONENT: The deterministic portion of the model is chosen in exactly the same manner as the regression models of the preceding chapters except that some of the independent variables might be time series variables or might be trigonometric functions of time (such as $\sin t$ or $\cos t$). It is helpful to think of the deterministic component as consisting of the trend (T_t), cyclical (C_t), and seasonal (S_t) effects described in Section 10.2.

For example, we may want to model the number of new housing starts, y_t, as a function of the prime interest rate, x_t. Then, one model for the mean of y_t is

$$E(y_t) = \beta_0 + \beta_1 x_t$$

for which the mean number of new housing starts is a multiple β_1 of the prime interest rate, plus a constant β_0. Another possibility is a second-order relationship,

$$E(y_t) = \beta_0 + \beta_1 x_t + \beta_2 x_t^2$$

which permits the *rate* of increase in the mean number of housing starts to increase or decrease with the prime interest rate.

Yet another possibility is to model the mean number of new housing starts as a function of both the prime interest rate and the year, t. Thus, the model

$$E(y_t) = \beta_0 + \beta_1 x_t + \beta_2 t + \beta_3 x_t t$$

implies that the mean number of housing starts increases linearly in x_t, the prime interest rate, but the rate of increase depends on the year t. If we wanted to adjust for seasonal (cyclical) effects due to t, we might introduce time into the model using trigonometric functions of t. This topic will be explained subsequently in greater detail.

Another important type of model for $E(y_t)$ is the **lagged independent variable model**. *Lagging* means that we are pairing observations on a dependent variable and independent variable at two different points in time, with the time corresponding to the independent variable lagging behind the time for the dependent variable. Suppose, for example, we believe that the monthly mean number of new housing starts is a function of the *previous* month's prime interest rate. Thus, we model y_t as a linear function of the lagged independent variable, prime interest rate, x_{t-1},

$$E(y_t) = \beta_0 + \beta_1 x_{t-1}$$

or, alternatively, as the second-order function,

$$E(y_t) = \beta_0 + \beta_1 x_{t-1} + \beta_2 x_{t-1}^2$$

For this example, the independent variable, prime interest rate x_t, is lagged 1 month behind the response y_t.

Many time series have distinct seasonal patterns. Retail sales are usually highest around Christmas, spring, and fall, with relative lulls in the winter and summer periods. Energy usage is highest in summer and winter, and lowest in spring and fall. Teenage unemployment rises in the summer months when schools are not in session, and falls near Christmas when many businesses hire part-time help.

When a time series' seasonality is exhibited in a relatively consistent pattern from year to year, we can model the pattern using trigonometric terms in the model for $E(y_t)$. For example, the model of a monthly series with mean $E(y_t)$ might be

$$E(y_t) = \beta_0 + \beta_1 \left(\cos \frac{2\pi}{12} t \right) + \beta_2 \left(\sin \frac{2\pi}{12} t \right)$$

This model would appear as shown in Figure 10.12. Note that the model is **cyclic**, with a **period** of 12 months. That is, the mean $E(y_t)$ completes a cycle every 12 months and then repeats the same cycle over the next 12 months. Thus, the **expected peaks and valleys** of the series remain the same from year to year. The coefficients β_1 and β_2 determine the **amplitude** and **phase shift** of the model. The amplitude is the magnitude of the seasonal effect, whereas the phase shift locates the peaks and valleys in time. For example, if we assume month 1 is January, the mean of the time series depicted in Figure 10.12 has a peak each April and a valley each October.

If the data are monthly or quarterly, we can treat the season as a qualitative independent variable (see Example 10.3), and write the model

$$E(y_t) = \beta_0 + \beta_1 S_1 + \beta_2 S_2 + \beta_3 S_3$$

FIGURE 10.12

A seasonal time series model

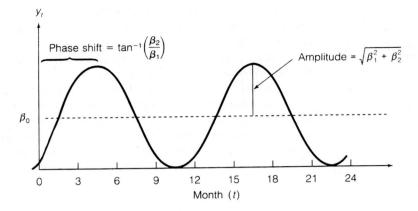

where

$$S_1 = \begin{cases} 1 & \text{if season is spring (II)} \\ 0 & \text{otherwise} \end{cases} \qquad S_2 = \begin{cases} 1 & \text{if season is summer (III)} \\ 0 & \text{otherwise} \end{cases}$$

$$S_3 = \begin{cases} 1 & \text{if season is fall (IV)} \\ 0 & \text{otherwise} \end{cases}$$

Thus, S_1, S_2, and S_3 are dummy variables that describe the four levels of season, letting winter (I) be the base level. The β coefficients determine the mean value of y_t for each season, as shown in Figure 10.13. Note that for the dummy variable model and the trigonometric model, we assume the seasonal effects are approximately the same from year to year. If they tend to increase or decrease with time, an interaction of the seasonal effect with time may be necessary. (An example will be given in Section 10.10.)

FIGURE 10.13

Seasonal model for quarterly data using dummy variables

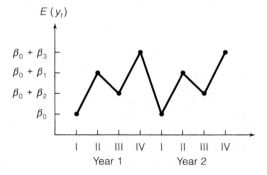

The appropriate form of the deterministic time series model will depend on both theory and data. Economic theory often provides several plausible models relating the mean response to one or more independent variables. The data can then be used to determine which, if any, of the models is best supported. The process is often an iterative one, beginning with preliminary models based on theoretical notions, using data to refine and modify these notions, collecting additional data to test the modified theories, and so forth.

CHOOSING THE RESIDUAL COMPONENT: The appropriate form of the residual component, R_t will depend on the pattern of autocorrelation in the residuals (see Sections 10.5 and 10.6). The autoregressive model of Section 10.5 is very useful for this aspect of time series modeling. The general form of an autoregressive model of order p is

$$R_t = \phi_1 R_{t-1} + \phi_2 R_{t-2} + \cdots + \phi_p R_{t-p} + \varepsilon_t$$

where ε_t is white noise (uncorrelated error). Recall that the name *autoregressive* comes from the fact that R_t is regressed on its own past values. As the order p is increased, more complex autocorrelation functions can be modeled. There are several other types of models that can be used for the random component, but the autoregressive model is very flexible and receives more application in business forecasting than the other models.

The simplest autoregressive error model is the **first-order autoregressive model**

$$R_t = \phi R_{t-1} + \varepsilon_t$$

Recall that the autocorrelation between residuals at two different points in time diminishes as the distance between the time points increases. Since many business and economic time series exhibit this property, the first-order autoregressive model is a popular choice for the residual component.

To summarize, we describe a general approach for constructing a time series:

1. Construct a regression model for the trend, seasonal, and cyclical components of $E(y_t)$. This model may be a polynomial in t for the trend (usually a straight-line or quadratic model) with trigonometric terms or dummy variables for the seasonal (cyclical) effects. The model may also include other time series variables as independent variables. For example, last year's rate of inflation may be used as a predictor of this year's Gross Domestic Product (GDP).

2. Next, construct a model for the random component (residual effect) of the model. A model that is widely used in practice is the first-order autoregressive error model

$$R_t = \phi R_{t-1} + \varepsilon_t$$

When the pattern of autocorrelation is more complex, use the general pth-order autoregressive model

$$R_t = \phi_1 R_{t-1} + \phi_2 R_{t-2} + \cdots + \phi_p R_{t-p} + \varepsilon_t$$

3. Combine the two components so that the model can be used for forecasting:

$$y_t = E(y_t) + R_t$$

Prediction intervals are calculated to measure the reliability of the forecasts. In the following two sections, we will demonstrate how time series models are fitted to data and used for forecasting. In Section 10.10, we will present an example in which we fit a seasonal time series model to a set of data.

EXERCISES

10.20. Suppose you are interested in buying stock in the Pepsi Company (PepsiCo). Your broker has advised you that your best strategy is to sell the stock at the first substantial jump in price. Hence, you are interested in a short-term investment. Before buying, you would like to model the closing price of PepsiCo, y_t, over time (in days), t.
  a. Write a first-order model for the deterministic portion of the model, $E(y_t)$.
  b. If a plot of the daily closing prices for the past month reveals a quadratic trend, write a plausible model for $E(y_t)$.
  c. Since the closing price of PepsiCo on day $(t + 1)$ is very highly correlated with the closing price on day t, your broker suggests that the random error components of the model are not white noise. Given this information, postulate a model for the error term, R_t.

10.21. An economist wishes to model the Gross Domestic Product (GDP) over time (in years) and also as a function of certain personal consumption expenditures. Let t = time in years and let
  y_t = GDP at time t
  x_{1t} = Durable goods at time t
  x_{2t} = Nondurable goods at time t
  x_{3t} = Services at time t
  a. The economist believes that y_t is linearly related to the independent variables x_{1t}, x_{2t}, x_{3t}, and t. Write the first-order model for $E(y_t)$.
  b. Rewrite the model if interaction between the independent variables and time is present.
  c. Postulate a model for the random error component, R_t. Explain why this model is appropriate.

10.22. Airlines sometimes overbook flights because of "no-show" passengers, i.e., passengers who have purchased a ticket but fail to board the flight. An airline supervisor wishes to be able to predict, for a flight from Miami to New York, the monthly accumulation of no-show passengers during the upcoming year, using data from the past 3 years. Let y_t = Number of no-shows during month t.
  a. Using dummy variables, propose a model for $E(y_t)$ that will take into account the seasonal (fall, winter,

spring, summer) variation that may be present in the data.
  b. Postulate a model for the error term R_t.
  c. Write the full time series model for y_t (include random error terms).
  d. Suppose the airline supervisor believes that the seasonal variation in the data is not constant from year to year, in other words, that there exists interaction between time and season. Rewrite the full model with the interaction terms added.

10.23. A farmer is interested in modeling the daily price of hogs at a livestock market. The farmer knows that the price varies over time (days) and also is reasonably confident that a seasonal effect is present.
  a. Write a seasonal time series model with trigonometric terms for $E(y_t)$, where y_t = Selling price (in dollars) of hogs on day t.
  b. Interpret the β parameters.
  c. Include in the model an interaction between time and the trigonometric components. What does the presence of interaction signify?
  d. Is it reasonable to assume that the random error component of the model, R_t, is white noise? Explain. Postulate a more appropriate model for R_t.

10.24. Numerous studies have been conducted to examine the relationship between seniority and productivity in business. A problem encountered in such studies is that individual output is often difficult to measure. G. A. Krohn developed a technique for estimating the experience–productivity relationship when such a measure is available (*Journal of Business & Economic Statistics*, Oct. 1983). Krohn modeled the batting average of a major league baseball player in year $t (y_t)$ as a function of the player's age in year $t (x_t)$ and an autoregressive error term (R_t).
  a. Write a model for $E(y_t)$ that hypothesizes, as did Krohn, a curvilinear relationship with x_t.
  b. Write a first-order autoregressive model for R_t.
  c. Use the models from parts **a** and **b** to write the full time series autoregressive model for y_t.

10.8 Fitting Time Series Models with Autoregressive Errors

We have proposed a general form for a time series model:

$$y_t = E(y_t) + R_t$$

where

$$E(y_t) = \beta_0 + \beta_1 x_1 + \cdots + \beta_k x_k$$

and, using an autoregressive model for R_t,

$$R_t = \phi R_{t-1} + \phi_2 R_{t-2} + \cdots + \phi_p R_{t-p} + \varepsilon_t$$

We now want to develop estimators for the parameters $\beta_0, \beta_1, \ldots, \beta_k$ of the regression model, and for the parameters $\phi_1, \phi_2, \ldots, \phi_p$ of the autoregressive model. The ultimate objective is to use the model to obtain forecasts (predictions) of future values of y_t, as well as to make inferences about the structure of the model itself.

We will introduce the techniques of fitting a time series model with a simple example. Refer to the data in Table 10.5, the annual sales for a firm in each of its 35 years of operation. Recall that the objective is to forecast future sales in years 36–40. In Section 10.4, we used a simple straight-line model for the mean sales

$$E(y_t) = \beta_0 + \beta_1 t$$

to make the forecasts.

The SAS printout showing the least squares estimates of β_0 and β_1 is reproduced in Figure 10.14. Although the model is useful for predicting annual sales (*p*-value for $H_0 : \beta_1 = 0$ is less than .0001), the Durbin–Watson statistic is $d = .821$, which is less than the tabulated value, $d_L = 1.40$ (Table 8 of Appendix C), for $\alpha = .05, n = 35$, and $k = 1$ independent variable. Thus, there is evidence that the residuals are

FIGURE 10.14

SAS printout for model of annual sales revenue

Dependent Variable: SALES

Analysis of Variance

Source	DF	Sum of Squares	Mean Square	F Value	Pr > F
Model	1	65875	65875	1615.72	<.0001
Error	33	1345.45355	40.77132		
Corrected Total	34	67221			

Root MSE	6.38524	R-Square	0.9800	
Dependent Mean	77.72286	Adj R-Sq	0.9794	
Coeff Var	8.21540			

Parameter Estimates

Variable	DF	Parameter Estimate	Standard Error	t Value	Pr > |t|
Intercept	1	0.40151	2.20571	0.18	0.8567
T	1	4.29563	0.10687	40.20	<.0001

Durbin-Watson D	0.821
Number of Observations	35
1st Order Autocorrelation	0.590

FIGURE 10.15

MINITAB residual plot for
annual sales model

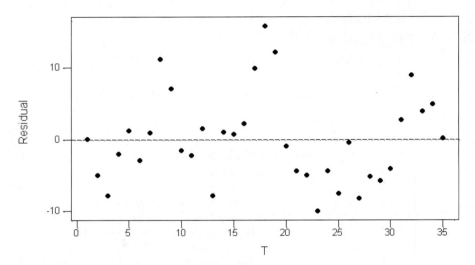

Residuals Versus T

(response is SALES)

positively correlated. The MINITAB plot of the least squares residuals over time,
in Figure 10.15, shows the pattern of positive autocorrelation. The residuals tend to
cluster in positive and negative runs; if the residual at time t is positive, the residual
at time $(t + 1)$ tends to be positive.

What are the consequences of fitting the least squares model when autocorrelated
residuals are present? Although *the least squares estimators of β_0 and β_1 remain
unbiased* even if the residuals are autocorrelated, i.e., $E(\hat{\beta}_0) = \beta_0$ and $E(\hat{\beta}_1) = \beta_1$,
the *standard errors given by least squares theory are usually smaller than the true
standard errors* when the residuals are positively autocorrelated. Consequently, t
values computed by the methods of Chapter 4 (which apply when the errors are
uncorrelated) will usually be inflated and will lead to a higher Type I error rate (α)
than the value of α selected for a test. Thus, the application of standard least squares
techniques to time series often produces misleading statistical test results that result
in overoptimistic evaluations of a model's predictive ability. There is a second reason
for seeking methods that specifically take into account the autocorrelated residuals.
If we can successfully model the residual autocorrelation, we should achieve a
smaller MSE and correspondingly narrower prediction intervals than those given by
the least squares model.

To account for the autocorrelated residual, we postulate a first-order autoregressive model,

$$R_t = \phi R_{t-1} + \varepsilon_t$$

Thus, we use the pair of models

$$y_t = \beta_0 + \beta_1 t + R_t$$
$$R_t = \phi R_{t-1} + \varepsilon_t$$

to describe the yearly sales of the firm. To estimate the parameters of the time series model (β_0, β_1, and ϕ), a modification of the least squares method is required. To do this, we use a *transformation* that is much like the variance-stabilizing transformations discussed in Chapter 8.

First, we multiply the model

$$y_t = \beta_0 + \beta_1 t + R_t \tag{10.1}$$

by ϕ at time $(t-1)$ to obtain

$$\phi y_{t-1} = \phi \beta_0 + \phi \beta_1 (t-1) + \phi R_{t-1} \tag{10.2}$$

Taking the difference between equations (10.1) and (10.2), we have

$$y_t - \phi y_{t-1} = \beta_0 (1 - \phi) + \beta_1 [t - \phi(t-1)] + (R_t - \phi R_{t-1})$$

or, since $R_t = \phi R_{t-1} + \varepsilon_t$, then

$$y_t^* = \beta_0^* + \beta_1 t^* + \varepsilon_t$$

where $y_t^* = y_t - \phi y_{t-1}$, $t^* = t - \phi(t-1)$, and $\beta_0^* = \beta_0(1 - \phi)$. Thus, we can use the transformed dependent variable y_t^* and transformed independent variable t^* to obtain least squares estimates of β_0^* and β_1. The residual ε_t is uncorrelated, so that the assumptions necessary for the least squares estimators are all satisfied. The estimator of the original intercept, β_0, can be calculated by

$$\hat{\beta}_0 = \frac{\hat{\beta}_0^*}{1 - \phi}$$

This transformed model appears to solve the problem of first-order autoregressive residuals. However, making the transformation requires knowing the value of the parameter ϕ. Also, we lose the initial observation, since the values of y_t^* and t^* can be calculated only for $t \geq 2$. The methods for estimating ϕ and adjustments for the values at $t = 1$ will not be detailed here. [See Anderson (1972) or Fuller (1978)]. Instead, we will present output from the SAS computer package, which both performs the transformation and estimates the model parameters, β_0, β_1, and ϕ.

The SAS printout of the straight-line, autoregressive time series model fit to the sales data is shown in Figure 10.16. The estimates of β_0 and β_1 in the deterministic component of the model (highlighted at the bottom of the printout) are $\hat{\beta}_0 = .4058$ and $\hat{\beta}_1 = 4.2959$. The estimate of the first-order autoregressive parameter ϕ (highlighted in the middle of the printout) is $-.589624$. However, the SAS time series model is defined so that ϕ takes the *opposite* sign from the value specified in our model. Consequently, you must multiply the estimate shown on the SAS printout by (-1) to obtain the estimate of ϕ for our model: $\hat{\varphi} = (-1)(-.589264) = .589264$. Therefore, the fitted models are:

$$\hat{y}_t = .4058 + 4.2959t + \hat{R}_t, \quad \hat{R}_t = .589624 \hat{R}_{t-1}$$

FIGURE 10.16

SAS printout for straight-line
model with autoregressive errors

The AUTOREG Procedure

Dependent Variable   SALES

Ordinary Least Squares Estimates

SSE	1345.45355	DFE	33
MSE	40.77132	Root MSE	6.38524
SBC	234.156237	AIC	231.045541
Regress R-Square	0.9800	Total R-Square	0.9800
Durbin-Watson	0.8207		

Variable	DF	Estimate	Standard Error	t Value	Approx Pr > |t|
Intercept	1	0.4015	2.2057	0.18	0.8567
T	1	4.2956	0.1069	40.20	<.0001

Estimates of Autocorrelations

Lag	Covariance	Correlation	-1 9 8 7 6 5 4 3 2 1 0 1 2 3 4 5 6 7 8 9 1
0	38.4415	1.000000	|********************|
1	22.6661	0.589624	|************ |

Preliminary MSE   25.0771

Estimates of Autoregressive Parameters

Lag	Coefficient	Standard Error	t Value
1	-0.589624	0.142779	-4.13

Yule-Walker Estimates

SSE	877.685377	DFE	32
MSE	27.42767	Root MSE	5.23714
SBC	223.18683	AIC	218.520786
Regress R-Square	0.9412	Total R-Square	0.9869
Durbin-Watson	1.8217		

Variable	DF	Estimate	Standard Error	t Value	Approx Pr > |t|
Intercept	1	0.4058	3.9970	0.10	0.9198
T	1	4.2959	0.1898	22.63	<.0001

Note, also, that there are two R^2 values shown at the bottom of the SAS printout, Figure 10.16. The quantity labeled as **Regress R-Square** is not the value of R^2 based on the original time series variable, y_t. Instead, it is based on the values of the transformed variable, y_t^*. When we refer to R^2 in this chapter, we will always mean the value of R^2 based on the original time series variable. This value, which usually will be larger than the R^2 for the transformed time series variable, is given on the printout as **Total R-Square.** Thus, the time series autoregressive model yields

$$MSE = 27.42767$$

and

$$R^2 = .9869.$$

A comparison of the least squares (Figure 10.7) and autoregressive (Figure 10.16) computer printouts is given in Table 10.6. Note that the autoregressive model reduces

TABLE 10.6 Comparison of Least Squares and Time Series Results

	LEAST SQUARES	AUTOREGRESSIVE
R^2	.980	.987
MSE	40.77	27.43
$\hat{\beta}_0$	.4015	.4058
$\hat{\beta}_1$	4.2956	4.2959
Standard error $(\hat{\beta}_0)$	2.2057	3.9970
Standard error $(\hat{\beta}_1)$	.1069	.1898
t statistic for $H_0 : \beta_1 = 0$	40.20	22.63
	$(p < .0001)$	$(p < .0001)$
$\hat{\phi}$	—	.5896
t statistic for $H_0 : \phi = 0$	—	4.13

MSE and increases R^2. The values of the estimators β_0 and β_1 change very little, but the estimated standard errors are considerably increased, thereby decreasing the t value for testing $H_0: \beta_1 = 0$. The implication that the linear relationship between sales y_t and year t is of significant predictive value is the same using either method. However, you can see that the underestimation of standard errors by using least squares in the presence of residual autocorrelation could result in the inclusion of unimportant independent variables in the model, since the t values will usually be inflated.

Is there evidence of positive autocorreled residuals? An approximate t test* of the hypothesis $H_0: \phi = 0$ yields a $t = 4.13$. With 32 df, this value is significant at less than $\alpha = .01$. Thus, the result of the Durbin–Watson d test is confirmed: There is adequate evidence of positive residual autocorrelation.[†]Furthermore, the first-order autoregressive model appears to describe this residual correlation well.

The steps for fitting a time series model to a set of data are summarized in the box. Once the model is estimated, the model can be used to forecast future values of the time series y_t.

EXERCISES

10.25. The Gross Domestic Product (GDP) is a measure of total U.S. output and is, therefore, an important indicator of the U.S. economy. The quarterly GDP values (in billions of dollars) from 1997 to 2001 are given in the table on p. 508. Let y_t be the GDP in quarter t, $t = 1, 2, 3, \ldots , 20$.
  a. Hypothesize a time series model for quarterly GDP that includes a straight-line long-term trend and autocorrelated residuals.

b. The SAS printout for the time series model $y_t = \beta_0 + \beta_1 t + \phi R_{t-1} + \varepsilon_t$ is shown on p. 509. Write the least squares prediction equation.
  c. Interpret the estimates of the model parameters, β_0, β_1, and ϕ.
  d. Interpret the values of R^2 and S.

---

*An explanation of this t test has been omitted. Consult the references at the end of the chapter for details of this test.
[†]This result is to be expected since it can be shown (proof omitted) that $\hat{\phi} \approx 1 - d/2$, where d is the value of the Durbin–Watson statistic.

Steps for Fitting Time Series Models

1. Use the least squares approach to obtain initial estimates of the β parameters. Do *not* use the t or F tests to assess the importance of the parameters, since the estimates of their standard errors may be biased (often underestimated).

2. Analyze the residuals to determine whether they are autocorrelated. The Durbin–Watson test is one technique for making this determination.

3. If there is evidence of autocorrelation, construct a model for the residuals. The autoregressive model is one useful model. Consult the references at the end of the chapter for more types of residual models and for methods of identifying the most suitable model.

4. Reestimate the β parameters, taking the residual model into account. This involves a simple transformation if an autoregressive model is used; several statistical software packages have computer routines to accomplish this.

 GDP

YEAR	QUARTER	GDP
1997	I	8124
	II	8280
	III	8391
	IV	8479
1998	I	8635
	II	8722
	III	8829
	IV	8975
1999	I	9093
	II	9161
	III	9297
	IV	9522
2000	I	9669
	II	9858
	III	9938
	IV	10028
2001	I	10142
	II	10203
	III	10225
	IV	10480

Source: Standard & Poor's Statistical Service: Current Statistics, Mar. 2002.

10.26. Refer to Exercise 10.8 (p. 490).

    a. Hypothesize a time series model for annual acreage of wheat harvested, y_t, that takes into account the residual autocorrelation.

    b. Fit the autoregressive time series model, part **a**. Interpret the estimates of the model parameters.

10.27. The Dow Jones Industrial Average (DJIA) is a widely followed stock market indicator. The values of the DJIA from 1971 to 2000 are given in the table on p. 509. Suppose we want to model the yearly DJIA, y_t, as a function of t, where t is the number of years since 1970 (i.e., $t = 1$ for 1971, $t = 2$ for 1972, ..., $t = 30$ for 2000).

    a. Construct a scatterplot of the data. Do you observe a long-term trend?

    b. Propose a time series model that includes a long-term quadratic trend and autocorrelated residuals.

    c. The SAS printout for the time series model, part **b**, is shown on p. 510. Identify and interpret (i) the estimates of the model parameters, (ii) the value of R^2, and (iii) the test for a quadratic long-term trend.

SAS Output for Exercise 10.25

```
                        The AUTOREG Procedure

                   Dependent Variable      GDP


                   Ordinary Least Squares Estimates

        SSE             61911.0549   DFE                     18
        MSE                   3440   Root MSE          58.64728
        SBC             223.503441   AIC             221.511977
        Regress R-Square    0.9940   Total R-Square      0.9940
        Durbin-Watson       0.9474


                                  Standard              Approx
        Variable    DF   Estimate     Error   t Value   Pr > |t|

        Intercept    1       8000   27.2435    293.65   <.0001
        T            1   124.0534    2.2742     54.55   <.0001

                    Estimates of Autocorrelations

Lag   Covariance    Correlation   -1 9 8 7 6 5 4 3 2 1 0 1 2 3 4 5 6 7 8 9 1

 0       3095.6      1.000000     |                    |********************|
 1       1629.2      0.526311     |                    |***********         |


            Preliminary MSE      2238.1


          Estimates of Autoregressive Parameters

                                 Standard
             Lag    Coefficient     Error    t Value

              1      -0.526311    0.206226     -2.55


                    Yule-Walker Estimates

        SSE             44761.1241   DFE                     17
        MSE                   2633   Root MSE          51.31284
        SBC             220.336349   AIC             217.349152
        Regress R-Square    0.9860   Total R-Square      0.9957
        Durbin-Watson       1.6799


                                 Standard              Approx
     Variable    DF   Estimate      Error   t Value   Pr > |t|

     Intercept    1       8000    44.1537    181.19   <.0001
     T            1   124.0390     3.5907     34.54   <.0001
```

Data for Exercise 10.27

💿 DJIA

YEAR	DJIA
1971	885
1972	951
1973	924
1974	759
1975	802
1976	975
1977	835
1978	805
1979	839
1980	964
1981	899
1982	1047
1983	1259

(continued)

YEAR	DJIA
1984	1212
1985	1547
1986	1896
1987	2276
1988	2061
1989	2508
1990	2679
1991	3169
1992	3296
1993	3540
1994	3793
1995	4534
1996	5780
1997	7438
1998	8548
1999	10483
2000	10710

Source: Standard & Poor's Statistical Service: Current Statistics, Mar. 2002.

SAS Output for Exercise 10.27

The AUTOREG Procedure

Dependent Variable   DJIA

Ordinary Least Squares Estimates

SSE	12853249.7	DFE	27
MSE	476046	Root MSE	689.96107
SBC	484.3772	AIC	480.173607
Regress R-Square	0.9468	Total R-Square	0.9468
Durbin-Watson	0.3259		

Variable	DF	Estimate	Standard Error	t Value	Approx Pr > |t|
Intercept	1	2066	404.5799	5.11	<.0001
T	1	-366.9565	60.1620	-6.10	<.0001
TSQ	1	20.7359	1.8831	11.01	<.0001

Estimates of Autocorrelations

Lag	Covariance	Correlation	-1 9 8 7 6 5 4 3 2 1 0 1 2 3 4 5 6 7 8 9 1
0	428442	1.000000	|********************|
1	330670	0.771796	|*************** |

Preliminary MSE   173232

Estimates of Autoregressive Parameters

Lag	Coefficient	Standard Error	t Value
1	-0.771796	0.124704	-6.19

Yule-Walker Estimates

SSE	4022306.13	DFE	26
MSE	154704	Root MSE	393.32440
SBC	453.831682	AIC	448.226892
Regress R-Square	0.8744	Total R-Square	0.9834
Durbin-Watson	1.3435		

Variable	DF	Estimate	Standard Error	t Value	Approx Pr > |t|
Intercept	1	1690	647.7797	2.61	0.0149
T	1	-349.5683	93.5466	-3.74	0.0009
TSQ	1	21.0921	2.8900	7.30	<.0001

10.28. Refer to Exercise 10.15 (p. 492) and the study on the long-term effects of the Employment Retirement Income Security Act (ERISA). Ledolter and Power also fitted quarterly time series models for the number of pension plan terminations and the number of profit-sharing plan terminations from the first quarter of 1956 through the third quarter of 1982 ($n = 107$ quarters). To account for residual correlation, they fitted straight-line autoregressive models of the form, $y_t = \beta_0 + \beta_1 t + \phi R_{t-1} + \varepsilon_t$. The results were as follows:

Pension plan: $\hat{y}_t = 3.54 + .039t + .40\hat{R}_{t-1}$

Profit-sharing plan: $\hat{y}_t = 3.45 + .038t + .22\hat{R}_{t-1}$

a. Interpret the estimates of the model parameters for pension plan terminations.

b. Interpret the estimates of the model parameters for profit-sharing plan terminations.

10.9 Forecasting with Time Series Autoregressive Models

Often, the ultimate objective of fitting a time series model is to forecast future values of the series. We will demonstrate the techniques for the simple model

$$y_t = \beta_0 + \beta_1 x_t + R_t$$

with the first-order autoregressive residual

$$R_t = \phi R_{t-1} + \varepsilon_t$$

Suppose we use the data $(y_1, x_1), (y_2, x_2), \ldots, (y_n, x_n)$ to obtain estimates of β_0, β_1, and ϕ, using the method presented in Section 10.8. We now want to forecast the value of y_{n+1}. From the model,

$$y_{n+1} = \beta_0 + \beta_1 x_{n+1} + R_{n+1}$$

where

$$R_{n+1} = \phi R_n + \varepsilon_{n+1}$$

Combining these, we obtain

$$y_{n+1} = \beta_0 + \beta_1 x_{n+1} + \phi R_n + \varepsilon_{n+1}$$

From this equation, we obtain the forecast of y_{n+1}, denoted F_{n+1}, by estimating each of the unknown quantities and setting ε_{n+1} to its expected value of 0:*

$$F_{n+1} = \hat{\beta}_0 + \hat{\beta}_1 x_{n+1} + \hat{\phi} \hat{R}_n$$

where $\hat{\beta}_0, \hat{\beta}_1$, and $\hat{\phi}$ are the estimates based on the time series model-fitting approach presented in Section 10.8. The estimate $\hat{R}_n$ of the residual R_n is obtained by noting that

$$R_n = y_n - (\beta_0 + \beta_1 x_n)$$

so that

$$\hat{R}_n = y_n - (\hat{\beta}_0 + \hat{\beta}_1 x_n)$$

---

*Note that the forecast requires the value of x_{n+1}. When x_t is itself a time series, the future value x_{n+1} will generally be unknown and must also be estimated. Often, $x_t = t$ (as in Example 10.4). In this case, the future time period (e.g., $t = n + 1$) is known and no estimate is required.

The two-step-ahead forecast of y_{n+2} is similarly obtained. The true value of y_{n+2} is

$$y_{n+2} = \beta_0 + \beta_1 x_{n+2} + R_{n+2}$$
$$= \beta_0 + \beta_1 x_{n+2} + \phi R_{n+1} + \varepsilon_{n+2}$$

and the forecast at $t = n + 2$ is

$$F_{n+2} = \hat{\beta}_0 + \hat{\beta}_1 x_{n+2} + \hat{\phi} \hat{R}_{n+1}$$

The residual R_{n+1} (and all future residuals) can now be obtained from the recursive relation

$$R_{n+1} = \phi R_n + \varepsilon_{n+1}$$

so that

$$\hat{R}_{n+1} = \hat{\phi} \hat{R}_n$$

Thus, the forecasting of future y values is an iterative process, with each new forecast making use of the previous residual to obtain the estimated residual for the future time period. The general forecasting procedure using time series models with first-order autoregressive residuals is outlined in the next box.

EXAMPLE   10.4

Suppose we want to forecast the sales of the company for the data in Table 10.5. In Section 10.8, we fit the regression–autoregression pair of models

$$y_t = \beta_0 + \beta_1 t + R_t \quad R_t = \phi R_{t-1} + \varepsilon_t$$

Using 35 years of sales data, we obtained the estimated models

$$\hat{y}_t = .4058 + 4.2959t + \hat{R}_t \quad \hat{R}_t = .5896 \hat{R}_{t-1}$$

Combining these, we have

$$\hat{y}_t = .4058 + 4.2959t + .5896 \hat{R}_{t-1}$$

a. Use the fitted model to forecast sales in years $t = 36, 37$, and 38.
b. Find approximate 95% prediction intervals for the forecasts.

Solution

a. The forecast for the 36th year requires an estimate of the last residual R_{35},

$$\hat{R}_{35} = y_{35} - [\hat{\beta}_0 + \hat{\beta}_1(35)]$$
$$= 150.9 - [.4058 + 4.2959(35)]$$
$$= .1377$$

Forecasting Using Time Series Models with First-Order Autoregressive Residuals

$$y_t = \beta_0 + \beta_1 x_{1t} + \beta_2 x_{2t} + \cdots + \beta_k x_{kt} + R_t$$
$$R_t = \phi R_{t-1} + \varepsilon_t$$

STEP 1 Use a statistical software package to obtain the estimated model

$$\hat{y}_t = \hat{\beta}_0 + \hat{\beta}_1 x_{1t} + \hat{\beta}_2 x_{2t} + \cdots + \hat{\beta}_k x_{kt} + \hat{R}_t, t = 1, 2, \ldots, n$$
$$\hat{R}_t = \hat{\phi} \hat{R}_{t-1}$$

STEP 2 Compute the estimated residual for the last time period in the data (i.e., $t = n$) as follows:

$$\hat{R}_n = y_n - \hat{y}_n$$
$$= y_n - (\hat{\beta}_0 + \hat{\beta}_1 x_{1n} + \hat{\beta}_2 x_{2n} + \cdots + \hat{\beta}_k x_{kn})$$

STEP 3 The forecast of the value y_{n+1} (i.e., the one-step-ahead forecast) is

$$F_{n+1} = \hat{\beta}_0 + \hat{\beta}_1 x_{1n+1} + \hat{\beta}_2 x_{2,n+1} + \cdots + \hat{\beta}_k x_{k,n+1} + \hat{\phi} \hat{R}_n$$

where $\hat{R}_n$ is obtained from step 2.

STEP 4 The forecast of the value y_{n+2} (i.e., the two-step-ahead forecast) is

$$F_{n+2} = \hat{\beta}_0 + \hat{\beta}_1 x_{1,n+2} + \hat{\beta}_2 x_{2,n+2} + \cdots + \hat{\beta}_k x_{k,n+2} + (\hat{\phi})^2 \hat{R}_n$$

where $\hat{R}_{n+1}$ is obtained from step 3.

In general, the m-step-ahead forecast is

$$F_{n+m} = \hat{\beta}_0 + \hat{\beta}_1 x_{1,n+m} + \hat{\beta}_2 x_{2,n+m} + \cdots + \hat{\beta}_k x_{k,n+m} + (\hat{\phi})^m \hat{R}_n$$

Then the one-step-ahead forecast (i.e., the sales forecast for year 36) is

$$F_{36} = \hat{\beta}_0 + \hat{\beta}_1 (36) + \hat{\phi} \hat{R}_{35}$$
$$= .4058 + 4.2959(36) + (.5896)(.1377)$$
$$= 155.14$$

Using the formula in the box, the two-step-ahead forecast (i.e., the sales forecast for year 37) is

$$F_{37} = \hat{\beta}_0 + \hat{\beta}_1 (37) + (\hat{\phi})^2 \hat{R}_{35}$$
$$= .4058 + 4.2959(37) + (.5896)^2 (.1377)$$
$$= 159.40$$

Similarly, the three-step-ahead forecast (i.e., the sales forecast for year 38) is

$$F_{38} = \hat{\beta}_0 + \hat{\beta}_1(38) + (\hat{\phi})^3 \hat{R}_{35}$$

$$= .4058 + 4.2959(38) + (.5896)^3(.1377)$$

$$= 163.68$$

Some statistical software packages (e.g., SAS) have options for computing forecasts using the autoregressive model. The three forecasted values, F_{36}, F_{37}, and F_{38}, are shown (shaded) at the bottom of the SAS printout, Figure 10.17, in the FORECAST column.

We can proceed in this manner to generate sales forecasts as far into the future as desired. However, the potential for error increases as the distance into the future increases. Forecast errors are traceable to three primary causes, as follows:

FIGURE 10.17

SAS printout of forecasts of annual sales revenue using straight-line model with autoregressive errors

T	SALES	FORECAST	LCL95	UCL95
1	4.8	4.702	-10.644	12.515
2	4.0	9.056	-3.980	16.489
3	5.5	10.347	-2.509	20.469
4	15.6	12.994	0.308	24.456
5	23.1	20.712	8.186	28.451
6	23.3	26.897	14.521	32.456
7	31.4	28.778	16.542	36.472
8	46.0	35.317	23.210	40.500
9	46.1	45.689	33.699	44.542
10	41.9	47.511	35.627	48.601
11	45.5	46.797	35.009	52.678
12	53.5	50.683	38.977	56.776
13	48.4	57.163	45.527	60.898
14	61.6	55.919	44.341	65.047
15	65.6	65.465	53.933	69.225
16	71.4	69.586	58.086	73.435
17	83.4	74.769	63.289	77.678
18	93.6	83.607	72.134	81.956
19	94.2	91.384	79.904	86.270
20	85.4	93.501	82.001	90.619
21	86.2	90.075	78.543	95.001
22	89.9	92.310	80.733	99.414
23	89.2	96.254	84.619	103.858
24	99.1	97.605	85.899	108.327
25	100.3	105.205	93.416	112.821
26	111.7	107.675	95.792	117.335
27	108.2	116.160	104.170	121.869
28	115.5	115.859	103.752	126.418
29	119.2	121.927	109.690	130.982
30	125.2	125.871	113.495	135.558
31	136.3	131.172	118.645	140.145
32	146.8	139.480	126.793	144.742
33	146.1	147.434	134.577	149.347
34	151.4	148.784	135.748	153.959
35	150.9	153.672	140.449	158.577
36	.	155.140	141.720	163.201
37	.	159.403	144.397	167.830
38	.	163.679	148.034	172.463

1. The form of the model may change at some future time. This is an especially difficult source of error to quantify, since we will not usually know when or if the model changes, or the extent of the change. The possibility of a change in the model structure is the primary reason we have consistently urged you to avoid predictions outside the observed range of the independent variables. However, time series forecasting leaves us little choice—by definition, the forecast will be a prediction at a future time.

2. A second source of forecast error is the uncorrelated residual, ε_t, with variance σ^2. For a first-order autoregressive residual, the forecast variance of the one-step-ahead prediction is σ^2, whereas that for the two-step-ahead prediction is $\sigma^2(1 + \phi^2)$, and, in general, for m steps ahead, the forecast variance[†] is $\sigma^2(1+\phi^2+\phi^4+\cdots+\phi^{2(m-1)})$. Thus, the forecast variance increases as the distance is increased. These variances allow us to form approximate 95% prediction intervals for the forecasts (see the next box).

3. A third source of variability is that attributable to the error of estimating the model parameters. This is generally of less consequence than the others, and is usually ignored in forming prediction intervals.

b. To obtain a prediction interval, we first estimate σ^2 by the MSE, the mean square for error from the time series regression analysis. For the sales data, we form an *approximate* 95% prediction interval for the sales in year 36:

$$F_{36} \pm 2\sqrt{\text{MSE}}$$

$$155.1 \pm 2\sqrt{27.42767}$$

$$155.1 \pm 10.1$$

or (144.8, 165.4). Thus, we forecast that the sales in year 36 will be between $145,000 and $165,000.

The approximate 95% prediction interval for year 37 is

$$F_{37} \pm 2\sqrt{\text{MSE}(1 + \phi^2)}$$

$$159.4 \pm 2\sqrt{27.42767[1 + (.5896)^2]}$$

$$159.4 \pm 11.9$$

or (147.5, 171.3). Note that this interval is wider than that for the one-step-ahead forecast. The intervals will continue to widen as we attempt to forecast farther ahead.

The formulas for computing *exact* 95% prediction intervals using the time series autoregressive model are complex and beyond the scope of this text. However, we can use statistical software to obtain them. The exact 95% prediction intervals for years 36–38 are shown at the bottom of the SAS printout, Figure 10.17, in the **LCL95** and **UCL95** columns. Note that the exact prediction intervals are wider than the approximate intervals. We again stress that the accuracy of these forecasts and intervals depends on the assumption that the model structure does not change during the forecasting period. If, for example, the company merges with another company during year 37, the structure of the sales model will almost surely change, and therefore, prediction intervals past year 37 are probably useless. ◆

[†]See Fuller (1976).

> **Approximate 95% Forecasting Limits Using Time Series Models with First-Order Autoregressive Residuals**
>
> *One-Step-Ahead Forecast:*
>
> $$\hat{y}_{n+1} \pm 2\sqrt{\text{MSE}}$$
>
> *Two-Step-Ahead Forecast:*
>
> $$\hat{y}_{n+2} \pm 2\sqrt{\text{MSE}(1 + \hat{\phi}^2)}$$
>
> *Three-Step-Ahead Forecast:*
>
> $$\hat{y}_{n+3} \pm 2\sqrt{\text{MSE}(1 + \hat{\phi}^2 + \hat{\phi}^4)}$$
>
> $$\vdots$$
>
> *m-Step-Ahead Forecast:*
>
> $$\hat{y}_{n+m} \pm 2\sqrt{\text{MSE}(1 + \hat{\phi}^2 + \hat{\phi}^4 + \cdots + \hat{\phi}^{2(m-1)})}$$
>
> [*Note*: MSE estimates σ^2, the variance of the uncorrelated residual ε_t.]

It is important to note that the forecasting procedure makes explicit use of the residual autocorrelation. The result is a better forecast than would be obtained using the standard least squares procedure of Chapter 4 (which ignores residual correlation). Generally, this is reflected by narrower prediction intervals for the time series forecasts than for the least squares prediction.[‡] The end result, then, of using a time series model when autocorrelation is present is that you obtain more reliable estimates of the β coefficients, smaller residual variance, and more accurate prediction intervals for future values of the time series.

EXERCISES

10.29. The quarterly time series model $y_t = \beta_0 + \beta_1 t + \beta_2 t^2 + \phi R_{t-1} + \varepsilon_t$ was fit to data collected for $n = 48$ quarters, with the following results:

$$\hat{y}_t = 220 + 17t - .3t^2 + .82\hat{R}_{t-1}$$

$$y_{48} = 350 \quad \text{MSE} = 10.5$$

a. Calculate forecasts for y_t for $t = 49, t = 50$, and $t = 51$.

b. Construct approximate 95% prediction intervals for the forecasts obtained in part **a**.

10.30. The annual time series model $y_t = \beta_0 + \beta_1 t + \phi R_{t-1} + \varepsilon_t$ was fit to data collected for $n = 30$ years with the

---

[‡]When n is large, approximate 95% prediction intervals obtained from the standard least squares procedure reduce to $\hat{y}_t \pm 2\sqrt{\text{MSE}}$ for *all* future values of the time series. These intervals may actually be narrower than the more accurate prediction intervals produced from the time series analysis.

following results:

$$\hat{y}_t = 10 + 2.5t + .64\hat{R}_{t-1}$$

$$y_{30} = 82 \quad MSE = 4.3$$

a. Calculate forecasts for y_t for $t = 31, t = 32$, and $t = 33$.

b. Construct approximate 95% prediction intervals for the forecasts obtained in part **a**.

10.31. Use the fitted time series model of Exercise 10.25 (p. 507) to forecast GDP for the four quarters of 2002 and calculate approximate 95% forecast limits. Go to your library (or search the Internet) to find the actual GDP values for 2002. Do the forecast intervals contain the actual 2002 GDP values?

10.32. Use the fitted time series model of Exercise 10.26 (p. 508) to forecast annual volume of wheat harvest for 2005. Place approximate 95% confidence bounds on the forecast.

10.33. Use the fitted time series model of Exercise 10.27 (p. 508) to forecast the DJIA for the years 2001 and 2002. Calculate approximate 95% prediction intervals for the forecasts. Do these intervals contain the actual 2001–2002 DJIA values? (Go to your library or search the Internet to find the actual DJIA values in 2001 and 2002.) If not, give a plausible explanation.

10.34. Refer to the gold price series, 1971–2001, Exercise 10.7 (p. 483). The data is saved in the GOLD file.

a. Hypothesize a deterministic model for $E(y_t)$ based on a plot of the time series.

b. Do you expect the random error term of the model, part **a**, to be uncorrected? Explain.

c. Hypothesize a model for the correlated error term, R_t.

d. Combine the two models, parts **a** and **c**, to form a time series forecasting model.

e. Fit the time series model, part **d**, to the data.

f. Use the fitted time series model from part **e** to forecast the gold price in 2002. Place an approximate 95% prediction interval around the forecast.

10.35. Refer to Exercise 10.28 (p. 510). The values of MSE for the quarterly time series models of retirement plan terminations are as follows:

Pension plan termination: $\quad$ MSE = .0440
Profit-sharing plan termination: $\quad$ MSE = .0402

a. Forecast the number of pension plan terminations for the fourth quarter of 1982 (i.e., $t = 108$). Assume that $y_{107} = 7.5$. [*Hint:* Remember that the forecasted number of pension plan terminations is $e^{\hat{y}_{108}}$.]

b. Place approximate 95% confidence bounds on the forecast obtained in part **a**. [*Hint:* First, calculate upper and lower confidence limits for y_{108}, then take antilogarithms.]

c. Repeat parts **a** and **b** for the number of profit-sharing plan terminations in the fourth quarter of 1982. Assume that $y_{107} = 7.6$.

10.10
Seasonal Time Series Models: An Example

We have used a simple regression model to illustrate the methods of model estimation and forecasting when the residuals are autocorrelated. In this section, we present a more realistic example that requires a seasonal model for $E(y_t)$, as well as an autoregressive model for the residual.

Critical water shortages have dire consequences for both business and private sectors of communities. Forecasting water usage for months in advance is essential to avoid such shortages. Suppose a community has monthly water usage records over the past 15 years. A plot of the last 6 years of the time series, y_t, is shown in Figure 10.18. Note that both an increasing trend and a seasonal pattern appear prominent in the data. The water usage seems to peak during the summer months and decline during the winter months. Thus, we might propose the following model:

$$E(y_t) = \beta_0 + \beta_1 t + \beta_2 \left(\cos \frac{2\pi}{12} t \right) + \beta_3 \left(\sin \frac{2\pi}{12} t \right)$$

Since the amplitude of the seasonal effect (that is, the magnitude of the peaks and valleys) appears to increase with time, we include in the model an interaction

FIGURE 10.18

Water usage time series

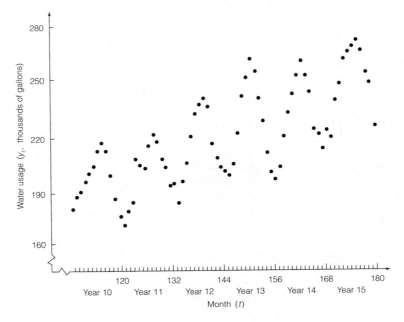

between time and trigonometric components, to obtain

$$E(y_t) = \beta_0 + \beta_1 t + \beta_2 \left(\cos \frac{2\pi}{12} t \right) + \beta_3 \left(\sin \frac{2\pi}{12} t \right) + \beta_4 t \left(\cos \frac{2\pi}{12} t \right) + \beta_5 t \left(\sin \frac{2\pi}{12} t \right)$$

The model for the random component R_t must allow for short-term cyclic effects. For example, in an especially hot summer, if the water usage, y_t, exceeds the expected usage, $E(y_t)$, for July, we would expect the same thing to happen in August. Thus, we propose a first-order autoregressive model* for the random component:

$$R_t = \phi R_{t-1} + \varepsilon_t$$

We now fit the models to the time series y_t, where y_t is expressed in thousands of gallons. The SAS printout is shown in Figure 10.19. The estimated models are given by

$$\hat{y}_t = 100.0832 + .8263t - 10.8011 \left(\cos \frac{2\pi}{12} t \right) - 7.0858 \left(\sin \frac{2\pi}{12} t \right)$$

$$- .0556t \left(\cos \frac{2\pi}{12} t \right) - .0296t \left(\sin \frac{2\pi}{12} t \right) + \hat{R}_t$$

$$\hat{R}_t = .6617 \hat{R}_{t-1}$$

with MSE = 23.135. The R^2 value of .99 indicates that the models provide a good fit to the data.

---

*A more complex time series model may be more appropriate. We use the simple first-order autoregressive model so you can follow the modeling process more easily.

FIGURE 10.19

SAS printout for time series model of water usage

The AUTOREG Procedure

Dependent Variable   USAGE

Estimates of Autoregressive Parameters

Lag	Coefficient	Standard Error	t Value
1	−0.661679	0.055886	−11.84

Yule-Walker Estimates

SSE	4025.513	DFE	174
MSE	23.135	Root MSE	4.810
SBC	1023.591	AIC	1002.941
Regress R-Square	0.9431	Total R-Square	0.9900
Durbin-Watson	0.5216		

Variable	DF	Estimate	Standard Error	t Value	Approx Pr > |t|
Intercept	1	100.0832	2.0761	48.21	<.0001
T	1	0.8263	0.0198	41.74	<.0001
COS	1	−10.8011	1.8559	−5.82	<.0001
SIN	1	−7.0858	1.8957	−3.74	.0003
COS_T	1	−0.0556	0.0177	−3.14	.0020
SIN T	1	−0.0296	0.0182	−1.63	.1049

We now use the models to forecast water usage for the next 12 months. The forecast for the first month is obtained as follows. The last residual value (obtained from a portion of the printout not shown) is $\hat{R}_{180} = -1.3247$. Then the formula for the one-step-ahead forecast is

$$F_{181} = \hat{\beta}_0 + \hat{\beta}_1(181) + \hat{\beta}_2 \left(\cos \frac{2\pi}{12} 181 \right) + \hat{\beta}_3 \left(\sin \frac{2\pi}{12} 181 \right)$$

$$+ \hat{\beta}_4(181) \left(\cos \frac{2\pi}{12} 181 \right) + \hat{\beta}_5(181) \left(\sin \frac{2\pi}{12} 181 \right) + \hat{\phi}\hat{R}_{180}$$

Substituting the values of $\hat{\beta}_0, \hat{\beta}_1, \ldots, \hat{\beta}_5$, and $\hat{\phi}$ shown in Figure 10.20, we obtain $F_{181} = 238.0$. Approximate 95% prediction bounds on this forecast are given by $\pm 2\sqrt{\text{MSE}} = \pm 2\sqrt{23.135} = \pm 9.6$.[†] That is, we expect our forecast for 1 month ahead to be within 9,600 gallons of the actual water usage. This forecasting process is then repeated for the next 11 months. The forecasts and their bounds are shown in Figure 10.20. Also shown are the actual values of water usage during year 16. Note that the forecast prediction intervals widen as we attempt to forecast farther

---

[†]We are ignoring the errors in the parameter estimates in calculating the forecast reliability. These errors should be small for a series of this length.

FIGURE 10.20

Forecasts of water usage

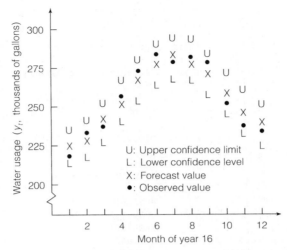

into the future. This property of the prediction intervals makes long-term forecasts very unreliable.

The variety and complexity of time series modeling techniques are overwhelming. We have barely scratched the surface here. However, if we have convinced you that time series modeling is a useful and powerful tool for business forecasting, we have accomplished our purpose. The successful construction of time series models requires much experience, and entire texts are devoted to the subject (see the references at the end of the chapter).

We conclude with a warning: Many oversimplified forecasting methods have been proposed. They usually consist of graphical extensions of a trend or seasonal pattern to future time periods. Although such pictorial techniques are easy to understand and therefore are intuitively appealing, they should be avoided. There is no measure of reliability for these forecasts, and thus the risk associated with making decisions based on them is very high.

10.11
Forecasting Using Lagged Values of the Dependent Variable (Optional)

In Section 10.7, we discussed a variety of choices for the deterministic component, $E(y_t)$, of the time series models. All these models were functions of independent variables, such as t, x_t, x_{t-1}, and seasonal dummy variables. Often, the forecast of y_t can be improved by adding *lagged values of the dependent variable* to the model. For example, since the price y_t of a stock on day t is highly correlated with the price on the previous day, (i.e., on day $t - 1$), a useful model for $E(y_t)$ is

$$E(y_t) = \beta_0 + \beta_1 y_{t-1}$$

Models with lagged values of y_t tend to violate the standard regression assumptions outlined in Section 4.2; thus, they must be fitted using specialized methods.

Box and Jenkins (1977) developed a method of analyzing time series models based on past values of y_t and past values of the random error ε_t. The general model, denoted **ARMA(p, q)**, takes the form

$$y_t + \phi_1 y_{t-1} + \phi_2 y_{t-2} + \cdots + \phi_p y_{t-p} = \varepsilon_t + \theta_1 \varepsilon_{t-1} + \theta_2 \varepsilon_{t-2} + \cdots + \theta_q \varepsilon_{t-q}$$

Note that the left side of the equation is a **pth-order autoregressive model** for y_t (see Section 10.5), whereas the right side of the equation is a **qth-order moving average model** for the random error ε_t (see Section 10.6).

The analysis of an ARMA(p, q) model is divided into three stages: (1) identification, (2) estimation, and (3) forecasting. In the identification stage, the values of p and q are determined from the sample data. That is, the order of both the autoregressive portion and the moving average portion of the model are identified.*For example, the analyst may find the best fit to be an ARMA model with $p = 2$ and $q = 0$. Substituting $p = 2$ and $q = 0$ into the previous equation, we obtain the ARMA(2, 0) model

$$y_t + \phi_1 y_{t-1} + \phi_2 y_{t-2} = \varepsilon_t$$

Note that since $q = 0$, there is no moving average component to the model.

Once the model is identified, the second stage involves obtaining estimates of the model's parameters. In the case of the ARMA(2, 0) model, we require estimates of the autoregressive parameters ϕ_1 and ϕ_2. Tests for model adequacy are conducted, and, if the model is deemed adequate, the estimated model is used to forecast future values of y_t in the third stage.

Analysis of ARMA(p, q) models for y_t requires a level of expertise that is beyond the scope of this text. Even with this level of expertise, the analyst cannot hope to proceed without the aid of a sophisticated computer program. Procedures for identifying, estimating, and forecasting with ARMA(p, q) models are available in SAS, SPSS, and MINITAB. Before attempting to run these procedures, however, you should consult the references provided at the end of this chapter.

Summary

Time series are often modeled as a combination of four components: **secular, seasonal, cyclical**, and **residual**. Both descriptive and inferential techniques are available for **estimating** the time series components and **forecasting** future values of the time series. The **moving average method** is a smoothing technique that uses estimates of the secular and seasonal components to forecast future values of a time series. However, the method requires you to extrapolate the moving average into the future to obtain the forecasts. Two alternative smoothing techniques that lead to explicit forecasts are **exponential smoothing** and the **Holt–Winters model**. Exponential smoothing is an adaptive forecasting method for time series with little or no secular or seasonal trends. The Holt–Winters model is an extension of the exponential smoothing technique that allows for trend and seasonal components.

One type of **inferential time series model** employs a combination of the **deterministic component $E(y_t)$** of the typical multiple regression model with an autoregressive model for the **autocorrelated residual**. The deterministic portion of the model accounts for the trend and seasonal components, and the autocorrelated residual deals with the problem of correlated errors.

---

*This step involves a careful examination of a plot of the sample autocorrelations. Certain patterns in the plot (such as those shown in Figures 10.9–10.12) allow the analyst to identify p and q.

The forecaster should be very careful to **distinguish between descriptive and inferential time series models**. If descriptive models (e.g., smoothing techniques) are used to predict future values of the series, no assessment of forecast reliability is possible. Only when a probabilistic model (e.g., a time series autoregressive model) is constructed can a prediction interval be used to evaluate the reliability of the forecast. Even then, if the structure of the model changes at some future time, forecasts beyond that point are probably useless. Careful application of time series modeling and forecasting will usually be rewarded with a better understanding of the phenomenon and with useful forecasts that assist in planning future strategy.

SUPPLEMENTARY EXERCISES

10.36. The level at which commercial lending institutions set mortgage interest rates has a significant effect on the volume of buying, selling, and construction of residential and commercial real estate. The data in the table are the annual average mortgage interest rates for conventional, fixed-rate, 30-year loans for the period 1980–2000. Forecast the 2003 average mortgage interest rate using each of the methods listed here.

INTRATE30

YEAR	INTEREST RATE (%)
1980	14.30
1981	16.54
1982	16.83
1983	13.92
1984	13.71
1985	12.91
1986	11.33
1987	10.46
1988	10.86
1989	12.07
1990	11.78
1991	11.14
1992	9.29
1993	8.09
1994	8.28
1995	7.86
1996	7.76
1997	7.57
1998	6.92
1999	7.46
2000	8.08

Source: Statistical Abstract of the United States, U.S. Bureau of the Census, 2002.

a. A 3-point moving average
b. The exponential smoothing technique ($w = .2$)
c. The Holt–Winters model with trend ($w = .2$ and $v = .5$)

d. Simple linear regression (Obtain a 95% prediction interval.)
e. A straight-line, first-order autoregressive model (Obtain an approximate 95% prediction interval.)

10.37. The accompanying table records the monthly number of mortgage applications) for new home construction processed by the Federal Housing Administration (FHA) for the period October, 2000 to December, 2001.

MORTAPPS

YEAR	MONTH	NUMBER OF MORTGAGE APPLICATIONS
2000		
	October	97,304
	November	91,116
	December	76,549
2001		
	January	129,494
	February	149,722
	March	169,997
	April	158,187
	May	158,979
	June	143,264
	July	135,532
	August	145,612
	September	123,360
	October	178,253
	November	160,636
	December	107,242

Source: Federal Housing Administration, U.S. Department of Housing and Urban Development.

a. Calculate and plot a 3-point moving average for the mortgage applications time series. Can you detect the secular trend? Does there appear to be a seasonal pattern?

b. Use the moving average from part **a** to forecast mortgage applications in January 2001.

c. Calculate and plot the exponentially smoothed series using $w = .6$.

d. Obtain the forecast for January 2001 using the exponential smoothing technique.

e. Obtain the forecast for January 2001 using the Holt–Winters model with trend and seasonal components and smoothing constants $w = .6$, $v = .7$, and $u = .5$.

f. Propose a time series model for monthly mortgage applications that accounts for secular trend, seasonal variation, and residual autocorrelation.

g. Fit the time series model specified in part **f**, using an available software package.

h. Use the time series model to forecast mortgage applications in January 2001. Obtain an approximate 95% prediction interval for the forecast.

10.38. Refer to the data on average monthly occupancies of hotels and motels in the cities of Atlanta and Phoenix, Exercise 10.14 (p. 492). The data are saved in the ROOMOCC file. In part **a** of Exercise 10.14, you hypothesized a model for mean occupancy (y_r) that accounted for seasonal variation in the series.

a. Modify the model of part **a** of Exercise 10.14 to account for first-order residual correlation.

b. Fit the model in part **a** to the data for each city. Interpret the results.

c. Would you recommend using the model to forecast monthly occupancy rates in year 3? Explain.

10.39. The accompanying table shows U.S. beer production for the years 1973–2000. Suppose you are interested in forecasting U.S. beer production in 2002.

🔘 BEER

YEAR	t	BEER PRODUCTION, y_t (millions of barrels)
1973	1	149
1974	2	156
1975	3	161
1976	4	164
1977	5	171
1978	6	179
1979	7	184
1980	8	194
1981	9	194
1982	10	196
1983	11	196
1984	12	193
1985	13	194
1986	14	197
1987	15	195
1988	16	197

(continued)

1989	17	199
1990	18	202
1991	19	204
1992	20	201
1993	21	202
1994	22	203
1995	23	200
1996	24	200
1997	25	199
1998	26	198
1999	27	198
2000	28	200

Source: 2001 Brewer's Almanac, U.S. Beer Institute.

a. Construct a time series plot for the data. Do you detect a long-term trend?

b. Hypothesize a model for y_t that incorporates the trend.

c. Fit the model to the data using the method of least squares.

d. Plot the least squares model from part **a** and extend the line to forecast y_{30}, the U.S. beer production (in millions of barrels) in 2002. How reliable do you think this forecast is?

e. Calculate and plot the residuals for the model obtained in part **a**. Is there visual evidence of residual autocorrelation?

f. How could you test to determine whether residual autocorrelation exists? If you have access to a computer package, carry out the test. Use $\alpha = .05$.

g. Hypothesize a time series model that will account for the residual autocorrelation. Fit the model to the data and interpret the results.

h. Compute a 95% prediction interval for y_{30}, the U.S. beer production in 2002. Why is this forecast preferred to that of part **b**?

10.40. Suppose you were to fit the time series model

$$E(y_t) = \beta_0 + \beta_1 t + \beta_2 t^2$$

to quarterly time series data collected over a 10-year period ($n = 40$ quarters).

a. Set up the test of hypothesis for positively autocorrelated residuals. Specify H_0, H_a, the test statistic, and the rejection region. Use $\alpha = .05$.

b. Suppose the Durbin–Watson d statistic is calculated to be 1.14. What is the appropriate conclusion?

10.41. Suppose a CPA firm wants to model its monthly income, y_t. The firm is growing at an increasing rate, so that the mean income will be modeled as a second-order function of t. In addition, the mean monthly income increases significantly each year from January through April because of processing tax returns.

a. Write a model for $E(y_t)$ to reflect both the second-order function of time, t, and the January–April jump in mean income.

b. Suppose the size of the January–April jump grows each year. How could this information be included in the model? Assume that 5 years of monthly data are available.

10.42. Refer to the data on annual OPEC oil imports, Exercise 10.4 (p. 482). The data are saved in the OPECOIL file.

a. Plot the time series.

b. Hypothesize a straight-line autoregressive time series model for annual amount of imported crude oil, y_t.

c. Fit the proposed model to the data. Interpret the results.

d. From the output, write the modified least squares prediction equation for y_t.

e. Forecast the amount of foreign crude oil imported into the United States from OPEC in 2003. Place approximate 95% prediction bounds on the forecast value.

REFERENCES

ANDERSON, T. W. *The Statistical Analysis of Time Series*. New York: Wiley, 1972.

ANSLEY, C. F., KOHN, R., and SHIVELY, T. S. "Computing p-values for the generalized Durbin–Watson and other invariant test statistics." *Journal of Econometrics*, Vol. 54, 1992.

BAILLIE, R. T., and BOLLERSLEV, T. "Prediction in dynamic models with time-dependent conditional variances." *Journal of Econometrics*, Vol. 52, 1992.

BOX, G. E. P., and JENKINS, G. M. *Time Series Analysis: Forecasting and Control*, Revised Edition, San Francisco: Holden-Day, 1977.

CHIPMAN, J.S. "Efficiency of least squares estimation of linear trend when residuals are autocorrelated." *Econometrica*, Vol. 47, 1979.

COCHRANE, D., and ORCUTT, G. H. "Application of least squares regression to relationships containing autocorrelated error terms." *Journal of the American Statistical Association*, vol. 44, 1949, 32–61.

ENGLE, R. F., LILIEN, D. M., and ROBINS, R. P., "Estimating time varying risk in the term structure: The ARCH-M model." *Econometrica*. Vol. 55, 1987.

FULLER, W. *Introduction to Time Series*, New York: Wiley, 1978.

FULLER, W. A. *Introduction to Statistical Time Series*, New York: Wiley, 1976.

GALLANT, A. R., and GOEBEL, J. J. "Nonlinear regression with autoregressive errors." *Journal of the American Statistical Association*, Vol. 71, 1976.

GODFREY, L. G. "Testing against general autoregressive and moving average error models when the regressors include lagged dependent variables," *Econometrica*. Vol. 46, 1978, 1293–1301.

GREENE, W. H. *Econometric Analysis*, 2nd ed. New York: Macmillan, 1993.

HAMILTON, J. D. *Time Series Analysis*, Princeton: Princeton University Press, 1994.

HARVEY, A. *The Econometric Analysis of Time Series*, 2nd ed. Cambridge: MIT Press, 1990.

JOHNSTON, J. *Econometric Methods*, 2nd ed. New York: McGraw-Hill, Inc., 1972.

JONES, R. H. "Maximum likelihood fitting of ARMA models to time series with missing observations." *Technometrics*, Vol. 22, 1980.

JUDGE, G. G., GRIFFITHS, W. E., HILL, R. C., and LEE, T. C. *The Theory and Practice of Econometrics*, 2nd ed. New York: Wiley, 1985.

MADDALA, G. S. *Econometrics*, New York: McGraw-Hill, 1977.

MAKRIDAKIS, S. et al. *The Forecasting Accuracy of Major Time Series Methods*. New York: Wiley, 1984.

MCLEOD, A. I., and LI, W. K. "Diagnostic checking ARMA time series models using squared-residual autocorrelations." *Journal of Time Series Analysis*, Vol. 4, 1983.

NELSON, D. B. "Stationarity and persistence in the GARCH(1,1) model." *Econometric Theory*, Vol. 6, 1990.

NELSON, D. B., and CAO, C. Q. "Inequality constraints in the univariate GARCH model." *Journal of Business & Economic Statistics*, Vol. 10, 1992.

PARK, R. E., and MITCHELL, B. M. "Estimating the autocorrelated error model with trended data." *Journal of Econometrics*, Vol. 13, 1980.

SHIVELY, T. S. "Fast evaluation of the distribution of the Durbin–Watson and other invariant test statistics in time series regression. *Journal of the American Statistical Association*, Vol. 85, 1990.

THEIL, H. *Principles of Econometrics*, New York: Wiley, 1971.

WHITE, K. J. "The Durbin–Watson test for autocorrelation in nonlinear models." *Review of Economics and Statistics*, Vol. 74, 1992.

Principles of Experimental Design

CONTENTS

OBJECTIVE

To present an overview of experiments designed to compare two or more population means; to explain the statistical principles of experimental design

---

11.1 Introduction

In Chapter 7, we learned that a regression analysis of observational data has some limitations. In particular, establishing a cause-and-effect relationship between an independent variable x and the response y is difficult since the values of other relevant independent variables—both those in the model and those omitted from the model—are not controlled. Recall that experimental data are data collected with the values of the x's set in advance of observing y (i.e., the values of the x's are controlled). With experimental data, we usually select the x's so that we can compare the mean responses, $E(y)$, for several different combinations of the x values.

The procedure for selecting sample data with the x's set in advance is called the **design of the experiment**. The statistical procedure for comparing the population means is called an **analysis of variance**. The objective of this chapter is to introduce some key aspects of experimental design. The analysis of the data from such experiments using an analysis of variance is the topic of Chapter 12.

11.2 Experimental Design Terminology

The study of experimental design originated with R.A. Fisher in the early 1900s in England. During these early years, it was associated solely with agricultural experimentation. The need for experimental design in agriculture was very clear: It takes a full year to obtain a single observation on the yield of a new variety of most crops. Consequently, the need to save time and money led to a study of ways to obtain more information using smaller samples. Similar motivations led to its subsequent acceptance and wide use in all fields of scientific experimentation. Despite this fact, the terminology associated with experimental design clearly indicates its early association with the biological sciences.

We will call the process of collecting sample data an **experiment** and the (*dependent*) variable to be measured, the **response** y. The planning of the sampling procedure is called the **design** of the experiment. The object upon which the response measurement y is taken is called an **experimental unit**.

Definition 11.1

The process of collecting sample data is called an **experiment**.

Definition 11.2

The plan for collecting the sample is called the **design of the experiment**.

Definition 11.3

The variable measured in the experiment is called the **response variable**.

Definition 11.4

The object upon which the response y is measured is called an **experimental unit**.

Independent variables that may be related to a response variable y are called **factors**. The value—that is, the intensity setting—assumed by a factor in an experiment is called a **level**. The combinations of levels of the factors for which the response will be observed are called **treatments**.

Definition 11.5

The independent variables, quantitative or qualitative, that are related to a response variable y are called **factors**.

Definition 11.6

The intensity setting of a factor (i.e., the value assumed by a factor in an experiment) is called a **level**.

> **Definition 11.7**
> A **treatment** is a particular combination of levels of the factors involved in an experiment.

EXAMPLE   11.1

A designed experiment.   A marketing study is conducted to investigate the effects of brand and shelf location on weekly coffee sales. Coffee sales are recorded for each of two brands (brand A and brand B) at each of three shelf locations (bottom, middle, and top). The $2 \times 3 = 6$ combinations of brand and shelf location were varied each week for a period of 18 weeks. Figure 11.1 is a layout of the design. For this experiment, identify

a. the experimental unit
b. the response, y
c. the factors
d. the factor levels
e. the treatments

FIGURE 11.1

Layout for designed experiment of Example 11.1

		SHELF LOCATION	
	Bottom	Middle	Top
A	Week 1 9 14	Week 2 7 16	Week 4 12 17
BRAND			
B	Week 5 10 13	Week 3 8 18	Week 6 11 15

Solution

a. Since the data will be collected each week for a period of 18 weeks, the experimental unit is 1 week.

b. The variable of interest, i.e., the response, is $y =$ weekly coffee sales. Note that weekly coffee sales is a quantitative variable.

c. Since we are interested in investigating the effect of brand and shelf location on sales, brand and shelf location are the factors. Note that both factors are qualitative variables, although, in general, they may be quantitative or qualitative.

d. For this experiment, brand is measured at two levels (A and B) and shelf location at three levels (bottom, middle, and top).

e. Since coffee sales are recorded for each of the six brand–shelf location combinations (brand A, bottom), (brand A, middle), (brand A, top), (brand B, bottom), (brand B, middle), and (brand B, top), then the experiment involves six treatments (see Figure 11.1). The term *treatments* is used to describe the factor level combinations to be included in an experiment because many experiments involve "treating" or doing something to alter the nature of the experimental unit. Thus, we might view the six brand–shelf location combinations as treatments on the experimental units in the marketing study involving coffee sales. ◆

Now that you understand some of the terminology, it is helpful to think of the design of an experiment in four steps.

STEP 1   Select the factors to be included in the experiment, and identify the parameters that are the object of the study. Usually, the target parameters are the population means associated with the factor level combinations (i.e., treatments).

STEP 2   Choose the treatments (the factor level combinations to be included in the experiment).

STEP 3   Determine the number of observations (sample size) to be made for each treatment. [This will usually depend on the standard error(s) that you desire.]

STEP 4   Plan how the treatments will be assigned to the experimental units. That is, decide on which design to use.

By following these steps, you can control the quantity of information in an experiment. We shall explain how this is done in Section 11.3.

11.3
Controlling the Information in an Experiment

The problem of acquiring good experimental data is analogous to the problem faced by a communications engineer. The receipt of any signal, verbal or otherwise, depends on the volume of the signal and the amount of background noise. The greater the volume of the signal, the greater will be the amount of information transmitted to the receiver. Conversely, the amount of information transmitted is reduced when the background noise is great. These intuitive thoughts about the factors that affect the information in an experiment are supported by the following fact: The standard errors of most estimators of the target parameters are proportional to σ (a measure of data variation or noise) and inversely proportional to the sample size (a measure of the volume of the signal). To illustrate, take the simple case where we wish to estimate a population mean μ by the sample mean $\bar{y}$. The standard error of the sampling distribution of $\bar{y}$ is

$$\sigma_{\bar{y}} = \frac{\sigma}{\sqrt{n}} \quad \text{(see Section 1.7)}$$

For a fixed sample size n, the smaller the value of σ, which measures the **variability (noise)** in the population of measurements, the smaller will be the standard error $\sigma_{\bar{y}}$. Similarly, by increasing the sample size n (**volume of the signal**) in a given experiment, you decrease $\sigma_{\bar{y}}$.

The first three steps in the design of an experiment—selecting the factors and treatments to be included in an experiment and specifying the sample sizes—determine the volume of the signal. You must select the treatments so that the observed values of y provide information on the parameters of interest. Then the larger the treatment sample sizes, the greater will be the quantity of information in the experiment. We present an example of a volume-increasing experiment in Section 11.5.

Is it possible to observe y and obtain no information on a parameter of interest? The answer is yes. To illustrate, suppose that you attempt to fit a first-order model

$$E(y) = \beta_0 + \beta_1 x$$

to a set of $n = 10$ data points, all of which were observed for a single value of x, say, $x = 5$. The data points might appear as shown in Figure 11.2. Clearly, there is no possibility of fitting a line to these data points. The only way to obtain information on β_0 and β_1 is to observe y for *different* values of x. Consequently, the $n = 10$ data points in this example contain absolutely no information on the parameters β_0 and β_1.

FIGURE 11.2

Data set with $n = 10$ responses, all at $x = 5$

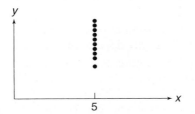

Step 4 in the design of an experiment provides an opportunity to reduce the noise (or experimental error) in an experiment. As we illustrate in Section 11.4, known sources of data variation can be reduced or eliminated by **blocking**—that is, observing all treatments within relatively homogeneous **blocks** of experimental material. When the treatments are compared within each block, any background noise produced by the block is canceled, or eliminated, allowing us to obtain better estimates of treatment differences.

Summary of Steps in Experimental Design

Volume-increasing:	1. Select the factors.
	2. Choose the treatments (factor level combinations).
	3. Determine the sample size for each treatment.
Noise-reducing:	4. Assign the treatments to the experimental units.

In summary, it is useful to think of experimental designs as being either "noise reducers" or "volume increasers." We will learn, however, that most designs are multifunctional. That is, they tend to both reduce the noise and increase the volume of the signal at the same time. Nevertheless, we will find that specific designs lean heavily toward one or the other objective.

11.4 Noise-Reducing Designs

Noise reduction in an experimental design, i.e., the removal of extraneous experimental variation, can be accomplished by an appropriate assignment of treatments to the experimental units. The idea is to compare treatments within blocks of relatively homogeneous experimental units. The most common design of this type is called a **randomized block design**.

To illustrate, suppose we want to compare the mean performance times of female long-distance runners using three different training liquids (e.g., fructose drinks,

TABLE 11.1   Completely Randomized
Design with $p = 3$ Treatments

RUNNER	TREATMENT (LIQUID) ASSIGNED
1	B
2	A
3	B
4	C
5	C
6	A
7	B
8	C
9	A
10	A
11	C
12	A
13	B
14	C
15	B

glucose drinks, and water) 1 hour prior to running a race. Thus, we want to compare the three means μ_A, μ_B, and μ_C, where μ_i is the mean processing time for liquid i. One way to design the experiment is to select 15 female runners (where the runners are the experimental units) and randomly assign one of the three liquids (treatments) to each runner. A diagram of this design, called a **completely randomized design** (since the treatments are randomly assigned to the experimental units), is shown in Table 11.1.

Definition 11.8

A **completely randomized design** to compare p treatments is one in which the treatments are randomly assigned to the experimental units.

This design has the obvious disadvantage that the performance times would vary greatly from runner depending on the fitness level of the athlete, the athlete's age, etc. A better design—one that contains more information on the mean performance times—would be to use only five runners and require each athlete to run three long-distance races, drinking a different liquid before each race. This *randomized block* procedure acknowledges the fact that performance time in a long-distance race varies substantially from runner to runner. By comparing the three performance times for each runner, we eliminate runner-to-runner variation from the comparison.

The randomized block design that we have just described is diagrammed in Figure 11.3. The figure shows that there are five runners. Each runner can be viewed as a **block** of three experimental units—one corresponding to the use of each of the training liquids, A, B, and C. The blocks are said to be **randomized** because the treatments (liquids) are randomly assigned to the experimental units within a block. For our example, the liquids drunk prior to a race would be assigned in random

FIGURE 11.3

Diagram for a randomized block design containing $b = 5$ blocks and $p = 3$ treatments

Blocks (Runners)   Treatments (Liquids)

1 | B | A | C |

2 | A | C | B |

3 | B | C | A |

4 | A | B | C |

5 | A | C | B |

order to avoid bias introduced by other unknown and unmeasured variables that may affect a runner's performance time.

In general, a randomized block design to compare p treatments will contain b relatively homogeneous blocks, with each block containing p experimental units. Each treatment appears once in every block, with the p treatments randomly assigned to the experimental units within each block.

Definition 11.9

A **randomized block design** to compare p treatments involves b blocks, each containing p relatively homogeneous experimental units. The p treatments are randomly assigned to the experimental units within each block, with one experimental unit assigned per treatment.

EXAMPLE 11.2

Suppose you want to compare the abilities of four real estate appraisers, A, B, C, and D. One way to make the comparison would be to randomly allocate a number of pieces of real estate—say, 40—ten to each of the four appraisers. Each appraiser would appraise the property, and you would record y, the difference between the appraised and selling prices expressed as a percentage of the selling price. Thus, y measures the appraisal error expressed as a percentage of selling price, and the treatment allocation to experimental units that we have described is a completely randomized design.

a. Discuss the problems with using a completely randomized design for this experiment.

b. Explain how you could employ a randomized block design.

Solution

a. The problem with using a completely randomized design for the appraisal experiment is that the comparison of mean percentage errors will be influenced by the nature of the properties. Some properties will be easier to appraise than others, and the variation in percentage errors that can be attributed to this fact will make it more difficult to compare the treatment means.

b. To eliminate the effect of property-to-property variability in comparing appraiser means, you could select only 10 properties and require each appraiser to appraise the value of each of the 10 properties. Although in this case there is probably no need for randomization, it might be desirable to randomly assign the order (in time) of the appraisals. This randomized block design, consisting of $p = 4$ treatments and $b = 10$ blocks would appear as shown in Figure 11.4. ◆

FIGURE 11.4

Diagram for a randomized block design: Example 11.2

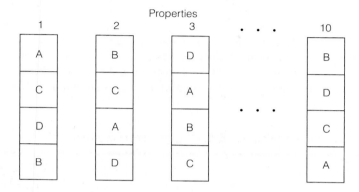

Each experimental design can be represented by a general linear model relating the response y to the factors (treatments, blocks, etc.) in the experiment. When the factors are qualitative in nature (as is often the case), the model includes dummy variables. For example, consider the completely randomized design portrayed in Table 11.1. Since the experiment involves three treatments (liquids), we require two dummy variables. The model for this completely randomized design would appear as follows:

$$y = \beta_0 + \beta_1 x_1 + \beta_2 x_2 + \varepsilon$$

where

$$x_1 = \begin{cases} 1 & \text{if liquid A} \\ 0 & \text{if not} \end{cases} \qquad x_2 = \begin{cases} 1 & \text{if liquid B} \\ 0 & \text{if not} \end{cases}$$

We have arbitrarily selected program C as the base level. From our discussion of dummy-variable models in Chapter 5, we know that the mean responses for the three liquids are

$$\mu_A = \beta_0 + \beta_1$$
$$\mu_B = \beta_0 + \beta_2$$
$$\mu_C = \beta_0$$

Recall that $\beta_1 = \mu_A - \mu_C$ and $\beta_2 = \mu_B - \mu_C$. Thus, to estimate the differences between the treatment means, we require estimates of β_1 and β_2.

Similarly, we can write the model for the randomized block design in Figure 11.3 as follows:

$$y = \beta_0 + \underbrace{\beta_1 x_1 + \beta_2 x_2}_{\text{Treatment effects}} + \underbrace{\beta_3 x_3 + \beta_4 x_4 + \beta_5 x_5 + \beta_6 x_6}_{\text{Block effects}} + \varepsilon$$

where

$$x_1 = \begin{cases} 1 & \text{if liquid A} \\ 0 & \text{if not} \end{cases} \qquad x_2 = \begin{cases} 1 & \text{if liquid B} \\ 0 & \text{if not} \end{cases} \qquad x_3 = \begin{cases} 1 & \text{if runner 1} \\ 0 & \text{if not} \end{cases}$$

$$x_4 = \begin{cases} 1 & \text{if runner 2} \\ 0 & \text{if not} \end{cases} \qquad x_5 = \begin{cases} 1 & \text{if runner 3} \\ 0 & \text{if not} \end{cases} \qquad x_6 = \begin{cases} 1 & \text{if runner 4} \\ 0 & \text{if not} \end{cases}$$

In addition to the treatment terms, the model includes four dummy variables representing the five blocks (runners). Note that we have arbitrarily selected runner 5 as the base level. Using this model, we can write each response y in the experiment of Figure 11.3 as a function of β's, as shown in Table 11.2.

TABLE 11.2   The Response for the Randomized Block Design Shown in Figure 11.3

BLOCKS (RUNNERS)	TREATMENTS (LIQUIDS)		
	$A(x_1 = 1, x_2 = 0)$	$B(x_1 = 0, x_2 = 1)$	$C(x_1 = 0, x_2 = 0)$
1 $(x_3 = 1, x_4 = x_5 = x_6 = 0)$	$y_{A1} = \beta_0 + \beta_1 + \beta_3 + \varepsilon_{A1}$	$y_{B1} = \beta_0 + \beta_2 + \beta_3 + \varepsilon_{B1}$	$y_{C1} = \beta_0 + \beta_3 + \varepsilon_{C1}$
2 $(x_4 = 1, x_3 = x_5 = x_6 = 0)$	$y_{A2} = \beta_0 + \beta_1 + \beta_4 + \varepsilon_{A2}$	$y_{B2} = \beta_0 + \beta_2 + \beta_4 + \varepsilon_{B2}$	$y_{C2} = \beta_0 + \beta_4 + \varepsilon_{C2}$
3 $(x_5 = 1, x_3 = x_4 = x_6 = 0)$	$y_{A3} = \beta_0 + \beta_1 + \beta_5 + \varepsilon_{A3}$	$y_{B3} = \beta_0 + \beta_2 + \beta_5 + \varepsilon_{B3}$	$y_{C3} = \beta_0 + \beta_5 + \varepsilon_{C3}$
4 $(x_6 = 1, x_3 = x_4 = x_5 = 0)$	$y_{A4} = \beta_0 + \beta_1 + \beta_6 + \varepsilon_{A4}$	$y_{B4} = \beta_0 + \beta_2 + \beta_6 + \varepsilon_{B4}$	$y_{C4} = \beta_0 + \beta_6 + \varepsilon_{C4}$
5 $(x_3 = x_4 = x_5 = x_6 = 0)$	$y_{A5} = \beta_0 + \beta_1 + \varepsilon_{A5}$	$y_{B5} = \beta_0 + \beta_2 + \varepsilon_{B5}$	$y_{C5} = \beta_0 + \varepsilon_{C5}$

For example, to obtain the model for the response y for treatment A in block 1 (denoted y_{A1}), we substitute $x_1 = 1$, $x_2 = 0$, $x_3 = 1$, $x_4 = 0$, $x_5 = 0$, and $x_6 = 0$ into the equation. The resulting model is

$$y_{A1} = \beta_0 + \beta_1 + \beta_3 + \varepsilon_{A1}$$

Now we will use Table 11.2 to illustrate how a randomized block design reduces experimental noise. Since each treatment appears in each of the five blocks, there are five measured responses per treatment. Averaging the five responses for treatment A shown in Table 11.2, we obtain

$$\bar{y}_A = \frac{y_{A1} + y_{A2} + y_{A3} + y_{A4} + y_{A5}}{5}$$

$$= [(\beta_0 + \beta_1 + \beta_3 + \varepsilon_{A1}) + (\beta_0 + \beta_1 + \beta_4 + \varepsilon_{A2}) + (\beta_0 + \beta_1 + \beta_5 + \varepsilon_{A3})$$

$$+ (\beta_0 + \beta_1 + \beta_6 + \varepsilon_{A4}) + (\beta_0 + \beta_1 + \varepsilon_{A5})]/5$$

$$= \frac{5\beta_0 + 5\beta_1 + (\beta_3 + \beta_4 + \beta_5 + \beta_6) + (\varepsilon_{A1} + \varepsilon_{A2} + \varepsilon_{A3} + \varepsilon_{A4} + \varepsilon_{A5})}{5}$$

$$= \beta_0 + \beta_1 + \frac{(\beta_3 + \beta_4 + \beta_5 + \beta_6)}{5} + \bar{\varepsilon}_A$$

Similarly, the mean responses for treatments B and C are obtained:

$$\overline{y}_B = \frac{y_{B1} + y_{B2} + y_{B3} + y_{B4} + y_{B5}}{5}$$

$$= \beta_0 + \beta_2 + \frac{(\beta_3 + \beta_4 + \beta_5 + \beta_6)}{5} + \overline{\varepsilon}_B$$

$$\overline{y}_C = \frac{y_{C1} + y_{C2} + y_{C3} + y_{C4} + y_{C5}}{5}$$

$$= \beta_0 + \frac{(\beta_3 + \beta_4 + \beta_5 + \beta_6)}{5} + \overline{\varepsilon}_C$$

Since the objective is to compare treatment means, we are interested in the differences $\overline{y}_A - \overline{y}_B, \overline{y}_A - \overline{y}_C,$ and $\overline{y}_B - \overline{y}_C$, which are calculated as follows:

$$\overline{y}_A - \overline{y}_B = [\beta_0 + \beta_1 + (\beta_3 + \beta_4 + \beta_5 + \beta_6)/5 + \overline{\varepsilon}_A]$$
$$- [\beta_0 + \beta_2 + (\beta_3 + \beta_4 + \beta_5 + \beta_6)/5 + \overline{\varepsilon}_B]$$
$$= (\beta_1 - \beta_2) + (\overline{\varepsilon}_A - \overline{\varepsilon}_B)$$

$$\overline{y}_A - \overline{y}_C = [\beta_0 + \beta_1 + (\beta_3 + \beta_4 + \beta_5 + \beta_6)/5 + \overline{\varepsilon}_A]$$
$$- [\beta_0 + (\beta_3 + \beta_4 + \beta_5 + \beta_6)/5 + \overline{\varepsilon}_C]$$
$$= \beta_1 + (\overline{\varepsilon}_A - \overline{\varepsilon}_C)$$

$$\overline{y}_B - \overline{y}_C = [\beta_0 + \beta_2 + (\beta_3 + \beta_4 + \beta_5 + \beta_6)/5 + \overline{\varepsilon}_B]$$
$$- [\beta_0 + (\beta_3 + \beta_4 + \beta_5 + \beta_6)/5 + \overline{\varepsilon}_C]$$
$$= \beta_2 + (\overline{\varepsilon}_B - \overline{\varepsilon}_C)$$

For each pairwise comparison, the block β's (β_3, β_4, β_5, and β_6) cancel out, leaving only the treatment β's (β_1 and β_2). That is, the experimental noise resulting from differences between blocks is eliminated when treatment means are compared. The quantities $\overline{\varepsilon}_A - \overline{\varepsilon}_B, \overline{\varepsilon}_A - \overline{\varepsilon}_C,$ and $\overline{\varepsilon}_B - \overline{\varepsilon}_C$ are the errors of estimation and represent the noise that tends to obscure the true differences between the treatment means.

What would occur if we employed the completely randomized design of Table 11.1 rather than the randomized block design? Since each runner is assigned to drink a single liquid, each treatment does not appear in each block. Consequently, when we compare the treatment means, the runner-to-runner variation (i.e., the block effects) will not cancel. For example, the difference between $\overline{y}_A$ and $\overline{y}_C$ would be

$$\overline{y}_A - \overline{y}_C = \beta_1 + \underbrace{(\text{Block } \beta\text{'s that do not cancel}) + (\overline{\varepsilon}_A - \overline{\varepsilon}_C)}_{\text{Error of estimation}}$$

Thus, for the completely randomized design, the error of estimation will be increased by an amount involving the block effects (β_3, β_4, β_5, and β_6) that do not cancel. These effects, which inflate the error of estimation, cancel out for the randomized block design, thereby reducing the noise in the experiment.

EXAMPLE 11.3

Refer to Example 11.2 and the randomized block design employed to compare the mean percentage error rates for the four appraisers. The design is illustrated in Figure 11.4.

a. Write the model for the randomized block design.

b. Interpret the β parameters of the model, part **a**.

c. How can we use the model, part **a**, to test for differences among the mean percentage error rates of the four appraisers?

Solution

a. The experiment involves a qualitative factor (Appraisers) at four levels, which represent the treatments. The blocks for the experiment are the 10 properties. Therefore, the model is

$$E(y) = \beta_0 + \underbrace{\beta_1 x_1 + \beta_2 x_2 + \beta_3 x_3}_{\text{Treatments (Appraisers)}} + \underbrace{\beta_4 x_4 + \beta_5 x_5 + \cdots + \beta_{12} x_{12}}_{\text{Blocks (Properties)}}$$

where

$$x_1 = \begin{cases} 1 & \text{if appraiser A} \\ 0 & \text{if not} \end{cases} \qquad x_2 = \begin{cases} 1 & \text{if appraiser B} \\ 0 & \text{if not} \end{cases} \qquad x_3 = \begin{cases} 1 & \text{if appraiser C} \\ 0 & \text{if not} \end{cases}$$

$$x_4 = \begin{cases} 1 & \text{if property 1} \\ 0 & \text{if not} \end{cases} \qquad x_5 = \begin{cases} 1 & \text{if property 2} \\ 0 & \text{if not,} \end{cases} \quad \cdots \quad x_{12} = \begin{cases} 1 & \text{if property 9} \\ 0 & \text{if not} \end{cases}$$

b. Note that we have arbitrarily selected appraiser D and property 10 as the base levels. Following our discussion in Section 5.8, the interpretations of the β's are

$\beta_1 = \mu_A - \mu_D$ for a given property

$\beta_2 = \mu_B - \mu_D$ for a given property

$\beta_3 = \mu_C - \mu_D$ for a given property

$\beta_4 = \mu_1 - \mu_{10}$ for a given appraiser

$\beta_5 = \mu_2 - \mu_{10}$ for a given appraiser

$\vdots$

$\beta_{12} = \mu_9 - \mu_{10}$ for a given appraiser

c. One way to determine whether the means for the four appraisers differ is to test the null hypothesis

$H_0 : \mu_A = \mu_B = \mu_C = \mu_D$

From our β interpretations in part **b**, this hypothesis is equivalent to testing

$H_0 : \beta_1 = \beta_2 = \beta_3 = 0$

To test this hypothesis, we drop the treatment β's (β_1, β_2, and β_3) from the complete model and fit the reduced model

$$E(y) = \beta_0 + \beta_4 x_4 + \beta_5 x_5 + \cdots + \beta_{12} x_{12}$$

Then we conduct the nested model partial F test (see Section 4.13), where

$$F = \frac{(\text{SSE}_{\text{Reduced}} - \text{SSE}_{\text{Complete}})/3}{\text{MSE}_{\text{Complete}}} \qquad \blacklozenge$$

The randomized block design represents one of the simplest types of noise-reducing designs. Other, more complex designs that employ the principle of blocking remove trends or variation in two or more directions. The **Latin square design** is useful when you want to eliminate two sources of variation, i.e., when you want to block in two directions. **Latin cube designs** allow you to block in three directions. A further variation in blocking occurs when the block contains fewer experimental units than the number of treatments. By properly assigning the treatments to a specified number of blocks, you can still obtain an estimate of the difference between a pair of treatments free of block effects. These are known as **incomplete block designs**. Consult the references for details on how to set up these more complex block designs.

EXERCISES

11.1. What two factors affect the quantity of information in an experiment?

11.2. How do block designs increase the quantity of information in an experiment?

11.3. *Applied Animal Behaviour Science* (October, 2000) published a study of the taste preferences of caged cockatiels. A sample of birds bred at the University of California, Davis, were randomly divided into three experimental groups. Group 1 was fed purified water in bottles on both sides of the cage. Group 2 was fed water on one side and a liquid sucrose (sweet) mixture on the opposite side of the cage. Group 3 was fed water on one side and a liquid sodium chloride (sour) mixture on the opposite side of the cage. One variable of interest to the researchers was total consumption of liquid by each cockatiel.
   a. What is the experimental unit for this study?
   b. Is the study a designed experiment? What type of design is employed?
   c. What are the factors in the study?
   d. Give the levels of each factor.
   e. How many treatments are in the study? Identify them.
   f. What is the response variable?
   g. Write the regression model for the designed experiment.

11.4. A commonly used index to estimate the reliability of a building subjected to lateral loads is the drift ratio. Sophisticated computer programs such as STAAD-III have been developed to estimate the drift ratio based on variables such as beam stiffness, column stiffness, story height, moment of inertia, etc. Civil engineers at SUNY, Buffalo, and the University of Central Florida performed an experiment to compare drift ratio estimates using STAAD-III with the estimates produced by a new, simpler micro-computer program called DRIFT (*Microcomputers in Civil Engineering*, 1993). Data for a 21-story building were used as input to the programs. Two runs were made with STAAD-III: Run 1 considered axial deformation of the building columns, and run 2 neglected this information. The goal of the analysis was to compare the mean drift ratios (where drift is measured as lateral displacement) estimated by the three computer runs.
   a. Identify the treatments in the experiment.
   b. Because lateral displacement will vary greatly across building levels (floors), a randomized block design will be used to reduce the level-to-level variation in drift. Explain, diagrammatically, the set-up of the design if all 21 levels are to be included in the study.
   c. Write the linear model for the randomized block design.

11.5. Refer to the randomized block design of Examples 11.2 and 11.3.
   a. Write the model for each observation of percentage appraisal error y for appraiser B. Sum the observations to obtain the average for appraiser B.
   b. Repeat part **a** for appraiser D.
   c. Show that $(\bar{y}_B - \bar{y}_D) = \beta_2 + (\bar{\varepsilon}_B - \bar{\varepsilon}_D)$. Note that the β's for blocks cancel when computing this difference.

11.5
Volume-Increasing Designs

In this section, we focus on how the proper choice of the treatments associated with *two or more factors* can increase the "volume" of information extracted from the experiment. The volume-increasing designs we will discuss are commonly known as **factorial designs** because they involve careful selection of the combinations of **factor levels** (i.e., treatments) in the experiment.

Consider a utility company that charges its customers a lower rate for using electricity during off-peak (less demanded) hours. The company is experimenting with several time-of-day pricing schedules. Two factors (i.e., independent variables) that the company can manipulate to form the schedule are price ratio, x_1, measured as the ratio of peak to off-peak prices, and peak period length, x_2, measured in hours. Suppose the utility company wants to investigate pricing ratio at two levels, 200% and 400%, and peak period length at two levels, 6 and 9 hours. The company will measure customer satisfaction, y, for several different schedules (i.e., combinations of x_1 and x_2) with the goal of comparing the mean satisfaction levels of the schedules. How should the company select the treatments for the experiment?

One method of selecting the combined levels of price ratio and peak period length to be assigned to the experimental units (customers) would be to use the "one-at-a-time" approach. According to this procedure, one independent variable is varied while the remaining independent variables are held constant. This process is repeated only once for each of the independent variables in the experiment. This plan would *appear* to be extremely logical and consistent with the concept of blocking introduced in Section 11.4—that is, making comparisons within relatively homogeneous conditions—but this is not the case, as we will demonstrate.

FIGURE 11.5

One-at-a-time approach to selecting treatments

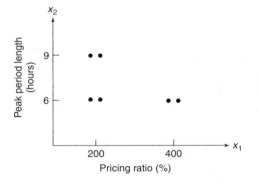

The one-at-a-time approach applied to price ratio (x_1) and peak period length (x_2) is illustrated in Figure 11.5. When length is held constant at $x_2 = 6$ hours, we will observe the response y at a ratio of $x_1 = 200\%$ and $x_1 = 400\%$, thus yielding one pair of y values to estimate the average change in customer satisfaction as a result of changing the pricing ratio (x_1). Also, when pricing ratio is held constant at $x_1 = 200\%$, we observe the response y at a peak period length of $x_2 = 9$ hours. This observation, along with the one at (200%, 6 hours), allows us to estimate the average change in customer satisfaction due to a change in peak period length (x_2). The three treatments just described, (200%, 6 hours), (400%, 6 hours), and (200%, 9 hours), are indicated as points in Figure 11.5. The figure shows two measurements (points) for each treatment. This is necessary to obtain an estimate of the standard deviation of the differences of interest.

A second method of selecting the factor level combinations would be to choose the same three treatments as implied by the one-at-a-time approach and then to choose the fourth treatment at (400%, 9 hours) as shown in Figure 11.6. In other words, we have varied both variables, x_1 and x_2, at the same time.

FIGURE 11.6

Selecting all possible treatments

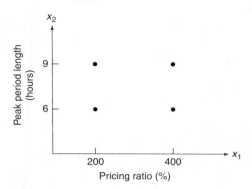

Which of the two designs yields more information about the treatment differences? Surprisingly, the design of Figure 11.6, with only four observations, yields more accurate information than the one-at-a-time approach with its six observations. First, note that both designs yield two estimates of the difference between the mean response y at $x_1 = 200\%$ and $x_1 = 400\%$ when peak period length (x_2) is held constant, and both yield two estimates of the difference between the mean response y at $x_2 = 6$ hours and $x_2 = 9$ hours when pricing ratio (x_1) is held constant. But what if the difference between the mean response y at $x_1 = 200\%$ and at $x_1 = 400\%$ depends on which level of x_2 is held fixed? That is, what if pricing ratio (x_1) and peak period length (x_2) *interact*? Then, we require estimates of the mean difference $(\mu_{200} - \mu_{400})$ when $x_2 = 6$ and the mean difference $(\mu_{200} - \mu_{400})$ when $x_2 = 9$. Estimates of both these differences are obtainable from the second design, Figure 11.6. However, since no estimate of the mean response for $x_1 = 400$ and $x_2 = 9$ is available from the one-at-a-time method, the interaction will go undetected for this design!

The importance of interaction between independent variables was emphasized in Section 4.9 and Chapter 5. If interaction is present, we cannot study the effect of one variable (or factor) on the response y independent of the other variable. Consequently, we require experimental designs that provide information on factor interaction.

Designs that accomplish this objective are called **factorial experiments**. A **complete factorial experiment** is one that includes all possible combinations of the levels of the factors as treatments. For the experiment on time-of-day pricing, we have two levels of pricing ratio (200% and 400%) and two levels of peak period length (6 and 9 hours). Hence, a complete factorial experiment will include $(2 \times 2) = 4$ treatments, as shown in Figure 11.6, and is called a **2 × 2 factorial design**.

Definition 11.10

A **factorial design** is a method for selecting the treatments (that is, the factor level combinations) to be included in an experiment. A complete factorial experiment is one in which the treatments consist of all factor level combinations.

If we were to include a third factor, say, season, at four levels, then a complete factorial experiment would include all $2 \times 2 \times 4 = 16$ combinations of pricing ratio, peak period length, and season. The resulting collection of data would be called a **$2 \times 2 \times 4$ factorial design**.

EXAMPLE   11.4

Suppose you plan to conduct an experiment to compare the yield strengths of nickel alloy tensile specimens charged in a sulfuric acid solution. In particular, you want to investigate the effect on mean strength of three factors: nickel composition at three levels (A_1, A_2, and A_3), charging time at three levels (B_1, B_2, and B_3), and alloy type at two levels (C_1 and C_2). Consider a complete factorial experiment. Identify the treatments for this $3 \times 3 \times 2$ factorial design.

Solution

The complete factorial experiment includes all possible combinations of nickel composition, charging time, and alloy type. We therefore would include the following treatments: $A_1B_1C_1$, $A_1B_1C_2$, $A_1B_2C_1$, $A_1B_2C_2$, $A_1B_3C_1$, $A_1B_3C_2$, $A_2B_1C_1$, $A_2B_1C_2$, $A_2B_2C_1$, $A_2B_2C_2$, $A_2B_3C_1$, $A_2B_3C_2$, $A_3B_1C_1$, $A_3B_1C_2$, $A_3B_2C_1$, $A_3B_2C_2$, $A_3B_3C_1$, $A_3B_3C_2$. These 18 treatments are diagrammed in Figure 11.7.   ◆

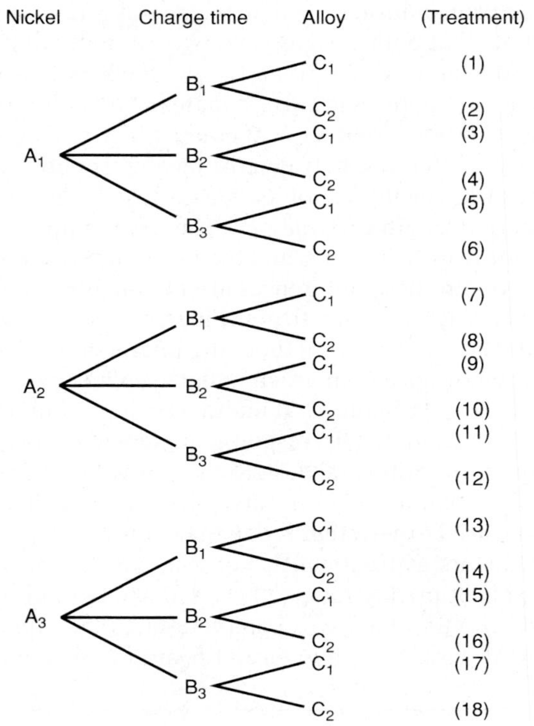

FIGURE 11.7

The 18 treatments for the
$3 \times 3 \times 2$ factorial of Example 11.4

The linear statistical model for a factorial design includes terms for each of the factors in the experiment—called **main effects**—and terms for factor interactions. For example, the model for the 2×2 factorial for the time-of-day pricing experiment includes a first-order term for the quantitative factor, pricing ratio (x_1); a first-order term for the quantitative factor, peak period length (x_2); and an interaction term:

$$y = \beta_0 + \underbrace{\beta_1 x_1 + \beta_2 x_2}_{\text{Main effects}} + \underbrace{\beta_3 x_1 x_2}_{\text{Interaction}} + \varepsilon$$

In general, the regression model for a complete factorial design for k factors contains terms for the following:

The main effects for each of the k factors

Two-way interaction terms for all pairs of factors

Three-way interaction terms for all combinations of three factors

$$\vdots$$

k-way interaction terms of all combinations of k factors

If the factors are qualitative, then we set up dummy variables and proceed as in the next example.

EXAMPLE   11.5

Write the model for the $3 \times 3 \times 2$ factorial experiment of Example 11.4.

Solution

Since the factors are qualitative, we set up dummy variables as follows:

$$x_1 = \begin{cases} 1 & \text{if nickel } A_1 \\ 0 & \text{if not} \end{cases} \qquad x_2 = \begin{cases} 1 & \text{if nickel } A_2 \\ 0 & \text{if not} \end{cases}$$

$$x_3 = \begin{cases} 1 & \text{if charge } B_1 \\ 0 & \text{if not} \end{cases} \qquad x_4 = \begin{cases} 1 & \text{if charge } B_2 \\ 0 & \text{if not} \end{cases}$$

$$x_5 = \begin{cases} 1 & \text{if alloy } C_1 \\ 0 & \text{if alloy } C_2 \end{cases}$$

Then the appropriate model is

$$y = \beta_0 + \underbrace{\beta_1 x_1 + \beta_2 x_2}_{\text{Nickel main effects}} + \underbrace{\beta_3 x_3 + \beta_4 x_4}_{\text{Charge main effects}} + \underbrace{\beta_5 x_5}_{\text{Alloy main effect}}$$

$$+ \underbrace{\beta_6 x_1 x_3 + \beta_7 x_1 x_4 + \beta_8 x_2 x_3 + \beta_9 x_2 x_4}_{\text{Nickel} \times \text{Charge}} + \underbrace{\beta_{10} x_1 x_5 + \beta_{11} x_2 x_5}_{\text{Nickel} \times \text{Alloy}}$$

$$+ \underbrace{\beta_{12} x_3 x_5 + \beta_{13} x_4 x_5}_{\text{Charge} \times \text{Alloy}}$$

$$+ \underbrace{\beta_{14} x_1 x_3 x_5 + \beta_{15} x_1 x_4 x_5 + \beta_{16} x_2 x_3 x_5 + \beta_{17} x_2 x_4 x_5}_{\text{Nickel} \times \text{Charge} \times \text{Alloy}}$$

◆

Note that the number of parameters in the model for the $3 \times 3 \times 2$ factorial design of Example 11.5 is 18, which is equal to the number of treatments contained in the experiment. This is always the case for a complete factorial experiment.

Consequently, if we fit the complete model to a single replication of the factorial treatments (i.e., one y observation measured per treatment), we will have no degrees of freedom available for estimating the error variance, σ^2. One way to solve this problem is to add additional data points to the sample. Researchers usually accomplish this by **replicating** the complete set of factorial treatments. That is, we collect two or more observed y values for each treatment in the experiment. This provides sufficient degrees of freedom for estimating σ^2.

One potential disadvantage of a complete factorial experiment is that it may require a large number of treatments. For example, an experiment involving 10 factors each at two levels would require $2^{10} = 1,024$ treatments! This might occur in an exploratory study where we are attempting to determine which of a large set of factors affect the response y. Several volume-increasing designs are available that employ only a fraction of the total number of treatments in a complete factorial experiment. For this reason, they are called **fractional factorial experiments**. Fractional factorials permit the estimation of the β parameters of lower-order terms (e.g., main effects and two-way interactions); however, β estimates of certain higher-order terms (e.g., three-way and four-way interactions) will be the same as some lower-order terms, thus confounding the results of the experiment. Consequently, a great deal of expertise is required to run and interpret fractional factorial experiments. Consult the references for details on fractional factorials and other more complex, volume-increasing designs.

EXERCISES

11.6. In what sense does a factorial experiment increase the quantity of information in an experiment?

11.7. In *Teaching of Psychology* (August, 1998), a study investigated whether final exam performance is affected by whether students take a practice test. Students in an introductory psychology class at Pennsylvania State University were initially divided into three groups based on their class standing: low, medium, or high. Within each group, the students were randomly assigned either to attend a review session or to take a practice test before the final exam. Six groups were formed: low, review; low, practice exam; medium, review; medium, practice exam; high, review; and high, practice exam: One goal of the study was to compare the mean final exam scores of the six groups of students.
  a. What is the experimental unit for this study?
  b. Is the study a designed experiment? What type of design is employed?
  c. What are the factors in the study?
  d. Give the levels of each factor.
  e. How many treatments are in the study? Identify them.
  f. What is the response variable?

11.8. Many cognitively demanding jobs (e.g., air traffic controller, radar/sonar operator) require efficient processing of visual information. Researchers at Georgia Tech investigated the variables that affect the reaction time of subjects performing a visual search task. (*Human Factors*, June 1993.) College students were trained on computers using one of two methods: continuously consistent or adjusted consistent. Each student was then assigned to one of six different practice sessions. Finally, the consistency of the search task was manipulated at four degrees: 100%, 67%, 50%, or 33%. The goal of the researcher was to compare the mean reaction times of students assigned to each of the (training method) × (practice session) × (task consistency) = $2 \times 6 \times 4 = 48$ experimental conditions.
  a. List the factors involved in the experiment.
  b. For each factor, state whether it is quantitative or qualitative.
  c. How many treatments are involved in this experiment? List them.

11.9. Consider a factorial design with two factors, A and B, each at three levels. Suppose we select the following treatment (factor level) combinations to be included in the experiment: A_1B_1, A_2B_1, A_3B_1, A_1B_2, and A_1B_3.

a. Is this a complete factorial experiment? Explain.

b. Explain why it is impossible to investigate AB interaction in this experiment.

11.10. Write the complete factorial model for a 2×3 factorial experiment where both factors are qualitative.

11.11. Write the complete factorial model for a $2 \times 3 \times 3$ factorial experiment where the factor at two levels is quantitative and the other two factors are qualitative.

11.12. Suppose you wish to investigate the effect of three factors on a response y. Explain why a factorial selection of treatments is better than varying each factor, one at a time, while holding the remaining two factors constant.

11.13. Why is the randomized block design a poor design to use to investigate the effect of two qualitative factors on a response y?

11.6 Selecting the Sample Size

We demonstrated how to select the sample size for estimating a single population mean or comparing two population means in Sections 1.8 and 1.10. We now show you how this problem can be solved for designed experiments.

As mentioned in Section 11.3, a measure of the quantity of information in an experiment that is pertinent to a particular population parameter is the standard error of the estimator of the parameter. A more practical measure is the half-width of the parameter's confidence interval, which will, of course, be a function of the standard error. For example, the half-width of a confidence interval for a population mean (given in Section 1.8) is

$$(t_{\alpha/2})s_{\bar{y}} = t_{\alpha/2}\left(\frac{s}{\sqrt{n}}\right)$$

Similarly, the half-width of a confidence interval for the slope β_1 of a straight-line model relating y to x (given in Section 3.6) is

$$(t_{\alpha/2})s_{\hat{\beta}_1} = t_{\alpha/2}\left(\frac{s}{\sqrt{SS_{xx}}}\right) = t_{\alpha/2}\left(\sqrt{\frac{SSE}{n-2}}\right)\left(\frac{1}{\sqrt{SS_{xx}}}\right)$$

In both cases, the half-width is a function of the total number of data points in the experiment; each interval half-width gets smaller as the total number of data points n increases. The same is true for a confidence interval for a parameter β_i of a general linear model, for a confidence interval for $E(y)$, and for a prediction interval for y. Since each designed experiment can be represented by a linear model, this result can be used to select, approximately, the number of replications (i.e., the number of observations measured for each treatment) in the experiment.

For example, consider a designed experiment consisting of three treatments, A, B, and C. Suppose we want to estimate $(\mu_B - \mu_C)$, the difference between the treatment means for B and C. From our knowledge of linear models for designed experiments, we know this difference will be represented by one of the β parameters in the model, say, β_2. The confidence interval for β_2 for a single replication of the experiment is

$$\hat{\beta}_2 \pm (t_{\alpha/2})s_{\hat{\beta}_2}$$

If we repeat exactly the same experiment r times (we call this r **replications**), it can be shown (proof omitted) that the confidence interval for β_2 will be

$$\hat{\beta}_2 \pm B \quad \text{where } B = t_{\alpha/2}\left(\frac{s_{\hat{\beta}_2}}{\sqrt{r}}\right)$$

To find r, we first set the half-width of the interval to the largest value, B, we are willing to tolerate. Then we approximate $(t_{\alpha/2})$ and $s_{\hat{\beta}_2}$ and solve for the number of replications r.

EXAMPLE   11.6

Consider a 2×2 factorial experiment to investigate the effect of two factors on the light output y of flashbulbs used in cameras. The two factors (and their levels) are: x_1 = Amount of foil contained in the bulb (100 and 200 milligrams) and x_2 = Speed of sealing machine (1.2 and 1.3 revolutions per minute). The complete model for the 2×2 factorial experiment is

$$E(y) = \beta_0 + \beta_1 x_1 + \beta_2 x_2 + \beta_3 x_1 x_2$$

How many replicates of the 2×2 factorial are required to estimate β_3, the interaction β, to within .3 of its true value using a 95% confidence interval?

Solution

To solve for the number of replicates, r, we want to solve the equation

$$t_{\alpha/2}\left(\frac{s_{\hat{\beta}_3}}{\sqrt{r}}\right) = B$$

You can see that we need to have an estimate of $s_{\hat{\beta}_3}$, the standard error of $\hat{\beta}_3$ for a single replication. Suppose it is known from a previous experiment conducted by the manufacturer of the flashbulbs that $s_{\hat{\beta}_3} \approx .2$. For a 95% confidence interval, $\alpha = .05$ and $\alpha/2 = .025$. Since we want the half-width of the interval to be $B = .3$, we have

$$t_{.025}\left(\frac{.2}{\sqrt{r}}\right) = .3$$

The degrees of freedom for $t_{.025}$ will depend on the sample size $n = (2 \times 2)r = 4r$; consequently, we must approximate its value. In fact, since the model includes four parameters, the degrees of freedom for t will be df(Error) $= n - 4 = 4r - 4 = 4(r - 1)$. At minimum, we require two replicates; hence, we will have at least $4(2 - 1) = 4$ df. In Table 2 of Appendix C, we find $t_{.025} = 2.776$ for df $= 4$. We will use this conservative estimate of t in our calculations.

Substituting $t = 2.776$ into the equation, we have

$$\frac{2.776(.2)}{\sqrt{r}} = .3$$

$$\sqrt{r} = \frac{(2.776)(.2)}{.3} = 1.85$$

$$r = 3.42$$

TABLE 11.3   **2 × 2 Factorial, with Four Replicates**

		AMOUNT OF FOIL, x_1	
		100	200
MACHINE SPEED,	1.2	4 observations on y	4 observations on y
x_2	1.3	4 observations on y	4 observations on y

Since we can run either three or four replications (but not 3.42), we should choose four replications to be reasonably certain that we will be able to estimate the interaction parameter, β_3, to within .3 of its true value. The 2 × 2 factorial with four replicates would be laid out as shown in Table 11.3. ◆

EXERCISES

11.14. Why is replication important in a complete factorial experiment?

11.15. Consider a 2 × 2 factorial. How many replications are required to estimate the interaction β to within two units with a 90% confidence interval? Assume that the standard error of the estimate of the interaction β (based on a single replication) is approximately 3.

11.16. For a randomized block design with b blocks, the estimated standard error of the estimated difference between any two treatment means is $s\sqrt{2/b}$. Use this formula to determine the number of blocks required to estimate $(\mu_A - \mu_B)$, the difference between two treatment means, to within 10 units using a 95% confidence interval. Assume $s \approx 15$.

11.7
The Importance of Randomization

All the basic designs presented in this chapter involve randomization of some sort. In a completely randomized design and a basic factorial experiment, the treatments are randomly assigned to the experimental units. In a randomized block design, the blocks are randomly selected and the treatments within each block are assigned in random order. Why randomize? The answer is related to the assumptions we make about the random error ε in the linear model. Recall (Section 4.2) our assumption that ε follows a normal distribution with mean 0 and constant variance σ^2 for fixed settings of the independent variables (i.e., for each of the treatments). Further, we assume that the random errors associated with repeated observations are independent of each other in a probabilistic sense.

Experimenters rarely know all of the important variables in a process, nor do they know the true functional form of the model. Hence, the functional form chosen to fit the true relation is only an approximation, and the variables included in the experiment form only a subset of the total. The random error, ε, is thus a composite error caused by the failure to include all of the important factors as well as the error in approximating the function.

Although many unmeasured and important independent variables affecting the response y do not vary in a completely random manner during the conduct of a designed experiment, we hope their behavior is such that their cumulative effect varies in a random manner and satisfies the assumptions upon which our inferential procedures are based. *The randomization in a designed experiment has the effect of randomly assigning these error effects to the treatments and assists in satisfying the assumptions on ε.*

Summary

Regression analysis based on observational data has at least one limitation: Even when the independent variables in a model are highly significant, we cannot infer a cause-and-effect relationship between the x's and y. The focus of this chapter was on data collected from a designed experiment in which the values of the independent variables are set in advance of observing y. By controlling the values of the x's, we hope to increase the amount of information extracted from the data.

Experimental design is a plan (or strategy) for collecting the experimental data. The goal is to increase the amount of information by controlling two factors:

1. **Volume** of the signal contained in the data.
2. **Noise** or random variation in the data that is measured by σ^2.

The first three steps in designing an experiment are as follows:

STEP 1   Select the **factors** (i.e., the independent variables) to be investigated.

STEP 2   Choose the factor level combinations (**treatments**).

STEP 3   Determine the number of observations for each treatment (i.e., the number of **replications** of the experiment).

These steps affect the volume of the signal contained in the data because they enable us to shift the information in the experiment so that it focuses on the parameter(s) of interest. An example of a volume-increasing design is a **factorial experiment**, in which all possible treatments (factor level combinations) are selected. With factorial designs, we shift the focus of the experiment to an investigation of factor interaction.

The fourth step in designing an experiment is

STEP 4   Decide how to assign the treatments to the experimental units.

Two basic methods of assigning treatments to the experimental units are the **completely randomized design** and the **randomized block design**. The latter is a noise-reducing design; by assigning treatments to relatively homogeneous blocks of experimental units, we can reduce the variation of treatment differences. The net effect of this action is to reduce experimental noise, measured by the variance of the random error ε that appears in the linear model.

The choice of design, noise-reducing or volume-increasing, will depend on your experimental objectives. In practice, researchers will attempt to employ both principles of design to increase the quantity of information in the experiment. For example, the treatments of a 2×2 factorial could be laid out in blocks to eliminate or reduce an unwanted source of variation. (An example of such a design is given in the next chapter.)

This chapter introduced the key principles of experimental design and presented some basic methods of collecting data in a designed experiment. Other, more complex designs, although beyond the scope of this text, may be more appropriate for your research problem. Consult the references listed at the end of the chapter to learn more about these designs. In Chapter 12, we demonstrate how to analyze experimental data using an **analysis of variance**.

SUPPLEMENTARY EXERCISES

11.17. How do you measure the quantity of information in a sample that is pertinent to a particular population parameter?

11.18. What steps in the design of an experiment affect the volume of the signal pertinent to a particular population parameter?

11.19. In what step in the design of an experiment can you possibly reduce the variation produced by extraneous and uncontrolled variables?

11.20. Explain the difference between a completely randomized design and a randomized block design. When is a randomized block design more advantageous?

11.21. Consider a two-factor factorial experiment where one factor is set at two levels and the other factor is set at four levels. How many treatments are included in the experiment? List them.

11.22. Write the complete factorial model for a $2 \times 2 \times 4$ factorial experiment where both factors at two levels are quantitative and the third factor at four levels is qualitative. If you conduct one replication of this experiment, how many degrees of freedom will be available for estimating σ^2?

11.23. Refer to Exercise 11.22. Write the model for y assuming that you wish to enter main-effect terms for the factors, but no terms for factor interactions. How many degrees of freedom will be available for estimating σ^2?

11.24. Retail store audits are periodic audits of a sample of retail sales to monitor inventory and purchases of a particular product. Such audits are often used by marketing researchers to estimate market share. A study was conducted to compare market shares of beer brands estimated by two different auditing methods.

a. Identify the treatments in the experiment.

b. Because of brand-to-brand variation in estimated market share, a randomized block design will be used. Explain how the treatments might be assigned to the experimental units if 10 beer brands are to be included in the study.

c. Write the linear model for the randomized block design.

11.25. Researchers investigated the effect of gender (male or female) and weight (light or heavy) on the length of time required by firefighters to perform a particular firefighting task (*Human Factors*, 1982). Eight firefighters were selected in each of the four gender—weight categories. Each firefighter was required to perform a certain task. The time (in minutes) needed to perform the task was recorded for each.

a. List the factors involved in the experiment.

b. For each factor, state whether it is quantitative or qualitative.

c. How many treatments are involved in this experiment? List them.

REFERENCES

Box G. E. P., Hunter, W. G., and Hunter, J. S. *Statistics for Experimenters*. New York: Wiley, 1957.

Cochran, W. G., and Cox, G. M. *Experimental Designs*, 2nd ed. New York: Wiley, 1957.

Davies, O. L. *The Design and Analysis of Industrial Experiments*. New York: Hafner, 1956.

Kirk, R. E. *Experimental Design*, 2nd ed. Belmont, Calif.: Brooks/Cole, 1982.

Mendenhall, W. *Introduction to Linear Models and the Design and Analysis of Experiments*. Belmont, Calif.: Wadsworth, 1968.

Neter, J., Kutner, M. Nachtsheim, C. and Wasserman, W. *Applied Linear Statistical Models*, 4th ed. Homewood, Ill.: Richard D. Irwin, 1996.

Winer, B. J. *Statistical Principles in Experimental Design*. New York: McGraw-Hill, 1962.

The Analysis of Variance for Designed Experiments

CONTENTS

OBJECTIVE

To present a method for analyzing data collected from designed experiments for comparing two or more population means; to define the relationship of the analysis of variance to regression analysis and to identify their common features

---

12.1 Introduction

Once the data for a designed experiment have been collected, we will want to use the sample information to make inferences about the population means associated with the various treatments. The method used to compare the treatment means is traditionally known as **analysis of variance**, or **ANOVA**. The analysis of variance procedure provides a set of formulas that enable us to compute test statistics and confidence intervals required to make these inferences.

The formulas—one set for each experimental design—were developed in the early 1900s, well before the invention of computers. The formulas are easy to use, although the calculations can become quite tedious. However, you will recall from Chapter 11 that a linear model is associated with each experimental design. Consequently, the same inferences derived from the ANOVA calculation formulas can be obtained by properly analyzing the model using a regression analysis and the computer.

In this chapter, the main focus is on the regression approach to analyzing data from a designed experiment. Several common experimental designs—some of which were presented in Chapter 11—are analyzed. We also provide the ANOVA calculation formulas for each design and show their relationship to regression. First, we provide the logic behind an analysis of variance and these formulas in Section 12.2.

12.2
The Logic Behind an Analysis of Variance

The concept behind an analysis of variance can be explained using the following simple example.

Consider an experiment with a single factor at two levels (that is, two treatments). Suppose we want to decide whether the two treatment means differ based on the means of two independent random samples, each containing $n_1 = n_2 = 5$ measurements, and that the y values appear as in Figure 12.1. Note that the five circles on the left are plots of the y values for sample 1 and the five solid dots on the right are plots of the y values for sample 2. Also, observe the horizontal lines that pass through the means for the two samples $\bar{y}_1$ and $\bar{y}_2$. Do you think the plots provide sufficient evidence to indicate a difference between the corresponding population means?

FIGURE 12.1

Plots of data for two samples

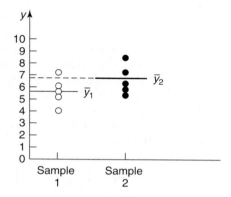

If you are uncertain whether the population means differ for the data in Figure 12.1, examine the situation for two different samples in Figure 12.2a. We think that you will agree that for these data, it appears that the population means differ. Examine a third case in Figure 12.2b. For these data, it appears that there is little or no difference between the population means.

What elements of Figures 12.1 and 12.2 did we intuitively use to decide whether the data indicate a difference between the population means? The answer to the question is that we visually compared the distance (the variation) *between* the sample means to the variation *within* the y values for each of the two samples. Since

FIGURE 12.2

Plots of data for two cases

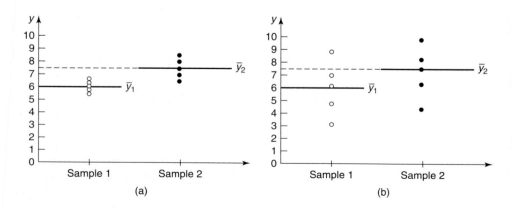

the difference between the sample means in Figure 12.2a is large relative to the within-sample variation, we inferred that the population means differ. Conversely, in Figure 12.2b, the variation between the sample means is small relative to the within-sample variation, and therefore there is little evidence to imply that the means are significantly different.

The variation within samples is measured by the pooled s^2 that we computed for the independent random samples t test of Section 1.9, namely,

$$\textit{Within-sample variation:}\quad s^2 = \frac{\sum_{i=1}^{n_1}(y_{i1} - \bar{y}_1)^2 + \sum_{i=1}^{n_2}(y_{i2} - \bar{y}_2)^2}{n_1 + n_2 - 2}$$

$$= \frac{\text{SSE}}{n_1 + n_2 - 2}$$

where y_{i1} is the ith observation in sample 1 and y_{i2} is the ith observation in sample 2. The quantity in the numerator of s^2 is often denoted **SSE**, the **sum of squared errors**. As with regression analysis, SSE measures unexplained variability. But in this case, it measures variability *unexplained* by the differences between the sample means.

A measure of the between-sample variation is given by the weighted sum of squares of deviations of the individual sample means about the mean for all 10 observations, $\bar{y}$, divided by the number of samples minus 1, i.e.,

$$\textit{Between-sample variation:}\quad \frac{n_1(\bar{y}_1 - \bar{y})^2 + n_2(\bar{y}_2 - \bar{y})^2}{2 - 1} = \frac{\text{SST}}{1}$$

The quantity in the numerator is often denoted **SST**, the **sum of squares for treatments**, since it measures the variability *explained* by the differences between the sample means of the two treatments.

For this experimental design, SSE and SST sum to a known total, namely,

$$\text{SS(Total)} = \sum (y_i - \bar{y})^2$$

[*Note*: SS(Total) is equivalent to SS_{yy} in regression.] Also, the ratio

$$F = \frac{\text{Between-sample variation}}{\text{Within-sample variation}}$$

$$= \frac{\text{SST}/1}{\text{SSE}/(n_1 + n_2 - 2)}$$

has an F distribution with $v_1 = 1$ and $v_2 = n_1 + n_2 - 2$ degrees of freedom (df) and therefore can be used to test the null hypothesis of no difference between the treatment means. The additivity property of the sums of squares led early researchers to view this analysis as a **partitioning** of $\text{SS(Total)} = \Sigma(y_i - \bar{y})^2$ into sources corresponding to the factors included in the experiment and to SSE. The simple formulas for computing the sums of squares, the additivity property, and the form of the test statistic made it natural for this procedure to be called **analysis**

of variance. We demonstrate the analysis of variance procedure and its relation to regression for several common experimental designs in Sections 12.3–12.6.

12.3
One-Factor Completely Randomized Designs

Recall (Section 11.2) the first two steps in designing an experiment: (1) decide on the factors to be investigated and (2) select the factor level combinations (treatments) to be included in the experiment. For example, suppose you wish to compare the length of time to assemble a device in a manufacturing operation for workers who have completed one of three training programs, A, B, and C. Then this experiment involves a single factor, training program, at three levels, A, B, and C. Since training program is the only factor, these levels (A, B, and C) represent the treatments. Now we must decide the sample size for each treatment (step 3) and figure out how to assign the treatments to the experimental units, namely, the specific workers (step 4).

As we learned in Chapter 11, the most common assignment of treatments to experimental units is called a **completely randomized design**. To illustrate, suppose we wish to obtain equal amounts of information on the mean assembly times for the three training procedures; i.e., we decide to assign equal numbers of workers to each of the three training programs. Also, suppose we use the procedure of Section 1.8 (Example 1.13) to select the sample size and determine the number of workers in each of the three samples to be $n_1 = n_2 = n_3 = 10$. Then a completely randomized design is one in which the $n_1 + n_2 + n_3 = 30$ workers are **randomly assigned**, 10 to each of the three treatments. *A random assignment is one in which any one assignment is as probable as any other*. This eliminates the possibility of bias that might occur if the workers were assigned in some systematic manner. For example, a systematic assignment might accidentally assign most of the manually dexterous workers to training program A, thus underestimating the true mean assembly time corresponding to A.

Example 12.1 illustrates how a **random number table** can be used to assign the 30 workers to the three treatments.

EXAMPLE 12.1

Use the random number table, Table 7 in Appendix C, to assign $n = 30$ experimental units to three treatment groups.

Solution

The first step is to number the 30 workers from 1 to 30. We will then use Table 7 in Appendix C to select two-digit numbers, discarding those that are larger than 30 or are identical, until we have a total of 20 two-digit numbers. We will then have 20 of the integers between 1 and 30 arranged in random order. The workers who have been assigned the first 10 numbers in the sequence are assigned to training program A, the second group of 10 workers are assigned to B, and the remaining workers are assigned to C.

To illustrate, suppose we start with the two-digit random number in row 5, column 6 of Table 7 and proceed down the column, selecting only two-digit numbers (the first two digits) less than or equal to 30 and deleting those that repeat. The first 20 are: 20, 18, 13, 16, 19, 04, 14, 06, 30, 25, 27, 17, 24, 21, 22, 02, 15, 05, 09, 08. The workers with the first 10 numbers are assigned to program A, the second 10 to B, and the remaining 10 to C. So the workers are assigned to the training program as shown in Table 12.1. ◆

TABLE 12.1 **Random Assignment of Workers to Treatments**		
A	B	C
20, 18, 13, 16, 19, 4, 14, 6, 30, 25	27, 17, 24, 21, 22, 2, 15, 5, 9, 8	1, 3, 7, 10, 11, 12, 23, 26, 28, 29

EXAMPLE   12.2

Suppose a beverage bottler wished to compare the effect of three different advertising displays on the sales of a beverage in supermarkets. Identify the experimental units you would use for the experiment, and explain how you would employ a completely randomized design to collect the sales data.

Solution

Presumably, the bottler has a list of supermarkets that market the beverage in a number of different cities. If we decide to measure the sales increase (or decrease) as the monthly dollar increase in sales (over the previous month) for a given supermarket, then the experimental unit is a 1-month unit of time in a specific supermarket. Thus, we would randomly select a 1-month period of time for each of $n_1 + n_2 + n_3$ supermarkets and assign n_1 supermarkets to receive display D_1, n_2 to receive D_2, and n_3 to receive D_3.   ◆

In some experimental situations, we are unable to assign the treatment to the experimental units randomly because of the nature of the experimental units themselves. For example, suppose we want to compare the mean annual salaries of professors in three College of Liberal Arts departments: chemistry, mathematics and sociology. Then the treatments—chemistry, mathematics, and sociology—cannot be "assigned" to the professors (experimental units). A professor is a member of the chemistry, mathematics, or sociology (or some other) department and cannot be arbitrarily assigned one of the treatments. Rather, we view the treatments (departments) as populations from which we will select independent random samples of experimental units (professors). A completely randomized design involves a comparison of the means for a number, say, p, of treatments, based on independent random samples of $n_1, n_2, \ldots, n_p$ observations, drawn from populations associated with treatments $1, 2, \ldots, p$, respectively. We repeat our definition of a completely randomized design (given in Section 11.4) with this modification. The general layout for a completely randomized design is shown in Figure 12.3.

FIGURE 12.3

Layout for a completely randomized design

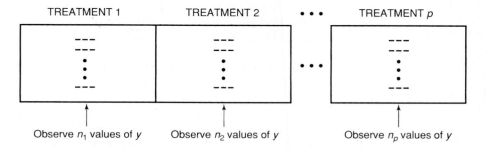

> **Definition 12.1**
>
> A **completely randomized design** to compare p treatment means is one in which the treatments are randomly assigned to the experimental units, or in which independent random samples are drawn from each of the p populations.

After collecting the data from a completely randomized design, we want to make inferences about p population means where μ_i is the mean of the population of measurements associated with treatment i, for $i = 1, 2, \ldots, p$. The null hypothesis to be tested is that the p treatment means are equal, i.e., $H_0 : \mu_1 = \mu_2 = \cdots = \mu_p$, and the alternative hypothesis we wish to detect is that at least two of the treatment means differ. The appropriate linear model for the response y is

$$E(y) = \beta_0 + \beta_1 x_1 + \beta_2 x_2 + \cdots + \beta_{p-1} x_{p-1}$$

where

$$x_1 = \begin{cases} 1 & \text{if treatment 2} \\ 0 & \text{if not} \end{cases} \quad x_2 = \begin{cases} 1 & \text{if treatment 3} \\ 0 & \text{if not} \end{cases} \cdots x_{p-1} = \begin{cases} 1 & \text{if treatment } p \\ 0 & \text{if not} \end{cases}$$

and (arbitrarily) treatment 1 is the base level. Recall that this 0–1 system of coding implies that

$$\begin{aligned} \beta_0 &= \mu_1 \\ \beta_1 &= \mu_2 - \mu_1 \\ \beta_2 &= \mu_3 - \mu_1 \\ \vdots \quad &\quad \vdots \\ \beta_{p-1} &= \mu_p - \mu_1 \end{aligned}$$

The null hypothesis that the p population means are equal is equivalent to the null hypothesis that all the treatment differences equal 0, i.e.,

$$H_0 : \beta_1 = \beta_2 = \cdots = \beta_{p-1} = 0$$

To test this hypothesis using regression, we use the technique of Section 4.13; that is, we compare the sum of squares for error, SSE_R, for the nested *reduced* model

$$E(y) = \beta_0$$

to the sum of squares for error, SSE_C, for the *complete* model

$$E(y) = \beta_0 + \beta_1 x_1 + \beta_2 x_2 + \cdots + \beta_{p-1} x_{p-1}$$

using the F statistic

$$F = \frac{(SSE_R - SSE_C)/\text{Number of } \beta \text{ parameters in } H_0}{SSE_C/[n - (\text{Number of } \beta \text{ parameters in the complete model})]}$$

$$= \frac{(SSE_R - SSE_C)/(p-1)}{SSE_C/(n-p)}$$

$$= \frac{(SSE_R - SSE_C)/(p-1)}{MSE_C}$$

where F is based on $\nu_1 = (p-1)$ and $\nu_2 = (n-p)$ df. If F exceeds the upper critical value, F_α, we reject H_0 and conclude that at least one of the treatment differences, $\beta_1, \beta_2, \ldots, \beta_{p-1}$, differs from zero; i.e., we conclude that at least two treatment means differ.

EXAMPLE   12.3

Show that the F statistic for testing the equality of treatment means in a completely randomized design is equivalent to a global F test of the complete model.

Solution

Since the reduced model contains only the β_0 term, the least squares estimate of β_0 is $\bar{y}$, and it follows that

$$SSE_R = \sum(y - \bar{y})^2 = SS_{yy}$$

We called this quantity the sum of squares for total in Chapter 4. The difference $(SSE_R - SSE_C)$ is simply $(SS_{yy} - SSE)$ for the complete model. Since in regression $(SS_{yy} - SSE) = SS \text{ (Model)}$, and the complete model has $(p-1)$ terms (excluding β_0),

$$F = \frac{(SSE_R - SSE_C)/(p-1)}{MSE_C} = \frac{SS \text{ (Model)}/(p-1)}{MSE} = \frac{MS \text{ (Model)}}{MSE}$$

Thus, it follows that the test statistic for testing the null hypothesis,

$$H_0 : \mu_1 = \mu_2 = \cdots = \mu_p$$

in a completely randomized design is the same as the F statistic for testing the global utility of the complete model for this design.   ◆

The regression approach to analyzing data from a completely randomized design is summarized in the next box. Note that the test requires several assumptions about the distributions of the response y for the p treatments and that these *assumptions are necessary regardless of the sizes of the samples.* (We have more to say about these assumptions in Section 12.9.)

EXAMPLE   12.4

Sociologists often conduct experiments to investigate the relationship between socioeconomic status and college performance. Socioeconomic status is generally partitioned into three groups: lower class, middle class, and upper class. Consider the problem of comparing the mean grade point average of those college freshmen associated with the lower class, those associated with the middle class, and those

---

Model and F Test for a Completely Randomized Design with p Treatments

Complete model: $E(y) = \beta_0 + \beta_1 x_1 + \beta_2 x_2 + \cdots + \beta_{p-1} x_{p-1}$

where $\quad x_1 = \begin{cases} 1 & \text{if treatment 2} \\ 0 & \text{if not} \end{cases} \qquad x_2 = \begin{cases} 1 & \text{if treatment 3} \\ 0 & \text{if not} \end{cases} \quad , \ldots ,$

$$x_{p-1} = \begin{cases} 1 & \text{if treatment } p \\ 0 & \text{if not} \end{cases}$$

$H_0: \beta_1 = \beta_2 = \cdots = \beta_{p-1} = 0$ (i.e., $H_0 : \mu_1 = \mu_2 = \cdots = \mu_p$)

H_a: At least one of the β parameters listed in H_0 differs from 0
(i.e., H_a: At least two means differ)

Test statistic: $\quad F = \dfrac{\text{MS(Model)}}{\text{MSE}}$

Rejection region: $\quad F > F_\alpha$, where the distribution of F is based on
$\quad \nu_1 = p - 1$ and $\nu_2 = (n - p)$ degrees of freedom.

Assumptions: 1. All p population probability distributions corresponding to
the p treatments are normal.
2. The population variances of the p treatments are equal.

---

associated with the upper class. The grade point averages (GPAs) for random samples of seven college freshmen associated with each of the three socioeconomic classes were selected from a university's files at the end of the academic year. The data are recorded in Table 12.2. Do the data provide sufficient evidence to indicate a difference among the mean freshmen GPAs for the three socioeconomic classes? Test using $\alpha = .05$.

Solution

This experiment involves a single factor, socioeconomic class, at three levels. Thus, we have a completely randomized design with $p = 3$ treatments. Let μ_L, μ_M, and μ_U represent the mean GPAs for students in the lower, middle, and upper socioeconomic classes, respectively. Then we want to test

$H_0 : \mu_L = \mu_M = \mu_U$

against

H_a: At least two of the three treatment means differ.

The appropriate linear model for $p = 3$ treatments is

Complete model: $E(y) = \beta_0 + \beta_1 x_1 + \beta_2 x_2$

GPA3

TABLE 12.2   **Grade Point Averages for Three Socioeconomic Groups**

	LOWER CLASS	MIDDLE CLASS	UPPER CLASS
	2.87	3.23	2.25
	2.16	3.45	3.13
	3.14	2.78	2.44
	2.51	3.77	2.54
	1.80	2.97	3.27
	3.01	3.53	2.81
	2.16	3.01	1.36
Sample means:	$\bar{y}_1 = 2.521$	$\bar{y}_2 = 3.249$	$\bar{y}_3 = 2.543$

where

$$x_1 = \begin{cases} 1 & \text{if middle socioeconomic class} \\ 0 & \text{if not} \end{cases}$$

$$x_2 = \begin{cases} 1 & \text{if upper socioeconomic class} \\ 0 & \text{if not} \end{cases}$$

Thus, we want to test $H_0 : \beta_1 = \beta_2 = 0$.

The SAS regression analysis for the complete model is shown in Figure 12.4. The F statistic for testing the overall adequacy of the model (shaded on the printout) is $F = 4.58$, where the distribution of F is based on $v_1 = (p - 1) = 3 - 1 = 2$ and $v_2 = (n - p) = 21 - 3 = 18$ df. For $\alpha = .05$, the critical value (obtained from Table 4 of Appendix C) is $F_{.05} = 3.55$ (see Figure 12.5).

Since the computed value of F, 4.58, exceeds the critical value, $F_{.05} = 3.55$, we reject H_0 and conclude (at the $\alpha = .05$ level of significance) that the mean GPA for college freshmen differs in at least two of the three socioeconomic classes. We can arrive at the same conclusion by noting that $\alpha = .05$ is greater than the p-value (.0247) shaded on the printout. ◆

The analysis of the data in Example 12.4 can also be accomplished using ANOVA computing formulas. In Section 12.2, we learned that an analysis of variance partitions SS(Total) $= \Sigma(y - \bar{y})^2$ into two components, SSE and SST (see Figure 12.6).

Recall that the quantity SST denotes the sum of squares for treatments and measures the variation explained by the differences between the treatment means. The sum of squares for error, SSE, is a measure of the unexplained variability, obtained by calculating a pooled measure of the variability *within* the p samples. If the treatment means truly differ, then SSE should be substantially smaller than SST. We compare the two sources of variability by forming an F statistic:

$$F = \frac{\text{SST}/(p - 1)}{\text{SSE}/(n - p)} = \frac{\text{MST}}{\text{MSE}}$$

where n is the total number of measurements. The numerator of the F statistic, MST = SST/$(p - 1)$, denotes **mean square for treatments** and is based on $(p - 1)$

FIGURE 12.4

SAS regression printout for the completely
randomized design of Example 12.4

Dependent Variable: GPA

Analysis of Variance

Source	DF	Sum of Squares	Mean Square	F Value	Pr > F
Model	2	2.39687	1.19843	4.58	0.0247
Error	18	4.71111	0.26173		
Corrected Total	20	7.10798			

Root MSE	0.51159	R-Square	0.3372
Dependent Mean	2.77095	Adj R-Sq	0.2636
Coeff Var	18.46275		

Parameter Estimates

Variable	DF	Parameter Estimate	Standard Error	t Value	Pr > |t|
Intercept	1	2.52143	0.19336	13.04	<.0001
X1	1	0.72714	0.27346	2.66	0.0160
X2	1	0.02143	0.27346	0.08	0.9384

FIGURE 12.5

Rejection region for
Example 12.4; numerator df = 2,
denominator df = 18, $\alpha = .05$

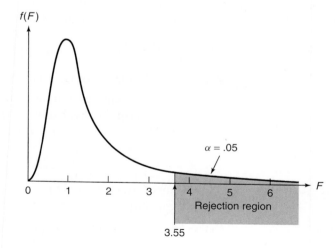

FIGURE 12.6

Partitioning of SS(Total) for a
completely randomized design

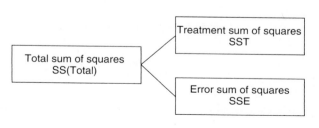

degrees of freedom—one for each of the p treatments minus one for the estimation of the overall mean. The denominator of the F statistic, MSE $=$ SSE/$(n - p)$, denotes **mean square for error** and is based on $(n - p)$ degrees of freedom—one for each of the n measurements minus one for each of the p treatment means being estimated. We have already demonstrated that this F statistic is identical to the global F value for the regression model specified previously.

For completeness, we provide the computing formulas for an analysis of variance in the box.

ANOVA Computing Formulas for a Completely Randomized Design

Sum of all n measurements $= \displaystyle\sum_{i=1}^{n} y_i$

Mean of all n measurements $= \bar{y}$

Sum of squares of all n measurements $= \displaystyle\sum_{i=1}^{n} y_i^2$

$$\text{CM} = \text{Correction for mean}$$

$$= \frac{(\text{Total of all observations})^2}{\text{Total number of observations}} = \frac{\left(\displaystyle\sum_{i=1}^{n} y_i\right)^2}{n}$$

$$\text{SS(Total)} = \text{Total sum of squares}$$
$$= (\text{Sum of squares of all observations}) - \text{CM}$$
$$= \sum_{i=1}^{n} y_i^2 - \text{CM}$$

$$\text{SST} = \text{Sum of squares for treatments}$$
$$= \left(\begin{array}{c}\text{Sum of squares of treatment totals with}\\ \text{each square divided by the number of}\\ \text{observations for that treatment}\end{array}\right) - \text{CM}$$
$$= \frac{T_1^2}{n_1} + \frac{T_2^2}{n_2} + \cdots + \frac{T_p^2}{n_p} - \text{CM}$$

$$\text{SSE} = \text{Sum of squares for error}$$
$$= \text{SS(Total)} - \text{SST}$$

$$\text{MST} = \text{Mean square for treatments}$$
$$= \frac{\text{SST}}{p - 1}$$

$$\text{MSE} = \text{Mean square for error}$$
$$= \frac{\text{SSE}}{n - p}$$

$$F = \frac{\text{MST}}{\text{MSE}}$$

EXAMPLE   12.5

Refer to Example 12.4. Analyze the data of Table 12.2 using the ANOVA approach. Use $\alpha = .05$.

Solution

Rather than performing the tedious calculations by hand (we leave this for the student as an exercise), we use a statistical software package with an ANOVA routine. All three of the software packages discussed in this text (SAS, MINITAB, and SPSS) have procedures that automatically compute the ANOVA sums of squares and the ANOVA F statistic.

The MINITAB ANOVA printout is shown in Figure 12.7. The value of the test statistic (shaded on the printout) is $F = 4.58$. Note that this is identical to the F value obtained using the regression approach in Example 12.4. The p-value of the test (also shaded) is $p = .025$. (Likewise, this quantity is identical to that in Example 12.4.) Since $\alpha = .05$ exceeds this p-value, we have sufficient evidence to conclude that the treatments differ. ◆

FIGURE 12.7

MINITAB ANOVA printout for the completely randomized design, Example 12.5

One-way ANOVA: GPA versus GROUP

```
Analysis of Variance for GPA
Source     DF       SS       MS        F         P
GROUP       2    2.397    1.198     4.58     0.025
Error      18    4.711    0.262
Total      20    7.108

                                  Individual 95% CIs for Mean
                                  Based on Pooled StDev
Level      N     Mean    StDev    --------+---------+---------+--------
1          7   2.5214   0.5041    (-------*--------)
2          7   3.2486   0.3526                    (-------*-------)
3          7   2.5429   0.6377    (-------*-------)
                                  --------+---------+---------+--------
Pooled StDev =   0.5116               2.50      3.00      3.50
```

The results of an analysis of variance are often summarized in tabular form. The general form of an ANOVA table for a completely randomized design is shown in the next box. The column head **SOURCE** refers to the source of variation, and for each source, **df** refers to the degrees of freedom, **SS** to the sum of squares, **MS** to the mean square, and F to the F statistic comparing the treatment mean square to the error mean square. Table 12.3 is the ANOVA summary table corresponding to the analysis of variance data for Example 12.5, obtained from the MINITAB printout.

ANOVA Summary Table for a Completely Randomized Design

SOURCE	df	SS	MS	F
Treatments	$p - 1$	SST	$\text{MST} = \dfrac{\text{SST}}{p-1}$	$F = \dfrac{\text{MST}}{\text{MSE}}$
Error	$n - p$	SSE	$\text{MSE} = \dfrac{\text{SSE}}{n-p}$	
Total	$n - 1$	SS(Total)		

TABLE 12.3   ANOVA Summary Table for Example 12.5

SOURCE	df	SS	MS	F
Socioeconomic class	2	2.40	1.198	4.58
Error	18	4.71	.262	
Total	20	7.11		

Because the completely randomized design involves the selection of independent random samples, we can find a confidence interval for a single treatment mean using the method of Section 1.8 or for the difference between two treatment means using the methods of Section 1.10. The estimate of σ^2 will be based on the pooled sum of squares within all p samples; that is,

$$MSE = s^2 = \frac{SSE}{n - p}$$

This is the same quantity that is used as the denominator for the analysis of variance F test. The formulas for the confidence intervals of Sections 1.8 and 1.10 are reproduced in the box.

Confidence Intervals for Means: Completely Randomized Design

Single treatment mean (say, treatment i): $\bar{y}_i \pm t_{\alpha/2}\left(\dfrac{s}{\sqrt{n_i}}\right)$

Difference between two treatment means (say, treatments i and j):

$$(\bar{y}_i - \bar{y}_j) \pm t_{\alpha/2}s\sqrt{\frac{1}{n_i} + \frac{1}{n_j}}$$

where $\bar{y}_i$ is the sample mean response for population (treatment) i, $s = \sqrt{MSE}$, and $t_{\alpha/2}$ is the tabulated value of t (Table 2 of Appendix C) that locates $\alpha/2$ in the upper tail of the t distribution with $(n - p)$ df (the degrees of freedom associated with error in the ANOVA).

EXAMPLE 12.6

Refer to Example 12.4. Find a 95% confidence interval for μ_L, the mean GPA of freshmen from the lower socioeconomic class.

Solution

From Table 12.3, MSE $= .262$. Then

$$s = \sqrt{MSE} = \sqrt{.262} = .512$$

The sample mean GPA for freshmen students from the lower class is

$$\bar{y}_L = \frac{17.64}{7} = 2.521$$

where 17.64 is the total of GPAs for the lower socioeconomic class obtained from Table 12.2. The tabulated value of $t_{.025}$ for 18 df (the same as for MSE) is 2.101. Therefore, a 95% confidence interval for μ_L, the mean GPA of college freshmen in the lower class, is

$$\bar{y}_L \pm (t_{\alpha/2})\frac{s}{\sqrt{n}} = 2.521 \pm (2.101)\frac{.512}{\sqrt{7}}$$

$$= 2.521 \pm .407$$

or (2.114, 2.928). This interval is shown graphically (shaded) on the MINITAB printout, Figure 12.7.

Note that this confidence interval is relatively wide—probably too wide to be of any practical value (considering that GPA is measured on a 4-point scale). The interval is this wide because of the large amount of variation within each socioeconomic class. For example, the GPA for freshmen in the lower class varies from 1.8 to 3.01. The more variable the data, the larger the value of s in the confidence interval and the wider the confidence interval. Consequently, if you want to obtain a more accurate estimate of treatment means with a narrower confidence interval, you will have to select larger samples of freshmen from within each socioeconomic class. ◆

Although we can use the formula given in the box to compare two treatment means in ANOVA, unless the two treatments are selected a priori (i.e., prior to conducting the ANOVA), we must apply one of the methods for comparing means presented in Sections 12.7 and 12.8 to obtain valid results.

EXERCISES

12.1. Refer to the completely randomized design of Exercise 11.3 (p. 537). Recall that the researchers want to compare the mean liquid consumptions of cockatiels in three feeding groups. Use the random number table (Table 7 in Appendix C) to randomly assign the cockatiels to the three groups. Assume there are 15 cockatiels in the study.

12.2. A partially completed ANOVA table for a completely randomized design is shown here.

SOURCE	df	SS	MS	F
Treatments	4	24.7	—	—
Error	—	—	—	
Total	34	62.4		

a. Complete the ANOVA table.

b. How many treatments are involved in the experiment?

c. Do the data provide sufficient evidence to indicate a difference among the treatment means? Test using $\alpha = .10$.

12.3. The data for a completely randomized design with two treatments are shown in the accompanying table.

EX12_3

TREATMENT 1	TREATMENT 2
10	12
7	8
8	13
11	10
10	10
9	11
9	

a. Give the linear model appropriate for analyzing the data using regression.

b. Fit the model, part **a**, to the data and conduct the analysis. [*Hint*: You do not need a computer to fit the model. Use the formulas provided in Chapter 3.]

12.4. Refer to Exercise 12.3.

a. Calculate MST for the data using the ANOVA formulas. What type of variability is measured by this quantity?

b. Calculate MSE for the data using the ANOVA formulas. What type of variability is measured by this quantity?

c. How many degrees of freedom are associated with MST?

d. How many degrees of freedom are associated with MSE?

e. Compute the test statistic appropriate for testing $H_0: \mu_1 = \mu_2$ against the alternative that the two treatment means differ, using a significance level of $\alpha = .05$. (Compare the value to the test statistic obtained using regression in part **b**, Exercise 12.3.)

f. Summarize the results from parts **a**–**e** in an ANOVA table.

g. Specify the rejection region, using a significance level of $\alpha = .05$.

h. State the proper conclusion.

12.5. Exercises 12.3 and 12.4 involve a test of the null hypothesis $H_0: \mu_1 = \mu_2$ based on independent random sampling (recall the definition of a completely randomized design). This test was conducted in Section 1.10 using a Student's t statistic.

a. Use the Student's t test to test the hypothesis $H_0: \mu_1 = \mu_2$ against the alternative hypothesis $H_a: \mu_1 \neq \mu_2$. Test using $\alpha = .05$.

b. It can be shown (proof omitted) that an F statistic with $v_1 = 1$ numerator degree of freedom and v_2 denominator degrees of freedom is equal to t^2, where t is a Student's t statistic based on v_2 degrees of freedom. Square the value of t calculated in part **a**, and show that it is equal to the value of F calculated in Exercises 12.3b and 12.4e.

c. Is the analysis of variance F test for comparing two population means a one- or a two-tailed test of $H_0: \mu_1 = \mu_2$? [*Hint*: Although the t test can be used to test either for $H_a: \mu_1 < \mu_2$, or for $H_a: \mu_1 < \mu_2$, the alternative hypothesis for the F test is H_a: The two means are different.]

12.6. The *Journal of Testing and Evaluation* (July 1992) published an investigation of the mean compression strength of corrugated fiberboard shipping containers. Comparisons were made for boxes of five different sizes: A, B, C, D, and E. Twenty identical boxes of each size were tested and the peak compression strength (pounds) recorded for each box. The accompanying figure shows the sample means for the five box types as well as the variation around each sample mean.

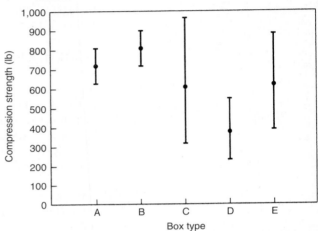

Source: Singh, S. P., et al. "Compression of single-wall corrugated shipping containers using fixed and floating test platens." *Journal of Testing and Evaluation*, Vol. 20, No. 4, July 1992, p. 319 (Figure 3).

a. Explain why the data are collected as a completely randomized design.

b. Refer to box types B and D. Based on the graph, does it appear that the mean compression strengths of these two box types are significantly different? Explain.

c. Based on the graph, does it appear that the mean compression strengths of all five box types are significantly different? Explain.

12.7. Robotics researchers investigated whether robots could be trained to behave like ants in an ant colony (*Nature*, Aug. 2000). Robots were trained and randomly assigned to "colonies" (i.e., groups) consisting of 3, 6, 9, or 12 robots. The robots were assigned the task of foraging for "food" and to recruit another robot when they identified a resource-rich area. One goal of the experiment was to compare the mean energy expended (per robot) of the four different colony sizes.

a. What type of experimental design was employed?

b. Identify the treatments and the dependent variable.

c. Set up the null and alternative hypotheses of the test.

d. The following ANOVA results were reported: $F = 7.70$, numerator df = 3, denominator df = 56, p-value $< .001$. Conduct the test at a significance level of $\alpha = .05$ and interpret the result.

12.8. An article in the *American Journal of Political Science* (Jan. 1998) examined the attitudes of three groups of professionals that influence U.S. policy. Random samples of 100 scientists, 100 journalists, and 100 government officials were asked about the safety of nuclear power plants. Responses were made on a seven-point scale, where 1 = very unsafe and 7 = very safe. The mean safety scores for the groups are: scientists, 4.1; journalists, 3.7; government officials, 4.2.

a. Identify the response variable for this study.

b. How many treatments are included in this study? Describe them.

c. Specify the null and alternative hypotheses that should be used to investigate whether there are differences in the attitudes of scientists, journalists, and government officials regarding the safety of nuclear power plants.

d. The MSE for the sample data is 2.355. At least how large must MST be in order to reject the null hypothesis of the test of part **a** using $\alpha = .05$?

e. If the MST $= 11.280$, what is the approximate p-value of the test of part **a**?

12.9. *The Archives of Disease in Childhood* (Apr. 2000) published a study of whether height influences a child's progression through elementary school. Within each grade, Australian school children were divided into equal thirds (tertiles) based on age (youngest third, middle third, and oldest third). The researchers compared the average heights of the three groups

using an analysis of variance. (All height measurements were standardized using z-scores.) A summary of the results for all grades combined, by gender, is shown in the table at the bottom of the page.

a. What is the null hypothesis for the ANOVA of the boys' data?

b. Write the linear model appropriate for analyzing the data.

c. Interpret the results of the test, part **a**. Use $\alpha = .05$.

d. Repeat parts **a–c** for the girls' data.

e. Summarize the results of the hypothesis tests in the words of the problem.

12.10. Objectivity is an essential characteristic of auditing. A study was conducted to investigate whether prior involvement in audit program design impairs an external auditor's objectivity in making decisions about that program (*Accounting and Finance*, Nov. 1993). A sample of 45 auditors was randomly divided into three equal-size groups, designated A/R, A/P, and Control. The A/R group designed an audit program for accounts receivable and evaluated an audit program for accounts payable designed by someone else. The A/P group did the reverse. Finally, the control group merely evaluated the audit programs for both accounts. All 45 auditors were then requested to allocate an additional 15 hours to investigate suspected irregularities in either one or both audit programs. The objective of the experiment was to compare the mean number of hours allocated to the accounts receivable for the three groups.

a. What type of design is used in this study?

b. Identify the treatments for this design.

c. A partial ANOVA table is shown on p. 564. Complete the table.

d. Based on the results of the analysis, what inference can the researchers make?

	SAMPLE SIZE	YOUNGEST TERTILE MEAN HEIGHT	MIDDLE TERTILE MEAN HEIGHT	OLDEST TERTILE MEAN HEIGHT	F VALUE	p-VALUE
Boys	1439	0.33	0.33	0.16	4.57	0.01
Girls	1409	0.27	0.18	0.21	0.85	0.43

Source: Wake, M., Coghlan, D., and Hesketh, K. "Does height influence progression through primary school grades?" *The Archives of Disease in Childhood*, Vol. 82, Apr. 2000 (Table 2).

ANOVA Table for Exercise 12.10

SOURCE	df	SS	MS	F	p-VALUE
Groups	—	71.51	—	—	.01
Error	—	—	7.65		
Total	—	392.98			

12.11. The *Journal of Hazardous Materials* (July 1995) published the results of a study of the chemical properties of three different types of hazardous organic solvents used to clean metal parts: aromatics, chloroalkanes, and esters. One variable studied was sorption rate, measured as mole percentage. Independent samples of solvents from each type were tested and their sorption rates were recorded, as shown in the table. A MINITAB analysis of variance of the data is provided.

◉ SORPRATE

AROMATICS		CHLOROALKANES		ESTERS		
1.06	.95	1.58	1.12	.29	.43	.06
.79	.65	1.45	.91	.06	.51	.09
.82	1.15	.57	.83	.44	.10	.17
.89	1.12	1.16	.43	.61	.34	.60
1.05				.55	.53	.17

Source: Reprinted from *Journal of Hazardous Materials*, Vol. 42, No. 2, J. D. Ortego *et al.*, "A review of polymeric geosynthetics used in hazardous waste facilities." p. 142 (Table 9), July 1995, Elsevier Science-NL, Sara Burgerhartstraat 25, 1055 KV Amsterdam, The Netherlands.

Minitab Output for Exercise 12.11

One-way ANOVA: SORPRATE versus SOLVENT

```
Analysis of Variance for SORPRATE
Source    DF       SS       MS       F       P
SOLVENT    2   3.3054   1.6527   24.51   0.000
Error     29   1.9553   0.0674
Total     31   5.2608
                              Individual 95% CIs For Mean
                              Based on Pooled StDev
Level      N     Mean    StDev  ----+---------+---------+---------+--
1          9   0.9422   0.1683                      (----*-----)
2          8   1.0063   0.4010                        (------*-----)
3         15   0.3300   0.2076  (----*----)
                              ----+---------+---------+---------+--
Pooled StDev =  0.2597           0.30      0.60      0.90      1.20
```

a. Construct an ANOVA table from the MINITAB printout.

b. Is there evidence of differences among the mean sorption rates of the three organic solvent types? Test using $\alpha = .10$.

12.12. Studies conducted at the University of Melbourne (Australia) indicate that there may be a difference between the pain thresholds of blondes and brunettes.

Men and women of various ages were divided into four categories according to hair color: light blond, dark blond, light brunette, and dark brunette. The purpose of the experiment was to determine whether hair color is related to the amount of pain evoked by common types of mishaps and assorted types of trauma. Each person in the experiment was given a pain threshold score based on his or her performance in a pain sensitivity test (the higher the score, the higher the person's pain tolerance). SAS was used to conduct the analysis of variance of the data listed in the table. The SAS printout is also provided.

◉ HAIRPAIN

LIGHT BLOND	DARK BLOND	LIGHT BRUNETTE	DARK BRUNETTE
62	63	42	32
60	57	50	39
71	52	41	51
55	41	37	30
48	43		35

SAS output for Exercise 12.12

```
                          The ANOVA Procedure
Dependent Variable: PAIN
                                    Sum of
Source              DF          Squares      Mean Square    F Value    Pr > F
Model                3      1360.726316      453.575439       6.79     0.0041
Error               15      1001.800000       66.786667
Corrected Total     18      2362.526316

                 R-Square     Coeff Var     Root MSE     PAIN Mean
                 0.575962      17.08184     8.172311      47.84211

Source              DF        Anova SS      Mean Square    F Value    Pr > F
COLOR                3      1360.726316      453.575439       6.79     0.0041
```

a. Based on the given information, what type of experimental design appears to have been employed?

b. Using the SAS printout, conduct a test to determine whether the mean pain thresholds differ among people possessing the four types of hair color. Use $\alpha = .05$.

c. What is the observed significance level for the test in part **b**? Interpret it.

d. What assumptions must be met in order to ensure the validity of the inferences you made in part **b**?

12.13. Psychologists at Lancaster University (United Kingdom) evaluated three methods of name retrieval in a controlled setting (*Journal of Experimental Psychology-Applied*, June 2000.) A sample of 139 students was randomly divided into three groups, and each group of students used a different method to learn the names of the other students in the group. Group 1 used the "simple name game," where the first student states his/her full name, the second student announces his/her name and the name of the first student, the third student says his/her name and the names of the first two students, etc. Group 2 used the "elaborate name game," a modification of the simple name game where the students not only state their names but also their favorite activity (e.g., sports). Group 3 used "pairwise introductions," where students are divided into pairs and each student must introduce the other member of the pair. One year later, all subjects were sent pictures of the students in their group and asked to state the full name of each. The researchers measured the percentage of names recalled for each student respondent. The data (simulated based on summary statistics provided in the table below. Conduct an analysis of variance to determine whether the mean percentages of names recalled differ for the three name retrieval methods. Use $\alpha = .05$.

NAMEGAME

SIMPLE NAME GAME

24	43	38	65	35	15	44	44	18	27	0	38	50	31
7	46	33	31	0	29	0	0	52	0	29	42	39	26
51	0	42	20	37	51	0	30	43	30	99	39	35	19
24	34	3	60	0	29	40	40						

ELABORATE NAME GAME

39	71	9	86	26	45	0	38	5	53	29	0	62	0
1	35	10	6	33	48	9	26	83	33	12	5	0	0
25	36	39	1	37	2	13	26	7	35	3	8	55	50

PAIRWISE INTRODUCTIONS

5	21	22	3	32	29	32	0	4	41	0	27	5	9
66	54	1	15	0	26	1	30	2	13	0	2	17	14
5	29	0	45	35	7	11	4	9	23	4	0	8	2
18	0	5	21	14									

Source: Morris, P.E., and Fritz, C.O. "The name game: Using retrieval practice to improve the learning of names." *Journal of Experimental Psychology-Applied*, Vol. 6, No. 2, June 2000 (data simulated from Figure 1).

12.14. What do people infer from facial expressions of emotion? This was the research question of interest in an article published in the *Journal of Nonverbal Behavior* (Fall 1996). A sample of 36 introductory psychology students was randomly divided into six groups. Each group was assigned to view one of six slides showing a person making a facial expression.*

The six expressions were (1) angry, (2) disgusted, (3) fearful, (4) happy, (5) sad, and (6) neutral faces. After viewing the slides, the students rated the degree of dominance they inferred from the facial expression (on a scale ranging from -15 to $+15$). The data (simulated from summary information provided in the article) are listed in the table. Conduct an analysis of variance to determine whether the mean dominance ratings differ among the six facial expressions. Use $\alpha = .10$.

FACES

ANGRY	DISGUSTED	FEARFUL	HAPPY	SAD	NEUTRAL
2.10	.40	.82	1.71	.74	1.69
.64	.73	-2.93	$-.04$	-1.26	$-.60$
.47	$-.07$	$-.74$	1.04	-2.27	$-.55$
.37	$-.25$	.79	1.44	$-.39$	.27
1.62	.89	$-.77$	1.37	-2.65	$-.57$
$-.08$	1.93	-1.60	.59	$-.44$	-2.16

12.15. Do you experience episodes of excessive eating accompanied by being overweight? If so, you may suffer from binge eating disorder. Cognitive-behavioral therapy (CBT), in which patients are taught how to make changes in specific behavior patterns (e.g., exercise, eat only low-fat foods), can be effective in treating the disorder. A group of Stanford University researchers investigated the effectiveness of interpersonal therapy (IPT) as a second level of treatment for binge eaters (*Journal of Consulting and Clinical Psychology*, June 1995). The researchers employed a design that randomly assigned a sample of 41 overweight individuals diagnosed with binge eating disorder to either a treatment group (30 subjects) or a control group (11 subjects). Subjects in the treatment group received 12 weeks of cognitive-behavior therapy, then were subdivided into two groups. Those who responded successfully to CBT (17 subjects) were assigned to a weight loss therapy (WLT) program for the next 12 weeks. Those CBT subjects who did not respond to treatment (13 subjects) received 12 weeks of IPT. The subjects in the control group received no therapy of any type. Thus, the study ultimately consisted of three groups of overweight binge eaters: the CBT-WLT group, the CBT-IPT group, and the control group. One outcome (response) variable measured for each subject was the number of binge eating episodes per week, x. Summary statistics for each of the three groups at the end of the 24-week period are shown in the table. The data

---

*In the actual experiment, each group viewed all six facial expression slides and the design employed was a Latin Square (beyond the scope of this text).

were analyzed as a completely randomized design with three treatments (CBT-WLT, CBT-IPT, and Control). Although the ANOVA tables were not provided in the article, sufficient information is provided in the table to reconstruct them.

	CBT-WLT	CBT-IPT	CONTROL
Sample size	17	13	11
Mean number of binges per week	.18	1.92	2.91
Standard deviation	0.4	1.7	2.0
Group Totals	3	25	32

Source: Agras, W. S., *et al.* "Does interpersonal therapy help patients with binge eating disorder who fail to respond to cognitive-behavioral therapy?" *Journal of Consulting and Clinical Psychology*, Vol. 63, No. 3, June 1995, p. 358 (Table 1).

a. Use the sum of the group totals and the total sample size to compute CM.
b. Use the individual group totals and sample sizes to compute SST.
c. Compute SSE using the pooled sum of squares formula:

$$SSE = \sum (y_{i1} - \bar{y}_1)^2 + \sum (y_{i2} - \bar{y}_2)^2$$
$$+ \sum (y_{i3} - \bar{y}_3)^2$$
$$= (n_1 - 1)s_1^2 + (n_2 - 1)s_2^2 + (n_3 - 1)s_3^2$$

d. Find SS(Total).
e. Construct an ANOVA table for the data.
f. Do the data provide sufficient evidence to indicate differences in the mean number of binges per week among the three groups? Test using $\alpha = .05$.
g. Give the linear model appropriate for analyzing the data using regression.
h. Use the information in the table to find the least squares prediction equation.

i. Comment on the use of a completely randomized design for this study. Have subjects been randomly and independently assigned to each group? How might this impact the results of the study?

12.16. Refer to the *Marine Technology* (Jan. 1995) study of major ocean oil spills by tanker vessels, Exercise 1.25 (p. 25). The spillage amounts (thousands of metric tons) and cause of accident for 48 tankers are saved in the OILSPILL file. (*Note*: Delete the two tankers with oil spills of unknown causes.) Conduct an analysis of variance (at $\alpha = .01$) to compare the mean spillage amounts for the four accident types: (1) collision, (2) grounding, (3) fire/explosion, and (4) hull failure. Interpret your results.

12.17. When marketing its products in a foreign country, should a company use its own salespeople or salespeople from the target market country? To answer this question, a study was designed to investigate the effect of salesperson nationality on buyer attitudes (*Journal of Business Research*, Vol. 22, 1991). A sample of U.S. MBA students was divided into two groups and shown a videotape of an advertisement for forklift trucks made in India. For group 1, an Indian sales representative made the presentation; for group 2, a U.S. sales representative made the presentation. After viewing the tape, the subjects were asked whether the salesperson was trustworthy (measured on a 5-point scale). The mean scores were compared using an ANOVA.
a. The ANOVA resulted in an F value of 2.32, with an observed significance level of .13. Is there evidence of a difference between the mean trustworthiness scores of the two groups of MBA students? Use $\alpha = .10$.
b. The sample mean scores for the two groups are $\bar{y}_1 = 3.12$ and $\bar{y}_2 = 3.49$. Suppose you were to test $H_0 : \mu_1 = \mu_2$ against $H_a : \mu_1 < \mu_2$ at $\alpha = .10$. Use the result, part **a**, to make the proper conclusion. [*Hint*: Use Exercise 12.5 and the fact that the p-value for a two-tailed t test is double the p-value for a one-tailed test.]

12.4 Randomized Block Designs

Randomized block design is a commonly used noise-reducing design. Recall (Definition 10.9) that a randomized block design employs groups of homogeneous experimental units (matched as closely as possible) to compare the means of the populations associated with p treatments. The general layout of a randomized block design is shown in Figure 12.8. Note that there are b blocks of relatively homogeneous experimental units. Since each treatment must be represented in each block, the blocks each contain p experimental units. Although Figure 12.8 shows the p treatments in order within the blocks, in practice they would be assigned to the experimental units in random order (hence the name **randomized block design**).

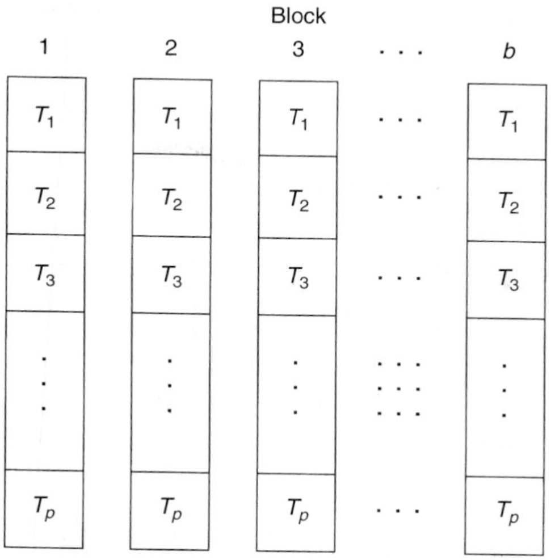

FIGURE 12.8

General form of a randomized block design (treatment is denoted by T_p)

Block

The complete model for a randomized block design contains $(p - 1)$ dummy variables for treatments and $(b - 1)$ dummy variables for blocks. Therefore, the total number of terms in the model, excluding β_0, is $(p - 1) + (b - 1) = p + b - 2$, as shown here.

Complete model:

$$E(y) = \beta_0 + \underbrace{\beta_1 x_1 + \beta_2 x_2 + \cdots + \beta_{p-1} x_{p-1}}_{\text{Treatment effects}} + \underbrace{\beta_p x_p + \cdots + \beta_{p+b-2} x_{p+b-2}}_{\text{Block effects}}$$

where

$$x_1 = \begin{cases} 1 & \text{if treatment 2} \\ 0 & \text{if not} \end{cases},$$

$$x_2 = \begin{cases} 1 & \text{if treatment 3} \\ 0 & \text{if not} \end{cases}, \ldots, x_{p-1} = \begin{cases} 1 & \text{if treatment } p \\ 0 & \text{if not} \end{cases}$$

$$x_p = \begin{cases} 1 & \text{if block 2} \\ 0 & \text{if not} \end{cases}, x_{p+1} = \begin{cases} 1 & \text{if block 3} \\ 0 & \text{if not} \end{cases}, \ldots, x_{p+b-2} = \begin{cases} 1 & \text{if block } b \\ 0 & \text{if not} \end{cases}$$

Note that the model does *not* include terms for treatment–block interaction. The reasons are twofold. First, the addition of these terms would leave 0 degrees of freedom for estimating σ^2. Second, the failure of the mean difference between a pair of treatments to remain the same from block to block is, by definition, experimental error. In other words, in a randomized block design, treatment–block interaction and experimental error are synonymous.

The primary objective of the analysis is to compare the p treatment means, $\mu_1, \mu_2, \ldots, \mu_p$. That is, we want to test the null hypothesis

$$H_0: \mu_1 = \mu_2 = \mu_3 = \cdots = \mu_p$$

Recall (Section 11.3) that this is equivalent to testing whether all the treatment parameters in the complete model are equal to 0, i.e.,

$$H_0: \beta_1 = \beta_2 = \cdots = \beta_{p-1} = 0$$

To perform this test using regression, we drop the treatment terms and fit the reduced model:

Reduced model for testing treatments

$$E(y) = \beta_0 + \beta_p x_p + \underbrace{\beta_{p+1} x_{p+1} + \cdots + \beta_{p+b-2} x_{p+b-2}}_{\text{Block effects}}$$

Then we compare the SSEs for the two models, SSE_R and SSE_C, using the "partial" F statistic:

$$F = \frac{(SSE_R - SSE_C)/\text{Number of } \beta\text{'s tested}}{MSE_C} = \frac{(SSE_R - SSE_C)/(p-1)}{MSE_C}$$

A significant F value implies that the treatment means differ.

Occasionally, experimenters want to determine whether blocking was effective in removing the extraneous source of variation, i.e., whether there is evidence of a difference among block means. In fact, if there are no differences among block means, the experimenter will lose information by blocking because blocking reduces the number of degrees of freedom associated with the estimated variance of the model, s^2. If blocking is *not* effective in reducing the variability, then the block parameters in the complete model will all equal 0 (i.e., there will be no differences among block means). Thus, we want to test

$$H_0: \beta_p = \beta_{p+1} = \cdots = \beta_{p+b-2} = 0$$

Another reduced model, with the block β's dropped, is fitted:

Reduced model for testing blocks

$$E(y) = \beta_0 + \underbrace{\beta_1 x_1 + \beta_2 x_2 + \cdots + \beta_{p-1} x_{p-1}}_{\text{Treatment effects}}$$

The SSE for this second reduced model is compared to the SSE for the complete model in the usual fashion. A significant F test implies that blocking is effective in removing (or reducing) the targeted extraneous source of variation.

These two tests are summarized in the following box.

Models and ANOVA F Tests for a Randomized Block Design with p Treatments and b Blocks

Complete model:

$$E(y) = \beta_0 + \overbrace{\beta_1 x_1 + \cdots + \beta_{p-1} x_{p-1}}^{(p-1)\text{treatment terms}} + \overbrace{\beta_p x_p + \cdots + \beta_{p+b-2} x_{p+b-2}}^{(b-1)\text{block terms}}$$

where

$$x_1 = \begin{cases} 1 & \text{if treatment 2} \\ 0 & \text{if not} \end{cases} \quad \cdots \quad x_{p-1} = \begin{cases} 1 & \text{if treatment } p \\ 0 & \text{if not} \end{cases}$$

$$x_p = \begin{cases} 1 & \text{if block 2} \\ 0 & \text{if not} \end{cases} \quad \cdots \quad x_{p+b-2} = \begin{cases} 1 & \text{if block } b \\ 0 & \text{if not} \end{cases}$$

TEST FOR COMPARING TREATMENT MEANS

$H_0: \beta_1 = \beta_2 = \cdots = \beta_{p-1} = 0$
(i.e., H_0: The p treatment means are equal)

H_a: At least one of the β parameters listed in H_0 differs from 0
(i.e., H_a: At least two treatment means differ)

Reduced model: $E(y) = \beta_0 + \beta_p x_p + \cdots + \beta_{p+b-2} x_{p+b-2}$

Test statistic: $F = \dfrac{(\text{SSE}_R - \text{SSE}_C)/(p-1)}{\text{SSE}_C/(n-p-b+1)}$

$$= \dfrac{(\text{SSE}_R - \text{SSE}_C)/(p-1)}{\text{MSE}_C}$$

where

$$\text{SSE}_R = \text{SSE for reduced model}$$
$$\text{SSE}_C = \text{SSE for complete model}$$
$$\text{MSE}_C = \text{MSE for complete model}$$

Rejection region: $F > F_\alpha$ where F is based on $v_1 = (p-1)$ and
$v_2 = (n - p - b + 1)$ degrees of freedom

TEST FOR COMPARING BLOCK MEANS

$H_0: \beta_p = \beta_{p+1} = \cdots = \beta_{p+b-2} = 0$
(i.e., H_0: The b block means are equal.)

H_a: At least one of the β parameters listed in H_0 differs from 0.
(i.e., H_a: At least two block means differ.)

Reduced model: $E(y) = \beta_0 + \beta_1 x_1 + \beta_2 x_2 + \cdots + \beta_{p-1} x_{p-1}$

Test statistic: $F = \dfrac{(\text{SSE}_R - \text{SSE}_C)/(b-1)}{\text{SSE}_C/(n-p-b+1)}$

$$= \dfrac{(\text{SSE}_R - \text{SSE}_C)/(b-1)}{\text{MSE}_C}$$

where

$$\text{SSE}_R = \text{SSE for reduced model}$$
$$\text{SSE}_C = \text{SSE for complete model}$$
$$\text{MSE}_C = \text{MSE for complete model}$$

Rejection region: $F > F_\alpha$ where F is based on $v_1 = (b-1)$ and $v_2 = (n-p-b+1)$ degrees of freedom

Assumptions:

1. The probability distribution of the difference between any pair of treatment observations within a block is approximately normal.
2. The variance of the difference is constant and the same for all pairs of observations.

EXAMPLE 12.7

Prior to submitting a bid for a construction job, cost engineers prepare a detailed analysis of the estimated labor and materials costs required to complete the job. This estimate will depend on the engineer who performs the analysis. An overly large estimate will reduce the chance of acceptance of a company's bid price, whereas an estimate that is too low will reduce the profit or even cause the company to lose money on the job. A company that employs three job cost engineers wanted to compare the mean level of the engineers' estimates. This was done by having each engineer estimate the cost of the same four jobs. The data (in hundreds of thousands of dollars) are shown in Table 12.4.

COST ENGINEERS

TABLE 12.4   **Data for the Randomized Block Design of Example 12.7**

		JOB				
		1	2	3	4	TREATMENT MEANS
	1	4.6	6.2	5.0	6.6	5.60
ENGINEER	2	4.9	6.3	5.4	6.8	5.85
	3	4.4	5.9	5.4	6.3	5.50
BLOCK MEANS		4.63	6.13	5.27	6.57	

a. Perform an analysis of variance on the data, and test to determine whether there is sufficient evidence to indicate differences among treatment means. Test using $\alpha = .05$.
b. Test to determine whether blocking on jobs was successful in reducing the job-to-job variation in the estimates. Use $\alpha = .05$.

Solution

a. The data for this experiment were collected according to a randomized block design because estimates of the same job were expected to be more nearly alike than estimates between jobs. Thus, the experiment involves three treatments (engineers) and four blocks (jobs).

The complete model for this design is

$$E(y) = \beta_0 + \underbrace{\beta_1 x_1 + \beta_2 x_2}_{\text{Treatments (engineers)}} + \underbrace{\beta_3 x_3 + \beta_4 x_4 + \beta_5 x_5}_{\text{Blocks (jobs)}}$$

where

$$y = \text{Cost estimate}$$

$$x_1 = \begin{cases} 1 & \text{if engineer 2} \\ 0 & \text{if not} \end{cases} \qquad x_2 = \begin{cases} 1 & \text{if engineer 3} \\ 0 & \text{if not} \end{cases}$$

Base level = Engineer 1

$$x_3 = \begin{cases} 1 & \text{if block 2} \\ 0 & \text{if not} \end{cases} \quad x_4 = \begin{cases} 1 & \text{if block 3} \\ 0 & \text{if not} \end{cases} \quad x_5 = \begin{cases} 1 & \text{if block 4} \\ 0 & \text{if not} \end{cases}$$

Base level = Block 1

The SAS printout for the complete model is shown in Figure 12.9. Note that $\text{SSE}_C = .18667$ and $\text{MSE}_C = .03111$ (shaded on the printout).

b. To test for differences among the treatment means, we will test

$$H_0: \mu_1 = \mu_2 = \mu_3$$

where μ_i = mean cost estimate of engineer i. This is equivalent to testing

$$H_0: \beta_1 = \beta_2 = 0$$

in the complete model. We fit the reduced model

$$E(y) = \beta_0 + \underbrace{\beta_3 x_3 + \beta_4 x_4 + \beta_5 x_5}_{\text{Blocks (jobs)}}$$

The SAS printout for this reduced model is shown in Figure 12.10. Note that $\text{SSE}_R = .44667$ (shaded on the printout). The remaining elements of the test follow.

Test statistic:

$$F = \frac{(\text{SSE}_R - \text{SSE}_C)/(p-1)}{\text{MSE}_C} = \frac{(.44667 - .18667)/2}{.03111} = 4.18$$

Rejection region: $F > 5.14$, where $F_{.05} = 5.14$ (from Table 4, Appendix C) is based on $v_1 = (p-1) = 2$ df and $v_2 = (n - p - b + 1) = 6$ df.

Conclusion: Since $F = 4.18$ is less than the critical value, 5.14, there is insufficient evidence, at the $\alpha = .05$ level of significance, to indicate differences among the mean estimates for the three cost engineers.

As an option, SAS will conduct this nested model F test. The test statistic, $F = 4.18$, is highlighted on the middle of the SAS complete model printout, Figure 12.9. The p-value of the test (also highlighted) is $p = .0730$. Since this value

FIGURE 12.9

SAS regression printout for randomized block design complete model, Example 12.7

```
                              Dependent Variable: COST

                              Analysis of Variance

                                      Sum of          Mean
Source                  DF           Squares         Square      F Value    Pr > F

Model                    5           7.02333        1.40467       45.15     0.0001
Error                    6           0.18667        0.03111
Corrected Total         11           7.21000


                Root MSE              0.17638    R-Square      0.9741
                Dependent Mean        5.65000    Adj R-Sq      0.9525
                Coeff Var             3.12183

                              Parameter Estimates

                        Parameter      Standard
Variable      DF        Estimate         Error      t Value    Pr > |t|

Intercept      1         4.58333       0.12472       36.75      <.0001
X1             1         0.25000       0.12472        2.00      0.0919
X2             1        -0.10000       0.12472       -0.80      0.4533
X3             1         1.50000       0.14402       10.42      <.0001
X4             1         0.63333       0.14402        4.40      0.0046
X5             1         1.93333       0.14402       13.42      <.0001


           Test ENGINEER Results for Dependent Variable COST

                                      Mean
Source                  DF           Square      F Value     Pr > F

Numerator                2           0.13000       4.18      0.0730
Denominator              6           0.03111


             Test JOBS Results for Dependent Variable COST

                                      Mean
Source                  DF           Square      F Value     Pr > F

Numerator                3           2.25444      72.46      <.0001
Denominator              6           0.03111
```

exceeds $\alpha = .05$, our conclusion is confirmed—there is insufficient evidence to reject H_0.

b. To test for the effectiveness of blocking on jobs, we test

$$H_0: \beta_3 = \beta_4 = \beta_5 = 0$$

in the complete model specified in part **a**. The reduced model is

$$E(y) = \beta_0 + \underbrace{\beta_1 x_1 + \beta_2 x_2}_{\text{Treatments (engineers)}}$$

FIGURE 12.10

SAS regression printout for randomized block
design reduced model for testing treatments

```
                        Dependent Variable: COST
                         Analysis of Variance

                            Sum of         Mean
Source              DF      Squares       Square    F Value   Pr > F
Model                3      6.76333       2.25444    40.38    <.0001
Error                8      0.44667       0.05583
Corrected Total     11      7.21000

            Root MSE          0.23629    R-Square   0.9380
            Dependent Mean    5.65000    Adj R-Sq   0.9148
            Coeff Var         4.18214

                        Parameter Estimates

                    Parameter      Standard
Variable      DF    Estimate         Error    t Value   Pr > |t|
Intercept      1     4.63333        0.13642     33.96    <.0001
X3             1     1.50000        0.19293      7.77    <.0001
X4             1     0.63333        0.19293      3.28     0.0111
X5             1     1.93333        0.19293     10.02    <.0001
```

FIGURE 12.11

SAS regression printout for
randomized block design reduced
model for testing blocks

```
                        Dependent Variable: COST
                         Analysis of Variance

                            Sum of         Mean
Source              DF      Squares       Square    F Value   Pr > F
Model                2      0.26000       0.13000     0.17    0.8477
Error                9      6.95000       0.77222
Corrected Total     11      7.21000

            Root MSE          0.87876    R-Square    0.0361
            Dependent Mean    5.65000    Adj R-Sq   -0.1781
            Coeff Var        15.55331

                        Parameter Estimates

                    Parameter      Standard
Variable      DF    Estimate         Error    t Value   Pr > |t|
Intercept      1     5.60000        0.43938     12.75    <.0001
X1             1     0.25000        0.62138      0.40     0.6968
X2             1    -0.10000        0.62138     -0.16     0.8757
```

The SAS printout for this second reduced model is shown in Figure 12.11. Note that
$SSE_R = 6.95$ (shaded on the printout). The elements of the test follow.

Test statistic:

$$F = \frac{(SSE_R - SSE_C)/(b-1)}{MSE_C} = \frac{(6.95 - .18667)/3}{.03111} = 72.46$$

Rejection region: $F > 4.76$, where $F_{.05} = 4.76$ (from Table 4, Appendix C) is based on $v_1 = (b-1) = 3$ df and $v_2 = (n-p-b+1) = 6$ df.

Conclusion: Since $F = 72.46$ exceeds the critical value 4.76, there is sufficient evidence (at $\alpha = .05$) to indicate differences among the block (job) means. It appears that blocking on jobs was effective in reducing the job-to-job variation in cost estimates.

We also requested SAS to perform this nested model F test for blocks. The results, $F = 72.46$ and p-value $<.0001$, are shaded at the bottom of the SAS complete model printout, Figure 12.9. The small p-value confirms our conclusion; there is sufficient evidence at ($\alpha = .05$) to reject H_0.   ◆

Caution: The result of the test for the equality of block means must be interpreted with care, especially when the calculated value of the F test statistic does not fall in the rejection region. This does not necessarily imply that the block means are the same, i.e., that blocking is unimportant. Reaching this conclusion would be equivalent to accepting the null hypothesis, a practice we have carefully avoided because of the unknown probability of committing a Type II error (that is, of accepting H_0 when H_a is true). In other words, even when a test for block differences is inconclusive, we may still want to use the randomized block design in similar future experiments. If the experimenter believes that the experimental units are more homogeneous within blocks than among blocks, he or she should use the randomized block design regardless of whether the test comparing the block means shows them to be different.

The traditional analysis of variance approach to analyzing the data collected from a randomized block design is similar to the completely randomized design. The partitioning of SS(Total) for the randomized block design is most easily seen by examining Figure 12.12. Note that SS(Total) is now partitioned into *three* parts:

$$SS(\text{Total}) = SSB + SST + SSE$$

The formulas for calculating SST and SSB follow the same pattern as the formula for calculating SST for the completely randomized design.

From these quantities, we obtain mean square for treatments, MST, mean square for blocks, MSB, and mean square for error, MSE, as shown in the next box. The

FIGURE 12.12

Partitioning of the total sum of squares for the randomized block design

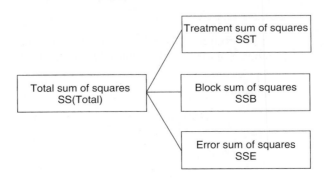

test statistics are

$$F = \frac{\text{MST}}{\text{MSE}} \text{ for testing treatments}$$

$$F = \frac{\text{MSB}}{\text{MSE}} \text{ for testing blocks}$$

These F values are equivalent to the "partial" F statistics of the regression approach.

ANOVA Computing Formulas for a Randomized Block Design

$$\sum_{i=1}^{n} y_i = \text{Sum of all } n \text{ measurements}$$

$$\sum_{i=1}^{n} y_i^2 = \text{Sum of squares of all } n \text{ measurements}$$

$$\text{CM} = \text{Correction for mean}$$

$$= \frac{(\text{Total of all measurements})^2}{\text{Total number of measurements}} = \frac{\left(\sum_{i=1}^{n} y_i\right)^2}{n}$$

$$\text{SS(Total)} = \text{Total sum of squares}$$

$$= (\text{Sum of squares of all measurements}) - \text{CM}$$

$$= \sum_{i=1}^{n} y_i^2 - \text{CM}$$

$$\text{SST} = \text{Sum of squares for treatments}$$

$$= \left(\begin{array}{c} \text{Sum of squares of treatment totals with} \\ \text{each square divided by } b, \text{ the number of} \\ \text{measurements for that treatment} \end{array} \right) - \text{CM}$$

$$= \frac{T_1^2}{b} + \frac{T_2^2}{b} + \cdots + \frac{T_p^2}{b} - \text{CM}$$

$$\text{SSB} = \text{Sum of squares for blocks}$$

$$= \left(\begin{array}{c} \text{Sum of squares for block totals with} \\ \text{each square divided by } p, \text{ the number} \\ \text{of measurements in that block} \end{array} \right) - \text{CM}$$

$$= \frac{B_1^2}{p} + \frac{B_2^2}{p} + \cdots + \frac{B_b^2}{p} - \text{CM}$$

$$\text{SSE} = \text{Sum of squares for error} = \text{SS(Total)} - \text{SST} - \text{SSB}$$

$$\text{MST} = \text{Mean square for treatments} = \frac{\text{SST}}{p-1}$$

$$MSB = \text{Mean square for blocks} = \frac{SSB}{b-1}$$

$$MSE = \text{Mean square for error} = \frac{SSE}{n-p-b+1}$$

$$F = \frac{MST}{MSE} \text{for testing treatments}$$

$$F = \frac{MSB}{MSE} \text{for testing blocks}$$

EXAMPLE 12.8

Refer to Example 12.7. Perform an analysis of variance of the data in Table 12.4 using the ANOVA approach.

Solution

Rather than perform the calculations by hand (again, we leave this as an exercise for the student), we utilize a statistical software package. The SPSS printout of the ANOVA is displayed in Figure 12.13. The F value for testing treatments, $F = 4.179$, and the F value for testing blocks, $F = 72.464$, are both shaded on the printout. Note that these values are identical to the F values computed using the regression approach, Example 12.8. The p-values of the tests (also shaded) lead to the same conclusions reached in Example 12.7. For example, the p-value for the test of treatment differences, $p = .073$, exceeds $\alpha = .05$; thus, there is insufficient evidence of differences among the treatment means. ◆

FIGURE 12.13

SPSS ANOVA printout for randomized block design

Tests of Between-Subjects Effects

Dependent Variable: COST

Source	Type III Sum of Squares	df	Mean Square	F	Sig.
Corrected Model	7.023a	5	1.405	45.150	.000
Intercept	383.070	1	383.070	12312.964	.000
ENGINEER	.260	2	.130	4.179	.073
JOB	6.763	3	2.254	72.464	.000
Error	.187	6	3.111E-02		
Total	390.280	12			
Corrected Total	7.210	11			

a. R Squared = .974 (Adjusted R Squared = .953)

As with a completely randomized design, the sources of variation and their respective degrees of freedom, sums of squares, and mean squares for a randomized block design are shown in an ANOVA summary table. The general format of such a table for a randomized block design is shown in the next box; the ANOVA table for the data of Table 12.4 is shown in Table 12.5. (These quantities are shaded on the SPSS printout, Figure 12.13.) Note that the degrees of freedom for the three sources of variation, treatments, blocks, and error, sum to the degrees of freedom for SS(Total). Similarly, the sums of squares for the three sources will always sum to SS(Total).

General Format of ANOVA Table for a Randomized Block Design

SOURCE	df	SS	MS	F
Treatments	$p - 1$	SST	$MST = \dfrac{SST}{p-1}$	$F = \dfrac{MST}{MSE}$
Blocks	$b - 1$	SSB	$MSB = \dfrac{SSB}{b-1}$	$F = \dfrac{MSB}{MSE}$
Error	$n - p - b + 1$	SSE	$MSE = \dfrac{SSE}{n-p-b+1}$	
Total	$n - 1$	SS(Total)		

TABLE 12.5 **ANOVA Summary Table for Example 12.8**

SOURCE	df	SS	MS	F
Treatments (Engineers)	2	.260	.130	4.18
Blocks (Jobs)	3	6.763	2.254	72.46
Error	6	.187	.031	
Total	11	7.210		

Confidence intervals for the difference between a pair of treatment means or block means for a randomized block design are shown in the following box.

Confidence Intervals for the Difference ($\mu_i - \mu_j$) Between a Pair of Treatment Means or Block Means

Treatment means: $(\overline{T}_i - \overline{T}_j) \pm t_{\alpha/2}s\sqrt{\dfrac{2}{b}}$

Block means: $(\overline{B}_i - \overline{B}_j) \pm t_{\alpha/2}s\sqrt{\dfrac{2}{p}}$

where

$$b = \text{Number of blocks}$$
$$p = \text{Number of treatments}$$
$$s = \sqrt{\text{MSE}}$$
$$\overline{T}_i = \text{Sample mean for treatment } i$$
$$\overline{B}_i = \text{Sample mean for block } i$$

and $t_{\alpha/2}$ is based on $(n - p - b + 1)$ degrees of freedom

EXAMPLE   12.9

Refer to Example 12.7. Find a 90% confidence interval for the difference between the mean level of estimates for engineers 1 and 2.

Solution

From Example 12.7, we know that $b = 4$, $\overline{T}_1 = 5.60$, $\overline{T}_2 = 5.85$, and $s^2 = \text{MSE}_C = .03111$. The degrees of freedom associated with s^2 (and, therefore, with $t_{\alpha/2}$) is 6. Therefore, $s = \sqrt{s^2} = \sqrt{.03111} = .176$ and $t_{\alpha/2} = t_{.05} = 1.943$. Substituting these values into the formula for the confidence interval for $(\mu_1 - \mu_2)$, we obtain

$$(\overline{T}_1 - \overline{T}_2) \pm (t_{\alpha/2})s\sqrt{\frac{2}{b}}$$

$$(5.60 - 5.85) \pm (1.943)(.176)\sqrt{\frac{2}{4}}$$

$$-.25 \pm .24$$

or, $-.49$ to $-.01$. Since each unit represents $100,000, we estimate the difference between the mean level of job estimates for estimators 1 and 2 to be enclosed by the interval, $-$49, 000$ to $-$1, 000$. [*Note:* At first glance, this result may appear to contradict the result of the F test for comparing treatment means. However, the observed significance level of the F test (.07) implies that significant differences exist between the means at $\alpha = .10$, which is consistent with the fact that 0 is not within the 90% confidence interval.]   ◆

There is one very important point to note when you block the treatments in an experiment. Recall from Section 11.3 that the block effects cancel. This fact enables us to calculate confidence intervals for the difference between treatment means using the formulas given in the box. But, if a sample treatment mean is used to estimate *a single treatment mean*, the block effects do not cancel. *Therefore, the only way that you can obtain an unbiased estimate of a single treatment mean (and corresponding confidence interval) in a blocked design is to randomly select the blocks from a large collection (population) of blocks and to treat the block effect as a second random component, in addition to random error.* Designs that contain two or more random components are called *nested designs* and are beyond the scope of this text. For more information on this topic, see the references at the end of this chapter.

| EXERCISES

12.18. The analysis of variance for a randomized block design produced the ANOVA table entries shown here.

SOURCE	df	SS	MS	F
Treatments	3	27.1	—	—
Blocks	5	—	14.90	—
Error	—	33.4	—	
Total	—	—		

The sample means for the four treatments are as follows:

$$\overline{y}_A = 9.7 \quad \overline{y}_B = 12.1 \quad \overline{y}_C = 6.2 \quad \overline{y}_D = 9.3$$

a. Complete the ANOVA table.
b. Do the data provide sufficient evidence to indicate a difference among the treatment means? Test using $\alpha = .01$.

c. Do the data provide sufficient evidence to indicate that blocking was a useful design strategy to employ for this experiment? Explain.

d. Find a 95% confidence interval for $(\mu_A - \mu_B)$.

e. Find a 95% confidence interval for $(\mu_B - \mu_D)$.

12.19. Do managers make inferences about an employee based on the employee's absenteeism? This question was researched in the *Journal of Management* (Vol. 18, 1992). A sample of 33 managers, all of whom were active members of a local association of human resources administrators, participated in the study. Each manager was presented with a booklet describing the absenteeism of three hypothetical employees. One employee was described as having an "excellent" record of attendance, the second an "average" record, and the third was described as being "absence-prone." The managers then estimated the total number of days each employee was absent during a 2-year period. The goal of the study was to compare the mean number of absences for the three hypothetical employees.

a. Identify the treatments and the blocks for this randomized block design.

b. Identify the response variable.

c. Specify the null hypothesis to be tested by the ANOVA.

d. Explain why a randomized block design is used in this study rather than a completely randomized design.

12.20. Rugel's pawpaw (yellow squirrel banana) is an endangered species of a dwarf shrub. Biologists from Stetson University conducted an experiment to determine the effects of fire on the shrub's growth (*Florida Scientist*, Spring 1997). Twelve experimental plots of land were selected in a pasture where the shrub is abundant. Within each plot, three pawpaws were randomly selected and treated as follows: one shrub was subjected to fire, another to clipping, and the third was left unmanipulated (a control). After five months, the number of flowers produced by each of the 36 shrubs was determined. The objective of the study was to compare the mean number of flowers produced by pawpaws for the three treatments (fire, clipping, and control).

a. Identify the type of experimental design employed, including the treatments, response variable, and experimental units.

b. Illustrate the layout of the design using a graphic similar to Figure 12.8.

c. Give the linear model appropriate for analyzing the data.

d. The ANOVA of the data resulted in a test statistic of $F = 5.42$ for treatments with p-value $= .009$. Interpret this result.

12.21. Using decoys is a common method of hunting waterfowl. A study in the *Journal of Wildlife Management* (July 1995) compared the effectiveness of three different decoy types—taxidermy-mounted decoys, plastic shell decoys, and full-bodied plastic decoys—in attracting Canada geese to sunken pit blinds. In order to account for an extraneous source of variation, three pit blinds were used as blocks in the experiment. Thus, a randomized block design with three treatments (decoy types) and three blocks (pit blinds) was employed. The response variable was the percentage of a goose flock to approach within 46 meters of the pit blind on a given day. The data are given in the table on p. 580.* A MINITAB printout of the analysis is shown below.

MINITAB output for Exercise 12.21

Two-way ANOVA: PERCENT versus DECOY, BLIND

Analysis of Variance for PERCENT

Source	DF	SS	MS	F	P
DECOY	2	30.1	15.0	0.61	0.589
BLIND	2	44.1	22.1	0.89	0.479
Error	4	99.3	24.8		
Total	8	173.6			

*The actual design employed in the study was more complex than the randomized block design shown here. In the actual study, each number in the table represents the mean daily percentage of goose flocks attracted to the blind, averaged over 13–17 days.

a. Find and interpret the F statistic for comparing the response means of the three decoy types.

b. Perform the analysis, part a, by fitting and comparing the appropriate linear models. Verify that the results agree.

Data for Exercise 12.21

 DECOY

BLIND	SHELL	FULL-BODIED	TAXIDERMY-MOUNTED
1	7.3	13.6	17.8
2	12.6	10.4	17.0
3	16.4	23.4	13.6

Source: Harrey, W.F., Hindman, L.J., and Rhodes, W.E. "Vulnerability of Canada geese to taxidermy-mounted decoys." *Journal of Wildlife Management*, Vol. 59, No. 3, July 1995, p. 475 (Table 1).

12.22. A commonly used index to estimate the reliability of a building subjected to lateral loads is the drift ratio. Sophisticated computer programs such as STAAD-III have been developed to estimate the drift ratio based on variables such as beam stiffness, column stiffness, story height, moment of inertia, and so on. Civil engineers at the State University of New York at Buffalo and the University of Central Florida performed an experiment to compare drift ratio estimates using STAAD-III with the estimates produced by a new, simpler microcomputer program called DRIFT (*Microcomputers in Civil Engineering*, 1993). Data for a 21-story building were used as input to the programs. Two runs were made with STAAD-III: Run 1 considered axial deformation of the building columns, and run 2 neglected this information. The goal of the analysis is to compare the mean drift ratios (where drift is measured as lateral displacement) estimated by the three computer runs (the two STAAD-III runs and DRIFT). The lateral displacements (in inches) estimated by the three programs are recorded in the next table for each of five building levels (1, 5, 10, 15, and 21). A MINITAB printout of the analysis of variance for the data is also shown in the next column.

a. Identify the treatments in the experiment.

b. Because lateral displacement will vary greatly across building levels (floors), a randomized block design will be used to reduce the level-to-level variation in drift. Explain, diagrammatically, the set-up of the design if all 21 levels are to be included in the study.

 STAAD

LEVEL	STAAD-III(1)	STAAD-III(2)	DRIFT
1	.17	.16	.16
5	1.35	1.26	1.27
10	3.04	2.76	2.77
15	4.54	3.98	3.99
21	5.94	4.99	5.00

Source: Valles, R. E., et al. "Simplified drift evaluation of wall-frame structures." *Microcomputers in Civil Engineering*, Vol. 8, 1993, p. 242 (Table 2).

MINITAB output for Exercise 12.22

Two-way ANOVA: DRIFT versus PROGRAM, LEVEL

```
Analysis of Variance for DRIFT
Source      DF       SS        MS        F       P
PROGRAM      2   0.4664    0.2332     4.79   0.043
LEVEL        4  52.1812   13.0453   267.74   0.000
Error        8   0.3898    0.0487
Total       14  53.0374
```

c. Using the information in the printout, compare the mean drift ratios estimated by the three programs.

12.23. Plant therapists believe that plants can reduce the stress levels of humans. A Kansas State University study was conducted to investigate this phenomenon. Two weeks before final exams, 10 undergraduate students took part in an experiment to determine what effect the presence of a live plant, a photo of a plant, or absence of a plant has on the student's ability to relax while isolated in a dimly lit room. Each student participated in three sessions—one with a live plant, one with a plant photo, and one with no plant (control).[†] During

PLANTS

STUDENT	LIVE PLANT	PLANT PHOTO	NO PLANT (CONTROL)
1	91.4	93.5	96.6
2	94.9	96.6	90.5
3	97.0	95.8	95.4
4	93.7	96.2	96.7
5	96.0	96.6	93.5
6	96.7	95.5	94.8
7	95.2	94.6	95.7
8	96.0	97.2	96.2
9	95.6	94.8	96.0
10	95.6	92.6	96.6

Source: Elizabeth Schreiber, Department of Statistics, Kansas State University, Manhattan, Kansas.

[†]The experiment is simplified for this exercise. The actual experiment involved 30 students who participated in 12 sessions.

SPSS Output for Exercise 12.23

Tests of Between-Subjects Effects

Dependent Variable: FINGTEMP

Source	Type III Sum of Squares	df	Mean Square	F	Sig.
Corrected Model	18.537[a]	11	1.685	.523	.863
Intercept	272176.875	1	272176.875	84413.380	.000
PLANT	.122	2	6.100E-02	.019	.981
STUDENT	18.415	9	2.046	.635	.754
Error	58.038	18	3.224		
Total	272253.450	30			
Corrected Total	76.575	29			

a. R Squared = .242 (Adjusted R Squared = -.221)

each session, finger temperature was measured at 1-minute intervals for 20 minutes. Since increasing finger temperature indicates an increased level of relaxation, the maximum temperature (in degrees) was used as the response variable. The data for the experiment, provided in the table on p. 580, were analyzed using the ANOVA procedure of SPSS. Use the SPSS printout above to make the proper inference.

12.24. A study was conducted to investigate the effect of prompting in a walking program (*Health Psychology*, Mar. 1995). Five groups of walkers—27 in each group—agreed to participate by walking for 20 minutes at least one day per week over a 24-week period. The participants were prompted to walk each week via telephone calls, but different prompting schemes were used for each group. Walkers in the control group received no prompting phone calls; walkers in the "frequent/low" group received a call once a week with low structure (i.e., "just touching base"); walkers in the "frequent/high" group received a call once a week with high structure (i.e., goals are set); walkers in the "infrequent/low" group received a call once every 3 weeks with low structure; and walkers in the "infrequent/high" group received a call once every 3 weeks with high structure. The table below lists the number of participants in each group who actually walked the minimum requirement each week for weeks 1, 4, 8, 12, 16, and 24. The data were subjected to an analysis of variance for a randomized block design, with the five walker groups representing the treatments and the six time periods (weeks) representing the blocks.

a. What is the purpose of blocking on weeks in this study?

b. Construct an ANOVA summary table using the printout.

c. Is there sufficient evidence of a difference in the mean number of walkers per week among the five walker groups? Use $\alpha = .05$.

⊙WALKERS

WEEK	CONTROL	FREQUENT/LOW	FREQUENT/HIGH	INFREQUENT/LOW	INFREQUENT/HIGH
1	7	23	25	21	19
4	2	19	25	10	12
8	2	18	19	9	9
12	2	7	20	8	2
16	2	18	18	8	7
24	1	17	17	7	6

Source: Lombard, D. N., *et al.* "Walking to meet health guidelines: The effect of prompting frequency and prompt structure." *Health Psychology*, Vol. 14, No. 2, Mar.1995, p. 167 (Table 2).

12.25. Eight amateur boxers participated in an experiment to investigate the effect of massage on boxing performance (*British Journal of Sports Medicine*, Apr. 2000). The punching power of each boxer (measured in Newtons) was recorded in the round following each of four different interventions: (M1) in round 1 following a pre-about sports massage, (R1) in round 1 following a pre-bout period of rest, (M5) in round 5 following a sports massage between rounds, and (R5) in round 5 following a period of rest between rounds. Based on information provided in the article, the data in the table were obtained. The main goal of the experiment is to compare the punching power means of the four interventions.

a. Give the complete model appropriate for this design.

b. Give the reduced model appropriate for testing for differences in the punching power means of the four interventions.

c. Give the reduced model appropriate for determining whether blocking by boxers was effective in removing an unwanted source of variability.

 BOXING

BOXER	INTERVENTION			
	M1	R1	M5	R5
1	1243	1244	1291	1262
2	1147	1053	1169	1177
3	1247	1375	1309	1321
4	1274	1235	1290	1285
5	1177	1139	1233	1238
6	1336	1313	1366	1362
7	1238	1279	1275	1261
8	1261	1152	1289	1266

Source: Hemmings, B., Smith, M., Graydon, J., and Dyson, R. "Effects of massage on physiological restoration, perceived recovery, and repeated sports performance." *British Journal of Sports Medicine*, Vol. 34, No. 2, Apr. 2000 (adapted from Table 3).

12.26. Refer to Exercise 12.25. The models of parts **a**, **b**, and **c** were fit to data in the table using MINITAB. The MINITAB printouts are displayed below and on p. 583.

a. Construct an ANOVA summary table

b. Is there evidence of differences in the punching power means of the four interventions? Use $\alpha = .05$.

MINITAB printout for complete model of Exercise 12.25a

```
The regression equation is
POWER = 1260 + 18.0 B1 - 106 B2 + 71.0 B3 + 29.0 B4 - 45.3 B5 + 102 B6
        + 21.3 B7 - 31.1 I1 + 6.3 I2 - 47.7 I3

Predictor      Coef      SE Coef        T          P
Constant    1260.16        20.84     60.48      0.000
B1            18.00        25.13      0.72      0.482
B2          -105.50        25.13     -4.20      0.000
B3            71.00        25.13      2.83      0.010
B4            29.00        25.13      1.15      0.261
B5           -45.25        25.13     -1.80      0.086
B6           102.25        25.13      4.07      0.001
B7            21.25        25.13      0.85      0.407
I1           -31.12        17.77     -1.75      0.094
I2             6.25        17.77      0.35      0.729
I3           -47.75        17.77     -2.69      0.014

S = 35.54      R-Sq = 83.4%      R-Sq(adj) = 75.4%

Analysis of Variance

Source            DF          SS         MS         F          P
Regression        10      132798      13280     10.51      0.000
Residual Error    21       26525       1263
Total             31      159323
```

MINITAB printout for complete model of Exercise 12.25b

```
The regression equation is
POWER = 1242 + 18.0 B1 - 106 B2 + 71.0 B3 + 29.0 B4 - 45.3 B5 + 102 B6
        + 21.3 B7
```

Predictor	Coef	SE Coef	T	P
Constant	1242.00	20.99	59.18	0.000
B1	18.00	29.68	0.61	0.550
B2	-105.50	29.68	-3.55	0.002
B3	71.00	29.68	2.39	0.025
B4	29.00	29.68	0.98	0.338
B5	-45.25	29.68	-1.52	0.140
B6	102.25	29.68	3.45	0.002
B7	21.25	29.68	0.72	0.481

```
S = 41.97      R-Sq = 73.5%     R-Sq(adj) = 65.7%
```

Analysis of Variance

Source	DF	SS	MS	F	P
Regression	7	117044	16721	9.49	0.000
Residual Error	24	42279	1762		
Total	31	159323			

MINITAB printout for complete model of Exercise 12.25c

```
The regression equation is
POWER = 1271 - 31.1 I1 + 6.3 I2 - 47.7 I3
```

Predictor	Coef	SE Coef	T	P
Constant	1271.50	25.32	50.22	0.000
I1	-31.12	35.80	-0.87	0.392
I2	6.25	35.80	0.17	0.863
I3	-47.75	35.80	-1.33	0.193

```
S = 71.61      R-Sq = 9.9%     R-Sq(adj) = 0.2%
```

Analysis of Variance

Source	DF	SS	MS	F	P
Regression	3	15754	5251	1.02	0.397
Residual Error	28	143569	5127		
Total	31	159323			

c. Is there evidence of a difference among the punching power means of the boxers? That is, is there evidence that blocking by boxers was effective in removing an unwanted source of variability? Use $\alpha = .05$.

12.27. A species of Caribbean mosquito is known to be resistant against certain insecticides. The effectiveness of five different types of insecticides—temephos, malathion, fenitrothion, fenthion, and chlorpyrifos—in controlling this mosquito species was investigated in the *Journal of the American Mosquito Control Association* (Mar. 1995). Mosquito larvae were collected from each of seven Caribbean locations. In a laboratory, the larvae from each location were divided into five batches and each batch was exposed to one of the five insecticides. The dosage of insecticide required to kill 50% of the larvae was recorded and divided by the known dosage for a susceptible mosquito strain. The resulting value is called the resistance ratio. (The higher the ratio, the more resistant the mosquito species is to the insecticide relative to the susceptible mosquito strain.) The resistance ratios for the study are listed in the table below. The researchers want to compare the mean resistance ratios of the five insecticides.

a. Explain why the experimental design is a randomized block design. Identify the treatments and the blocks.

b. Conduct a complete analysis of the data. Are any of the insecticides more effective than any of the others?

⊛ MOSQUITO

	INSECTICIDE				
LOCATION	TEMEPHOS	MALATHION	FENITROTHION	FENTHION	CHLORPYRIFOS
Anguilla	4.6	1.2	1.5	1.8	1.5
Antigua	9.2	2.9	2.0	7.0	2.0
Dominica	7.8	1.4	2.4	4.2	4.1
Guyana	1.7	1.9	2.2	1.5	1.8
Jamaica	3.4	3.7	2.0	1.5	7.1
St. Lucia	6.7	2.7	2.7	4.8	8.7
Suriname	1.4	1.9	2.0	2.1	1.7

Source: Rawlins, S. C., and Oh Hing Wan, J. "Resistance in some Caribbean population of *Aedes aegypti* to several insecticides." *Journal of the American Mosquito Control Association*, Vol. 11, No. 1, Mar. 1995 (Table 1).

12.5 Two-Factor Factorial Experiments

In Section 11.4, we learned that factorial experiments are volume-increasing designs conducted to investigate the effect of two or more independent variables (factors) on the mean value of the response y. In this section, we focus on the analysis of two-factor factorial experiments.

Suppose, for example, we want to relate the mean number of defects on a finished item—say, a new desk top—to two factors, type of nozzle for the varnish spray gun and length of spraying time. Suppose further that we want to investigate the mean number of defects per desk for three types (three levels) of nozzles (N_1, N_2, and N_3) and for two lengths (two levels) of spraying time (S_1 and S_2). If we choose the treatments for the experiment to include all combinations of the three levels of nozzle type with the two levels of spraying time (i.e., we observe the number of defects for the factor level combinations N_1S_1, N_1S_2, N_2S_1, N_2S_2, N_3S_1, N_3S_2), our design is called a **complete 3 × 2 factorial experiment**. Note that the design will contain $3 \times 2 = 6$ treatments.

Factorial experiments, you will recall, are useful methods for selecting treatments because they permit us to make inferences about factor interactions. The complete model for the 3×2 factorial experiment contains $(3 - 1) = 2$ main effect terms for nozzles, $(2 - 1) = 1$ main effect term for spray time, and $(3 - 1)(2 - 1) = 2$ nozzle–spray time interaction terms:

$$E(y) = \beta_0 + \underbrace{\beta_1 x_1 + \beta_2 x_2}_{\substack{\text{Main effects} \\ \text{Nozzle}}} + \underbrace{\beta_3 x_3}_{\substack{\text{Main effect} \\ \text{Spray time}}} + \underbrace{\beta_4 x_1 x_2 + \beta_5 x_1 x_3}_{\substack{\text{Interaction} \\ \text{Nozzle} \times \text{Spray time}}}$$

The independent variables (factors) in the model can be either quantitative or qualitative. If they are quantitative, the main effects are represented by terms such as x, x^2, x^3, etc.; if qualitative, the main effects are represented by dummy variables. In our 3×2 factorial experiment, nozzle type is qualitative and spraying time is quantitative; hence, the x variables in the model are defined as follows:

$$x_1 = \begin{cases} 1 & \text{if nozzle } N_1 \\ 0 & \text{if not} \end{cases} \qquad x_2 = \begin{cases} 1 & \text{if nozzle } N_2 \\ 0 & \text{if not} \end{cases} \qquad \text{Base level} = N_3$$

$$x_3 = \text{Length of spraying time (in minutes)}$$

Note that the model for the 3×2 factorial contains a total of $3 \times 2 = 6$ β parameters. If we observe only a single value of the response y for each of the $3 \times 2 = 6$ treatments, then $n = 6$ and df(Error) for the complete model is $(n - 6) = 0$. Consequently, for a factorial experiment, *the number r of observations per factor level combination (i.e., the number of replications of the factorial experiment) must always be 2 or more*. Otherwise, no degrees of freedom are available for estimating σ^2.

To test for factor interaction, we drop the interaction terms and fit the reduced model:

$$E(y) = \beta_0 + \underbrace{\beta_1 x_1 + \beta_2 x_2}_{\substack{\text{Main effects} \\ \text{Nozzle}}} + \underbrace{\beta_3 x_3}_{\substack{\text{Main effect} \\ \text{Spray time}}}$$

The null hypothesis of no interaction, $H_0 : \beta_4 = \beta_5 = 0$, is tested by comparing the SSEs for the two models in a partial F statistic. This test for interaction is summarized, in general, in the box.

Tests for factor main effects are conducted in a similar manner. The main effect terms of interest are dropped from the complete model and the reduced model is fitted. The SSEs for the two models are compared in the usual fashion.

Before we work through a numerical example of an analysis of variance for a factorial experiment, we must understand the practical significance of the tests for factor interaction and factor main effects. We illustrate these concepts in Example 12.10.

Models and ANOVA F Test for Interaction in a Two-Factor Factorial Experiment with Factor A at a Levels and Factor B at b Levels
Complete model:

$$E(y) = \beta_0 + \overbrace{\beta_1 x_1 + \cdots + \beta_{a-1} x_{a-1}}^{\text{Main effect } A \text{ terms}} + \overbrace{\beta_a x_a + \cdots + \beta_{a+b-2} x_{a+b-2}}^{\text{Main effect } B \text{ terms}}$$

$$+ \overbrace{\beta_{a+b-1} x_1 x_a + \beta_{a+b} x_1 x_{a+1} + \cdots + \beta_{ab-1} x_{a-1} x_{a+b-2}}^{AB \text{ interaction terms}}$$

where*

$$x_1 = \begin{cases} 1 & \text{if level 2 of factor } A \\ 0 & \text{if not} \end{cases} \cdots$$

$$x_{a-1} = \begin{cases} 1 & \text{if level } a \text{ of factor } A \\ 0 & \text{if not} \end{cases}$$

$$x_a = \begin{cases} 1 & \text{if level 2 of factor } B \\ 0 & \text{if not} \end{cases} \cdots$$

$$x_{a+b-2} = \begin{cases} 1 & \text{if level } b \text{ of factor } B \\ 0 & \text{if not} \end{cases}$$

$H_0: \beta_{a+b-1} = \beta_{a+b} = \cdots = \beta_{ab-1} = 0$
   (i.e., H_0: No interaction between factors A and B.)

H_a: At least one of the β parameters listed in H_0 differs from 0.
   (i.e., H_a: Factors A and B interact.)

Reduced model:

$$E(y) = \beta_0 + \overbrace{\beta_1 x_1 + \cdots + \beta_{a-1} x_{a-1}}^{\text{Main effect A terms}} + \overbrace{\beta_a x_a + \cdots + \beta_{a+b-2} x_{a+b-2}}^{\text{Main effect B terms}}$$

Test statistic: $F = \dfrac{(\text{SSE}_R - \text{SSE}_C)/[(a-1)(b-1)]}{\text{SSE}_C/[ab(r-1)]}$

$ = \dfrac{(\text{SSE}_R - \text{SSE}_C)/[(a-1)(b-1)]}{\text{MSE}_C}$

where

$\text{SSE}_R = $ SSE for reduced model

$\text{SSE}_C = $ SSE for complete model

$\text{MSE}_C = $ MSE for complete model

$r = $ Number of replications (i.e., number of y measurements per cell of the $a \times b$ factorial)

Rejection region: $F > F_\alpha$, where F is based on $v_1 = (a-1)(b-1)$ and $v_2 = ab(r-1)$ df

Assumptions: 1. The population probability distribution of the observations for any factor level combination is approximately normal.

2. The variance of the probability distribution is constant and the same for all factor level combinations.

Note: The independent variables, $x_1, x_2, \ldots, x_{a+b-2}$, are defined for an experiment in which both factors represent *qualitative* variables. When a factor is *quantitative*, you may choose to represent the main effects with quantitative terms such as x, x^2, x^3, and so forth.

EXAMPLE 12.10

A company that stamps gaskets out of sheets of rubber, plastic, and other materials wants to compare the mean number of gaskets produced per hour for two different types of stamping machines. Practically, the manufacturer wants to determine whether one machine is more productive than the other. Even more important is determining whether one machine is more productive in making rubber gaskets while the other is more productive in making plastic gaskets. To answer these questions, the manufacturer decides to conduct a 2×3 factorial experiment using three types of gasket material, B_1, B_2, and B_3, with each of the two types of stamping machines, A_1 and A_2. Each machine is operated for three 1-hour time periods for each of the gasket materials, with the eighteen 1-hour time periods assigned to the six machine–material combinations in random order. (The purpose of the randomization is to eliminate the possibility that uncontrolled environmental factors might bias the results.) Suppose we have calculated and plotted the six treatment means. Two hypothetical plots of the six means are shown in Figures 12.14a and 12.14b. The three means for stamping machine A_1 are connected by solid line segments and the corresponding three means for machine A_2 by dashed line segments. What do these plots imply about the productivity of the two stamping machines?

FIGURE 12.14
Hypothetical plot of the means for the six machine–material combinations

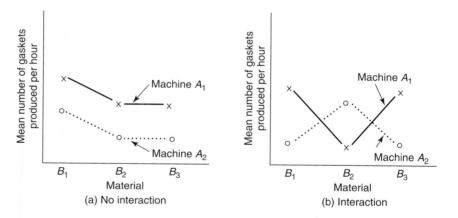

(a) No interaction   (b) Interaction

Solution

Figure 12.14a suggests that machine A_1 produces a larger number of gaskets per hour, regardless of the gasket material, and is therefore superior to machine A_2. On the average, machine A_1 stamps more cork (B_1) gaskets per hour than rubber or plastic, but the *difference* in the mean numbers of gaskets produced by the two machines remains approximately the same, regardless of the gasket material. Thus, the difference in the mean number of gaskets produced by the two machines is *independent* of the gasket material used in the stamping process.

In contrast to Figure 12.14a, Figure 12.14b shows the productivity for machine A_1 to be greater than that for machine A_2, when the gasket material is cork (B_1) or plastic (B_3). But the means are reversed for rubber (B_2) gasket material. For this material, machine A_2 produces, on the average, more gaskets per hour than machine A_1. Thus, Figure 12.14b illustrates a situation where the mean value of the response variable *depends* on the combination of the factor levels. When this situation occurs, we say that the factors *interact*. Thus, one of the most important objectives of a factorial experiment is to detect factor interaction if it exists.   ◆

Definition 12.2

In a factorial experiment, when the difference in the mean levels of factor A depends on the different levels of factor B, we say that the factors A and B **interact**. If the difference is independent of the levels of B, then there is **no interaction** between A and B.

Tests for main effects are relevant only when no interaction exists between factors. Generally, the test for interaction is performed first. *If there is evidence of factor interaction, then we will not perform the tests on the main effects.* Rather, we will want to focus attention on the individual cell (treatment) means, perhaps locating one that is the largest or the smallest.

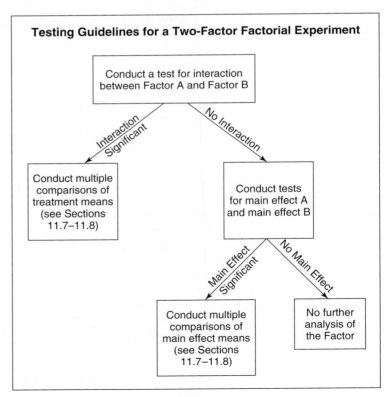

EXAMPLE 12.11

A manufacturer, whose daily supply of raw materials is variable and limited, can use the material to produce two different products in various proportions. The profit per unit of raw material obtained by producing each of the two products depends on the length of a product's manufacturing run and, hence, on the amount of raw material assigned to it. Other factors, such as worker productivity and machine breakdown, affect the profit per unit as well, but their net effect on profit is random and uncontrollable. The manufacturer has conducted an experiment to investigate the effect of the level of supply of raw materials, S, and the ratio of its assignment, R, to the two product manufacturing lines on the profit y per unit of raw material. The ultimate goal would be to be able to choose the best ratio R to match each day's supply of raw materials, S. The levels of supply of the raw material chosen for the experiment were 15, 18, and 21 tons; the levels of the ratio of allocation to the two product lines were $\frac{1}{2}$, 1, and 2. The response was the profit (in dollars) per unit of raw material supply obtained from a single day's production. Three replications of a complete 3×3 factorial experiment were conducted in a random sequence (i.e., a completely randomized design). The data for the 27 days are shown in Table 12.6.

a. Write the complete model for the experiment.

b. Do the data present sufficient evidence to indicate an interaction between supply S and ratio R? Use $\alpha = .05$.

c. Based on the result, part **b**, should we perform tests for main effects?

◉ RAWMATERIAL

TABLE 12.6 Data for Example 12.11

		RAW MATERIAL SUPPLY (S), tons		
		15	18	21
RATIO OF RAW MATERIAL ALLOCATION (R)	$\frac{1}{2}$	23, 20, 21	22, 19, 20	19, 18, 21
	1	22, 20, 19	24, 25, 22	20, 19, 22
	2	18, 18, 16	21, 23, 20	20, 22, 24

Solution

a. Both factors, supply and ratio, are quantitative. Accordingly, when the factors in a factorial experiment are quantitative, the main effects can be represented by terms such as x, x^2, x^3, and so forth. Since each factor has three levels, we require two main effects, x and x^2, for each factor. (In general, the number of main effect terms will be one less than the number of levels for a factor.) Consequently, the complete factorial model for this 3×3 factorial experiment is

$$E(y) = \beta_0 + \underbrace{\beta_1 x_1 + \beta_2 x_1^2}_{\text{Supply main effects}} \quad \underbrace{+ \beta_3 x_2 + \beta_4 x_2^2}_{\text{Ratio main effects}}$$

$$+ \underbrace{\beta_5 x_1 x_2 + \beta_6 x_1 x_2^2 + \beta_7 x_1^2 x_2 + \beta_8 x_1^2 x_2^2}_{\text{Supply} \times \text{Ratio interaction}}$$

where

$$x_1 = \text{Supply of raw material (in tons)}$$
$$x_2 = \text{Ratio of allocation}$$

Note that the interaction terms for the model are constructed by taking the products of the various main effect terms, one from each factor. For example, we included terms involving the products of x_1 with x_2 and x_2^2. The remaining interaction terms were formed by multiplying x_1^2 by x_2 and by x_2^2.

b. To test the null hypothesis that supply and ratio do not interact, we must test the null hypothesis that the interaction terms are not needed in the linear model of part **a**:

$H_0: \beta_5 = \beta_6 = \beta_7 = \beta_8 = 0$

This requires that we fit the reduced model

$$E(y) = \beta_0 + \beta_1 x_1 + \beta_2 x_1^2 + \beta_3 x_2 + \beta_4 x_2^2$$

and perform the partial F test outlined in Section 4.13. The test statistic is

$$F = \frac{(\text{SSE}_R - \text{SSE}_C)/4}{\text{MSE}_C}$$

where

$$\text{SSE}_R = \text{SSE for reduced model}$$
$$\text{SSE}_C = \text{SSE for complete model}$$
$$\text{MSE}_C = \text{MSE for complete model}$$

The complete model of part **a** and the reduced model here were fitted to the data in Table 12.6 using SAS. The SAS printouts are displayed in Figures 12.15a and 12.15b. The pertinent quantities, shaded on the printouts, are

$\text{SSE}_C = 43.33333$ (see Figure 12.15a)

$\text{MSE}_C = 2.40741$ (see Figure 12.15a)

$\text{SSE}_R = 89.55556$ (see Figure 12.15b)

Substituting these values into the formula for the test statistic, we obtain

$$F = \frac{(\text{SSE}_R - \text{SSE}_C)/4}{\text{MSE}_C} = \frac{(89.55556 - 43.33333)/4}{2.40741} = 4.80$$

This "partial" F value is shaded at the bottom of the SAS printout, Figure 12.15a, as is the p-value of the test, .0082. Since $\alpha = .05$ exceeds the p-value, we reject H_0 and conclude that supply and ratio interact.

c. The presence of interaction tells you that the mean profit depends on the particular combination of levels of supply S and ratio R. Consequently, there is little point in checking to see whether the means differ for the three levels of supply or whether they differ for the three levels of ratio (i.e., we will not perform

FIGURE 12.15a

SAS regression printout for complete factorial model

Dependent Variable: PROFIT

Analysis of Variance

Source	DF	Sum of Squares	Mean Square	F Value	Pr > F
Model	8	74.66667	9.33333	3.88	0.0081
Error	18	43.33333	2.40741		
Corrected Total	26	118.00000			

Root MSE	1.55158	R-Square	0.6328	
Dependent Mean	20.66667	Adj R-Sq	0.4696	
Coeff Var	7.50766			

Parameter Estimates

Variable	DF	Parameter Estimate	Standard Error	t Value	Pr > |t|
Intercept	1	245.33333	130.49665	1.88	0.0764
SUPPLY	1	-25.07407	14.71842	-1.70	0.1057
SUPPSQ	1	0.67901	0.40837	1.66	0.1137
RATIO	1	-534.33333	252.45535	-2.12	0.0485
RATSQ	1	192.66667	97.17011	1.98	0.0629
RAT_SUPP	1	60.55556	28.47387	2.13	0.0475
S_RATSQ	1	-22.14815	10.95960	-2.02	0.0584
R_SUPPSQ	1	-1.66667	0.79003	-2.11	0.0492
RSQ_SSQ	1	0.61728	0.30408	2.03	0.0574

Test INTERACT Results for Dependent Variable PROFIT

Source	DF	Mean Square	F Value	Pr > F
Numerator	4	11.55556	4.80	0.0082
Denominator	18	2.40741		

Test HIGHORDR Results for Dependent Variable PROFIT

Source	DF	Mean Square	F Value	Pr > F
Numerator	3	3.71958	1.55	0.2373
Denominator	18	2.40741		

FIGURE 12.15b

SAS regression printout for reduced factorial model

Dependent Variable: PROFIT

Analysis of Variance

Source	DF	Sum of Squares	Mean Square	F Value	Pr > F
Model	4	28.44444	7.11111	1.75	0.1757
Error	22	89.55556	4.07071		
Corrected Total	26	118.00000			

Root MSE	2.01760	R-Square	0.2411	
Dependent Mean	20.66667	Adj R-Sq	0.1031	
Coeff Var	9.76258			

Parameter Estimates

Variable	DF	Parameter Estimate	Standard Error	t Value	Pr > |t|
Intercept	1	-43.48148	29.32961	-1.48	0.1524
SUPPLY	1	6.81481	3.29854	2.07	0.0508
SUPPSQ	1	-0.18519	0.09152	-2.02	0.0553
RATIO	1	5.66667	4.35851	1.30	0.2070
RATSQ	1	-2.29630	1.67759	-1.37	0.1849

the tests for main effects). For example, the supply level that gave the highest mean profit (over all levels of R) might not be the same supply—ratio level combination that produces the largest mean profit per unit of raw material. ◆

The traditional analysis of variance approach to analyzing a complete two-factor factorial with factor A at a levels and factor B at b levels utilizes the fact that the total sum of squares, SS(Total), can be partitioned into four parts, SS(A), SS(B), SS(AB), and SSE (see Figure 12.16). The first two sums of squares, SS(A) and SS(B), are called **main effect sums of squares** to distinguish them from the **interaction sum of squares**. SS(AB).

FIGURE 12.16

Partitioning of the total sum of squares for a complete two-factor factorial experiment

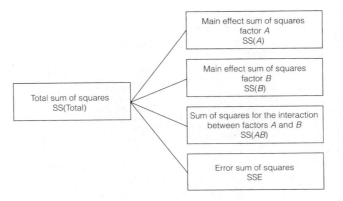

Since the sums of squares and the degrees of freedom for the analysis of variance are additive, the ANOVA table appears as shown in the following box.

ANOVA Table for an $a \times b$ Factorial Design with r Observations per Cell

SOURCE	df	SS	MS	F
Main effects A	$(a-1)$	SS(A)	MS(A) = SS(A)/$(a-1)$	MS(A) / MSE
Main effects B	$(b-1)$	SS(B)	MS(B) = SS(B)/$(b-1)$	MS(B) / MSE
AB interaction	$(a-1)(b-1)$	SS(AB)	MS(AB) = SS(AB)/$[(a-1)(b-1)]$	MS(AB) / MSE
Error	$ab(r-1)$	SSE	MSE = SSE/$[ab(r-1)]$	
Total	$abr-1$	SS(Total)		

(Note: $n = abr$).

Note that the F statistics for testing factor main effects and factor interaction are obtained by dividing the appropriate mean square by MSE. The numerator df for the test of interest will equal the df of the source of variation being tested; the denominator df will always equal df(Error). These F tests are equivalent to the F tests obtained by fitting complete and reduced models in regression.[*]

---

*The ANOVA F tests for main effects shown in the ANOVA summary table are equivalent to those of the regression approach only when the reduced model includes interaction terms. Since we usually test for main effects only after determining that interaction is nonsignificant, some statisticians favor dropping the interaction terms from both the complete and reduced models prior to conducting the main effect tests. For example, to test for main effect A, the complete model includes terms for main effects A

For completeness, the formulas for calculating the ANOVA sums of squares for a complete two-factor factorial experiment are given in the next box.

ANOVA Computing Formulas for a Two-Factor Factorial Experiment

CM = Correction for the mean

$$= \frac{(\text{Total of all } n \text{ measurements})^2}{n}$$

$$= \frac{\left(\sum_{i=1}^{n} y_i\right)^2}{n}$$

$SS(\text{Total})$ = Total sum of squares

= Sum of squares of all n measurements − CM

$$= \sum_{i=1}^{n} y_i^2 - CM$$

$SS(A)$ = Sum of squares for main effects, independent variable 1

$$= \left(\begin{array}{c} \text{Sum of squares of the totals } A_1, A_2, \ldots, A_a \\ \text{divided by the number of measurements} \\ \text{in a single total, namely, } br \end{array} \right) - CM$$

$$= \frac{\sum_{i=1}^{a} A_i^2}{br} - CM$$

$SS(B)$ = Sum of squares for main effects, independent variable 2

$$= \left(\begin{array}{c} \text{Sum of squares of the totals } B_1, B_2, \ldots, B_b \\ \text{divided by the number of measurements} \\ \text{in a single total, namely, } ar \end{array} \right) - CM$$

$$= \frac{\sum_{i=1}^{b} B_i^2}{ar} - CM$$

---

and B, whereas the reduced model includes terms for main effect B only. To obtain the equivalent result using the ANOVA approach, the sums of squares for AB interaction and error are "pooled" and a new MSE is computed, where

$$MSE = \frac{SS(AB) + SSE}{n - a - b + 1}$$

SS(AB) = Sum of squares for AB interaction

$$= \left(\begin{array}{c} \text{Sum of squares of the cell totals} \\ AB_{11}, AB_{12}, \ldots, AB_{ab} \text{ divided by} \\ \text{the number of measurements} \\ \text{in a single total, namely, } r \end{array} \right) - \text{SS}(A) - \text{SS}(B) - \text{CM}$$

$$= \frac{\displaystyle\sum_{i=1}^{b} \sum_{i=1}^{a} AB_{ij}^2}{r} - \text{SS}(A) - \text{SS}(B) - \text{CM}$$

where

a = Number of levels of independent variable 1

b = Number of levels of independent variable 2

r = Number of measurements for each pair of levels of independent variables 1 and 2

n = Total number of measurements

   = $a \times b \times r$

A_i = Total of all measurements of independent variable 1 at level $i \, (i = 1, 2, \ldots, a)$

B_j = Total of all measurements of independent variable 2 at level j $(j = 1, 2, \ldots, b)$

AB_{ij} = Total of all measurements at the ith level of independent variable 1 and at the jth level of independent variable 2 $(i = 1, 2, \ldots, a;$ $j = 1, 2, \ldots, b)$

EXAMPLE 12.12

Refer to Example 12.11.

a. Construct an ANOVA summary table for the analysis.

b. Conduct the test for supply × ratio interaction using the traditional analysis of variance approach.

Solution

a. Although the formulas given in the previous box are straightforward, they can become quite tedious to use. Therefore, we use a statistical software package to conduct the ANOVA. A SAS printout of the ANOVA is displayed in Figure 12.17. The value of SS(Total), given in the SAS printout under **Sum of Squares** in the **Corrected Total** row, is SS(Total) = 118. The sums of squares, mean squares, and F values for the factors S, R, and $S \times R$ interaction are given under the **Anova SS, Mean Square**, and **F Value** columns, respectively, in the bottom portion of the printout. These values are shown in Table 12.7.

FIGURE 12.17

SAS ANOVA printout for complete factorial design

The ANOVA Procedure

Dependent Variable: PROFIT

Source	DF	Sum of Squares	Mean Square	F Value	Pr > F
Model	8	74.6666667	9.3333333	3.88	0.0081
Error	18	43.3333333	2.4074074		
Corrected Total	26	118.0000000			

R-Square	Coeff Var	Root MSE	PROFIT Mean
0.632768	7.507656	1.551582	20.66667

Source	DF	Anova SS	Mean Square	F Value	Pr > F
SUPPLY	2	20.22222222	10.11111111	4.20	0.0318
RATIO	2	8.22222222	4.11111111	1.71	0.2094
SUPPLY*RATIO	4	46.22222222	11.55555556	4.80	0.0082

TABLE 12.7   ANOVA Table for Example 12.12

SOURCE	df	SS	MS
Supply	2	20.22	10.11
Ratio	2	8.22	4.11
Supply × Ratio interaction	4	46.22	11.56
Error	18	43.33	2.41
Total	26	118.00	

b. To test the hypothesis that supply and ratio interact, we use the test statistic

$$F = \frac{MS(SR)}{MSE} = \frac{11.56}{2.41} = 4.80 \text{ (shaded on the SAS printout, Figure 12.17)}$$

The p-value of the test (also shaded on the SAS printout) is .0082. (Both of these values are identical to the values obtained using regression in Example 12.11.) Since the p-value is less than the selected value of $\alpha = .05$, we conclude that supply and ratio interact. ◆

Confidence intervals for a single treatment mean and for the difference between two treatment means in a factorial experiment are provided in the following boxes.

$100(1 - \alpha)\%$ Confidence Interval for the Mean of a Single Treatment: Factorial Experiment

$$\bar{y}_{ij} \pm (t_{\alpha/2}) \left(\frac{s}{\sqrt{r}} \right)$$

where

$\bar{y}_{ij}$ = Sample mean for the treatment identified by level i of the first factor and level j of the second factor

r = Number of measurements per treatment

$s = \sqrt{\text{MSE}}$

and $t_{\alpha/2}$ is based on $ab(r - 1)$ df.

$100(1 - \alpha)\%$ Confidence Interval for the Difference Between a Pair of Treatment Means: Factorial Experiment

$$(\bar{y}_1 - \bar{y}_2) \pm (t_{\alpha/2}) s \sqrt{\frac{2}{r}}$$

where

$\bar{y}_1$ = Sample mean of the r measurements for the first treatment

$\bar{y}_2$ = Sample mean of the r measurements for the second treatment

$s = \sqrt{\text{MSE}}$ and $t_{\alpha/2}$ is based on $ab(r - 1)$ df.

EXAMPLE   12.13

Refer to Examples 12.11 and 12.12.

a. Find a 95% confidence interval to estimate the mean profit per unit of raw materials when $S = 18$ tons and the ratio of production is $R = 1$.

b. Find a 95% confidence interval to estimate the difference in mean profit per unit of raw materials when $(S = 18, R = \frac{1}{2})$ and $(S = 18, R = 1)$.

Solution

a. A 95% confidence interval for the mean $E(y)$ when supply $S = 18$ and $R = 1$ is

$$\bar{y}_{18,1} \pm (t_{.025}) \left(\frac{s}{\sqrt{r}} \right)$$

where $\bar{y}_{18,1}$ is the mean of the $r = 3$ values of y for $S = 18$, $R = 1$ (obtained from Table 12.6), and $t_{.025} = 2.101$ is based on 18 df. Substituting, we obtain

$$\frac{71}{3} \pm (2.101) \left(\frac{1.55}{\sqrt{3}} \right)$$

$$23.67 \pm 1.88$$

Therefore, our interval estimate for the mean profit per unit of raw material where $S = 18$ and $R = 1$ is \$21.79 to \$25.55.

b. A 95% confidence interval for the difference in mean profit per unit of raw material for two different combinations of levels of S and R is

$$(\bar{y}_1 - \bar{y}_2) \pm (t_{.025})s\sqrt{\frac{2}{r}}$$

where $\bar{y}_1$ and $\bar{y}_2$ represent the means of the $r = 3$ replications for the factor level combinations $(S = 18, R = \frac{1}{2})$ and $(S = 18, R = 1)$, respectively. From Table 12.7, the sums of the three measurements for these two treatments are 61 and 71. Substituting, we obtain

$$\left(\frac{61}{3} - \frac{71}{3} \right) \pm (2.101)(1.55)\sqrt{\frac{2}{3}}$$

$$-3.33 \pm 2.66$$

Therefore, the interval estimate for the difference in mean profit per unit of raw material for the two factor level combinations is $(-\$5.99, -\$.67)$. The negative values indicate that we estimate the mean for $(S = 18, R = \frac{1}{2})$ to be less than the mean for $(S = 18, R = 1)$ by between \$.67 and \$5.99. ◆

Throughout this chapter, we have presented two methods for analyzing data from a designed experiment: the regression approach and the traditional ANOVA approach. In a factorial experiment, the two methods yield identical results when both factors are qualitative; however, regression will provide more information when at least one of the factors is quantitative. For example, the analysis of variance in Example 12.12 enables us to estimate the mean profit per unit of supply for *only* the nine combinations of supply–ratio levels. It will not permit us to estimate the mean response for some other combination of levels of the independent variables not included among the nine used in the factorial experiment. Alternatively, the prediction equation obtained from the regression analysis in Example 12.11 enables us to estimate the mean profit per unit of supply when $(S = 17, R = 1)$. We could not obtain this estimate from the analysis of variance in Example 12.12.

The prediction equation found by regression analysis also contributes other information not provided by traditional analysis of variance. For example, we might wish to estimate the rate of change in the mean profit, $E(y)$, for unit changes in S, R, or both for specific values of S and R. Or, we might want to determine whether the third- and fourth-order terms in the complete model of Example 12.11 really contribute additional information for the prediction of profit, y.

We illustrate some of these applications in the next two examples.

EXAMPLE   12.14

Do the data provide sufficient information to indicate that third- and fourth-order terms in the complete factorial model given in Example 12.11 contribute information for the prediction of y? Use $\alpha = .05$.

Solution

If the response to the question is yes, then at least one of the parameters, β_6, β_7, or β_8, of the complete factorial model differs from 0 (i.e., they are needed in the model). Consequently, the null hypothesis is

$$H_0 : \beta_6 = \beta_7 = \beta_8 = 0$$

and the alternative hypothesis is

H_a: At least one of the three β's is nonzero.

To test this hypothesis, we compute the drop in SSE between the appropriate reduced and complete model.

For this application the complete model is the complete factorial model of Example 12.11:

$$\textit{Complete model}: \quad E(y) = \beta_0 + \beta_1 x_1 + \beta_2 x_1^2 + \beta_3 x_2 + \beta_4 x_2^2 + \beta_5 x_1 x_2$$
$$+ \beta_6 x_1 x_2^2 + \beta_7 x_1^2 x_2 + \beta_8 x_1^2 x_2^2$$

The reduced model is this complete model minus the third- and fourth-order terms; i.e., the reduced model is the second-order model shown here:

$$\textit{Reduced model}: E(y) = \beta_0 + \beta_1 x_1 + \beta_2 x_1^2 + \beta_3 x_2 + \beta_4 x_2^2 + \beta_5 x_1 x_2$$

Recall (from Figure 12.15a) that the SSE and MSE for the complete model are $SSE_C = 43.3333$ and $MSE_C = 2.4704$. A SAS printout of the regression analysis of the reduced model is shown in Figure 12.18. The SSE for the reduced model (shaded) is $SSE_R = 54.49206$.

Consequently, the test statistic required to conduct the nested model F test is

Test statistic:

$$F = \frac{(SSE_R - SSE_C)/(\text{Number of } \beta\text{'s tested})}{MSE_C} = \frac{(54.49206 - 43.3333)/3}{2.4704}$$
$$= 1.55$$

This "partial" F value can also be obtained using SAS options and is given at the bottom of the SAS complete model printout, Figure 12.15a, as well as the p-value of the test, .2373.

Conclusion: Since $\alpha = .05$ is less than p-value $= .2373$, we cannot reject the null hypothesis that $\beta_6 = \beta_7 = \beta_8 = 0$. That is, there is insufficient evidence (at $\alpha = .05$) to indicate that the third- and fourth-order terms associated with β_6, β_7, and β_8 contribute information for the prediction of y. Since the complete factorial model contributes no more information about y than the reduced (second-order) model, we recommend using the reduced model in practice.   ◆

FIGURE 12.18

SAS regression printout for reduced
(second-order) factorial model

```
                    Dependent Variable: PROFIT
                       Analysis of Variance

                            Sum of          Mean
Source              DF      Squares         Square    F Value   Pr > F

Model                5     63.50794       12.70159      4.89    0.0040
Error               21     54.49206        2.59486
Corrected Total     26    118.00000


          Root MSE              1.61086    R-Square    0.5382
          Dependent Mean       20.66667    Adj R-Sq    0.4283
          Coeff Var             7.79447


                       Parameter Estimates

                    Parameter     Standard
Variable      DF    Estimate      Error       t Value    Pr > |t|

Intercept      1    -27.81481     23.80152      -1.17     0.2557
SUPPLY         1      5.94444      2.64418       2.25     0.0354
RATIO          1     -7.76190      5.04523      -1.54     0.1389
RAT_SUPP       1      0.74603      0.20295       3.68     0.0014
SUPPSQ         1     -0.18519      0.07307      -2.53     0.0193
RATSQ          1     -2.29630      1.33939      -1.71     0.1012
```

EXAMPLE 12.15

Use the second-order model of Example 12.14 and find a 95% confidence interval for the mean profit per unit supply of raw material when $S = 17$ and $R = 1$.

Solution

The portion of the SAS printout for the second-order model with 95% confidence intervals for $E(y)$ is shown in Figure 12.19.

The confidence interval for $E(y)$ when $S = 17$ and $R = 1$ is given in the last row of the printout. You can see that the interval is (20.9687, 23.7244). Thus, we estimate (with confidence coefficient equal to .95) that the mean profit per unit of supply will lie between $20.97 and $23.72 when $S = 17$ tons and $R = 1$. Beyond this immediate result, this example illustrates the power and versatility of a regression analysis. In particular, there is no way to obtain this estimate from the analysis of variance in Example 12.12. However, a computerized regression package can be easily programmed to include the confidence interval automatically. ◆

EXERCISES

12.28. The analysis of variance for a 3×2 factorial experiment, with four observations per treatment, produced the ANOVA summary table entries shown here.
a. Complete the ANOVA summary table.
b. Test for interaction between factor A and factor B. Use $\alpha = .05$.

SOURCE	df	SS	MS	F
A	—	100	—	—
B	1	—	—	—
AB	2	—	2.5	—
Error	—	—	2.0	
Total	—	700		

FIGURE 12.19

SAS printout of confidence intervals for reduced
(second-order) factorial model

Obs	SUPPLY	RATIO	Dep Var PROFIT	Predicted Value	Std Error Mean Predict	95% CL Mean		Residual
1	15	0.5	23.0000	20.8254	0.8033	19.1549	22.4959	2.1746
2	15	0.5	20.0000	20.8254	0.8033	19.1549	22.4959	-0.8254
3	15	0.5	21.0000	20.8254	0.8033	19.1549	22.4959	0.1746
4	18	0.5	22.0000	21.4444	0.6932	20.0029	22.8860	0.5556
5	18	0.5	19.0000	21.4444	0.6932	20.0029	22.8860	-2.4444
6	18	0.5	20.0000	21.4444	0.6932	20.0029	22.8860	-1.4444
7	21	0.5	19.0000	18.7302	0.8033	17.0596	20.4007	0.2698
8	21	0.5	18.0000	18.7302	0.8033	17.0596	20.4007	-0.7302
9	21	0.5	21.0000	18.7302	0.8033	17.0596	20.4007	2.2698
10	15	1	22.0000	20.8175	0.7006	19.3605	22.2744	1.1825
11	15	1	20.0000	20.8175	0.7006	19.3605	22.2744	-0.8175
12	15	1	19.0000	20.8175	0.7006	19.3605	22.2744	-1.8175
13	18	1	24.0000	22.5556	0.6932	21.1140	23.9971	1.4444
14	18	1	25.0000	22.5556	0.6932	21.1140	23.9971	2.4444
15	18	1	22.0000	22.5556	0.6932	21.1140	23.9971	-0.5556
16	21	1	20.0000	20.9603	0.7006	19.5034	22.4173	-0.9603
17	21	1	19.0000	20.9603	0.7006	19.5034	22.4173	-1.9603
18	21	1	22.0000	20.9603	0.7006	19.5034	22.4173	1.0397
19	15	2	18.0000	17.3571	0.8590	15.5707	19.1436	0.6429
20	15	2	18.0000	17.3571	0.8590	15.5707	19.1436	0.6429
21	15	2	16.0000	17.3571	0.8590	15.5707	19.1436	-1.3571
22	18	2	21.0000	21.3333	0.6932	19.8917	22.7749	-0.3333
23	18	2	23.0000	21.3333	0.6932	19.8917	22.7749	1.6667
24	18	2	20.0000	21.3333	0.6932	19.8917	22.7749	-1.3333
25	21	2	20.0000	21.9762	0.8590	20.1897	23.7627	-1.9762
26	21	2	22.0000	21.9762	0.8590	20.1897	23.7627	0.0238
27	21	2	24.0000	21.9762	0.8590	20.1897	23.7627	2.0238
28	17	1	.	22.3466	0.6625	20.9687	23.7244	.

c. Test for differences in main effect means for factor *A*. Use $\alpha = .05$.

d. Test for differences in main effect means for factor *B*. Use $\alpha = .05$.

12.29. A coagulation—microfiltration process for removing bacteria from water was investigated in *Environmental Science & Engineering* (Sept. 1, 2000). Chemical engineers at Seoul National University performed a designed experiment to estimate the effect of both the level of the coagulant and acidity (pH) level on the coagulation efficiency of the process. Six levels of coagulant (5, 10, 20, 50, 100, and 200 milligrams per liter) and six pH levels (4.0, 5.0, 6.0, 7.0, 8.0, and 9.0) were employed. Water specimens collected from the Han River in Seoul, Korea, were placed in jars, and each jar randomly assigned to receive one of the $6 \times 6 = 36$ combinations of coagulant level and pH level.

a. What type of experimental design was applied in this study?

b. Give the factors, factor levels, and treatments for the study.

12.30. Many temperate-zone animal species exhibit physiological and morphological changes when the hours of daylight begin to decrease during autumn months. A study was conducted to investigate the "short day" traits of collared lemmings (*The Journal of Experimental Zoology*, Sept. 1993). A total of 124 lemmings were bred in a colony maintained with a photoperiod of 22 hours of light per day. At weaning (19 days of age), the lemmings were weighed and randomly assigned to live under one of two photoperiods: (1) 16 hours or less of light per day, and (2) more than 16 hours light per day. (Each group was assigned the same number of males and females.) After 10 weeks, the lemmings were weighed again. The response variable of interest was the gain in body weight (measured in grams) over the 10-week experimental period. The researchers analyzed the data using an ANOVA for a 2×2 factorial design, where the two factors are photoperiod (at two levels) and gender (at two levels).

a. Construct an ANOVA table for the experiment, listing the sources of variation and associated degrees of freedom.

b. Give the models that will enable the researchers to test for photoperiod by gender interaction.

c. The *F* test for interaction was not significant. Interpret this result practically.

d. The p-values for testing for photoperiod and gender main effects were both smaller than .001. Interpret these results practically.

12.31. Parapsychologists define "lucky" people as individuals who report that seemingly chance events consistently tend to work out in their favor. A team of British psychologists designed a study to examine the effects of luckiness and competition on performance in a guessing task (*The Journal of Parapsychology*, Mar. 1997). Each in a sample of 56 college students was classified as lucky, unlucky, or uncertain based on their responses to a Luckiness Questionnaire. In addition, the participants were randomly assigned to either a competitive or noncompetitive condition. All students were then asked to guess the outcomes of 50 flips of a coin. The response variable measured was percentage of coin-flips correctly guessed.
   a. An ANOVA for a 2×3 factorial design was conducted on the data. Identify the factors and their levels for this design.
   b. The results of the ANOVA are summarized in the table. Fully interpret the results.

SOURCE	df	F	p-VALUE
Luckiness (L)	2	1.39	.26
Competition (C)	1	2.84	.10
L × C	2	0.72	.72
Error	50		
Total	55		

12.32. A study published in *Teaching Psychology* (May, 1998) examined how external clues influence student performance. Introductory psychology students were randomly assigned to one of four different midterm examinations. Form 1 was printed on blue paper and contained difficult questions, while form 2 was also printed on blue paper but contained simple questions. Form 3 was printed on red paper, with difficult questions; form 4 was printed on red paper with simple questions. The researchers were interested in the impact that Color (red or blue) and Question (simple or difficult) had on mean exam score.
   a. What experimental design was employed in this study? Identify the factors and treatments.
   b. Give the complete model appropriate for analyzing the data for this experiment.
   c. The researchers conducted an ANOVA and found a significant interaction between Color and Question (p-value $< .03$). Interpret this result.
   d. The sample mean scores (percentage correct) for the four exam forms are listed below. Plot the four means on a graph to illustrate the Color × Question interaction.

FORM	COLOR	QUESTION	MEAN SCORE
1	Blue	Difficult	53.3
2	Blue	Simple	80.0
3	Red	Difficult	39.3
4	Red	Simple	73.6

12.33. The chemical element antimony is sometimes added to tin–lead solder to replace the more expensive tin and to reduce the cost of soldering. A factorial experiment was conducted to determine how antimony affects the strength of the tin–lead solder joint (*Journal of Materials Science*, May 1986). Tin–lead solder specimens were prepared using one of four possible cooling methods (water-quenched, WQ; oil-quenched, OQ; air-blown, AB; and furnace-cooled, FC) and with one of four possible amounts of antimony (0%, 3%, 5%, and 10%) added to the composition. Three solder joints were randomly assigned to each of the $4 \times 4 = 16$ treatments and the shear strength of each measured. The experimental results, shown in the accompanying table, were subjected to an ANOVA using MINITAB. The printout is shown on p. 602.

TINLEAD

AMOUNT OF ANTIMONY % weight	COOLING METHOD	SHEAR STRENGTH, MPa		
0	WQ	17.6	19.5	18.3
0	OQ	20.0	24.3	21.9
0	AB	18.3	19.8	22.9
0	FC	19.4	19.8	20.3
3	WQ	18.6	19.5	19.0
3	OQ	20.0	20.9	20.4
3	AB	21.7	22.9	22.1
3	FC	19.0	20.9	19.9
5	WQ	22.3	19.5	20.5
5	OQ	20.9	22.9	20.6
5	AB	22.9	19.7	21.6
5	FC	19.6	16.4	20.5
10	WQ	15.2	17.1	16.6
10	OQ	16.4	19.0	18.1
10	AB	15.8	17.3	17.1
10	FC	16.4	17.6	17.6

Source: Tomlinson, W. J., and Cooper, G. A. "Fracture mechanism of brass/Sn-Pb-Sb solder joints and the effect of production variables on the joint strength." *Journal of Materials Science*, Vol. 21, No. 5, May 1986, p. 1731 (Table II). Copyright 1986 Chapman and Hall.

MINITAB output for Exercise 12.33

Two-way ANOVA: STRENGTH versus ANTIMONY, METHOD

```
Analysis of Variance for STRENGTH
Source         DF        SS        MS        F        P
ANTIMONY        3    104.19     34.73    20.12    0.000
METHOD          3     28.63      9.54     5.53    0.004
Interaction     9     25.13      2.79     1.62    0.152
Error          32     55.25      1.73
Total          47    213.20
```

a. Construct an ANOVA summary table for the experiment.

b. Conduct a test to determine whether the two factors, amount of antimony and cooling method, interact. Use $\alpha = .01$.

c. Interpret the result obtained in part b.

d. If appropriate, conduct the tests for main effects. Use $\alpha = .01$.

12.34. What is the optimal method of directing newcomers to a specific location in a complex building? Researchers at Ball State University (Indiana) investigated this "wayfinding" problem and reported their results in *Human Factors* (Mar. 1993). Subjects met in a starting room on a multilevel building and were asked to locate the "goal" room as quickly as possible. (Some of the subjects were provided directional aids, whereas others were not.) Upon reaching their destination, the subjects returned to the starting room and were given a second room to locate. (One of the goal rooms was located in the east end of the building, the other in the west end.) The experimentally controlled variables in the study were aid type at three levels (signs, map, no aid) and room order at two levels (east/west, west/east). Subjects were randomly assigned to each of the $3 \times 2 = 6$ experimental conditions and the travel time (in seconds) recorded. The results of the analysis of the east room data for this 3×2 factorial design are provided in the table. Interpret the results.

SOURCE	df	MS	F	p-VALUE
Aid type	2	511,323.06	76.67	$p < .0001$
Room order	1	13,005.08	1.95	$p > .10$
Aid × Order	2	8,573.13	1.29	$p < .10$
Error	46	6,668.94		

Source: Butler, D. L., *et al.* "Wayfinding by newcomers in a complex building." *Human Factors*, Vol. 35, No. 1, Mar. 1993, p. 163 (Table 2).

12.35. Do women enjoy the thrill of a close basketball game as much as men? To answer this question, male and female undergraduate students were recruited to participate in an experiment (*Journal of Sport & Social Issues*, Feb. 1997). The students watched one of eight live televised games of a recent NCAA basketball tournament. (None of the games involved a home team to which the students could be considered emotionally committed.) The "suspense" of each game was classified into one of four categories according to the closeness of scores at the game's conclusions: minimal (15 point or greater differential), moderate (10–14 point differential), substantial (5–9 point differential), and extreme (1–4 point differential). After the game, each student rated his or her enjoyment on an 11-point scale ranging from 0 (not at all) to 10 (extremely). The enjoyment rating data were analyzed as a 4×2 factorial design, with suspense (four levels) and gender (two levels) as the two factors. The $4 \times 2 = 8$ treatment means are shown in the accompanying table.

	GENDER	
SUSPENSE	MALE	FEMALE
Minimal	1.77	2.73
Moderate	5.38	4.34
Substantial	7.16	7.52
Extreme	7.59	4.92

Source: Gan, Su-lin, *et al.* "The thrill of a close game: Who enjoys it and who doesn't?" *Journal of Sport & Social Issues*, Vol. 21, No. 1, Feb. 1997, pp. 59–60.

a. Plot the treatment means in a graph similar to Figure 12.23. Does the pattern of means suggest interaction between suspense and gender? Explain.

b. The ANOVA F-test for interaction yielded the following results: numerator df = 3, denominator

$df = 68$, $F = 4.42$, p-value $= .007$. What can you infer from these results?

c. Based on the test, part **b**, is the difference between the mean enjoyment levels of males and females the same, regardless of the suspense level of the game?

12.36. *Time-of-day pricing* is a plan by which customers are charged a lower rate for using electricity during off-peak (less demanded) hours. One experiment (reported in the *Journal of Consumer Research*, June 1982) was conducted to measure customer satisfaction with several time-of-day pricing schemes. The experiment consisted of two factors, price ratio (the ratio of peak to off-peak prices) and peak period length, each at three levels. The $3 \times 3 = 9$ combinations of price ratio and peak period length represent the nine time-of-day pricing schemes. For each pricing scheme, customers were randomly selected and asked to rate satisfaction with the plan using an index from 10 to 38, with 38 indicating extreme satisfaction. Suppose four customers were sampled for each pricing scheme. The table gives the satisfaction scores for these customers. [*Note*: The data are based on mean scores provided in the *Journal of Consumer Research* article.]

💿 TIMEOFDAY

		PRICING RATIO					
		2:1		4:1		8:1	
PEAK PERIOD LENGTH	6 hours	25	28	31	29	24	28
		26	27	26	27	25	26
	9 hours	26	27	25	24	33	28
		29	30	30	26	25	27
	12 hours	22	20	33	27	30	31
		25	21	25	27	26	27

a. Use a statistical software package to conduct an analysis of variance of the data. Report the results in an ANOVA table.

b. Compute the nine customer satisfaction index means.

c. Plot the nine means from part **b** on a graph similar to Figure 11.14. Does it appear that the two factors, price ratio and peak period length, interact? Explain.

d. Do the data provide sufficient evidence of interaction between price ratio and peak period length? Test using $\alpha = .05$.

e. Do the data provide sufficient evidence that mean customer satisfaction differs for the three peak period lengths? Test using $\alpha = .05$.

f. When is the test of part **e** appropriate?

g. Find a 90% confidence interval for the mean customer satisfaction rating of a pricing scheme with a peak period length of 9 hours and pricing ratio of 2:1.

h. Find a 95% confidence interval for the difference between the mean customer satisfaction ratings of pricing schemes 9 hours, 8:1 ratio and 6 hours, 8:1 ratio. Interpret the interval.

12.37. A computer lab at the University of Oklahoma is open 24 hours a day, seven days a week. In *Production and Inventory Management Journal* (3rd Quarter, 1999), S. Barman investigated whether computer usage differed significantly (1) among the seven days of the week and (2) among the 24 hours of the day. Using student log-on records, data on hourly student loads (number of users per hour) were collected during a 7-week period. A factorial ANOVA was employed with the results presented in the table.

SOURCE	df	SS	MS	F	p-VALUE
Day	6	18732.13	3122.02	68.39	.0001
Time	23	164629.86	7157.82	156.80	.0001
Day × Time	138	7685.22	55.69	1.22	.0527

Source: Barman, S. "A statistical analysis of the attendance pattern of a computer laboratory." *Production and Inventory Management Journal*, 3rd Quarter, 1999, pp. 26–30.

a. Is this an observational or a designed experiment? Explain.

b. What are the two factors of the experiment and how many levels of each factor are used?

c. This is an $a \times b$ factorial experiment. What are a and b?

d. Specify the null and alternative hypotheses that should be used to test for an interaction effect between the two factors of the study.

e. Conduct the test of part **d** using $\alpha = .05$. Interpret your result in the context of the problem.

f. If appropriate, conduct main effects tests for both day and time. Use $\alpha = .05$. Interpret your results in the context of the problem.

12.38. In the *Journal of Nutrition* (July 1995), University of Georgia researchers examined the impact of a vitamin B supplement (nicotinamide) on the kidney. The experimental "subjects" were 28 Zucker rats—a species that tends to develop kidney problems. Half of the rats were classified as obese and half as lean. Within each group, half were randomly assigned to receive a vitamin B–supplemented diet and half were not. Thus, a 2×2 factorial experiment was conducted with seven rats assigned to each of the four combinations of size (lean or obese) and diet (supplemental or not). One of the response variables measured was weight (in grams) of the kidney at the end of a 20-week feeding period. The data (simulated from summary information provided in the journal article) are shown in the table.

⊙ VITAMINB

		DIET			
		Regular		Vitamin B Supplement	
RAT SIZE	LEAN	1.62 1.80 1.71 1.81	1.47 1.37 1.71	1.51 1.65 1.45 1.44	1.63 1.35 1.66
	OBESE	2.35 2.97 2.54 2.93	2.84 2.05 2.82	2.93 2.72 2.99 2.19	2.63 2.61 2.64

a. Conduct an analysis of variance on the data. Summarize the results in an ANOVA table.

b. Conduct the appropriate ANOVA F-tests at $\alpha = .01$. Interpret the results.

12.39. *Teaching of Psychology* (Aug. 1998) published a study on whether a practice test helps students prepare for a final exam. Students in an introductory psychology class were grouped according to their class standing and whether they attended a review session or took a practice test prior to the final exam. The experimental design was a 3×2 factorial design, with Class Standing at 3 levels (low, medium, or high) and Exam Preparation at 2 levels (practice exam or review session). There were 22 students in each of the $3 \times 2 = 6$ treatment groups. After completing the final exam, each student rated her or his exam preparation on an 11-point scale ranging from 0 (not helpful at all) to 10 (extremely helpful). The data for this experiment (simulated from summary statistics provided in the article) are saved in the PRACEXAM file. The first five and last five observations in the data set are listed below. Conduct a complete analysis of variance of the helpfulness ratings data, including (if warranted) multiple comparisons of means. Do your findings support the research conclusion that "students at all levels of academic ability benefit from a . . . practice exam"?

⊙ PRACEXAM

EXAM PREPARATION	CLASS STANDING	HELPFULNESS RATING
PRACTICE	LOW	6
PRACTICE	LOW	7
PRACTICE	LOW	7
PRACTICE	LOW	5
PRACTICE	LOW	3
⋮	⋮	⋮
REVIEW	HI	5
REVIEW	HI	2
REVIEW	HI	5
REVIEW	HI	4
REVIEW	HI	3

Source: Balch, W. R. "Practice versus review exams and final exam performance." *Teaching of Psychology*, Vol. 25, No. 3, Aug. 1998 (adapted from Table 1).

12.40. As part of a study on the rate of combustion of artificial graphite in humid air flow, researchers conducted an experiment to investigate oxygen diffusivity through a water vapor mixture. A 3×9 factorial experiment was conducted with mole fraction of water (H_2O) at three levels and temperature of the nitrogen–water mixture at nine levels. The data are shown here.

⊙ WATERVAPOR

TEMPERATURE	MOLE FRACTION OF H_2O		
°K	.0022	.017	.08
1,000	1.68	1.69	1.72
1,100	1.98	1.99	2.02
1,200	2.30	2.31	2.35
1,300	2.64	2.65	2.70
1,400	3.00	3.01	3.06
1,500	3.38	3.39	3.45
1,600	3.78	3.79	3.85
1,700	4.19	4.21	4.27
1,800	4.63	4.64	4.71

Source: Matsui, K., Tsuji, H., and Makino, A. "The effects of water vapor concentration on the rate of combustion of an artificial graphite in humid air flow." *Combustion and Flame*, Vol. 50, 1983, pp. 107–118. Copyright 1983 by The Combustion Institute. Reprinted by permission of Elsevier Science Publishing Co., Inc.

a. Explain why the traditional analysis of variance (using the ANOVA formulas) is inappropriate for the analysis of these data.

b. Plot the data to determine if a first- or second-order model for mean oxygen diffusivity, $E(y)$, is more appropriate.

c. Write an interaction model relating mean oxygen diffusivity, $E(y)$, to temperature x_1 and mole fraction x_2.

d. Suppose that temperature and mole fraction of H_2O do not interact. What does this imply about the relationship between $E(y)$ and x_1 and x_2?

e. Do the data provide sufficient information to indicate that temperature and mole fraction of H_2O interact? Use the MINITAB printout below to conduct the test at $\alpha = .05$.

f. Give the least squares prediction equation for $E(y)$.

g. Substitute into the prediction equation to predict the mean diffusivity when the temperature of the process is $1,300°K$ and the mole fraction of water is .017.

h. Locate the 95% confidence interval for mean diffusivity when the temperature of the process is $1,300°K$ and the mole fraction of water is .017 shown on the MINITAB printout. Interpret the result.

MINITAB output for Exercise 12.40

```
The regression equation is
OXYDIFF = - 2.10 + 0.00368 TEMP - 0.24 MOLE + 0.00073 TEMPMOLE

Predictor        Coef      SE Coef         T        P
Constant     -2.09528      0.09035    -23.19    0.000
TEMP       0.00368411   0.00006347     58.05    0.000
MOLE           -0.238        1.913     -0.12    0.902
TEMPMOLE     0.000733     0.001344      0.55    0.591


S = 0.06081      R-Sq = 99.7%      R-Sq(adj) = 99.6%

Analysis of Variance

Source             DF          SS         MS         F        P
Regression          3     24.7733     8.2578   2233.31    0.000
Residual Error     23      0.0850     0.0037
Total              26     24.8583


Predicted Values for New Observations

New Obs      Fit     SE Fit         95.0% CI            95.0% PI
1         2.7062     0.0139   ( 2.6774,  2.7350)  ( 2.5772,  2.8352)

Values of Predictors for New Observations

New Obs     TEMP      MOLE   TEMPMOLE
1           1300    0.0170       22.1
```

12.6
More Complex Factorial Designs (Optional)

In this optional section, we present some useful factorial designs that are more complex than the basic two-factor factorial of Section 12.5. These designs fall under the general category of a **k-way classification of data**. A k-way classification of data arises when we run all combinations of the levels of k independent variables. These independent variables can be factors or blocks.

For example, consider a replicated $2 \times 3 \times 3$ factorial experiment in which the $2 \times 3 \times 3 = 18$ treatments are assigned to the experimental units according to a completely randomized design. Since every combination of the three factors (a total of 18) is examined, the design is often called a three-way classification of data. Similarly, a k-way classification of data would result if we randomly assign the treatments of a $(k-1)$-factor factorial experiment to the experimental units of a randomized block design. For example, if we assigned the $2 \times 3 = 6$ treatments of a complete 2×3 factorial experiment to blocks containing six experimental units each, the data would be arranged in a three-way classification, i.e., according to the two factors and the blocks.

The formulas required for calculating the sums of squares for main effects and interactions for an analysis of variance for a k-way classification of data are complicated and are, therefore, not given here. If you are interested in the computational formulas, see the references. As with the designs in the previous three sections, we provide the appropriate linear model for these more complex designs and use either regression or the standard ANOVA output of a statistical software package to analyze the data.

EXAMPLE 12.16   Consider a $2 \times 3 \times 3$ factorial experiment with qualitative factors and $r = 3$ experimental units randomly assigned to each treatment.

a. Write the appropriate linear model for the design.

b. Indicate the sources of variation and their associated degrees of freedom in a partial ANOVA table.

Solution   **a.** Denote the three qualitative factors as A, B, and C, with A at two levels, and B and C at three levels. Then the linear model for the experiment will contain one parameter corresponding to main effects for A, two each for B and C, $(1)(2) = 2$ each for the AB and AC interactions, $(2)(2) = 4$ for the BC interaction, and $(1)(2)(2) = 4$ for the three-way ABC interaction. Three-way interaction terms measure the failure of two-way interaction effects to remain the same from one level to another level of the third factor.

$$E(y) = \beta_0 + \underbrace{\beta_1 x_1}_{\substack{\text{Main effect} \\ A}} + \underbrace{\beta_2 x_2 + \beta_3 x_3}_{\substack{\text{Main effects} \\ B}} + \underbrace{\beta_4 x_4 + \beta_5 x_5}_{\substack{\text{Main effects} \\ C}}$$

$$+ \underbrace{\beta_6 x_1 x_2 + \beta_7 x_1 x_3}_{A \times B \text{ interaction}} + \underbrace{\beta_8 x_1 x_4 + \beta_9 x_1 x_5}_{A \times C \text{ interaction}}$$

$$+ \underbrace{\beta_{10} x_2 x_4 + \beta_{11} x_2 x_5 + \beta_{12} x_3 x_4 + \beta_{13} x_3 x_5}_{B \times C \text{ interaction}}$$

$$+ \underbrace{\beta_{14} x_1 x_2 x_4 + \beta_{15} x_1 x_3 x_4 + \beta_{16} x_1 x_2 x_5 + \beta_{17} x_1 x_3 x_5}_{A \times B \times C \text{ interaction}}$$

where

$$x_1 = \begin{cases} 1 & \text{if level 1 of } A \\ 0 & \text{if level 2 of } A \end{cases} \qquad x_2 = \begin{cases} 1 & \text{if level 1 of } B \\ 0 & \text{if not} \end{cases}$$

$$x_3 = \begin{cases} 1 & \text{if level 2 of B} \\ 0 & \text{if not} \end{cases} \qquad x_4 = \begin{cases} 1 & \text{if level 1 of C} \\ 0 & \text{if not} \end{cases}$$

$$x_5 = \begin{cases} 1 & \text{if level 2 of C} \\ 0 & \text{if not} \end{cases}$$

TABLE 12.8 Table of Sources and Degrees of Freedom for Example 12.16

SOURCE	df
Main effect A	1
Main effect B	2
Main effect C	2
AB interaction	2
AC interaction	2
BC interaction	4
ABC interaction	4
Error	36
Total	53

b. The sources of variation and the respective degrees of freedom corresponding to these sets of parameters are shown in Table 12.8.

The degrees of freedom for SS(Total) will always equal $(n - 1)$—that is, n minus 1 degree of freedom for β_0. Since the degrees of freedom for all sources must sum to the degrees of freedom for SS(Total), it follows that the degrees of freedom for error will equal the degrees of freedom for SS(Total), minus the sum of the degrees of freedom for main effects and interactions, i.e., $(n - 1) - 17$. Our experiment will contain three observations for each of the $2 \times 3 \times 3 = 18$ treatments; therefore, $n = (18)(3) = 54$, and the degrees of freedom for error will equal $53 - 17 = 36$.

If data for this experiment were analyzed on a computer, the computer printout would show the analysis of variance table that we have constructed and would include the associated mean squares, values of the F test statistics, and their observed significance levels. Each F statistic would represent the ratio of the source mean square to MSE $= s^2$. ◆

EXAMPLE 12.17

A transistor manufacturer conducted an experiment to investigate the effects of three factors on productivity (measured in thousands of dollars of items produced) per 40-hour week. The factors were as follows:

a. Length of work week (two levels): five consecutive 8-hour days or four consecutive 10-hour days

b. Shift (two levels): day or evening shift

c. Number of coffee breaks (three levels): 0, 1, or 2

The experiment was conducted over a 24-week period with the $2 \times 2 \times 3 = 12$ treatments assigned randomly to the 24 weeks. The data for this completely randomized design are shown in Table 12.9. Perform an analysis of variance for the data.

Solution

The data were subjected to an analysis of variance. The SAS printout is shown in Figure 12.20. Pertinent sections of the SAS printout are boxed and numbered, as follows:

1. The value of SS(Total), shown in the **Corrected Total** row of box 1, is 1,091.833333. The number of degrees of freedom associated with this quantity is $(n - 1) = (24 - 1) = 23$. Box 1 gives the partitioning (the analysis of variance) of this quantity into two sources of variation. The first source, **Model**, corresponds to the 11 parameters (all except β_0) in the model. The second source is **Error**. The degrees of freedom, sums of squares, and mean squares for these quantities are

TRANSISTOR1

TABLE 12.9   **Data for Example 12.17**

		DAY SHIFT			NIGHT SHIFT		
		COFFEE BREAKS			COFFEE BREAKS		
		0	1	2	0	1	2
LENGTH	4 days	94	105	96	90	102	103
OF		97	106	91	89	97	98
WORK	5 days	96	100	82	81	90	94
WEEK		92	103	88	84	92	96

FIGURE 12.20

SPSS ANOVA printout for $2 \times 2 \times 3$ factorial

The ANOVA Procedure

Dependent Variable: PRODUCT

Source	DF	Sum of Squares	Mean Square	F Value	Pr > F
① Model	11	1009.833333	91.803030	13.43	<.0001
Error	12	82.000000	6.833333		
Corrected Total	23	1091.833333			

R-Square	Coeff Var	Root MSE	PRODUCT Mean
④ 0.924897	2.768647	③ 2.614065	94.41667

Source	DF	Anova SS	Mean Square	F Value	Pr > F
SHIFT	1	48.1666667	48.1666667	7.05	0.0210
DAYS	1	204.1666667	204.1666667	29.88	0.0001
② SHIFT*DAYS	1	8.1666667	8.1666667	1.20	0.2958
BREAKS	2	334.0833333	167.0416667	24.45	<.0001
SHIFT*BREAKS	2	385.5833333	192.7916667	28.21	<.0001
BREAKS*DAYS	2	8.0833333	4.0416667	0.59	0.5689
SHIFT*BREAKS*DAYS	2	21.5833333	10.7916667	1.58	0.2461

shown in their respective columns. For example, MSE = 6.833333. The *F* statistic for testing

$$H_0 : \beta_1 = \beta_2 = \ldots = \beta_{11} = 0$$

is based on $v_1 = 11$ and $v_2 = 12$ degrees of freedom and is shown on the printout as $F = 13.43$. The observed significance level, shown under **Pr > F**, is less than .0001. This small observed significance level presents ample evidence to indicate that at least one of the three independent variables—shifts, number of days in a work week, or number of coffee breaks per day—contributes information for the prediction of mean productivity.

2. To determine which sets of parameters are actually contributing information for the prediction of *y*, we examine the breakdown (box 2) of SS(Model) into components corresponding to the sets of parameters for main effects **SHIFT, DAYS**, and **BREAKS**, and parameters for two-way interactions, **SHIFT*DAYS**, **SHIFT*BREAKS**, and **DAYS*BREAKS**. The last **Model** source of variation corresponds to the set of all three-way **SHIFT*DAYS*BREAKS** parameters. Note that the degrees of freedom for these sources sum to 11, the number of degrees of freedom for **Model**. Similarly, the sum of the component sums of squares is equal to SS(Model). Box 2 does not give the mean squares associated with the sources, but it does give the F values associated with testing hypotheses concerning the set of parameters associated with each source. Box 2 also gives the observed significance levels of these tests. You can see that there is ample evidence to indicate the presence of a **SHIFT*BREAKS** interaction. The F tests associated with all three main effect parameter sets are also statistically significant at the $\alpha = .05$ level of significance. The practical implication of these results is that there is evidence to indicate that all three independent variables, shift, number of work days per week, and number of coffee breaks per day, contribute information for the prediction of productivity. The presence of a **SHIFT*BREAKS** interaction means that the effect of the number of breaks on productivity is not the same from shift to shift. Thus, the specific number of coffee breaks that might achieve maximum productivity on one shift might be different from the number of breaks that would achieve maximum productivity on the other shift.

3. Box 3 gives the value of $s = \sqrt{\text{MSE}} = 2.614065$. This value would be used to construct a confidence interval to compare the difference between 2 of the 12 treatment means. The confidence interval for the difference between a pair of means, $(\mu_i - \mu_j)$, would be

$$(\overline{y}_i - \overline{y}_j) \pm t_{\alpha/2} s \sqrt{\frac{2}{r}}$$

where r is the number of replications of the factorial experiment within a completely randomized design. There were $r = 2$ observations for each of the 12 treatments (factor level combinations) in this example.

4. Box 4 gives the value of R^2, a measure of how well the model fits the experimental data. It is of value primarily when the number of degrees of freedom for error is large—say, at least 5 or 6. The larger the number of degrees of freedom for error, the greater will be its practical importance. The value of R^2 for this analysis, .924897, indicates that the model provides a fairly good fit to the data. It also suggests that the model could be improved by adding new predictor variables or, possibly, by including higher-order terms in the variables originally included in the model. ◆

EXAMPLE   12.18

In a manufacturing process, a plastic rod is produced by heating a granular plastic to a molten state and then extruding it under pressure through a nozzle. An experiment was conducted to investigate the effect of two factors, extrusion temperature (°F) and pressure (pounds per square inch), on the rate of extrusion (inches per second) of the molded rod. A complete 2×2 factorial experiment (that is, with each factor at two levels) was conducted. Three batches of granular plastic were used for the experiment, with each batch (viewed as a block) divided into four equal parts. The four portions of granular plastic for a given batch were randomly assigned to the four treatments; this was repeated for each of the three batches, resulting in a 2×2 factorial experiment laid out in three blocks. The data are shown in Table 12.10. Perform an analysis of variance for these data.

RODMOLD

TABLE 12.10   **Data for Example 12.18**

		BATCH (BLOCK)				
		1		2		3
		PRESSURE		PRESSURE		PRESSURE
		40	60	40	60	40	60
TEMPERATURE	200°	1.35	1.74	1.31	1.67	1.40	1.86
	300°	2.48	3.63	2.29	3.30	2.14	3.27

Solution

This experiment consists of a three-way classification of the data corresponding to batches (blocks), pressure, and temperature. The analysis of variance for this 2×2 factorial experiment (four treatments) laid out in a randomized block design (three blocks) yields the sources and degrees of freedom shown in Table 12.11.

The linear model for the experiment is

$$E(y) = \beta_0 + \overbrace{\beta_1 x_1}^{\substack{\text{Main} \\ \text{effect} \\ P}} + \overbrace{\beta_2 x_2}^{\substack{\text{Main} \\ \text{effect} \\ T}} + \overbrace{\beta_3 x_1 x_2}^{\substack{\text{PT} \\ \text{inter-} \\ \text{action}}} + \overbrace{\beta_4 x_3 + \beta_5 x_4}^{\substack{\text{Block} \\ \text{terms}}}$$

where

$$x_1 = \text{Pressure} \qquad x_2 = \text{Temperature}$$

$$x_3 = \begin{cases} 1 & \text{if block 2} \\ 0 & \text{otherwise} \end{cases} \qquad x_4 = \begin{cases} 1 & \text{if block 3} \\ 0 & \text{otherwise} \end{cases}$$

The SPSS printout for the analysis of variance is shown in Figure 12.21. The F test for the overall model (shaded at the top of the printout) is highly significant (p-value = .000). Thus, there is ample evidence to indicate differences among the block means or the treatment means or both. Proceeding to the breakdown of the model sources, you can see that the values of the F statistics for pressure, temperature, and the temperature × pressure interaction are all highly significant

---

TABLE 12.11   **Table of Sources and Degrees of Freedom for Example 12.18**

SOURCE	df
Pressure (P)	1
Temperature (T)	1
Blocks	2
Pressure × Temperature interaction	1
Error	6
Total	11

FIGURE 12.21

SPSS ANOVA printout for Example 12.18

Tests of Between-Subjects Effects

Dependent Variable: RATE

Source	Type III Sum of Squares	df	Mean Square	F	Sig.
Corrected Model	7.149ᵃ	5	1.430	83.226	.000
Intercept	58.256	1	58.256	3390.818	.000
PRESSURE	1.687	1	1.687	98.222	.000
TEMP	5.044	1	5.044	293.590	.000
PRESSURE * TEMP	.361	1	.361	20.985	.004
BATCH	5.732E-02	2	2.866E-02	1.668	.265
Error	.103	6	1.718E-02		
Total	65.509	12			
Corrected Total	7.252	11			

a. R Squared = .986 (Adjusted R Squared = .974)

(that is, their observed significance levels are very small). Therefore, all of the terms ($\beta_1 x_1$, $\beta_2 x_2$, and $\beta_3 x_1 x_2$) contribute information for the prediction of y.

The treatments in the experiment were assigned according to a randomized block design. Thus, we expected the extrusion of the plastic to vary from batch to batch. However, the F test for testing differences among block (batch) means is not statistically significant (p-value = .2654); there is insufficient evidence to indicate a difference in the mean extrusion of the plastic from batch to batch. Blocking does not appear to have increased the amount of information in the experiment. ◆

Many other complex designs, such as fractional factorials, Latin square designs, and incomplete blocks designs, fall under the general k-way classification of data. Consult the references for the layout of these designs and the linear models appropriate for analyzing them.

EXERCISES

12.41. An experiment was conducted to investigate the effects of three factors—paper stock, bleaching compound, and coating type—on the whiteness of fine bond paper. Three paper stocks (factor A), four types of bleaches (factor B), and two types of coatings (factor C) were used for the experiment. Six paper specimens were prepared for each of the $3 \times 4 \times 2$ stock–bleach–coating combinations and a measure of whiteness was recorded.

a. Construct an analysis of variance table showing the sources of variation and the respective degrees of freedom.

b. Suppose MSE = .14, MS(AB) = .39, and the mean square for all interactions combined is .73. Do the data provide sufficient evidence to indicate any interactions among the three factors? Test using $\alpha = .05$.

c. Do the data present sufficient evidence to indicate an AB interaction? Test using $\alpha = .05$. From a practical point of view, what is the significance of an AB interaction?

d. Suppose SS(A) = 2.35, SS(B) = 2.71, and SS(C) = .72. Find SS(Total). Then find R^2 and interpret its value.

12.42. In increasingly severe oil-well environments, oil producers are interested in high-strength nickel alloys that are corrosion-resistant. Since nickel alloys are especially susceptible to hydrogen embrittlement, an experiment was conducted to compare the yield strengths of nickel alloy tensile specimens cathodically charged in a 4% sulfuric acid solution saturated with carbon disulfide, a hydrogen recombination poison. Two alloys were combined: inconel alloy (75% nickel composition) and incoloy (30% nickel composition). The alloys were tested under two material conditions (cold rolled and cold drawn), each at three different charging times (0, 25, and 50 days). Thus, a $2 \times 2 \times 3$ factorial experiment was conducted, with alloy type at two levels, material condition at two levels, and charging time at three levels. Two hydrogen-charged tensile specimens were prepared for each of the $2 \times 2 \times 3 = 12$ factor level combinations. Their yield strengths (kilograms per square inch) are recorded in the table. The SAS analysis of variance printout for the data is also shown.

a. Is there evidence of any interactions among the three factors? Test using $\alpha = .05$. [*Note:* This means that you must test all the interaction parameters. The drop in SSE appropriate for the test would be the sum of all interaction sums of squares.]

b. Now examine the F tests shown on the printout for the individual interactions. Which, if any, of the interactions are statistically significant at the .05 level of significance?

NICKEL

		ALLOY TYPE						
		INCONEL				INCOLOY		
		COLD ROLLED		COLD DRAWN		COLD ROLLED		COLD DRAWN
CHARGING TIME	0 days	53.4	52.6	47.1	49.3	50.6	49.9	30.9	31.4
	25 days	55.2	55.7	50.8	51.4	51.6	53.2	31.7	33.3
	50 days	51.0	50.5	45.2	44.0	50.5	50.2	29.7	28.1

SAS output for Exercise 12.42

The ANOVA Procedure

Dependent Variable: YIELD

Source	DF	Sum of Squares	Mean Square	F Value	Pr > F
Model	11	1931.734583	175.612235	258.73	<.0001
Error	12	8.145000	0.678750		
Corrected Total	23	1939.879583			

R-Square	Coeff Var	Root MSE	YIELD Mean
0.995801	1.801942	0.823863	45.72083

Source	DF	Anova SS	Mean Square	F Value	Pr > F
ALLOY	1	552.0004167	552.0004167	813.26	<.0001
MATCOND	1	956.3437500	956.3437500	1408.98	<.0001
ALLOY*MATCOND	1	339.7537500	339.7537500	500.56	<.0001
TIME	2	71.0408333	35.5204167	52.33	<.0001
ALLOY*TIME	2	7.9858333	3.9929167	5.88	0.0166
MATCOND*TIME	2	4.1725000	2.0862500	3.07	0.0836
ALLOY*MATCOND*TIME	2	0.4375000	0.2187500	0.32	0.7306

12.43. Refer to Exercise 12.42. Since charging time is a quantitative factor, we could plot the strength y versus charging time x_1 for each of the four combinations of alloy type and material condition. This suggests that a prediction equation relating mean strength $E(y)$ to charging time x_1 may be useful. Consider the model

$$E(y) = \beta_0 + \beta_1 x_1 + \beta_2 x_1^2 + \beta_3 x_2 + \beta_4 x_3 + \beta_5 x_2 x_3$$
$$+ \beta_6 x_1 x_2 + \beta_7 x_1 x_3 + \beta_8 x_1 x_2 x_3$$
$$+ \beta_9 x_1^2 x_2 + \beta_{10} x_1^2 x_3 + \beta_{11} x_1^2 x_2 x_3$$

where

$x_1 = $ Charging time

$$x_2 = \begin{cases} 1 & \text{if inconel alloy} \\ 0 & \text{if incoloy alloy} \end{cases} \qquad x_3 = \begin{cases} 1 & \text{if cold rolled} \\ 0 & \text{if cold drawn} \end{cases}$$

a. Using the model terms, give the relationship between mean strength $E(y)$ and charging time x_1 for cold-drawn incoloy alloy.

b. Using the model terms, give the relationship between mean strength $E(y)$ and charging time x_1 for cold-drawn inconel alloy.

c. Using the model terms, give the relationship between mean strength $E(y)$ and charging time x_1 for cold-rolled inconel alloy.

d. Fit the model to the data and find the least-squares prediction equation.

e. Refer to part **d**. Find the prediction equations for each of the four combinations of alloy type and material condition.

f. Refer to part **d**. Plot the data points for each of the four combinations of alloy type and material condition. Graph the respective prediction equations.

12.44. Refer to Exercises 12.42–12.43. If the relationship between mean strength $E(y)$ and charging time x_1 is the same for all four combinations of alloy type and material condition, the appropriate model for $E(y)$ is

$$E(y) = \beta_0 + \beta_1 x_1 + \beta_2 x_1^2$$

Fit the model to the data. Use the regression results, together with the information from Exercise 12.43, to decide whether the data provide sufficient evidence to indicate differences among the second-order models relating $E(y)$ to x_1 for the four categories of alloy type and material condition. Test using $\alpha = .05$.

12.45. A study was conducted to evaluate the use of computer-assisted instruction (CAI) in teaching an introductory FORTRAN programming course (GE 102) in the College of Engineering at Oregon State University (*Engineering Education*, Feb. 1986). One of the objectives was to investigate the effect of four factors on a student's final exam score (y) in the course. The factors and their respective levels are as follows.

Group (3 levels):
    Control group (student receives no CAI)
    Guided CAI
    Self-paced CAI
Math background (4 levels):
    High school algebra
    Trigonometry
    Differential calculus
    Integral calculus
Computer background (3 levels):
    None (little or no exposure)
    Some (one programming language)
    Extensive (more than one language)
Grade in prerequisite course, GE 101 (3 levels):
    A, B, or C

a. How many treatments are associated with this four-way factorial experiment?

b. Write the complete factorial model for this experiment. [*Hint*: Use dummy variables to represent the factors.]

c. What hypothesis would you test to determine whether interaction exists among the four factors?

12.46. A 2×2 factorial experiment was conducted for each of 3 weeks to determine the effect of two factors, temperature and pressure, on the yield of a chemical. Temperature was set at $300°$ and $500°$. The pressure maintained in the reactor was set at 100 and 200 pounds per square inch. Four days were randomly selected within each week and the four factor level combinations were randomly assigned to them.

a. What type of design was used for this experiment?

b. Construct an analysis of variance table showing all sources and their respective degrees of freedom.

12.7
Follow-Up Analysis: Tukey's Multiple Comparisons of Means

Many practical experiments are conducted to determine the largest (or the smallest) mean in a set. For example, suppose a chemist has developed five chemical solutions for removing a corrosive substance from metal. The chemist would want to determine which solution will remove the greatest amount of corrosive substance in a single application. Similarly, a production engineer might want to determine which among six machines or which among three foremen achieves the highest mean productivity per hour. A stockbroker might want to choose one stock, from among four, that yields the highest mean return, and so on.

Once differences among, say, five treatment means have been detected in an ANOVA, choosing the treatment with the largest mean might appear to be a simple matter. We could, for example, obtain the sample means $\bar{y}_1, \bar{y}_2, \ldots, \bar{y}_5$, and compare them by constructing a $(1 - \alpha)100\%$ confidence interval for the difference between each pair of treatment means. However, there is a problem associated with this procedure: **A confidence interval for $\mu_i - \mu_j$, with its corresponding value of α, is valid only when the two treatments (i and j) to be compared are selected prior to experimentation**. After you have looked at the data, you cannot use a confidence interval to compare the treatments for the largest and smallest sample means because they will always be farther apart, on the average, than any pair of treatments selected at random. Furthermore, **if you construct a series of confidence intervals, each with a chance α of indicating a difference between a pair of means if no difference exists, then the risk of making at least one Type I error in the series of inferences will be larger than the value of α specified for a single interval**.

There are a number of procedures for comparing and ranking a group of treatment means as part of a **follow-up** (or **post-hoc**) **analysis** to the ANOVA. The one that we present in this section, known as **Tukey's method for multiple comparisons**, utilizes the Studentized range

$$q = \frac{\bar{y}_{\max} - \bar{y}_{\min}}{s/\sqrt{n}}$$

(where $\bar{y}_{\max}$ and $\bar{y}_{\min}$ are the largest and smallest sample means, respectively) to determine whether the difference in any pair of sample means implies a difference in the corresponding treatment means. The logic behind this **multiple comparisons procedure** is that if we determine a critical value for the difference between the largest and smallest sample means, $|\bar{y}_{\max} - \bar{y}_{\min}|$, one that implies a difference in their respective treatment means, then any other pair of sample means that differ by as much as or more than this critical value would also imply a difference in the corresponding treatment means. Tukey's (1949) procedure selects this critical distance, ω, so that the probability of making one or more Type I errors (concluding that a difference exists between a pair of treatment means if, in fact, they are identical) is α. Therefore, the risk of making a Type I error applies to the whole procedure, i.e., to the comparisons of all pairs of means in the experiment, rather than to a single comparison. Consequently, the value of α selected by the researchers is called an **experimentwise error rate** (in contrast to a **comparisonwise error rate**).

Tukey's procedure relies on the assumption that the p sample means are based on independent random samples, *each containing an equal number n_t of observations.* Then if $s = \sqrt{\text{MSE}}$ is the computed standard deviation for the analysis, the distance ω is

$$\omega = q_\alpha(p, v)\frac{s}{\sqrt{n_t}}$$

The tabulated statistic $q_\alpha(p, v)$ is the critical value of the Studentized range, the value that locates α in the upper tail of the q distribution. This critical value depends on α, the number of treatment means involved in the comparison, and v, the number of degrees of freedom associated with MSE, as shown in the box. Values of $q_\alpha(p, v)$ for $\alpha = .05$ and $\alpha = .01$ are given in Tables 11 and 12, respectively, of Appendix C.

EXAMPLE   12.19   Refer to the ANOVA for the completely randomized design, Examples 12.4 and 12.5. Recall that we rejected the null hypothesis of no differences among the mean

GPAs for the three socioeconomic groups of college freshmen. Use Tukey's method to compare the three treatment means.

Solution

STEP 1   For this follow-up analysis, we will select an experimentwise error rate of $\alpha = .05$.

STEP 2   From previous examples, we have $(p - 3)$ treatments, $v = 18$ df for error, $s = \sqrt{MSE} = .512$, and $n_t = 7$ observations per treatment. The critical value of the Studentized range (obtained from Table 11, Appendix C) is $q_{.05}(3, 18) = 3.61$. Substituting these values into the formula for ω, we obtain

$$\omega = q_{.05}(3, 18)\left(\frac{s}{\sqrt{n_t}}\right) = 3.61\left(\frac{.512}{\sqrt{7}}\right) = .698$$

Tukey's Multiple Comparisons Procedure: Equal Sample Sizes

1. Select the desired experimentwise error rate, α.
2. Calculate

$$\omega = q_\alpha(p, v)\frac{s}{\sqrt{n_t}}$$

   where

$$p = \text{Number of sample means (i.e., number of treatments)}$$
$$s = \sqrt{MSE}$$
$$v = \text{Number of degrees of freedom associated with MSE}$$
$$n_t = \text{Number of observations in each of the } p \text{ samples (i.e., number}$$
$$\text{of observations per treatment)}$$
$$q_\alpha(p, v) = \text{Critical value of the Studentized range (Tables 11 and 12}$$
$$\text{of Appendix C)}$$

3. Calculate and rank the p sample means.
4. Place a bar over those pairs of treatment means that differ by less than ω. A pair of treatments not connected by an overbar (i.e., differing by more than ω) implies a difference in the corresponding population means.

Note: The confidence level associated with all inferences drawn from the analysis is $(1 - \alpha)$.

STEP 3   The sample means for the three socioeconomic groups (obtained from Table 12.12) are

$$\bar{y}_L = 2.521 \quad \bar{y}_M = 3.249 \quad \bar{y}_U = 2.543$$

STEP 4   Based on the critical difference $\omega = .70$, the three treatment means are ranked as follows:

Sample means:	2.521	2.543	3.249
Treatments:	Lower	Upper	Middle

From this information, we infer that the mean freshman GPA for the middle class is significantly larger than the means for the other two classes, since $\bar{y}_M$ exceeds both $\bar{y}_L$ and $\bar{y}_U$ by more than the critical value. However, the lower and upper classes are connected by a horizontal line since $|\bar{y}_L - \bar{y}_U|$ is less than ω. This indicates that the means for these treatments are not significantly different.

In summary, the Tukey analysis reveals that the mean GPA for the middle class of students is significantly larger than the mean GPAs of either the upper or lower classes, but that the means of the upper and lower classes are not significantly different. These inferences are made with an overall confidence level of $(1-\alpha) = .95$.   ◆

As Example 12.19 illustrates, Tukey's multiple comparisons of means procedure involves quite a few calculations. Most analysts utilize statistical software packages to conduct Tukey's method. The SAS, MINITAB, and SPSS printouts of the Tukey analysis for Example 12.19 are shown in Figures 12.22a, 12.22b, and 12.22c, respectively. Optionally, SAS presents the results in one of two forms. In the top printout, Figure 12.22a, SAS lists the treatment means vertically in descending order. Treatment means connected by the same letter (A, B, C, etc.) in the left column are *not* significantly different. You can see from Figure 12.22a that the middle class has a different letter (A) than the upper and lower classes (assigned the letter B). In the bottom printout of Figure 12.22a, SAS lists the Tukey confidence intervals for $(\mu_i - \mu_j)$, for all possible treatment pairs, i and j. Intervals that include 0 imply that the two treatments compared are not significantly different. The only interval at the bottom of Figure 12.22a that includes 0 is the one involving the upper and lower classes; hence, the GPA means for theses two treatments are not significantly different. All the confidence intervals involving the middle class indicate that the middle class mean GPA is larger than either the upper or lower class mean.

Both MINITAB and SPSS present the Tukey comparisons in the form of confidence intervals for pairs of treatment means. Figures 12.22b and 12.22c (top) show the lower and upper endpoints of a confidence interval for $(\mu_1 - \mu_2)$, $(\mu_1 - \mu_3)$, and $(\mu_2 - \mu_3)$, where "1" represents the lower class, "2" represents the middle class, and "3" represents the upper class. SPSS, like SAS, also produces a list of the treatment means arranged in subsets. The bottom of Figure 12.22c shows the means for treatments 1 and 3 (lower and upper classes) in the same subset, implying that these two means are not significantly different. The mean for treatment 2 (middle class) is in a different subset; hence, its treatment mean is significantly different than the others.

EXAMPLE   12.20

Refer to Example 12.17. In a simpler experiment, the transistor manufacturer investigated the effects of just two factors on productivity (measured in thousands of dollars of items produced) per 40-hour week. The factors were:

Length of work week (two levels):   five consecutive 8-hour days or four consecutive 10-hour days

Number of coffee breaks (three levels): 0, 1, or 2

FIGURE 12.22a

SAS printout of Tukey's multiple comparisons of
means, Example 12.19

```
                        The ANOVA Procedure

              Tukey's Studentized Range (HSD) Test for GPA

NOTE: This test controls the Type I experimentwise error rate, but it generally has a higher
                     Type II error rate than REGWQ.


         Alpha                                      0.05
         Error Degrees of Freedom                     18
         Error Mean Square                        0.261729
         Critical Value of Studentized Range      3.60930
         Minimum Significant Difference            0.6979


     Means with the same letter are not significantly different.


       Tukey Grouping        Mean     N    CLASS

              A             3.2486     7    MIDDLE

              B             2.5429     7    UPPER
              B
              B             2.5214     7    LOWER

                        The ANOVA Procedure

              Tukey's Studentized Range (HSD) Test for GPA

      NOTE: This test controls the Type I experimentwise error rate.


         Alpha                                      0.05
         Error Degrees of Freedom                     18
         Error Mean Square                        0.261729
         Critical Value of Studentized Range      3.60930
         Minimum Significant Difference            0.6979


   Comparisons significant at the 0.05 level are indicated by ***.


                         Difference      Simultaneous
             CLASS        Between        95% Confidence
           Comparison      Means            Limits

       MIDDLE - UPPER      0.7057      0.0078   1.4036   ***
       MIDDLE - LOWER      0.7271      0.0292   1.4251   ***
       UPPER  - MIDDLE    -0.7057     -1.4036  -0.0078   ***
       UPPER  - LOWER      0.0214     -0.6765   0.7193
       LOWER  - MIDDLE    -0.7271     -1.4251  -0.0292   ***
       LOWER  - UPPER     -0.0214     -0.7193   0.6765
```

The experiment was conducted over a 12-week period with the $2 \times 3 = 6$ treatments assigned in a random manner to the 12 weeks. The data for this two-factor factorial experiment are shown in Table 12.12 (p. 619).

a. Perform an analysis of variance for the data.

b. Compare the six population means using Tukey's multiple comparisons procedure. Use $\alpha = .05$.

Solution

a. The SAS printout of the ANOVA for the 2×3 factorial is shown in Figure 12.23. Note that the test for interaction between the two factors, length (L) and breaks (B), is significant at $\alpha = .01$. (The p-value, .0051, is shaded on the printout.) Since interaction implies that the level of length (L) that yields the highest mean

FIGURE 12.22*b*

MINITAB printout of Tukey's multiple comparisons of means, Example 12.19

```
Tukey's pairwise comparisons

    Family error rate = 0.0500
Individual error rate = 0.0200

Critical value = 3.61

Intervals for (column level mean) - (row level mean)

                     1                2

      2           -1.4252
                  -0.0291

      3           -0.7195          0.0077
                   0.6766          1.4038
```

FIGURE 12.22*c*

SPSS printout of Tukey's multiple comparisons of means, Example 12.19

Post Hoc Tests

Multiple Comparisons

Dependent Variable: GPA

Tukey HSD

(I) GROUP	(J) GROUP	Mean Difference (I-J)	Std. Error	Sig.	95% Confidence Interval Lower Bound	95% Confidence Interval Upper Bound
1	2	-.7271*	.27346	.040	-1.4251	-.0292
	3	-.0214	.27346	.997	-.7193	.6765
2	1	.7271*	.27346	.040	.0292	1.4251
	3	.7057*	.27346	.047	.0078	1.4036
3	1	.0214	.27346	.997	-.6765	.7193
	2	-.7057*	.27346	.047	-1.4036	-.0078

*. The mean difference is significant at the .05 level.

Homogeneous Subsets

GPA

Tukey HSD[a]

GROUP	N	Subset for alpha = .05 — 1	Subset for alpha = .05 — 2
1	7	2.5214	
3	7	2.5429	
2	7		3.2486
Sig.		.997	1.000

Means for groups in homogeneous subsets are displayed.

a. Uses Harmonic Mean Sample Size = 7.000.

productivity may differ across different levels of breaks (B), we ignore the tests for main effects and focus our investigation on the individual treatment means.

b. The sample means for the six factor level combinations are highlighted in the middle of the SAS printout, Figure 12.23. Since the sample means represent measures of productivity in the manufacture of transistors, we want to find the

TRANSISTOR2

TABLE 12.12   **Data for Example 12.20**

		COFFEE BREAKS		
		0	1	2
LENGTH OF	4 days	101 102	104 107	95 92
WORK WEEK	5 days	95 93	109 110	83 87

FIGURE 12.23

SAS ANOVA printout for Example 12.20

The GLM Procedure

Dependent Variable: PRODUCT

Source	DF	Sum of Squares	Mean Square	F Value	Pr > F
Model	5	811.6666667	162.3333333	48.70	<.0001
Error	6	20.0000000	3.3333333		
Corrected Total	11	831.6666667			

R-Square	Coeff Var	Root MSE	PRODUCT Mean
0.975952	1.859839	1.825742	98.16667

Source	DF	Type III SS	Mean Square	F Value	Pr > F
DAYS	1	48.0000000	48.0000000	14.40	0.0090
BREAKS	2	667.1666667	333.5833333	100.07	<.0001
DAYS*BREAKS	2	96.5000000	48.2500000	14.47	0.0051

Least Squares Means
Adjustment for Multiple Comparisons: Tukey

DAYS	BREAKS	PRODUCT LSMEAN	LSMEAN Number
4	0	101.500000	1
4	1	105.500000	2
4	2	93.500000	3
5	0	94.000000	4
5	1	109.500000	5
5	2	85.000000	6

Least Squares Means for effect DAYS*BREAKS
Pr > |t| for H0: LSMean(i)=LSMean(j)

Dependent Variable: PRODUCT

i/j	1	2	3	4	5	6
1		0.3573	0.0329	0.0437	0.0329	0.0008
2	0.3573		0.0046	0.0057	0.3573	0.0002
3	0.0329	0.0046		0.9997	0.0010	0.0250
4	0.0437	0.0057	0.9997		0.0012	0.0192
5	0.0329	0.3573	0.0010	0.0012		<.0001
6	0.0008	0.0002	0.0250	0.0192	<.0001	

length of work week and number of coffee breaks that yield the highest mean productivity.

In the presence of interaction, SAS displays the results of the Tukey multiple comparisons by listing the p-values for comparing all possible treatment mean pairs. These p-values are shown at the bottom of Figure 12.23. First we demonstrate how to conduct the multiple comparisons using the formulas in the box. Then we explain(in notes) how to use p-values reported in the SAS to rank the means.

The first step in the ranking procedure is to calculate ω for $p = 6$ (we are ranking six treatment means), $n_t = 2$ (two observations per treatment), $\alpha = .05$, and $s = \sqrt{\text{MSE}} = \sqrt{3.33} = 1.83$ (where MSE is shaded in Figure 12.23). Since MSE is based on $\nu = 6$ degrees of freedom, we have

$$q_{.05}(6, 6) = 5.63$$

and

$$\omega = q_{.05}(6, 6)\left(\frac{s}{\sqrt{n_t}}\right)$$
$$= (5.63)\left(\frac{1.83}{\sqrt{2}}\right)$$
$$= 7.27$$

Therefore, population means corresponding to pairs of sample means that differ by more than $\omega = 7.27$ will be judged to be different. The six sample means are ranked as follows:

Sample means	85.0	93.5	94.0	101.5	105.5	109.5
Treatments (Length, Breaks)	(5, 2)	(4, 2)	(5, 0)	(4, 0)	(4, 1)	(5, 1)
Number on SAS printout:	6	3	4	1	2	5

Using $\omega = 7.27$ as a yardstick to determine differences between pairs of treatments, we have placed connecting bars over those means that *do not* significantly differ. The following conclusions can be drawn:

1. There is evidence of a difference between the population mean of the treatment corresponding to a 5-day work week with two coffee breaks (with the smallest sample mean of 85.0) and every other treatment mean. Therefore, we can conclude that the 5-day, 2-break work week yields the lowest mean productivity among all length-break combinations.

   [*Note*: This inference can also be derived from the p-values shown under the mean 6 column at the bottom of the SAS printout, Figure 12.23. Each p-value (obtained using Tukey's adjustment) is used to compare the (5,2) treatment mean with each of the other treatment means. Since all the p-values are less than our selected experiment-wise error rate of $\alpha = .05$, the (5,2) treatment mean is significantly different than each of the other means.]

2. The population mean of the treatment corresponding to a 5-day, 1-break work week (with the largest sample mean of 109.5) is significantly larger than the treatments corresponding to the four smallest sample means. However, there is no evidence of a difference between the 5-day, 1-break treatment mean and the 4-day, 1-break treatment mean (with a sample mean of 105.5).

[*Note*: This inference is supported by the Tukey-adjusted *p*-values shown under the mean 5 column —the column for the (5,1) treatment—in Figure 12.23. The only *p*-value that is *not* smaller than .05 is the one comparing mean 5 to mean 2, where mean 2 represents the (4,1) treatment.]

3. There is no evidence of a difference between the 4-day, 1-break treatment mean (with a sample mean of 105.5) and the 4-day, 0-break treatment mean (with a sample mean of 101.5). Both of these treatments, though, have significantly larger means than the treatments corresponding to the three smallest sample means.

[*Note*: This inference is supported by the Tukey-adjusted *p*-values shown under the mean 2 column —the column for the (4,1) treatment —in Figure 12.23. The *p*-value comparing mean 2 to mean 1, where mean 1 represents the (4,0) treatment, exceeds $\alpha = .05$.]

4. There is no evidence of a difference between the treatments corresponding to the sample means 93.5 and 94.0, i.e., between the (4,2) and (5,0) treatment means.

[*Note*: This inference can also be obtained by observing that the Tukey-adjusted *p*-value shown in Figure 12.23 under the mean 4 column —the column for the (5,0) treatment —and in the mean 3 row —the row for the (4,2) treatment —is greater than $\alpha = .05$.]

In summary, the treatment means appear to fall into four groups, as follows:

	TREATMENTS (LENGTH, BREAKS)
Group 1 (lowest mean productivity)	$(5, 2)$
Group 2	$(4, 2)$ and $(5, 0)$
Group 3	$(4, 0)$ and $(4, 1)$
Group 4 (highest mean productivity)	$(4, 1)$ and $(5, 1)$

Notice that it is unclear where we should place the treatment corresponding to a 4-day, 1-break work week because of the overlapping bars above its sample mean, 105.5. That is, although there is sufficient evidence to indicate that treatments (4, 0) and (5, 1) differ, neither has been shown to differ significantly from treatment (4, 1). Tukey's method guarantees that the probability of making one or more Type I errors in these pairwise comparisons is only $\alpha = .05$. ◆

Remember that Tukey's multiple comparisons procedure requires the sample sizes associated with the treatments to be equal. This, of course, will be satisfied for the randomized block designs and factorial experiments described in Sections 12.4 and 12.5, respectively. The sample sizes, however, may not be equal in a completely randomized design (Section 12.3). In this case a modification of Tukey's method (sometimes called the Tukey—Kramer method) is necessary, as described in the accompanying box. The technique requires that the critical difference ω_{ij} be calculated for each pair of treatments (i, j) in the experiment and pairwise comparisons made based on the appropriate value of ω_{ij}. However, when Tukey's method is used with unequal sample sizes, the value of α selected a priori by the researcher only approximates the true experimentwise error rate. In fact, when applied to unequal sample sizes, the procedure has been found to be more conservative, i.e., less likely to detect differences between pairs of treatment means when they exist, than in the case of equal sample sizes. For this reason, researchers sometimes look to alternative methods of multiple comparisons when the sample sizes are unequal. Two of these methods are presented in optional Section 12.8.

Tukey's Approximate Multiple Comparisons Procedure for Unequal Sample Sizes

1. Calculate for each treatment pair (i, j)

$$\omega_{ij} = q_\alpha(p, v) \frac{s}{\sqrt{2}} \sqrt{\frac{1}{n_i} + \frac{1}{n_j}}$$

where

$$p = \text{Number of sample means}$$

$$s = \sqrt{\text{MSE}}$$

$$v = \text{Number of degrees of freedom associated with MSE}$$

$$n_i = \text{Number of observations in sample for treatment } i$$

$$n_j = \text{Number of observations in sample for treatment } j$$

$$q_\alpha(p, v) = \text{Critical value of the Studentized range}$$
$$\text{(Tables 11 and 12 of Appendix C)}$$

2. Rank the p sample means and place a bar over any treatment pair (i, j) that differs by less than ω_{ij}. Any pair of sample means not connected by an overbar (i.e., differing by more than ω) implies a difference in the corresponding population means.

Note: This procedure is approximate, i.e., the value of α selected by the researcher approximates the true probability of making at least one Type I error.

In general, multiple comparisons of treatment means should be performed only as a follow-up analysis to the ANOVA, i.e., only after we have conducted the appropriate analysis of variance F test(s) and determined that sufficient evidence exists of differences among the treatment means. Be wary of conducting multiple comparisons when the ANOVA F test indicates no evidence of a difference among a small number of treatment means—this may lead to confusing and contradictory results.*

> **Warning**
>
> In practice, it is advisable to avoid conducting multiple comparisons of a small number of treatment means when the corresponding ANOVA F test is nonsignificant; otherwise, confusing and contradictory results may occur.

EXERCISES

12.47. Refer to the *Nature* (Aug. 2000) study of robots trained to behave like ants, Exercise 12.7 (p. 562). Multiple comparisons of mean energy expended for the four colony sizes were conducted using an experimentwise error rate of .05. The results are summarized below.

Sample mean:	.97	.95	.93	.80
Group size:	3	6	9	12

a. How many pairwise comparisons are conducted in this analysis?

b. Interpret the results shown in the table.

12.48. Refer to the *American Journal of Political Science* (Jan. 1998) study of the attitudes of three groups of professionals (scientists, journalists, and federal government policymakers) regarding the safety of nuclear power plants, Exercise 12.8 (p. 563). The mean safety scores for the groups were:

Government officials	4.2
Scientists	4.1
Journalists	3.7

a. Determine the number of pairwise comparisons of treatment means that can be made in this study.

b. Using an experimentwise error rate of $\alpha = .05$, Tukey's minimum significant difference for comparing means is .23. Use this information to conduct a multiple comparisons of the safety score means. Fully interpret the results.

12.49. In business, the prevailing theory is that companies can be categorized into one of four types based on their strategic profile: *reactors*, which are dominated by industry forces; *defenders*, which specialize in lowering costs for established products while maintaining quality; *prospectors*, which develop new/improved products; and *analyzers*, which operate in two product areas—one stable, and the other dynamic. The *American Business Review* (Jan. 1990) reported on a study that proposes a fifth organization type, *balancers*, which operate in three product spheres—one stable and two dynamic. Each firm in a sample of 78 glassware firms was categorized as one of these five types, and the level of performance (process research and development ratio) of each was measured.

a. A completely randomized design ANOVA of the data resulted in a significant (at $\alpha = .05$) F value for treatments (organization types). Interpret this result.

b. Multiple comparisons of the five mean performance levels (using Tukey's procedure at $\alpha = .05$) are summarized in the following table. Interpret the results.

Mean	.138	.235	.820	.826	.911
Type	Reactor	Prospector	Defender	Analyzer	Balancer

Source: Wright, P., *et al.* "Business performance and conduct of organization types: A study of select special-purpose and laboratory glassware firms." *American Business Review*, Jan. 1990, p. 95 (Table 4).

12.50. Refer to the *Accounting and Finance* (Nov. 1993) study of auditors' objectivity, Exercise 12.10 (p. 563). The means of the three groups of auditors were ranked using a Tukey multiple comparisons procedure at $\alpha = .05$, as shown here:

Mean number of hours allocated	6.7	7.6	9.7
Group	A/R	C	A/P

---

*When a large number of treatments are to be compared, a borderline, nonsignificant F value (e.g., $.05 < p\text{-value} < .10$) may mask differences between some of the means. In this situation, it is better to ignore the F test and proceed directly to a multiple comparisons procedure.

At the beginning of the study, the researchers theorized that the A/R group would allocate the least audit effort to receivables and that the A/P group would allocate the most. Formally stated, the researchers believed that $\mu_{AR} < \mu_C < \mu_{AP}$. Do the results support this theory? Explain.

12.51. Refer to the *Human Factors* (Apr. 1990) study of the performance of a computerized speech recognizer, Exercise 4.58 (p. 231). Accuracy was measured at three levels (90%, 95%, and 99%) and vocabulary size at three levels (75%, 87.5%, and 100%). The data on task completion times (minutes) were subjected to an analysis of variance for a 3×3 factorial design. The F test for accuracy–vocabulary interaction resulted in a p-value less than .0003.

a. Interpret the result of the test for interaction.

b. As a follow-up to the test for interaction, the mean task completion times for the three levels of accuracy were compared under each level of vocabulary. Do you agree with this method of analysis? Explain.

c. Refer to part **b.** Tukey's multiple comparisons method was used to compare the three accuracy means within each level of vocabulary at an experimentwise error rate of $\alpha = .05$. The results are summarized in the table. Interpret these results.

	ACCURACY LEVEL		
VOCABULARY SIZE	99%	95%	90%
75%	15.49	19.29	22.19
87.5%	12.77	14.31	16.48
100%	8.67	9.68	11.88

Source: Casali, S. P., Williges, B. H., and Dryden, R. D. "Effects of recognition accuracy and vocabulary size of a speech recognition system on task performance and user acceptance." *Human Factors*, Vol. 32, No. 2, Apr. 1990, p. 190 (Figure 2).

12.52. Refer to the *Journal of Nonverbal Behavior* study of students' evaluations of facial expressions, Exercise 12.14 (p. 565). Use a statistical software package with Tukey's method to rank the dominance rating means of the six facial expressions. (Use an experimentwise error rate of $\alpha = .10$.)

12.53. The *Journal of Computer Information Systems* (Spring 1993) published the results of a study of end-user computing. Data on the ratings of 18 specific end-user computing (EUC) policies were obtained for each of 82 managers. (Managers rated policies on a 5-point scale, where 1 = no value and 5 = necessity.) The goal was to compare the mean ratings of the 18 EUC policies; thus, a randomized block design with 18 treatments (policies) and 82 blocks (managers) was used. Since the ANOVA F test for treatments was significant at $\alpha = .01$, a follow-up analysis was conducted. The mean ratings for the 18 EUC policies are reported in the table. Using an overall significance level of $\alpha = .05$, the Tukey critical difference for comparing the 18 means was determined to be $\omega = .32$.

a. Determine the pairs of EUC policy means that are significantly different.

b. According to the researchers, the group of policies receiving the highest rated values have mean ratings of 4.0 and above. Do you agree with this assessment?

EUC POLICY	MEAN RATING
1. Organizational value	2.439
2. Training	2.683
3. Goals	2.854
4. Justify applications	3.098
5. Relation with MIS	3.293
6. Hardware movement	3.366
7. Accountability	3.390
8. Justify data	3.561
9. Ownership of files	3.756
10. In-house software	3.854
11. Copyright infringement	3.878
12. Compatibility	4.000
13. Document files	4.000
14. Role of networking	4.049
15. Data confidentiality	4.073
16. Data security	4.219
17. Hardware standards	4.293
18. Software purchases	4.317

Source: Mitchell, R. B., and Neal, R. "Status of planning and control systems in the end-user computing environment." *Journal of Computer Information Systems*, Vol. 33, No. 3, Spring 1993, p. 29 (Table 4).

12.54. Refer to Exercise 12.20 (p. 579). The three treatment means were compared using Tukey's method at $\alpha = .05$. Interpret the results shown below.

Mean number of flowers:	1.17	10.58	17.08
Treatment:	Control	Clipping	Burning

12.55. Refer to Exercise 12.33 (p. 601). Use Tukey's multiple comparisons procedure to compare the mean shear strengths for the four antimony amounts. Identify the means that appear to differ. Use $\alpha = .01$.

12.56. Refer to Exercise 12.36 (p. 603). Use Tukey's multiple comparisons procedure to compare the mean satisfaction scores for the three peak period lengths under each of the three pricing ratios. Identify the means that appear to differ under each pricing ratio. Use $\alpha = .01$.

12.8

Other Multiple Comparisons Methods (Optional)

In this optional section, we present two alternatives to Tukey's method of multiple comparisons of treatment means. The choice of methods will depend on the type of experimental design used and the particular error rate that the researcher wants to control.

SCHEFFÉ METHOD: Recall that Tukey's method of multiple comparisons is designed to control the experimentwise error rate, i.e., the probability of making at least one Type I error in the comparison of *all pairs* of treatment means in the experiment. Therefore, Tukey's method should be applied when you are interested in pairwise comparisons only.

Scheffé (1953) developed a more general procedure for comparing all possible linear combinations of the treatment means, called **contrasts**.

> ### Definition 12.3
> A **contrast** L is a linear combination of the p treatment means in a designed experiment, i.e.,
>
> $$L = \sum_{i=1}^{p} c_i \mu_i$$
>
> where the constants $c_1, c_2, \ldots, c_p$ sum to 0, i.e., $\sum_{i=1}^{p} c_i = 0$.

For example, in an experiment with four treatments (A, B, C, D), you might want to compare the following contrasts, where μ_i represents the population mean for treatment i:

$$L_1 = \frac{\mu_A + \mu_B}{2} - \frac{\mu_C + \mu_D}{2}$$

$$L_2 = \mu_A - \mu_D$$

$$L_3 = \frac{\mu_B + \mu_C + \mu_D}{3} - \mu_A$$

The contrast L_2 involves a comparison of a pair of treatment means, whereas L_1 and L_3 are more complex comparisons of the treatments. Thus, pairwise comparisons are special cases of general contrasts.

As in Tukey's method, the value of α selected by the researcher using Scheffé's method applies to the procedure as a whole, i.e., to the comparisons of all possible contrasts (not just those considered by the researcher). Unlike Tukey's method, however, the probability of at least one Type I error, α, is exact regardless of whether the sample sizes are equal. For this reason, some researchers prefer Scheffé's method to Tukey's method in the case of unequal samples, even if only pairwise comparisons of treatment means are made. The Scheffé method for general contrasts is outlined in the box.

In the special case of all pairwise comparisons in an experiment with four treatments, the relevant contrasts reduce to $L_1 = \mu_A - \mu_B$, $L_2 = \mu_A - \mu_C$, $L_3 = \mu_A - \mu_D$, and so forth. Notice that for each of these contrasts $\sum c_i^2/n_i$ reduces to $(1/n_i + 1/n_j)$, where n_i and n_j are the sizes of treatments i and j, respectively. [For example, for contrast L_1, $c_1 = 1$, $c_2 = -1$, $c_3 = c_4 = 0$, and $\sum c_i^2/n_i = (1/n_1 + 1/n_2)$.] Consequently, the formula for S in the general contrast method can be simplified and pairwise comparisons made using the technique of Section 12.7. The Scheffé method for pairwise comparisons of treatment means is shown in the box.

Scheffé's Multiple Comparisons Procedure for General Contrasts

1. For each contrast $L = \sum_{i=1}^{p} c_i \mu_i$, calculate

$$\hat{L} = \sum_{i=1}^{p} c_i \bar{y}_i$$

and

$$S = \sqrt{(p-1)(F_\alpha)(\text{MSE}) \sum_{i=1}^{p} \left(\frac{c_i^2}{n_i} \right)}$$

where

$\quad p =$ Number of sample (treatment) means

$\quad \text{MSE} =$ Mean squared error

$\quad n_i =$ Number of observations in sample for treatment i

$\quad \bar{y}_i =$ Sample mean for treatment i

$\quad F_\alpha =$ Critical value of F distribution with $p - 1$ numerator df and ν denominator df (Tables 3, 4, 5, and 6 of Appendix C)

$\quad \nu =$ Number of degrees of freedom associated with MSE

2. Calculate the confidence interval $\hat{L} \pm S$ for each contrast. The confidence coefficient, $1 - \alpha$, applies to the procedure as a whole, i.e., to the entire set of confidence intervals for all possible contrasts.

EXAMPLE 12.21   Refer to the completely randomized design for comparing the mean GPAs of three socioeconomic classes of college freshmen, Examples 12.4 and 12.19. In Example 12.19, we used Tukey's method to rank the three treatment means.

Conduct the multiple comparisons using Scheffé's method at $\alpha = .05$. Use the fact that MSE $= .262$, $p = 3$ treatments, $v = $ df(Error) $= 18$, and the three treatment means are $\overline{y}_{\text{Lower}} = 2.521$, $\overline{y}_{\text{Middle}} = 3.249$, and $\overline{y}_{\text{Upper}} = 2.543$.

Solution

Recall that in this completely randomized design, there are seven observations per treatment; thus, $n_i = n_j = 7$ for all treatment pairs (i, j).

Since the values of p, F, and MSE are fixed, the critical difference S_{ij} will be the same for all treatment pairs (i, j). Substituting in the appropriate values into the formula given in the box, the critical difference is:

$$S_{ij} = \sqrt{(p-1)(F_{.05})\text{MSE}\left(\frac{1}{n_i} + \frac{1}{n_i}\right)}$$

$$= \sqrt{(2)(3.55)(.262)\left(\frac{1}{7} + \frac{1}{7}\right)} = .729$$

Treatment means differing by more than $S = .729$ will imply a significant difference between the corresponding population means. Now, the difference between the largest sample mean and the smallest sample mean, $(3.249 - 2.521) = .728$, is less than $S = .729$. Consequently, we obtain the rankings shown below:

Sample means:	2.521	2.543	3.249
Treatments:	Lower	Upper	Middle

Thus, Scheffé's method fails to detect any significant differences among the GPA means of the three socioeconomic classes at an experiment-wise error rate of $\alpha = .05$.

[*Note*: This inference is confirmed by the results shown on the SAS printout of the Scheffé analysis, Figure 12.24. All three classes have the same "Scheffé grouping" letter (A).] ◆

FIGURE 12.24

SAS printout of Scheffé's multiple comparisons of GPA means, Example 12.21

The ANOVA Procedure

Scheffé's Test for GPA

NOTE: This test controls the Type I experimentwise error rate.

Alpha	0.05
Error Degrees of Freedom	18
Error Mean Square	0.261729
Critical Value of F	3.55456
Minimum Significant Difference	0.7291

Means with the same letter are not significantly different.

Scheffe Grouping	Mean	N	CLASS
A	3.2486	7	MIDDLE
A			
A	2.5429	7	UPPER
A			
A	2.5214	7	LOWER

Scheffé's Multiple Comparisons Procedure for Pairwise Comparisons of Treatment Means

1. Calculate Scheffé's critical difference for each pair of treatments (i, j):

$$S_{ij} = \sqrt{(p-1)(F_\alpha)(\text{MSE})\left(\frac{1}{n_i} + \frac{1}{n_j}\right)}$$

where

$$
\begin{aligned}
p &= \text{Number of sample (treatment) means} \\
\text{MSE} &= \text{Mean squared error} \\
n_i &= \text{Number of observations in sample} \\
&\quad \text{for treatment } i \\
n_j &= \text{Number of observations in sample} \\
&\quad \text{for treatment } j \\
F_\alpha &= \text{Critical value of } F \text{ distribution with } p-1 \\
&\quad \text{numerator df and } v \text{ denominator df} \\
&\quad \text{(Tables 3, 4, 5, and 6 of Appendix C)} \\
v &= \text{Number of degrees of freedom} \\
&\quad \text{associated with MSE}
\end{aligned}
$$

2. Rank the p sample means and place a bar over any treatment pair (i, j) that differs by less than S_{ij}. Any pair of sample means not connected by an overbar implies a difference in the corresponding population means.

Note that in Example 12.21, the Scheffé method produced a critical difference of $S = .729$—a value larger than Tukey's critical difference of $\omega = .698$ (Example 12.19). This implies that Tukey's method produces narrower confidence intervals than Scheffé's method for differences in pairs of treatment means. Therefore, if only pairwise comparisons of treatments are to be made, Tukey's is the preferred method as long as the sample sizes are equal. The Scheffé method, on the other hand, yields narrower confidence intervals (i.e., smaller critical differences) for situations in which the goal of the researchers is to make comparisons of general contrasts.

BONFERRONI APPROACH: As noted previously, Tukey's multiple comparisons procedure is approximate in the case of unequal sample sizes. That is, the value of α selected a priori only approximates the true probability of making at least one Type I error. The Bonferroni approach is an exact method that is applicable in either the equal or the unequal sample size case (see Miller, 1981). Furthermore, Bonferroni's procedure covers all possible comparisons of treatments, including pairwise comparisons, general contrasts, or combinations of pairwise comparisons and more complex contrasts.

The Bonferroni approach is based on the following result (proof omitted): If g comparisons are to be made, each with confidence coefficient $1 - \alpha/g$, then the overall probability of making one or more Type I errors (i.e., the experimentwise error rate) is at most α. That is, the set of intervals constructed using the Bonferroni method yields an overall confidence level of at least $1 - \alpha$. For example, if you want to construct $g = 2$ confidence intervals with an experimentwise error rate of at most $\alpha = .05$, then each individual interval must be constructed using a confidence level of $1 - .05/2 = .975$.

The Bonferroni approach for general contrasts is shown in the next box. When applied only to pairwise comparisons of treatments, the Bonferroni approach can be carried out as shown in the next box.

Bonferroni Multiple Comparisons Procedure for General Contrasts

1. For each contrast $L = \sum_{i=1}^{p} c_i \mu_i$, calculate

$$\hat{L} = \sum_{i=1}^{p} c_i \bar{y}_i$$

and

$$B = t_{\alpha/(2g)} s \sqrt{\sum_{i=1}^{p} \left(\frac{c_i^2}{n_i} \right)}$$

where

$p =$ Number of sample (treatment) means
$g =$ Number of contrasts
$s = \sqrt{\text{MSE}}$
$v =$ Number of degrees of freedom associated with MSE
$n_i =$ Number of observations in sample for treatment i
$\bar{y}_i =$ Sample mean for treatment i
$t_{\alpha/(2g)} =$ Critical value of t distribution with v df and tail area $\alpha/(2g)$

2. Calculate the confidence interval $\hat{L} \pm B$ for each contrast. The confidence coefficient for the procedure as a whole, i.e., for the entire set of confidence intervals, is *at least* $(1 - \alpha)$.

EXAMPLE 12.22

Refer to Example 12.21. Use Bonferroni's method to perform pairwise comparisons of the three treatment means. Use $\alpha = .05$.

Solution

From Example 12.21, we have $p = 3$, $s = \sqrt{.262} = .512$, $v = 18$, and $n_i = n_j = 7$ for all treatment pairs (i, j). For $p = 3$ means, the number of pairwise comparisons to be made is

$$g = \frac{p(p-1)}{2} = \frac{3(2)}{2} = 3$$

Thus, we need to find the critical value, $t_{\alpha/(2g)} = t_{.05/[2(3)]} = t_{.0083}$, for the t distribution based on $v = 18$ df. This value, although not shown in Table 2 in Appendix C, is approximately 2.64.* Substituting $t_{.0083} \approx 2.64$ into the equation for Bonferroni's critical difference B_{ij}, we have

$$B_{ij} \approx (t_{.0083})s\sqrt{\frac{1}{n_i} + \frac{1}{n_j}} = (2.64)(.512)\sqrt{\frac{1}{7} + \frac{1}{7}} = .722$$

for any treatment pair (i, j).

Using the value $B_{ij} = .722$ to detect significant differences between treatment means, we obtain the following results:

Sample means:	2.521	2.543	3.249
Treatments:	Lower	Upper	Middle

You can see that the middle class mean GPA is significantly larger than the mean for the lower class. However, there is no significant difference between the pair of means for the lower and upper classes and no significant difference between the pair of means for the upper and middle classes. In other words, it is unclear whether

FIGURE 12.25

SAS printout of Bonferroni's multiple comparisons of GPA means, Example 12.22

```
                    The ANOVA Procedure

                Bonferroni (Dunn) t Tests for GPA

NOTE: This test controls the Type I experimentwise error rate, but it generally has a higher
                    Type II error rate than REGWQ.


            Alpha                               0.05
            Error Degrees of Freedom              18
            Error Mean Square               0.261729
            Critical Value of t             2.63914
            Minimum Significant Difference   0.7217


     Means with the same letter are not significantly different.


        Bon Grouping         Mean      N    CLASS

                     A      3.2486     7    MIDDLE
                     A
                B    A      2.5429     7    UPPER
                B
                B           2.5214     7    LOWER
```

*We obtained the value using the SAS probability generating function for a Student's t distribution.

the upper class mean GPA should be grouped with the treatment with the largest mean (middle class) or the treatment with the smallest mean (lower class). [*Note*: These inferences are supported by the results shown in the SAS printout of the Bonferroni analysis, Figure 12.25. The middle and upper class treatments have the same "Bonferroni grouping" letter (A), while the upper and lower class treatments have the same letter (B).] All inferences derived from this analysis can be made at an overall confidence level of at least $(1 - \alpha) = .95$. ◆

Bonferroni Multiple Comparisons Procedure for Pairwise Comparisons of Treatment Means

1. Calculate for each treatment pair (i, j)

$$B_{ij} = t_{\alpha/(2g)} s \sqrt{\frac{1}{n_i} + \frac{1}{n_j}}$$

where

$p =$ Number of sample (treatment) means in the experiment

$g =$ Number of pairwise comparisons

    [*Note* : If all pairwise comparisons are to be made,

    then $g = p(p - 1)/2$]

$s = \sqrt{\text{MSE}}$

$v =$ Number of degrees of freedom associated with MSE

$n_i =$ Number of observations in sample for treatment i

$n_j =$ Number of observations in sample for treatment j

$t_{\alpha/(2g)} =$ Critical value of t distribution with v df and tail area

    $\alpha/(2g)$(Table 2 in Appendix C)

2. Rank the sample means and place a bar over any treatment pair (i, j) whose sample means differ by less than B_{ij}. Any pair of means not connected by an overbar implies a difference in the corresponding population means.

Note: The level of confidence associated with all inferences drawn from the analysis is at least $(1 - \alpha)$.

When applied to pairwise comparisons of treatments, the Bonferroni method, like the Scheffé procedure, produces wider confidence intervals (reflected by the magnitude of the critical difference) than Tukey's method. (In Example 12.22. Bonferroni's critical difference is $B \approx .722$ compared to Tukey's $\omega = .698$.) Therefore, if only pairwise comparisons are of interest, Tukey's procedure is again superior. However, if the sample sizes are unequal or more complex contrasts are to be compared, the Bonferroni technique may be preferred. Unlike the Tukey and Scheffé methods, however, Bonferroni's procedure requires that you know in advance how many contrasts are to be compared. Also, the value needed to calculate the critical difference B, $t_{\alpha/(2g)}$, may not be available in the t tables provided in most texts, and you will have to estimate it.

In this section, we have presented two alternatives to Tukey's multiple comparisons procedure. The technique you select will depend on several factors, including the sample sizes and the type of comparisons to be made. Keep in mind, however, that many other methods of making multiple comparisons are available, and one or more of these techniques may be more appropriate to use in your particular application. Consult the references given at the end of this chapter for details on other techniques.

EXERCISES

12.57. Refer to *The Archives of Disease in Childhood* (Apr. 2000) comparison of the average heights of the three groups of Australian school children based on age, Exercise 12.9 (p. 563). The three height means for boys were ranked using the Bonferroni method at $\alpha = .05$. The results are summarized below. (Recall that all height measurements were standardized using z-scores.)

Sample mean:	0.16	0.33	0.33
Age group:	Oldest	Youngest	Middle

a. Is there a significant difference between the standardized height means for the oldest and youngest boys?

b. Is there a significant difference between the standardized height means for the oldest and middle-aged boys?

c. Is there a significant difference between the standardized height means for the youngest and middle-aged boys?

d. What is the experimentwise error rate for the inferences made in parts **a**–**c**? Interpret this value.

e. The researchers did not perform a Bonferroni analysis of the height means for the three groups of girls. Explain why not.

12.58. Refer to the *Journal of Hazardous Materials* study of mean sorption rates for three types of organic solvents, Exercise 12.11 (p. 564). SAS was used to produce Bonferroni confidence intervals for all pairs of treatment means. The SAS printout is shown above right.

a. Find and interpret the experimentwise error rate shown on the printout.

b. Locate and interpret the confidence interval for the difference between the mean sorption rates of aromatics and esters.

c. Use the confidence intervals to determine which pairs of treatment means are significantly different.

SAS Output for Exercise 12.58

The ANOVA Procedure

Bonferroni (Dunn) t Tests for SORPRATE

Note: This test controls the Type I experimentwise error rate, but it generally has a higher Type II error rate than Tukey's for all pairwise comparisons.

Alpha	0.05
Error Degrees of Freedom	29
Error Mean Square	0.067426
Critical Value of t	2.54091

Comparisons significant at the 0.05 level are indicated by***

SOLVENT Comparison	Difference Between Means	Simultaneous 95% Confidence Limits	
CHLOR - AROMA	0.06403	−0.25657 0.38462	
CHLOR - ESTER	0.67625	0.38740 0.96510	***
AROMA - CHLOR	−0.06403	−0.38462 0.25657	
AROMA - ESTER	0.61222	0.33403 0.89041	***
ESTER - CHLOR	−0.67625	−0.96510 −0.38740	***
ESTER - AROMA	−0.61222	−0.89041 −0.33403	***

12.59. Does recalling a traumatic dental experience increase your level of anxiety at the dentist's office? In a study published in *Psychological Reports* (Aug. 1997), researchers at Wittenberg University randomly assigned 74 undergraduate psychology students to one of three experimental conditions. Subjects in the "Slide" condition viewed 10 slides of scenes from a dental office. Subjects in the "Questionnaire" condition completed a full dental history questionnaire; one of the questions asked them to describe their worst dental experience. Subjects in the "Control" condition received no actual treatment. All students then completed the Dental Fear Scale, with scores ranging from 27 (no fear) to 135 (extreme fear). The sample dental fear means for the Slide, Questionnaire, and Control groups were reported as 43.1, 53.8, and 41.8, respectively.

a. A completely randomized design ANOVA was carried out on the data, with the following results: $F = 4.43$, p-value $< .05$. Interpret these results.

b. According to the article, a Bonferroni ranking of the three dental fear means (at $\alpha = .05$) "indicated a significant difference between the mean scores on the Dental Fear Scale for the Control and Questionnaire groups, but not for the means between the Control and Slide groups." Summarize these results in a chart that connects means that are not significantly different.

12.60. Refer to the *Journal of Consulting and Clinical Psychology* study of binge eaters, Exercise 12.15 (p. 565). You found evidence of a difference among the treatment means for the three treatments, CBT-WT, CBT-IPT, and Control. Conduct a Bonferroni analysis to rank the three treatment means using $\alpha = .03$.

12.61. Refer to the *Marine Technology study of major ocean* spills by tankers, Exercise 12.16 (p. 566). Use a statistical software package with Scheffé's method to rank the spillage amount means of the four accident types. (Use an experimentwise error rate of $\alpha = .01$.)

12.62. A field experiment was conducted at a not-for-profit research and development organization to examine the expectations, attitudes, and decisions of employees with regard to training programs (*Academy of Management Journal*, Sept. 1987). In particular, the study was aimed at determining how managers' evaluations of a training program were affected by the prior information they received and by the degree of choice they had in entering the program. These two factors, prior information and degree of choice, were each varied at two levels. The prior information managers received about the training program was either a realistic preview of the program and its benefits or a traditional announcement that tended to exaggerate the workshop's benefits. Degree of choice was either low (mandatory attendance) or high (little pressure from supervisors to attend). Twenty-one managers were randomly assigned to each of the $2 \times 2 = 4$ experimental conditions; thus, a 2×2 factorial design was employed. At the end of the training program, each manager was asked to rate his or her satisfaction with the workshop on a 7-point scale ($1 =$ no satisfaction, $7 =$ extremely satisfied). The ratings were subjected to an analysis of variance, with the results shown in the partial ANOVA summary table.

SOURCE	df	SS	MS	F
Prior information (P)	1	—	1.55	—
Degree of choice (D)	1	—	22.26	—
PD interaction	1	—	.61	—
Error	80	—	1.43	
Total	83	—		

Source: Hicks, W. D., and Klimoski, R. J. "Entry into training programs and its effects on training outcomes: A field experiment." *Academy of Management Journal*, Vol. 30, No. 3, Sept. 1987, p. 548.

a. Complete the ANOVA summary table.

b. Conduct the appropriate ANOVA F tests (use $\alpha = .05$). Interpret the results.

c. The sample mean satisfaction ratings of managers for the four combinations of prior information and degree of choice are shown in the table below. Use Tukey's method to rank the four means. Use. $\alpha = .06$.

d. Use the Scheffé method to perform all pairwise comparisons of the four treatment means. Use $\alpha = .05$.

e. Use the Bonferroni approach to perform all pairwise comparisons of the four treatment means. Use $\alpha = .05$.

f. Compare the results, parts **c–e**.

		PRIOR INFORMATION	
		REALISTIC PREVIEW	TRADITIONAL ANNOUNCEMENT
DEGREE OF CHOICE	High	6.20	6.06
	Low	5.33	4.82

Source: Hicks, W. D., and Klimoski, R. J. "Entry into training programs and its effects on training outcomes: A field experiment." *Academy of Management Journal*, Vol. 30, No. 3, Sept. 1987, p. 548.

12.9
Checking ANOVA Assumptions

For each of the experiments and designs discussed in this chapter, we listed in the relevant boxes the assumptions underlying the analysis in the terminology of ANOVA. For example, in the box on page 555, the assumptions for a completely randomized design are that (1) the p probability distributions of the response y corresponding to the p treatments are normal and (2) the population variances of the p treatments are equal. Similarly, for randomized block designs and factorial designs, the data for the treatments must come from normal probability distributions with equal variances.

These assumptions are equivalent to those required for a regression analysis (see Section 4.2). The reason, of course, is that the probabilistic model for the response y that underlies each design is the familiar general linear regression model of Chapter 4. A brief overview of the techniques available for checking the ANOVA assumptions follows.

DETECTING NONNORMAL POPULATIONS

1. For each treatment, construct a histogram, stem-and-leaf display, or normal probability plot for the response, y. Look for highly skewed distributions. [*Note*: For relatively large samples (e.g., 20 or more observations per treatment), ANOVA, like regression, is robust with respect to the normality assumption. That is, slight departures from normality will have little impact on the validity of the inferences derived from the analysis. [If the sample size for each treatment is small, then these graphs will probably be of limited use.]

2. Formal statistical tests of normality (such as the **Shapiro–Wilk test** or **Kol-mogorov–Smirnov test**) are also available. The null hypothesis is that the probability distribution of the response y is normal. These tests, however, are sensitive to slight departures from normality. Since in most scientific applications the normality assumption will not be satisfied exactly, these tests will likely result in a rejection of the null hypothesis and, consequently, are of limited use in practice. Consult the references for more information on these formal tests.

3. If the distribution of the response departs greatly from normality, a **normalizing transformation** may be necessary. For example, for highly skewed distributions, transformations on the response y such as $\log(y)$ or $\sqrt{y}$ tend to "normalize" the data since these functions "pull" the observations in the tail of the distribution back toward the mean.

DETECTING UNEQUAL VARIANCES

1. For each treatment, construct a box plot or **frequency** (dot) plot for y and look for differences in spread (variability). If the variability of the response in each plot is about the same, then the assumption of equal variances is likely to be satisfied. [*Note*: ANOVA is robust with respect to unequal variances for **balanced designs**, i.e., designs with equal sample sizes for each treatment.]

2. When the sample sizes are small for each treatment, only a few points are graphed on the frequency plots, making it difficult to detect differences in

variation. In this situation, you may want to use one of several formal statistical tests of homogeneity of variances that are available. For p treatments, the null hypothesis is $H_0 : \sigma_1^2 = \sigma_2^2 = \ldots = \sigma_p^2$, where σ_i^2 is the population variance of the response y corresponding to the ith treatment. If all p populations are approximately normal, **Bartlett's test for homogeneity of variances** can be applied. Bartlett's test works well when the data come from normal (or near normal) distributions. The results, however, can be misleading for nonnormal data. In situations where the response is clearly not normally distributed, **Levene's test** is more appropriate. The elements of these tests are shown in the accompanying boxes. Note that Bartlett's test statistic depends on whether the sample sizes are equal or unequal.

Bartlett's Test of Homogeneity of Variance

H_0: $\sigma_1^2 = \sigma_2^2 = \ldots = \sigma_p^2$

H_a: At least two variances differ.

Test statistic (equal sample sizes):

$$B = \frac{(n-1)\left[p \ln \bar{s}^2 - \sum \ln s_i^2 \right]}{1 + \dfrac{p+1}{3p(n-1)}}$$

where

$n = n_1 = n_2 = \cdots = n_p$

$s_i^2 = $ Sample variance for sample i

$\bar{s}^2 = $ Average of the p sample variances $= \left(\sum s_i^2 \right) \Big/ p$

$\ln x = $ Natural logarithm (i.e., log to the base e) of the quantity x

Test statistic (unequal sample sizes):

$$B = \frac{\left[\sum (n_i - 1) \right] \ln \bar{s}^2 - \sum (n_i - 1) \ln s_i^2}{1 + \dfrac{1}{3(p-1)} \left\{ \sum \dfrac{1}{(n_i - 1)} - \dfrac{1}{\sum (n_i - 1)} \right\}}$$

where

$n_i = $ Sample size for sample i

$s_i^2 = $ Sample variance for sample i

$\bar{s}^2 = $ Weighted average of the p sample variances $= \dfrac{\sum (n_i - 1) s_i^2}{\sum (n_i - 1)}$

$\ln x = $ Natural logarithm (i.e., log to the base e) of the quantity x

> *Rejection region* : $B > \chi_\alpha^2$, where χ_α^2, locates an area α in the upper tail of a χ^2 distribution with $(p-1)$ degrees of freedom
>
> *Assumptions* :   1. Independent random samples are selected from the p populations.
> 2. All p populations are normally distributed.

Levene's Test of Homogeneity of Variance

H_0: $\sigma_1^2 = \sigma_2^2 = \ldots = \sigma_p^2$

H_a: At least two variances differ

Test statistic: $F = MST/MSE$

where *MST* and *MSE* are obtained from an ANOVA with p treatments conducted on the transformed response variable $y_i^* = |y_i - Med_p|$, and Med_p is the median of the response y values for treatment p.

> *Rejection region*: $F > F_\alpha$, where F_α locates an area α in the upper tail of an F-distribution with $v_1 = (p-1)$ df and $v_2 = (n-p)$ df.
>
> *Assumptions* :   1. Independent random samples are selected from the p treatment populations.
> 2. The response variable y is a continuous random variable.

---

3. When unequal variances are detected, use one of the **variance-stabilizing transformations** of the response y discussed in Section 8.3.

EXAMPLE   12.23

Refer to the ANOVA for the completely randomized design, Example 12.4. Recall that we found differences among the mean GPAs for the three socioeconomic classes of college freshmen. Check to see if the ANOVA assumptions are satisfied for this analysis.

Solution

First, we'll check the assumption of normality. For this design, there are only seven observations per treatment (class); consequently, constructing graphs (e.g., histograms or stem-and-leaf plots) for each treatment will not be very informative. Alternatively, we can combine the data for the three treatments and form a histogram for all 21 observations in the data set. A MINITAB histogram for the response variable, GPA, is shown in Figure 12.26. Clearly, the data fall into an approximately mound-shaped distribution. This result is supported by the MINITAB normal probability plot for GPA displayed in Figure 12.27. The results of a test for normality of the data is also shown (highlighted) in Figure 12.27. Since the p-value of the test exceeds .10, there is insufficient evidence (at $\alpha = .05$) to conclude that the data are nonnormal. Consequently, it appears that the GPAs come from a normal distribution.

FIGURE 12.26

MINITAB histogram for all the GPAs in the completely randomized design

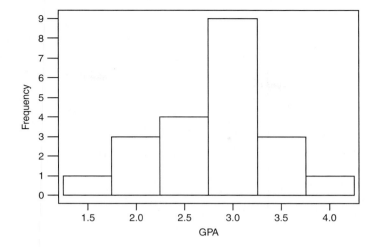

FIGURE 12.27

MINITAB normal probability plot and normality test for all the GPAs in the completely randomized design

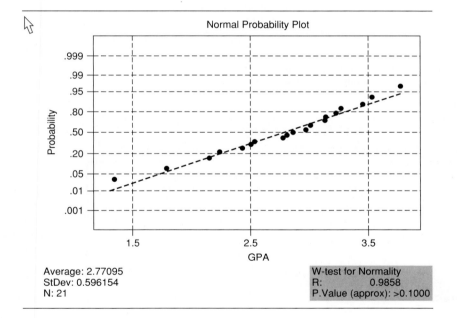

Average: 2.77095
StDev: 0.596154
N: 21

W-test for Normality
R: 0.9858
P.Value (approx): >0.1000

Next, we check the assumption of equal variances. MINITAB dot plots for GPA are displayed in Figure 12.28. Note that the variability of the response in each plot is about the same; thus, the assumption of equal variances appears to be satisfied. To formally test the hypothesis, H_0: $\sigma_1^2 = \sigma_2^2 = \sigma_3^2$, we conduct both Bartlett's and Levene's test for homogeneity of variances. Rather than use the computing formulas shown in the boxes, we resort to a statistical software package. The MINITAB printout of the test results are shown in Figure 12.29. The p-values for both tests are shaded on the printout. Since both p-values exceed at $\alpha = .05$, there is insufficient evidence to reject the null hypothesis of equal variances. Therefore, it appears that the assumption of equal variance is satisfied also. ◆

FIGURE 12.28

MINITAB dot plots for the GPAs
in the completely randomized
design, by treatment group

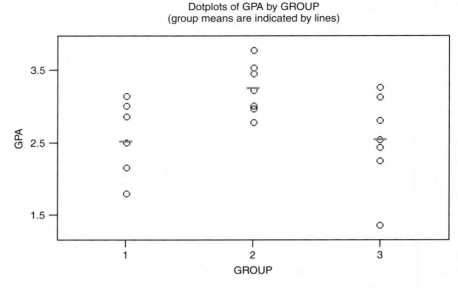

FIGURE 12.29

MINITAB printout of tests for homogeneity of
GPA variances for the completely randomized
design

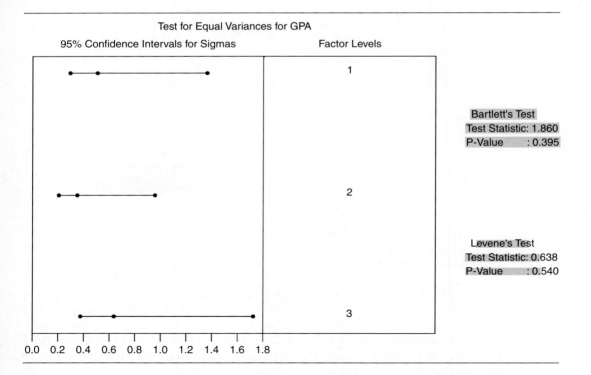

In most applications, the assumptions will not be satisfied exactly. These analysis of variance procedures are flexible, however, in the sense that slight departures from the assumptions will not significantly affect the analysis or the validity of the resulting inferences. On the other hand, gross violations of the assumptions (e.g., a nonconstant variance) will cast doubt on the validity of the inferences. Therefore, you should make it standard practice to verify that the assumptions are (approximately) satisfied.

EXERCISES

12.63. Check the assumptions for the completely randomized design ANOVA of Exercise 12.11 (p. 564).

12.64. Check the assumptions for the completely randomized design ANOVA of Exercise 12.13 (p. 564).

12.65. Check the assumptions for the completely randomized design ANOVA of Exercise 12.16 (p. 566).

12.66. Check the assumptions for the factorial design ANOVA of Exercise 12.38 (p. 603).

12.67. Check the assumptions for the factorial design ANOVA of Exercise 12.39 (p. 604).

Summary

In this chapter, we demonstrated how to analyze data collected from a designed experiment using either regression analysis or **analysis of variance (ANOVA)**. An analysis of variance partitions the total sum of squares, SS(Total), into SSE and, depending on the design, sums of squares for treatments (SST), blocks (SSB), and factor main effects and interaction. Tests for treatment means, block means, factor interactions, and so forth are obtained by calculating the ratio of the appropriate mean square to MSE and conducting an F test. These F tests can also be conducted by fitting the appropriate linear models to the data using regression. The models differ depending on the specific design employed.

Throughout this chapter we showed that an ANOVA and a regression analysis yield equivalent results. But an analysis of variance possesses both advantages and disadvantages compared with a regression analysis. One major advantage of an analysis of variance is that it is easy to perform on a hand calculator. Second, when y is affected by more than one source of random variation, an analysis of variance permits us to separate these sources and to estimate the variances of their respective random components. (A discussion of these models is not included in this text.) The disadvantages are its restrictions and limitations, as follows:

1. **The set of analysis of variance formulas appropriate for a particular experimental design applies only to that design**. If the data collected are observational (i.e., the independent variables are uncontrolled), an analysis of variance is inappropriate. No deviations from the design are permitted. Consequently, the method is of value only for special types of designed experiments.

2. **In contrast to a regression analysis, the analysis of variance formulas change from one design to another**. (There is a pattern, as indicated in Section 12.3, but the pattern is usually not apparent to a beginner.)

3. **The ANOVA formulas for a factorial experiment and many other experimental designs can be used only when the sample sizes are equal**. However, the regression approach applies to both equal and unequal sample sizes for the various factor level combinations.

4. **An analysis of variance does not give you a prediction equation**. This is a great handicap when one (or more) of the independent variables is quantitative.

5. **Although a linear model is always implied in an analysis of variance, it is rarely presented or discussed when analyses of data have been performed**. Consequently, the thrust of an analysis of variance is often counter (although it need not be) to the notion of modeling, which is the modern quantitative way of analyzing real-world phenomena.

In conclusion, perhaps the most important point for you to note in this chapter is the following: **If your data can be modeled using a linear model that contains a single random error component (which is the model used throughout this text), then a regression analysis can do everything that an analysis of variance can do and it can do more!** But you will need a computer to do it. Regression analysis is programmed into nearly all statistical software packages, and it can be used to analyze data obtained from both designed and undesigned experiments. Consequently, a beginner may be advised to stick to regression analyses for analyzing the relationship between a set of independent variables and a response y if software is available to perform the computations.

SUPPLEMENTARY EXERCISES

[*Note*: Exercises marked with an asterisk (*) are from the optional sections in this chapter.]

12.68. Vanadium (V) is an essential trace element found in living organisms. An experiment was conducted to compare the concentrations of V in biological materials using isotope dilution mass spectrometry (*Analytical Chemistry*, Nov. 1985). The accompanying table gives the quantities of V (measured in nanograms per gram) in dried samples of oyster tissue, citrus leaves, bovine liver, and human serum. The data were used to fit the linear model

$$E(y) = \beta_0 + \beta_1 x_1 + \beta_2 x_2 + \beta_3 x_3$$

where

$$x_1 = \begin{cases} 1 & \text{if oyster} \\ & \text{tissue} \\ 0 & \text{if not} \end{cases} \quad x_2 = \begin{cases} 1 & \text{if citrus} \\ & \text{leaves} \\ 0 & \text{if not} \end{cases}$$

$$x_3 = \begin{cases} 1 & \text{if bovine liver} \\ 0 & \text{if not} \end{cases}$$

The SAS printout is shown on p. 641.

💿 VANADIUM

OYSTER TISSUE	CITRUS LEAVES	BOVINE LIVER	HUMAN SERUM	
2.35	2.32	.39	.10	.16
1.30	3.07	.54	.17	.16
.34	4.09	.30	.14	

a. Identify the treatments in this experiment.
b. Is there sufficient evidence (at $\alpha = .05$) to indicate that the mean V concentrations differ among the four biological materials?
c. Use the β estimates of the model to estimate the mean V concentrations in each of the four biological materials.

SAS output for Exercise 12.68

```
                        Dependent Variable: V

                        Analysis of Variance

                              Sum of          Mean
Source              DF       Squares        Square     F Value    Pr > F

Model                3      18.86832       6.28944      17.31     0.0003
Error               10       3.63252       0.36325
Corrected Total     13      22.50084


           Root MSE              0.60270    R-Square      0.8386
           Dependent Mean        1.10214    Adj R-Sq      0.7901
           Coeff Var            54.68474


                        Parameter Estimates

                      Parameter       Standard
     Variable    DF    Estimate         Error     t Value    Pr > |t|

     Intercept    1     0.14600        0.26954       0.54      0.5999
     X1           1     1.18400        0.44015       2.69      0.0227
     X2           1     3.01400        0.44015       6.85      <.0001
     X3           1     0.26400        0.44015       0.60      0.5620
```

12.69. Bell Communications Research (Bellcore) publishes, on average, 150 technical memos (TMs) each month. Since not all TMs are relevant to all employees, Bellcore recognizes that "information filtering" will naturally occur—each employee will retain only the information that is relevant to him or her. Bellcore researchers conducted a study of four automated information filtering methods (*Communications of the Association for Computing Machinery*, Dec. 1992). These methods are (1) keyword match-word profile, (2) latent semantic indexing (LSI)-word profile, (3) keyword match-document profile, and (4) LSI-document profile. A sample of TM abstracts were filtered by each method, and the relevance of the filtered abstracts was rated on a 7-point scale by a panel of Bellcore employees. In addition, a subset of TM abstracts were randomly selected and rated by the panel. The mean relevance ratings of the five filtering methods (four automated methods plus the random selection method) were compared using an analysis of variance for a completely randomized design.
a. Identify the treatments in the experiment.
b. Identify the response variable.
c. The ANOVA resulted in a test statistic of $F = 117.5$, based on 4 numerator degrees of freedom and 132 denominator degrees of freedom. Interpret this result at a significance level of $\alpha = .05$.

12.70. Epidemiologists have theorized that the risk of coronary heart disease can be reduced by an increased consumption of fish. One study, begun in 1960, monitored the diet and health of a random sample of middle-aged Dutch men. The men were divided into five groups according to the numbers of grams of fish consumed per day: 0, 1–14, 15–29, 30–44, and 45 or more. One of the many variables measured on each subject was intake of polysaccharides (a substance linked to coronary heart disease.) An analysis of variance on the levels of polysaccharides (measured as a percentage of energy) in the five groups of men resulted in the partial ANOVA table given here.

SOURCE	df	SS	MS	F
Groups	—	534.97	—	—
Error	—	23,659.45	—	
Total	851	24,194.42		

Source: Kromhout, D., Bosschieter, E. B., and Coulander, C. D. L. "The inverse relation between fish consumption and 20-year mortality from coronary heart disease." *New England Journal of Medicine*, May 9, 1985, Vol. 312, No. 19, pp. 1205–1209. Reprinted by permission.

a. Give the total number of Dutch men included in this portion of the study.
b. Complete the ANOVA summary table.
c. Is there sufficient evidence of differences among the mean levels of polysaccharides in the five groups of men? Test using $\alpha = .01$.

d. The mean levels of polysaccharides found in the five groups of Dutch men are provided in the accompanying table. Use Tukey's method to rank the group means. Use $\alpha = .05$.

FISH CONSUMPTION grams/day	SAMPLE SIZE	MEAN LEVEL OF POLY- SACCHARIDES
0	159	27.0
1–14	283	27.0
15–29	215	26.6
30–44	116	25.7
45 or more	79	24.4

12.71. A fast-food chain that specializes in Mexican food (tacos, burritos, etc.) is opening a new franchise in a university town. An important consideration in determining where the franchise will be located is traffic density. Five possible locations (each near a major intersection) are under consideration by the chain. To compare the average density of traffic at the possible sites, company employees are placed at each of the five locations to count the number of cars passing each location daily for a period of 10 randomly selected days. (At location IV, the counter assigned to record traffic density became ill and could obtain data for only eight of the days.) The results are listed in the table. Assuming the samples of days were independently selected, conduct a complete analysis of the data using an available statistical software package.

⊚ TEXMEX

LOCATION				
I	II	III	IV	V
344	412	237	518	367
382	441	390	501	445
353	607	365	577	480
395	531	355	642	323
207	486	217	489	366
312	508	268	475	325
407	337	117	532	316
421	419	273	540	381
366	499	288		407
222	387	351		339

12.72. A nuclear power plant that uses water from the surrounding bay for cooling its condensers is required by the Environmental Protection Agency (EPA) to determine whether discharging its heated water into the bay has a detrimental effect on the plant life in the water. The EPA requests that the power plant investigate three strategically chosen locations, called *stations*. Stations 1 and 2 are located near the plant's discharge tubes, whereas station 3 is located farther out in the bay. During one randomly selected day in each of six months, a diver descends to each of the stations, randomly samples a square meter area of the bottom, and counts the number of blades of the different types of grasses present. The results for one important grass type are listed in the table.

⊚ NUCLEAR

MONTH	STATION		
	1	2	3
April	32	40	30
May	28	31	53
June	25	22	61
July	37	30	56
August	20	26	48
September	18	21	30

The goal of the study is to determine whether the mean number of blades found per square meter per month differs for at least two of the three stations. Use an available statistical software package to conduct an ANOVA of the data.

12.73. *Industrial Marketing Management* (1993) published the results of a study of humor in trade magazine advertisements. A sample of 665 ads were categorized according to nationality (U.S., British, or German) and industry (29 categories, ranging from accounting to travel). Then a panel of judges determined the degree of humor in each ad using a 5-point scale (where 1 = not at all humorous and 5 = very humorous). The data were analyzed using a 3×29 factorial ANOVA, where the factors were nationality (at 3 levels) and industry (at 29 levels). The results of the ANOVA are reported in the accompanying table.*

SOURCE	df	SS	MS	F	p-VALUE
Nationality (N)	2	1.44	.72	2.40	.087
Industry (I)	28	48.00	1.71	5.72	.000
$N \times I$	49	20.28	.41	1.38	.046

Source: McCullough, L. S., and Taylor, R. K. "Humor in American, British, and German ads." *Industrial Marketing Management*, Vol. 22, 1993, p. 21 (Table 1).

---

*As a result of missing data, the number of degrees of freedom for Nationality × Industry interaction is less than $2 \times 28 = 56$.

a. Using $\alpha = .05$, interpret the ANOVA results. Comment on the order in which the ANOVA F tests should be conducted and whether any of the tests should be ignored.

b. According to the researchers, "British ads were more likely to be humorous than German or U.S. ads in the Graphics industry. German ads were least humorous in the Grocery and Mining industries, but funnier than U.S. ads in the Medical industry and funnier than British ads in the Packaging industry." Do these inferences agree or conflict with the conclusions reached in part **a**? Explain.

12.74. *Science* (Jan. 1, 1999) reported on the ability of 7-month-old infants to learn an unfamiliar language. In one experiment, 16 infants were trained in an artificial language. Then, each infant was presented with two 3-word sentences that consisted entirely of new words (e.g., "wo fe wo"). One sentence was consistent (i.e., constructed from the same grammar as in the training session) and one sentence was inconsistent (i.e., constructed from grammar in which the infant was not trained). The variable measured in each trial was the time (in seconds) the infant spent listening to the speaker, with the goal to compare the mean listening times of consistent and inconsistent sentences.

a. The data were analyzed as a randomized block design with the 16 infants representing the blocks and the two sentence types (consistent and inconsistent) representing the treatments. Do you agree with this data analysis method? Explain.

b. Refer to part **a**. The test statistic for testing treatments was $F = 25.7$ with an associated observed significance level of $p < .001$. Interpret this result.

c. Explain why the data could also be analyzed as a paired difference experiment, with a test statistic of $t = 5.07$.

d. The mean listening times and standard deviations for the two treatments are given here. Use this information to calculate the F statistic for comparing the treatment means in an ANOVA for a completely randomized design. Explain why this test statistic provides weaker evidence of a difference between treatment means than the test in part **b**.

	CONSISTENT SENTENCES	INCONSISTENT SENTENCES
Mean	6.3	9.0
Standard dev.	2.6	2.16

e. Explain why there is no need to control the experimentwise error rate when ranking the treatment means for this experiment.

12.75. An experiment was conducted to examine the effects of alcohol on the marital interactions of husbands and wives (*Journal of Abnormal Psychology*, Nov. 1998). A total of 135 couples participated in the experiment. The husband in each couple was classified as aggressive (60 husbands) or nonaggressive (75 husbands), based on an interview and his response to a questionnaire. Before the marital interactions of the couples were observed, each husband was randomly assigned to three groups: receive no alcohol, receive several alcoholic mixed drinks, or receive placebos (nonalcoholic drinks disguised as mixed drinks). Consequently, a 2×3 factorial design was employed, with husband's aggressiveness at 2 levels (aggressive or nonaggressive) and husband's alcohol condition at 3 levels (no alcohol, alcohol, and placebo). The response variable observed during the marital interaction was severity of conflict (measured on a 100-point scale).

a. A partial ANOVA table is shown below. Fill in the missing degrees of freedom.

SOURCE	df	F	p-VALUE
Aggressiveness (A)	—	16.43	< .001
Alcohol Condition (C)	—	6.00	< .01
A × C	—		—
Error	129		
Total	—		

b. Interpret the p-value of the F-test for Aggressiveness.

c. Interpret the p-value of the F-test for Alcohol Condition.

d. The F-test for interaction was omitted from the article. Discuss the dangers of making inferences based on the tests, parts **a** and **b**, without knowing the result of the interaction test.

12.76. A company conducted an experiment to determine the effects of three types of incentive pay plans on worker productivity for both union and non-union workers. The company used plants in adjacent towns; one was unionized and the other was not. One-third of the production workers in each plant were assigned to each incentive plan. Then six workers were randomly selected from each group, and their productivity (in number of items produced) was measured for a 1-week period. The six productivity measures for the 2×3 factor combinations are listed in the table on p. 644. Conduct the analysis for the company using an available statistical software package.

PAYPLAN

		INCENTIVE PLAN					
		A		B		C	
	Union	337	328	346	373	317	341
		362	319	351	338	335	329
UNION		305	344	355	365	310	315
AFFILIATION		359	346	371	377	350	336
	Nonunion	345	396	352	401	349	351
		381	373	399	378	374	340

12.77. The steam explosion of peat yields fermentable carbohydrates that have a number of potentially important industrial uses. A study of the steam explosion process was initiated to determine the optimum conditions for the release of fermentable carbohydrate (*Biotechnology and Bioengineering*, Feb. 1986). Triplicate samples of peat were treated for .5, 1.0, 2.0, 3.0, and 5.0 minutes at $170°$, $200°$, and $215°C$, in the steam explosion process. Thus, the experiment consists of two factors—temperature at three levels and treatment time at five levels. The accompanying table gives the percentage of carbohydrate solubilized for each of the $3 \times 5 = 15$ peat samples.

PEAT

TEMPERATURE °C	TIME minutes	CARBOHYDRATE SOLUBILIZED %
170	.5	1.3
170	1.0	1.8
170	2.0	3.2
170	3.0	4.9
170	5.0	11.7
200	.5	9.2
200	1.0	17.3
200	2.0	18.1
200	3.0	18.1
200	5.0	18.8
215	.5	12.4
215	1.0	20.4
215	2.0	17.3
215	3.0	16.0
215	5.0	15.3

Source: Forsberg, C. W., *et al*. "The release of fermentable carbohydrate from peat by steam explosion and its use in the microbial production of solvents." *Biotechnology and Bioengineering*, Vol. 28, No. 2, Feb. 1986, p. 179 (Table 1). Copyright 1986.

a. What type of experimental design was used?

b. Explain why the traditional analysis of variance formulas are inappropriate for the analysis of these data.

c. Write a second-order model relating mean amount of carbohydrate solubilized, $E(y)$, to temperature (x_1) and time (x_2).

d. Explain how you could test the hypothesis that the two factors, temperature (x_1) and time (x_2), interact.

e. Fit the model and perform the test for interaction.

12.78. In the late 1970s and 1980s, environmental scientists hypothesized that *acid rain* could be a serious environmental problem. It is formed by the combination of water vapor in clouds with nitrous oxide and sulfur dioxide, which are among the emissions products of coal and oil combustion. The pH of rain in central and northern Florida consistently ranges from 4.5 to 5, indicating an acidic condition. (On the 14-point pH scale, any value below 7 is considered acidic, whereas values above 7 are considered alkaline.) To determine the effects of acid rain on soil pH in a natural ecosystem, engineers at the University of Florida's Institute of Food and Agricultural Sciences irrigated experimental plots near Gainesville, Florida, with rainwater at two pH levels, 3.7 and 4.5. The acidity of the soil was then measured at three different depths, 0–15, 15–30, and 30–46 centimeters. Tests were conducted during three different time periods. The resulting soil pH values are shown in the table on p. 645. Suppose we treat the experiment as a 2×3 factorial laid out in three blocks, where the factors are acid rain at two pH levels and soil depth at three levels, and the blocks are the three time periods. The SAS printout for the analysis of variance is provided on p. 645.

a. Is there evidence of an interaction between pH level of acid rain and soil depth? Test using $\alpha = .05$.

b. Conduct a test to determine whether blocking over time was effective in removing an extraneous source of variation. Use $\alpha = .05$.

⬤ ACIDRAIN

		APRIL 3 ACID RAIN pH 3.7	APRIL 3 ACID RAIN pH 4.5	JUNE 16 ACID RAIN pH 3.7	JUNE 16 ACID RAIN pH 4.5	JUNE 30 ACID RAIN pH 3.7	JUNE 30 ACID RAIN pH 4.5
SOIL DEPTH, cm	0–15	5.33	5.33	5.47	5.47	5.20	5.13
	15–30	5.27	5.03	5.50	5.53	5.33	5.20
	30–46	5.37	5.40	5.80	5.60	5.33	5.17

Source: "Acid rain linked to growth of coal-fired power." *Florida Agricultural Research*, 83, Vol. 2, No. 1, Winter 1983.

SAS output for Exercise 12.78

The ANOVA Procedure

Dependent Variable: SOILPH

Source	DF	Sum of Squares	Mean Square	F Value	Pr > F
Model	7	0.48685556	0.06955079	6.99	0.0034
Error	10	0.09952222	0.00995222		
Corrected Total	17	0.58637778			

R-Square	Coeff Var	Root MSE	SOILPH Mean
0.830276	1.861595	0.099761	5.358889

Source	DF	Anova SS	Mean Square	F Value	Pr > F
DEPTH	2	0.06714444	0.03357222	3.37	0.0759
RAINPH	1	0.03042222	0.03042222	3.06	0.1110
DEPTH*RAINPH	2	0.00781111	0.00390556	0.39	0.6854
DATE	2	0.38147778	0.19073889	19.17	0.0004

12.79. Most people are right-handed due to the propensity of the left hemisphere of the brain to control sequential movement. Similarly, the fact that some tasks are performed better with the left hand is likely due to the superiority of the right hemisphere of the brain to process the necessary information. Does such cerebral specialization in spatial processing occur in adults with Down syndrome? A 2×2 factorial experiment was conducted to answer this question (*American Journal on Mental Retardation*, May 1995). A sample of adults with Down syndrome was compared to a control group of normal individuals of a similar age. Thus, one factor was Group at two levels (Down syndrome and control) and the second factor was the Handedness (left or right) of the subject. All the subjects performed a task that typically yields a left-hand advantage. The response variable was "laterality index," measured on a -100 to 100 point scale. (A large positive index indicates a right-hand advantage, while a large negative index indicates a left-hand advantage.)

a. Identify the treatments in this experiment.

b. Construct a graph that would support a finding of no interaction between the two factors.

c. Construct a graph that would support a finding of interaction between the two factors.

d. The *F*-test for factor interaction yielded an observed significance level of $p < .05$. Interpret this result.

e. Multiple comparisons of all pairs of treatment means yielded the rankings shown here. Interpret the results.

Mean laterality index: Group/Handed:	-30 Down/ Left	-4 Control/ Right	$-.5$ Control/ Left	$+.5$ Down/ Right

f. The experiment-wise error rate for the analysis, part **e**, was $\alpha = .05$. Interpret this value.

12.80. Turnover among truck drivers is a major problem for both carriers and shippers. Since knowledge of driver-related job attitudes is valuable for predicting and controlling future turnover, a study of the work-related attitudes of truck drivers was conducted (*Transportation Journal*, Fall 1993). The two factors considered in the study were career stage and time spent on road. Career stage was set at three levels: early (less than 2 years), mid-career (between 3 and 10 years), and late (more than 10 years). Road time was dichotomized as short (gone for one weekend or less) and long (gone for longer than one weekend). Data were collected on job satisfaction for drivers sampled in each of the $3 \times 2 = 6$ combinations of career stage and road time. [Job satisfaction was measured on a 5-point scale, where 1 = really dislike and 5 = really like.]

a. Identify the response variable for this experiment.

b. Identify the factors for this experiment.

c. Identify the treatments for this experiment.

d. The ANOVA table for the analysis is shown here. Fully interpret the results.

SOURCE	F VALUE	p-VALUE
Career stage (*CS*)	26.67	$p \leq .001$
Road time (*RT*)	.19	$p > .05$
$CS \times RT$	1.59	$p < .05$

Source: McElroy, J. C., *et al*. "Career stage, time spent on the road, and truckload driver attitudes." *Transportation Journal*, Vol. 33, No. 1, Fall 1993, p. 10 (Table 2).

e. The researchers theorized that the impact of road time on job satisfaction may be different depending on the career stage of the driver. Do the results support this theory?

f. The researchers also theorized that career stage affects the job satisfaction of truck drivers. Do the results support this theory?

*g Since career stage was found to be the only significant factor in the 3×2 factorial ANOVA, the mean job satisfaction levels of the three career stages (early, middle, and late) were compared using the Bonferroni method. Find the adjusted α level to use in the analysis, if the researchers desire an overall significance level of $\alpha = .09$.

*h The sample mean job satisfaction levels for the three career stages are given here. Assuming equal sample sizes for each stage and a Bonferroni critical difference of $B = .06$, rank the means.

Mean job satisfaction	3.47	3.38	3.36
Career stage	Early	Middle	Late

12.81. The traditional retail store audit is one of the most widely used marketing research tools among consumer package goods companies. It involves periodic audits of a sample of retail outlets to monitor inventory and purchases of a particular product. V. K. Prasad, W. R. Casper, and R. J. Schieffer conducted a study to compare market data yielded by retail store audits with alternative, less costly auditing procedures—weekend

selldown audits and store purchases audits (*Journal of Marketing*, Winter 1984). The market shares of six major brands of beer distributed in eastern cities were estimated using each of the three store audit methods. The data are provided in the table.

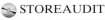 STOREAUDIT

BRAND	TRADITIONAL STORE AUDIT	WEEKEND SELLDOWN AUDIT	STORE PURCHASES AUDIT
1	18.0	19.0	20.7
2	15.3	17.3	14.0
3	8.9	8.5	10.1
4	6.5	4.9	6.1
5	5.3	6.1	4.6
6	3.4	3.0	3.1

Source: Prasad, V. K., Casper, W. R., and Schieffer, R. J. "Alternatives to the traditional retail store audit: A field study." *Journal of Marketing*, Winter 1984, 48, pp. 54–61. Reprinted by permission of the American Marketing Association.

a. Construct an ANOVA summary table for the data.

b. Is there sufficient evidence to indicate a difference in the mean estimates of beer-brand market shares produced by the three auditing methods? Test using $\alpha = .05$.

c. Estimate the difference between the mean estimates of beer-brand market shares produced by the traditional store audit and the weekend selldown audit using a 95% confidence interval.

12.82. The percentage of water removed from paper as it passes through a dryer depends on the temperature of the dryer and the speed of the paper passing through it. A laboratory experiment was conducted to investigate the relationship between dryer temperature T at three levels (100°, 120°, and 140°F) and exposure time E (which is related to speed) also at three levels (10, 20, and 30 seconds). Four paper specimens were prepared for each of the $3 \times 3 = 9$ conditions. The data (percentage of water removed) are shown in the accompanying table. Carry out a complete analysis of the data using an available statistical software package.

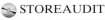 DRYPAPER

		TEMPERATURE (*T*)					
		100		120		140	
EXPOSURE TIME (*E*)	10	24	26	33	33	45	49
		21	25	36	32	44	45
	20	39	34	51	50	67	64
		37	40	47	52	68	65
	30	58	55	75	71	89	87
		56	53	70	73	86	83

12.83. The *Accounting Review* (Jan. 1991) reported on a study of the effect of two factors, confirmation of accounts receivable and verification of sales transactions, on account misstatement risk by auditors. Both factors were held at the same two levels: completed or not completed. Thus, the experimental design is a 2×2 factorial design.

    a. Identify the factors, factor levels, and treatments for this experiment.

    b. Explain what factor interaction means for this experiment.

    c. A graph of the hypothetical mean misstatement risks for each of the $2 \times 2 = 4$ treatments is shown here. In this hypothetical case, does it appear that interaction exists?

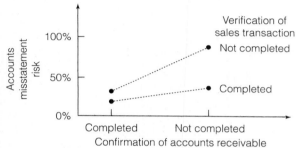

Source: Brown, C. E. and Solomon, I. "Configural information processing in auditing: The role of domain-specific knowledge." *The Accounting Review*. Vol. 66, No. 1, Jan 1991. p. 105 (Figure 1).

12.84. A production manager who supervises an assembly operation wants to investigate the effect of the incoming rate (parts per minute) x_1 of components and room temperature x_2 on the productivity (number of items produced per minute) y. The component parts approach the worker on a belt and return to the worker if not selected on the first trip past the assembly point. It is thought that an increase in the arrival rate of components has a positive effect on the assembly rate, up to a point, after which increases may annoy the assembler and reduce productivity. Similarly, it is suspected that lowering the room temperature is beneficial to a point, after which reductions may reduce productivity. The experimenter used the same assembly position for each worker. Thirty-two workers were used for the experiment, two each assigned to the 16 factor level combinations of a 4×4 factorial experiment. The data, in parts per minute averaged over a 5-minute period, are shown in the table at the bottom of the page.

    a. Perform an analysis of variance for the data. Display the computed quantities in an ANOVA table.

    b. Write the linear model implied by the analysis of variance. [*Hint:* For a quantitative variable recorded at four levels, main effects include terms for x, x^2, and x^3.]

    c. Do the data provide sufficient evidence to indicate differences among the mean responses for the 16 treatments of the 4×4 factorial experiment? Test using $\alpha = .05$.

    d. Do the data provide sufficient evidence to indicate an interaction between arrival rate x_1 and room temperature x_2 on worker productivity? Test using $\alpha = .05$.

    e. Find the value of R^2 that would be obtained if you were to fit the linear model in part **b** to the data.

    f. Explain why a regression analysis would be a useful addition to the inferential methods used in parts **a–e**.

12.85. A second-order model would be a reasonable choice to model the data of Exercise 12.84. To simplify the analysis, we will code the arrival rate and temperature values as follows:

$$x_1 = \frac{\text{Arrival rate} - 55}{5} \qquad x_2 = \frac{\text{Temperature} - 72.5}{2.5}$$

A printout of the SAS regression analysis is shown on p. 648.

    a. Write the second-order model for the response. Note the difference between this model and the ANOVA model in Exercise 12.84, part **a**.

    b. Give the prediction relating the response y to the coded independent variables x_1 and x_2.

🔘 ASSEMBLY1

		RATE OF INCOMING COMPONENTS (x_1), PARTS PER MINUTE							
		40		50		60		70	
	65	24.0,	23.8	25.6,	25.4	29.2,	29.4	28.4,	27.6
ROOM	70	25.0,	26.0	28.8,	28.8	31.6,	32.0	30.2,	30.0
TEMPERATURE	75	25.6,	25.0	27.6,	28.0	29.8,	28.6	28.0,	27.0
(x_2),° F	80	24.0,	24.6	27.6,	26.2	27.6,	28.6	26.0,	24.4

SAS output for Exercise 12.85

Dependent Variable: PPM5

Analysis of Variance

Source	DF	Sum of Squares	Mean Square	F Value	Pr > F
Model	5	130.80680	26.16136	27.73	<.0001
Error	26	24.53320	0.94358		
Corrected Total	31	155.34000			

Root MSE	0.97138	R-Square	0.8421	
Dependent Mean	27.32500	Adj R-Sq	0.8117	
Coeff Var	3.55492			

Parameter Estimates

Variable	DF	Parameter Estimate	Standard Error	t Value	Pr > |t|
Intercept	1	29.85625	0.34876	85.61	<.0001
X1	1	0.56000	0.07679	7.29	<.0001
X2	1	-0.16250	0.07679	-2.12	0.0441
X1X2	1	-0.11350	0.03434	-3.30	0.0028
X1SQ	1	-0.27500	0.04293	-6.41	<.0001
X2SQ	1	-0.23125	0.04293	-5.39	<.0001

c. Why does the SSE given in the computer printout differ from the SSE obtained in Exercise 12.84?

d. Find the value of R^2 appropriate for your second-order model and interpret its value.

e. Do the data provide sufficient evidence to indicate that the complete factorial model provides more information for predicting y than a second-order model?

*12.86 Suppose you want to investigate the effect of two factors—arrival rate of product components, A, and temperature of the room, T—on the length of time, y, required by individual workers to perform a product assembly operation. Each factor will be held at two levels: arrival rate at .5 and 1.0 component per second, and temperature at 70° and 80°F. Thus, a 2×2 factorial experiment will be employed. To block out worker-to-worker variability, each of 10 randomly selected workers will be required to assemble the product under all four experimental conditions. Therefore, the four treatments (working conditions) will be assigned to the experimental units (workers) using a randomized block design, where the workers represent the blocks. The appropriate complete model for the randomized block design is

Treatment effects (main effects and interaction terms for arrival rate and temperature)

$$\text{Complete model}: E(y) = \beta_0 + \overbrace{\beta_1 x_1 + \beta_2 x_2 + \beta_3 x_1 x_2}$$

Block (worker) effects

$$+ \overbrace{\beta_4 x_3 + \beta_5 x_4 + \cdots + \beta_{12} x_{11}}$$

where

x_1 = Arrival rate x_2 = Temperature

$$x_3 = \begin{cases} 1 & \text{if worker 1} \\ 0 & \text{if not} \end{cases}$$

$$x_4 = \begin{cases} 1 & \text{if worker 2} \\ 0 & \text{if not} \end{cases}$$

$$\vdots$$

$$x_{11} = \begin{cases} 1 & \text{if worker 9} \\ 0 & \text{if not} \end{cases}$$

ASSEMBLY2

				WORKER									
				1	2	3	4	5	6	7	8	9	10
	70°F	ARRIVAL	.5	1.7	1.3	1.7	2.0	2.0	2.3	2.0	2.8	1.5	1.6
ROOM		RATE	1.0	.8	.8	1.5	1.2	1.2	1.7	1.1	1.5	.5	1.0
TEMPERATURE		(component	.5	1.3	1.5	2.3	1.6	2.2	2.1	1.8	2.4	1.3	1.8
	80°F	per second)	1.0	1.8	1.5	2.3	2.0	2.7	2.2	2.3	2.6	1.3	1.8

[Note that $x_3 = x_4 = \ldots = x_{11} = 0$ if worker (block) 10 is the assembler.] The assembly time data for the 2×2 factorial experiment with a randomized block design are given in the table above.

a. Use regression to determine if a difference exists among the four treatment means?

b. Does the effect of a change in arrival rate on assembly time depend on temperature (i.e., do arrival rate and temperature interact)?

c. Estimate the mean loss (or gain) in assembly time as arrival rate is increased from .5 to 1.0 component per second and temperature is held at 70°F. What inference can you make based on this estimate?

12.87. Ducks inhabiting the Great Salt Lake marshes feed on a variety of animals, including water boatmen, brine shrimp, beetles, and snails. The changes in the availability of these animal species for ducks during the summer was investigated (*Wetlands*, March 1995). The goal was to compare the mean amount (measured as biomass) of a particular duck food species across four different summer time periods: (1) July 9–23, (2) July 24–Aug. 8, (3) Aug. 9–23, and (4) Aug. 24–31. Ten stations in the marshes were randomly selected, and the biomass density in a water specimen collected from each was measured. Biomass measurements (milligrams per square meter) were collected during each of the four summer time periods at each station, with stations treated as a blocking factor. Thus, the data were analyzed as a randomized block design.

a. Fill in the missing degrees of freedom in the randomized block ANOVA table shown here.

SOURCE	df	F	p-VALUE
Time Period	—	11.25	.0001
Station	—	—	—
Error	—		
Total	39		

b. The F value (and corresponding p-value) shown in the ANOVA table, part **a**, were computed from an analysis of biomass of water boatmen nymphs (a common duck food). Interpret these results.

c. A multiple comparisons of time period means was conducted using an experimentwise error rate of .05. The results are summarized below. Identify the time period(s) with the largest and smallest mean biomass.

Mean biomass (mg/m^2):	19	54.5	90	148
Time period:	8/24–8/31	8/9–8/23	7/24–8/8	7/9–7/23

*12.88 A $2 \times 2 \times 2 \times 2 = 2^4$ factorial experiment was conducted to investigate the effect of four factors on the light output, y, of flashbulbs. Two observations were taken for each of the factorial treatments. The factors are amount of foil contained in a bulb (100 and 120 milligrams); speed of sealing machine (1.2 and 1.3 revolutions per minute); shift (day or night); and machine operator (A or B). The data for the two replications of the 2^4 factorial experiment are shown in the table on p. 650. To simplify computations, define

$$x_1 = \frac{\text{Amount of foil} - 110}{10}$$

$$x_2 = \frac{\text{Speed of machine} - 1.25}{.05}$$

so that x_1 and x_2 will take values -1 and $+1$. Also, define

$$x_3 = \begin{cases} -1 & \text{if night shift} \\ 1 & \text{if day shift} \end{cases} \quad x_4 = \begin{cases} -1 & \text{if machine operator B} \\ 1 & \text{if machine operator A} \end{cases}$$

a. Write the complete factorial model for y as a function of x_1, x_2, x_3, and x_4.

b. How many degrees of freedom will be available for estimating σ^2?

FLASHBULBS

		AMOUNT OF FOIL			
		100 milligrams		120 milligrams	
		SPEED OF MACHINE			
		1.2 rpm	1.3 rpm	1.2 rpm	1.3 rpm
DAY	Operator B	6; 5	5; 4	16; 14	13; 14
SHIFT	Operator A	7; 5	6; 5	16; 17	16; 15
NIGHT	Operator B	8; 6	7; 5	15; 14	17; 14
SHIFT	Operator A	5; 4	4; 3	15; 13	13; 14

c. Do the data provide sufficient evidence (at $\alpha = .05$) to indicate that any of the factors contribute information for the prediction of y?

d. Identify the factors that appear to affect the amount of light y in the flashbulbs.

12.89. A trade-off study regarding the inspection and test of transformer parts was conducted by the quality department of a major defense contractor. The investigation was structured to examine the effects of varying inspection levels and incoming test times to detect early part failure or fatigue. The levels of inspection selected were full military inspection (*A*), reduced military specification level (*B*), and commercial grade (*C*). Operational burn-in test times chosen for this study were at 1-hour increments from 1 hour to 9 hours. The response was failures per thousand pieces obtained from samples taken from lot sizes inspected to a specified level and burned-in over a prescribed time length. Three replications were randomly sequenced under each condition, making this a complete 3×9 factorial experiment (a total of 81 observations). The data are shown in the table below. Analyze the data and interpret the results.

BURNIN

BURN-IN, hours	INSPECTION LEVELS								
	Full Military Specification, A			Reduced Military Specification, B			Commercial, C		
1	7.60	7.50	7.67	7.70	7.10	7.20	6.16	6.13	6.21
2	6.54	7.46	6.84	5.85	6.15	6.15	6.21	5.50	5.64
3	6.53	5.85	6.38	5.30	5.60	5.80	5.41	5.45	5.35
4	5.66	5.98	5.37	5.38	5.27	5.29	5.68	5.47	5.84
5	5.00	5.27	5.39	4.85	4.99	4.98	5.65	6.00	6.15
6	4.20	3.60	4.20	4.50	4.56	4.50	6.70	6.72	6.54
7	3.66	3.92	4.22	3.97	3.90	3.84	7.90	7.47	7.70
8	3.76	3.68	3.80	4.37	3.86	4.46	8.40	8.60	7.90
9	3.46	3.55	3.45	5.25	5.63	5.25	8.82	9.76	9.52

Source: Danny La Nuez, former graduate student, College of Business Administration, University of South Florida.

REFERENCES

Box, G. E. P., Hunter, W. G., and Hunter, J. S. *Statistics for Experimenters*. New York: Wiley, 1978.

Cochran, W. G., and Cox, G. M. *Experimental Designs*, 2nd ed. New York: Wiley, 1957.

Hicks, C. R. *Fundamental Concepts in the Design of Experiments*, 3rd ed. New York: CBC College Publishing, 1982.

Hochberg, Y., and Tamhane, A. C. *Multiple Comparison Procedures*. New York: Wiley, 1987

Hsu, J. C. *Multiple Comparisons, Theory and Methods*. New York: Chapman & Hall, 1996.

Johnson, R., and Wichern, D. *Applied Multivariate Statistical Methods*, 3rd ed. Upper Saddle River, N. J.: Prentice Hall, 1992.

Kirk, R. E. *Experimental Design*, 2nd ed. Belmont, Calif.: Brooks/Cole, 1982.

Kramer, C. Y. "Extension of multiple range tests to group means with unequal number of replications." *Biometrics*, Vol. 12, 1956, pp. 307–310.

Levene, H. *Contributions to Probability and Statistics*. Stanford, Calif.: Stanford University Press, 1960, pp. 278–292.

Mendenhall, W. *Introduction to Linear Models and the Design and Analysis of Experiments*. Belmont, Calif.: Wadsworth, 1968.

Miller, R. G. *Simultaneous Statistical Inference*, 2nd ed. New York: Springer-Verlag, 1981.

Montgomery, D. C. *Design and Analysis of Experiments*, 3rd ed. New York: John Wiley & Sons, 1991.

Neter, J., Kutner, M., Nachtsheim, C., and Wasserman, W. *Applied Linear Statistical Models*, 4th ed. Homewood, Ill.: Richard D. Irwin, 1996.

Scheffe, H. "A method for judging all contrasts in the analysis of variance." *Biometrika*, Vol. 40, 1953, pp. 87–104.

Scheffe, H. *The Analysis of Variance*. New York: Wiley, 1959.

Searle, S. R., Casella, G., and McCulloch, C. E. *Variance Components*. New York: Wiley, 1992.

Tukey, J. W. "Comparing individual means in the analysis of variance." *Biometrics*, Vol. 5, 1949, pp. 99–114.

Uusipaikka, E. "Exact simultaneous confidence intervals for multiple comparisons among three or four mean values," *Journal of the American Statistical Association*, Vol. 80, 1985, pp. 196–201.

Winer, B. J. *Statistical Principals in Experimental Design*, 2nd ed. New York: McGraw-Hill, 1971.

Modeling the Sale Prices of Residential Properties in Four Neighborhoods

CONTENTS

OBJECTIVE

To demonstrate how regression analysis can be used to model and compare the relationship between the sale prices and assessed values of residential properties for four different city neighborhoods

| 13.1 The Problem

This case study concerns a problem of interest to real estate appraisers, tax assessors, real estate investors, and home buyers—namely, the relationship between the appraised value of a property and its sale price. The sale price for any given property will vary depending on the price set by the seller, the strength of appeal of the property to a specific buyer, and the state of the money and real estate markets. Therefore, we can think of the sale price of a specific property as possessing a relative frequency distribution. The mean of this distribution might be regarded as a measure of the fair value of the property. Presumably, this is the value that a property appraiser or a tax assessor would like to attach to a given property.

The purpose of this case study is to examine the relationship between the mean sale price $E(y)$ of a property and the following independent variables:

1. Appraised land value of the property
2. Appraised value of the improvements on the property
3. Neighborhood in which the property is listed

The objectives of the study are twofold:

1. To determine whether the data indicate that appraised values of land and improvements are related to sale prices. That is, do the data supply sufficient evidence to indicate that these variables contribute information for the prediction of sale price?

2. To acquire the prediction equation relating appraised value of land and improvements to sale price and to determine whether this relationship is the same for a variety of neighborhoods. In other words, do the appraisers use the same appraisal criteria for various types of neighborhoods?

13.2
The Data
TAMSALES4

The data for the study were supplied by the property appraiser's office of Hillsborough County, Florida, and consist of the appraised land and improvement values and sale prices for residential properties sold in the city of Tampa, Florida, during 2000. Four neighborhoods (Carrollwood Village, Tampa Palms, Town & Country, and Davis Isles), each relatively homogeneous but differing sociologically and in property types and values, were identified within the city and surrounding area. The subset of sales and appraisal data pertinent to these four neighborhoods—a total of 675 observations—was used to develop a prediction equation relating sale prices to appraised land and improvement values. The data (recorded in thousands of dollars) are saved in the TAMSALES4 file and are fully described in Appendix G.

13.3
The Theoretical Model

FIGURE 13.1

The theoretical relationship between mean sale price and appraised value x

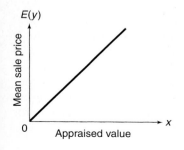

If the mean sale price $E(y)$ of a property, were, in fact, equal to its appraised value, x, the relationship between $E(y)$ and x would be a straight line with slope equal to 1, as shown in Figure 13.1. But does this situation exist in reality? The property appraiser's data could be several years old and consequently may represent (because of inflation) only a percentage of the actual mean sale price. Also, experience has shown that the sale price–appraisal relationship is sometimes curvilinear in nature. One reason is that appraisers have a tendency to overappraise or underappraise properties in specific price ranges, say, very low-priced or very high-priced properties. In fact, it is common for realtors and real estate appraisers to model the natural logarithm of sales price, $\ln(y)$, as a function of appraised value, x. Recall (Section 7.7) that modeling $\ln(y)$ as a linear function of x introduces a curvilinear relationship between y and x.

To gain insight into the sales-appraisal relationship, we used MINITAB to construct a scatterplot of sales price versus total appraised value for all 675 observations in the data set. The plotted points are shown in Figure 13.2, as well as a graph of a straight line fit to the data. Despite the concerns mentioned above, it appears that a linear model will fit the data well. Consequently, we use y = sale price (in thousands of dollars) as the dependent variable and consider only straight-line models. Later in this case study, we will compare the linear model to a model with $\ln(y)$ as the dependent variable.

FIGURE 13.2
MINITAB scatterplot of
sales-appraisal data

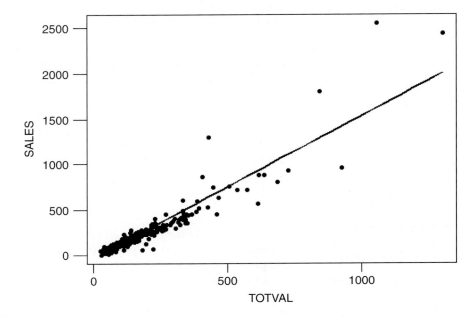

<hr/>

13.4

**The Hypothesized
Regression Models**

We want to relate sale price y to three independent variables: the qualitative factor, neighborhood (four levels), and the two quantitative factors, appraised land value and appraised improvement value. We consider the following four models as candidates for this relationship.

Model 1 is a first-order model that will trace a response plane for mean of sale price, $E(y)$, as a function of $x_1 =$ appraised land value (in thousands of dollars) and $x_2 =$ appraised improvement value (in thousands of dollars). This model will assume that the response planes are identical for all four neighborhoods, i.e., that a first-order model is appropriate for relating y to x_1 and x_2 and that the relationship between the sale price and the appraised value of a property is the same for all neighborhoods. This model is:

Model 1
First-order model, identical
for all neighborhoods

$$
E(y) = \beta_0 + \overbrace{\beta_1 x_1}^{\substack{\text{Appraised land} \\ \text{value}}} + \overbrace{\beta_2 x_2}^{\substack{\text{Appraised} \\ \text{improvement value}}}
$$

In Model 1, we are assuming that the change in sale price y for every \$1,000 (1-unit) increase in appraised land value x_1 (represented by β_1) is constant for a fixed appraised improvements value, x_2. Likewise, the change in y for every \$1,000 increase in x_2 (represented by β_2) is constant for fixed x_1.

Model 2 will assume that the relationship between $E(y)$ and x_1 and x_2 is first-order (a planar response surface), but that the planes' y-intercepts differ depending on the neighborhood. This model would be appropriate if the appraiser's procedure for establishing appraised values produced a relationship between mean sale price and x_1 and x_2 that differed in at least two neighborhoods, but the differences remained constant for different values of x_1 and x_2. Model 2 is

Model 2
First-order model, constant differences between neighborhoods

$$E(y) = \beta_0 + \overbrace{\beta_1 x_1}^{\substack{\text{Appraised land}\\\text{value}}} + \overbrace{\beta_2 x_2}^{\substack{\text{Appraised}\\\text{improvement value}}} + \overbrace{\beta_3 x_3 + \beta_4 x_4 + \beta_5 x_5}^{\substack{\text{Main effect terms}\\\text{for neighborhoods}}}$$

where

$$x_1 = \text{Appraised land value}$$

$$x_2 = \text{Appraised improvement value}$$

$$x_3 = \begin{cases} 1 & \text{if Carrollwood Village neighborhood} \\ 0 & \text{if not} \end{cases}$$

$$x_4 = \begin{cases} 1 & \text{if Davis Isles neighborhood} \\ 0 & \text{if not} \end{cases}$$

$$x_5 = \begin{cases} 1 & \text{if Tampa Palms neighborhood} \\ 0 & \text{if not} \end{cases}$$

The fourth neighborhood, Town & Country, was chosen as the base level. Consequently, the model will predict $E(y)$ for this neighborhood when $x_3 = x_4 = x_5 = 0$. Although it allows for neighborhood differences, Model 2 assumes that change in sale price y for every \$1,000 increase in either x_1 or x_2 does not depend on neighborhood.

Model 3 is similar to Model 2 except that we will add interaction terms between the neighborhood dummy variables and x_1 and between the neighborhood dummy variables and x_2. These interaction terms allow the change in y for increases in x_1 or x_2 to vary depending on the neighborhood. The equation of Model 3 is

Model 3
First-order model, no restrictions on neighborhood differences

$$E(y) = \beta_0 + \overbrace{\beta_1 x_1}^{\substack{\text{Appraised land}\\\text{value}}} + \overbrace{\beta_2 x_2}^{\substack{\text{Appraised improvement}\\\text{value}}}$$

Main effect terms for neighborhoods

$$+ \overbrace{\beta_3 x_3 + \beta_4 x_4 + \beta_5 x_5}$$

Interaction, appraised land by neighborhood

$$+ \overbrace{\beta_6 x_1 x_3 + \beta_7 x_1 x_4 + \beta_8 x_1 x_5}$$

Interaction, appraised improvement by neighborhood

$$+ \overbrace{\beta_9 x_2 x_3 + \beta_{10} x_2 x_4 + \beta_{11} x_2 x_5}$$

Note that for Model 3, the change in sale price y for every \$1,000 increase in appraised land value x_1 (holding x_2 fixed) is (β_1) in Town and Country and $(\beta_1 + \beta_6)$ in Carrollwood Village.

Model 4 differs from the previous three models by the addition of terms for x_1, x_2-interaction. Thus, Model 4 is a second-order (interaction) model that will trace (geometrically) a second-order response surface, one for each neighborhood. The interaction model follows:

Model 4

Interaction model in x_1 and x_2 that differs from one neighborhood to another

$$
E(y) = \overbrace{\beta_0 + \beta_1 x_1 + \beta_2 x_2 + \beta_3 x_1 x_2}^{\text{Interaction model in } x_1 \text{ and } x_2} + \overbrace{\beta_4 x_3 + \beta_5 x_4 + \beta_6 x_5}^{\text{Main effect terms for neighborhoods}}
$$
$$
\left. \begin{array}{l} + \beta_7 x_1 x_3 + \beta_8 x_1 x_4 + \beta_9 x_1 x_5 + \beta_{10} x_2 x_3 \\ + \beta_{11} x_2 x_4 + \beta_{12} x_2 x_5 + \beta_{13} x_1 x_2 x_3 \\ + \beta_{14} x_1 x_2 x_4 + \beta_{15} x_1 x_2 x_5 \end{array} \right\} \begin{array}{l} \text{Interaction terms: } x_1, \\ x_2, \text{ and } x_1 x_2 \text{ terms by} \\ \text{neighborhood} \end{array}
$$

Unlike Models 1–3, Model 4 allows the change in y for increases in x_1 to depend on x_2, and vice versa. For example, the change in sale price for a \$1,000 increase in appraised land value in the base level neighborhood (Town & Country) is $(\beta_1 + \beta_3 x_2)$. Model 4 also allows for these sale price changes to vary from neighborhood to neighborhood (due to the neighborhood interaction terms).

We will fit Models 1–4 to the data. Then, we will compare the models using the nested model F test outlined in Section 4.13. Conservatively, we conduct each test at $\alpha = .01$.

13.5 Model Comparisons

The SAS printouts for Models 1–4 are shown in Figures 13.3–13.6, respectively. These printouts, yield the values of MSE, R_{adj}^2, and s listed in Table 13.1.

FIGURE 13.3
SAS regression printout for Model 1

```
                        Dependent Variable: SALES

                          Analysis of Variance

                                Sum of          Mean
Source               DF        Squares        Square     F Value    Pr > F

Model                 2       22718514      11359257     2865.48    <.0001
Error               672        2663923    3964.17090
Corrected Total     674       25382436


              Root MSE              62.96166    R-Square     0.8950
              Dependent Mean       158.65704    Adj R-Sq     0.8947
              Coeff Var             39.68413


                          Parameter Estimates

                         Parameter      Standard
Variable        DF        Estimate         Error    t Value    Pr > |t|

Intercept        1       -18.14077       3.81517      -4.75      <.0001
LAND             1         2.06956       0.05132      40.32      <.0001
IMP              1         1.13826       0.04411      25.81      <.0001
```

FIGURE 13.4

SAS regression printout for Model 2

```
                            Dependent Variable: SALES
                              Analysis of Variance

                                    Sum of            Mean
Source                    DF        Squares         Square    F Value   Pr > F

Model                      5       22852371        4570474    1208.53   <.0001
Error                    669        2530065     3781.86145
Corrected Total          674       25382436


              Root MSE               61.49684    R-Square      0.9003
              Dependent Mean        158.65704    Adj R-Sq      0.8996
              Coeff Var              38.76087


                              Parameter Estimates

                              Parameter       Standard
       Variable      DF        Estimate          Error    t Value    Pr > |t|

       Intercept      1       -16.14281        3.91237      -4.13     <.0001
       LAND           1         2.12696        0.06363      33.43     <.0001
       IMP            1         1.22741        0.05093      24.10     <.0001
       CWD            1        -7.10714        6.52843      -1.09      0.2767
       DAV            1       -30.78334        9.13640      -3.37      0.0008
       TAM            1       -44.79203        8.30959      -5.39     <.0001

            Test NBHDS Results for Dependent Variable SALES

                                            Mean
         Source                 DF        Square     F Value   Pr > F

         Numerator               3         44619      11.80     <.0001
         Denominator           669    3781.86145
```

TABLE 13.1 **Summary of Regressions of the Models**

MODEL	MSE	R_{adj}^2	S
1	3964	.895	63.0
2	3782	.900	61.5
3	2949	.922	54.3
4	2396	.936	48.9

Test # 1

Model 1 versus Model 2

To test the hypothesis that a single first-order model is appropriate for all neighborhoods, we wish to test the null hypothesis that the neighborhood parameters in Model 2 are all equal to 0, i.e.,

$$H_0: \beta_3 = \beta_4 = \beta_5 = 0$$

That is, we want to compare the complete model, Model 2, to the reduced model, Model 1. The test statistic is

$$F = \frac{(SSE_R - SSE_C)/\text{Number of } \beta \text{ parameters in } H_0}{MSE_C}$$

$$= \frac{(SSE_1 - SSE_2)/3}{MSE_2}$$

FIGURE 13.5

SAS regression printout for
Model 3

Dependent Variable: SALES

Analysis of Variance

Source	DF	Sum of Squares	Mean Square	F Value	Pr > F
Model	11	23427072	2129734	722.12	<.0001
Error	663	1955364	2949.26740		
Corrected Total	674	25382436			

Root MSE		54.30716	R-Square	0.9230	
Dependent Mean		158.65704	Adj R-Sq	0.9217	
Coeff Var		34.22928			

Parameter Estimates

Variable	DF	Parameter Estimate	Standard Error	t Value	Pr > |t|
Intercept	1	2.88102	9.90309	0.29	0.7712
LAND	1	1.16832	0.44459	2.63	0.0088
IMP	1	1.17633	0.16907	6.96	<.0001
CWD	1	20.51098	16.69340	1.23	0.2196
DAV	1	-94.70878	13.17841	-7.19	<.0001
TAM	1	17.37181	14.66131	1.18	0.2365
LAN_CWD	1	0.59055	0.57725	1.02	0.3067
LAN_DAV	1	0.77172	0.44883	1.72	0.0860
LAN_TAM	1	-0.52816	0.57218	-0.92	0.3563
IMP_CWD	1	-0.31284	0.24825	-1.26	0.2080
IMP_DAV	1	0.63365	0.18150	3.49	0.0005
IMP_TAM	1	0.02616	0.20883	0.13	0.9004

Test QN_NBHDS Results for Dependent Variable SALES

Source	DF	Mean Square	F Value	Pr > F
Numerator	6	95784	32.48	<.0001
Denominator	663	2949.26740		

Although the information is available to compute this value by hand, we utilize the SAS option to conduct the nested model F test. The test statistic value, shaded at the bottom of Figure 13.4, is $F = 11.80$. The p-value of the test, also shaded, is less than .0001. Since $\alpha = .01$ exceeds this p-value, we have evidence to indicate that the addition of the neighborhood dummy variables in Model 2 contributes significantly to the prediction of y. The practical implication of this result is that the appraiser is not assigning appraised values to properties in such a way that the first-order relationship between (sales), y, and appraised values x_1 and x_2 is the same for all neighborhoods.

Test # 2
Model 2 versus Model 3

Can the prediction equation be improved by the addition of interactions between neighborhood and x_1 and neighborhood and x_2? That is, do the data provide sufficient evidence to indicate that Model 3 is a better predictor of sale price than Model 2? To answer this question, we will test the null hypothesis that the parameters associated with all neighborhood interaction terms in Model 3 equal 0. Thus, Model 2 is now the reduced model and Model 3 is the complete model.

Checking the equation of Model 3, you will see that there are six neighborhood interaction terms and that the parameters included in H_0 will be

$$H_0: \beta_6 = \beta_7 = \beta_8 = \beta_9 = \beta_{10} = \beta_{11} = 0$$

FIGURE 13.6

SAS regression printout for Model 4

Dependent Variable: SALES

Analysis of Variance

Source	DF	Sum of Squares	Mean Square	F Value	Pr > F
Model	15	23803546	1586903	662.34	<.0001
Error	659	1578890	2395.88786		
Corrected Total	674	25382436			

Root MSE	48.94781	R-Square	0.9378	
Dependent Mean	158.65704	Adj R-Sq	0.9364	
Coeff Var	30.85133			

Parameter Estimates

Variable	DF	Parameter Estimate	Standard Error	t Value	Pr > |t|
Intercept	1	4.36822	24.73371	0.18	0.8599
LAND	1	1.09236	1.24446	0.88	0.3804
IMP	1	1.14927	0.44648	2.57	0.0103
CWD	1	9.16488	34.87611	0.26	0.7928
DAV	1	13.50801	27.40135	0.49	0.6222
TAM	1	-1.23297	29.26886	-0.04	0.9664
LAN_CWD	1	0.84012	1.34194	0.63	0.5315
LAN_DAV	1	0.06131	1.24729	0.05	0.9608
LAN_TAM	1	-0.13275	1.30625	-0.10	0.9191
IMP_CWD	1	-0.16222	0.54600	-0.30	0.7665
IMP_DAV	1	-0.01202	0.45366	-0.03	0.9789
IMP_TAM	1	0.12073	0.46248	0.26	0.7941
LAN_IMP	1	0.00131	0.02028	0.06	0.9486
L_I_CWD	1	-0.00352	0.02084	-0.17	0.8658
L_I_DAV	1	0.00164	0.02028	0.08	0.9355
L_I_TAM	1	-0.00222	0.02029	-0.11	0.9129

Test LAND_IMP Results for Dependent Variable SALES

Source	DF	Mean Square	F Value	Pr > F
Numerator	4	94119	39.28	<.0001
Denominator	659	2395.88786		

To test H_0, we require the test statistic

$$F = \frac{(\text{SSE}_R - \text{SSE}_C)/\text{Number of } \beta \text{ parameters in } H_0}{\text{MSE}_C} = \frac{(\text{SSE}_2 - \text{SSE}_3)/6}{\text{MSE}_3}$$

This value, $F = 32.48$, is shaded at the bottom of the SAS printout for Model 3, Figure 13.5. The p-value of the test (also shaded) is less than .0001. Thus, there is sufficient evidence (at $\alpha = .01$) to indicate that the neighborhood interaction terms of Model 3 contribute information for the prediction of y. Practically, this test implies that the rate of change of sale price y with either appraised value, x_1 or x_2, differs for each of the four neighborhoods.

Test # 3

Model 3 versus Model 4

We have already shown that the first-order prediction equations vary among neighborhoods. To determine whether the (second-order) interaction terms involving the appraised values, x_1 and x_2, contribute significantly to the prediction of y, we test the hypothesis that the four parameters involving x_1x_2 in Model 4 all equal 0. The null hypothesis is

$$H_0: \beta_3 = \beta_{13} = \beta_{14} = \beta_{15} = 0$$

and the alternative hypothesis is that at least one of these parameters does not equal 0. Using Model 4 as the complete model and Model 3 as the reduced model, the test statistic required is:

$$F = \frac{(\text{SSE}_R - \text{SSE}_C)/\text{Number of } \beta \text{ parameters in } H_0}{\text{MSE}_C} = \frac{(\text{SSE}_3 - \text{SSE}_4)/4}{\text{MSE}_4}$$

This value (shaded at the bottom of the SAS printout for Model 4, Figure 13.6) is $F = 39.28$. Again, the p-value of the test (shaded) is less than .0001. This small p-value supports the alternative hypothesis that the x_1x_2 interaction terms of Model 4 contribute significantly to the prediction of y.

The results of the preceding tests suggest that Model 4 is the best of the four models for modeling sale price y. The global F value for testing

$$H_0: \beta_1 = \beta_2 = \cdots = \beta_{15} = 0$$

is highly significant ($F = 662.34$, p-value $< .0001$); the R^2 value indicates that the model explains almost 94% of the variability in sale price. You may notice that several of the t tests involving the individual β parameters in Model 4 are nonsignificant. Be careful not to conclude that these terms should be dropped from the model.

Whenever a model includes a large number of interactions (as in Model 4) and/or squared terms, several t tests will often be nonsignificant even if the global F test is highly significant. This result is due partly to the unavoidable intercorrelations among the main effects for a variable, its interactions, and its squared terms (see the discussion on multicollinearity in Section 7.5). We warned in Chapter 4 of the dangers of conducting a series of t tests to determine model adequacy. For a model with a large number of β's, such as Model 4, you should avoid conducting any t tests at all and rely on the global F test and partial F tests to determine the important terms for predicting y.

Before we proceed to estimation and prediction with Model 4, we conduct one final test. Recall our discussion of the theoretical relationship between sale price and appraised value in Section 13.2. Will a model with the natural log of y as the dependent variable outperform Model 4? To check this, we fit a model identical to Model 4—call it Model 5—but with $y^* = \ln(y)$ as the dependent variable. A portion of the SAS printout for Model 5 is displayed in Figure 13.7. From the printout, we obtain $R^2 = .89$, $s = .222$, and global $F = 35.34$ (p-value $< .0001$). Model 5 is clearly "statistically" useful for predicting sale price based on the global F test. However, we must be careful not to judge which of the two models is better based on the values of R^2 and s shown on the printouts because the dependent variable is not the same for each model. (Recall the "warning" given at the end of Section 4.11.) An informal procedure for comparing the two models requires that we obtain predicted values for $\ln(y)$ using the prediction equation for Model 5, transform these predicted

FIGURE 13.7

Portion of SAS regression printout for Model 5

```
                    Dependent Variable: LNSALES

                        Analysis of Variance

                         Sum of          Mean
Source           DF      Squares         Square      F Value    Pr > F
Model            15     265.84292       17.72286     359.34    <.0001
Error           659      32.50250        0.04932
Corrected Total 674     298.34542


      Root MSE              0.22208     R-Square     0.8911
      Dependent Mean        4.78920     Adj R-Sq     0.8886
      Coeff Var             4.63717
```

log values back to sale prices using the equation, $e^{\widehat{ln(y)}}$, then recalculate the R^2 and s values for Model 5 using the formulas given in Chapter 4. We used the programming language of SAS to compute these values; they are compared to the values of R^2 and s for Model 4 in Table 13.2.

You can see that Model 4 outperforms Model 5 with respect to both statistics. Model 4 has a slightly higher R^2 value and a lower model standard deviation. These results support our decision to build a model with sale price y rather than $\ln(y)$ as the dependent variable.

TABLE 13.2   Comparison of R^2 and s for Models 4 and 5

	R^2	s
Model 4:	.938 (from Figure 13.6)	48.9 (from Figure 13.6)
Model 5:	.914 (using transformation)	57.4 (using transformation)

13.6 Interpreting the Prediction Equation

Substituting the estimates of the Model 4 parameters (Figure 13.6) into the prediction equation we have

$$\hat{y} = 4.368 + 1.092x_1 + 1.149x_2 + .0013x_1x_2 + 9.165x_3 + 13.508x_4$$
$$- 1.233x_5 + .840x_1x_3 + .061x_1x_4 - .133x_1x_5 - .162x_2x_3$$
$$- .012x_2x_4 + .121x_2x_5 - .0035x_1x_2x_3 + .0016x_1x_2x_4$$
$$- .6022x_1x_2x_5$$

We have noted that the model yields four response surfaces, one for each neighborhood. One way to interpret the prediction equation is to first find the equation of the response surface for each neighborhood. Substituting the appropriate values of the neighborhood dummy variables, x_3, x_4, and x_5, into the equation and combining like terms, we obtain the following:

Carrollwood Village: $(x_3 = 1, x_4 = x_5 = 0)$

$$\hat{y} = (4.368 + 9.165) + (1.092 + .840)x_1 + (1.149 - .162)x_2$$
$$+(.0013 - .0035)x_1x_2$$
$$= 13.533 + 1.932x_1 + .987x_2 - .0022x_1x_2$$

Davis Isles: $(x_3 = 0, x_4 = 1, x_5 = 0)$

$$\hat{y} = (4.368 + 13.508) + (1.092 + .061)x_1 + (1.149 - .012)x_2$$
$$+(.0013 + .0016)x_1x_2$$
$$= 17.876 + 4.429x_1 + 1.137x_2 + .0029x_1x_2$$

Tampa Palms: $(x_3 = x_4 = 0, x_5 = 1)$

$$\hat{y} = (4.638 - 1.233) + (1.092 - .133)x_1 + (1.149 + .121)x_2$$
$$+(.0013 - .0022)x_1x_2$$
$$= 3.135 + 4.235x_1 + 1.27x_2 - .0009x_1x_2$$

Town & Country: $(x_3 = x_4 = x_5 = 0)$

$$\hat{y} = 4.368 + 1.092x_1 + 1.149x_2 + .0013x_1x_2$$

Note that each equation is in the form of an interaction model involving appraised land value x_1 and appraised improvements x_2. To interpret the β estimates of each interaction equation, we hold one independent variable fixed, say, x_1, and focus on the slope of the line relating y to x_2. For example, holding appraised land value constant at \$30,000 $(x_1 = 30)$, the slope of the line relating y to x_2 for Town & Country (the base level neighborhood) is

$$\hat{\beta}_2 + \hat{\beta}_3x_1 = 1.149 + .0013(30) = 1.188$$

Thus, for residential properties in Town & Country with appraised land value of \$30,000, the sale price will increase \$1,188 for every \$1,000 increase in appraised improvements.

Similar interpretations can be made for the slopes for other combinations of neighborhoods and appraised land value x_1. The estimated slopes for several of these combinations are computed and shown in Table 13.3. Because of the interaction terms in the model, the increases in sale price for a \$1,000 increase in appraised improvements, x_2, differ for each neighborhood and for different levels of land value, x_1.

Some trends are evident from Table 13.3. For fixed appraised land value, x_1, the increase in sales price for every \$1,000 increase in appraised improvements is smallest for Carrollwood Village. For the Carrollwood Village and Tampa Palms neighborhoods, the slope decreases as appraised land value increases; but for the Davis Isles and Town & Country neighborhoods, the slope increases as appraised land value increases.

TABLE 13.3   Estimated Dollar Increase in Sale Price for $1,000 Increase in Appraised Improvements

		NEIGHBORHOOD			
		CARROLLWOOD VILLAGE	DAVIS ISLES	TAMPA PALMS	TOWN & COUNTRY
APPRAISED LAND VALUE	$30,000	1,026	1,224	1,243	1,188
	$50,000	877	1,282	1,225	1,214
	$70,000	833	1,340	1,207	1,240

Another way to describe the prediction equation would be to graph predicted sale price for each neighborhood as a function of appraised improvements value x_2 for different levels of appraised land value x_1. The lines (one line for each neighborhood) are shown in Figure 13.8 for appraised land value $x_1 = \$30,000$. Similar sets of lines are shown in Figures 13.9–13.11 for $x_1 = \$50,000, \$60,000$, and $\$100,000$, respectively. Note that not all figures have lines corresponding to all four neighborhoods. This is because the appraised land values for sales in some neighborhoods might not have been as low as $30,000 or as high as $100,000. Some lines are shortened for similar reasons; i.e., the appraised improvements values for some neighborhoods might not span the range shown on the x_2-axis.

Before we examine the curves in Figures 13.8–13.11, we want some information concerning the nature of the neighborhoods. The SAS printout in Figure 13.12 gives the means, standard deviations, and other descriptive statistics for the sale price y and the appraised values x_1 and x_2 for each of the four neighborhoods. The mean sale prices confirm what the authors know to be true, i.e., that neighborhoods Davis Isles and Tampa Palms, are two of the relatively expensive residential areas in the city. Most of the inhabitants are older, established professional or business people. In contrast, Carrollwood Village and Town & Country are moderately priced residential areas inhabited primarily by either young married couples who are starting their careers. or older, retired couples.

In Figures 13.8–13.11, the estimated mean sale price $\hat{y}$ increases as the appraised improvements value x_2 increases, but the slope is not always the same. The more expensive neighborhoods tend to have the steeper slope, and their estimated lines tend to be above the lines of the less expensive neighborhoods.

There are some other interesting features to note. For example, in Figure 13.10, observe the different predicted sale prices for an appraised improvement value of $100,000. You can see that properties in Carrollwood Village (a moderately-priced neighborhood) sell at a higher predicted price than those in the more expensive Tampa Palms neighborhood when the appraised land value is $x_1 = \$60,000$. This would suggest that the appraised values of properties in this price range in Tampa Palms are too high compared with similar properties in Carrollwood Village. Perhaps a low appraised property value in Tampa Palms corresponds to a very small lot; this might have a strong depressive effect on sale prices.

FIGURE 13.8
SAS graph of predicted sale price for land value
of $30,000

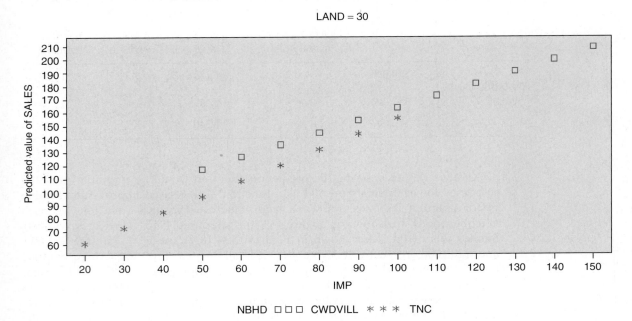

FIGURE 13.9
SAS graph of predicted sale price for land value
of $50,000

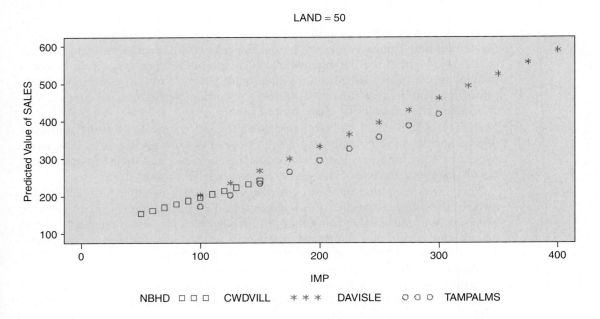

FIGURE 13.10

SAS graph of predicted sale price for land value
of $60,000

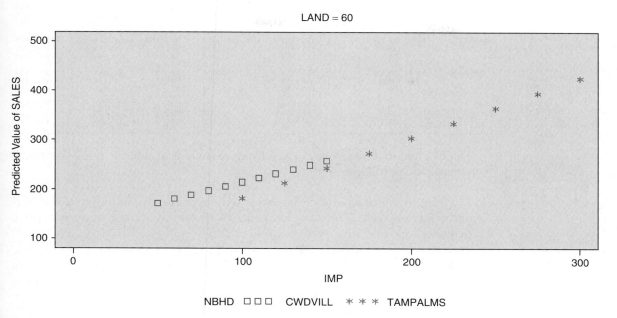

FIGURE 13.11

SAS graph of predicted sale price for land value
of $100,000

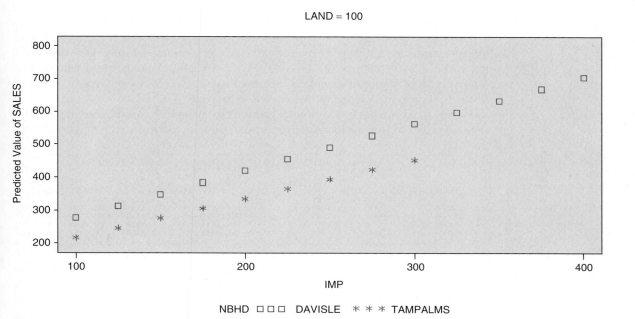

FIGURE 13.12

SAS descriptive statistics printout, by neighborhood

The MEANS Procedure

NBHD	N Obs	Variable	Mean	Std Dev	N	Minimum	Maximum
CWDVILL	138	SALES	158.648	63.924	138	34.300	407.000
		LAND	27.766	18.819	138	0.100	83.475
		IMP	100.081	38.117	138	40.376	279.491
DAVISLE	92	SALES	368.877	421.261	92	65.800	2550.000
		LAND	127.421	120.849	92	0.100	758.331
		IMP	117.958	112.664	92	23.485	567.349
TAMPALMS	92	SALES	236.558	139.461	92	59.000	957.500
		LAND	52.055	33.115	92	0.100	258.790
		IMP	152.169	97.302	92	54.166	670.606
TNC	353	SALES	83.570	28.679	353	5.500	195.000
		LAND	17.094	6.905	353	0.100	42.294
		IMP	51.615	18.159	353	14.049	140.755

The two major points to be derived from an analysis of the sale price–appraised value lines are as follows:

1. The rate at which sale price increases with appraised value differs for different neighborhoods. This increase tends to be largest for the more expensive neighborhoods.
2. The lines for Carrollwood Village frequently lie above those for Tampa Palms, indicating that properties in the higher-priced Tampa Palms neighborhood are being underappraised relative to sale price compared with properties in the lower-priced Carrollwood Village neighborhood.

13.7
Predicting the Sale Price of a Property

How well do appraised land value x_1 and appraised improvements value x_2 predict residential property sale price? Recall that from Model 4 (Figure 13.6), we obtained $R^2 = .938$, indicating that the model accounts for approximately 94% of the sample variability in the sale price values, y. This seems to indicate that the model provides a reasonably good fit to the data, but note that $s = 48.9$ (see Figure 13.6). Our interpretation is that approximately 95% of the predicted sale price values will fall within $(2s)(\$1,000) = (2)(48.9)(\$1,000) = \$97,800$ of their actual values. This relatively large standard deviation may lead to large errors of prediction for some residential properties if the model is used in practice.

Figure 13.13 is a portion of a SAS printout showing 95% prediction intervals for the sale prices of four residential properties, one selected from each neighborhood.

FIGURE 13.13

A portion of the SAS printout showing 95% prediction intervals for sale price

Output Statistics

Obs	NBRHOOD	LAND	IMP	Dep Var SALES	Predicted Value	Std Error Mean Predict	95% CL Predict	
1	CWDVILL	40	100	.	180.6758	5.9591	83.8537	277.497
2	DAVISLE	180	250	.	642.4968	9.8019	544.4762	740.517
3	TAMPALMS	75	200	.	315.4166	7.9332	218.0500	412.783
4	TNC	20	60	.	96.7404	3.2550	0.4157	193.065

Note the large widths of the intervals. These wide prediction intervals cast doubt on whether the prediction equation could be of practical value in predicting property sale prices. We feel certain that a much more accurate predictor of sale price could be developed by relating y to the variables that describe the property (such as location, square footage, and number of bedrooms) and those that describe the market (mortgage interest rates, availability of money, and so forth).

13.8
Conclusions

The results of the regression analyses described in Section 13.5 indicate that the relationships between property sale prices and appraised values are not consistent from one neighborhood to another. Further, the widths of the prediction intervals in Section 13.7 are rather sizable, indicating that there is room for improvement in the methods used to determine appraised property values.

EXERCISES

13.1. Explain why the tests of model adequacy conducted in Section 13.5 give no assurance that Model 4 will be a successful predictor of sale price in the future.

13.2. Use the data-splitting technique of Section 5.11 to assess the external validity of Model 4.

TAMSALES7

13.3. Recall that the data for this case study are described in Appendix G. The full data set, named TAMSALES7,

contains sale price information for the four neighborhoods compared in this case study, as well as sale price information for three additional neighborhoods (Avila, Temple Terrace, and Ybor city). Use the full data set to build a regression model for sale price of a residential property. Part of your analysis will involve a comparison of the seven neighborhoods.

An Analysis of Rain Levels in California

CONTENTS

OBJECTIVE

To illustrate how a residual analysis can be used to detect an important omitted variable in regression

---

14.1 The Problem

For this case study, we focus on an application of regression analysis in the science of geography. Writing in the journal *Geography* (July 1980), P. J. Taylor sought to describe the method of multiple regression to the research geographer "in a completely nontechnical manner." Taylor chose to investigate the variation in average annual precipitation in California—"a typical research problem that would be tackled using multiple regression analysis." In this chapter, we use Taylor's data to build a model for average annual precipitation, y. Then we examine the residuals, the deviations between the predicted and the actual precipitation levels, to detect (as Taylor did) an important independent variable omitted from the regression model.

14.2 The Data

The state of California operates numerous meteorological stations. One of the many functions of each station is to monitor rainfall on a daily basis. This information is then used to produce an average annual precipitation level for each station.

Table 14.1 lists average annual precipitation levels (in inches) for a sample of 30 meteorological stations scattered throughout the state. (These are the data analyzed by Taylor.) In addition to average annual precipitation (y), the table lists three

⊙ CALIRAIN

TABLE 14.1   Data for 30 Meteorological Stations in California

STATION	AVERAGE ANNUAL PRECIPITATION y, inches	ALTITUDE x_1, feet	LATITUDE x_2, degrees	DISTANCE FROM COAST x_3, miles
1. Eureka	39.57	43	40.8	1
2. Red Bluff	23.27	41	40.2	97
3. Thermal	18.20	4,152	33.8	70
4. Fort Bragg	37.48	74	39.4	1
5. Soda Springs	49.26	6,752	39.3	150
6. San Francisco	21.82	52	37.8	5
7. Sacramento	18.07	25	38.5	80
8. San Jose	14.17	95	37.4	28
9. Giant Forest	42.63	6,360	36.6	145
10. Salinas	13.85	74	36.7	12
11. Fresno	9.44	331	36.7	114
12. Pt. Piedras	19.33	57	35.7	1
13. Paso Robles	15.67	740	35.7	31
14. Bakersfield	6.00	489	35.4	75
15. Bishop	5.73	4,108	37.3	198
16. Mineral	47.82	4,850	40.4	142
17. Santa Barbara	17.95	120	34.4	1
18. Susanville	18.20	4,152	40.3	198
19. Tule Lake	10.03	4,036	41.9	140
20. Needles	4.63	913	34.8	192
21. Burbank	14.74	699	34.2	47
22. Los Angeles	15.02	312	34.1	16
23. Long Beach	12.36	50	33.8	12
24. Los Banos	8.26	125	37.8	74
25. Blythe	4.05	268	33.6	155
26. San Diego	9.94	19	32.7	5
27. Daggett	4.25	2,105	34.09	85
28. Death Valley	1.66	−178	36.5	194
29. Crescent City	74.87	35	41.7	1
30. Colusa	15.95	60	39.2	91

independent variables that are believed (by California geographers) to have the most impact on the amount of rainfall at each station, as follows:

1. Altitude of the station (x_1, feet)
2. Latitude of the station (x_2, degrees)
3. Distance of the station from the Pacific coast (x_3, miles)

14.3
A Model for Average Annual Precipitation

As an initial attempt in explaining the average annual precipitation in California, Taylor considered the following first-order model:

Model 1:
$$E(y) = \beta_0 + \beta_1 x_1 + \beta_2 x_2 + \beta_3 x_3$$

Model 1 assumes that the relationship between average annual precipitation y and each independent variable is linear, and the effect of each x on y is independent of the other x's (i.e., no interaction).

The model is fit to the data of Table 14.1, resulting in the SPSS printout shown in Figure 14.1. The key numbers on the printout are shaded and interpreted as follows.

Global $F = 13.016$ (p-value $= .000$): At any significant level $\alpha > .0001$, we reject the null hypothesis H_0: $\beta_1 = \beta_2 = \beta_3 = 0$. Thus, there is sufficient evidence to indicate that the model is "statistically" useful for predicting average annual precipitation, y.

$R^2_{\text{adj}} = .554$: After accounting for sample size and number of β parameters in the model, approximately 55% of the sample variation in average annual precipitation levels is explained by the first-order model with altitude (x_1), latitude (x_2), and distance from Pacific coast (x_3).

$s = 11.09799$: Approximately 95% of the actual average annual precipitation levels at the stations will fall within $2s = 22.2$ inches of the values predicted by the first-order model.

FIGURE 14.1

SPSS regression printout for Model 1

Model Summary[b]

Model	R	R Square	Adjusted R Square	Std. Error of the Estimate
1	.775[a]	.600	.554	11.09799

a. Predictors: (Constant), DISTANCE, LATITUDE, ALTITUDE

b. Dependent Variable: RAIN

ANOVA[b]

Model		Sum of Squares	df	Mean Square	F	Sig.
1	Regression	4809.356	3	1603.119	13.016	.000[a]
	Residual	3202.298	26	123.165		
	Total	8011.654	29			

a. Predictors: (Constant), DISTANCE, LATITUDE, ALTITUDE

b. Dependent Variable: RAIN

Coefficients[a]

Model		Unstandardized Coefficients		Standardized Coefficients	t	Sig.
		B	Std. Error	Beta		
1	(Constant)	-102.357	29.205		-3.505	.002
	ALTITUDE	4.091E-03	.001	.516	3.358	.002
	LATITUDE	3.451	.795	.554	4.342	.000
	DISTANCE	-.143	.036	-.596	-3.931	.001

a. Dependent Variable: RAIN

$\hat{\beta}_1 = .00409$: Holding latitude (x_2) and distance from coast (x_3) constant, we estimate average annual precipitation (y) of a station to increase .0041 inch for every 1-foot increase in the station's altitude (x_1).

$\hat{\beta}_2 = 3.451$: Holding altitude (x_1) and distance from coast (x_3) constant, we estimate average annual precipitation (y) of a station to increase 3.45 inches for every 1-degree increase in the station's latitude (x_2).

$\hat{\beta}_3 = -.143$: Holding altitude (x_1) and latitude (x_2) constant, we estimate average annual precipitation (y) of a station to decrease .143 inch for every 1-mile increase in the station's distance from the Pacific coast (x_3).

Note also that t tests for the three independent variables in the model are all highly significant (p-value $< .01$). Therefore, it appears that the first-order model is adequate for predicting a meteorological station's average annual precipitation.

Can we be certain, without further analysis, that additional independent variables or higher-order terms will not improve the prediction equation? The answer, of course, is no. In the next section, we use a residual analysis to help guide us to a better model.

14.4
A Residual Analysis of the Model

The residuals of Model 1 are analyzed using the graphs discussed in Chapter 8. The SPSS printout in Figure 14.2 shows both a histogram and a normal probability plot for the standardized residuals. Both graphs appear to support the regression assumption of normally distributed errors.

The SPSS printouts shown in Figures 14.3 and are plots of the residuals versus predicted rainfall, $\hat{y}$, and against each of the independent variables. Other than one or two unusual observations (outliers), the plots exhibit no distinctive patterns or trends. Consequently, transformations on the independent variables for the purposes of improving the fit of the model or for stabilizing the error variance do not seem to be necessary.

On the surface, the residual plots seem to imply that no adjustments to the first-order model can be made to improve the prediction equation. Taylor, however, used his knowledge of geography and regression to examine Figure 14.3 more closely. He found that the residuals shown in Figure 14.3 actually exhibit a fairly consistent pattern. Taylor noticed that stations located on the west-facing slopes of the California mountains invariably had positive residuals whereas stations on the leeward side of the mountains had negative residuals.

To see what Taylor observed more clearly, we plotted the residuals of Model 1 against $\hat{y}$ using either "W" or "L" as a plotting symbol. Stations numbered 1, 4, 5, 6, 9, 12, 16, 17, 21, 22, 23, 26, and 29 in Table 14.1 were assigned a "W" since they all are located on west-facing slopes, whereas the remaining stations were assigned an "L" since they are leeward-facing. The revised residual plot, with a "W" or "L" assigned to each point, is shown in the SPSS printout, Figure 14.4. You can see that with few exceptions, the "W" points have positive residuals (implying that the least squares model underpredicted the level of precipitation), whereas the "L" points have negative residuals (implying that the least squares model overpredicted the level of precipitation). In Taylor's words, the results shown in Figure 14.4 "suggest a very clear shadow effect of the mountains, for which California is known." Thus, it appears we

FIGURE 14.2

SPSS histogram and normal
probability plot of Model 1
residuals

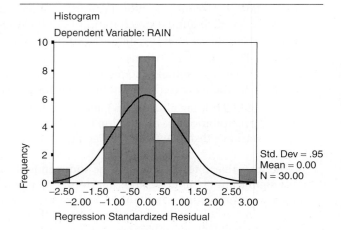

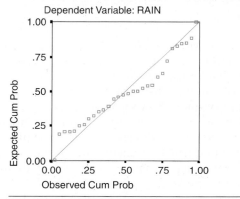

can improve the fit of the model by adding a variable that represents the
shadow effect.

14.5
Adjustments
to the Model

To account for the shadow effect of the California mountains, consider the dummy
variable

$$\text{Shadow: } x_4 = \begin{cases} 1 & \text{if station on the leeward side} \\ 0 & \text{if station on the westward side} \end{cases}$$

The model with the shadow effect takes the form

Model 2: $\qquad\qquad\qquad E(y) = \beta_0 + \beta_1 x_1 + \beta_2 x_2 + \beta_3 x_3 + \beta_4 x_4$

Model 2, like Model 1, allows for straight-line relationships between precipitation
and altitude (x_1), precipitation and latitude (x_2), and precipitation and distance
from coast (x_3). The y-intercepts of these lines, however, will depend on the shadow
effect (i.e., whether the station is leeward or westward).

FIGURE 14.3

SPSS residual plots for Model 1

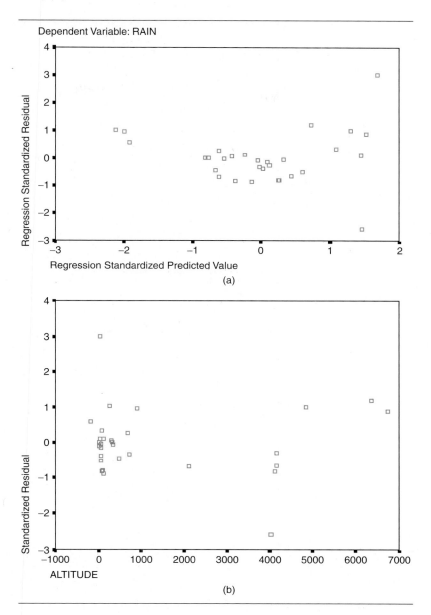

The SAS printout for Model 2 is shown in Figure 14.5. Note that the adjusted R^2 for Model 2 is .6963—an increase of about 15% from Model 1. This implies that the shadow-effect model (Model 2) explains about 15% more of the sample variation in average annual precipitation than the no-shadow-effect model (Model 1).

Is this increase a statistically significant one? To answer this question, we test the contribution for the shadow effect by testing

$$H_0: \beta_4 = 0 \quad \text{against} \quad H_a: \beta_4 \neq 0$$

FIGURE 14.3
continued

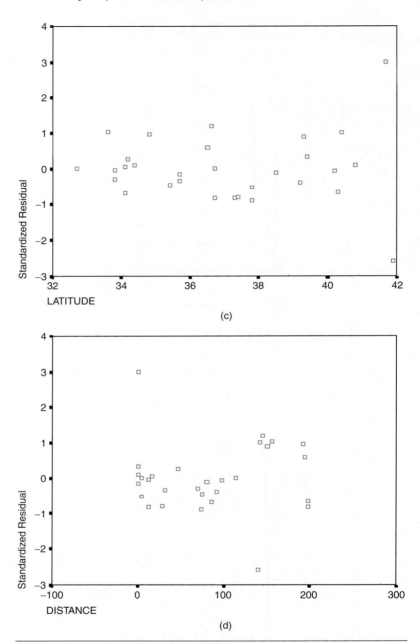

(c)

(d)

The test statistic, shaded on Figure 14.5, is $t = -3.63$ and the two-tailed p-value (also shaded) is $p = .0013$. Thus, there is sufficient evidence (at $\alpha = .01$) to conclude that $\beta_4 \neq 0$; i.e., the shadow-effect term contributes to the prediction of average annual precipitation.

FIGURE 14.4

SPSS plot of residuals for Model 1
with shadow effect

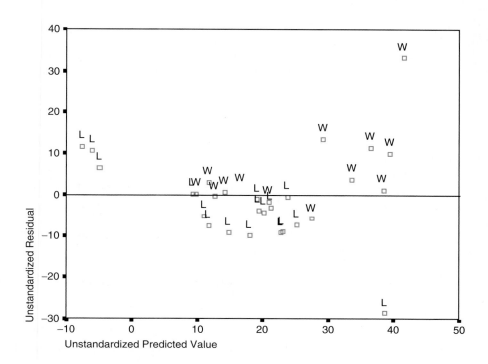

FIGURE 14.5

SAS regression printout for Model 2

```
                        Dependent Variable: RAIN

                         Analysis of Variance

                                Sum of          Mean
Source               DF       Squares         Square    F Value    Pr > F

Model                 4    5913.81738     1478.45434      17.62    <.0001
Error                25    2097.83621       83.91345
Corrected Total      29    8011.65359


           Root MSE                9.16043    R-Square      0.7382
           Dependent Mean         19.80733    Adj R-Sq      0.6963
           Coeff Var              46.24766


                        Parameter Estimates

                     Parameter      Standard
Variable        DF    Estimate         Error    t Value    Pr > |t|

Intercept        1   -97.89872      24.13791      -4.06      0.0004
ALTITUDE         1     0.00221       0.00113       1.95      0.0627
LATITUDE         1     3.45376       0.65609       5.26      <.0001
DISTANCE         1    -0.05365       0.03879      -1.38      0.1789
SHADOW           1   -15.85771       4.37100      -3.63      0.0013
```

Can Model 2 be improved by adding interaction terms? Consider Model 3:

Model 3: $E(y) = \beta_0 + \beta_1 x_1 + \beta_2 x_2 + \beta_3 x_3 + \beta_4 x_4 + \beta_5 x_1 x_4 + \beta_6 x_2 x_4 + \beta_7 x_3 x_4$

Note that Model 3 includes interactions between the shadow effect (x_4) and each of the quantitative independent variables. This model allows the slopes of the lines relating y to x_1, y to x_2, and y to x_3 to depend on the shadow effect (x_4). The SAS printout for Model 3 is given in Figure 14.6.

FIGURE 14.6

SAS printout for Model 3

Dependent Variable: RAIN

Analysis of Variance

Source	DF	Sum of Squares	Mean Square	F Value	Pr > F
Model	7	6921.64039	988.80577	19.96	<.0001
Error	22	1090.01319	49.54605		
Corrected Total	29	8011.65359			

Root MSE	7.03890	R-Square	0.8639	
Dependent Mean	19.80733	Adj R-Sq	0.8207	
Coeff Var	35.53682			

Parameter Estimates

Variable	DF	Parameter Estimate	Standard Error	t Value	Pr > |t|
Intercept	1	-160.70358	25.78066	-6.23	<.0001
ALTITUDE	1	0.00453	0.00418	1.08	0.2897
LATITUDE	1	5.14128	0.69811	7.36	<.0001
DISTANCE	1	-0.13008	0.17571	-0.74	0.4669
SHADOW	1	127.14457	37.57712	3.38	0.0027
SHAD_ALT	1	-0.00372	0.00433	-0.86	0.3992
SHAD_LAT	1	-3.78713	1.01911	-3.72	0.0012
SHAD_DIS	1	0.07079	0.17842	0.40	0.6954

Test SHAD_INTERACT Results for Dependent Variable RAIN

Source	DF	Mean Square	F Value	Pr > F
Numerator	3	335.94100	6.78	0.0021
Denominator	22	49.54605		

To determine whether these interaction terms are important, we test

H_0: $\beta_5 = \beta_6 = \beta_7 = 0$

H_a: At least one of the β's $\neq 0$

The test is carried out by comparing Models 2 and 3 with the nested model partial F test of Section 4.13. The F test statistic, shaded at the bottom of Figure 14.6, is $F = 6.78$ and the associated p-value (also shaded) is $p = .0021$.

Consequently, there is sufficient evidence (at $\alpha = .01$) to reject H_0 and conclude that at least one of the interaction β's is nonzero. This implies that Model 3, with the interaction terms, is a better predictor of average annual precipitation than Model 2.

The improvement of Model 3 over the other two models can be seen practically by examining R^2_{adj} and s on the printout, Figure 13.7. For Model 3, $R^2_{adj} = .8207$, an increase of about 12% from Model 2 and 27% from Model 1. The standard deviation of Model 3 is $s = 7.04$, compared to $s = 9.16$ for Model 2 and $s = 11.1$ for Model 1. Thus, in practice, we expect Model 3 to predict average annual precipitation of a meteorological station to within about 14 inches of its true value. (This is compared to a bound on the error of prediction of about 22 inches for Model 1 and 18 inches for Model 2.) Clearly, a model that incorporates the shadow effect and its interactions with altitude, latitude, and distance from coast is a more useful predictor of average annual precipitation, y.

14.6 Conclusions

We have demonstrated how a residual analysis can help the analyst find important independent variables that were originally omitted from the regression model. This technique, however, requires substantial knowledge of the problem, data, and potentially important predictor variables. Without knowledge of the presence of a shadow effect in California, the geographer Taylor could not have enhanced the residual plot, Figure 14.3, and consequently would not have seen its potential for improving the fit of the model.

EXERCISES

14.1. Conduct an outlier analysis of the residuals for Model 1. Identify any influential observations, and suggest how to handle these observations. (The data are saved in the CALIRAIN file.)

14.2. Determine whether interactions between the quantitative variables, altitude (x_1), latitude (x_2), and distance from coast (x_3) will improve the fit of the model.

REFERENCES

TAYLOR, P. J. "A pedagogic application of multiple regression analysis." *Geography*, July 1980, Vol. 65, pp. 203–212.

Reluctance to Transmit Bad News: The MUM Effect

CONTENTS

OBJECTIVE

To present a designed experiment that investigates the effects of two manipulated factors on the reluctance of human subjects to transmit bad news to others

---

15.1
The Problem

In a 1970 experiment, psychologists S. Rosen and A. Tesser found that people were reluctant to transmit bad news to peers in a nonprofessional setting. Rosen and Tesser termed this phenomenon the "MUM effect."* Since that time, numerous studies have investigated the impact of the MUM effect in a professional setting, e.g., on doctor–patient relationships, organizational functioning, and group psychotherapy. The consensus: The reluctance to transmit bad news continues to be a major professional concern.

Why do people keep mum when given an opportunity to transmit bad news to others? Two theories have emerged from this research. The first maintains that the MUM effect is an *aversion to private discomfort*. To avoid discomforts such as empathy with the victim's distress or guilt feelings for their own good fortune, would-be communicators of bad news keep mum. The second theory is that the MUM effect is a *public display*. People experience little or no discomfort when transmitting bad news, but keep mum to avoid an unfavorable impression or to pay homage to a social norm.

---

*Rosen, S., and Tesser, A. "On reluctance to communicate undesirable information: The MUM effect." *Journal of Communication*, Vol. 22, 1970, pp. 124–141.

The subject of this case study is an article by C. F. Bond and E. L. Anderson (*Journal of Experimental Social Psychology*, Vol. 23, 1987). Bond and Anderson conducted a controlled experiment to determine which of the two explanations for the MUM effect is more plausible. "If the MUM effect is an aversion to private discomfort," they state, "subjects should show the effect whether or not they are visible [to the victim]. If the effect is a public display, it should be stronger if the subject is visible than if the subject cannot be seen."

15.2
The Design

Forty undergraduates (25 males and 15 females) at Duke University participated in the experiment to fulfill an introductory psychology course requirement. Each subject was asked to administer an IQ test to another student and then provide the test taker with his or her percentile score. Unknown to the subject, the test taker was a confederate student working with the researchers.

The experiment manipulated two factors, *subject visibility* and *confederate success*, each at two levels. Subject visibility was manipulated by giving written instructions to each subject. Some subjects were told that they were *visible* to the test taker through a glass plate and the others were told that they were *not visible* through a one-way mirror. Confederate success was manipulated by supplying the subject with one of two bogus answer keys. With one answer key, the confederate would always seem to succeed at the test, placing him or her in the top 20% of all Duke undergraduates; when the other answer key was used, the confederate would always seem to fail, ranking in the bottom 20%.

Ten subjects were randomly assigned to each of the $2 \times 2 = 4$ experimental conditions; thus, a 2×2 factorial design with 10 replications was employed. The design is diagrammed in Table 15.1. For convenience, we use the letters NS, NF, VS, and VF to represent the four treatments.

TABLE 15.1   **2 × 2 Factorial design**

| | | CONFEDERATE SUCCESS | |
		Success	Failure
SUBJECT VISIBILITY	Visible	Subject 1 2 ⋮ (VS) 10	Subject 21 22 ⋮ (VF) 30
	Not Visible	Subject 11 12 ⋮ (NS) 20	Subject 31 32 ⋮ (NF) 40

One of several behavioral variables that were measured during the experiment was *latency to feedback*, defined as time (in seconds) between the end of the test and delivery of feedback (i.e., the percentile score) from the subject to the test taker. This case focuses on an analysis of variance of the dependent variable, latency to feedback. The longer it takes the subject to deliver the score, presumably the greater the MUM effect. The experimental data are saved in the MUM file.* With an analysis of this data, the researchers hope to determine whether either one of the two factors, subject visibility or confederate success, has an impact on the MUM effect, and, if so, whether the factors are independent.

15.3
Analysis of Variance Models and Results

Since both factors, subject visibility and confederate success, are qualitative, the complete model for this 2×2 factorial experiment is written as follows.

$$\text{Complete model:} \quad E(y) = \beta_0 + \underbrace{\beta_1 x_1}_{\substack{\text{Visibility} \\ \text{main} \\ \text{effect}}} + \underbrace{\beta_2 x_2}_{\substack{\text{Success} \\ \text{main} \\ \text{effect}}} + \underbrace{\beta_3 x_1 x_2}_{\substack{\text{Visibility} \times \text{Success} \\ \text{interaction}}}$$

where

$$y = \text{Latency to feedback}$$

$$x_1 = \begin{cases} 1 & \text{if subject visible} \\ 0 & \text{if not} \end{cases} \qquad x_2 = \begin{cases} 1 & \text{if confederate success} \\ 0 & \text{if confederate failure} \end{cases}$$

To test for factor interaction, we can compare the complete model to the reduced model

$$\text{Reduced model:} \quad E(y) = \beta_0 + \beta_1 x_1 + \beta_2 x_2$$

using the partial F test, or, equivalently, we can conduct a t test on the interaction parameter, β_3. Either way, the null hypothesis to be tested is

$$H_0: \quad \beta_3 = 0$$

Alternatively, we can use the ANOVA routine of a statistical software package to conduct the test.

The SAS ANOVA printout is shown in Figure 15.1. From the highlighted portions of the printout, we form the ANOVA table displayed in Table 15.2. The F statistic for testing the visibility-success interaction reported in the table is $F = 63.40$ with a p-value less than .0001. Therefore, we reject H_0 at $\alpha = .05$ and conclude that the two factors, subject visibility and confederate success, interact.

Practically, this result implies that the effect of confederate success on mean latency to feedback, $E(y)$, depends on whether the subject is visible. Similarly, the effect of subject visibility on $E(y)$ depends on the success or failure of the confederate student. In other words, we cannot examine the effect of one factor on latency to feedback without knowing the level of the second factor. Consequently, we ignore the F test for factor main effects and focus on the nature of the differences among the means of the $2 \times 2 = 4$ experimental conditions.

---

*The data are simulated based on summary results reported in the journal article.

FIGURE 15.1
SAS ANOVA printout for the 2×2 factorial

```
                          The GLM Procedure
Dependent Variable: FEEDBACK

     Source                    DF      Sum of
                                       Squares     Mean Square    F Value    Pr > F
     Model                      3    37371.27500   12457.09167     37.90    <.0001

     Error                     36    11833.70000     328.71389

     Corrected Total           39    49204.97500


              R-Square    Coeff Var      Root MSE      FEEDBACK Mean
              0.759502    19.03961      18.13047          95.22500


     Source              DF     Type III SS    Mean Square    F Value    Pr > F
     SUBJECT              1      8381.02500     8381.02500     25.50    <.0001
     CONFED               1      8151.02500     8151.02500     24.80    <.0001
     SUBJECT*CONFED       1     20839.22500    20839.22500     63.40    <.0001
```

TABLE 15.2   ANOVA Table for the 2×2 Factorial Experiment

SOURCE	df	SS	MS	F	p-VALUE
Subject visibility	1	8,381.025	8,381.025	25.50	<.0001
Confederate success	1	8,151.025	8,151.025	24.80	<.0001
Visibility × Success	1	20,839.225	20,839.225	63.40	<.0001
Error	36	11,833.700	328.714		
Total	39	49,204.975			

15.4
Follow-up Analysis

The sample latency to feedback means (in seconds) for each of the four experimental conditions are highlighted on the SAS printout, Figure 15.2. These four means are listed in Table 15.3 and plotted using MINITAB in Figure 15.3.

We will conduct a follow-up analysis of the ANOVA by ranking the four treatment means. Since a balanced design is employed ($n = 10$ subjects per treatment), Tukey's method of multiple comparisons of means will be used.

TABLE 15.3   Sample Means for the Four Experimental Conditions

		CONFEDERATE SUCCESS	
		Success	Failure
SUBJECT	Visible	72.6	146.8
	Not visible	89.3	72.2

FIGURE 15.2

SAS printout of Tukey multiple
comparisons of feedback means

```
                    The GLM Procedure
                   Least Squares Means
           Adjustment for Multiple Comparisons: Tukey

                                   FEEDBACK      LSMEAN
             SUBJECT    CONFED        LSMEAN      Number

             NotVis     Failure     72.200000        1
             NotVis     Success     89.300000        2
             Visibl     Failure    146.800000        3
             Visibl     Success     72.600000        4


           Least Squares Means for effect SUBJECT*CONFED
                Pr > |t| for H0: LSMean(i)=LSMean(j)

                   Dependent Variable: FEEDBACK

     i/j           1             2             3             4

      1                       0.1696        <.0001        1.0000
      2         0.1696                      <.0001        0.1858
      3         <.0001        <.0001                      <.0001
      4         1.0000        0.1858        <.0001
```

Rather than use the formula for calculating Tukey's critical difference, ω (given in Section 12.7), we use the Tukey option for factorial experiments available in SAS. The results are shown at the bottom of the SAS printout, Figure 15.2. Tukey-adjusted p-values that are less than $\alpha = .05$ are highlighted on the printout. These identify the significantly different pairs of treatment means: (3,2), (3,1) and (4,3), where 1, 2, 3, and 4 correspond to treatments NF, NS, VF, and VS, respectively. The implication is that mean 3 (the VF treatment mean) is significantly different from each of the other treatment means; but, none of the other treatment means are significantly different. The rankings are summarized in the traditional fashion in Table 15.4.

FIGURE 15.3

MINITAB plot of sample means
for the 2×2 factorial

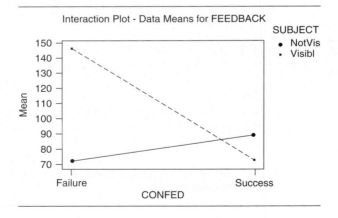

Another way to view the results in Table 15.4 is to notice that there is no significant difference between the two nonvisible subject means for confederate success and failure (i.e., treatments NF and NS). However, there is a significant difference between the two visible subject means for confederate success and failure

TABLE 15.4   **Summary of Tukey's Rankings of the Treatment Means**

Sample mean:	72.2	72.6	89.3	146.8
Treatment:	NF	VS	NS	VF
(Mean #)	(1)	(4)	(2)	(3)

(i.e., treatments VF and VS). Thus, only when the subject is visible can we conclude that confederate success has an effect on the mean latency to feedback. Furthermore, since $\bar{y}_{VF} = 146.8$ is over twice as large as $\bar{y}_{VS} = 72.6$, the researchers conclude that "subjects appear reluctant to transmit bad news—but only when they are visible to the news recipient."

15.5 Conclusions

In their discussion of these results, the researchers conclude:

In this experiment, subjects were required to give a test taker either success or failure feedback. While doing so, they presumed themselves to be visible to the test taker or visible to no one. Subjects who were visible took twice as long to deliver failure feedback as success feedback; those who were not visible delivered failure and success feedback with equal speed.

These results are not consistent with the discomfort explanation as originally conceived. We had imagined that subjects might empathize with another's failure, that mere observation of the failure would be sufficient to arouse vicarious distress. We found no behavioral evidence of such discomfort. . . . We also imagined that subjects would be reluctant to induce discomfort by announcing a poor intelligence performance, and that they would defer the announcement while checking the IQ test score. We found evidence of this deferral—but only when the subject could be seen. In private, subjects seemed blithe to others' misfortune—as quick to relay bad as good news. As the latency results suggest, there is no inherent discomfort in the transmission of bad news.

EXERCISES

15.1. Use Table 15.2 to determine SSEs for the complete and reduced ANOVA models. Then use these values to obtain the F statistic for testing interaction.

15.2. The journal article made no mention of an analysis of the ANOVA residuals. Discuss the potential problems of an ANOVA when no residual analysis is conducted. Carry out this analysis of the data in the MUM file.

15.3. A second dependent variable measured in the study was *gaze*, defined as the proportion of time the subject was looking toward the confederate test taker on the other side of the glass plate. Gaze was measured at four points in time using videotape segments: early in the test, late in the test, during the wait for feedback, and after the feedback. Construct a complete model for analyzing gaze as a function of subject visibility, confederate success, and videotape segment. Identify the important tests to conduct.

REFERENCES

BOND, C. F., and ANDERSON, E. L. "The reluctance to transmit bad news: Private discomfort or public display?" *Journal of Experimental Social Psychology*, Vol. 23, 1987, pp. 176–187.

An Investigation of Factors Affecting the Sale Price of Condominium Units Sold at Public Auction

OBJECTIVE

To show how regression analysis can be used to develop a model relating sale price of condominium units to a set of independent variables and, particularly, to show how the model can be used to reveal some interesting relationships among these variables

CONTENTS

16.1 The Problem

This chapter contains a partial investigation of the factors that affect the sale price of oceanside condominium units. It represents an extension of an analysis of the same data by Herman Kelting (1979).

The sales data were obtained for a new oceanside condominium complex consisting of two adjacent and connected eight-floor buildings. The complex contains 200 units of equal size (approximately 500 square feet each). The locations of the buildings relative to the ocean, the swimming pool, the parking lot, etc., are shown in Figure 16.1. There are several features of the complex that you should note. The units facing south, called *ocean-view*, face the beach and ocean. In addition, units in building 1 have a good view of the pool. Units to the rear of the building, called *bay-view*, face the parking lot and an area of land that ultimately borders a bay. The view from the upper floors of these units is primarily of wooded, sandy terrain. The bay is very distant and barely visible.

The only elevator in the complex is located at the east end of building 1, as are the office and the game room. People moving to or from the higher-floor units in building 2 would likely use the elevator and move through the passages to their units. Thus, units on the higher floors and at a greater distance from the elevator would be less convenient; they would require greater effort in moving baggage, groceries,

FIGURE 16.1

Layout of condominium complex

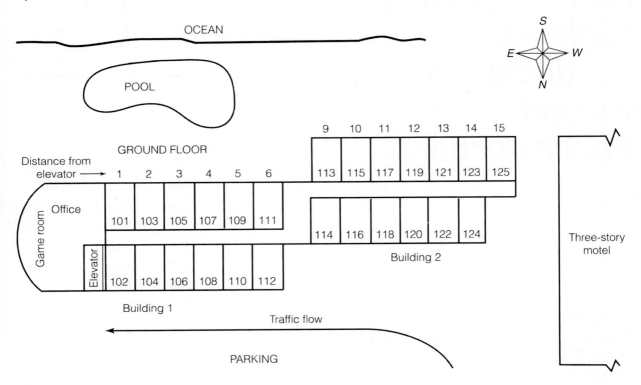

etc., and would be farther from the game room, the office, and the swimming pool. These units also possess an advantage: There would be the least amount of traffic through the hallways in the area, and hence they are the most private.

Lower-floor oceanside units are most suited to active people; they open onto the beach, ocean, and pool. They are within easy reach of the game room and they are easily reached from the parking area.

Checking Figure 16.1, you will see that some of the units in the center of the complex, units numbered _11 and _14, have part of their view blocked. We would expect this to be a disadvantage. We will show you later that this expectation is true for the ocean-view units and that these units sold at a lower price than adjacent ocean-view units.

The condominium complex was completed during a recession; sales were slow and the developer was forced to sell most of the units at auction approximately 18 months after opening. Many unsold units were furnished by the developer and rented prior to the auction.

This condominium complex was particularly suited to our study. The single elevator located at one end of the complex produces a remarkably high level of both inconvenience and privacy for the people occupying units on the top floors in building 2. Consequently, the data provide a good opportunity to investigate the relationship between sale price, height of the unit (floor number), distance of the unit from the elevator, and presence or absence of an ocean view. The presence or absence of furniture in each of the units also enables us to investigate the effect of the availability of furniture on sale price. Finally, the auction data are completely buyer-specified and hence consumer-oriented in contrast to most other real estate sales data, which are, to a high degree, seller- and broker-specified.

16.2 The Data

In addition to the sale price (measured in hundreds of dollars) the following data were recorded for each of the 106 units sold at the auction:

1. *Floor height*   The floor location of the unit; this variable, x_1, could take values $1, 2, \ldots, 8$.

2. *Distance from elevator*   This distance, measured along the length of the complex, was expressed in number of condominium units. An additional two units of distance was added to the units in building 2 to account for the walking distance in the connecting area between the two buildings. Thus, the distance of unit 105 from the elevator would be 3, and the distance between unit 113 and the elevator would be 9. This variable, x_2, could take values $1, 2, \ldots, 15$.

3. *View of ocean*   The presence or absence of an ocean view was recorded for each unit and entered into the model with a dummy variable, x_3, where $x_3 = 1$ if the unit possessed an ocean view and $x_3 = 0$ if not. Note that units not possessing an ocean view would face the parking lot.

4. *End unit*   We expected the partial reduction of view of end units on the ocean side (numbers ending in 11) to reduce their sale price. The ocean view of these end units is partially blocked by building 2. This qualitative variable was entered into the model with a dummy variable, x_4, where $x_4 = 1$ if the unit has a unit number ending in 11 and $x_4 = 0$ if not.

5. *Furniture*   The presence or absence of furniture was recorded for each unit. This qualitative variable was entered into the model using a single dummy variable, x_5, where $x_5 = 1$ if the unit was furnished and $x_5 = 0$ if not.

CONDO

The raw data used in this analysis are saved in the CONDO file.

16.3 The Models

This case involves five independent variables, two quantitative (floor height x_1 and distance from elevator x_2) and three qualitative (view of ocean, end unit, and furniture). We postulated four models relating mean sale price to these five factors. The models, numbered 1–4, are developed in sequence, Model 1 being the simplest and Model 4, the most complex. Each of Models 2 and 3 contains all the terms of the preceding models along with new terms that we think will improve their predictive ability. Thus, Model 2 contains all the terms contained in Model 1 plus some new terms, and hence it should predict mean sale price as well as or better than Model 1. Similarly, Model 3 should predict as well as or better than Model 2. Model 4 does not contain all the terms contained in Model 3, but that is only because we have entered floor height into Model 4 as a qualitative independent variable.

Consequently, Model 4 contains all the predictive ability of Model 3, and it could be an improvement over Model 3 if our theory is correct. The logic employed in this sequential model-building procedure will be explained in the following discussion.

The simplest theory that we might postulate is that the five factors affect the price in an independent manner and that the effect of the two quantitative factors on sale price is linear. Thus, we envision a set of planes, each identical except for their y-intercepts. We would expect sale price planes for ocean-view units to be higher than those with a bay view, those corresponding to end units ($_11$) would be lower than for non–end units, and those with furniture would be higher than those without.

Model 1
First-order, main effects

$$E(y) = \beta_0 + \beta_1 x_1 + \beta_2 x_2 + \beta_3 x_3 + \beta_4 x_4 + \beta_5 x_5$$

where

$x_1 =$ Floor height $(x_1 = 1, 2, \ldots, 8)$

$x_2 =$ Distance from elevator $(x_2 = 1, 2, \ldots, 15)$

$$x_3 = \begin{cases} 1 & \text{if an ocean view} \\ 0 & \text{if not} \end{cases} \qquad x_4 = \begin{cases} 1 & \text{if an end unit} \\ 0 & \text{if not} \end{cases} \qquad x_5 = \begin{cases} 1 & \text{if furnished} \\ 0 & \text{if not} \end{cases}$$

The second theory that we considered was that the effects on sale price of floor height and distance from elevator might not be linear. Consequently, we constructed Model 2, which is similar to Model 1 except that second-order terms are included for x_1 and x_2. This model envisions a single second-order response surface for $E(y)$ in x_1 and x_2 that possesses identically the same shape, regardless of the view, whether the unit is an end unit, and whether the unit is furnished. Expressed in other terminology, Model 2 assumes that there is no interaction between any of the qualitative factors (view of ocean, end unit, and furniture) and the quantitative factors (floor height and distance from elevator).

Model 2
Second-order, main effects

$$\text{Second-order model in Floor and Distance}$$

$$E(y) = \beta_0 + \overbrace{\beta_1 x_1 + \beta_2 x_2 + \beta_3 x_1 x_2 + \beta_4 x_1^2 + \beta_5 x_2^2}$$

$$+ \underbrace{\beta_6 x_3}_{\text{View of ocean}} + \underbrace{\beta_7 x_4}_{\text{End unit}} + \underbrace{\beta_8 x_5}_{\text{Furniture}}$$

Model 2 may possess a serious shortcoming. It assumes that the shape of the second-order response surface relating mean sale price $E(y)$ to x_1 and x_2 is identical for ocean-view and bay-view units. Since we think that there is a strong possibility that completely different preference patterns may govern the purchase of these two groups of units, we will construct a model that provides for two completely different second-order response surfaces—one for ocean-view units and one for bay-view units. Further, we will assume that the effects of the two qualitative factors, end unit and furniture, are additive; i.e., their presence or absence will simply shift the mean sale price response surface up or down by a fixed amount. Thus, Model 3 is given as follows:

$$\text{Second-order model in floor and distance}$$

$$E(y) = \beta_0 \ + \ \overbrace{\beta_1 x_1 + \beta_2 x_2 + \beta_3 x_1 x_2 + \beta_4 x_1^2 + \beta_5 x_2^2}$$

$$\overset{\text{View of ocean}}{} \qquad \overset{\text{End unit}}{} \qquad \overset{\text{Furniture}}{}$$

$$+ \ \overbrace{\beta_6 x_3} \qquad + \qquad \overbrace{\beta_7 x_4} \qquad + \qquad \overbrace{\beta_8 x_5}$$

$$+ \quad \underset{\substack{\text{Interaction of the second-order model} \\ \text{with view of ocean}}}{}$$

$$+ \ \overbrace{\beta_9 x_1 x_3 + \beta_{10} x_2 x_3 + \beta_{11} x_1 x_2 x_3 + \beta_{12} x_1^2 x_3 + \beta_{13} x_2^2 x_3}$$

Model 3
Second-order,
view
interactions

As a fourth possibility, we constructed a model similar to Model 3 but entered floor height as a qualitative factor at eight levels. This requires seven dummy variables:

$$x_6 = \begin{cases} 1 & \text{if first floor} \\ 0 & \text{if not} \end{cases}$$

$$x_7 = \begin{cases} 1 & \text{if second floor} \\ 0 & \text{if not} \end{cases}$$

$$\vdots$$

$$x_{12} = \begin{cases} 1 & \text{if seventh floor} \\ 0 & \text{if not} \end{cases}$$

Thus, Model 4 is:

$$E(y) = \ \beta_0 + \ \underbrace{\beta_1 x_2 + \beta_2 x_2^2}_{\text{Second-order in distance}}$$

$$+ \ \underbrace{\beta_3 x_3}_{\text{View}} \quad + \underbrace{\beta_4 x_4}_{\text{End unit}} \quad + \underbrace{\beta_5 x_5}_{\text{Furnished}}$$

$$+ \ \underbrace{\beta_6 x_6 + \beta_7 x_7 + \beta_8 x_8 + \beta_9 x_9 + \beta_{10} x_{10} + \beta_{11} x_{11} + \beta_{12} x_{12}}_{\text{Floor main effects}}$$

Model 4
Floor height
dummy
variable model

$$+ \ \underbrace{\beta_{13} x_2 x_6 + \beta_{14} x_2 x_7 + \cdots + \beta_{19} x_2 x_{12} + \beta_{20} x_2^2 x_6 + \beta_{21} x_2^2 x_7 + \cdots}_{}$$

$$\underbrace{+ \ \beta_{26} x_2^2 x_{12}}_{\text{Distance-Floor interactions}}$$

$$+ \ \underbrace{\beta_{27} x_2 x_3 + \beta_{28} x_2^2 x_3}_{\text{Distance-View interactions}} \quad + \quad \underbrace{\beta_{29} x_3 x_6 + \beta_{30} x_3 x_7 + \cdots + \beta_{35} x_3 x_{12}}_{\text{Floor-View interactions}}$$

$$+ \quad \beta_{36}x_2x_3x_6 + \beta_{37}x_2x_3x_7 + \cdots \beta_{42}x_2x_3x_{12}$$

$$+ \quad \underbrace{\beta_{43}x_2^2x_3x_6 + \beta_{44}x_2^2x_3x_7 + \cdots \beta_{49}x_2^2x_3x_{12}}_{\text{Distance-View-Floor interactions}}$$

The reasons for entering floor height as a qualitative factor are twofold:

1. Higher-floor units have better views but less accessibility to the outdoors. This latter characteristic could be a particularly undesirable feature for these units.
2. The views of some lower-floor bayside units were blocked by a nearby three-story motel.

If our supposition is correct and if these features would have a depressive effect on the sale price of these units, then the relationship between floor height and mean sale price would not be second-order (a smooth curvilinear relationship). Entering floor height as a qualitative factor would permit a better fit to this irregular relationship and improve the prediction equation. Thus, Model 4 is identical to Model 3 except that Model 4 contains seven main effect terms for floor height (in contrast to two for Model 3), and it also contains the corresponding interactions of these variables with the other variables included in Model 3*. We will subsequently show that there was no evidence to indicate that Model 4 contributes more information for the prediction of *y* than Model 3.

16.4
The Regression
Analyses

This section gives the regression analyses for the four models described in Section 16.3. You will see that our approach is to build the model in a sequential manner. In each case, we use a partial *F* test to see whether a particular model contributes more information for the prediction of sale price than its predecessor.

This procedure is more conservative than a step-down procedure. In a step-down approach you would assume Model 4 to be the appropriate model, then test and possibly delete terms. But deleting terms can be particularly risky because, in doing so, you are tacitly accepting the null hypothesis. Thus, you risk deleting important terms from the model and do so with an unknown probability of committing a Type II error.

Do not be unduly influenced by the individual *t* tests associated with an analysis. As you will see, it is possible for a set of terms to contribute information for the prediction of *y* when none of their respective *t* values are statistically significant. This is because the *t* test focuses on the contribution of a single term, given that all the other terms are retained in the model. Therefore, if a set of terms contributes overlapping information, it is possible that none of the terms individually would be statistically significant, even when the set as a whole contributes information for the prediction of *y*.

The SAS regression analysis computer printouts for fitting Models 1, 2, 3, and 4 to the data are shown in Figures 16.2, 16.3, 16.4, and 16.5, respectively. A summary containing the key results (e.g., R_{adj}^2 and *s*) for these models is provided in Table 16.1.

---

*Some of the terms in Model 4 were not estimable because sales were not consummated for some combinations of the independent variables. This is why the SAS printout in Figure 16.5 shows only 41 df for the model.

FIGURE 16.2

SAS regression printout for Model 1

```
                    Dependent Variable: PRICE100

                        Analysis of Variance

                            Sum of          Mean
Source              DF      Squares        Square    F Value    Pr > F

Model                5        23534    4706.82958      47.98    <.0001
Error              100   9810.07851      98.10079
Corrected Total    105        33344


            Root MSE              9.90458    R-Square    0.7058
            Dependent Mean     191.81132    Adj R-Sq    0.6911
            Coeff Var            5.16371


                        Parameter Estimates

                    Parameter    Standard
Variable      DF     Estimate       Error    t Value    Pr > |t|

Intercept      1    177.70349     4.16842      42.63     <.0001
FLOOR          1     -0.71514     0.53077      -1.35     0.1809
DIST           1     -0.87325     0.24495      -3.57     0.0006
VIEW           1     31.27285     2.23121      14.02     <.0001
END            1    -17.80782     3.98195      -4.47     <.0001
FURNISH        1      9.98376     2.05150       4.87     <.0001
```

TABLE 16.1   A Summary of the Regressions for Models 1, 2, 3, and 4

MODEL	df(MODEL)	SSE	df(ERROR)	MSE	R^2_{adj}	s
1	5	9810	100	98.1	.691	9.90
2	8	9033	97	93.1	.707	9.65
3	13	7844	92	85.3	.732	9.23
4	41	5207	64	81.4	.744	9.02

Examining the SAS printout for the first-order model (Model 1) in Figure 16.2, you can see that the value of the F statistic for testing the null hypothesis

$$H_0: \quad \beta_1 = \beta_2 = \cdots = \beta_5 = 0$$

is 47.98. This is statistically significant at a level of $\alpha = .0001$. Consequently, there is ample evidence to indicate that the overall model contributes information for the prediction of y. At least one of the five factors contributes information for the prediction of sale price.

If you examine the t tests for the individual parameters, you will see that they are all statistically significant except the test for β_1, the parameter associated with floor height x_1 (p-value = .1809). The failure of floor height x_1 to reveal itself as an important information contributor goes against our intuition, and it demonstrates the pitfalls that can attend an unwary attempt to interpret the results of t tests in a regression analysis. Intuitively, we would expect floor height to be an important factor. You might argue that units on the higher floors possess a better view and hence should command a higher mean sale price. Or, you might argue that units on the lower floors have greater accessibility to the pool and ocean and, consequently, should be in greater demand. Why, then, is the t test for floor height not statistically significant? The answer is that both of the preceding arguments are correct, one for the oceanside and one for the bayside. Thus, you will subsequently see that there is an interaction between floor height and view of ocean. Ocean-view units on the lower floors sell at higher prices than ocean-view units on the higher floors. In contrast, bay-view units on the higher floors command higher prices than bay-view units on the lower floors. These two contrasting effects tend to cancel (because we have not included interaction terms in the model) and thereby give the false impression that floor height is not an important variable for predicting mean sale price.

But, of course, we are looking ahead. Our next step is to determine whether Model 2 is better than Model 1.

Are floor height x_1 and distance from elevator x_2 related to sale price in a curvilinear manner; i.e., should we be using a second-order response surface instead of a first-order surface to relate $E(y)$ to x_1 and x_2? To answer this question, we will examine the drop in SSE from Model 1 to Model 2. The null hypothesis "Model 2 contributes no more information for the prediction of y than Model 1" is equivalent to testing

$$H_0: \quad \beta_3 = \beta_4 = \beta_5 = 0$$

where β_3, β_4, and β_5 appear in Model 2. The F statistic for the test, based on 3 and 97 df, is

$$F = \frac{(\text{SSE}_1 - \text{SSE}_2)/\#\beta\text{'s in } H_0}{\text{MSE}_2} = \frac{(9810 - 9033)/3}{93.1}$$
$$= 2.78$$

This value is shown (shaded) at the bottom of Figure 16.3 as well as the corresponding p-value of the test. Since $\alpha = .05$ exceeds p-value $= .0452$, we reject H_0. There is evidence to indicate that Model 2 contributes more information for the prediction of y than Model 1. This tells us that there is evidence of curvature in the response surfaces relating mean sale price, $E(y)$, to floor height x_1 and distance from elevator x_2.

You will recall that the difference between Models 2 and 3 is that Model 3 allows for two differently shaped second-order surfaces, one for ocean-view units and another for bay-view units; Model 2 employs a single surface to represent both types of units. Consequently, we wish to test the null hypothesis that "a single second-order surface adequately characterizes the relationship between $E(y)$, floor height x_1, and distance from elevator x_2 for both ocean-view and bay-view units" [i.e., Model 2 adequately models $E(y)$] against the alternative hypothesis that you need two different second-order surfaces [i.e, you need Model 3]. Thus,

$$H_0: \quad \beta_9 = \beta_{10} = \cdots = \beta_{13} = 0$$

FIGURE 16.3

SAS regression printout for Model 2

```
                    Dependent Variable: PRICE100

                         Analysis of Variance

                              Sum of          Mean
Source                DF     Squares         Square     F Value    Pr > F

Model                  8       24311     3038.84678      32.63    <.0001
Error                 97   9033.45215       93.12837
Corrected Total      105       33344


           Root MSE              9.65030   R-Square     0.7291
           Dependent Mean      191.81132   Adj R-Sq     0.7067
           Coeff Var             5.03114


                         Parameter Estimates

                     Parameter     Standard
Variable      DF      Estimate        Error     t Value    Pr > |t|

Intercept      1     194.59573      7.66067       25.40    <.0001
FLOOR          1      -6.83830      2.45493       -2.79     0.0064
DIST           1      -2.64122      1.22794       -2.15     0.0340
FLR_DIS        1       0.04384      0.13564        0.32     0.7472
FLOORSQ        1       0.58394      0.23786        2.45     0.0159
DISTSQ         1       0.11426      0.07714        1.48     0.1418
VIEW           1      30.42124      2.19627       13.85    <.0001
END            1     -16.80585      4.10416       -4.09    <.0001
FURNISH        1      11.27207      2.05390        5.49    <.0001
```

```
Test SECOND_ORDER Results for Dependent Variable PRICE100

                              Mean
Source                DF     Square     F Value    Pr > F

Numerator              3   258.87545       2.78    0.0452
Denominator           97    93.12837
```

where $\beta_9, \beta_{10}, \ldots, \beta_{13}$ are parameters in Model 3. The F statistic for this test, based on 5 and 92 df, is

$$F = \frac{(\text{SSE}_2 - \text{SSE}_3)/\#\beta\text{'s in } H_0}{\text{MSE}_3}$$

$$= \frac{(9033 - 7844)/5}{85.3}$$

$$= 2.79$$

This value and its associated p-value are shown (shaded) at the bottom of Figure 16.4. Since $\alpha = .05$ exceeds p-value $= .0216$, we reject H_0 and conclude that there is evidence to indicate that we need two different second-order surfaces to relate $E(y)$ to x_1 and x_2, one each for ocean-view and bay-view units.

FIGURE 16.4

SAS regression printout for Model 3

Dependent Variable: PRICE100

Analysis of Variance

Source	DF	Sum of Squares	Mean Square	F Value	Pr > F
Model	13	25500	1961.56185	23.01	<.0001
Error	92	7843.92233	85.26003		
Corrected Total	105	33344			

Root MSE	9.23364	R-Square	0.7648	
Dependent Mean	191.81132	Adj R-Sq	0.7315	
Coeff Var	4.81392			

Parameter Estimates

Variable	DF	Parameter Estimate	Standard Error	t Value	Pr > |t|
Intercept	1	142.05564	27.95942	5.08	<.0001
FLOOR	1	8.79374	9.41542	0.93	0.3528
DIST	1	-4.00545	3.03226	-1.32	0.1898
FLR_DIS	1	0.32600	0.34146	0.95	0.3422
FLOORSQ	1	-0.58321	0.80662	-0.72	0.4715
DISTSQ	1	0.11069	0.16598	0.67	0.5065
VIEW	1	84.52959	28.85730	2.93	0.0043
END	1	-16.30753	4.00143	-4.08	<.0001
FURNISH	1	12.56003	2.04377	6.15	<.0001
FLR_VIEW	1	-14.11343	9.74820	-1.45	0.1511
DIS_VIEW	1	0.36326	3.34173	0.11	0.9137
F_D_VIEW	1	-0.29131	0.37258	-0.78	0.4363
FLRSQ_VU	1	0.95473	0.84502	1.13	0.2615
DISSQ_VU	1	0.06503	0.19068	0.34	0.7338

Test VIEW_INTERACT Results for
Dependent Variable PRICE100

Source	DF	Mean Square	F Value	Pr > F
Numerator	5	237.90596	2.79	0.0216
Denominator	92	85.26003		

Finally, we question whether Model 4 will provide an improvement over Model 3; i.e., will we gain information for predicting y by entering floor height into the model as a qualitative factor at eight levels? Although Models 3 and 4 are not "nested" models, we can compare them using a "partial" F test. The F statistic to test the null hypothesis "Model 4 contributes no more information for predicting y than does Model 3" compares the drop in SSE from Model 3 to Model 4 with s_4^2. This F statistic, based on 28 df (the difference in the numbers of parameters in Models 4 and 3) and 64 df, is

$$F = \frac{(\text{SSE}_3 - \text{SSE}_4)/\#\beta\text{'s in } H_0}{\text{MSE}_4}$$

$$= \frac{(7844 - 5207)/28}{81.4}$$

$$= 1.16$$

FIGURE 16.5

Portion of SAS regression printout for Model 4

```
              Dependent Variable: PRICE100

                    Analysis of Variance

                           Sum of        Mean
Source              DF     Squares       Square     F Value    Pr > F

Model               41       28137    686.27992       8.44    <.0001
Error               64  5206.74955     81.35546
Corrected Total    105       33344


        Root MSE            9.01973   R-Square     0.8438
        Dependent Mean    191.81132   Adj R-Sq     0.7438
        Coeff Var           4.70240
```

Since the p-value of the test is not shown on Figure 16.5, we will conduct the test by finding the rejection region. Checking Table 4 in Appendix C, you will find that the value for $F_{.05}$, based on 28 and 64 df, is approximately 1.65. Since the computed F is less than this value, there is not sufficient evidence to indicate that Model 4 is a significant improvement over Model 3.

Having checked the four theories of Section 16.3, we have evidence to indicate that Model 3 is the best of the four models. The R^2_{adj} for the model implies that about 73% of the sample variation in sale prices can be explained. We will examine the prediction equation for Model 3 and see what it tells us about the relationship between the mean sale price $E(y)$ and the five factors used in our study; but first, it is important that we examine the residuals for Model 3 to determine whether the standard least squares assumptions about the random error term are satisfied.

16.5
An Analysis of the Residuals from Model 3

The four standard least squares assumptions about the random error term ε are (from Chapter 4) the following:

1. The mean is 0.
2. The variance (σ^2) is constant for all settings of the independent variables.
3. The errors follow a normal distribution.
4. The errors are independent.

If one or more of these assumptions are violated, any inferences derived from the Model 3 regression analysis are suspect. It is unlikely that assumption 1 (0 mean) is violated because the method of least squares guarantees that the mean of the residuals is 0. The same can be said for assumption 4 (independent errors) since the sale price data are not a time series. However, verifying assumptions 2 and 3 requires a thorough examination of the residuals from Model 3.

Recall that we can check for heteroscedastic errors (i.e., errors with unequal variances) by plotting the residuals against the predicted values. This residual plot is shown in Figure 16.6. If the variances were not constant, we would expect to see a

FIGURE 16.6
SAS plot of standardized residuals from Model 3

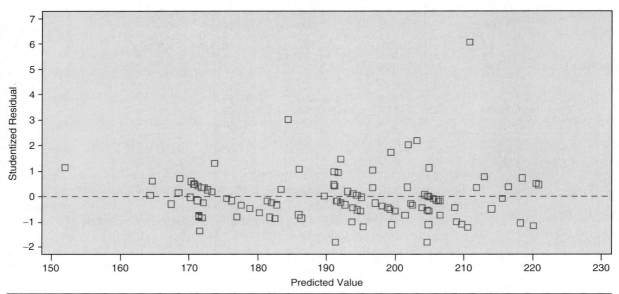

PRICE100 = 142.06 +8.7937 FLOOR −4.0055 DIST +0.326 FLR_DIS −0.5832 FLOORSQ +0.1107 DISTSQ +84.53 VIEW −16.308 END +12.56 FURNISH −14.113 FLR_VIEW +0.3633 DIS_VIEW −0.2913F_D_VIEW +0.9547 FLRSQ_VU +0.065 DISSQ_VU

cone-shaped pattern (since the response is sale price) in Figure 16.6, with the spread of the residuals increasing as $\hat{y}$ increases. Note, however, that except for one point that appears at the top of the graph, the residuals appear to be randomly scattered around 0. Therefore, assumption 2 (constant variance) appears to be satisfied.

To check the normality assumption (assumption 3), we have generated a SAS histogram of the residuals in Figure 16.7. It is very evident that the distribution of residuals is not normal, but skewed to the right. At this point, we could opt to use a transformation on the response (similar to the variance-stabilizing transformations discussed in Section 8.3) to normalize the residuals. However, a nonnormal error distribution is often due to the presence of a single outlier. If this outlier is eliminated (or corrected), the normality assumption may then be satisfied.

Figure 16.6 is a plot of the *standardized* residuals. In Section 8.4, we defined outliers to have standardized residuals that exceed 3 (in absolute value) and suspect outliers as standardized residuals that fall between 2 and 3 (in absolute value). Therefore, we can use Figure 16.6 to detect outliers. Note that there is one outlier and one suspect outlier, both with large *positive* standardized residuals (approximately 7 and 3, respectively). Should we automatically eliminate these two observations from the analysis and refit Model 3? Although many analysts adopt such an approach, we should carefully examine the observations before deciding to eliminate them. We may discover a correctable recording (or coding) error, or we may find that the outliers are very influential and are due to an inadequate model (in which case, it is the model that needs fixing, not the data).

FIGURE 16.7

SAS histogram of residuals from Model 3

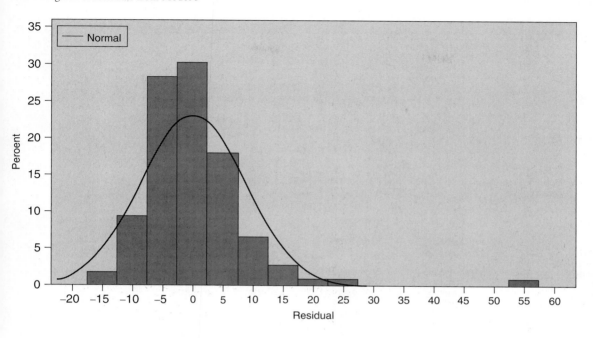

TABLE 16.2   Influence Diagnostics for Two Outliers in Model 3

OBSERVATION	RESPONSE y	PREDICTED VALUE $\hat{y}$	RESIDUAL $y - \hat{y}$	LEVERAGE h	COOK'S DISTANCE D
35	265	210.77	54.23	.0605	.169
49	210	184.44	25.56	.1607	.125

An examination of the SAS printout of the Model 3 residuals (not shown) reveals that the two observations in question are identified by observation numbers 35 and 49 (where the observations are numbered from 1 to 106). The sale prices, floor heights, and so forth, for these two data points were found to be recorded and coded correctly. To determine how influential these outliers are on the analysis, influence diagnostics were generated using SAS. The results are summarized in Table 16.2.

Based on the "rules of thumb" given in Section 8.5, neither observation has strong influence on the analysis. Both leverage (h) values fall below $2(k + 1)/n = 2(14)/106 = .264$, indicating that the observations are not influential with respect to their x values; and both Cook's D values fall below .96 [the 50th percentile of an F distribution with $v_1 = k + 1 = 14$ and $v_2 = n - (k + 1) = 106 - 14 = 92$ degrees of freedom], implying that they do not exhibit strong overall influence on the regression results (e.g., the β estimates). Consequently, if we remove these outliers from the data and refit Model 3, the least squares prediction equation will not be greatly affected and the normality assumption will probably be more nearly satisfied.

FIGURE 16.8
SAS regression printout for Model 3 with outliers
removed

Dependent Variable: PRICE100

Analysis of Variance

Source	DF	Sum of Squares	Mean Square	F Value	Pr > F
Model	13	23655	1819.64590	41.77	<.0001
Error	90	3921.13212	43.56813		
Corrected Total	103	27577			

Root MSE	6.60062	R-Square	0.8578	
Dependent Mean	190.93269	Adj R-Sq	0.8373	
Coeff Var	3.45704			

Parameter Estimates

Variable	DF	Parameter Estimate	Standard Error	t Value	Pr > |t|
Intercept	1	151.73426	20.09147	7.55	<.0001
FLOOR	1	3.76115	6.84884	0.55	0.5843
DIST	1	-1.93178	2.22898	-0.87	0.3884
FLR_DIS	1	0.25242	0.24477	1.03	0.3052
FLOORSQ	1	-0.15321	0.58763	-0.26	0.7949
DISTSQ	1	0.00420	0.12159	0.03	0.9725
VIEW	1	77.52403	20.73006	3.74	0.0003
END	1	-14.99258	2.86477	-5.23	<.0001
FURNISH	1	10.89332	1.47165	7.40	<.0001
FLR_VIEW	1	-10.91058	7.08792	-1.54	0.1272
DIS_VIEW	1	-1.79122	2.44305	-0.73	0.4653
F_D_VIEW	1	-0.24675	0.26701	-0.92	0.3579
FLRSQ_VU	1	0.72313	0.61542	1.18	0.2431
DISSQ_VU	1	0.20033	0.13878	1.44	0.1523

The SAS printout for the refitted model is shown in Figure 16.8. Note that df(Error) is reduced from 92 to 90 (since we eliminated the two outlying observations), and the β estimates remain relatively unchanged. However, the model standard deviation is decreased from 9.23 to 6.60 and the R^2_{adj} is increased from .73 to .84, implying that the refitted model will yield more accurate predictions of sale price. A residual plot for the refitted model is shown in Figure 16.9 and a histogram of the residuals in Figure 16.10. The residual plot (Figure 16.9) reveals no evidence of outliers, and the histogram of the residuals (Figure 16.10) is now approximately normal.

16.6
What the Model 3 Regression Analysis Tells Us

We have settled on Model 3 (with two observations deleted) as our choice to relate mean sale price $E(y)$ to five factors: the two quantitative factors, floor height x_1 and distance from elevator x_2; and the three qualitative factors, view of ocean, end unit, and furniture. This model postulates two different second-order surfaces relating mean sale price $E(y)$ to x_1 and x_2, one for ocean-view units and one for bay-view units. The effect of each of the two qualitative factors, end unit (numbered _11) and furniture, is to produce a change in mean sale price that is identical for

FIGURE 16.9

SAS plot of standardized residuals from Model 3
with outliers removed

PRICE100 = 151.73 +3.7611 FLOOR −1.9318 DIST +0.2524 FLR_DIS −0.1532 FLOORSQ +0.0042 DISTSQ
+77.524 VIEW −14.993 END +10.893 FURNISH −10.911 FLR_VIEW −1.7912 DIS_VIEW
−0.2467 F_D_VIEW +0.7231 FLRSQ_VU +0.2003 DISSQ_VU

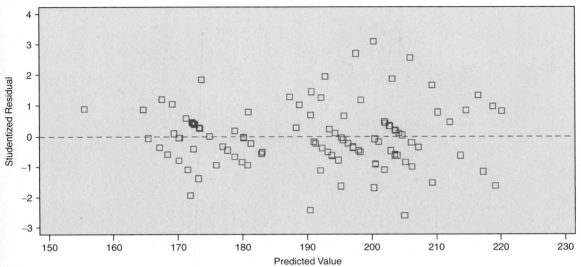

all combinations of values of x_1 and x_2. In other words, assigning a value of 1 to one of the dummy variables increases (or decreases) the estimated mean sale price by a fixed amount. The net effect is to push the second-order surface upward or downward, with the direction depending on the level of the specific qualitative factor. The estimated increase (or decrease) in mean sale price because of a given qualitative factor is given by the estimated value of the β parameter associated with its dummy variable.

For example, the prediction equation (with rounded values given for the parameter estimates) obtained from Figure 16.8 is

$$\hat{y} = 151.7 + 3.76x_1 - 1.93x_2$$
$$+.25x_1x_2 - .15x_1^2 + .004x_2^2$$
$$+77.52x_3 - 14.99x_4 + 10.89x_5$$
$$-10.91x_1x_3 - 1.79x_2x_3 - .25x_1x_2x_3$$
$$+.72x_1^2x_3 + .20x_2^2x_3$$

Since the dummy variables for end unit and furniture are, respectively, x_4 and x_5, the estimated changes in mean sale price for these qualitative factors are

End unit ($x_4 = 1$): $\hat{\beta}_7 \times (\$100) = -\$1,499$
Furnished ($x_5 = 1$): $\hat{\beta}_8 \times (\$100) = +\$1,089$

FIGURE 16.10

SAS histogram of residuals from Model 3 with
outliers removed

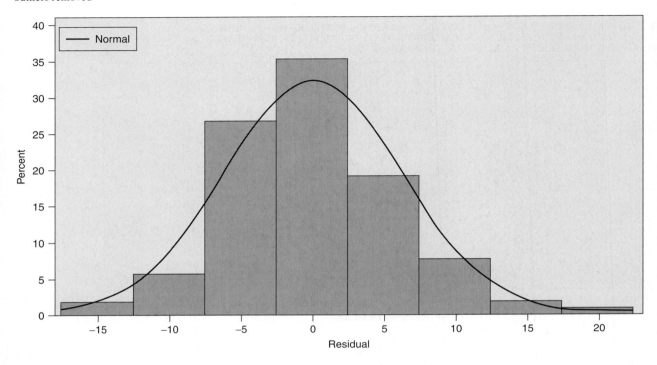

Thus, if you substitute $x_4 = 1$ into the prediction equation, the estimated mean decrease in sale price for an end unit is \$1,499, regardless of the view, floor, and whether it is furnished.

The effect of floor height x_1 and distance from elevator x_2 can be determined by plotting $\hat{y}$ as a function of one of the variables for given values of the other. For example, suppose we wish to examine the relationship between $\hat{y}$, x_1, and x_2 for bay-view ($x_3 = 0$), non-end units ($x_4 = 0$) with no furniture ($x_5 = 0$). The prediction curve relating $\hat{y}$ to distance from elevator x_2 can be graphed for each floor by first setting $x_1 = 1$, then $x_1 = 2, \ldots, x_1 = 8$. The graphs of these curves are shown in Figure 16.11. The floor heights are indicated by the symbols at the bottom of the graph. In Figure 16.11, we can also see some interesting patterns in the estimated mean sale prices:

1. The higher the floor of a bay-view unit, the higher will be the mean sale price. Low floors look out onto the parking lot and, all other variables held constant, are least preferred.

2. The relationship is curvilinear and is not the same for each floor.

3. Units on the first floor near the office have a higher estimated mean sale price than second- or third-floor units located in the west end of the complex. Perhaps the reason for this is that these units are close to the pool and the game room, and these advantages outweigh the disadvantage of a poor view.

FIGURE 16.11

SAS graph of predicted price versus distance
(bay-view units)

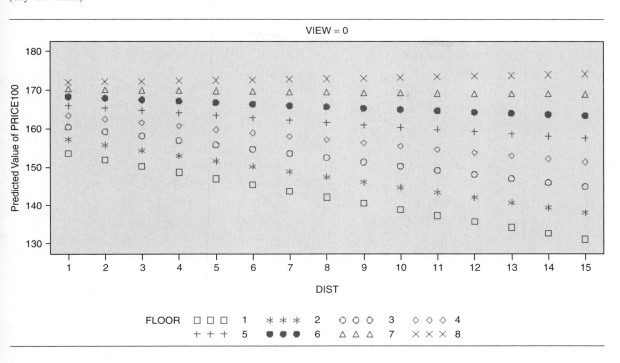

4. The mean sale price decreases as the distance from the elevator and center of activity increases for the lower floors, but the decrease is less as you move upward, floor to floor. Finally, note that the estimated mean sale price increases substantially for units on the highest floor that are farthest away from the elevator. These units are subjected to the least human traffic and are, therefore, the most private. Consequently, a possible explanation for their high price is that buyers place a higher value on the privacy provided by the units than the negative value that they assign to their inconvenience. One additional explanation for the generally higher estimated sale price for units at the ends of the complex may be that they possess more windows.

A similar set of curves is shown in Figure 16.12 for ocean-view units ($x_3 = 1$). You will note some amazing differences between these curves and those for the bay-view units in Figure 16.11 (these differences explain why we needed two separate second-order surfaces to describe these two sets of units). The preference for floors is completely reversed on the ocean side of the complex: the lower the floor, the higher the estimated mean sale price. Apparently, people selecting the ocean-view units are primarily concerned with accessibility to the ocean, pool, beach, and game room. Note that the estimated mean sale price is highest near the elevator. It drops and then rises as you reach the units farthest from the elevator. An explanation for this phenomenon is similar to the one that we used for the bayside units. Units near the elevator are more accessible and nearer to the recreational facilities. Those

FIGURE 16.12
SAS graph of predicted price versus distance
(ocean-view units)

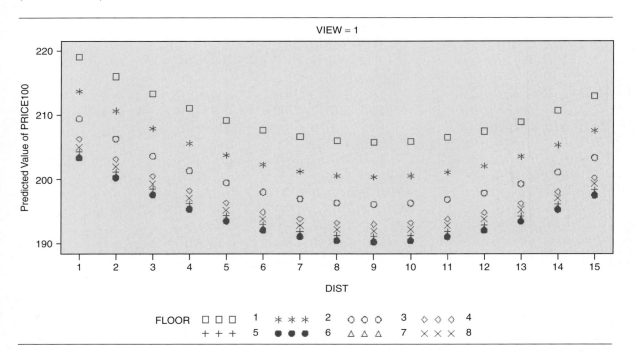

farthest from the elevators afford the greatest privacy. Units near the center of the complex offer reduced amounts of both accessibility *and* privacy. Notice that units adjacent to the elevator command a higher estimated mean sale price than those near the west end of the complex, suggesting that accessibility has a greater influence on price than privacy.

Rather than examine the graphs of $\hat{y}$ as a function of distance from elevator x_2, you may want to see how $\hat{y}$ behaves as a function of floor height x_1 for units located at various distances from the elevator. These estimated mean sale price curves are shown for bay-view units in Figure 16.13 and for ocean-view units in Figure 16.14. To avoid congestion in the graphs, we have shown only the curves for distances $x_2 = 1$, 5, 10 and 15. The symbols representing these distances are shown at the bottom of the graphs. We leave it to you and to the real estate experts to deduce the practical implications of these curves.

16.7
Comparing the Mean Sale Price for Two Types of Units (Optional)

[*Note*: This section requires an understanding of the mechanics of a multiple regression analysis presented in Appendix A.]

Comparing the mean sale price for two types of units might seem a useless endeavor, considering that all the units have been sold and that we will never be able to sell the units in the same economic environment that existed at the time the data were collected. Nevertheless, this information might be useful to a real estate appraiser or to a developer who is pricing units in a new and similar condominium

FIGURE 16.13

SAS graph of predicted price versus floor height
(bay-view units)

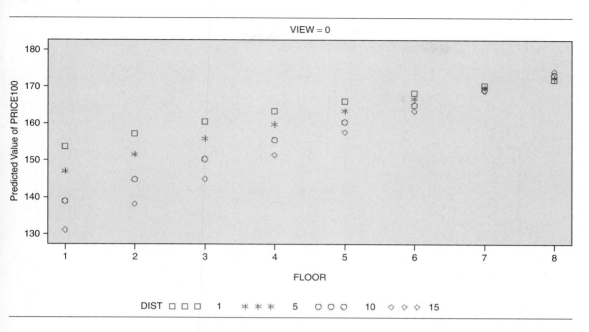

complex. We will assume that a comparison is useful and show you how it can be accomplished.

Suppose you want to estimate the difference in mean sale price for units in two different locations and with or without furniture. For example, suppose you wish to estimate the difference in mean sale price between the first-floor ocean-view and bay-view units located at the east end of building 1 (units 101 and 102 in Figure 16.1). Both these units afford a maximum of accessibility, but they possess different views. We assume that both are furnished. The estimate of the mean sale price $E(y)$ for the first-floor, bay-view unit will be the value of $\hat{y}$ when $x_1 = 1, x_2 = 1, x_3 = 0, x_4 = 0, x_5 = 1$. Similarly, the estimated value of $E(y)$ for the first-floor, ocean-view unit is obtained by substituting $x_1 = 1, x_2 = 1, x_3 = 1, x_4 = 0$, and $x_5 = 1$ into the prediction equation.

We will let $\hat{y}_o$ and $\hat{y}_b$ represent the estimated mean sale prices for the first-floor ocean-view and bay-view units, respectively. Then the estimator of the difference in mean sale prices for the two units is

$$\ell = \hat{y}_o - \hat{y}_b$$

We represent this estimator by the symbol ℓ, because it is a linear function of the parameter estimators $\hat{\beta}_0, \hat{\beta}_1, \ldots, \hat{\beta}_{13}$; i.e.,

$$\hat{y}_o = \hat{\beta}_0 + \hat{\beta}_1(1) + \hat{\beta}_2(1) + \hat{\beta}_3(1)(1) + \hat{\beta}_4(1)^2 + \hat{\beta}_5(1)^2$$
$$+ \hat{\beta}_6(1) + \hat{\beta}_7(0) + \hat{\beta}_8(1) + \hat{\beta}_9(1)(1) + \hat{\beta}_{10}(1)(1)$$
$$+ \hat{\beta}_{11}(1)(1)(1) + \hat{\beta}_{12}(1)^2(1) + \hat{\beta}_{13}(1)^2(1)$$

FIGURE 16.14

SAS graph of predicted price versus floor height
(ocean-view units)

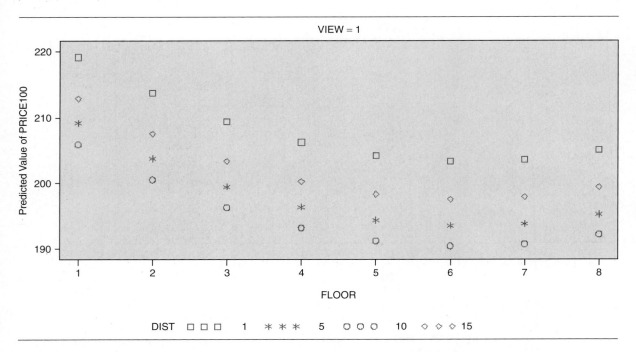

$$\hat{y}_b = \hat{\beta}_0 + \hat{\beta}_1(1) + \hat{\beta}_2(1) + \hat{\beta}_3(1)(1) + \hat{\beta}_4(1)^2 + \hat{\beta}_5(1)^2$$
$$+ \hat{\beta}_6(0) + \hat{\beta}_7(0) + \hat{\beta}_8(1) + \hat{\beta}_9(1)(0) + \hat{\beta}_{10}(1)(0)$$
$$+ \hat{\beta}_{11}(1)(1)(0) + \hat{\beta}_{12}(1)^2(0) + \hat{\beta}_{13}(1)^2(0)$$

then

$$\ell = \hat{y}_0 - \hat{y}_b = \hat{\beta}_6 + \hat{\beta}_9 + \hat{\beta}_{10} + \hat{\beta}_{11} + \hat{\beta}_{12} + \hat{\beta}_{13}$$

A 95% confidence interval for the mean value of a linear function of the estimators $\hat{\beta}_0, \hat{\beta}_1, \ldots, \hat{\beta}_k$, given in Section A.7 of Appendix A, is

$$\ell \pm (t_{.025})s\sqrt{\mathbf{a'(X'X)}^{-1}\mathbf{a}}$$

where in our case, $\ell = \hat{y}_0 - \hat{y}_b$ is the estimate of the difference in mean values for the two units, $E(y_0) - E(y_b)$; s is the least squares estimate of the standard deviation from the regression analysis of Model 3 (Figure 16.8); and $\mathbf{(X'X)}^{-1}$, the inverse matrix for the Model 3 regression analysis, is shown in Figure 16.15. The **a** matrix is a column matrix containing elements $a_0, a_1, a_2, \ldots a_{13}$, where $a_0, a_1, a_2, \ldots, a_{13}$ are the coefficients of $\hat{\beta}_0, \hat{\beta}_1, \ldots, \hat{\beta}_{13}$ in the linear function ℓ, i.e.,

$$\ell = a_0\hat{\beta}_0 + a_1\hat{\beta}_1 + \cdots + a_{13}\hat{\beta}_{13}$$

Since our linear function is

$$\ell = \hat{\beta}_6 + \hat{\beta}_9 + \hat{\beta}_{10} + \hat{\beta}_{11} + \hat{\beta}_{12} + \hat{\beta}_{13}$$

it follows that $a_6 = a_9 = a_{10} = a_{11} = a_{12} = a_{13} = 1$ and $a_0 = a_1 = a_2 = a_3 = a_4 = a_5 = a_7 = a_8 = 0$.

Substituting the values of $\hat{\beta}_6, \hat{\beta}_9, \hat{\beta}_{10}, \ldots, \hat{\beta}_{13}$ (given in Figure 16.8) into ℓ, we have

$$\ell = \hat{y}_o - \hat{y}_b = \hat{\beta}_6 + \hat{\beta}_9 + \hat{\beta}_{10} + \hat{\beta}_{11} + \hat{\beta}_{12} + \hat{\beta}_{13}$$
$$= 77.52 - 10.91 - 1.79 - .25 + .72 + .20 = 65.49$$

or, $6,549.

The value of $t_{.025}$ needed for the confidence interval is approximately equal to 1.96 (because of the large number of degrees of freedom), and the value of s, given in Figure 16.8, is $s = 6.60$. Finally, the matrix product $a'(X'X)^{-1}a$ can be obtained by multiplying the a matrix (described in the preceding paragraph) and the $(X'X)^{-1}$ matrix given in Figure 16.15. It can be shown (proof omitted) that this matrix product is the sum of the elements of the $(X'X)^{-1}$ matrix highlighted in Figure 16.15. Substituting these values into the formula for the confidence interval, we obtain

$$\overbrace{\hat{y}_o - \hat{y}_b}^{\ell} \pm t_{.025}s\sqrt{a'(X'X)^{-1}a} = 65.49 \pm (1.96)(6.60)\sqrt{4.55}$$
$$= 65.49 \pm 27.58 = (37.91, 93.07)$$

Therefore, we estimate the difference in the mean sale prices of first-floor ocean-view and bay-view units (units 101 and 102) to lie within the interval $3,791 to $9,307.

You can use the technique described above to compare the mean sale prices for any pair of units.

16.8
Conclusions

You may be able to propose a better model for mean sale price than Model 3, but we think that Model 3 provides a good fit to the data. Further, it reveals some interesting information on the preferences of buyers of oceanside condominium units.

Lower floors are preferred on the ocean side; the closer the units lie to the elevator and pool, the higher the estimated price. Some preference is given to the privacy of units located in the upper floor west-end.

Higher floors are preferred on the bay-view side (the side facing away from the ocean), with maximum preference given to units near the elevator (convenient and close to activities) and, to a lesser degree, to the privacy afforded by the west-end units.

EXERCISES

16.1. Of the 200 units in the condominium complex, 106 were sold at auction and the remainder were sold (some more than once) at the developer's fixed price. This case study analyzed the data for the 106 units sold at auction. The data in the CONDO file (described in Appendix J). Includes all 200 units. Fit Models 1, 2, and 3 to the data for all 200 units.

FIGURE 16.15

SAS printout of $(\mathbf{X'X}^{-1})$ matrix for Model 3

Variable	Intercept	FLOOR	DIST	FLR_DIS	FLOORSQ
Intercept	9.2651939926	-2.992278509	-0.256509984	0.0170883115	0.2368571042
FLOOR	-2.992278509	1.0766279902	-0.005854674	0.0023211328	-0.090368025
DIST	-0.256509984	-0.005854674	0.1140368456	-0.008968213	0.004048291
FLR_DIS	0.0170883115	0.0023211328	-0.008968213	0.0013751281	-0.000755996
FLOORSQ	0.2368571042	-0.090368025	0.004048291	-0.000755996	0.0079255903
DISTSQ	0.0133588561	-0.001568449	-0.004451619	0.0000311222	0.0001270988
VIEW	-9.261052131	2.9935687455	0.2560234919	-0.017075652	-0.23705016
END	-0.003509088	-0.001093121	0.000412168	-0.000010725	0.0001635619
FURNISH	-0.036216143	-0.011281739	0.0042538497	-0.000110692	0.0016880681
FLR_VIEW	2.9939400548	-1.0761104	0.0056595128	-0.002316054	0.090290579
DIS_VIEW	0.2609271215	0.0072306621	-0.114555671	0.008981714	-0.004254178
F_D_VIEW	-0.016875095	-0.002254713	0.0089431694	-0.001374476	0.0007460576
FLRSQ_VU	-0.237301892	0.0902294686	-0.003996047	0.0007546363	-0.007904858
DISSQ_VU	-0.013735053	0.0014512596	0.0044958064	-0.000032272	-0.000109564

Variable	DISTSQ	VIEW	END	FURNISH	FLR_VIEW
Intercept	0.0133588561	-9.261052131	-0.003509088	-0.036216143	2.9939400548
FLOOR	-0.001568449	2.9935687455	-0.001093121	-0.011281739	-1.0761104
DIST	-0.004451619	0.2560234919	0.000412168	0.0042538497	0.0056595128
FLR_DIS	0.0000311222	-0.017075652	-0.000010725	-0.000110692	-0.002316054
FLOORSQ	0.0001270988	-0.23705016	0.0001635619	0.0016880681	0.090290579
DISTSQ	0.0003393423	-0.013335485	-0.000019801	-0.00020436	0.0015778251
VIEW	-0.013335485	9.8635213822	-0.003460162	0.0305310915	-3.162628076
END	-0.000019801	-0.003460162	0.188369971	0.0048165173	0.0123690474
FURNISH	-0.00020436	0.0305310915	0.0048165173	0.0497096845	0.0090011283
FLR_VIEW	0.0015778251	-3.162628076	0.0123690474	0.0090011283	1.1531049299
DIS_VIEW	0.0044765442	-0.325308972	-0.020075044	-0.010316741	-0.004258676
F_D_VIEW	-0.000029919	0.0221320773	-0.000318277	-0.000181965	0.0015446723
FLRSQ_VU	-0.000129609	0.2497183294	-0.000886821	-0.001077559	-0.097391471
DISSQ_VU	-0.000341465	0.0159493901	0.0015437548	0.0007207221	-0.001457361

Variable	DIS_VIEW	F_D_VIEW	FLRSQ_VU	DISSQ_VU
Intercept	0.2609271215	-0.016875095	-0.237301892	-0.013735053
FLOOR	0.0072306621	-0.002254713	0.0902294686	0.0014512596
DIST	-0.114555671	0.0089431694	-0.003996047	0.0044958064
FLR_DIS	0.008981714	-0.001374476	0.0007546363	-0.000032272
FLOORSQ	-0.004254178	0.0007460576	-0.007904858	-0.000109564
DISTSQ	0.0044765442	-0.000029919	-0.000129609	-0.000341465
VIEW	-0.325308972	0.0221320773	0.2497183294	0.0159493901
END	-0.020075044	-0.000318277	-0.000886821	0.0015437548
FURNISH	-0.010316741	-0.000181965	-0.001077559	0.0007207221
FLR_VIEW	-0.004258676	0.0015446723	-0.097391471	-0.001457361
DIS_VIEW	0.1369922565	-0.00972204	0.0044120185	-0.005677362
F_D_VIEW	-0.00972204	0.0016363944	-0.000836949	-0.000017251
FLRSQ_VU	0.0044120185	-0.000836949	0.0086931319	0.0001387393
DISSQ_VU	-0.005677362	-0.000017251	0.0001387393	0.0004420689

16.2. Postulate some models that you think might be an improvement over Model 3. For example, consider a qualitative variable for sales method (auction or fixed price). Fit these models to the CONDO data set. Test to see whether they do, in fact, contribute more information for predicting sale price than Model 3.

REFERENCES

KELTING, H. "Investigation of condominium sale prices in three market scenarios: Utility of stepwise, interactive, multiple regression analysis and implications for design and appraisal methodology." Unpublished paper, University of Florida, Gainesville, 1979.

Modeling Daily Peak Electricity Demands

CONTENTS

OBJECTIVE

To present a time series approach to modeling daily peak electricity demands on a power company and to show how to use the time series model for short-term forecasting

17.1 The Problem

To operate effectively, power companies must be able to predict daily peak demand for electricity. *Demand* (or *load*) is defined as the rate (measured in megawatts) at which electric energy is delivered to customers. Since demand is normally recorded on an hourly basis, daily peak demand refers to the maximum hourly demand in a 24-hour period. Power companies are continually developing and refining statistical models of daily peak demand.

Models of daily peak demand serve a twofold purpose. First, the models provide short-term *forecasts* that will assist in the economic planning and dispatching of electric energy. Second, models that relate peak demand to one or more weather variables provide estimates of historical peak demands under a set of alternative weather conditions. That is, since changing weather conditions represent the primary source of variation in peak demand, the model can be used to answer the often-asked question, "What would the peak daily demand have been had normal weather prevailed?" This second application, commonly referred to as *weather normalization*, is mainly an exercise in *backcasting* (i.e., adjusting historical data) rather than forecasting (Jacob, 1985).

Since peak demand is recorded over time (days), the dependent variable is a time series and one approach to modeling daily peak demand is to use a time series model. This chapter presents key results of a study designed to compare several alternative methods of modeling 1983 daily peak demand for the Florida Power Corporation (FPC). For this case study, we focus on two time series models and a multiple regression model proposed in the original FPC study. Then we demonstrate how to forecast daily peak demand using one of the time series models. (We leave the problem of backcasting as an exercise.)

17.2
The Data

The data for the study consist of daily observations on peak demand recorded by the FPC for the period beginning November 1, 1982, and ending October 31, 1983, and several factors that are known to influence demand. (For reasons of confidentiality, the data are not made available on CD that accompanies this text.) It is typically assumed that demand consists of two components, a non–weather-sensitive "base" demand that is not influenced by temperature changes and a weather-sensitive demand that is highly responsive to changes in temperature.

The principal factor that affects the usage of non–weather-sensitive appliances (such as refrigerators, generators, light, and computers) is the *day of the week*. Typically, Saturdays have lower peak demands than weekdays due to decreased commercial and industrial activity, whereas Sundays and holidays exhibit even lower peak demand levels as commercial and industrial activity declines even further.

The single most important factor affecting the usage of weather-sensitive appliances (such as heaters and air conditioners) is *temperature*. During the winter months, as temperatures drop below comfortable levels, customers begin to operate their electric heating units, thereby increasing the level of demand placed on the system. Similarly, during the summer months, as temperatures climb above comfortable levels, the use of air conditioning drives demand upward. Since the FPC serves 32 counties along west-central and northern Florida, it was necessary to obtain temperature conditions from multiple weather stations. This was accomplished by identifying three primary weather stations within the FPC service area and recording the temperature value at the hour of peak demand each day at each station. A weighted average of these three daily temperatures was used to represent coincident temperature (i.e., temperature at the hour of peak demand) for the entire FPC service area, where the weights were proportional to the percentage of total electricity sales attributable to the weather zones surrounding each of the three weather stations.

To summarize, the independent variable (y_t) and the independent variables recorded for each of the 365 days of the November 1982–October 1983 year were as follows:

Dependent Variable:

y_t = Peak demand (in megawatts) observed on day t

Independent Variables:

Day of the week: Weekday, Saturday, or Sunday/holiday

Temperature: Coincident temperature (in degrees), i.e., the temperature recorded at the hour of the peak demand on day t, calculated as a weighted average of three daily temperatures

17.3
The Models

In any modeling procedure, it is often helpful to graph the data in a scatterplot. Figure 17.1 shows a graph of the daily peak demand (y_t) from November 1982 through October 1983. The effects of seasonal weather on peak demand are readily apparent from the figure. One way to account for this seasonal variation is to include dummy variables for months or trigonometric terms in the model (refer to Section 10.6). However, since temperature is such a strong indicator of the weather, the FPC chose a simpler model with temperature as the sole seasonal weather variable.

FIGURE 17.1

Daily peak megawatt demands, November 1982–October 1983

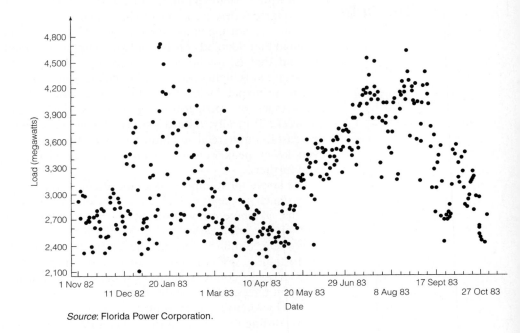

Source: Florida Power Corporation.

Figure 17.2 is a scatterplot of daily peak demands versus coincident temperature. Note the nonlinear relationship that exists between the two variables. During the cool winter months, peak demand is inversely related to temperature; lower temperatures cause increased usage of heating equipment, which in turn causes higher peak demands. In contrast, the summer months reveal a positive relationship between peak demand and temperature; higher temperatures yield higher peak demands because of greater usage of air conditioners. You might think that a second-order (quadratic) model would be a good choice to account for the U-shaped distribution of peak demands shown in Figure 17.2. The FPC, however, rejected such a model for two reasons:

1. A quadratic model yields a symmetrical shape (i.e., a parabola) and would, therefore, not allow independent estimates of the winter and summer peak demand–temperature relationship.

2. In theory, there exists a mild temperature range where peak demand is assumed to consist solely of the non–weather-sensitive base demand component. For this range, a temperature change will not spur any additional heating or cooling

FIGURE 17.2

Daily peak demand versus temperature, November 1982—October 1983

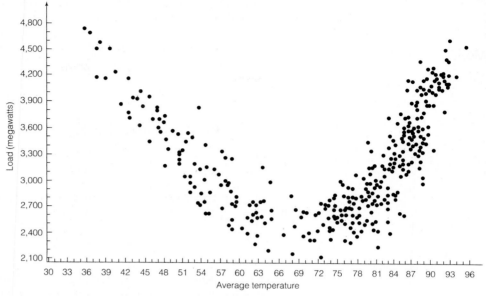

Source: Florida Power Corporation.

and, consequently, has no impact on demand. The lack of linearity in the bottom portion of the U-shaped parabola fitted by the quadratic model would yield overestimates of peak demand at the extremes of the mild temperature range and underestimates for temperatures in the middle of this range (see Figure 17.3).

FIGURE 17.3

Theoretical relationship between daily peak demand and temperature

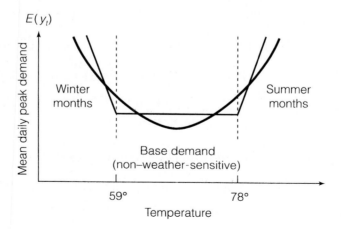

The solution was to model daily peak demand with a piecewise linear regression model (see Section 9.2). This approach has the advantage of allowing the peak demand–temperature relationship to vary between some prespecified temperature ranges, as well as providing a mechanism for joining the separate pieces.

Using the piecewise linear specification as the basic model structure, the following model of daily peak demand was proposed:

Model 1

$$y_t = \beta_0 + \underbrace{\beta_1(x_{1t} - 59)x_{2t} + \beta_2(x_{1t} - 78)x_{3t}}_{\text{Temperature}} + \underbrace{\beta_3 x_{4t} + \beta_4 x_{5t}}_{\text{Day of the week}} + \varepsilon_t$$

where

$x_{1t} = $ Coincident temperature on day t

$$x_{2t} = \begin{cases} 1 & \text{if } x_{1t} < 59 \\ 0 & \text{if not} \end{cases} \qquad x_{3t} = \begin{cases} 1 & \text{if } x_{1t} > 78 \\ 0 & \text{if not} \end{cases}$$

$$x_{4t} = \begin{cases} 1 & \text{if Saturday} \\ 0 & \text{if not} \end{cases} \qquad x_{5t} = \begin{cases} 1 & \text{if Sunday/holiday} \\ 0 & \text{if not} \end{cases} \qquad \text{(Base level} = \text{Weekday)}$$

$\varepsilon_t = $ Uncorrelated error term

Model 1 proposes three different straight-line relationships between peak demand (y_t) and coincident temperature x_{1t}, one for each of the three temperature ranges corresponding to winter months (less than 59°), non–weather-sensitive months (between 59° and 78°), and summer months (greater than 78°).* The model also allows for variations in demand because of day of the week (Saturday, Sunday/holiday, or weekday). Since interaction between temperature and day of the week is omitted, the model is assuming that the differences between mean peak demand for weekdays and weekends/holidays is constant for the winter-sensitive, summer-sensitive, and non–weather-sensitive months.

We will illustrate the mechanics of the piecewise linear terms by finding the equations of the three demand–temperature lines for weekdays (i.e., $x_{4t} = x_{5t} = 0$). Substituting $x_{4t} = 0$ and $x_{5t} = 0$ into the model, we have

Winter-sensitive months $(x_{1t} < 59°, \ x_{2t} = 1, \ x_{3t} = 0)$:

$$\begin{aligned} E(y_t) &= \beta_0 + \beta_1(x_{1t} - 59)(1) + \beta_2(x_{1t} - 78)(0) + \beta_3(0) + \beta_4(0) \\ &= \beta_0 + \beta_1(x_{1t} - 59) \\ &= (\beta_0 - 59\beta_1) + \beta_1 x_{1t} \end{aligned}$$

Summer-sensitive months $(x_{1t} > 78°, \ x_{2t} = 0, \ x_{3t} = 1)$:

$$\begin{aligned} E(y_t) &= \beta_0 + \beta_1(x_{1t} - 59)(0) + \beta_2(x_{1t} - 78)(1) + \beta_3(0) + \beta_4(0) \\ &= \beta_0 + \beta_2(x_{1t} - 78) \\ &= (\beta_0 - 78\beta_2) + \beta_2 x_{1t} \end{aligned}$$

*The temperature values, 59° and 78°, identify where the winter- and summer-sensitive portions of demand join the base demand component. These "knot values" were determined from visual inspection of the graph in Figure 17.2.

Non−weather-sensitive months $(59° \le x_{1t} \le 78°, \; x_{2t} = x_{3t} = 0)$:

$$E(y_t) = \beta_0 + \beta_1(x_{1t} - 59)(0) + \beta_2(x_{1t} - 78)(0) + \beta_3(0) + \beta_4(0)$$
$$= \beta_0$$

Note that the slope of the demand−temperature line for winter-sensitive months (when $x_{1t} < 59$) is β_1 (which we expect to be negative), whereas the slope for summer-sensitive months (when $x_{1t} > 78$) is β_2 (which we expect to be positive). The intercept term β_0 represents the mean daily peak demand observed in the non−weather-sensitive period (when $59 \le x_{1t} \le 78$). Notice also that the peak demand during non−weather-sensitive days does not depend on temperature (x_{1t}).

Model 1 is a multiple regression model that relies on the standard regression assumptions of independent errors (ε_t uncorrelated). This may be a serious shortcoming in view of the fact that the data are in the form of a time series. To account for possible autocorrelated residuals, two time series models were proposed:

Model 2
$$y_t = \beta_0 + \beta_1(x_{1t} - 59)x_{2t} + \beta_2(x_{1t} - 78)x_{3t} + \beta_3 x_{4t} + \beta_4 x_{5t} + R_t$$
$$R_t = \phi_1 R_{t-1} + \varepsilon_t$$

Model 2 proposes a regression−autoregression pair of models for daily peak demand (y_t). The deterministic component, $E(y_t)$, is identical to the deterministic component of Model 1; however, a first-order autoregressive model is chosen for the random error component. Recall (from Section 9.5) that a first-order autoregressive model is appropriate when the correlation between residuals diminishes as the distance between time periods (in this case, days) increases.

Model 3
$$y_t = \beta_0 + \beta_1(x_{1t} - 59)x_{2t} + \beta_2(x_{1t} - 78)x_{3t} + \beta_3 x_{4t} + \beta_4 x_{5t} + R_t$$
$$R_t = \phi_1 R_{t-1} + \phi_2 R_{t-2} + \phi_5 R_{t-5} + \phi_7 R_{t-7} + \varepsilon_t$$

Model 3 extends the first-order autoregressive error model of Model 2 to a seventh-order autoregressive model with lags at 1, 2, 5, and 7. In theory, the peak demand on day t will be highly correlated with the peak demand on day $t + 1$. However, there also may be significant correlation between demand 2 days, 5 days, and/or 1 week (7 days) apart. This more general error model is proposed to account for any residual correlation that may occur as a result of the week-to-week variation in peak demand, in addition to the day-to-day variation.

17.4
The Regression and Autoregression Analyses

The multiple regression computer printout for Model 1 is shown in Figure 17.4, and a plot of the least squares fit is shown in Figure 17.5. The model appears to provide a good fit to the data, with $R^2 = .8307$ and $F = 441.73$ (significant at $p = .0001$). The value, $s = 245.585$ implies that we can expect to predict daily peak demand accurate to within $2s \approx 491$ megawatts of its true value. However, we must be careful not to conclude at this point that the model is useful for predicting peak demand. Recall that in the presence of autocorrelated residuals, the standard errors of the regression coefficients are underestimated, thereby inflating the corresponding t statistics for testing $H_0: \beta_i = 0$. At worst, this could lead to the false conclusion

FIGURE 17.4

SAS least squares regression printout for daily peak demand, Model 1

The REG Procedure
Dependent Variable: LOAD

Analysis of Variance

Source	DF	Sum of Squares	Mean Square	F Value	Pr > F
Model	4	106565982	26641496	441.73	<.0001
Error	360	21712247	60311.8		
Corrected Total	364	128278229			

Root MSE	245.585	R-Square	0.8307
Dependent Mean	3191.863	Adj R-Sq	0.8289
Coeff Var	7.694		

Parameter Estimates

Variable	DF	Parameter Estimate	Standard Error	t Value	Pr > |t|
Intercept	1	2670.171	21.2518	125.64	<.0001
AVTW	1	-82.040	2.9419	-27.89	<.0001
AVTS	1	114.443	3.0505	37.52	<.0001
SAT	1	-164.932	37.9902	-4.34	0.0001
SUN	1	-285.114	35.3283	-8.07	<.0001

Durbin-Watson D	0.705
Number of Observations	365
1st Order Autocorrelation	0.648

that a β parameter is significantly different from 0; at best, the results, although significant, give an overoptimistic view of the predictive ability of the model.

To determine whether the residuals of the multiple regression model are positively autocorrelated, we conduct the Durbin–Watson test:

H_0: Uncorrelated residuals

H_a: Positive residual correlation

Recall that the Durbin–Watson test is designed specifically for detecting first-order autocorrelation in the residuals, R_t. Thus, we can write the null and alternative hypotheses as

H_0: $\phi_1 = 0$

H_a: $\phi_1 > 0$

where $R_t = \phi_1 R_{t-1} + \varepsilon_t$, and $\varepsilon_t =$ uncorrelated error (white noise).

The test statistic, shaded at the bottom of the printout in Figure 17.4, is $d = .705$. Recall that small values of d lead us to reject H_0: $\phi_1 = 0$ in favor of the alternative H_a: $\phi_1 > 0$. For $\alpha = .01$, $n = 365$, and $k = 4$ (the number of β parameters in the

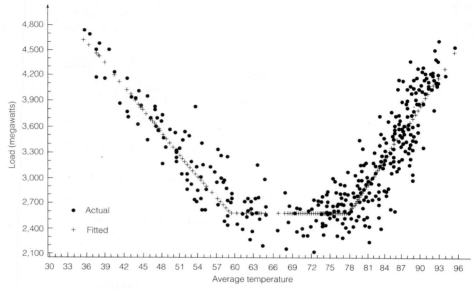

FIGURE 17.5

Daily peak demand versus temperature: Actual versus fitted piecewise linear model

Source: Florida Power Corporation.

Source: Florida Power Corporation.

model, excluding β_0), the lower bound on the critical value (obtained from Table 9 of Appendix C) is approximately $d_L = 1.46$. Since the value of the test statistic, $d = .705$, falls well below the lower bound, there is strong evidence at $\alpha = .01$ of positive (first-order) autocorrelated residuals. Thus, we need to incorporate terms for residual autocorrelation into the model.

The time series printouts for Models 2 and 3 are shown in Figures 17.6 and 17.7, respectively. A summary of the results for all three models is given in Table 17.1.

TABLE 17.1   **Summary of Results for Models 1, 2, and 3**

	MODEL 1	MODEL 2	MODEL 3
R^2	.8307	.9225	.9351
MSE	60,312	27,687	23,398
s	245.585	166.4	153.0

The addition of the first-order autoregressive error term in Model 2 yielded a drastic improvement to the fit of the model. The value of R^2 increased from .83 for Model 1 to .92, and the standard deviation s decreased from 245.6 to 166.4. These results support the conclusion reached by the Durbin–Watson test—namely, that the first-order autoregressive lag parameter ϕ_1 is significantly different from 0.

Does the more general autoregressive error model (Model 3) provide a better approximation to the pattern of correlation in the residuals than the first-order autoregressive model (Model 2)? To test this hypothesis, we would need to test $H_0: \phi_2 = \phi_5 = \phi_7 = 0$. Although we omit discussion of tests on autoregressive

FIGURE 17.6

SAS printout for autoregressive
time series model of daily peak
demand, Model 2

The AUTOREG Procedure
Dependent Variable: LOAD

Estimates of Autocorrelations

Lag	Covariance	Correlation	-1 9 8 7 6 5 4 3 2 1 0 1 2 3 4 5 6 7 8 9 1
0	59485.6	1.00000	|********************|
1	38519.4	0.64754	|************** |

Preliminary MSE =   34542.75

Estimates of the Autoregressive Paramenters

Lag	Coefficient	Standard Error	t Value
1	-0.647554	0.039887	-16.23

Yule-Walker Estimates

SSE	9939789	DFE	358
MSE	27687.4	Root MSE	166.39
Regress R-Square	0.7626	Total R-Square	0.9225

Variable	DF	Estimate	Standard Error	t Value	Approx Pr > |t|
Intercept	1	2812.967	29.8791	94.15	<.0001
AVTW	1	-65.337	2.6639	-24.53	<.0001
AVTS	1	83.455	3.8531	21.66	<.0001
SAT	1	-130.828	22.4136	-5.84	<.0001
SUN	1	-275.551	21.3737	-12.89	<.0001

parameters in this text,* we can arrive at a decision from a pragmatic point of view by again comparing the values of R^2 and s for the two models. The more complex autoregressive model proposed by Model 3 yields a slight increase in R^2 (.935 compared to .923 for Model 2) and a slight decrease in the value of s (153.0 compared to 166.4 for Model 2). The additional lag parameters, although they may be statistically significant, may not be practically significant. The practical analyst may decide that the first-order autoregressive process proposed by Model 2 is the more desirable option since it is easier to use to forecast peak daily demand (and therefore more explainable to managers) while yielding approximate prediction errors (as measured by $2s$) that are only slightly larger than those for Model 3.

For the purposes of illustration, we use Model 2 to forecast daily peak demand in the following section.

---

*For details of tests on autoregressive parameters, see Fuller (1976).

FIGURE 17.7

SAS printout for 7th-order autoregressive time series model of daily peak demand, Model 3

The AUTOREG Procedure
Dependent Variable: LOAD

Estimates of Autocorrelations

Lag	Covariance	Correlation	-1 9 8 7 6 5 4 3 2 1 0 1 2 3 4 5 6 7 8 9 1
0	59485.6	1.00000	|********************|
1	38519.4	0.64754	|*************
2	35741.0	0.60083	|************
3	32868.2	0.55254	|***********
4	29917.9	0.50294	|**********
5	31340.9	0.52686	|***********
6	30061.9	0.50536	|**********
7	31508.0	0.52967	|***********

Preliminary MSE =   28841.4

Estimates of the Autoregressive Paramenters

Lag	Coefficient	Standard Error	t Value
1	-0.367936	0.049902	-7.37
2	-0.207028	0.051722	-4.00
3	0.000000	0.000000	
4	0.000000	0.000000	
5	-0.135264	0.049072	-2.76
6	0.000000	0.000000	
7	-0.153385	0.048430	-3.17

Yule-Walker Estimates

SSE	8329842	DFE	356	
MSE	23398.4	Root MSE	152.97	
Regress R-Square	0.8112	Total R-Square	0.9351	

Variable	DF	Estimate	Standard Error	t Value	Approx Pr > |t|
Intercept	1	2809.950	58.2346	48.25	<.0001
AVTW	1	-71.282	2.2621	-31.51	<.0001
AVTS	1	79.120	4.1806	18.93	<.0001
SAT	1	-150.524	23.4728	-6.41	<.0001
SUN	1	-262.273	21.6832	-12.10	<.0001

17.5 Forecasting Daily Peak Electricity Demand

Suppose the FPC decided to use Model 2 to forecast daily peak demand for the first seven days of November 1983. The estimated model,* obtained from Figure 17.6, is

---

*Remember that the estimate of ϕ_1 is obtained by multiplying the value reported on the SAS printout by (−1).

given by

$$\hat{y}_t = 2{,}812.967 - 65.337(x_{1t} - 59)x_{2t} + 83.455(x_{1t} - 78)x_{3t}$$
$$-130.828x_{4t} - 275.551x_{5t} + \hat{R}_t$$
$$\hat{R}_t = .6475\hat{R}_{t-1}$$

The forecast for November 1, 1983 ($t = 366$), requires an estimate of the residual R_{365}, where $\hat{R}_{365} = y_{365} - \hat{y}_{365}$. The last day of the November 1982–October 1983 time period ($t = 365$) was October 31, 1983, a Monday. On this day the peak demand was recorded as $y_{365} = 2{,}752$ megawatts and the coincident temperature as $x_{1,365} = 77°$. Substituting the appropriate values of the dummy variables into the equation for $\hat{y}_t$ (i.e., $x_{2t} = 0$, $x_{3t} = 0$, $x_{4t} = 0$, and $x_{5t} = 0$),
we have

$$\hat{R}_{365} = y_{365} - \hat{y}_{365}$$
$$= 2{,}752 - [2{,}812.967 - 65.337(77 - 59)(0) + 83.455(77 - 78)(0)$$
$$-130.828(0) - 275.551(0)]$$
$$= 2{,}752 - 2{,}812.967 = -60.967$$

Then the formula for calculating the forecast for Tuesday, November 1, 1983, is

$$\hat{y}_{366} = 2{,}812.967 - 65.337(x_{1,366} - 59)x_{2,366} + 83.455(x_{1,366} - 78)x_{3,366}$$
$$-130.828x_{4,366} - 275.551x_{5,366} + \hat{R}_{366}$$

where

$$\hat{R}_{366} = \hat{\phi}_1 \hat{R}_{365} = (.6475)(-60.967) = -39.476$$

Note that the forecast requires an estimate of coincident temperature on that day, $\hat{x}_{1,366}$. If the FPC wants to forecast demand under normal weather conditions, then this estimate can be obtained from historical data for that day. Or, the FPC may choose to rely on a meteorologist's weather forecast for that day. For this example, assume that $\hat{x}_{1,366} = 76°$ (the actual temperature recorded by the FPC). Then $x_{2,366} = x_{3,366} = 0$ (since $59 \leq \hat{x}_{1,366} \leq 78$) and $x_{4,366} = x_{5,366} = 0$ (since the target day is a Tuesday). Substituting these values and the value of $\hat{R}_{366}$ into the equation, we have

$$\hat{y}_{366} = 2{,}812.967 - 65.337(76 - 59)(0) + 83.455(76 - 78)(0)$$
$$-130.828(0) - 275.551(0) - 39.476$$
$$= 2{,}773.49$$

Similarly, a forecast for Wednesday, November 2, 1983 (i.e., $t = 367$), can be obtained:

$$\hat{y}_{367} = 2{,}812.967 - 65.337(x_{1,367} - 59)x_{2,367} + 83.455(x_{1,367} - 78)x_{3,367}$$
$$-130.828x_{3,367} - 275.551x_{4,367} + \hat{R}_{367}$$

where $\hat{R}_{367} = \hat{\phi}_1 \hat{R}_{366} = (.6475)(-39.476) = -25.561$, and $x_{3,367} = x_{4,367} = 0$. For an estimated coincident temperature of $\hat{x}_{1,367} = 77°$ (again, this is the actual temperature recorded on that day), we have $x_{2,367} = 0$ and $x_{3,367} = 0$. Substituting these values into the prediction equation, we obtain

$$\hat{y}_{367} = 2{,}812.967 - 65.337(77 - 59)(0) + 83.455(77 - 78)(0)$$
$$-130.828(0) - 275.551(0) - 25.561$$
$$= 2{,}812.967 - 25.561$$
$$= 2{,}787.41$$

Approximate 95% prediction intervals for the two forecasts are calculated as follows:

Tuesday, Nov. 1, 1983:

$$\hat{y}_{366} \pm 1.96\sqrt{\text{MSE}}$$
$$= 2{,}773.49 \pm 1.96\sqrt{27{,}687.44}$$
$$= 2{,}773.49 \pm 326.14 \text{ or } (2{,}447.35, \ 3{,}099.63)$$

Wednesday, Nov. 2, 1983:

$$\hat{y}_{367} \pm 1.96\sqrt{\text{MSE}(1 + \hat{\phi}_1^2)}$$
$$= 2{,}787.41 \pm 1.96\sqrt{(27{,}687.44)[1 + (.6475)^2]}$$
$$= 2{,}787.41 \pm 388.53 \text{ or } (2{,}398.88, 3{,}175.94)$$

The forecasts, approximate 95% prediction intervals, and actual daily peak demands (recorded by the FPC) for the first seven days of November 1983 are given in Table 17.2. Note that actual peak demand y_t falls within the corresponding prediction interval for all seven days. Thus, the model appears to be useful for making short-term forecasts of daily peak demand. Of course, if the prediction intervals were extremely wide, this result would be of no practical value. For example, the

TABLE 17.2 **Forecasts and Actual Peak Demands for the First Seven Days of November 1983**

DATE	DAY t	FORECAST $\hat{y}_t$	APPROXIMATE 95% PREDICTION INTERVAL	ACTUAL DEMAND y_t	ACTUAL TEMPERATURE x_{1t}
Tues., Nov. 1	366	2,773.49	(2,447.35, 3,099.63)	2,799	76
Wed., Nov. 2	367	2,787.41	(2,398.88, 3,175.94)	2,784	77
Thurs., Nov. 3	368	2,796.42	(2,384.53, 3,208.31)	2,845	77
Fri., Nov. 4	369	2,802.25	(2,380.92, 3,223.58)	2,701	76
Sat., Nov. 5	370	2,675.20	(2,249.97, 3,100.43)	2,512	72
Sun., Nov. 6	371	2,532.92	(2,106.07, 2,959.77)	2,419	71
Mon., Nov. 7	372	2,810.06	(2,382.59, 3,237.53)	2,749	68

forecast error $y_t - \hat{y}_t$, measured as a percentage of the actual value y_t, may be large even when y_t falls within the prediction interval. Various techniques, such as the percent forecast error, are available for evaluating the accuracy of forecasts. Consult the references given at the end of Chapter 10 for details on these techniques.

17.6 Conclusions

This case study presents a time series approach to modeling and forecasting daily peak demand observed at Florida Power Corporation. A graphical analysis of the data provided the means of identifying and formulating a piecewise linear regression model relating peak demand to temperature and day of the week. The multiple regression model, although providing a good fit to the data, exhibited strong signs of positive residual autocorrelation.

Two autoregressive time series models were proposed to account for the autocorrelated errors. Both models were shown to provide a drastic improvement in model adequacy. Either could be used to provide reliable short-term forecasts of daily peak demand or for weather normalization (i.e., estimating the peak demand if normal weather conditions had prevailed).

EXERCISES

17.1. All three models discussed in this case study make the underlying assumption that the peak demand–temperature relationship is independent of day of the week. Write a model that includes interaction between temperature and day of the week. Show the effect the interaction has on the straight-line relationships between peak demand and temperature. Explain how you could test the significance of the interaction terms.

17.2. Consider the problem of using Model 2 for weather normalization. Suppose the temperature on Saturday, March 5, 1983 (i.e., $t = 125$), was $x_{1,125} = 25°$, unusually cold for that day. Normally, temperatures range from $40°$ to $50°$ on March 5 in the FPC service area. Substitute $x_{1,125} = 45°$ into the prediction equation to obtain an estimate of the peak demand expected if normal weather conditions had prevailed on March 5, 1983. Calculate an approximate 95% prediction interval for the estimate. [*Hint:* Use $\hat{y}_{125} \pm 1.96\sqrt{\text{MSE}}$.]

REFERENCES

FULLER, W. A. *Introduction to Statistical Time Series*. New York: Wiley, 1976.

JACOB, M. F. "A time series approach to modeling daily peak electricity demands." Paper presented at the SAS Users Group International Annual Conference, Reno, Nevada, 1985.

The Mechanics of a Multiple Regression Analysis

CONTENTS

A.1 Introduction

The rationale behind a multiple regression analysis and the types of inferences it permits you to make are the subjects of Chapter 4. We noted that the method of least squares most often leads to a very difficult computational problem—namely, the solution of a set of $(k+1)$ simultaneous linear equations in the unknown values of the estimates $\hat{\beta}_0, \hat{\beta}_1, \ldots, \hat{\beta}_k$—and that the formulas for the estimated standard errors $s_{\hat{\beta}_0}, s_{\hat{\beta}_1}, \ldots, s_{\hat{\beta}_k}$ are too complicated to express as ordinary algebraic formulas. We circumvented both these problems easily. We relied on the least squares estimates, confidence intervals, tests, etc., provided by a standard regression analysis software package. Thus, Chapter 4 provides a basic working knowledge of the types of inferences you might wish to make from a multiple regression analysis and explains how to interpret the results. If we can do this, why would we wish to know the actual process performed by the computer?

There are several answers to this question:

1. Some multiple regression statistical software packages do not print all the information you may want. As one illustration, we noted in Chapter 4 that very often the objective of a regression analysis is to develop a prediction equation that can be used to estimate the mean value of y (say, mean profit or mean yield)

for given values of the predictor variables $x_1, x_2, \ldots, x_k$. Some software packages do not give the confidence interval for $E(y)$ or a prediction interval for y. Thus, you might need to know how to find the necessary quantities from the analysis and perform the computations yourself.

2. A multiple regression software package may possess the capability of computing some specific quantity that you desire, but you may find the instructions on how to "call" for this special calculation difficult to understand. It may be easier to identify the components required for your computation and do it yourself.

3. For some designed experiments, finding the least squares equations and solving them is a trivial operation. Understanding the process by which the least squares equations are generated and understanding how they are solved will help you understand how experimental design affects the results of a regression analysis. Thus, a knowledge of the computations involved in performing a regression analysis will help you to better understand the contents of Chapters 10 and 11.

To summarize, "knowing how it is done" is not essential for performing an ordinary regression analysis or interpreting its results. But "knowing how" helps, and it is essential for an understanding of many of the finer points associated with a multiple regression analysis. This appendix explains "how it is done" without getting into the unpleasant task of performing the computations for solving the least squares equations. This mechanical and tedious procedure can be left to a computer (the solutions are verifiable). We illustrate the procedure in Appendix B.

A.2
Matrices and Matrix Multiplication

Although it is very difficult to give the formulas for the multiple regression least squares estimators and for their estimated standard errors in ordinary algebra, it is easy to do so using **matrix algebra**. Thus, by arranging the data in particular rectangular patterns called **matrices** and by performing various operations with them, we can obtain the least squares estimates and their estimated standard errors. In this section and Sections A.3 and A.4, we will define what we mean by a matrix and explain various operations that can be performed with matrices. We will explain how to use this information to conduct a regression analysis in Section A.5.

Three matrices, **A**, **B**, and **C**, are shown here. Note that each matrix is a rectangular arrangement of numbers with one number in every row–column position.

$$\mathbf{A} = \begin{bmatrix} 2 & 3 \\ 0 & 1 \\ -1 & 6 \end{bmatrix} \qquad \mathbf{B} = \begin{bmatrix} 3 & 0 & 1 \\ -1 & 0 & 1 \\ 4 & 2 & 0 \end{bmatrix} \qquad \mathbf{C} = \begin{bmatrix} 1 \\ 2 \\ 1 \end{bmatrix}$$

> **Definition A.1**
> A **matrix** is a rectangular array of numbers.*

The numbers that appear in a matrix are called **elements** of the matrix. If a matrix contains r rows and c columns, there will be an element in each of the row–column

---

*For our purpose, we assume that the numbers are real.

positions of the matrix, and the matrix will have $r \times c$ elements. For example, the matrix $\mathbf{A}$ shown previously contains $r = 3$ rows, $c = 2$ columns, and $rc = (3)(2) = 6$ elements, one in each of the six row–column positions.

Definition A.2

A number in a particular row–column position is called an **element** of the matrix.

Notice that the matrices $\mathbf{A}$, $\mathbf{B}$, and $\mathbf{C}$ contain different numbers of rows and columns. The numbers of rows and columns give the **dimensions** of a matrix.

When we give a formula in matrix notation, the elements of a matrix will be represented by symbols. For example, if we have a matrix

$$\mathbf{A} = \begin{bmatrix} a_{11} & a_{12} & a_{13} \\ a_{21} & a_{22} & a_{23} \end{bmatrix}$$

the symbol a_{ij} will denote the element in the ith row and jth column of the matrix. The first subscript always identifies the row and the second identifies the column in which the element is located. For example, the element a_{12} is in the first row and second column of the matrix $\mathbf{A}$. The rows are always numbered from top to bottom, and the columns are always numbered from left to right.

Definition A.3

A matrix containing r rows and c columns is said to be an **$r \times c$ matrix** where r and c are the **dimensions** of the matrix.

Definition A.4

If $r = c$, a matrix is said to be a **square matrix**.

Matrices are usually identified by capital letters, such as $\mathbf{A}$, $\mathbf{B}$, $\mathbf{C}$, corresponding to the letters of the alphabet employed in ordinary algebra. The difference is that in ordinary algebra, a letter is used to denote a single real number, whereas in matrix algebra, *a letter denotes a rectangular array of numbers*. The operations of matrix algebra are very similar to those of ordinary algebra—you can add matrices, subtract them, multiply them, and so on. But since we are concerned only with the applications of matrix algebra to the solution of the least squares equations, we will define only the operations and types of matrices that are pertinent to that subject.

The most important operation for us is matrix multiplication, which requires **row–column multiplication**. To illustrate this process, suppose we wish to find the product $\mathbf{AB}$, where

$$\mathbf{A} = \begin{bmatrix} 2 & 1 \\ 4 & -1 \end{bmatrix} \qquad \mathbf{B} = \begin{bmatrix} 2 & 0 & 3 \\ -1 & 4 & 0 \end{bmatrix}$$

We will always multiply the rows of **A** (the matrix on the left) by the columns of **B** (the matrix on the right). The product formed by the first row of **A** times the first column of **B** is obtained by multiplying the elements in corresponding positions and summing these products. Thus, the first row, first column product, shown diagrammatically here, is

$$(2)(2) + (1)(-1) = 4 - 1 = 3$$

$$\mathbf{AB} = \begin{bmatrix} 2 & 1 \\ 4 & -1 \end{bmatrix} \begin{bmatrix} 2 & 0 & 3 \\ -1 & 4 & 0 \end{bmatrix} = \begin{bmatrix} 3 & & \end{bmatrix}$$

Similarly, the first row, second column product is

$$(2)(0) + (1)(4) = 0 + 4 = 4$$

So far we have

$$\mathbf{AB} = \begin{bmatrix} 3 & 4 & \end{bmatrix}$$

To find the complete matrix product **AB**, all we need to do is find each element in the **AB** matrix. Thus, we will define an element in the ith row, jth column of **AB** as the product of the ith row of **A** and the jth column of **B**. We complete the process in Example A.1.

EXAMPLE   A.1

Find the product **AB**, where

$$\mathbf{A} = \begin{bmatrix} 2 & 1 \\ 4 & -1 \end{bmatrix} \qquad \mathbf{B} = \begin{bmatrix} 2 & 0 & 3 \\ -1 & 4 & 0 \end{bmatrix}$$

Solution

If we represent the product **AB** as

$$\mathbf{C} = \begin{bmatrix} c_{11} & c_{12} & c_{13} \\ c_{21} & c_{22} & c_{23} \end{bmatrix}$$

we have already found $c_{11} = 3$ and $c_{12} = 4$. Similarly, the element c_{21}, the element in the second row, first column of **AB**, is the product of the second row of **A** and the first column of **B**:

$$(4)(2) + (-1)(-1) = 8 + 1 = 9$$

Proceeding in a similar manner to find the remaining elements of **AB**, we have

$$\mathbf{AB} = \begin{bmatrix} 2 & 1 \\ 4 & -1 \end{bmatrix} \begin{bmatrix} 2 & 0 & 3 \\ -1 & 4 & 0 \end{bmatrix} = \begin{bmatrix} 3 & 4 & 6 \\ 9 & -4 & 12 \end{bmatrix} \qquad \blacklozenge$$

Now, try to find the product **BA**, using matrices **A** and **B** from Example A.1. You will observe two very important differences between multiplication in matrix algebra and multiplication in ordinary algebra:

1. You cannot find the product **BA** because you cannot perform row–column multiplication. You can see that the dimensions do not match by placing the matrices side by side.

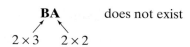

BA   does not exist

The number of elements (3) in a row of **B** (the matrix on the left) does not match the number of elements (2) in a column of **A** (the matrix on the right). Therefore, you cannot perform row–column multiplication, and the matrix product **BA** does not exist. The point is, not all matrices can be multiplied. You can find products for matrices **A** and **B** only when **A** is $r \times d$ and **B** is $d \times c$. That is:

Requirement for Matrix Multiplication

The two inner dimension numbers must be equal. The dimensions of the product will always be given by the outer dimension numbers:

Dimensions of AB Are $r \times c$

---

2. The second difference between ordinary and matrix multiplication is that in ordinary algebra, $ab = ba$. In matrix algebra, **AB** usually does not equal **BA**. In fact, as noted in item 1, **BA** may not even exist.

Definition A.5

The product **AB** of an $r \times d$ matrix **A** and a $d \times c$ matrix **B** is an $r \times c$ matrix **C**, where the element $c_{ij} (i = 1, 2, \ldots, r; j = 1, 2, \ldots, c)$ of **C** is the product of the ith row of **A** and the jth column of **B**.

EXAMPLE A.2

Given the matrices below, find **IA** and **IB**.

$$\mathbf{A} = \begin{bmatrix} 2 \\ 1 \\ 3 \end{bmatrix} \qquad \mathbf{B} = \begin{bmatrix} 3 & 0 \\ 1 & 2 \\ 4 & -1 \end{bmatrix} \qquad \mathbf{I} = \begin{bmatrix} 1 & 0 & 0 \\ 0 & 1 & 0 \\ 0 & 0 & 1 \end{bmatrix}$$

Solution

Notice that the product

exists and that it will be of dimensions 3×1. Performing the row–column multiplication yields

$$\mathbf{IA} = \begin{bmatrix} 1 & 0 & 0 \\ 0 & 1 & 0 \\ 0 & 0 & 1 \end{bmatrix} \begin{bmatrix} 2 \\ 1 \\ 3 \end{bmatrix} = \begin{bmatrix} 2 \\ 1 \\ 3 \end{bmatrix}$$

Similarly,

exists and is of dimensions 3×2. Performing the row—column multiplications, we obtain

$$\mathbf{IB} = \begin{bmatrix} 1 & 0 & 0 \\ 0 & 1 & 0 \\ 0 & 0 & 1 \end{bmatrix} \begin{bmatrix} 3 & 0 \\ 1 & 2 \\ 4 & -1 \end{bmatrix} = \begin{bmatrix} 3 & 0 \\ 1 & 2 \\ 4 & -1 \end{bmatrix}$$

Notice that the **I** matrix possesses a special property. We have **IA** = **A** and **IB** = **B**. We will comment further on this property in Section A.3.   ◆

EXERCISES

A.1. Given the matrices **A**, **B**, and **C**:

$$\mathbf{A} = \begin{bmatrix} 3 & 0 \\ -1 & 4 \end{bmatrix} \qquad \mathbf{B} = \begin{bmatrix} 2 & 1 \\ 0 & -1 \end{bmatrix}$$

$$\mathbf{C} = \begin{bmatrix} 1 & 0 & 3 \\ -2 & 1 & 2 \end{bmatrix}$$

a. Find **AB**.   **b**. Find **AC**.   **c**. Find **BA**.

A.2. Given the matrices **A**, **B**, and **C**:

$$\mathbf{A} = \begin{bmatrix} 3 & 1 & 3 \\ 2 & 0 & 4 \\ -4 & 1 & 2 \end{bmatrix} \qquad \mathbf{B} = \begin{bmatrix} 1 & 0 & 2 \end{bmatrix} \qquad \mathbf{C} = \begin{bmatrix} 3 \\ 0 \\ 2 \end{bmatrix}$$

a. Find **AC**.   **b**. Find **BC**.
c. Is it possible to find **AB**? Explain.

A.3. Suppose **A** is a 3×2 matrix and **B** is a 2×4 matrix.
   a. What are the dimensions of **AB**?
   b. Is it possible to find the product **BA**? Explain.

A.4. Suppose matrices **B** and **C** are of dimensions 1×3 and 3×1, respectively.
   a. What are the dimensions of the product **BC**?
   b. What are the dimensions of **CB**?
   c. If **B** and **C** are the matrices shown in Exercise A.2, find **CB**.

A.5. Given the matrices **A**, **B**, and **C**:

$$\mathbf{A} = \begin{bmatrix} 1 & 0 & 0 \\ 0 & 3 & 0 \\ 0 & 0 & 2 \end{bmatrix}$$

$$\mathbf{B} = \begin{bmatrix} 2 & 3 \\ -3 & 0 \\ 4 & -1 \end{bmatrix} \quad \mathbf{C} = [3 \quad 0 \quad 2]$$

   a. Find **AB**.   **b.** Find **CA**.   **c.** Find **CB**.

A.6. Given the matrices:

$$\mathbf{A} = [3 \quad 0 \quad -1 \quad 2] \quad \mathbf{B} = \begin{bmatrix} 2 \\ -1 \\ 0 \\ 3 \end{bmatrix}$$

   a. Find **AB**.   **b.** Find **BA**.

A.3
Identity Matrices and Matrix Inversion

In ordinary algebra, the number 1 is the identity element for the multiplication operation. That is, 1 is the element such that any other number, say, c, multiplied by the identity element is equal to c. Thus, $4(1) = 4$, $(-5)(1) = -5$, and so forth.

The corresponding identity element for multiplication in matrix algebra, identified by the symbol **I**, is a matrix such that

$$\mathbf{AI} = \mathbf{IA} = \mathbf{A} \quad \text{for any matrix } \mathbf{A}$$

The difference between identity elements in ordinary algebra and matrix algebra is that in ordinary algebra there is only one identity element, the number 1. In matrix algebra, the identity matrix must possess the correct dimensions for the product **IA** to exist. Consequently, there is an infinitely large number of identity matrices—all square and possessing the same pattern. The 1×1, 2×2, and 3×3 identity matrices are

$$\underset{1 \times 1}{\mathbf{I}} = [1] \qquad \underset{2 \times 2}{\mathbf{I}} = \begin{bmatrix} 1 & 0 \\ 0 & 1 \end{bmatrix} \qquad \underset{3 \times 3}{\mathbf{I}} = \begin{bmatrix} 1 & 0 & 0 \\ 0 & 1 & 0 \\ 0 & 0 & 1 \end{bmatrix}$$

In Example A.2, we demonstrated the fact that this matrix satisfies the property

$$\mathbf{IA} = \mathbf{A}$$

Definition A.6

If **A** is any matrix, then a matrix **I** is defined to be an **identity matrix** if $\mathbf{AI} = \mathbf{IA} = \mathbf{A}$. The matrices that satisfy this definition possess the pattern

$$\mathbf{I} = \begin{bmatrix} 1 & 0 & 0 & \ldots & 0 \\ 0 & 1 & 0 & \ldots & 0 \\ 0 & 0 & 1 & \ldots & 0 \\ \cdot & \cdot & \cdot & \ldots & \cdot \\ \cdot & \cdot & \cdot & \ldots & \cdot \\ \cdot & \cdot & \cdot & \ldots & \cdot \\ 0 & 0 & 0 & \ldots & 1 \end{bmatrix}$$

EXAMPLE　A.3

If **A** is the matrix shown here, find **IA** and **AI**.

$$\mathbf{A} = \begin{bmatrix} 3 & 4 & -1 \\ 1 & 0 & 2 \end{bmatrix}$$

Solution

$$\underset{2 \times 2 \quad 2 \times 3}{\overset{\mathbf{IA}}{\frown}} = \begin{bmatrix} 1 & 0 \\ 0 & 1 \end{bmatrix} \begin{bmatrix} 3 & 4 & -1 \\ 1 & 0 & 2 \end{bmatrix} = \begin{bmatrix} 3 & 4 & -1 \\ 1 & 0 & 2 \end{bmatrix} = \mathbf{A}$$

$$\underset{2 \times 3 \quad 3 \times 3}{\overset{\mathbf{AI}}{\frown}} = \begin{bmatrix} 3 & 4 & -1 \\ 1 & 0 & 2 \end{bmatrix} \begin{bmatrix} 1 & 0 & 0 \\ 0 & 1 & 0 \\ 0 & 0 & 1 \end{bmatrix} = \begin{bmatrix} 3 & 4 & -1 \\ 1 & 0 & 2 \end{bmatrix} = \mathbf{A}$$

Notice that the identity matrices used to find the products **IA** and **AI** were of different dimensions. This was necessary for the products to exist. ◆

The identity element assumes importance when we consider the process of division and its role in the solution of equations. In ordinary algebra, division is essentially multiplication using the reciprocals of elements. For example, the equation

$$2x = 6$$

can be solved by dividing both sides of the equation by 2, *or* it can be solved by *multiplying* both sides of the equation by $\frac{1}{2}$, which is the reciprocal of 2. Thus,

$$\left(\frac{1}{2}\right) 2x = \frac{1}{2}(6)$$

$$x = 3$$

What is the reciprocal of an element? It is the element such that the reciprocal times the element is equal to the identity element. Thus, the reciprocal of 3 is $\frac{1}{3}$ because

$$3\left(\frac{1}{3}\right) = 1$$

The identity matrix plays the same role in matrix algebra. Thus, the reciprocal of a matrix **A**, called the **inverse of A** and denoted by the symbol $\mathbf{A}^{-1}$, is a matrix such that $\mathbf{A}\mathbf{A}^{-1} = \mathbf{A}^{-1}\mathbf{A} = \mathbf{I}$.

Inverses are defined only for square matrices, but not all square matrices possess inverses. Those that do have inverses play an important role in solving the least squares equations and in other aspects of a regression analysis. We will show you one important application of the inverse matrix in Section A.4. The procedure for finding the inverse of a matrix is demonstrated in Appendix B.

Definition A.7

The square matrix $\mathbf{A}^{-1}$ is said to be the **inverse** of the square matrix $\mathbf{A}$ if

$$\mathbf{A}^{-1}\mathbf{A} = \mathbf{A}\mathbf{A}^{-1} = \mathbf{I}$$

The procedure for finding an inverse matrix is computationally quite tedious and is performed most often using a computer. There are several exceptions. For example, finding the inverse of one type of matrix, called a **diagonal matrix**, is easy. A diagonal matrix is one that has nonzero elements down the **main diagonal** (running from top left of the matrix to bottom right) and 0 elements elsewhere. Thus, the identity matrix is a diagonal matrix (with 1's along the main diagonal), as are the following matrices:

$$\mathbf{A} = \begin{bmatrix} 3 & 0 & 0 \\ 0 & 1 & 0 \\ 0 & 0 & 2 \end{bmatrix} \qquad \mathbf{B} = \begin{bmatrix} 5 & 0 & 0 & 0 \\ 0 & 2 & 0 & 0 \\ 0 & 0 & 1 & 0 \\ 0 & 0 & 0 & 5 \end{bmatrix}$$

Definition A.8

A **diagonal matrix** is one that contains nonzero elements on the main diagonal and 0 elements elsewhere.

You can verify that the inverse of

$$\mathbf{A} = \begin{bmatrix} 3 & 0 & 0 \\ 0 & 1 & 0 \\ 0 & 0 & 2 \end{bmatrix} \quad \text{is} \quad \mathbf{A}^{-1} = \begin{bmatrix} \frac{1}{3} & 0 & 0 \\ 0 & 1 & 0 \\ 0 & 0 & \frac{1}{2} \end{bmatrix}$$

That is, $\mathbf{A}\mathbf{A}^{-1} = \mathbf{I}$. In general, the inverse of a diagonal matrix is given by the following theorem, which is stated without proof:

Theorem A.1

The **inverse of a diagonal matrix**

$$\mathbf{D} = \begin{bmatrix} d_{11} & 0 & 0 & \cdots & 0 \\ 0 & d_{22} & 0 & \cdots & 0 \\ 0 & 0 & d_{33} & \cdots & 0 \\ \cdot & \cdot & \cdot & \cdots & \cdot \\ \cdot & \cdot & \cdot & \cdots & \cdot \\ \cdot & \cdot & \cdot & \cdots & \cdot \\ 0 & 0 & 0 & \cdots & d_{nn} \end{bmatrix} \quad \text{is } \mathbf{D}^{-1} = \begin{bmatrix} 1/d_{11} & 0 & 0 & \cdots & 0 \\ 0 & 1/d_{22} & 0 & \cdots & 0 \\ 0 & 0 & 1/d_{33} & \cdots & 0 \\ \cdot & \cdot & \cdot & \cdots & \cdot \\ \cdot & \cdot & \cdot & \cdots & \cdot \\ \cdot & \cdot & \cdot & \cdots & \cdot \\ 0 & 0 & 0 & \cdots & 1/d_{nn} \end{bmatrix}$$

A second type of matrix that is easy to invert is a 2×2 matrix. The following theorem shows how to find the inverse of this type of matrix.

Theorem A.2

The **inverse of a 2 × 2 matrix**

$$\mathbf{A} = \begin{bmatrix} a & b \\ c & d \end{bmatrix} \text{ is } \mathbf{A}^{-1} = \begin{bmatrix} \frac{d}{ad-bc} & \frac{-b}{ad-bc} \\ \frac{-c}{ad-bc} & \frac{a}{ad-bc} \end{bmatrix}$$

You can verify that the inverse of

$$\mathbf{A} = \begin{bmatrix} 1 & -2 \\ -2 & 6 \end{bmatrix} \text{ is } \mathbf{A}^{-1} = \begin{bmatrix} 3 & 1 \\ 1 & \frac{1}{2} \end{bmatrix}$$

We demonstrate another technique for finding $\mathbf{A}^{-1}$ in Appendix B.

EXERCISES

A.7. Let $\mathbf{A} = \begin{bmatrix} 3 & 0 & 2 \\ -1 & 1 & 4 \end{bmatrix}$.

  a. Give the identity matrix that will be used to obtain the product $\mathbf{IA}$.

  b. Show that $\mathbf{IA} = \mathbf{A}$.

  c. Give the identity matrix that will be used to find the product $\mathbf{AI}$.

  d. Show that $\mathbf{AI} = \mathbf{A}$.

A.8. Given the following matrices $\mathbf{A}$ and $\mathbf{B}$, show that $\mathbf{AB} = \mathbf{I}$, that $\mathbf{BA} = \mathbf{I}$, and consequently, verify that $\mathbf{B} = \mathbf{A}^{-1}$.

$$\mathbf{A} = \begin{bmatrix} 1 & 0 & 0 \\ 0 & 2 & 0 \\ 0 & 0 & 3 \end{bmatrix} \quad \mathbf{B} = \begin{bmatrix} 1 & 0 & 0 \\ 0 & \frac{1}{2} & 0 \\ 0 & 0 & \frac{1}{3} \end{bmatrix}$$

A.9. If

$$\mathbf{A} = \begin{bmatrix} 12 & 0 & 0 & 8 \\ 0 & 12 & 0 & 0 \\ 0 & 0 & 8 & 0 \\ 8 & 0 & 0 & 8 \end{bmatrix}$$

verify that

$$\mathbf{A}^{-1} = \begin{bmatrix} \frac{1}{4} & 0 & 0 & -\frac{1}{4} \\ 0 & \frac{1}{12} & 0 & 0 \\ 0 & 0 & \frac{1}{8} & 0 \\ -\frac{1}{4} & 0 & 0 & \frac{3}{8} \end{bmatrix}$$

A.10. If

$$\mathbf{A} = \begin{bmatrix} 3 & 0 & 0 \\ 0 & 5 & 0 \\ 0 & 0 & 7 \end{bmatrix}$$

show that

$$\mathbf{A}^{-1} = \begin{bmatrix} \frac{1}{3} & 0 & 0 \\ 0 & \frac{1}{5} & 0 \\ 0 & 0 & \frac{1}{7} \end{bmatrix}$$

A.11. Verify Theorem A.1.

A.12. Verify Theorem A.2.

A.13. Find the inverse of

$$\mathbf{A} = \begin{bmatrix} 2 & -1 \\ 2 & 3 \end{bmatrix}$$

Consider the following set of simultaneous linear equations in two unknowns:

$$2v_1 + v_2 = 7$$

$$v_1 - v_2 = 2$$

Note that the solution for these equations is $v_1 = 3$, $v_2 = 1$.

Now define the matrices

$$\mathbf{A} = \begin{bmatrix} 2 & 1 \\ 1 & -1 \end{bmatrix} \qquad \mathbf{V} = \begin{bmatrix} v_1 \\ v_2 \end{bmatrix} \qquad \mathbf{G} = \begin{bmatrix} 7 \\ 2 \end{bmatrix}$$

Thus, $\mathbf{A}$ is the matrix of coefficients of v_1 and v_2, $\mathbf{V}$ is a column matrix containing the unknowns (written in order, top to bottom), and $\mathbf{G}$ is a column matrix containing the numbers on the right-hand side of the equal signs.

Now, the given system of simultaneous equations can be rewritten as a **matrix equation:**

$$\mathbf{AV} = \mathbf{G}$$

By a matrix equation, we mean that the product matrix, $\mathbf{AV}$, is equal to the matrix $\mathbf{G}$. *Equality of matrices means that corresponding elements are equal.* You can see that this is true for the expression $\mathbf{AV} = \mathbf{G}$, since

$$\underset{2 \times 2 \qquad 2 \times 1}{\overset{\mathbf{AV}}{\overbrace{\qquad\qquad}}} = \begin{bmatrix} 2 & 1 \\ 1 & -1 \end{bmatrix} \begin{bmatrix} v_1 \\ v_2 \end{bmatrix} = \begin{bmatrix} (2v_1 + v_2) \\ (v_1 - v_2) \end{bmatrix} = \underset{2 \times 1}{\overset{\mathbf{G}}{}}$$

The matrix procedure for expressing a system of two simultaneous linear equations in two unknowns can be extended to express a set of k simultaneous equations in k unknowns. If the equations are written in the orderly pattern

$$a_{11}v_1 + a_{12}v_2 + \cdots + a_{1k}v_k = g_1$$
$$a_{21}v_1 + a_{22}v_2 + \cdots + a_{2k}v_k = g_2$$
$$\vdots \qquad \vdots \qquad\qquad \vdots \qquad \vdots$$
$$a_{k1}v_1 + a_{k2}v_2 + \cdots + a_{kk}v_k = g_k$$

then the set of simultaneous linear equations can be expressed as the matrix equation $\mathbf{AV} = \mathbf{G}$, where

$$\mathbf{A} = \begin{bmatrix} a_{11} & a_{12} & \cdots & a_{1k} \\ a_{21} & & \cdots & a_{2k} \\ \vdots & & & \vdots \\ a_{k1} & & \cdots & a_{kk} \end{bmatrix} \qquad \mathbf{V} = \begin{bmatrix} v_1 \\ v_2 \\ \vdots \\ v_k \end{bmatrix} \qquad \mathbf{G} = \begin{bmatrix} g_1 \\ g_2 \\ \vdots \\ g_k \end{bmatrix}$$

Now let us solve this system of simultaneous equations. (If they are uniquely solvable, it can be shown that $\mathbf{A}^{-1}$ exists.) Multiplying both sides of the matrix equation by $\mathbf{A}^{-1}$, we have

$$(\mathbf{A}^{-1})\mathbf{AV} = (\mathbf{A}^{-1})\mathbf{G}$$

But since $\mathbf{A}^{-1}\mathbf{A} = \mathbf{I}$, we have

$$(\mathbf{I})\mathbf{V} = \mathbf{A}^{-1}\mathbf{G}$$
$$\mathbf{V} = \mathbf{A}^{-1}\mathbf{G}$$

In other words, if we know $\mathbf{A}^{-1}$, we can find the solution to the set of simultaneous linear equations by obtaining the product $\mathbf{A}^{-1}\mathbf{G}$.

Matrix Solution to a Set of Simultaneous Linear Equations, $\mathbf{AV} = \mathbf{G}$

Solution: $\mathbf{V} = \mathbf{A}^{-1}\mathbf{G}$

EXAMPLE A.4

Apply the result from the box to find the solution to the set of simultaneous linear equations

$$2v_1 + v_2 = 7$$
$$v_1 - v_2 = 2$$

Solution

The first step is to obtain the inverse of the coefficient matrix,

$$\mathbf{A} = \begin{bmatrix} 2 & 1 \\ 1 & -1 \end{bmatrix}$$

namely,

$$\mathbf{A}^{-1} = \begin{bmatrix} \frac{1}{3} & \frac{1}{3} \\ \frac{1}{3} & -\frac{2}{3} \end{bmatrix}$$

(This matrix can be found using a packaged computer program for matrix inversion or, for this simple case, you could use the procedure explained in Appendix B.) As a check, note that

$$\mathbf{A}^{-1}\mathbf{A} = \begin{bmatrix} \frac{1}{3} & \frac{1}{3} \\ \frac{1}{3} & -\frac{2}{3} \end{bmatrix} \begin{bmatrix} 2 & 1 \\ 1 & -1 \end{bmatrix} = \begin{bmatrix} 1 & 0 \\ 0 & 1 \end{bmatrix} = \mathbf{I}$$

The second step is to obtain the product $\mathbf{A}^{-1}\mathbf{G}$. Thus,

$$\mathbf{V} = \mathbf{A}^{-1}\mathbf{G} = \begin{bmatrix} \frac{1}{3} & \frac{1}{3} \\ \frac{1}{3} & -\frac{2}{3} \end{bmatrix} \begin{bmatrix} 7 \\ 2 \end{bmatrix} = \begin{bmatrix} 3 \\ 1 \end{bmatrix}$$

Since

$$\mathbf{V} = \begin{bmatrix} v_1 \\ v_2 \end{bmatrix} = \begin{bmatrix} 3 \\ 1 \end{bmatrix}$$

it follows that $v_1 = 3$ and $v_2 = 1$. You can see that these values of v_1 and v_2 satisfy the simultaneous linear equations and are the values that we specified as a solution at the beginning of this section. ◆

EXERCISES

A.1. Suppose the simultaneous linear equations

$$3v_1 + v_2 = 5$$

$$v_1 - v_2 = 3$$

are expressed as a matrix equation,

$$AV = G$$

a. Find the matrices $\mathbf{A}$, $\mathbf{V}$, and $\mathbf{G}$.
b. Verify that

$$\mathbf{A}^{-1} = \begin{bmatrix} \frac{1}{4} & \frac{1}{4} \\ \frac{1}{4} & -\frac{3}{4} \end{bmatrix}$$

[*Note*: A procedure for finding $\mathbf{A}^{-1}$ is given in Appendix B.]

c. Solve the equations by finding $\mathbf{V} = \mathbf{A}^{-1}\mathbf{G}$.

A.2. For the simultaneous linear equations

$$10v_1 + 20v_3 - 60 = 0$$

$$20v_2 - 60 = 0$$

$$20v_1 + 68v_3 - 176 = 0$$

a. Find the matrices $\mathbf{A}$, $\mathbf{V}$, and $\mathbf{G}$.
b. Verify that

$$\mathbf{A}^{-1} = \begin{bmatrix} \frac{17}{70} & 0 & -\frac{1}{14} \\ 0 & \frac{1}{20} & 0 \\ -\frac{1}{14} & 0 & \frac{1}{28} \end{bmatrix}$$

c. Solve the equations by finding $\mathbf{V} = \mathbf{A}^{-1}\mathbf{G}$.

A.5
The Least Squares Equations and Their Solutions

To apply matrix algebra to a regression analysis, we must place the data in matrices in a particular pattern. We will suppose that the linear model is

$$y = \beta_0 + \beta_1 x_1 + \beta_2 x_2 + \cdots + \beta_k x_k + \varepsilon$$

where (from Chapter 4) $x_1, x_2, \ldots, x_k$ could actually represent the squares, cubes, cross products, or other functions of predictor variables, and ε is a random error. We will assume that we have collected n data points, i.e., n values of y and corresponding values of $x_1, x_2, \ldots, x_k$, and that these are denoted as shown in the table:

DATA POINT	y Value	x_1	x_2	$\cdots$	x_k
1	y_1	x_{11}	x_{21}		x_{k1}
2	y_2	x_{12}	x_{22}		x_{k2}
$\vdots$	$\vdots$	$\vdots$	$\vdots$		$\vdots$
n	y_n	x_{1n}	x_{2n}		x_{kn}

Then the two data matrices $\mathbf{Y}$ and $\mathbf{X}$ are as shown in the next box.

The Data Matrices Y and X and the $\hat{\beta}$ Matrix

$$\mathbf{Y} = \begin{bmatrix} y_1 \\ y_2 \\ y_3 \\ \vdots \\ y_n \end{bmatrix} \quad \mathbf{X} = \begin{bmatrix} 1 & x_{11} & x_{21} & \cdots & x_{k1} \\ 1 & x_{12} & x_{22} & \cdots & x_{k2} \\ 1 & x_{13} & x_{23} & \cdots & x_{k3} \\ \vdots & \vdots & \vdots & & \vdots \\ 1 & x_{1n} & x_{2n} & \cdots & x_{kn} \end{bmatrix} \quad \hat{\beta} = \begin{bmatrix} \hat{\beta}_0 \\ \hat{\beta}_1 \\ \hat{\beta}_2 \\ \vdots \\ \hat{\beta}_k \end{bmatrix}$$

Notice that the first column in the **X** matrix is a column of 1's. Thus, we are inserting a value of x, namely, x_0, as the coefficient of β_0, where x_0 is a variable always equal to 1. Therefore, there is one column in the **X** matrix for each β parameter. Also, remember that a particular data point is identified by specific rows of the **Y** and **X** matrices. For example, the y value y_3 for data point 3 is in the third row of the **Y** matrix, and the corresponding values of $x_1, x_2, \ldots, x_k$ appear in the third row of the **X** matrix.

The $\hat{\beta}$ matrix shown in the box contains the least squares estimates (which we are attempting to obtain) of the coefficients $\beta_0, \beta_1, \ldots, \beta_k$ of the linear model

$$y = \beta_0 + \beta_1 x_1 + \beta_2 x_2 + \cdots + \beta_k x_k + \varepsilon$$

To write the least squares equation, we need to define what we mean by the **transpose of a matrix**. If

$$\mathbf{Y} = \begin{bmatrix} 5 \\ 1 \\ 0 \\ 4 \\ 2 \end{bmatrix} \quad \mathbf{X} = \begin{bmatrix} 1 & 0 \\ 1 & 1 \\ 1 & 4 \\ 1 & 2 \\ 1 & 6 \end{bmatrix}$$

then the transpose matrices of the **Y** and **X** matrices, denoted as **Y′** and **X′**, respectively, are

$$\mathbf{Y'} = \begin{bmatrix} 5 & 1 & 0 & 4 & 2 \end{bmatrix} \quad \mathbf{X'} = \begin{bmatrix} 1 & 1 & 1 & 1 & 1 \\ 0 & 1 & 4 & 2 & 6 \end{bmatrix}$$

Definition A.9

The **transpose of a matrix A**, denoted as **A′**, is obtained by interchanging corresponding rows and columns of the **A** matrix. That is, the ith row of the **A** matrix becomes the ith column of the **A′** matrix.

Using the **Y** and **X** data matrices, their transposes, and the $\hat{\beta}$ matrix, we can write the least squares equations (proof omitted) as:

Least Squares Matrix Equation

$$(\mathbf{X'X})\hat{\boldsymbol{\beta}} = \mathbf{X'Y}$$

Thus, $(\mathbf{X'X})$ is the coefficient matrix of the least squares estimates $\hat{\beta}_0, \hat{\beta}_1, \ldots, \hat{\beta}_k$, and $\mathbf{X'Y}$ gives the matrix of constants that appear on the right-hand side of the equality signs. In the notation of Section A.4,

$$\mathbf{A} = \mathbf{X'X} \quad \mathbf{V} = \hat{\boldsymbol{\beta}} \quad \mathbf{G} = \mathbf{X'Y}$$

The solution, which follows from Section A.4, is

Least Squares Matrix Solution

$$\hat{\boldsymbol{\beta}} = (\mathbf{X'X})^{-1}\mathbf{X'Y}$$

Thus, to solve the least squares matrix equation, the computer calculates $(\mathbf{X'X})$, $(\mathbf{X'X})^{-1}$, $\mathbf{X'Y}$, and, finally, the product $(\mathbf{X'X})^{-1}\mathbf{X'Y}$. We will illustrate this process using the data for the advertising example from Section 3.3.

EXAMPLE A.5

Find the least squares line for the data given in Table A.1.

TABLE A.1

MONTH	ADVERTISING EXPENDITURE x, hundreds of dollars	SALES REVENUE y, thousands of dollars
1	1	1
2	2	1
3	3	2
4	4	2
5	5	4

Solution

The model is

$$y = \beta_0 + \beta_1 x_1 + \varepsilon$$

and the $\mathbf{Y}$, $\mathbf{X}$, and $\hat{\boldsymbol{\beta}}$ matrices are

$$\mathbf{Y} = \begin{bmatrix} 1 \\ 1 \\ 2 \\ 2 \\ 4 \end{bmatrix} \quad \mathbf{X} = \begin{matrix} & \begin{matrix} x_0 & x_1 \end{matrix} \\ & \begin{bmatrix} 1 & 1 \\ 1 & 2 \\ 1 & 3 \\ 1 & 4 \\ 1 & 5 \end{bmatrix} \end{matrix} \quad \hat{\boldsymbol{\beta}} = \begin{bmatrix} \hat{\beta}_0 \\ \hat{\beta}_1 \end{bmatrix}$$

Then,

$$\mathbf{X'X} = \begin{bmatrix} 1 & 1 & 1 & 1 & 1 \\ 1 & 2 & 3 & 4 & 5 \end{bmatrix} \begin{bmatrix} 1 & 1 \\ 1 & 2 \\ 1 & 3 \\ 1 & 4 \\ 1 & 5 \end{bmatrix} = \begin{bmatrix} 5 & 15 \\ 15 & 55 \end{bmatrix}$$

$$\mathbf{X'Y} = \begin{bmatrix} 1 & 1 & 1 & 1 & 1 \\ 1 & 2 & 3 & 4 & 5 \end{bmatrix} \begin{bmatrix} 1 \\ 1 \\ 2 \\ 2 \\ 4 \end{bmatrix} = \begin{bmatrix} 10 \\ 37 \end{bmatrix}$$

The last matrix that we need is $(\mathbf{X'X})^{-1}$. This matrix, which can be found by using Theorem A.2 (or by using the method of Appendix B), is

$$(\mathbf{X'X})^{-1} = \begin{bmatrix} 1.1 & -.3 \\ -.3 & .1 \end{bmatrix}$$

Then the solution to the least squares equation is

$$\hat{\boldsymbol{\beta}} = (\mathbf{X'X})^{-1}\mathbf{X'Y} = \begin{bmatrix} 1.1 & -.3 \\ -.3 & .1 \end{bmatrix} \begin{bmatrix} 10 \\ 37 \end{bmatrix} = \begin{bmatrix} -.1 \\ .7 \end{bmatrix}$$

Thus, $\hat{\beta}_0 = -.1$, $\hat{\beta}_1 = .7$, and the prediction equation is

$$\hat{y} = -.1 + .7x$$

You can verify that this is the same answer as obtained in Section 3.3.   ◆

EXAMPLE   A.6

Table A.2 contains data on monthly electrical power usage and home size for a sample of $n = 10$ homes. Find the least squares solution for fitting the monthly power usage y to size of home x for the model

$$y = \beta_0 + \beta_1 x + \beta_2 x^2 + \varepsilon$$

Solution

The $\mathbf{Y}$, $\mathbf{X}$, and $\hat{\boldsymbol{\beta}}$ matrices are as follows:

TABLE A.2   **Data for Power Usage Study**

SIZE OF HOME x, square feet	MONTHLY USAGE y, kilowatt-hours	SIZE OF HOME x, square feet	MONTHLY USAGE y, kilowatt-hours
1,290	1,182	1,840	1,711
1,350	1,172	1,980	1,804
1,470	1,264	2,230	1,840
1,600	1,493	2,400	1,956
1,710	1,571	2,930	1,954

$$Y = \begin{bmatrix} 1,182 \\ 1,172 \\ 1,264 \\ 1,493 \\ 1,571 \\ 1,711 \\ 1,804 \\ 1,840 \\ 1,956 \\ 1,954 \end{bmatrix}$$

$$X = \begin{matrix} x_0 & x & x^2 \\ \begin{bmatrix} 1 & 1,290 & 1,664,100 \\ 1 & 1,350 & 1,822,500 \\ 1 & 1,470 & 2,160,900 \\ 1 & 1,600 & 2,560,000 \\ 1 & 1,710 & 2,924,100 \\ 1 & 1,840 & 3,385,600 \\ 1 & 1,980 & 3,920,400 \\ 1 & 2,230 & 4,972,900 \\ 1 & 2,400 & 5,760,000 \\ 1 & 2,930 & 8,584,900 \end{bmatrix} \end{matrix}$$

Then

$$X'X = \begin{bmatrix} 10 & 18,800 & 37,755,400 \\ 18,800 & 37,755,400 & 8,093.9 \times 10^7 \\ 37,755,400 & 8,093.9 \times 10^7 & 1.843 \times 10^{14} \end{bmatrix}$$

$$X'Y = \begin{bmatrix} 15,947 \\ 31,283,250 \\ 6.53069 \times 10^{10} \end{bmatrix}$$

and (obtained using a statistical software package):

$$(X'X)^{-1} = \begin{bmatrix} 26.9156 & -.027027 & 6.3554 \times 10^{-6} \\ -.027027 & 2.75914 \times 10^{-5} & -6.5804 \times 10^{-9} \\ 6.3554 \times 10^{-6} & -6.5804 \times 10^{-9} & 1.5934 \times 10^{-12} \end{bmatrix}$$

Finally, performing the multiplication, we obtain

$$\hat{\beta} = (X'X)^{-1}X'Y$$

$$= \begin{bmatrix} 26.9156 & -.027027 & 6.3554 \times 10^{-6} \\ -.027027 & 2.75914 \times 10^{-5} & -6.5804 \times 10^{-9} \\ 6.3554 \times 10^{-6} & -6.5804 \times 10^{-9} & 1.5934 \times 10^{-12} \end{bmatrix} \begin{bmatrix} 15,947 \\ 31,283,250 \\ 6.53069 \times 10^{10} \end{bmatrix}$$

$$= \begin{bmatrix} -1,216.14389 \\ 2.39893 \\ -.00045 \end{bmatrix}$$

Thus,

$$\hat{\beta}_0 = -1,216.14389$$
$$\hat{\beta}_1 = 2.39893$$
$$\hat{\beta}_2 = -.00045$$

and the prediction equation is

$$\hat{y} = -1,216.14389 + 2.39893x - .00045x^2$$

The MINITAB printout for the regression analysis is shown in Figure A.1. Note that the β estimates we obtained agree with the shaded values.   ◆

FIGURE A.1

MINITAB regression printout for power usage model

```
The regression equation is
Y = - 1216 + 2.40 X -0.000450 XSQ

Predictor        Coef      SE Coef         T        P
Constant       -1216.1        242.8     -5.01    0.002
X               2.3989       0.2458      9.76    0.000
XSQ         -0.00045004   0.00005908     -7.62    0.000

S = 46.80        R-Sq = 98.2%      R-Sq(adj) = 97.7%

Analysis of Variance

Source           DF         SS         MS         F        P
Regression        2     831070     415535    189.71    0.000
Residual Error    7      15333       2190
Total             9     846402
```

EXERCISES

A.16. Use the method of least squares to fit a straight line to the five data points:

x	−2	−1	0	1	2
y	4	3	3	1	−1

a. Construct $\mathbf{Y}$ and $\mathbf{X}$ matrices for the data.
b. Find $\mathbf{X'X}$ and $\mathbf{X'Y}$.
c. Find the least squares estimates $\hat{\boldsymbol{\beta}} = (\mathbf{X'X})^{-1}\mathbf{X'Y}$. [*Note*: See Theorem A.1 for information on finding $(\mathbf{X'X})^{-1}$.]
d. Give the prediction equation.

A.17. Use the method of least squares to fit the model $E(y) = \beta_0 + \beta_1 x$ to the six data points:

x	1	2	3	4	5	6
y	1	2	2	3	5	6

a. Construct $\mathbf{Y}$ and $\mathbf{X}$ matrices for the data.
b. Find $\mathbf{X'X}$ and $\mathbf{X'Y}$.
c. Verify that

$$(\mathbf{X'X})^{-1} = \begin{bmatrix} \frac{13}{15} & -\frac{7}{35} \\ -\frac{7}{35} & \frac{2}{35} \end{bmatrix}$$

d. Find the $\hat{\boldsymbol{\beta}}$ matrix.

e. Give the prediction equation.

A.18. An experiment was conducted in which two y observations were collected for each of five values of x:

x	-2		-1		0		1		2	
y	1.1	1.3	2.0	2.1	2.7	2.8	3.4	3.6	4.1	4.0

Use the method of least squares to fit the second-order model, $E(y) = \beta_0 + \beta_1 x + \beta_2 x^2$, to the 10 data points.

a. Give the dimensions of the $\mathbf{Y}$ and $\mathbf{X}$ matrices.

b. Verify that

$$(\mathbf{X}'\mathbf{X})^{-1} = \begin{bmatrix} \frac{17}{70} & 0 & -\frac{1}{14} \\ 0 & \frac{1}{20} & 0 \\ -\frac{1}{14} & 0 & \frac{1}{28} \end{bmatrix}$$

c. Both $\mathbf{X}'\mathbf{X}$ and $(\mathbf{X}'\mathbf{X})^{-1}$ are symmetric matrices. What is a symmetric matrix?

d. Find the $\hat{\boldsymbol{\beta}}$ matrix and the least squares prediction equation.

e. Plot the data points and graph the prediction equation.

A.6
Calculating SSE and s^2

You will recall that the variances of the estimators of all the β parameters and of $\hat{y}$ depend on the value of σ^2, the variance of the random error ε that appears in the linear model. Since σ^2 will rarely be known in advance, we must use the sample data to estimate its value.

Matrix Formulas for SSE and s^2

$$\text{SSE} = \mathbf{Y}'\mathbf{Y} - \boldsymbol{\beta}'\mathbf{X}'\mathbf{Y}$$

$$s^2 = \frac{\text{SSE}}{n - \text{Number of } \beta \text{ parameters in model}}$$

We demonstrate the use of these formulas with the advertising–sales data of Example A.5.

EXAMPLE A.7

Find the SSE for the advertising–sales data of Example A.5.

Solution

From Example A.5,

$$\hat{\boldsymbol{\beta}} = \begin{bmatrix} -.1 \\ .7 \end{bmatrix} \quad \text{and} \quad \mathbf{X}'\mathbf{Y} = \begin{bmatrix} 10 \\ 37 \end{bmatrix}$$

Then,

$$\mathbf{Y}'\mathbf{Y} = \begin{bmatrix} 1 & 1 & 2 & 2 & 4 \end{bmatrix} \begin{bmatrix} 1 \\ 1 \\ 2 \\ 2 \\ 4 \end{bmatrix} = 26$$

and

$$\hat{\beta}'\mathbf{X}'\mathbf{Y} = \begin{bmatrix} -.1 & .7 \end{bmatrix} \begin{bmatrix} 10 \\ 37 \end{bmatrix} = 24.9$$

So

$$\text{SSE} = \mathbf{Y}'Y - \hat{\beta}'\mathbf{X}'\mathbf{Y} = 26 - 24.9 = 1.1$$

(Note that this is the same answer as that obtained in Section 3.3.) Finally,

$$s^2 = \frac{\text{SSE}}{n - \text{Number of } \beta \text{ parameters in model}} = \frac{1.1}{5 - 2} = .367$$

This estimate is needed to construct a confidence interval for β_1, to test a hypothesis concerning its value, or to construct a confidence interval for the mean sales for a given advertising expenditure.   ◆

A.7
Standard Errors of Estimators, Test Statistics, and Confidence Intervals for $\beta_0, \beta_1, \ldots, \beta_k$

This appendix is important because all the relevant information pertaining to the standard errors of the sampling distributions of $\hat{\beta}_0, \hat{\beta}_1, \ldots, \hat{\beta}_k$ (and hence of $\hat{\mathbf{Y}}$) is contained in $(\mathbf{X}'\mathbf{X})^{-1}$. Thus, if we denote the $(\mathbf{X}'\mathbf{X})^{-1}$ matrix as

$$(\mathbf{X}'\mathbf{X})^{-1} = \begin{bmatrix} c_{00} & c_{01} & \cdots & c_{0k} \\ c_{10} & c_{11} & \cdots & c_{1k} \\ c_{20} & c_{21} & \cdots & c_{2k} \\ \vdots & \vdots & \vdots & \vdots \\ c_{k0} & c_{k1} & \cdots & c_{kk} \end{bmatrix}$$

then it can be shown (proof omitted) that the standard errors of the sampling distributions of $\hat{\beta}_0, \hat{\beta}_1, \ldots, \hat{\beta}_k$ are

$$\sigma_{\hat{\beta}_0} = \sigma\sqrt{c_{00}}$$
$$\sigma_{\hat{\beta}_1} = \sigma\sqrt{c_{11}}$$
$$\sigma_{\hat{\beta}_2} = \sigma\sqrt{c_{22}}$$
$$\vdots$$
$$\sigma_{\hat{\beta}_k} = \sigma\sqrt{c_{kk}}$$

where σ is the standard deviation of the random error ε. In other words, the diagonal elements of $(\mathbf{X}'\mathbf{X})^{-1}$ give the values of $c_{00}, c_{11}, \ldots, c_{kk}$ that are required for finding the standard errors of the estimators $\hat{\beta}_0, \hat{\beta}_1, \ldots, \hat{\beta}_k$. The estimated values of the standard errors are obtained by replacing σ by s in the formulas for the standard errors. Thus, the estimated standard error of $\hat{\beta}_1$ is $s_{\hat{\beta}_1} = s\sqrt{c_{11}}$.

The confidence interval for a single β parameter, β_i, is given in the next box.

Confidence Interval for β_i

$$\hat{\beta}_i \pm t_{\alpha/2}(\text{Estimated standard error of } \hat{\beta}_i)$$

or

$$\hat{\beta}_i \pm (t_{\alpha/2})s\sqrt{c_{ii}}$$

where $t_{\alpha/2}$ is based on the number of degrees of freedom associated with s.

Similarly, the test statistic for testing the null hypothesis $H_0 : \beta_i = 0$ is as shown in the following box.

Test Statistic for $H_0: \beta_i = 0$

$$t = \frac{\hat{\beta}_i}{s\sqrt{c_{ii}}}$$

EXAMPLE   A.8

Refer to Example A.5 and find the estimated standard error for the sampling distribution of $\hat{\beta}_1$, the estimator of the slope of the line β_1. Then give a 95% confidence interval for β_1.

Solution

The $(\mathbf{X}'\mathbf{X})^{-1}$ matrix for the least squares solution of Example A.5 was

$$(\mathbf{X}'\mathbf{X})^{-1} = \begin{bmatrix} 1.1 & -.3 \\ -.3 & .1 \end{bmatrix}$$

Therefore, $c_{00} = 1.1, c_{11} = .1$, and the estimated standard error for $\hat{\beta}_1$ is

$$s_{\hat{\beta}_1} = s\sqrt{c_{11}} = \sqrt{.367}(\sqrt{.1}) = .192$$

The value for s, $\sqrt{.367}$, was obtained from Example A.7.
A 95% confidence interval for β_1 is

$$\hat{\beta}_1 \pm (t_{\alpha/2})s\sqrt{c_{11}}$$
$$.7 \pm (3.182)(.192) = (.09, 1.31)$$

The t value, $t_{.025}$, is based on $(n - 2) = 3$ df. Observe that this is the same confidence interval as the one obtained in Section 3.6.   ◆

EXAMPLE   A.9

Refer to Example A.6 and the least squares solution for fitting power usage y to the size of a home x using the model

$$y = \beta_0 + \beta_1 x + \beta_2 x^2 + \varepsilon$$

The MINITAB printout for the analysis is reproduced in Figure A.2.

a. Compute the estimated standard error for $\hat{\beta}_1$, and compare this result with the value shaded in Figure A.2.

b. Compute the value of the test statistic for testing $H_0: \beta_2 = 0$. Compare this with the value shaded in Figure A.2.

FIGURE A.2

MINITAB regression printout for power usage model

```
The regression equation is
Y = - 1216 + 2.40 X -0.000450 XSQ

Predictor          Coef      SE Coef          T          P
Constant        -1216.1        242.8      -5.01      0.002
X                2.3989       0.2458       9.76      0.000
XSQ         -0.00045004   0.00005908      -7.62      0.000

S = 46.80        R-Sq = 98.2%      R-Sq(adj) = 97.7%

Analysis of Variance

Source            DF          SS          MS          F          P
Regression         2      831070      415535     189.71      0.000
Residual Error     7       15333        2190
Total              9      846402
```

Solution

The fitted model is

$$\hat{y} = -1{,}216.14389 + 2.39893x - .00045x^2$$

The $(\mathbf{X'X})^{-1}$ matrix, obtained in Example A.6, is

$$(\mathbf{X'X})^{-1} = \begin{bmatrix} 26.9156 & -.027027 & 6.3554 \times 10^{-6} \\ -.027027 & 2.75914 \times 10^{-5} & -6.5804 \times 10^{-9} \\ 6.3554 \times 10^{-6} & -6.5804 \times 10^{-9} & 1.5934 \times 10^{-12} \end{bmatrix}$$

From $(\mathbf{X'X})^{-1}$, we know that

$c_{00} = 26.9156$
$c_{11} = 2.75914 \times 10^{-5}$
$c_{22} = 1.5934 \times 10^{-12}$

and from the printout, $s = 46.80$.

a. The estimated standard error of $\hat{\beta}_1$ is

$$s_{\hat{\beta}_1} = s\sqrt{c_{11}}$$

$$= (46.80)\sqrt{2.75914 \times 10^{-1}} = .2458$$

Notice that this agrees with the value of $s_{\hat{\beta}_1}$ shaded in the MINITAB printout (Figure A.2).

b. The value of the test statistic for testing $H_0: \beta_2 = 0$ is

$$t = \frac{\hat{\beta}_2}{s\sqrt{c_{22}}} = \frac{-.00045}{(46.80)\sqrt{1.5934 \times 10^{-12}}} = -7.62$$

Notice that this value of the t statistic agrees with the value -7.62 the shaded in the printout (Figure A.2).   ◆

EXERCISES

A.19. Do the data given in Exercise A.16 provide sufficient evidence to indicate that x contributes information for the prediction of y? Test $H_0: \beta_1 = 0$ against $H_a: \beta_1 \neq 0$ using $\alpha = .05$.

A.20. Find a 90% confidence interval for the slope of the line in Exercise A.19.

A.21. The term in the second-order model $E(y) = \beta_0 + \beta_1 x + \beta_2 x^2$ that controls the curvature in its graph is $\beta_2 x^2$. If $\beta_2 = 0$, $E(y)$ graphs as a straight line. Do the data given in Exercise A.18 provide sufficient evidence to indicate curvature in the model for $E(y)$? Test $H_0: \beta_2 = 0$ against $H_a: \beta_2 \neq 0$ using $\alpha = .10$.

A.8
A Confidence Interval for a Linear Function of the β Parameters; a Confidence Interval for $E(y)$

Suppose we were to postulate that the mean value of the productivity, y, of a company is related to the size of the company, x, and that the relationship could be modeled by the expression

$$E(y) = \beta_0 + \beta_1 x + \beta_2 x^2$$

A graph of $E(y)$ might appear as shown in Figure A.3.

We might have several reasons for collecting data on the productivity and size of a set of n companies and for finding the least squares prediction equation,

$$\hat{y} = \hat{\beta}_0 + \hat{\beta}_1 x + \hat{\beta}_2 x^2$$

FIGURE A.3
Graph of mean productivity $E(y)$

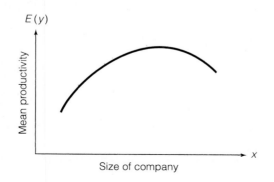

$E(y)$

Mean productivity

Size of company

x

For example, we might wish to estimate the mean productivity for a company of a given size (say, $x = 2$). That is, we might wish to estimate

$$E(y) = \beta_0 + \beta_1 x + \beta_2 x^2$$
$$= \beta_0 + 2\beta_1 + 4\beta_2 \quad \text{where} \quad x = 2$$

Or we might wish to estimate the marginal increase in productivity, the slope of a tangent to the curve, when $x = 2$ (see Figure A.4). The marginal productivity for y when $x = 2$ is the rate of change of $E(y)$ with respect to x, evaluated at $x = 2$.* The marginal productivity for a value of x, denoted by the symbol $dE(y)/dx$, can be shown (proof omitted) to be

$$\frac{dE(y)}{dx} = \beta_1 + 2\beta_2 x$$

FIGURE A.4

Marginal productivity

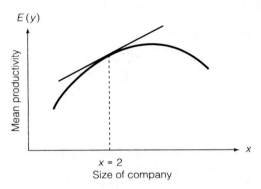

Therefore, the marginal productivity at $x = 2$ is

$$\frac{dE(y)}{dx} = \beta_1 + 2\beta_2(2) = \beta_1 + 4\beta_2$$

For $x = 2$, both $E(y)$ and the marginal productivity are *linear* functions of the unknown parameters $\beta_0, \beta_1, \beta_2$ in the model. The problem we pose in this section is that of finding confidence intervals for linear functions of β parameters or testing hypotheses concerning their values. The information necessary to solve this problem is rarely given in a standard multiple regression analysis computer printout, but we can find these confidence intervals or values of the appropriate test statistics from knowledge of $(\mathbf{X'X})^{-1}$.

For the model

$$y = \beta_0 + \beta_1 x_1 + \cdots + \beta_k x_k + \varepsilon$$

we can make an inference about a linear function of the β parameters, say,

$$a_0\beta_0 + a_1\beta_1 + \cdots + a_k\beta_k$$

---

*If you have had calculus, you can see that the marginal productivity for y given x is the first derivative of $E(y) = \beta_0 + \beta_1 x + \beta_2 x^2$ with respect to x.

where $a_0, a_1, \ldots, a_k$ are known constants. We will use the corresponding linear function of least squares estimates,

$$\ell = a_0\hat{\beta}_0 + a_1\hat{\beta}_1 + \cdots + a_k\hat{\beta}_k$$

as our best estimate of $a_0\beta_0 + a_1\beta_1 + \cdots + a_k\beta_k$.

Then, for the assumptions on the random error ε (stated in Section 4.2), the sampling distribution for the estimator l will be normal, with mean and standard error as given in the first box on page 795. This indicates that l is an unbiased estimator of

$$E(\ell) = a_0\beta_0 + a_1\beta_1 + \cdots + a_k\beta_k$$

and that its sampling distribution would appear as shown in Figure A.5.

FIGURE A.5
Sampling distribution for ℓ

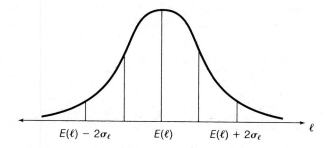

$$E(\ell) - 2\sigma_\ell \qquad E(\ell) \qquad E(\ell) + 2\sigma_\ell \qquad \ell$$

Mean and Standard Error of ℓ

$$E(\ell) = a_0\beta_0 + a_1\beta_1 + \cdots + a_k\beta_k$$

$$\sigma_l = \sqrt{\sigma^2 \mathbf{a}'(\mathbf{X}'\mathbf{X})^{-1}\mathbf{a}}$$

where σ^2 is the variance of ε, $(\mathbf{X}'\mathbf{X})^{-1}$ is the inverse matrix obtained in fitting the least squares model to the set of data, and

$$\mathbf{a} = \begin{bmatrix} a_0 \\ a_1 \\ a_2 \\ \vdots \\ a_k \end{bmatrix}$$

It can be demonstrated that a $100(1 - \alpha)\%$ confidence interval for $E(\ell)$ is as shown in the next box.

A $100(1 - \alpha)\%$ Confidence Interval for $E(\ell)$

$$\ell \pm t_{\alpha/2}\sqrt{s^2 \mathbf{a}'(\mathbf{X}'\mathbf{X})^{-1}\mathbf{a}}$$

where

$$E(\ell) = a_0\beta_0 + a_1\beta_1 + \cdots + a_k\beta_k$$

$$\ell = a_0\hat{\beta}_0 + a_1\hat{\beta}_1 + \cdots + a_k\hat{\beta}_k \quad \mathbf{a} = \begin{bmatrix} a_0 \\ a_1 \\ a_2 \\ \vdots \\ a_k \end{bmatrix}$$

s^2 and $(\mathbf{X}'\mathbf{X})^{-1}$ are obtained from the least squares procedure, and $t_{\alpha/2}$ is based on the number of degrees of freedom associated with s^2.

The linear function of the β parameters that is most often the focus of our attention is

$$E(y) = \beta_0 + \beta_1 x_1 + \cdots + \beta_k x_k$$

That is, we want to find a confidence interval for $E(y)$ for specific values of $x_1, x_2, \ldots, x_k$. For this special case,

$$\ell = \hat{y}$$

and the **a** matrix is

$$\mathbf{a} = \begin{bmatrix} 1 \\ x_1 \\ x_2 \\ \vdots \\ x_k \end{bmatrix}$$

where the symbols $x_1, x_2, \ldots, x_k$ in the **a** matrix indicate the specific numerical values assumed by these variables. Thus, the procedure for forming a confidence interval for $E(y)$ is as shown in the box.

EXAMPLE A.10   Refer to the data of Example A.5 for sales revenue y and advertising expenditure x. Find a 95% confidence interval for the mean sales revenue $E(y)$ when advertising expenditure is $x = 4$.

> **A $100(1 - \alpha)\%$ Confidence Interval for $E(y)$**
>
> $$\ell \pm t_{\alpha/2}\sqrt{s^2 \mathbf{a}'(\mathbf{X}'\mathbf{X})^{-1}\mathbf{a}}$$
>
> where
>
> $$E(y) = \beta_0 + \beta_1 x_1 + \beta_2 x_2 + \cdots + \beta_k x_k$$
>
> $$\ell = \hat{y} = \hat{\beta}_0 + \hat{\beta}_1 x_1 + \cdots + \hat{\beta}_k x_k \quad \mathbf{a} = \begin{bmatrix} 1 \\ x_1 \\ x_2 \\ \vdots \\ x_k \end{bmatrix}$$
>
> s^2 and $(\mathbf{X}'\mathbf{X})^{-1}$ are obtained from the least squares analysis, and $t_{\alpha/2}$ is based on the number of degrees of freedom associated with s^2, namely, $n - (k + 1)$.

Solution

The confidence interval for $E(y)$ for a given value of x is

$$\hat{y} \pm t_{\alpha/2}\sqrt{s^2 \mathbf{a}'(\mathbf{X}'\mathbf{X})^{-1}\mathbf{a}}$$

Consequently, we need to find and substitute the values of $\mathbf{a}'(\mathbf{X}'\mathbf{X})^{-1}\mathbf{a}$, $t_{\alpha/2}$, and $\hat{y}$ into this formula. Since we wish to estimate

$$\begin{aligned} E(y) &= \beta_0 + \beta_1 x \\ &= \beta_0 + \beta_1(4) \quad \text{when} \quad x = 4 \\ &= \beta_0 + 4\beta_1 \end{aligned}$$

it follows that the coefficients of β_0 and β_1 are $a_0 = 1$ and $a_1 = 4$, and thus,

$$\mathbf{a} = \begin{bmatrix} 1 \\ 4 \end{bmatrix}$$

From Examples A.5 and A.7, $\hat{y} = -.1 + .7x$,

$$(\mathbf{X}'\mathbf{X})^{-1} = \begin{bmatrix} 1.1 & -.3 \\ -.3 & .1 \end{bmatrix}$$

and $s^2 = .367$. Then,

$$\mathbf{a}'(\mathbf{X}'\mathbf{X})^{-1}\mathbf{a} = \begin{bmatrix} 1 & 4 \end{bmatrix} \begin{bmatrix} 1.1 & -.3 \\ -.3 & .1 \end{bmatrix} \begin{bmatrix} 1 \\ 4 \end{bmatrix}$$

We first calculate

$$\mathbf{a}'(\mathbf{X}'\mathbf{X})^{-1} = \begin{bmatrix} 1 & 4 \end{bmatrix} \begin{bmatrix} 1.1 & -.3 \\ -.3 & .1 \end{bmatrix} = \begin{bmatrix} -.1 & .1 \end{bmatrix}$$

Then,

$$\mathbf{a}'(\mathbf{X}'\mathbf{X})^{-1}\mathbf{a} = [-.1 \quad .1]\begin{bmatrix} 1 \\ 4 \end{bmatrix} = .3$$

The t value, $t_{.025}$, based on 3 df is 3.182. So, a 95% confidence interval for the mean sales revenue with an advertising expenditure of 4 is

$$\hat{y} \pm t_{\alpha/2}\sqrt{s^2\mathbf{a}'(\mathbf{X}'\mathbf{X})^{-1}\mathbf{a}}$$

Since $\hat{y} = -.1 + .7x = -.1 + (.7)(4) = 2.7$, the 95% confidence interval for $E(y)$ when $x = 4$ is

$$2.7 \pm (3.182)\sqrt{(.367)(.3)}$$
$$2.7 \pm 1.1$$

Notice that this is exactly the same result as obtained in Example 3.4.   ◆

EXAMPLE   A.11

An economist recorded a measure of productivity y and the size x for each of 100 companies producing cement. A regression model,

$$y = \beta_0 + \beta_1 x + \beta_2 x^2 + \varepsilon$$

fit to the $n = 100$ data points produced the following results:

$$\hat{y} = 2.6 + .7x - .2x^2$$

where x is coded to take values in the interval $-2 < x < 2,$[†] and

$$(\mathbf{X}'\mathbf{X})^{-1} = \begin{bmatrix} .0025 & .0005 & -.0070 \\ .0005 & .0055 & 0 \\ -.0070 & 0 & .0050 \end{bmatrix} \quad s = .14$$

Find a 95% confidence interval for the marginal increase in productivity given that the coded size of a plant is $x = 1.5$.

Solution

The mean value of y for a given value of x is

$$E(y) = \beta_0 + \beta_1 x + \beta_2 x^2$$

Therefore, the marginal increase in y for $x = 1.5$ is

$$\frac{dE(y)}{dx} = \beta_1 + 2\beta_2 x$$
$$= \beta_1 + 2(1.5)\beta_2$$

[†]We give a formula for *coding* observational data in Section 5.6.

Or,

$$E(l) = \beta_1 + 3\beta_2 \quad \text{when} \quad x = 1.5$$

Note from the prediction equation, $\hat{y} = 2.6 + .7x - .2x^2$, that $\hat{\beta}_1 = .7$ and $\hat{\beta}_2 = -.2$. Therefore,

$$l = \hat{\beta}_1 + 3\hat{\beta}_2 = .7 + 3(-.2) = .1$$

and

$$\mathbf{a} = \begin{bmatrix} a_0 \\ a_1 \\ a_2 \end{bmatrix} = \begin{bmatrix} 0 \\ 1 \\ 3 \end{bmatrix}$$

We next calculate

$$\mathbf{a}'(\mathbf{X}'\mathbf{X})^{-1}\mathbf{a} = \begin{bmatrix} 0 & 1 & 3 \end{bmatrix} \begin{bmatrix} .0025 & .0005 & -.0070 \\ .0005 & .0055 & 0 \\ -.0070 & 0 & .0050 \end{bmatrix} \begin{bmatrix} 0 \\ 1 \\ 3 \end{bmatrix} = .0505$$

Then, since s is based on $n - (k + 1) = 100 - 3 = 97$ df, $t_{.025} \approx 1.96$, and a 95% confidence interval for the marginal increase in productivity when $x = 1.5$ is

$$\ell \pm t_{.025} \sqrt{(s^2)\mathbf{a}'(\mathbf{X}'\mathbf{X})^{-1}\mathbf{a}}$$

or

$$.1 \pm (1.96)\sqrt{(.14)^2(.0505)}$$
$$.1 \pm .062$$

Thus, the marginal increase in productivity, the slope of the tangent to the curve

$$E(y) = \beta_0 + \beta_1 x + \beta_2 x^2$$

is estimated to lie in the interval $.1 \pm .062$ at $x = 1.5$. A graph of $\hat{y} = 2.6 + .7x - .2x^2$ is shown in Figure A.6.   ◆

A.9
A Prediction Interval for Some Value of y to Be Observed in the Future

We have indicated in Sections 3.9 and 4.12 that two of the most important applications of the least squares predictor $\hat{y}$ are estimating the mean value of y (the topic of the preceding section) and predicting a new value of y, yet unobserved, for specific values of $x_1, x_2, \ldots, x_k$. The difference between these two inferential problems (when each would be pertinent) was explained in Chapters 3 and 4, but we will give another example to make certain that the distinction is clear.

Suppose you are the manager of a manufacturing plant and that y, the daily profit, is a function of various process variables $x_1, x_2, \ldots, x_k$. Suppose you want to know

FIGURE A.6

A graph of $\hat{y} = 2.6 + .7x - .2x^2$

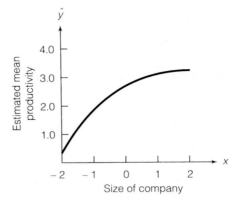

how much money you would make *in the long run* if the x's are set at specific values. For this case, you would be interested in finding a confidence interval for the mean profit per day, $E(y)$. In contrast, suppose you planned to operate the plant for just one more day! Then you would be interested in predicting the value of y, the profit associated with tomorrow's production.

We have indicated that the error of prediction is always larger than the error of estimating $E(y)$. You can see this by comparing the formula for the prediction interval (shown in the next box) with the formula for the confidence interval for $E(y)$ that was given in Section A.8.

A 100$(1 - \alpha)$% Prediction Interval for y

$$\hat{y} \pm t_{\alpha/2}\sqrt{s^2 + s^2\mathbf{a}'(\mathbf{X}'\mathbf{X})^{-1}\mathbf{a}} = \hat{y} \pm t_{\alpha/2}\sqrt{s^2[1 + \mathbf{a}'(\mathbf{X}'\mathbf{X})^{-1}\mathbf{a}]}$$

where

$$\hat{y} = \hat{\beta}_0 + \hat{\beta}_1 x_1 + \cdots + \hat{\beta}_k x_k$$

s^2 and $(\mathbf{X}'\mathbf{X})^{-1}$ are obtained from the least squares analysis,

$$\mathbf{a} = \begin{bmatrix} 1 \\ x_1 \\ x_2 \\ \vdots \\ x_k \end{bmatrix}$$

contains the numerical values of $x_1, x_2, \ldots, x_k$, and $t_{\alpha/2}$ is based on the number of degrees of freedom associated with s^2, namely, $n - (k + 1)$.

EXAMPLE   A.12

Refer to the sales–advertising expenditure example (Example A.10). Find a 95% prediction interval for the sales revenue next month, if it is known that next month's advertising expenditure will be $x = 4$.

Solution

The 95% prediction interval for sales revenue y is

$$\hat{y} \pm t_{\alpha/2}\sqrt{s^2[1 + \mathbf{a}'(\mathbf{X}'\mathbf{X})^{-1}\mathbf{a}]}$$

From Example A.10, when $x = 4$, $\hat{y} = -.1 + .7x = -.1 + (.7)(4) = 2.7$, $s^2 = .367$, $t_{.025} = 3.182$, and $\mathbf{a}'(\mathbf{X}'\mathbf{X})^{-1}\mathbf{a} = .3$. Then the 95% prediction interval for y is

$$2.7 \pm (3.182)\sqrt{(.367)(1 + .3)}$$

$$2.7 \pm 2.2$$

You will find that this is the same solution as obtained in Example 3.5. ◆

EXERCISES

A.22. Refer to Exercise A.16. Find a 90% confidence interval for $E(y)$ when $x = 1$. Interpret the interval.

A.23. Refer to Exercise A.16. Suppose you plan to observe y for $x = 1$. Find a 90% prediction interval for that value of y. Interpret the interval.

A.24. Refer to Exercise A.17. Find a 90% confidence interval for $E(y)$ when $x = 2$. Interpret the interval.

A.25. Refer to Exercise A.17. Find a 90% prediction interval for a value of y to be observed in the future when $x = 2$. Interpret the interval.

A.26. Refer to Exercise A.18. Find a 90% confidence interval for the mean value of y when $x = 1$. Interpret the interval.

A.27. Refer to Exercise A.18. Find a 90% prediction interval for a value of y to be observed in the future when $x = 1$.

A.28. The productivity (items produced per hour) per worker on a manufacturing assembly line is expected to increase as piecework pay rate (in dollars) increases; it is expected to stabilize after a certain pay rate has been reached. The productivity of five different workers was recorded for each of five piecework pay rates, $.80, $.90, $1.00, $1.10, $1.20, thus giving $n = 25$ data points.

A multiple regression analysis using a second-order model,

$$E(y) = \beta_0 + \beta_1 x + \beta_2 x^2$$

gave

$$\hat{y} = 2.08 + 8.42x - 1.65x^2$$

$$\text{SSE} = 26.62, \text{SS}_{yy} = 784.11, \text{ and}$$

$$(\mathbf{X}'\mathbf{X})^{-1} = \begin{bmatrix} .020 & -.010 & .015 \\ -.010 & .040 & -.006 \\ .015 & -.006 & .028 \end{bmatrix}$$

a. Find s^2.

b. Find a 95% confidence interval for the mean productivity when the pay rate is $1.10. Interpret this interval.

c. Find a 95% prediction interval for the production of an individual worker who is paid at a rate of $1.10 per piece. Interpret the interval.

d. Find R^2 and interpret the value.

Summary

Except for the tedious process of inverting a matrix (discussed in Appendix B), we have covered the major steps performed by a computer in fitting a linear statistical model to a set of data using the method of least squares. We have also explained how to find the confidence intervals, prediction intervals, and values of test statistics that would be pertinent in a regression analysis.

In addition to providing a better understanding of a multiple regression analysis, the most important contributions of this appendix are contained in Sections A.8 and A.9. If you want to make a specific inference concerning the mean value of y or any linear function of the β parameters and if you are unable to obtain the results from the computer package you are using, you will find the contents of Sections A.8 and A.9 very useful. Since you will almost always be able to find a computer program package to find $(\mathbf{X}'\mathbf{X})^{-1}$, you will be able to calculate the desired confidence interval(s) and so forth on your own.

SUPPLEMENTARY EXERCISES

A.29. Use the method of least squares to fit a straight line to the six data points:

x	−5	−3	−1	1	3	5
y	1.1	1.9	3.0	3.8	5.1	6.0

a. Construct **Y** and **X** matrices for the data.
b. Find $\mathbf{X}'\mathbf{X}$ and $\mathbf{X}'\mathbf{Y}$.
c. Find the least squares estimates,

$$\hat{\beta} = (\mathbf{X}'\mathbf{X})^{-1}\mathbf{X}'\mathbf{Y}$$

[*Note*: See Theorem A.1 for information on finding $(\mathbf{X}'\mathbf{X})^{-1}$.]
d. Give the prediction equation.
e. Find SSE and s^2.
f. Does the model contribute information for the prediction of y? Test $H_0: \beta_1 = 0$. Use $\alpha = .05$.
g. Find r^2 and interpret its value.
h. Find a 90% confidence interval for $E(y)$ when $x = .5$. Interpret the interval.

A.30. An experiment was conducted to investigate the effect of extrusion pressure P and temperature T on the strength y of a new type of plastic. Two plastic specimens were prepared for each of five combinations of pressure and temperature. The specimens were then tested in random order, and the breaking strength for each specimen was recorded. The independent variables were coded to simplify computations, i.e.,

$$x_1 = \frac{P - 200}{10} \quad x_2 = \frac{T - 400}{25}$$

The $n = 10$ data points are listed in the table.

y	x_1	x_2
5.2; 5.0	−2	2
.3; −.1	−1	−1
−1.2; −1.1	0	−2
2.2; 2.0	1	−1
6.2; 6.1	2	2

a. Give the **Y** and **X** matrices needed to fit the model $y = \beta_0 + \beta_1 x_1 + \beta_2 x_2 + \varepsilon$.
b. Find the least squares prediction equation.
c. Find SSE and s^2.
d. Does the model contribute information for the prediction of y? Test using $\alpha = .05$.
e. Find R^2 and interpret its value.
f. Test the null hypothesis that $\beta_1 = 0$. Use $\alpha = .05$. What is the practical implication of the test?
g. Find a 90% confidence interval for the mean strength of the plastic for $x_1 = -2$ and $x_2 = 2$.
h. Suppose a single specimen of the plastic is to be installed in the engine mount of a Douglas DC-10 aircraft. Find a 90% prediction interval for the strength of this specimen if $x_1 = -2$ and $x_2 = 2$.

A.31. Suppose we obtained two replications of the experiment described in Exercise A.17; i.e., two values of y were observed for each of the six values of x. The data are shown below.

x	1		2		3		4		5		6	
y	1.1	.5	1.8	2.0	2.0	2.9	3.8	3.4	4.1	5.0	5.0	5.8

a. Suppose (as in Exercise A.17) you wish to fit the model $E(y) = \beta_0 + \beta_1 x$. Construct **Y** and **X** matrices for the data. [*Hint*: Remember, the **Y** matrix must be of dimension 12×1.]

b. Find $\mathbf{X'X}$ and $\mathbf{X'Y}$.

c. Compare the $\mathbf{X'X}$ matrix for two replications of the experiment with the $\mathbf{X'X}$ matrix obtained for a single replication (part **b** of Exercise A.17). What is the relationship between the elements in the two matrices?

d. Observe the $(\mathbf{X'X})^{-1}$ matrix for a single replication (see part **c** of Exercise A.17). Verify that the $(\mathbf{X'X})^{-1}$ matrix for two replications contains elements that are equal to $\frac{1}{2}$ of the values of the corresponding elements in the $(\mathbf{X'X})^{-1}$ matrix for a single replication of the experiment. [*Hint*: Show that the product of the $(\mathbf{X'X})^{-1}$ matrix (for two replications) and the $\mathbf{X'X}$ matrix from part c equals the identity matrix **I**.]

e. Find the prediction equation.

f. Find SSE and s^2.

g. Do the data provide sufficient information to indicate that x contributes information for the prediction of y? Test using $\alpha = .05$.

h. Find r^2 and interpret its value.

A.32. Refer to Exercise A.31.

a. Find a 90% confidence interval for $E(y)$ when $x = 4.5$. Interpret the interval.

b. Suppose we wish to predict the value of y if, in the future, $x = 4.5$. Find a 90% prediction interval for y and interpret the interval.

A.33. Refer to Exercise A.31. Suppose you replicated the experiment described in Exercise A.17 three times; i.e., you collected three observations on y for each value of x. Then $n = 18$.

a. What would be the dimensions of the **Y** matrix?

b. Write the **X** matrix for three replications. Compare with the **X** matrices for one and for two replications. Note the pattern.

c. Examine the $\mathbf{X'X}$ matrices obtained for one and two replications of the experiment (obtained in Exercises A.17 and A.31, respectively). Deduce the values of the elements of the $\mathbf{X'X}$ matrix for three replications.

d. Look at your answer to Exercise A.31, part **d**. Deduce the values of the elements in the $(\mathbf{X'X})^{-1}$ matrix for three replications.

e. Suppose you wanted to find a 90% confidence interval for $E(y)$ when $x = 4.5$ based on three replications of the experiment. Find the value of $\mathbf{a'(X'X)}^{-1}\mathbf{a}$ that appears in the confidence interval and compare with the value of $\mathbf{a'(X'X)}^{-1}\mathbf{a}$ that would be obtained for a single replication of the experiment.

f. Approximately how much of a reduction in the width of the confidence interval is obtained by using three versus two replications? [*Note*: The values of s computed from the two sets of data will almost certainly be different.]

REFERENCES

DRAPER, N. and SMITH, H. *Applied Regression Analysis*. 3rd ed. New York: Wiley, 1998.

GRAYBILL, F. A. *Theory and Application of the Linear Model*. North Scituate, Mass.: Duxbury, 1976.

MENDENHALL, W. *Introduction to Linear Models and the Design and Analysis of Experiments*. Belmont, Calif.: Wadsworth, 1968.

NETER, J., KUTNER, M. H., NACHTSHEIM, C. J. and WASSERMAN, W. *Applied Linear Statistical Models*, 4th ed. Homewood, Ill.: Richard D. Irwin, 1996.

A Procedure for Inverting a Matrix

There are several different methods for inverting matrices. All are tedious and time-consuming. Consequently, in practice, you will invert almost all matrices using a computer. The purpose of this section is to present one method for inverting small (2×2 or 3×3) matrices manually, thus giving you an appreciation of the enormous computing problem involved in inverting large matrices (and, consequently, in fitting linear models containing many terms to a set of data). In particular, you will be able to understand why rounding errors creep into the inversion process and, consequently, why two different computer programs might invert the same matrix and produce inverse matrices with slightly different corresponding elements.

The procedure we will demonstrate to invert a matrix $\mathbf{A}$ requires us to perform a series of operations on the rows of the $\mathbf{A}$ matrix. For example, suppose

$$\mathbf{A} = \begin{bmatrix} 1 & -2 \\ -2 & 6 \end{bmatrix}$$

We will identify two different ways to operate on a row of a matrix:*

---

*We omit a third row operation, because it would add little and could be confusing.

1. We can multiply every element in one particular row by a constant, *c*. For example, we could operate on the first row of the **A** matrix by multiplying every element in the row by a constant, say, 2. Then the resulting row would be $[2 - 4]$.

2. We can operate on a row by multiplying another row of the matrix by a constant and then adding (or subtracting) the elements of that row to elements in corresponding positions in the row operated upon. For example, we could operate on the first row of the **A** matrix by multiplying the second row by a constant, say, 2:

$$2[-2 \quad 6] = [-4 \quad 12]$$

Then we add this row to row 1:

$$[(1 - 4)(-2 + 12)] = [-3 \quad 10]$$

Note one important point. We operated on the *first* row of the **A** matrix. Although we used the second row of the matrix to perform the operation, *the second row would remain unchanged.* Therefore, the row operation on the **A** matrix that we have just described would produce the new matrix,

$$\begin{bmatrix} -3 & 10 \\ -2 & 6 \end{bmatrix}$$

Matrix inversion using row operations is based on an elementary result from matrix algebra. It can be shown (proof omitted) that performing a series of row operations on a matrix **A** is equivalent to multiplying **A** by a matrix **B**; i.e., row operations produce a new matrix, **BA**. This result is used as follows: Place the **A** matrix and an identity matrix **I** of the same dimensions side by side. Then perform the same series of row operations on both **A** and **I** until the **A** matrix has been changed into the identity matrix **I**. This means that you have multiplied both **A** and **I** by some matrix **B** such that:

$$\mathbf{A} = \begin{bmatrix} \\ \\ \end{bmatrix} \qquad\qquad \mathbf{I} = \begin{bmatrix} 1 & 0 & 0 & \cdots & 0 \\ 0 & 1 & 0 & \cdots & 0 \\ 0 & 0 & 1 & \cdots & 0 \\ \vdots & \vdots & \vdots & & \vdots \\ 0 & 0 & 0 & \cdots & 1 \end{bmatrix}$$

$$\downarrow \qquad \leftarrow \text{Row operations change } \mathbf{A} \text{ to } \mathbf{I} \rightarrow \qquad \downarrow$$

$$\mathbf{I} = \begin{bmatrix} \\ \\ \end{bmatrix} \qquad\qquad \mathbf{B} = \begin{bmatrix} \\ \\ \end{bmatrix}$$

$$\mathbf{BA} = \mathbf{I} \text{ and } \mathbf{BI} = \mathbf{B}$$

Since $\mathbf{BA} = \mathbf{I}$, it follows that $\mathbf{B} = \mathbf{A}^{-1}$. Therefore, as the $\mathbf{A}$ matrix is transformed by row operations into the identity matrix $\mathbf{I}$, the identity matrix $\mathbf{I}$ is transformed into $\mathbf{A}^{-1}$, i.e.,

$$\mathbf{BI} = \mathbf{B} = \mathbf{A}^{-1}$$

We will show you how this procedure works with two examples.

EXAMPLE   B.1

Find the inverse of the matrix

$$\mathbf{A} = \begin{bmatrix} 1 & -2 \\ -2 & 6 \end{bmatrix}$$

Solution

Place the $\mathbf{A}$ matrix and a 2×2 identity matrix side by side and then perform the following series of row operations (we will indicate by an arrow the row operated upon in each operation):

$$\mathbf{A} = \begin{bmatrix} 1 & -2 \\ -2 & 6 \end{bmatrix} \qquad \mathbf{I} = \begin{bmatrix} 1 & 0 \\ 0 & 1 \end{bmatrix}$$

OPERATION 1:   Multiply the first row by 2 and add it to the second row:

$$\rightarrow \begin{bmatrix} 1 & -2 \\ 0 & 2 \end{bmatrix} \qquad \begin{bmatrix} 1 & 0 \\ 2 & 1 \end{bmatrix}$$

OPERATION 2:   Multiply the second row by $\frac{1}{2}$:

$$\rightarrow \begin{bmatrix} 1 & -2 \\ 0 & 1 \end{bmatrix} \qquad \begin{bmatrix} 1 & 0 \\ 1 & \frac{1}{2} \end{bmatrix}$$

OPERATION 3:   Multiply the second row by 2 and add it to the first row:

$$\rightarrow \begin{bmatrix} 1 & 0 \\ 0 & 1 \end{bmatrix} \qquad \begin{bmatrix} 3 & 1 \\ 1 & \frac{1}{2} \end{bmatrix}$$

Thus,

$$\mathbf{A}^{-1} = \begin{bmatrix} 3 & 1 \\ 1 & \frac{1}{2} \end{bmatrix}$$

(Note that our solution matches the one obtained using Theorem A.2.)

The final step in finding an inverse is to check your solution by finding the product $\mathbf{A}^{-1}\mathbf{A}$ to see whether it equals the identity matrix $\mathbf{I}$. To check:

$$\mathbf{A}^{-1}\mathbf{A} = \begin{bmatrix} 3 & 1 \\ 1 & \frac{1}{2} \end{bmatrix} \begin{bmatrix} 1 & -2 \\ -2 & 6 \end{bmatrix}$$

$$= \begin{bmatrix} 1 & 0 \\ 0 & 1 \end{bmatrix}$$

Since this product is equal to the identity matrix, it follows that our solution for $\mathbf{A}^{-1}$ is correct.   ◆

EXAMPLE   B.2

Find the inverse of the matrix

$$\mathbf{A} = \begin{bmatrix} 2 & 0 & 3 \\ 0 & 4 & 1 \\ 3 & 1 & 2 \end{bmatrix}$$

Solution

Place an identity matrix alongside the $\mathbf{A}$ matrix and perform the row operations:

OPERATION 1:   Multiply row 1 by $\frac{1}{2}$:

$$\rightarrow \begin{bmatrix} 1 & 0 & \frac{3}{2} \\ 0 & 4 & 1 \\ 3 & 1 & 2 \end{bmatrix} \qquad \begin{bmatrix} \frac{1}{2} & 0 & 0 \\ 0 & 1 & 0 \\ 0 & 0 & 1 \end{bmatrix}$$

OPERATION 2:   Multiply row 1 by 3 and subtract from row 3:

$$\rightarrow \begin{bmatrix} 1 & 0 & \frac{3}{2} \\ 0 & 4 & 1 \\ 0 & 1 & -\frac{5}{2} \end{bmatrix} \qquad \begin{bmatrix} \frac{1}{2} & 0 & 0 \\ 0 & 1 & 0 \\ -\frac{3}{2} & 0 & 1 \end{bmatrix}$$

OPERATION 3:   Multiply row 2 by $\frac{1}{4}$:

$$\rightarrow \begin{bmatrix} 1 & 0 & \frac{3}{2} \\ 0 & 1 & \frac{1}{4} \\ 0 & 1 & -\frac{5}{2} \end{bmatrix} \qquad \begin{bmatrix} \frac{1}{2} & 0 & 0 \\ 0 & \frac{1}{4} & 0 \\ -\frac{3}{2} & 0 & 1 \end{bmatrix}$$

OPERATION 4:   Subtract row 2 from row 3:

$$\rightarrow \begin{bmatrix} 1 & 0 & \frac{3}{2} \\ 0 & 1 & \frac{1}{4} \\ 0 & 0 & -\frac{11}{4} \end{bmatrix} \qquad \begin{bmatrix} \frac{1}{2} & 0 & 0 \\ 0 & \frac{1}{4} & 0 \\ -\frac{3}{2} & -\frac{1}{4} & 1 \end{bmatrix}$$

OPERATION 5:   Multiply row 3 by $-\frac{4}{11}$:

$$\rightarrow \begin{bmatrix} 1 & 0 & \frac{3}{2} \\ 0 & 1 & \frac{1}{4} \\ 0 & 0 & 1 \end{bmatrix} \qquad \begin{bmatrix} \frac{1}{2} & 0 & 0 \\ 0 & \frac{1}{4} & 0 \\ \frac{12}{22} & \frac{1}{11} & -\frac{4}{11} \end{bmatrix}$$

OPERATION 6:   Operate on row 2 by subtracting $\frac{1}{4}$ of row 3:

$$\rightarrow \begin{bmatrix} 1 & 0 & \frac{3}{2} \\ 0 & 1 & 0 \\ 0 & 0 & 1 \end{bmatrix} \quad \begin{bmatrix} \frac{1}{2} & 0 & 0 \\ -\frac{3}{22} & \frac{5}{22} & \frac{1}{11} \\ \frac{12}{22} & \frac{1}{11} & -\frac{4}{11} \end{bmatrix}$$

OPERATION 7:   Operate on row 1 by subtracting $\frac{3}{2}$ of row 3:

$$\rightarrow \begin{bmatrix} 1 & 0 & 0 \\ 0 & 1 & 0 \\ 0 & 0 & 1 \end{bmatrix} \quad \begin{bmatrix} -\frac{7}{22} & -\frac{3}{22} & \frac{6}{11} \\ -\frac{3}{22} & \frac{5}{22} & \frac{1}{11} \\ \frac{6}{11} & \frac{1}{11} & -\frac{4}{11} \end{bmatrix} = \mathbf{A}^{-1}$$

To check the solution, we find the product:

$$\mathbf{A}^{-1}\mathbf{A} = \begin{bmatrix} -\frac{7}{22} & -\frac{3}{22} & \frac{6}{11} \\ -\frac{3}{22} & \frac{5}{22} & \frac{1}{11} \\ \frac{6}{11} & \frac{1}{11} & -\frac{4}{11} \end{bmatrix} \begin{bmatrix} 2 & 0 & 3 \\ 0 & 4 & 1 \\ 3 & 1 & 2 \end{bmatrix}$$

$$= \begin{bmatrix} 1 & 0 & 0 \\ 0 & 1 & 0 \\ 0 & 0 & 1 \end{bmatrix}$$

Since the product $\mathbf{A}^{-1}\mathbf{A}$ is equal to the identity matrix, it follows that our solution for $\mathbf{A}^{-1}$ is correct.   ◆

Examples B.1 and B.2 indicate the strategy employed when performing row operations on the $\mathbf{A}$ matrix to change it into an identity matrix. Multiply the first row by a constant to change the element in the top left row into a 1. Then perform operations to change all elements in the first column into 0's. Then operate on the second row and change the second diagonal element into a 1. Then operate to change all elements in the second column beneath row 2 into 0's. Then operate on the diagonal element in row 3, etc. When all elements on the main diagonal are 1's and all below the main diagonal are 0's, perform row operations to change the last column to 0; then the next-to-last, etc., until you get back to the first column. The procedure for changing the off-diagonal elements to 0's is indicated diagrammatically as shown in Figure B.1.

The preceding instructions on how to invert a matrix using row operations suggest that the inversion of a large matrix would involve many multiplications, subtractions, and additions and, consequently, could produce large rounding errors in the calculations unless you carry a large number of significant figures in the calculations. This explains why two different multiple regression analysis computer programs may produce different estimates of the same β parameters, and it emphasizes the importance of carrying a large number of significant figures in all computations when inverting a matrix.

FIGURE B.1

Diagram of matrix inversion steps

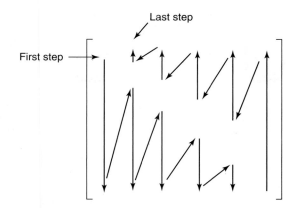

You can find other methods for inverting matrices in any linear algebra textbook. All work—exactly—in theory. It is only in the actual process of performing the calculations that the rounding errors occur.

EXERCISES

B.1. Invert the following matrices and check your answers to make certain that $A^{-1}A = AA^{-1} = I$:

a. $A = \begin{bmatrix} 3 & 2 \\ 4 & 5 \end{bmatrix}$

b. $A = \begin{bmatrix} 3 & 0 & -2 \\ 1 & 4 & 2 \\ 5 & 1 & 1 \end{bmatrix}$

c. $A = \begin{bmatrix} 1 & 0 & 1 \\ 0 & 2 & 1 \\ 1 & 1 & 3 \end{bmatrix}$

d. $A = \begin{bmatrix} 4 & 0 & 10 \\ 0 & 10 & 0 \\ 10 & 0 & 5 \end{bmatrix}$

[*Note*: No answers are given to these exercises. You will know whether your answer is correct if $A^{-1}A = I$.]

Useful Statistical Tables

CONTENTS

TABLE C.1   Normal Curve Areas

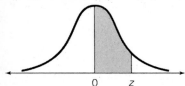

z	.00	.01	.02	.03	.04	.05	.06	.07	.08	.09
.0	.0000	.0040	.0080	.0120	.0160	.0199	.0239	.0279	.0319	.0359
.1	.0398	.0438	.0478	.0517	.0557	.0596	.0636	.0675	.0714	.0753
.2	.0793	.0832	.0871	.0910	.0948	.0987	.1026	.1064	.1103	.1141
.3	.1179	.1217	.1255	.1293	.1331	.1368	.1406	.1443	.1480	.1517
.4	.1554	.1591	.1628	.1664	.1700	.1736	.1772	.1808	.1844	.1879
.5	.1915	.1950	.1985	.2019	.2054	.2088	.2123	.2157	.2190	.2224
.6	.2257	.2291	.2324	.2357	.2389	.2422	.2454	.2486	.2517	.2549
.7	.2580	.2611	.2642	.2673	.2704	.2734	.2764	.2794	.2823	.2852
.8	.2881	.2910	.2939	.2967	.2995	.3023	.3051	.3078	.3106	.3133
.9	.3159	.3186	.3212	.3238	.3264	.3289	.3315	.3340	.3365	.3389
1.0	.3413	.3438	.3461	.3485	.3508	.3531	.3554	.3577	.3599	.3621
1.1	.3643	.3665	.3686	.3708	.3729	.3749	.3770	.3790	.3810	.3830
1.2	.3849	.3869	.3888	.3907	.3925	.3944	.3962	.3980	.3997	.4015
1.3	.4032	.4049	.4066	.4082	.4099	.4115	.4131	.4147	.4162	.4177
1.4	.4192	.4207	.4222	.4236	.4251	.4265	.4279	.4292	.4306	.4319
1.5	.4332	.4345	.4357	.4370	.4382	.4394	.4406	.4418	.4429	.4441
1.6	.4452	.4463	.4474	.4484	.4495	.4505	.4515	.4525	.4535	.4545
1.7	.4554	.4564	.4573	.4582	.4591	.4599	.4608	.4616	.4625	.4633
1.8	.4641	.4649	.4656	.4664	.4671	.4678	.4686	.4693	.4699	.4706
1.9	.4713	.4719	.4726	.4732	.4738	.4744	.4750	.4756	.4761	.4767
2.0	.4772	.4778	.4783	.4788	.4793	.4798	.4803	.4808	.4812	.4817
2.1	.4821	.4826	.4830	.4834	.4838	.4842	.4846	.4850	.4854	.4857
2.2	.4861	.4864	.4868	.4871	.4875	.4878	.4881	.4884	.4887	.4890
2.3	.4893	.4896	.4898	.4901	.4904	.4906	.4909	.4911	.4913	.4916
2.4	.4918	.4920	.4922	.4925	.4927	.4929	.4931	.4932	.4934	.4936
2.5	.4938	.4940	.4941	.4943	.4945	.4946	.4948	.4949	.4951	.4952
2.6	.4953	.4955	.4956	.4957	.4959	.4960	.4961	.4962	.4963	.4964
2.7	.4965	.4966	.4967	.4968	.4969	.4970	.4971	.4972	.4973	.4974
2.8	.4974	.4975	.4976	.4977	.4977	.4978	.4979	.4979	.4980	.4981
2.9	.4981	.4982	.4982	.4983	.4984	.4984	.4985	.4985	.4986	.4986
3.0	.4987	.4987	.4987	.4988	.4988	.4989	.4989	.4989	.4990	.4990

Source: Abridged from Table 1 of A. Hald, *Statistical Tables and Formulas* (New York: John Wiley & Sons, Inc.), 1952. Reproduced by permission of the publisher.

TABLE C.2   Critical Values for Student's *t*

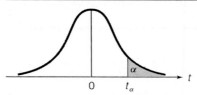

ν	$t_{.100}$	$t_{.050}$	$t_{.025}$	$t_{.010}$	$t_{.005}$	$t_{.001}$	$t_{.0005}$
1	3.078	6.314	12.706	31.821	63.657	318.31	636.62
2	1.886	2.920	4.303	6.965	9.925	22.326	31.598
3	1.638	2.353	3.182	4.541	5.841	10.213	12.924
4	1.533	2.132	2.776	3.747	4.604	7.173	8.610
5	1.476	2.015	2.571	3.365	4.032	5.893	6.869
6	1.440	1.943	2.447	3.143	3.707	5.208	5.959
7	1.415	1.895	2.365	2.998	3.499	4.785	5.408
8	1.397	1.860	2.306	2.896	3.355	4.501	5.041
9	1.383	1.833	2.262	2.821	3.250	4.297	4.781
10	1.372	1.812	2.228	2.764	3.169	4.144	4.587
11	1.363	1.796	2.201	2.718	3.106	4.025	4.437
12	1.356	1.782	2.179	2.681	3.055	3.930	4.318
13	1.350	1.771	2.160	2.650	3.012	3.852	4.221
14	1.345	1.761	2.145	2.624	2.977	3.787	4.140
15	1.341	1.753	2.131	2.602	2.947	3.733	4.073
16	1.337	1.746	2.120	2.583	2.921	3.686	4.015
17	1.333	1.740	2.110	2.567	2.898	3.646	3.965
18	1.330	1.734	2.101	2.552	2.878	3.610	3.922
19	1.328	1.729	2.093	2.539	2.861	3.579	3.883
20	1.325	1.725	2.086	2.528	2.845	3.552	3.850
21	1.323	1.721	2.080	2.518	2.831	3.527	3.819
22	1.321	1.717	2.074	2.508	2.819	3.505	3.792
23	1.319	1.714	2.069	2.500	2.807	3.485	3.767
24	1.318	1.711	2.064	2.492	2.797	3.467	3.745
25	1.316	1.708	2.060	2.485	2.787	3.450	3.725
26	1.315	1.706	2.056	2.479	2.779	3.435	3.707
27	1.314	1.703	2.052	2.473	2.771	3.421	3.690
28	1.313	1.701	2.048	2.467	2.763	3.408	3.674
29	1.311	1.699	2.045	2.462	2.756	3.396	3.659
30	1.310	1.697	2.042	2.457	2.750	3.385	3.646
40	1.303	1.684	2.021	2.423	2.704	3.307	3.551
60	1.296	1.671	2.000	2.390	2.660	3.232	3.460
120	1.289	1.658	1.980	2.358	2.617	3.160	3.373
∞	1.282	1.645	1.960	2.326	2.576	3.090	3.291

TABLE C.3   Critical Values for the F Statistic: $F_{.10}$

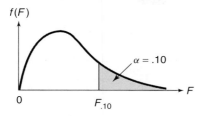

ν_2 \ ν_1	1	2	3	4	5	6	7	8	9
1	39.86	49.50	53.59	55.83	57.24	58.20	58.91	59.44	59.86
2	8.53	9.00	9.16	9.24	9.29	9.33	9.35	9.37	9.38
3	5.54	5.46	5.39	5.34	5.31	5.28	5.27	5.25	5.24
4	4.54	4.32	4.19	4.11	4.05	4.01	3.98	3.95	3.94
5	4.06	3.78	3.62	3.52	3.45	3.40	3.37	3.34	3.32
6	3.78	3.46	3.29	3.18	3.11	3.05	3.01	2.98	2.96
7	3.59	3.26	3.07	2.96	2.88	2.83	2.78	2.75	2.72
8	3.46	3.11	2.92	2.81	2.73	2.67	2.62	2.59	2.56
9	3.36	3.01	2.81	2.69	2.61	2.55	2.51	2.47	2.44
10	3.29	2.92	2.73	2.61	2.52	2.46	2.41	2.38	2.35
11	3.23	2.86	2.66	2.54	2.45	2.39	2.34	2.30	2.27
12	3.18	2.81	2.61	2.48	2.39	2.33	2.28	2.24	2.21
13	3.14	2.76	2.56	2.43	2.35	2.28	2.23	2.20	2.16
14	3.10	2.73	2.52	2.39	2.31	2.24	2.19	2.15	2.12
15	3.07	2.70	2.49	2.36	2.27	2.21	2.16	2.12	2.09
16	3.05	2.67	2.46	2.33	2.24	2.18	2.13	2.09	2.06
17	3.03	2.64	2.44	2.31	2.22	2.15	2.10	2.06	2.03
18	3.01	2.62	2.42	2.29	2.20	2.13	2.08	2.04	2.00
19	2.99	2.61	2.40	2.27	2.18	2.11	2.06	2.02	1.98
20	2.97	2.59	2.38	2.25	2.16	2.09	2.04	2.00	1.96
21	2.96	2.57	2.36	2.23	2.14	2.08	2.02	1.98	1.95
22	2.95	2.56	2.35	2.22	2.13	2.06	2.01	1.97	1.93
23	2.94	2.55	2.34	2.21	2.11	2.05	1.99	1.95	1.92
24	2.93	2.54	2.33	2.19	2.10	2.04	1.98	1.94	1.91
25	2.92	2.53	2.32	2.18	2.09	2.02	1.97	1.93	1.89
26	2.91	2.52	2.31	2.17	2.08	2.01	1.96	1.92	1.88
27	2.90	2.51	2.30	2.17	2.07	2.00	1.95	1.91	1.87
28	2.89	2.50	2.29	2.16	2.06	2.00	1.94	1.90	1.87
29	2.89	2.50	2.28	2.15	2.06	1.99	1.93	1.89	1.86
30	2.88	2.49	2.28	2.14	2.05	1.98	1.93	1.88	1.85
40	2.84	2.44	2.23	2.09	2.00	1.93	1.87	1.83	1.79
60	2.79	2.39	2.18	2.04	1.95	1.87	1.82	1.77	1.74
120	2.75	2.35	2.13	1.99	1.90	1.82	1.77	1.72	1.68
∞	2.71	2.30	2.08	1.94	1.85	1.77	1.72	1.67	1.63

NUMERATOR DEGREES OF FREEDOM (column headings 1–9)

DENOMINATOR DEGREES OF FREEDOM (row labels)

(continued overleaf)

TABLE C.3 (***Continued***)

ν_2 \ ν_1 (DENOMINATOR DEGREES OF FREEDOM)	NUMERATOR DEGREES OF FREEDOM									
	10	12	15	20	24	30	40	60	120	∞
1	60.19	60.71	61.22	61.74	62.00	62.26	62.53	62.79	63.06	63.33
2	9.39	9.41	9.42	9.44	9.45	9.46	9.47	9.47	9.48	9.49
3	5.23	5.22	5.20	5.18	5.18	5.17	5.16	5.15	5.14	5.13
4	3.92	3.90	3.87	3.84	3.83	3.82	3.80	3.79	3.78	3.76
5	3.30	3.27	3.24	3.21	3.19	3.17	3.16	3.14	3.12	3.10
6	2.94	2.90	2.87	2.84	2.82	2.80	2.78	2.76	2.74	2.72
7	2.70	2.67	2.63	2.59	2.58	2.56	2.54	2.51	2.49	2.47
8	2.54	2.50	2.46	2.42	2.40	2.38	2.36	2.34	2.32	2.29
9	2.42	2.38	2.34	2.30	2.28	2.25	2.23	2.21	2.18	2.16
10	2.32	2.28	2.24	2.20	2.18	2.16	2.13	2.11	2.08	2.06
11	2.25	2.21	2.17	2.12	2.10	2.08	2.05	2.03	2.00	1.97
12	2.19	2.15	2.10	2.06	2.04	2.01	1.99	1.96	1.93	1.90
13	2.14	2.10	2.05	2.01	1.98	1.96	1.93	1.90	1.88	1.85
14	2.10	2.05	2.01	1.96	1.94	1.91	1.89	1.86	1.83	1.80
15	2.06	2.02	1.97	1.92	1.90	1.87	1.85	1.82	1.79	1.76
16	2.03	1.99	1.94	1.89	1.87	1.84	1.81	1.78	1.75	1.72
17	2.00	1.96	1.91	1.86	1.84	1.81	1.78	1.75	1.72	1.69
18	1.98	1.93	1.89	1.84	1.81	1.78	1.75	1.72	1.69	1.66
19	1.96	1.91	1.86	1.81	1.79	1.76	1.73	1.70	1.67	1.63
20	1.94	1.89	1.84	1.79	1.77	1.74	1.71	1.68	1.64	1.61
21	1.92	1.87	1.83	1.78	1.75	1.72	1.69	1.66	1.62	1.59
22	1.90	1.86	1.81	1.76	1.73	1.70	1.67	1.64	1.60	1.57
23	1.89	1.84	1.80	1.74	1.72	1.69	1.66	1.62	1.59	1.55
24	1.88	1.83	1.78	1.73	1.70	1.67	1.64	1.61	1.57	1.53
25	1.87	1.82	1.77	1.72	1.69	1.66	1.63	1.59	1.56	1.52
26	1.86	1.81	1.76	1.71	1.68	1.65	1.61	1.58	1.54	1.50
27	1.85	1.80	1.75	1.70	1.67	1.64	1.60	1.57	1.53	1.49
28	1.84	1.79	1.74	1.69	1.66	1.63	1.59	1.56	1.52	1.48
29	1.83	1.78	1.73	1.68	1.65	1.62	1.58	1.55	1.51	1.47
30	1.82	1.77	1.72	1.67	1.64	1.61	1.57	1.54	1.50	1.46
40	1.76	1.71	1.66	1.61	1.57	1.54	1.51	1.47	1.42	1.38
60	1.71	1.66	1.60	1.54	1.51	1.48	1.44	1.40	1.35	1.29
120	1.65	1.60	1.55	1.48	1.45	1.41	1.37	1.32	1.26	1.19
∞	1.60	1.55	1.49	1.42	1.38	1.34	1.30	1.24	1.17	1.00

Source: From M. Merrington and C. M. Thompson, "Tables of percentage points of the inverted beta (F)-distribution," *Biometrika*; 1943, 33, 73–88. Reproduced by permission of the *Biometrika* Trustees.

TABLE C.4   Critical Values for the *F* Statistic: $F_{.05}$

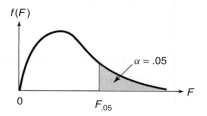

v_2 \ v_1	NUMERATOR DEGREES OF FREEDOM								
	1	2	3	4	5	6	7	8	9
1	161.4	199.5	215.7	224.6	230.2	234.0	236.8	238.9	240.5
2	18.51	19.00	19.16	19.25	19.30	19.33	19.35	19.37	19.38
3	10.13	9.55	9.28	9.12	9.01	8.94	8.89	8.85	8.81
4	7.71	6.94	6.59	6.39	6.26	6.16	6.09	6.04	6.00
5	6.61	5.79	5.41	5.19	5.05	4.95	4.88	4.82	4.77
6	5.99	5.14	4.76	4.53	4.39	4.28	4.21	4.15	4.10
7	5.59	4.74	4.35	4.12	3.97	3.87	3.79	3.73	3.68
8	5.32	4.46	4.07	3.84	3.69	3.58	3.50	3.44	3.39
9	5.12	4.26	3.86	3.63	3.48	3.37	3.29	3.23	3.18
10	4.96	4.10	3.71	3.48	3.33	3.22	3.14	3.07	3.02
11	4.84	3.98	3.59	3.36	3.20	3.09	3.01	2.95	2.90
12	4.75	3.89	3.49	3.26	3.11	3.00	2.91	2.85	2.80
13	4.67	3.81	3.41	3.18	3.03	2.92	2.83	2.77	2.71
14	4.60	3.74	3.34	3.11	2.96	2.85	2.76	2.70	2.65
15	4.54	3.68	3.29	3.06	2.90	2.79	2.71	2.64	2.59
16	4.49	3.63	3.24	3.01	2.85	2.74	2.66	2.59	2.54
17	4.45	3.59	3.20	2.96	2.81	2.70	2.61	2.55	2.49
18	4.41	3.55	3.16	2.93	2.77	2.66	2.58	2.51	2.46
19	4.38	3.52	3.13	2.90	2.74	2.63	2.54	2.48	2.42
20	4.35	3.49	3.10	2.87	2.71	2.60	2.51	2.45	2.39
21	4.32	3.47	3.07	2.84	2.68	2.57	2.49	2.42	2.37
22	4.30	3.44	3.05	2.82	2.66	2.55	2.46	2.40	2.34
23	4.28	3.42	3.03	2.80	2.64	2.53	2.44	2.37	2.32
24	4.26	3.40	3.01	2.78	2.62	2.51	2.42	2.36	2.30
25	4.24	3.39	2.99	2.76	2.60	2.49	2.40	2.34	2.28
26	4.23	3.37	2.98	2.74	2.59	2.47	2.39	2.32	2.27
27	4.21	3.35	2.96	2.73	2.57	2.46	2.37	2.31	2.25
28	4.20	3.34	2.95	2.71	2.56	2.45	2.36	2.29	2.24
29	4.18	3.33	2.93	2.70	2.55	2.43	2.35	2.28	2.22
30	4.17	3.32	2.92	2.69	2.53	2.42	2.33	2.27	2.21
40	4.08	3.23	2.84	2.61	2.45	2.34	2.25	2.18	2.12
60	4.00	3.15	2.76	2.53	2.37	2.25	2.17	2.10	2.04
120	3.92	3.07	2.68	2.45	2.29	2.17	2.09	2.02	1.96
∞	3.84	3.00	2.60	2.37	2.21	2.10	2.01	1.94	1.88

DENOMINATOR DEGREES OF FREEDOM

(continued overleaf)

TABLE C.4   (*Continued*)

v_2	v_1 NUMERATOR DEGREES OF FREEDOM									
	10	12	15	20	24	30	40	60	120	∞
1	241.9	243.9	245.9	248.0	249.1	250.1	251.1	252.2	253.3	254.3
2	19.40	19.41	19.43	19.45	19.45	19.46	19.47	19.48	19.49	19.50
3	8.79	8.74	8.70	8.66	8.64	8.62	8.59	8.57	8.55	8.53
4	5.96	5.91	5.86	5.80	5.77	5.75	5.72	5.69	5.66	5.63
5	4.74	4.68	4.62	4.56	4.53	4.50	4.46	4.43	4.40	4.36
6	4.06	4.00	3.94	3.87	3.84	3.81	3.77	3.74	3.70	3.67
7	3.64	3.57	3.51	3.44	3.41	3.38	3.34	3.30	3.27	3.23
8	3.35	3.28	3.22	3.15	3.12	3.08	3.04	3.01	2.97	2.93
9	3.14	3.07	3.01	2.94	2.90	2.86	2.83	2.79	2.75	2.71
10	2.98	2.91	2.85	2.77	2.74	2.70	2.66	2.62	2.58	2.54
11	2.85	2.79	2.72	2.65	2.61	2.57	2.53	2.49	2.45	2.40
12	2.75	2.69	2.62	2.54	2.51	2.47	2.43	2.38	2.34	2.30
13	2.67	2.60	2.53	2.46	2.42	2.38	2.34	2.30	2.25	2.21
14	2.60	2.53	2.46	2.39	2.35	2.31	2.27	2.22	2.18	2.13
15	2.54	2.48	2.40	2.33	2.29	2.25	2.20	2.16	2.11	2.07
16	2.49	2.42	2.35	2.28	2.24	2.19	2.15	2.11	2.06	2.01
17	2.45	2.38	2.31	2.23	2.19	2.15	2.10	2.06	2.01	1.96
18	2.41	2.34	2.27	2.19	2.15	2.11	2.06	2.02	1.97	1.92
19	2.38	2.31	2.23	2.16	2.11	2.07	2.03	1.98	1.93	1.88
20	2.35	2.28	2.20	2.12	2.08	2.04	1.99	1.95	1.90	1.84
21	2.32	2.25	2.18	2.10	2.05	2.01	1.96	1.92	1.87	1.81
22	2.30	2.23	2.15	2.07	2.03	1.98	1.94	1.89	1.84	1.78
23	2.27	2.20	2.13	2.05	2.01	1.96	1.91	1.86	1.81	1.76
24	2.25	2.18	2.11	2.03	1.98	1.94	1.89	1.84	1.79	1.73
25	2.24	2.16	2.09	2.01	1.96	1.92	1.87	1.82	1.77	1.71
26	2.22	2.15	2.07	1.99	1.95	1.90	1.85	1.80	1.75	1.69
27	2.20	2.13	2.06	1.97	1.93	1.88	1.84	1.79	1.73	1.67
28	2.19	2.12	2.04	1.96	1.91	1.87	1.82	1.77	1.71	1.65
29	2.18	2.10	2.03	1.94	1.90	1.85	1.81	1.75	1.70	1.64
30	2.16	2.09	2.01	1.93	1.89	1.84	1.79	1.74	1.68	1.62
40	2.08	2.00	1.92	1.84	1.79	1.74	1.69	1.64	1.58	1.51
60	1.99	1.92	1.84	1.75	1.70	1.65	1.59	1.53	1.47	1.39
120	1.91	1.83	1.75	1.66	1.61	1.55	1.50	1.43	1.35	1.25
∞	1.83	1.75	1.67	1.57	1.52	1.46	1.39	1.32	1.22	1.00

Source: From M. Merrington and C. M. Thompson, "Tables of percentage points of the inverted beta (*F*)-distribution," *Biometrika*, 1943, 33, 73–88. Reproduced by permission of the *Biometrika* Trustees.

The left side of the table is labeled vertically: DENOMINATOR DEGREES OF FREEDOM

TABLE C.5   Critical Values for the *F* Statistic: $F_{.025}$

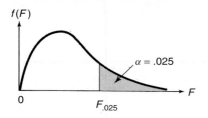

				NUMERATOR DEGREES OF FREEDOM					
ν_2	1	2	3	4	5	6	7	8	9
1	647.8	799.5	864.2	899.6	921.8	937.1	948.2	956.7	963.3
2	38.51	39.00	39.17	39.25	39.30	39.33	39.36	39.37	39.39
3	17.44	16.04	15.44	15.10	14.88	14.73	14.62	14.54	14.47
4	12.22	10.65	9.98	9.60	9.36	9.20	9.07	8.98	8.90
5	10.01	8.43	7.76	7.39	7.15	6.98	6.85	6.76	6.68
6	8.81	7.26	6.60	6.23	5.99	5.82	5.70	5.60	5.52
7	8.07	6.54	5.89	5.52	5.29	5.12	4.99	4.90	4.82
8	7.57	6.06	5.42	5.05	4.82	4.65	4.53	4.43	4.36
9	7.21	5.71	5.08	4.72	4.48	4.32	4.20	4.10	4.03
10	6.94	5.46	4.83	4.47	4.24	4.07	3.95	3.85	3.78
11	6.72	5.26	4.63	4.28	4.04	3.88	3.76	3.66	3.59
12	6.55	5.10	4.47	4.12	3.89	3.73	3.61	3.51	3.44
13	6.41	4.97	4.35	4.00	3.77	3.60	3.48	3.39	3.31
14	6.30	4.86	4.24	3.89	3.66	3.50	3.38	3.29	3.21
15	6.20	4.77	4.15	3.80	3.58	3.41	3.29	3.20	3.12
16	6.12	4.69	4.08	3.73	3.50	3.34	3.22	3.12	3.05
17	6.04	4.62	4.01	3.66	3.44	3.28	3.16	3.06	2.98
18	5.98	4.56	3.95	3.61	3.38	3.22	3.10	3.01	2.93
19	5.92	4.51	3.90	3.56	3.33	3.17	3.05	2.96	2.88
20	5.87	4.46	3.86	3.51	3.29	3.13	3.01	2.91	2.84
21	5.83	4.42	3.82	3.48	3.25	3.09	2.97	2.87	2.80
22	5.79	4.38	3.78	3.44	3.22	3.05	2.93	2.84	2.76
23	5.75	4.35	3.75	3.41	3.18	3.02	2.90	2.81	2.73
24	5.72	4.32	3.72	3.38	3.15	2.99	2.87	2.78	2.70
25	5.69	4.29	3.69	3.35	3.13	2.97	2.85	2.75	2.68
26	5.66	4.27	3.67	3.33	3.10	2.94	2.82	2.73	2.65
27	5.63	4.24	3.65	3.31	3.08	2.92	2.80	2.71	2.63
28	5.61	4.22	3.63	3.29	3.06	2.90	2.78	2.69	2.61
29	5.59	4.20	3.61	3.27	3.04	2.88	2.76	2.67	2.59
30	5.57	4.18	3.59	3.25	3.03	2.87	2.75	2.65	2.57
40	5.42	4.05	3.46	3.13	2.90	2.74	2.62	2.53	2.45
60	5.29	3.93	3.34	3.01	2.79	2.63	2.51	2.41	2.33
120	5.15	3.80	3.23	2.89	2.67	2.52	2.39	2.30	2.22
∞	5.02	3.69	3.12	2.79	2.57	2.41	2.29	2.19	2.11

DENOMINATOR DEGREES OF FREEDOM

(continued overleaf)

TABLE C.5   (*Continued*)

ν_2 \ ν_1	NUMERATOR DEGREES OF FREEDOM									
	10	12	15	20	24	30	40	60	120	∞
1	968.6	976.7	984.9	993.1	997.2	1001	1006	1010	1014	1018
2	39.40	39.41	39.43	39.45	39.46	39.46	39.47	39.48	39.49	39.50
3	14.42	14.34	14.25	14.17	14.12	14.08	14.04	13.99	13.95	13.90
4	8.84	8.75	8.66	8.56	8.51	8.46	8.41	8.36	8.31	8.26
5	6.62	6.52	6.43	6.33	6.28	6.23	6.18	6.12	6.07	6.02
6	5.46	5.37	5.27	5.17	5.12	5.07	5.01	4.96	4.90	4.85
7	4.76	4.67	4.57	4.47	4.42	4.36	4.31	4.25	4.20	4.14
8	4.30	4.20	4.10	4.00	3.95	3.89	3.84	3.78	3.73	3.67
9	3.96	3.87	3.77	3.67	3.61	3.56	3.51	3.45	3.39	3.33
10	3.72	3.62	3.52	3.42	3.37	3.31	3.26	3.20	3.14	3.08
11	3.53	3.43	3.33	3.23	3.17	3.12	3.06	3.00	2.94	2.88
12	3.37	3.28	3.18	3.07	3.02	2.96	2.91	2.85	2.79	2.72
13	3.25	3.15	3.05	2.95	2.89	2.84	2.78	2.72	2.66	2.60
14	3.15	3.05	2.95	2.84	2.79	2.73	2.67	2.61	2.55	2.49
15	3.06	2.96	2.86	2.76	2.70	2.64	2.59	2.52	2.46	2.40
16	2.99	2.89	2.79	2.68	2.63	2.57	2.51	2.45	2.38	2.32
17	2.92	2.82	2.72	2.62	2.56	2.50	2.44	2.38	2.32	2.25
18	2.87	2.77	2.67	2.56	2.50	2.44	2.38	2.32	2.26	2.19
19	2.82	2.72	2.62	2.51	2.45	2.39	2.33	2.27	2.20	2.13
20	2.77	2.68	2.57	2.46	2.41	2.35	2.29	2.22	2.16	2.09
21	2.73	2.64	2.53	2.42	2.37	2.31	2.25	2.18	2.11	2.04
22	2.70	2.60	2.50	2.39	2.33	2.27	2.21	2.14	2.08	2.00
23	2.67	2.57	2.47	2.36	2.30	2.24	2.18	2.11	2.04	1.97
24	2.64	2.54	2.44	2.33	2.27	2.21	2.15	2.08	2.01	1.94
25	2.61	2.51	2.41	2.30	2.24	2.18	2.12	2.05	1.98	1.91
26	2.59	2.49	2.39	2.28	2.22	2.16	2.09	2.03	1.95	1.88
27	2.57	2.47	2.36	2.25	2.19	2.13	2.07	2.00	1.93	1.85
28	2.55	2.45	2.34	2.23	2.17	2.11	2.05	1.98	1.91	1.83
29	2.53	2.43	2.32	2.21	2.15	2.09	2.03	1.96	1.89	1.81
30	2.51	2.41	2.31	2.20	2.14	2.07	2.01	1.94	1.87	1.79
40	2.39	2.29	2.18	2.07	2.01	1.94	1.88	1.80	1.72	1.64
60	2.27	2.17	2.06	1.94	1.88	1.82	1.74	1.67	1.58	1.48
120	2.16	2.05	1.94	1.82	1.76	1.69	1.61	1.53	1.43	1.31
∞	2.05	1.94	1.83	1.71	1.64	1.57	1.48	1.39	1.27	1.00

Left axis label: DENOMINATOR DEGREES OF FREEDOM

Source: From M. Merrington and C. M. Thompson, "Tables of percentage points of the inverted beta (*F*)-distribution," *Biometrika*, 1943, 33, 73–88. Reproduced by permission of the *Biometrika* Trustees.

TABLE C.6   **Critical Values for the *F* Statistic: $F_{.01}$**

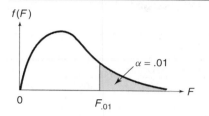

	v_1	NUMERATOR DEGREES OF FREEDOM								
v_2		1	2	3	4	5	6	7	8	9
	1	4,052	4,999.5	5,403	5,625	5,764	5,859	5,928	5,982	6,022
	2	98.50	99.00	99.17	99.25	99.30	99.33	99.36	99.37	99.39
	3	34.12	30.82	29.46	28.71	28.24	27.91	27.67	27.49	27.35
	4	21.20	18.00	16.69	15.98	15.52	15.21	14.98	14.80	14.66
	5	16.26	13.27	12.06	11.39	10.97	10.67	10.46	10.29	10.16
	6	13.75	10.92	9.78	9.15	8.75	8.47	8.26	8.10	7.98
	7	12.25	9.55	8.45	7.85	7.46	7.19	6.99	6.84	6.72
	8	11.26	8.65	7.59	7.01	6.63	6.37	6.18	6.03	5.91
	9	10.56	8.02	6.99	6.42	6.06	5.80	5.61	5.47	5.35
DENOMINATOR DEGREES OF FREEDOM	10	10.04	7.56	6.55	5.99	5.64	5.39	5.20	5.06	4.94
	11	9.65	7.21	6.22	5.67	5.32	5.07	4.89	4.74	4.63
	12	9.33	6.93	5.95	5.41	5.06	4.82	4.64	4.50	4.39
	13	9.07	6.70	5.74	5.21	4.86	4.62	4.44	4.30	4.19
	14	8.86	6.51	5.56	5.04	4.69	4.46	4.28	4.14	4.03
	15	8.68	6.36	5.42	4.89	4.56	4.32	4.14	4.00	3.89
	16	8.53	6.23	5.29	4.77	4.44	4.20	4.03	3.89	3.78
	17	8.40	6.11	5.18	4.67	4.34	4.10	3.93	3.79	3.68
	18	8.29	6.01	5.09	4.58	4.25	4.01	3.84	3.71	3.60
	19	8.18	5.93	5.01	4.50	4.17	3.94	3.77	3.63	3.52
	20	8.10	5.85	4.94	4.43	4.10	3.87	3.70	3.56	3.46
	21	8.02	5.78	4.87	4.37	4.04	3.81	3.64	3.51	3.40
	22	7.95	5.72	4.82	4.31	3.99	3.76	3.59	3.45	3.35
	23	7.88	5.66	4.76	4.26	3.94	3.71	3.54	3.41	3.30
	24	7.82	5.61	4.72	4.22	3.90	3.67	3.50	3.36	3.26
	25	7.77	5.57	4.68	4.18	3.85	3.63	3.46	3.32	3.22
	26	7.72	5.53	4.64	4.14	3.82	3.59	3.42	3.29	3.18
	27	7.68	5.49	4.60	4.11	3.78	3.56	3.39	3.26	3.15
	28	7.64	5.45	4.57	4.07	3.75	3.53	3.36	3.23	3.12
	29	7.60	5.42	4.54	4.04	3.73	3.50	3.33	3.20	3.09
	30	7.56	5.39	4.51	4.02	3.70	3.47	3.30	3.17	3.07
	40	7.31	5.18	4.31	3.83	3.51	3.29	3.12	2.99	2.89
	60	7.08	4.98	4.13	3.65	3.34	3.12	2.95	2.82	2.72
	120	6.85	4.79	3.95	3.48	3.17	2.96	2.79	2.66	2.56
	∞	6.63	4.61	3.78	3.32	3.02	2.80	2.64	2.51	2.41

(*continued overleaf*)

TABLE C.6   (*Continued*)

ν_2 \ ν_1	10	12	15	20	24	30	40	60	120	∞
1	6,056	6,106	6,157	6,209	6,235	6,261	6,287	6,313	6,339	6,366
2	99.40	99.42	99.43	99.45	99.46	99.47	99.47	99.48	99.49	99.50
3	27.23	27.05	26.87	26.69	26.60	26.50	26.41	26.32	26.22	26.13
4	14.55	14.37	14.20	14.02	13.93	13.84	13.75	13.65	13.56	13.46
5	10.05	9.89	9.72	9.55	9.47	9.38	9.29	9.20	9.11	9.02
6	7.87	7.72	7.56	7.40	7.31	7.23	7.14	7.06	6.97	6.88
7	6.62	6.47	6.31	6.16	6.07	5.99	5.91	5.82	5.74	5.65
8	5.81	5.67	5.52	5.36	5.28	5.20	5.12	5.03	4.95	4.86
9	5.26	5.11	4.96	4.81	4.73	4.65	4.57	4.48	4.40	4.31
10	4.85	4.71	4.56	4.41	4.33	4.25	4.17	4.08	4.00	3.91
11	4.54	4.40	4.25	4.10	4.02	3.94	3.86	3.78	3.69	3.60
12	4.30	4.16	4.01	3.86	3.78	3.70	3.62	3.54	3.45	3.36
13	4.10	3.96	3.82	3.66	3.59	3.51	3.43	3.34	3.25	3.17
14	3.94	3.80	3.66	3.51	3.43	3.35	3.27	3.18	3.09	3.00
15	3.80	3.67	3.52	3.37	3.29	3.21	3.13	3.05	2.96	2.87
16	3.69	3.55	3.41	3.26	3.18	3.10	3.02	2.93	2.84	2.75
17	3.59	3.46	3.31	3.16	3.08	3.00	2.92	2.83	2.75	2.65
18	3.51	3.37	3.23	3.08	3.00	2.92	2.84	2.75	2.66	2.57
19	3.43	3.30	3.15	3.00	2.92	2.84	2.76	2.67	2.58	2.49
20	3.37	3.23	3.09	2.94	2.86	2.78	2.69	2.61	2.52	2.42
21	3.31	3.17	3.03	2.88	2.80	2.72	2.64	2.55	2.46	2.36
22	3.26	3.12	2.98	2.83	2.75	2.67	2.58	2.50	2.40	2.31
23	3.21	3.07	2.93	2.78	2.70	2.62	2.54	2.45	2.35	2.26
24	3.17	3.03	2.89	2.74	2.66	2.58	2.49	2.40	2.31	2.21
25	3.13	2.99	2.85	2.70	2.62	2.54	2.45	2.36	2.27	2.17
26	3.09	2.96	2.81	2.66	2.58	2.50	2.42	2.33	2.23	2.13
27	3.06	2.93	2.78	2.63	2.55	2.47	2.38	2.29	2.20	2.10
28	3.03	2.90	2.75	2.60	2.52	2.44	2.35	2.26	2.17	2.06
29	3.00	2.87	2.73	2.57	2.49	2.41	2.33	2.23	2.14	2.03
30	2.98	2.84	2.70	2.55	2.47	2.39	2.30	2.21	2.11	2.01
40	2.80	2.66	2.52	2.37	2.29	2.20	2.11	2.02	1.92	1.80
60	2.63	2.50	2.35	2.20	2.12	2.03	1.94	1.84	1.73	1.60
120	2.47	2.34	2.19	2.03	1.95	1.86	1.76	1.66	1.53	1.38
∞	2.32	2.18	2.04	1.88	1.79	1.70	1.59	1.47	1.32	1.00

NUMERATOR DEGREES OF FREEDOM; DENOMINATOR DEGREES OF FREEDOM

Source: From M. Merrington and C. M. Thompson, "Tables of percentage points of the inverted beta (*F*)-distribution," *Biometrika*, 1943, 33, 73–88. Reproduced by permission of the *Biometrika* Trustees.

TABLE C.7   Random Numbers

ROW \ COLUMN	1	2	3	4	5	6	7	8	9	10	11	12	13	14
1	10480	15011	01536	02011	81647	91646	69179	14194	62590	36207	20969	99570	91291	90700
2	22368	46573	25595	85393	30995	89198	27982	53402	93965	34095	52666	19174	39615	99505
3	24130	48360	22527	97265	76393	64809	15179	24830	49340	32081	30680	19655	63348	58629
4	42167	93093	06243	61680	07856	16376	39440	53537	71341	57004	00849	74917	97758	16379
5	37570	39975	81837	16656	06121	91782	60468	81305	49684	60672	14110	06927	01263	54613
6	77921	06907	11008	42751	27756	53498	18602	70659	90655	15053	21916	81825	44394	42880
7	99562	72905	56420	69994	98872	31016	71194	18738	44013	48840	63213	21069	10634	12952
8	96301	91977	05463	07972	18876	20922	94595	56869	69014	60045	18425	84903	42508	32307
9	89579	14342	63661	10281	17453	18103	57740	84378	25331	12566	58678	44947	05585	56941
10	85475	36857	53342	53988	53060	59533	38867	62300	08158	17983	16439	11458	18593	64952
11	28918	69578	88231	33276	70997	79936	56865	05859	90106	31595	01547	85590	91610	78188
12	63553	40961	48235	03427	49626	69445	18663	72695	52180	20847	12234	90511	33703	90322
13	09429	93969	52636	92737	88974	33488	36320	17617	30015	08272	84115	27156	30613	74952
14	10365	61129	87529	85689	48237	52267	67689	93394	01511	26358	85104	20285	29975	89868
15	07119	97336	71048	08178	77233	13916	47564	81056	97735	85977	29372	74461	28551	90707
16	51085	12765	51821	51259	77452	16308	60756	92144	49442	53900	70960	63990	75601	40719
17	02368	21382	52404	60268	89368	19885	55322	44819	01188	65255	64835	44919	05944	55157
18	01011	54092	33362	94904	31273	04146	18594	29852	71585	85030	51132	01915	92747	64951
19	52162	53916	46369	58586	23216	14513	83149	98736	23495	64350	94738	17752	35156	35749
20	07056	97628	33787	09998	42698	06691	76988	13602	51851	46104	88916	19509	25625	58104
21	48663	91245	85828	14346	09172	30168	90229	04734	59193	22178	30421	61666	99904	32812
22	54164	58492	22421	74103	47070	25306	76468	26384	58151	06646	21524	15227	96909	44592
23	32639	32363	05597	24200	13363	38005	94342	28728	35806	06912	17012	64161	18296	22851
24	29334	27001	87637	87308	58731	00256	45834	15398	46557	41135	10367	07684	36188	18510
25	02488	33062	28834	07351	19731	92420	60952	61280	50001	67658	32586	86679	50720	94953
26	81525	72295	04839	96423	24878	82651	66566	14778	76797	14780	13300	87074	79666	95725
27	29676	20591	68086	26432	46901	20849	89768	81536	86645	12659	92259	57102	80428	25280
28	00742	57392	39064	66432	84673	40027	32832	61362	98947	96067	64760	64584	96096	98253
29	05366	04213	25669	26422	44407	44048	37937	63904	45766	66134	75470	66520	34693	90449
30	91921	26418	64117	94305	26766	25940	39972	22209	71500	64568	91402	42416	07844	69618
31	00582	04711	87917	77341	42206	35126	74087	99547	81817	42607	43808	76655	62028	76630
32	00725	69884	62797	56170	86324	88072	76222	36086	84637	93161	76038	65855	77919	88006
33	69011	65795	95876	55293	18988	27354	26575	08625	40801	59920	29841	80150	12777	48501
34	25976	57948	29888	88604	67917	48708	18912	82271	65424	69774	33611	54262	85963	03547
35	09763	83473	73577	12908	30883	18317	28290	35797	05998	41688	34952	37888	38917	88050
36	91576	42595	27958	30134	04024	86385	29880	99730	55536	84855	29080	09250	79656	73211
37	17955	56349	90999	49127	20044	59931	06115	20542	18059	02008	73708	83517	36103	42791
38	46503	18584	18845	49618	02304	51038	20655	58727	28168	15475	56942	53389	20562	87338
39	92157	89634	94824	78171	84610	82834	09922	25417	44137	48413	25555	21246	35509	20468
40	14577	62765	35605	81263	39667	47358	56873	56307	61607	49518	89656	20103	77490	18062
41	98427	07523	33362	64270	01638	92477	66969	98420	04880	45585	46565	04102	46880	45709
42	34914	63976	88720	82765	34476	17032	87589	40836	32427	70002	70663	88863	77775	69348
43	70060	28277	39475	46473	23219	53416	94970	25832	69975	94884	19661	72828	00102	66794
44	53976	54914	06990	67245	68350	82948	11398	42878	80287	88267	47363	46634	06541	97809
45	76072	29515	40980	07391	58745	25774	22987	80059	39911	96189	41151	14222	60697	59583
46	90725	52210	83974	29992	65831	38857	50490	83765	55657	14361	31720	57375	56228	41546
47	64364	67412	33339	31926	14883	24413	59744	92351	97473	89286	35931	04110	23726	51900
48	08962	00358	31662	25388	61642	34072	81249	35648	56891	69352	48373	45578	78547	81788
49	95012	68379	93526	70765	10592	04542	76463	54328	02349	17247	28865	14777	62730	92277
50	15664	10493	20492	38391	91132	21999	59516	81652	27195	48223	46751	22923	32261	85653
51	16408	81899	04153	53381	79401	21438	83035	92350	36693	31238	59649	91754	72772	02338

(continued overleaf)

TABLE C.7 **(Continued)**

ROW \ COLUMN	1	2	3	4	5	6	7	8	9	10	11	12	13	14
52	18629	81953	05520	91962	04739	13092	97662	24822	94730	06496	35090	04822	86774	98289
53	73115	35101	47498	87637	99016	71060	88824	71013	18735	20286	23153	72924	35165	43040
54	57491	16703	23167	49323	45021	33132	12544	41035	80780	45393	44812	12515	98931	91202
55	30405	83946	23792	14422	15059	45799	22716	19792	09983	74353	68668	30429	70735	25499
56	16631	35006	85900	98275	32388	52390	16815	69298	82732	38480	73817	32523	41961	44437
57	96773	20206	42559	78985	05300	22164	24369	54224	35083	19687	11052	91491	60383	19746
58	38935	64202	14349	82674	66523	44133	00697	35552	35970	19124	63318	29686	03387	59846
59	31624	76384	17403	53363	44167	64486	64758	75366	76554	31601	12614	33072	60332	92325
60	78919	19474	23632	27889	47914	02584	37680	20801	72152	39339	34806	08930	85001	87820
61	03931	33309	57047	74211	63445	17361	62825	39908	05607	91284	68833	25570	38818	46920
62	74426	33278	43972	10119	89917	15665	52872	73823	73144	88662	88970	74492	51805	99378
63	09066	00903	20795	95452	92648	45454	09552	88815	16553	51125	79375	97596	16296	66092
64	42238	12426	87025	14267	20979	04508	64535	31355	86064	29472	47689	05974	52468	16834
65	16153	08002	26504	41744	81959	65642	74240	56302	00033	67107	77510	70625	28725	34191
66	21457	40742	29820	96783	29400	21840	15035	34537	33310	06116	95240	15957	16572	06004
67	21581	57802	02050	89728	17937	37621	47075	42080	97403	48626	68995	43805	33386	21597
68	55612	78095	83197	33732	05810	24813	86902	60397	16489	03264	88525	42786	05269	92532
69	44657	66999	99324	51281	84463	60563	79312	93454	68876	25471	93911	25650	12682	73572
70	91340	84979	46949	81973	37949	61023	43997	15263	80644	43942	89203	71795	99533	50501
71	91227	21199	31935	27022	84067	05462	35216	14486	29891	68607	41867	14951	91696	85065
72	50001	38140	66321	19924	72163	09538	12151	06878	91903	18749	34405	56087	82790	70925
73	65390	05224	72958	28609	81406	39147	25549	48542	42627	45233	57202	94617	23772	07896
74	27504	96131	83944	41575	10573	08619	64482	73923	36152	05184	94142	25299	84387	34925
75	37169	94851	39117	89632	00959	16487	65536	49071	39782	17095	02330	74301	00275	48280
76	11508	70225	51111	38351	19444	66499	71945	05422	13442	78675	84081	66938	93654	59894
77	37449	30362	06694	54690	04052	53115	62757	95348	78662	11163	81651	50245	34971	52924
78	46515	70331	85922	38329	57015	15765	97161	17869	45349	61796	66345	81073	49106	79860
79	30986	81223	42416	58353	21532	30502	32305	86482	05174	07901	54339	58861	74818	46942
80	63798	64995	46583	09785	44160	78128	83991	42865	92520	83531	80377	35909	81250	54238
81	82486	84846	99254	67632	43218	50076	21361	64816	51202	88124	41870	52689	51275	83556
82	21885	32906	92431	09060	64297	51674	64126	62570	26123	05155	59194	52799	28225	85762
83	60336	98782	07408	53458	13564	59089	26445	29789	85205	41001	12535	12133	14645	23541
84	43937	46891	24010	25560	86355	33941	25786	54990	71899	15475	95434	98227	21824	19585
85	97656	63175	89303	16275	07100	92063	21942	18611	47348	20203	18534	03862	78095	50136
86	03299	01221	05418	38982	55758	92237	26759	86367	21216	98442	08303	56613	91511	75928
87	79626	06486	03574	17668	07785	76020	79924	25651	83325	88428	85076	72811	22717	50585
88	85636	68335	47539	03129	65651	11977	02510	26113	99447	68645	34327	15152	55230	93448
89	18039	14367	61337	06177	12143	46609	32989	74014	64708	00533	35398	58408	13261	47908
90	08362	15656	60627	36478	65648	16764	53412	09013	07832	41574	17639	82163	60859	75567
91	79556	29068	04142	16268	15387	12856	66227	38358	22478	73373	88732	09443	82558	05250
92	92608	82674	27072	32534	17075	27698	98204	63863	11951	34648	88022	56148	34925	57031
93	23982	25835	40055	67006	12293	02753	14827	23235	35071	99704	37543	11601	35503	85171
94	09915	96306	05908	97901	28395	14186	00821	80703	70426	75647	76310	88717	37890	40129
95	59037	33300	26695	62247	69927	76123	50842	43834	86654	70959	79725	93872	28117	19233
96	42488	78077	69882	61657	34136	79180	97526	43092	04098	73571	80799	76536	71255	64239
97	46764	86273	63003	93017	31204	36692	40202	35275	57306	55543	53203	18098	47625	88684
98	03237	45430	55417	63282	90816	17349	88298	90183	36600	78406	06216	95787	42579	90730
99	86591	81482	52667	61582	14972	90053	89534	76036	49199	43716	97548	04379	46370	28672
100	38534	01715	94964	87288	65680	43772	39560	12918	86537	62738	19636	51132	25739	56947

Source: Abridged from W. H. Beyer (ed.). *CRC Standard Mathematical Tables*, 24th edition. Cleveland: The Chemical Rubber Company, 1976.

TABLE C.8   Critical Values for the Durbin–Watson d Statistic ($\alpha = .05$)

n	$k = 1$		$k = 2$		$k = 3$		$k = 4$		$k = 5$	
	d_L	d_U	d_L	d_U	d_L	d_U	d_L	d_U	d_L	d_U
15	1.08	1.36	.95	1.54	.82	1.75	.69	1.97	.56	2.21
16	1.10	1.37	.98	1.54	.86	1.73	.74	1.93	.62	2.15
17	1.13	1.38	1.02	1.54	.90	1.71	.78	1.90	.67	2.10
18	1.16	1.39	1.05	1.53	.93	1.69	.82	1.87	.71	2.06
19	1.18	1.40	1.08	1.53	.97	1.68	.86	1.85	.75	2.02
20	1.20	1.41	1.10	1.54	1.00	1.68	.90	1.83	.79	1.99
21	1.22	1.42	1.13	1.54	1.03	1.67	.93	1.81	.83	1.96
22	1.24	1.43	1.15	1.54	1.05	1.66	.96	1.80	.86	1.94
23	1.26	1.44	1.17	1.54	1.08	1.66	.99	1.79	.90	1.92
24	1.27	1.45	1.19	1.55	1.10	1.66	1.01	1.78	.93	1.90
25	1.29	1.45	1.21	1.55	1.12	1.66	1.04	1.77	.95	1.89
26	1.30	1.46	1.22	1.55	1.14	1.65	1.06	1.76	.98	1.88
27	1.32	1.47	1.24	1.56	1.16	1.65	1.08	1.76	1.01	1.86
28	1.33	1.48	1.26	1.56	1.18	1.65	1.10	1.75	1.03	1.85
29	1.34	1.48	1.27	1.56	1.20	1.65	1.12	1.74	1.05	1.84
30	1.35	1.49	1.28	1.57	1.21	1.65	1.14	1.74	1.07	1.83
31	1.36	1.50	1.30	1.57	1.23	1.65	1.16	1.74	1.09	1.83
32	1.37	1.50	1.31	1.57	1.24	1.65	1.18	1.73	1.11	1.82
33	1.38	1.51	1.32	1.58	1.26	1.65	1.19	1.73	1.13	1.81
34	1.39	1.51	1.33	1.58	1.27	1.65	1.21	1.73	1.15	1.81
35	1.40	1.52	1.34	1.58	1.28	1.65	1.22	1.73	1.16	1.80
36	1.41	1.52	1.35	1.59	1.29	1.65	1.24	1.73	1.18	1.80
37	1.42	1.53	1.36	1.59	1.31	1.66	1.25	1.72	1.19	1.80
38	1.43	1.54	1.37	1.59	1.32	1.66	1.26	1.72	1.21	1.79
39	1.43	1.54	1.38	1.60	1.33	1.66	1.27	1.72	1.22	1.79
40	1.44	1.54	1.39	1.60	1.34	1.66	1.29	1.72	1.23	1.79
45	1.48	1.57	1.43	1.62	1.38	1.67	1.34	1.72	1.29	1.78
50	1.50	1.59	1.46	1.63	1.42	1.67	1.38	1.72	1.34	1.77
55	1.53	1.60	1.49	1.64	1.45	1.68	1.41	1.72	1.38	1.77
60	1.55	1.62	1.51	1.65	1.48	1.69	1.44	1.73	1.41	1.77
65	1.57	1.63	1.54	1.66	1.50	1.70	1.47	1.73	1.44	1.77
70	1.58	1.64	1.55	1.67	1.52	1.70	1.49	1.74	1.46	1.77
75	1.60	1.65	1.57	1.68	1.54	1.71	1.51	1.74	1.49	1.77
80	1.61	1.66	1.59	1.69	1.56	1.72	1.53	1.74	1.51	1.77
85	1.62	1.67	1.60	1.70	1.57	1.72	1.55	1.75	1.52	1.77
90	1.63	1.68	1.61	1.70	1.59	1.73	1.57	1.75	1.54	1.78
95	1.64	1.69	1.62	1.71	1.60	1.73	1.58	1.75	1.56	1.78
100	1.65	1.69	1.63	1.72	1.61	1.74	1.59	1.76	1.57	1.78

TABLE C.9 Critical Values for the Durbin-Watson d Statistic ($\alpha = .01$)

n	$k = 1$		$k = 2$		$k = 3$		$k = 4$		$k = 5$	
	d_L	d_U	d_L	d_U	d_L	d_U	d_L	d_U	d_L	d_U
15	.81	1.07	.70	1.25	.59	1.46	.49	1.70	.39	1.96
16	.84	1.09	.74	1.25	.63	1.44	.53	1.66	.44	1.90
17	.87	1.10	.77	1.25	.67	1.43	.57	1.63	.48	1.85
18	.90	1.12	.80	1.26	.71	1.42	.61	1.60	.52	1.80
19	.93	1.13	.83	1.26	.74	1.41	.65	1.58	.56	1.77
20	.95	1.15	.86	1.27	.77	1.41	.68	1.57	.60	1.74
21	.97	1.16	.89	1.27	.80	1.41	.72	1.55	.63	1.71
22	1.00	1.17	.91	1.28	.83	1.40	.75	1.54	.66	1.69
23	1.02	1.19	.94	1.29	.86	1.40	.77	1.53	.70	1.67
24	1.04	1.20	.96	1.30	.88	1.41	.80	1.53	.72	1.66
25	1.05	1.21	.98	1.30	.90	1.41	.83	1.52	.75	1.65
26	1.07	1.22	1.00	1.31	.93	1.41	.85	1.52	.78	1.64
27	1.09	1.23	1.02	1.32	.95	1.41	.88	1.51	.81	1.63
28	1.10	1.24	1.04	1.32	.97	1.41	.90	1.51	.83	1.62
29	1.12	1.25	1.05	1.33	.99	1.42	.92	1.51	.85	1.61
30	1.13	1.26	1.07	1.34	1.01	1.42	.94	1.51	.88	1.61
31	1.15	1.27	1.08	1.34	1.02	1.42	.96	1.51	.90	1.60
32	1.16	1.28	1.10	1.35	1.04	1.43	.98	1.51	.92	1.60
33	1.17	1.29	1.11	1.36	1.05	1.43	1.00	1.51	.94	1.59
34	1.18	1.30	1.13	1.36	1.07	1.43	1.01	1.51	.95	1.59
35	1.19	1.31	1.14	1.37	1.08	1.44	1.03	1.51	.97	1.59
36	1.21	1.32	1.15	1.38	1.10	1.44	1.04	1.51	.99	1.59
37	1.22	1.32	1.16	1.38	1.11	1.45	1.06	1.51	1.00	1.59
38	1.23	1.33	1.18	1.39	1.12	1.45	1.07	1.52	1.02	1.58
39	1.24	1.34	1.19	1.39	1.14	1.45	1.09	1.52	1.03	1.58
40	1.25	1.34	1.20	1.40	1.15	1.46	1.10	1.52	1.05	1.58
45	1.29	1.38	1.24	1.42	1.20	1.48	1.16	1.53	1.11	1.58
50	1.32	1.40	1.28	1.45	1.24	1.49	1.20	1.54	1.16	1.59
55	1.36	1.43	1.32	1.47	1.28	1.51	1.25	1.55	1.21	1.59
60	1.38	1.45	1.35	1.48	1.32	1.52	1.28	1.56	1.25	1.60
65	1.41	1.47	1.38	1.50	1.35	1.53	1.31	1.57	1.28	1.61
70	1.43	1.49	1.40	1.52	1.37	1.55	1.34	1.58	1.31	1.61
75	1.45	1.50	1.42	1.53	1.39	1.56	1.37	1.59	1.34	1.62
80	1.47	1.52	1.44	1.54	1.42	1.57	1.39	1.60	1.36	1.62
85	1.48	1.53	1.46	1.55	1.43	1.58	1.41	1.60	1.39	1.63
90	1.50	1.54	1.47	1.56	1.45	1.59	1.43	1.61	1.41	1.64
95	1.51	1.55	1.49	1.57	1.47	1.60	1.45	1.62	1.42	1.64
100	1.52	1.56	1.50	1.58	1.48	1.60	1.46	1.63	1.44	1.65

Source: From J. Durbin and G. S. Watson, "Testing for serial correlation in least squares regression, II," *Biometrika*, 1951, 30, 159–178. Reproduced by permission of the *Biometrika* Trustees.

TABLE C.10   Critical Values for the χ^2 Statistic

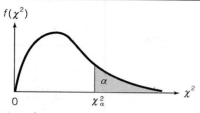

DEGREES OF FREEDOM	$\chi^2_{.995}$	$\chi^2_{.990}$	$\chi^2_{.975}$	$\chi^2_{.950}$	$\chi^2_{.900}$
1	.0000393	.0001571	.0009821	.0039321	.0157908
2	.0100251	.0201007	.0506356	.102587	.210720
3	.0717212	.114832	.215795	.351846	.584375
4	.206990	.297110	.484419	.710721	1.063623
5	.411740	.554300	.831211	1.145476	1.61031
6	.675727	.872085	1.237347	1.63539	2.20413
7	.989265	1.239043	1.68987	2.16735	2.83311
8	1.344419	1.646482	2.17973	2.73264	3.48954
9	1.734926	2.087912	2.70039	3.32511	4.16816
10	2.15585	2.55821	3.24697	3.94030	4.86518
11	2.60321	3.05347	3.81575	4.57481	5.57779
12	3.07382	3.57056	4.40379	5.22603	6.30380
13	3.56503	4.10691	5.00874	5.89186	7.04150
14	4.07468	4.66043	5.62872	6.57063	7.78953
15	4.60094	5.22935	6.26214	7.26094	8.54675
16	5.14224	5.81221	6.90766	7.96164	9.31223
17	5.69724	6.40776	7.56418	8.67176	10.0852
18	76.26481	7.01491	8.23075	9.39046	10.8649
19	6.84398	7.63273	8.90655	10.1170	11.6509
20	7.43386	8.26040	9.59083	10.8508	12.4426
21	8.03366	8.89720	10.28293	11.5913	13.2396
22	8.64272	9.54249	10.9823	12.3380	14.0415
23	9.26042	10.19567	11.6885	13.0905	14.8479
24	9.88623	10.8564	12.4011	13.8484	15.6587
25	10.5197	11.5240	13.1197	14.6114	16.4734
26	11.1603	12.1981	13.8439	15.3791	17.2919
27	11.8076	12.8786	14.5733	16.1513	18.1138
28	12.4613	13.5648	15.3079	16.9279	18.9392
29	13.1211	14.2565	16.0471	17.7083	19.7677
30	13.7867	14.9535	16.7908	18.4926	20.5992
40	20.7065	22.1643	24.4331	26.5093	29.0505
50	27.9907	29.7067	32.3574	34.7642	37.6886
60	35.5346	37.4848	40.4817	43.1879	46.4589
70	43.2752	45.4418	48.7576	51.7393	55.3290
80	51.1720	53.5400	57.1532	60.3915	64.2778
90	59.1963	61.7541	65.6466	69.1260	73.2912
100	67.3276	70.0648	74.2219	77.9295	82.3581
150	109.142	112.668	117.985	122.692	128.275
200	152.241	156.432	162.728	168.279	174.835
300	240.663	245.972	253.912	260.878	269.068
400	330.903	337.155	346.482	354.641	364.207
500	422.303	429.388	439.936	449.147	459.926

(continued overleaf)

TABLE C.10 (Continued)

DEGREES OF FREEDOM	$\chi^2_{.100}$	$\chi^2_{.050}$	$\chi^2_{.025}$	$\chi^2_{.010}$	$\chi^2_{.500}$
1	2.70554	3.84146	5.02389	6.63490	7.87944
2	4.60517	5.99147	7.37776	9.21034	10.5966
3	6.25139	7.81473	9.34840	11.3449	12.8381
4	7.77944	9.48773	11.1433	13.2767	14.8602
5	9.23635	11.0705	12.8325	15.0863	16.7496
6	10.6446	12.5916	14.4494	16.8119	18.5476
7	12.0170	14.0671	16.0128	18.4753	20.2777
8	13.3616	15.5073	17.5346	20.0902	21.9550
9	14.6837	16.9190	19.0228	21.6660	23.5893
10	15.9871	18.3070	20.4831	23.2093	25.1882
11	17.2750	19.6751	21.9200	24.7250	26.7569
12	18.5494	21.0261	23.3367	26.2170	28.2995
13	19.8119	22.3621	24.7356	27.6883	29.8194
14	21.0642	23.6848	26.1190	29.1413	31.3193
15	22.3072	24.9958	27.4884	30.5779	32.8013
16	23.5418	26.2962	28.8454	31.9999	34.2672
17	24.7690	27.5871	30.1910	33.4087	35.7185
18	25.9894	28.8693	31.5264	34.8053	37.1564
19	27.2036	30.1435	32.8523	36.1908	38.5822
20	28.4120	31.4104	34.1696	37.5662	39.9968
21	29.6151	32.6705	35.4789	38.9321	41.4010
22	30.8133	33.9244	36.7807	40.2894	42.7956
23	32.0069	35.1725	38.0757	41.6384	44.1813
24	33.1963	36.4151	39.3641	42.9798	45.5585
25	34.3816	37.6525	40.6465	44.3141	46.9278
26	36.5631	38.8852	41.9232	45.6417	48.2899
27	36.7412	40.1133	43.1944	46.9630	49.6449
28	37.9159	41.3372	44.4607	48.2782	50.9933
29	39.0875	42.5569	45.7222	49.5879	52.3356
30	40.2560	43.7729	46.9792	50.8922	53.6720
40	51.8050	55.7585	59.3417	63.6907	66.7659
50	63.1671	67.5048	71.4202	76.1539	79.4900
60	74.3970	79.0819	83.2976	88.3794	91.9517
70	85.5271	90.5312	95.0231	100.425	104.215
80	96.5782	101.879	106.629	112.329	116.321
90	107.565	113.145	118.136	124.116	128.299
100	118.498	124.342	129.561	135.807	140.169
150	172.581	179.581	185.800	193.208	198.360
200	226.021	233.994	241.058	249.445	255.264
300	331.789	341.395	349.874	359.906	366.844
400	436.649	447.632	457.305	468.724	476.606
500	540.930	553.127	563.852	576.493	585.207

TABLE C.11   Percentage Points of the Studentized Range $q(p, v)$, Upper 5%

v \ p	2	3	4	5	6	7	8	9	10	11
1	17.97	26.98	32.82	37.08	40.41	43.12	45.40	47.36	49.07	50.59
2	6.08	8.33	9.80	10.88	11.74	12.44	13.03	13.54	13.99	14.39
3	4.50	5.91	6.82	7.50	8.04	8.48	8.85	9.18	9.46	9.72
4	3.93	5.04	5.76	6.29	6.71	7.05	7.35	7.60	7.83	8.03
5	3.64	4.60	5.22	5.67	6.03	6.33	6.58	6.80	6.99	7.17
6	3.46	4.34	4.90	5.30	5.63	5.90	6.12	6.32	6.49	6.65
7	3.34	4.16	4.68	5.06	5.36	5.61	5.82	6.00	6.16	6.30
8	3.26	4.04	4.53	4.89	5.17	5.40	5.60	5.77	5.92	6.05
9	3.20	3.95	4.41	4.76	5.02	5.24	5.43	5.59	5.74	5.87
10	3.15	3.88	4.33	4.65	4.91	5.12	5.30	5.46	5.60	5.72
11	3.11	3.82	4.26	4.57	4.82	5.03	5.20	5.35	5.49	5.61
12	3.08	3.77	4.20	4.51	4.75	4.95	5.12	5.27	5.39	5.51
13	3.06	3.73	4.15	4.45	4.69	4.88	5.05	5.19	5.32	5.43
14	3.03	3.70	4.11	4.41	4.64	4.83	4.99	5.13	5.25	5.36
15	3.01	3.67	4.08	4.37	4.60	4.78	4.94	5.08	5.20	5.31
16	3.00	3.65	4.05	4.33	4.56	4.74	4.90	5.03	5.15	5.26
17	2.98	3.63	4.02	4.30	4.52	4.70	4.86	4.99	5.11	5.21
18	2.97	3.61	4.00	4.28	4.49	4.67	4.82	4.96	5.07	5.17
19	2.96	3.59	3.98	4.25	4.47	4.65	4.79	4.92	5.04	5.14
20	2.95	3.58	3.96	4.23	4.45	4.62	4.77	4.90	5.01	5.11
24	2.92	3.53	3.90	4.17	4.37	4.54	4.68	4.81	4.92	5.01
30	2.89	3.49	3.85	4.10	4.30	4.46	4.60	4.72	4.82	4.92
40	2.86	3.44	3.79	4.04	4.23	4.39	4.52	4.63	4.73	4.82
60	2.83	3.40	3.74	3.98	4.16	4.31	4.44	4.55	4.65	4.73
120	2.80	3.36	3.68	3.92	4.10	4.24	4.36	4.47	4.56	4.64
∞	2.77	3.31	3.63	3.86	4.03	4.17	4.29	4.39	4.47	4.55

(continued overleaf)

TABLE C.11   (*Continued*)

ν \ p	12	13	14	15	16	17	18	19	20
1	51.96	53.20	54.33	55.36	56.32	57.22	58.04	58.83	59.56
2	14.75	15.08	15.38	15.65	15.91	16.14	16.37	16.57	16.77
3	9.95	10.15	10.35	10.52	10.69	10.84	10.98	11.11	11.24
4	8.21	8.37	8.52	8.66	8.79	8.91	9.03	9.13	9.23
5	7.32	7.47	7.60	7.72	7.83	7.93	8.03	8.12	8.21
6	6.79	6.92	7.03	7.14	7.24	7.34	7.43	7.51	7.59
7	6.43	6.55	6.66	6.76	6.85	6.94	7.02	7.10	7.17
8	6.18	6.29	6.39	6.48	6.57	6.65	6.73	6.80	6.87
9	5.98	6.09	6.19	6.28	6.36	6.44	6.51	6.58	6.64
10	5.83	5.93	6.03	6.11	6.19	6.27	6.34	6.40	6.47
11	5.71	5.81	5.90	5.98	6.06	6.13	6.20	6.27	6.33
12	5.61	5.71	5.80	5.88	5.95	6.02	6.09	6.15	6.21
13	5.53	5.63	5.71	5.79	5.86	5.93	5.99	6.05	6.11
14	5.46	5.55	5.64	5.71	5.79	5.85	5.91	5.97	6.03
15	5.40	5.49	5.57	5.65	5.72	5.78	5.85	5.90	5.96
16	5.35	5.44	5.52	5.59	5.66	5.73	5.79	5.84	5.90
17	5.31	5.39	5.47	5.54	5.61	5.67	5.73	5.79	5.84
18	5.27	5.35	5.43	5.50	5.57	5.63	5.69	5.74	5.79
19	5.23	5.31	5.39	5.46	5.53	5.59	5.65	5.70	5.75
20	5.20	5.28	5.36	5.43	5.49	5.55	5.61	5.66	5.71
24	5.10	5.18	5.25	5.32	5.38	5.44	5.49	5.55	5.59
30	5.00	5.08	5.15	5.21	5.27	5.33	5.38	5.43	5.47
40	4.90	4.98	5.04	5.11	5.16	5.22	5.27	5.31	5.36
60	4.81	4.88	4.94	5.00	5.06	5.11	5.15	5.20	5.24
120	4.71	4.78	4.84	4.90	4.95	5.00	5.04	5.09	5.13
∞	4.62	4.68	4.74	4.80	4.85	4.89	4.93	4.97	5.01

Source: *Biometrika Tables for Statisticians*, Vol. I, 3rd ed., edited by E. S. Pearson and H. O. Hartley (Cambridge University Press, 1966). Reproduced by permission of Professor E. S. Pearson and the *Biometrika* Trustees.

TABLE C.12   Percentage Points of the Studentized Range $q(p, v)$, Upper 1%

v \ p	2	3	4	5	6	7	8	9	10	11
1	90.03	135.0	164.3	185.6	202.2	215.8	227.2	237.0	245.6	253.2
2	14.04	19.02	22.29	24.72	26.63	28.20	29.53	30.68	31.69	32.59
3	8.26	10.62	12.17	13.33	14.24	15.00	15.64	16.20	16.69	17.13
4	6.51	8.12	9.17	9.96	10.58	11.10	11.55	11.93	12.27	12.57
5	5.70	6.98	7.80	8.42	8.91	9.32	9.67	9.97	10.24	10.48
6	5.24	6.33	7.03	7.56	7.97	8.32	8.61	8.87	9.10	9.30
7	4.95	5.92	6.54	7.01	7.37	7.68	7.94	8.17	8.37	8.55
8	4.75	5.64	6.20	6.62	6.96	7.24	7.47	7.68	7.86	8.03
9	4.60	5.43	5.96	6.35	6.66	6.91	7.13	7.33	7.49	7.65
10	4.48	5.27	5.77	6.14	6.43	6.67	6.87	7.05	7.21	7.36
11	4.39	5.15	5.62	5.97	6.25	6.48	6.67	6.84	6.99	7.13
12	4.32	5.05	5.50	5.84	6.10	6.32	6.51	6.67	6.81	6.94
13	4.26	4.96	5.40	5.73	5.98	6.19	6.37	6.53	6.67	6.79
14	4.21	4.89	5.32	5.63	5.88	6.08	6.26	6.41	6.54	6.66
15	4.17	4.84	5.25	5.56	5.80	5.99	6.16	6.31	6.44	6.55
16	4.13	4.79	5.19	5.49	5.72	5.92	6.08	6.22	6.35	6.46
17	4.10	4.74	5.14	5.43	5.66	5.85	6.01	6.15	6.27	6.38
18	4.07	4.70	5.09	5.38	5.60	5.79	5.94	6.08	6.20	6.31
19	4.05	4.67	5.05	5.33	5.55	5.73	5.89	6.02	6.14	6.25
20	4.02	4.64	5.02	5.29	5.51	5.69	5.84	5.97	6.09	6.19
24	3.96	4.55	4.91	5.17	5.37	5.54	5.69	5.81	5.92	6.02
30	3.89	4.45	4.80	5.05	5.24	5.40	5.54	5.65	5.76	5.85
40	3.82	4.37	4.70	4.93	5.11	5.26	5.39	5.50	5.60	5.69
60	3.76	4.28	4.59	4.82	4.99	5.13	5.25	5.36	5.45	5.53
120	3.70	4.20	4.50	4.71	4.87	5.01	5.12	5.21	5.30	5.37
∞	3.64	4.12	4.40	4.60	4.76	4.88	4.99	5.08	5.16	5.23

(*continued overleaf*)

TABLE C.12   (*Continued*)

v \ p	12	13	14	15	16	17	18	19	20
1	260.0	266.2	271.8	277.0	281.8	286.3	290.0	294.3	298.0
2	33.40	34.13	34.81	35.43	36.00	36.53	37.03	37.50	37.95
3	17.53	17.89	18.22	18.52	18.81	19.07	19.32	19.55	19.77
4	12.84	13.09	13.32	13.53	13.73	13.91	14.08	14.24	14.40
5	10.70	10.89	11.08	11.24	11.40	11.55	11.68	11.81	11.93
6	9.48	9.65	9.81	9.95	10.08	10.21	10.32	10.43	10.54
7	8.71	8.86	9.00	9.12	9.24	9.35	9.46	9.55	9.65
8	8.18	8.31	8.44	8.55	8.66	8.76	8.85	8.94	9.03
9	7.78	7.91	8.03	8.13	8.23	8.33	8.41	8.49	8.57
10	7.49	7.60	7.71	7.81	7.91	7.99	8.08	8.15	8.23
11	7.25	7.36	7.46	7.56	7.65	7.73	7.81	7.88	7.95
12	7.06	7.17	7.26	7.36	7.44	7.52	7.59	7.66	7.73
13	6.90	7.01	7.10	7.19	7.27	7.35	7.42	7.48	7.55
14	6.77	6.87	6.96	7.05	7.13	7.20	7.27	7.33	7.39
15	6.66	6.76	6.84	6.93	7.00	7.07	7.14	7.20	7.26
16	6.56	6.66	6.74	6.82	6.90	6.97	7.03	7.09	7.15
17	6.48	6.57	6.66	6.73	6.81	6.87	6.94	7.00	7.05
18	6.41	6.50	6.58	6.65	6.72	6.79	6.85	6.91	6.97
19	6.34	6.43	6.51	6.58	6.65	6.72	6.78	6.84	6.89
20	6.28	6.37	6.45	6.52	6.59	6.65	6.71	6.77	6.82
24	6.11	6.19	6.26	6.33	6.39	6.45	6.51	6.56	6.61
30	5.93	6.01	6.08	6.14	6.20	6.26	6.31	6.36	6.41
40	5.76	5.83	5.90	5.96	6.02	6.07	6.12	6.16	6.21
60	5.60	5.67	5.73	5.78	5.84	5.89	5.93	5.97	6.01
120	5.44	5.50	5.56	5.61	5.66	5.71	5.75	5.79	5.83
∞	5.29	5.35	5.40	5.45	5.49	5.54	5.57	5.61	5.65

Source: *Biometrika Tables for Statisticians*, Vol. 1, 3rd ed., edited by E. S. Pearson and H. O. Hartley (Cambridge University Press, 1966). Reproduced by permission of Professor E. S. Pearson and the *Biometrika* Trustees.

SAS for Windows Tutorial

CONTENTS

D.1 SAS Windows Environment

Upon entering into a SAS session, you will see a screen similar to Figure D.1. The window at the bottom of the screen is the SAS Editor window. The SAS program commands for creating and analyzing data are specified in this window. The window at the top of the screen is the SAS Log window, which logs whether or not each command line has been successfully executed. Once a program is run, a third window appears—the SAS Output window. This window will show the results of the analysis. The SAS printouts shown throughout this text appear in the SAS Output window.

D.2 Creating a SAS Data Set Ready for Analysis

In the SAS Editor window, three basic types of instructions are utilized:

1. *DATA entry commands*: instructions on how the data will be entered
2. *Input data values*: the values of the variables in the data set

FIGURE D.1

Initial screen viewed by SAS user

FIGURE D.2

SAS commands for creating and printing a data set

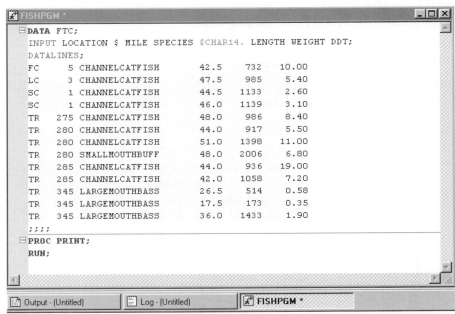

3. *Statistical procedural (PROC) commands*: instruction on what type of analysis is to be conducted on the data

The commands shown in Figure D.2 create and print a SAS data set, called FTC, that is a subset of the FISHDDT data file of Chapter 1. The names of the SAS variables (e.g., SPECIES) are listed on the INPUT command. (*Note*: Qualitative variable names are followed by a dollar sign.) The input data values must be typed (or copied) directly into the Editor window. If the data are saved in an external data file (as is the FISHDDT data set), you can access it using the INFILE command. This is illustrated in Figure D.3. (*Note*: The program in Figure D.3 also shows how to create interaction and squared terms using the standard symbol, *, for multiplication.)

FIGURE D.3

SAS commands for accessing/printing an external data file

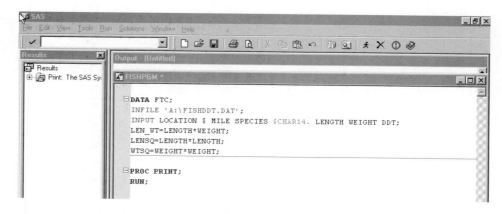

To submit the SAS program (and obtain results in the Output window), you must click on the Run button shown on the menu bar at the top of the SAS screen (see Figure D.3).

D.3
Using SAS Analyst

For SAS users who are not familiar with SAS procedure commands, SAS has available a "user-friendly" menu interface called SAS Analyst. In SAS Analyst, you do not need to know any SAS commands. You obtain results by simply clicking on the appropriate menu options.

After you have specified the SAS data set that you want to analyze (see Section D.2), enter into an Analyst session by clicking on the Solutions button on the SAS menu bar, then click on Analysis, then on Analyst (see Figure D.4). The resulting screen appears as in Figure D.5.

FIGURE D.4

Entering into a SAS analyst session

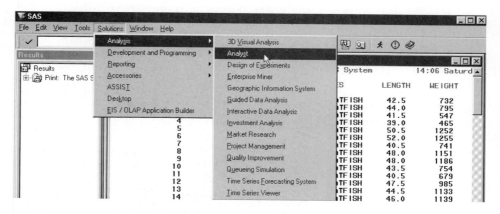

Notice that Figure D.5 shows an empty "data table." To put the data you want to analyze with SAS Analyst in the table, you must click on the File button on the SAS menu bar, then click on Open by SAS Name, then click on Work, and finally click on the name of the SAS data (e.g., FTC) that you created using SAS commands in the Editor window. Now the data table will contain the values of the data set (see Figure D.6).

FIGURE D.5
Initial screen in a SAS analyst session

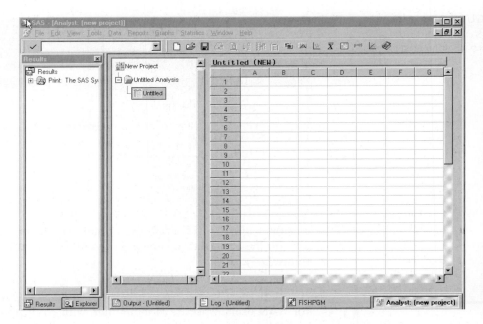

FIGURE D.6
Data table in SAS analyst

Ftc (Browse)

	LOCATION	MILE	SPECIES	LENGTH	WEIGHT	DDT
1	FC	5	CHANNELCATFISH	42.5	732	10
2	FC	5	CHANNELCATFISH	44	795	18
3	FC	5	CHANNELCATFISH	41.5	547	23
4	FC	5	CHANNELCATFISH	39	465	21
5	FC	5	CHANNELCATFISH	50.5	1252	50
6	FC	5	CHANNELCATFISH	52	1255	150
7	LC	3	CHANNELCATFISH	40.5	741	28
8	LC	3	CHANNELCATFISH	48	1151	7.7
9	LC	3	CHANNELCATFISH	48	1186	2
10	LC	3	CHANNELCATFISH	43.5	754	19
11	LC	3	CHANNELCATFISH	40.5	679	18
12	LC	3	CHANNELCATFISH	47.5	985	5.4
13	SC	1	CHANNELCATFISH	44.5	1133	2.6
14	SC	1	CHANNELCATFISH	46	1139	3.1
15	SC	1	CHANNELCATFISH	48	1186	3.5
16	SC	1	CHANNELCATFISH	45	984	9.1
17	SC	1	CHANNELCATFISH	43	965	7.8
18	SC	1	CHANNELCATFISH	45	1084	4.1
19	TR	275	CHANNELCATFISH	48	986	8.4
20	TR	275	CHANNELCATFISH	45	1023	15
21	TR	275	CHANNELCATFISH	49	1266	25
22	TR	275	CHANNELCATFISH	50	1086	5.6

Once you access the data in this fashion, you are ready to analyze it using the menu-driven features of SAS Analyst.

D.4
Listing Data

To obtain a listing (printout) of your data using SAS Analyst, click on the Reports button on the Analyst menu bar, then click on List Data (see Figure D.7). The resulting menu, or dialog box, appears as in Figure D.8.

FIGURE D.7

SAS analyst options for obtaining a list of your data

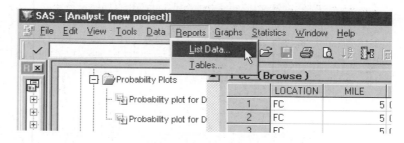

FIGURE D.8

List data dialog box

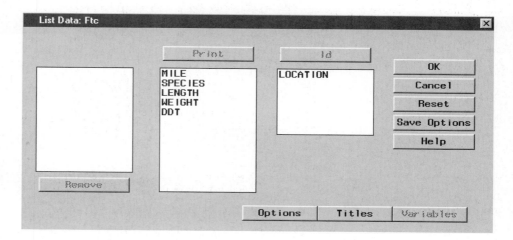

Enter the names of the variables you want to print in the Print box (you can do this by simply clicking on the variables), then click OK. The printout will show up on your screen.

D.5
Graphing Data

To obtain graphical descriptions of your data (e.g., bar charts, histograms, scatterplots, etc.) using SAS Analyst, click on the Graphs button on the Analyst menu bar (see Figure D.9). The resulting menu list appears as shown in Figure D.9. Four of the options covered in this text are Bar Chart, Histogram, (Normal) Probability Plot, and Scatterplot. Click on the graph of your choice to view the appropriate dialog box. For example, the dialog boxes for a vertical bar chart and a scatterplot are shown, respectively, in Figures D.10 and D.11. Make the appropriate variable selections and click OK to view the graph.

FIGURE D.9
SAS analyst options for graphing
your data

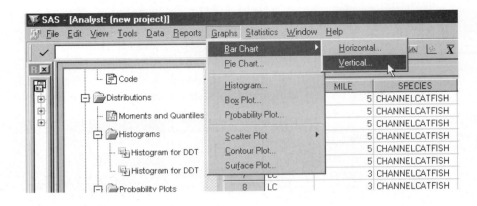

FIGURE D.10
Bar chart dialog box

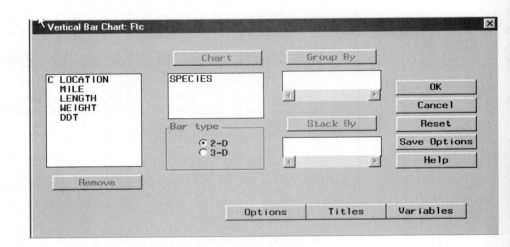

FIGURE D.11
Scatterplot dialog box

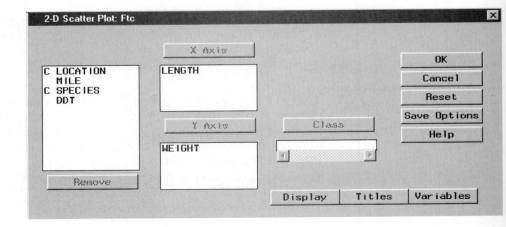

D.6
Descriptive Statistics and Correlations

To obtain numerical descriptive measures for a quantitative variable (e.g., mean, standard deviation, etc.) using SAS Analyst, click on the Statistics button on the Analyst menu bar, then click on Descriptive and finally click on Summary Statistics (see Figure D.12). The resulting dialog box appears in Figure D.13.

Select the quantitative variable you want to analyze and place it in the Analysis box. (As an option, you can obtain summary statistics on this quantitative variable for different levels of a qualitative variable by placing the qualitative variable in the

FIGURE D.12

SAS analyst options for obtaining descriptive statistics

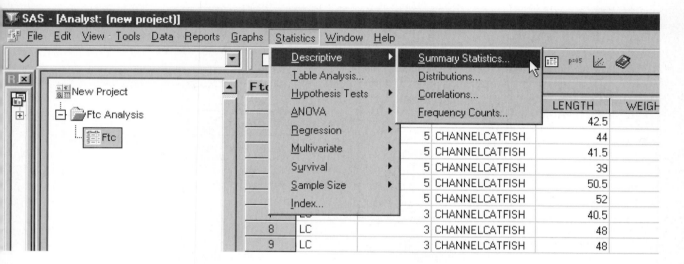

FIGURE D.13

Descriptive statistics dialog box

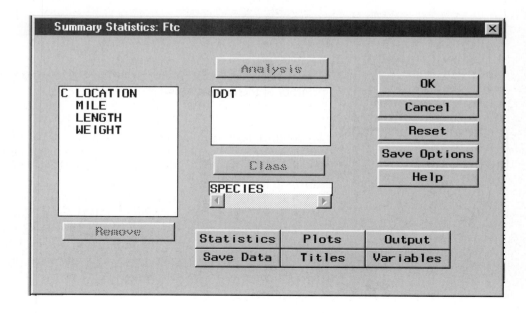

Class box. Also, you can control which particular descriptive statistics appear by clicking the Statistics button on the dialog box and making your selections.) Click on OK to view the descriptive statistics printout.

To obtain Pearson product moment correlations for pairs of quantitative variables using SAS Analyst, click on the Statistics button on the Analyst menu bar, then click on Descriptive and finally click on Correlation (see Figure D.12). The resulting dialog box appears in Figure D.14.

FIGURE D.14
Correlation dialog box

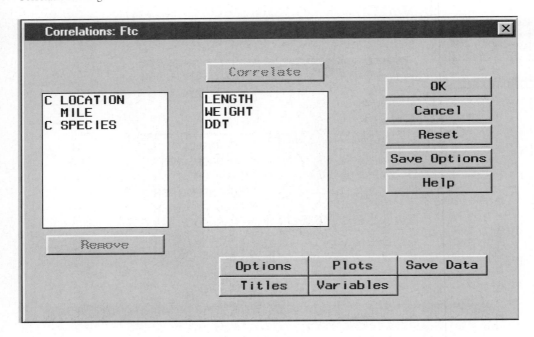

Enter the variables of interest in the Correlate box, then click OK to obtain a printout of the correlations.

D.7
Hypothesis Tests

To conduct tests of hypotheses on population parameters (e.g., mean, variance) for quantitative variables using SAS Analyst, click on the Statistics button on the Analyst menu bar, then click on Hypothesis Tests. The resulting menu appears as shown in Figure D.15.

Click on the test of interest to view the appropriate dialog box. For example, the dialog boxes for a One-Sample *t*-Test for a Mean and a Two-Sample *t*-Test for Means are shown, respectively, in Figures D.16 and D.17. Specify the quantitative variable to be tested, the null hypothesis value and form of the alternative hypothesis in the appropriate boxes. For a two-sample test, you must also specify the qualitative variable that represents the two samples (groups). Click OK to view the test results.

FIGURE D.15
SAS analyst options for
hypothesis testing

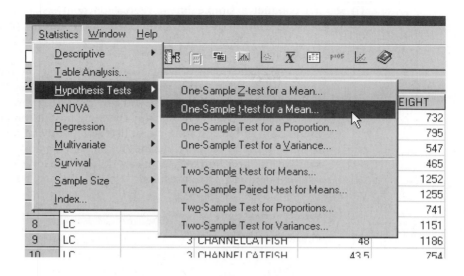

FIGURE D.16
One-sample *t*-test for mean dialog
box

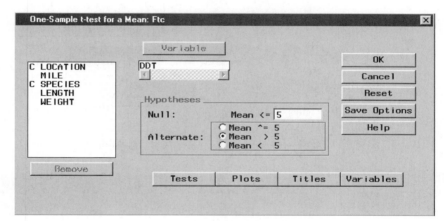

FIGURE D.17
Two-sample *t*-test for means
dialog box

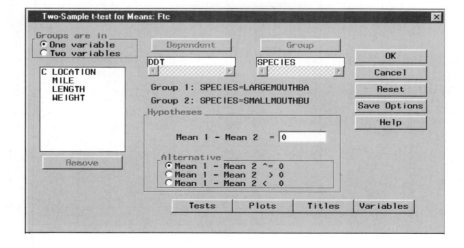

D.8
Simple Linear Regression

To conduct a simple linear regression analysis using SAS Analyst, click on the Statistics button on the Analyst menu bar, then click on Regression, and finally click on Simple, as shown in Figure D.18. The resulting dialog box appears as shown in Figure D.19.

Specify the quantitative dependent variable in the Dependent box and the quantitative independent variable in the Explanatory box. Be sure to select Linear in the Model box. Optionally, you can get SAS to produce confidence intervals for the model parameters by clicking the Statistics button and checking the appropriate menu item in the resulting menu list. Also, you can obtain prediction intervals and residual plots by clicking the Predictions button and Plots button, respectively, and making the appropriate selections on the resulting menus. Click OK to view the simple linear regression results.

FIGURE D.18

SAS analyst options for simple linear regression

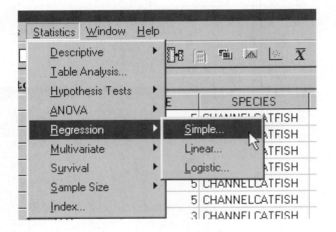

FIGURE D.19

Dialog box for simple linear regression

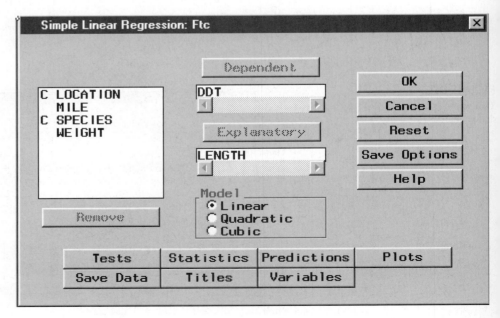

To conduct a multiple regression analysis of a general linear model using SAS Analyst, click on the Statistics button on the Analyst menu bar, then click on Regression, and finally click on Linear, as shown in Figure D.20. The resulting dialog box appears in Figure D.21.

FIGURE D.20

SAS analyst options for multiple regression

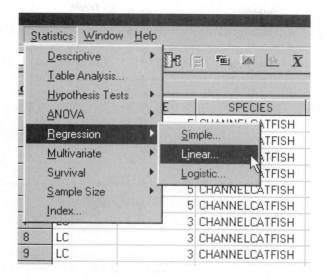

FIGURE D.21

Dialog box for multiple regression

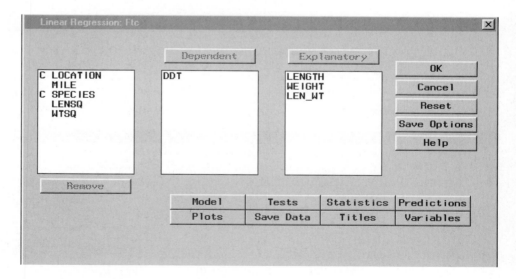

Specify the dependent variable in the Dependent box and the independent variables in the model in the Explanatory box. Optionally, you can get SAS to produce confidence intervals for the model parameters by clicking the Statistics button and checking the appropriate menu item in the resulting menu list. To produce variance inflation factors, click the Statistics button, then click the Tests button, and then select the Variance Inflation Factors option. Also, you can obtain prediction intervals and residual plots by clicking the Predictions button and Plots button, respectively, and making the appropriate selections on the resulting menus. When all the options you desire have been checked, click OK to view the multiple regression results.

[*Note*: If your model includes interaction and/or squared terms, you must create these higher-order variables in the DATA command lines in your SAS program before entering into a SAS Analyst session. See Figure D.3 for an example.]

As an alternative, you can fit general linear models using the ANOVA option available in SAS Analyst. To do this, click on the Statistics button on the Analyst menu bar, then click on ANOVA, and finally click on Linear Models, as shown in Figure D.22. The resulting dialog box appears in Figure D.23.

FIGURE D.22
SAS analyst options for general linear models

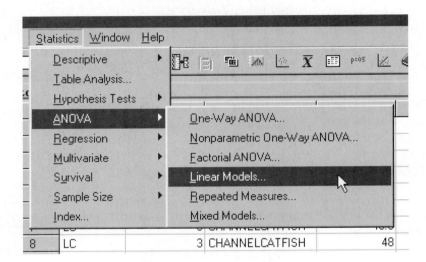

FIGURE D.23
Dialog box for general linear models

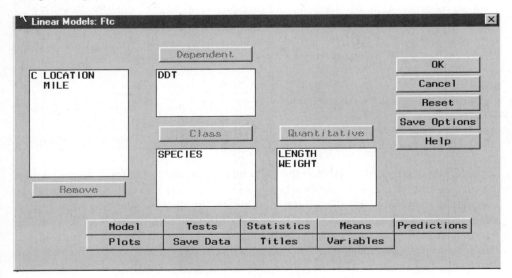

Specify the dependent variable in the Dependent box, the quantitative independent variables in the model in the Quantitative box, and the qualitative independent variables in the Class box. (*Note*: SAS will automatically create the appropriate number of dummy variables for each qualitative variable specified.) After making the variable selections, click the Model button to view the dialog box shown in Figure D.24.

FIGURE D.24

Dialog box for selecting model terms

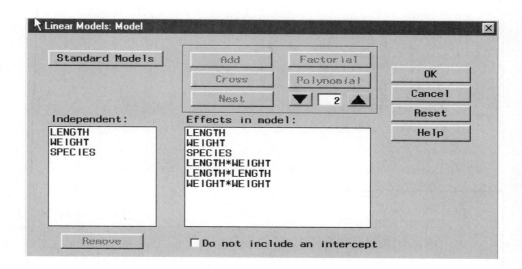

Specify the terms in the model using the Cross button (for interactions) and the Polynomial button (for higher-order terms). The model terms will appear in the Effects in Model box. Click OK to return to the Linear Models dialog box (Figure D.23). Click the Statistics button and check Calculate Parameter Estimates on the resulting menu to produce the estimates of the model parameters. Also, you can obtain prediction intervals and residual plots by clicking the Predictions button and Plots button, respectively, and by making the appropriate selections on the resulting menus. When all the options you desire have been checked, click OK to view the multiple regression results.

D.10
Stepwise Regression

To conduct a stepwise regression analysis using SAS Analyst, click on the Statistics button on the Analyst menu bar, then click on Regression, and click on Linear, as shown in Figure D.20. The resulting dialog box appears in Figure D.21. Specify the dependent variable in the Dependent box and the independent variables in the stepwise model in the Explanatory box. Now click on the Model button. The resulting menu appears as shown in Figure D.25.

For the stepwise regression method, choose Stepwise Selection. (The default method is Full Model.) For the all-possible-regressions-selection method, choose Mallows' Cp, R-Square, or Adjusted R-Square. As an option, you can select the value of α to use in the analysis by clicking on the Criteria button and specifying the value. (The default is $\alpha = .05$.) Click OK to view the stepwise regression results.

FIGURE D.25
Model menu selection for multiple regression

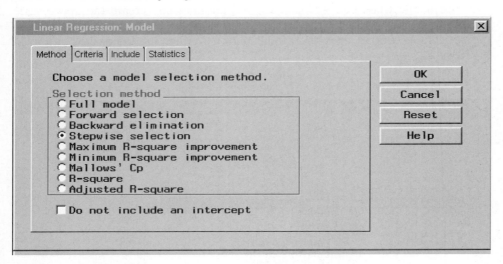

D.11
Residual Analysis
and Influence
Diagnostics

To conduct a residual analysis using SAS Analyst, click on the Statistics button on the Analyst menu bar, then click on Regression, and then click on Linear, as shown in Figure D.20. The resulting dialog box appears in Figure D.21. Specify the dependent variable in the Dependent box and the independent variables in the model in the Explanatory box. To produce residual scatterplots, click on the Plots button, then click on the Residual button. The resulting menu appears as shown in Figure D.26.

Select the residual scatterplot you want to produce by clicking on the open box next to the type of residual (vertical axis) and variable (horizontal axis) on the

FIGURE D.26
Residual plots menu selection for multiple
regression

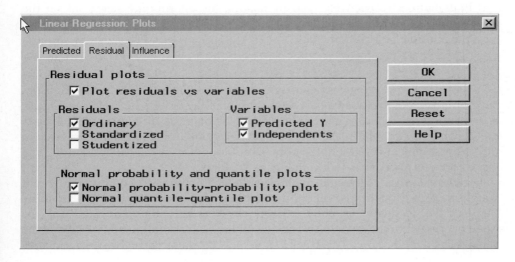

menu, as shown in Figure D.26. A normal probability plot of the residuals can also be produced by checking the appropriate box. Click OK to view the residual plots.

To produce influence diagnostics, click on the Save Data button in the Multiple Regression dialog box (Figure D.21). The resulting menu appears as shown in Figure D.27.

FIGURE D.27

Save data menu for multiple regression

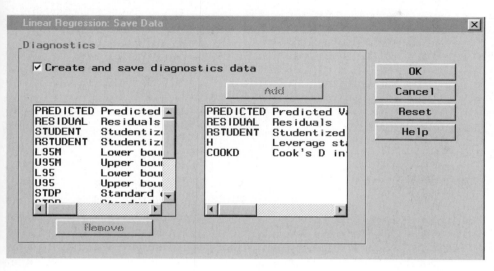

Click the box next to Create and Save Diagnostics Data, then select the diagnostics you want to produce by double clicking each of your choices. These diagnostics will appear in the Add box, as shown in Figure D.27. Click OK to return to the Multiple Regression dialog box (Figure D.21), then click OK to run the analyses.

To view the influence diagnostics, double click on the Diagnostics Table option, which now appears on the SAS Analyst main menu screen, as shown in Figure D.28.

D.12
Logistic Regression

To conduct a logistic regression analysis for a two-level dependent qualitative variable using SAS Analyst, click on the Statistics button on the Analyst menu bar, then click on Regression, and finally click on Logistic, as shown in Figure D.29. The resulting dialog box appears in Figure D.30.

Specify the dependent variable in the Dependent box and the level of the dependent variable that will be modeled as π in the Model Pr{} box. Qualitative independent variables in the model are specified in the Class box, while quantitative independent variables are specified in the Quantitative box. [*Note*: SAS will automatically create the appropriate number of dummy variables for each qualitative variable specified.] After making the variable selections, click the Model button to view the dialog box shown in Figure D.31.

Specify the terms in the model using the Cross button (for interactions) and the Polynomial button (for higher-order terms). The model terms will appear in the Effects in Model box. Click OK to return to the Logistic Regression dialog

FIGURE D.28
Select diagnostics table from main SAS analyst menu

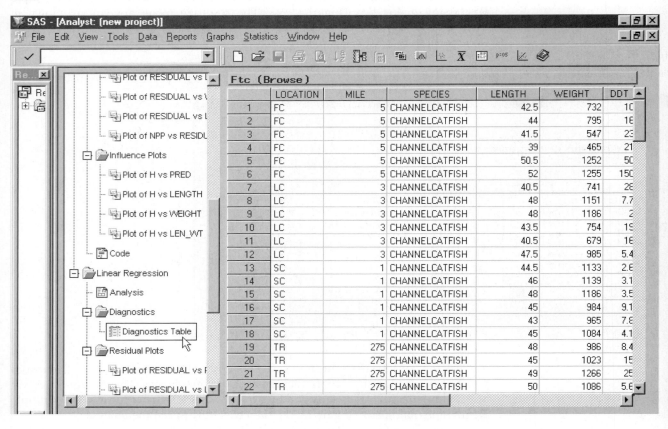

FIGURE D.29
SAS analyst options for logistic regression

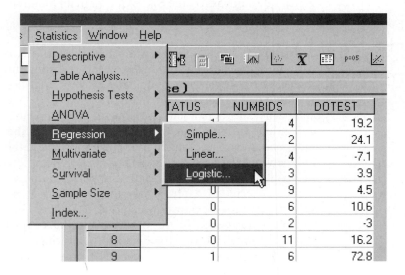

box (Figure D.30). Click the Statistics button and check Classification table on the resulting menu to produce a classification table for the analysis. Also, you can obtain prediction intervals for π by clicking the Predictions button and making the appropriate selections on the resulting menu. When all the options you desire have been checked, click OK to view the logistic regression results.

FIGURE D.30

Dialog box for logistic regression

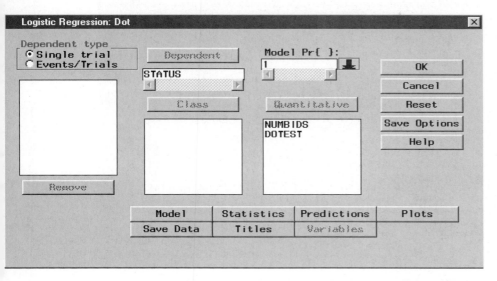

FIGURE D.31

Dialog box for selecting model terms in logistic regession

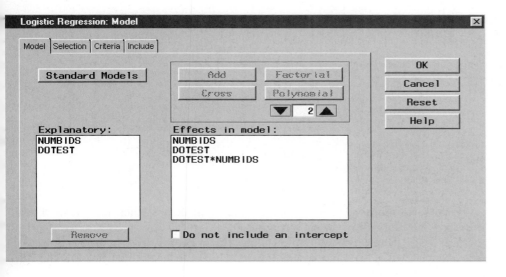

D.13
One-Way Analysis of Variance

To conduct a one-way ANOVA for a completely randomized design using SAS Analyst, click on the Statistics button on the Analyst menu bar, then click on ANOVA, and finally click on One-Way ANOVA, as shown in Figure D.32. The resulting dialog box appears in Figure D.33.

Specify the dependent variable in the Dependent box and the qualitative variable that represents the single factor in the Independent box.

FIGURE D.32
SAS analyst options for analysis of variance

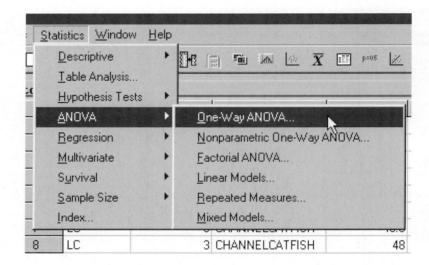

FIGURE D.33
Dialog box for one-way ANOVA

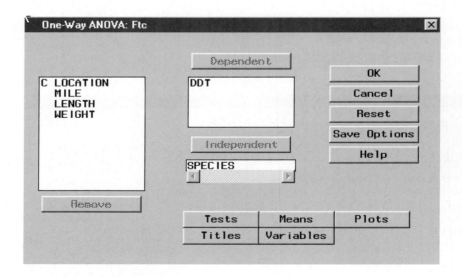

To perform multiple comparisons of treatment means, click the Means button to obtain the dialog box shown in Figure D.34. On this box, select the comparison method (e.g., Bonferroni's method) and the comparison-wise error rate (e.g., significance level of .05), and specify the main effect to be tested. Click OK to return to the One-Way ANOVA dialog box (Figure D.33).

FIGURE D.34

Dialog box for multiple comparisons of means

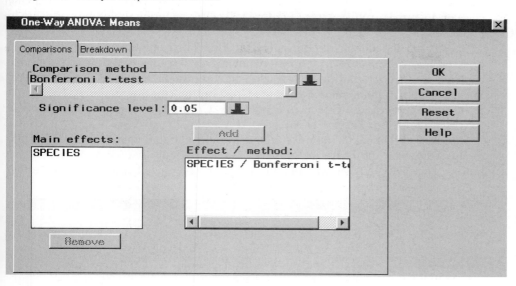

To perform a test of equality of variances, click the Tests button on the One-Way ANOVA dialog box to obtain the menu shown in Figure D.35. From this menu, select the test to be performed (e.g., Levene's test), then click OK to return to the One-Way ANOVA dialog box (Figure D.33). Click OK to view the ANOVA results.

FIGURE D.35

Menu for testing equality of variances

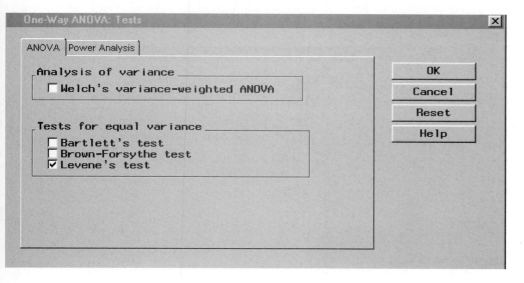

D.14
Analysis of Variance
for Factorial
Designs

To conduct an ANOVA for a factorial design using SAS Analyst, click on the Statistics button on the Analyst menu bar, then click on ANOVA, and finally click on Factorial ANOVA (see Figure D.32). The resulting dialog box appears in Figure D.36.

Specify the dependent variable in the Dependent box and the qualitative variables that represent the factors in the Independent box. To specify the factorial model, click on the Models button. The dialog box shown in Figure D.37 will appear. Specify the terms in the model using the Cross button (for interactions). The model terms will appear in the Effects in Model box. Click OK to return to the Factorial ANOVA dialog box (Figure D.36).

To run multiple comparisons of means for all treatment combinations, click the Means button on the Factorial ANOVA dialog box and then click the LS Means button. The dialog box shown in Figure D.38 appears. Specify the interaction effect

FIGURE D.36

Dialog box for factorial ANOVA

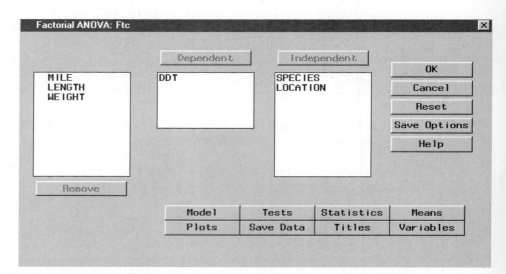

FIGURE D.37

Dialog box for factorial model selection

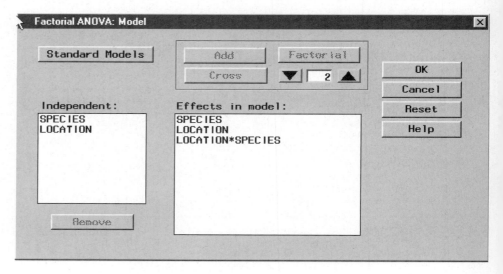

Dialog box for multiple comparisons of factorial means

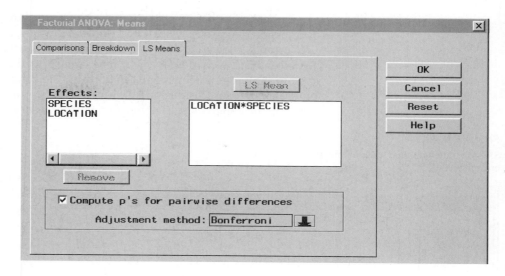

of interest in the LS Mean box and select the comparison method (e.g., Bonferroni's method), then click OK to return to the Factorial ANOVA dialog box (Figure D.36). Click OK to run and view the ANOVA results.

D.15
Time Series
Forecasting Models

To conduct the Durbin-Watson test for autocorrelated errors in a model for time series data using SAS Analyst, first specify the model to be fit. That is, click on the Statistics button on the Analyst menu bar, then click on Regression, and click on Linear, as shown in Figure D.20. Specify the dependent and independent variables in the model on the resulting dialog box (see Figure D.21). Once the model has been specified, click on the Statistics button on the Multiple Regression dialog box, then click on Tests to obtain the menu shown in Figure D.39.

FIGURE D.39
Linear regression tests menu

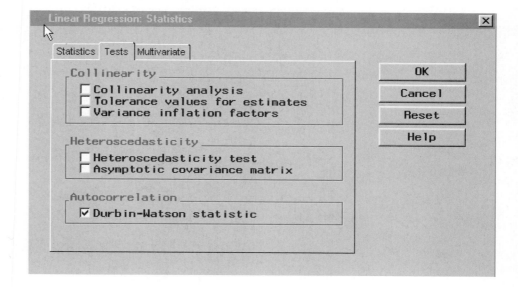

Check the Durbin–Watson Statistic box, then click OK to return to the Multiple Regression dialog box. Click OK to view the results.

Currently, you cannot fit a time series model with autoregressive errors using SAS Analyst. Consequently, you must return to the Editor window of SAS and type in the program commands to do this. The commands for fitting a time series autoregressive forecasting model with AR(1) errors are shown in Figure D.40. After entering the commands, click the Run button to submit the program.

FIGURE D.40
SAS program to fit time series autoregressive
error model

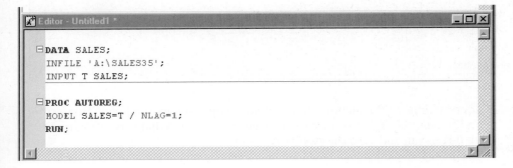

```
DATA SALES;
INFILE 'A:\SALES35';
INPUT T SALES;

PROC AUTOREG;
MODEL SALES=T / NLAG=1;
RUN;
```

SPSS for Windows Tutorial

CONTENTS

E.1 SPSS Windows Environment

Upon entering into an SPSS session, you will see a screen similar to Figure E.1. The main portion of the screen is an empty spreadsheet, with columns representing variables and rows representing observations (or cases). The very top of the screen is the SPSS main menu bar, with buttons for the different functions and procedures available in SPSS. Once you have entered data into the spreadsheet, you can analyze the data by clicking the appropriate menu buttons.

E.2 Creating an SPSS Spreadsheet Data File Ready for Analysis

You can create an SPSS data file by entering data directly into the spreadsheet. Figure E.2 shows the length-to-width ratio data for the BONES file analyzed in Chapter 1. The variables (columns) can be named by selecting the "Variable View" button at the bottom of the screen and typing in the name of each variable.

If the data are saved in an external data file (e.g., the FISHDDT data set of Chapter 1), you can access it using the options available in SPSS. Click the File button on the menu bar, then click Read Text Data, as shown in Figure E.3. The dialog box shown in Figure E.4 will appear.

FIGURE E.1
Initial screen viewed by SPSS user

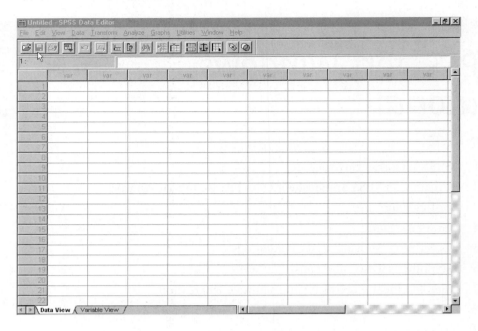

FIGURE E.2
Data entered into the SPSS
spreadsheet

	ratio	var	var	var	var
1	10.73				
2	9.57				
3	6.66				
4	9.89				
5	8.89				
6	9.29				
7	9.35				
8	8.17				
9	9.07				
10	9.94				
11	8.86				
12	8.93				
13	9.20				
14	8.07				
15	9.93				
16	8.80				
17	10.33				
18	8.37				
19	8.91				
20	10.02				
21	9.98				
22	6.85				

FIGURE E.3

SPSS options for reading data from an external file

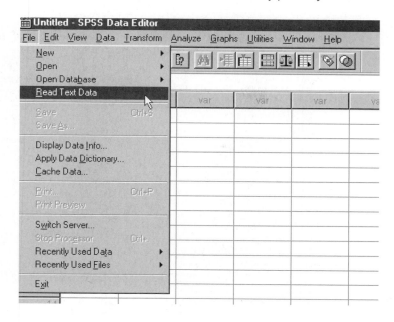

FIGURE E.4

Selecting the external data file

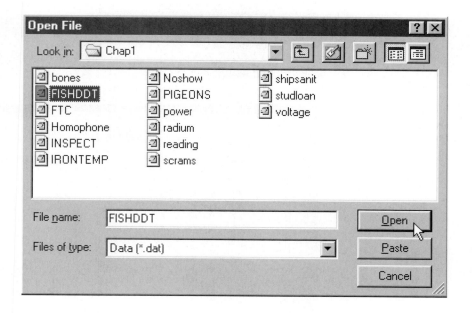

Specify the disk drive and folder that contains the data file, click on the data file, then click Open, as shown in Figure E.4. This will invoke the SPSS Text Import Wizard. The Text Import Wizard presents a series of six screen menus, the first of which is shown in Figure E.5. Make the appropriate selections on the screen, and click Next to go to the next screen. When finished, click Finish. The SPSS spreadsheet will reappear with the data from the external data file, as shown in Figure E.6.

FIGURE E.5

The SPSS text import wizard, screen 1

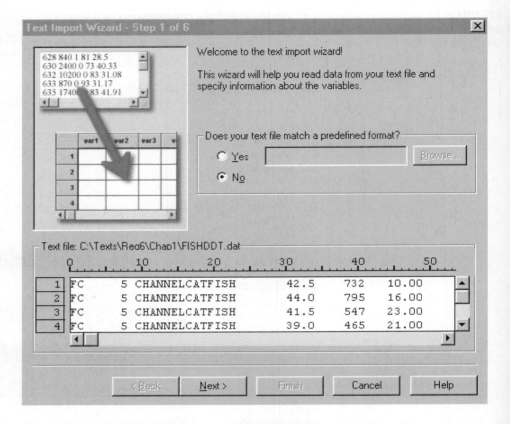

FIGURE E.6

The SPSS spreadsheet with the imported data

Reminder: The variables (columns) can be named by selecting the Variable View button at the bottom of the spreadsheet screen and typing in the name of each variable.

E.3
Listing Data

To obtain a listing (printout) of your data, click on the Analyze button on the SPSS main menu bar, then click on Reports, then on Report Summaries in Rows (see Figure E.7). The resulting menu, or dialog box, appears as in Figure E.8. Enter the names of the variables you want to print in the Data Columns box (you can do this by simply clicking on the variables), then click OK. The printout will show up on your screen.

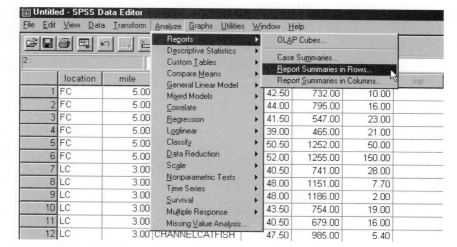

FIGURE E.7
SPSS menu options for obtaining a list of your data

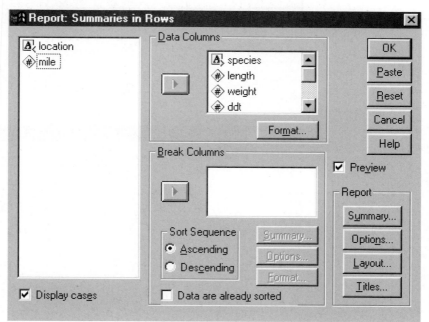

FIGURE E.8
Report data dialog box

E.4
Graphing Data

To obtain graphical descriptions of your data (e.g., bar charts, histograms, scatterplots, etc.), click on the Graphs button on the SPSS menu bar. The resulting menu list appears as shown in Figure E.9. Several of the options covered in this text are Bar (Chart), Scatter (plot), Histogram, and P-P (normal probability plot). Click on the graph of your choice to view the appropriate dialog box. For example, the dialog boxes for a histogram and scatterplot are shown, respectively, in Figures E.10 and E.11. Make the appropriate variable selections, and click OK to view the graph.

FIGURE E.9
SPSS menu options for graphing your data

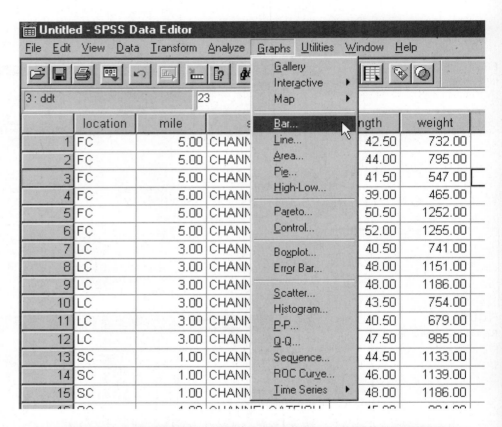

FIGURE E.10
Histogram dialog box

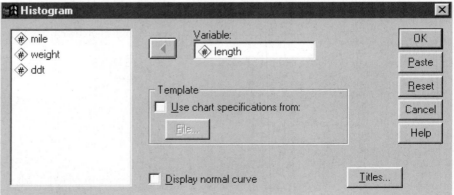

FIGURE E.11
Scatterplot dialog box

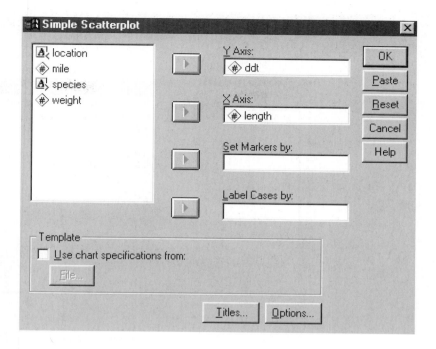

E.5
Descriptive
Statistics
and Correlations

To obtain numerical descriptive measures for a quantitative variable (e.g., mean, standard deviation, etc.), click on the Analyze button on the main menu bar, then click on Descriptive Statistics and finally click on Descriptives (see Figure E.12). The resulting dialog box appears in Figure E.13.

Select the quantitative variables you want to analyze and place them in the Variable(s) box. You can control which particular descriptive statistics appear by

FIGURE E.12
SPSS options for obtaining
descriptive statistics

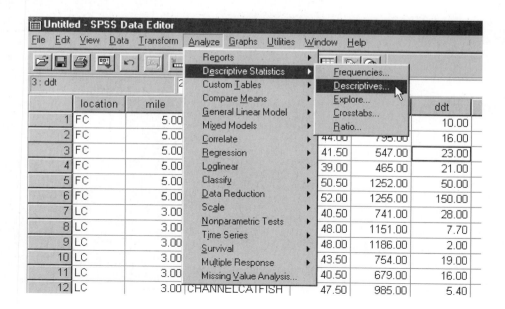

clicking the Options button on the dialog box and making your selections.) Click on OK to view the descriptive statistics printout.

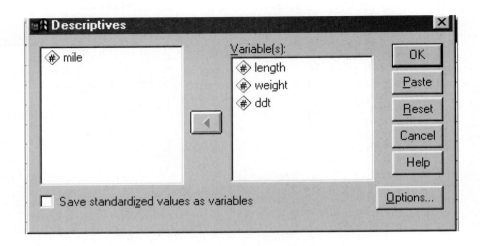

To obtain Pearson product moment correlations for pairs of quantitative variables, click on the Analyze button on the main menu bar, then click on Correlate (see Figure E.12), finally click on Bivariate. The resulting dialog box appears in Figure E.14. Enter the variables of interest in the Variables box, then click OK to obtain a printout of the correlations.

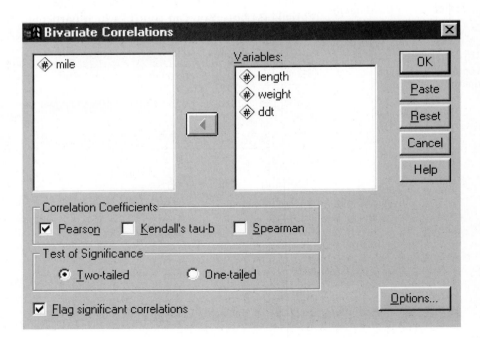

E.6
Hypothesis Tests
on Means

To conduct tests of hypotheses of population means for quantitative variables, click on the Analyze button on the SPSS menu bar, then click on Compare Means. The resulting menu appears as shown in Figure E.15.

Click on the test of interest to view the appropriate dialog box. For example, the dialog boxes for a One-Sample *t*-test and an Independent-Samples *t*-test are shown, respectively, in Figures E.16 and E.17. For a one-sample *t*-test, specify the quantitative variable to be tested in the Test Variable(s) box, and the null hypothesis value in the Test Value box. For a two-sample test, you will need to specify the qualitative variable that represents the two samples in the Grouping Variable box. Optionally, you can obtain a confidence interval for the mean (or difference in means) by clicking the Options box and making the appropriate selection. Click OK to view the test results.

FIGURE E.15

SPSS options for hypothesis tests of means

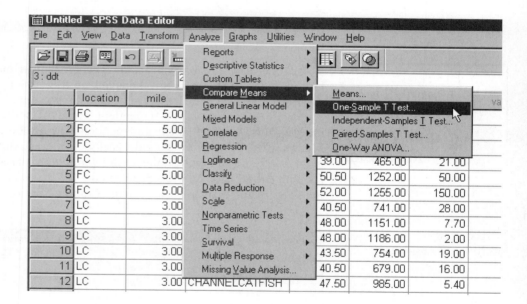

FIGURE E.16

One-sample *t*-test for mean dialog box

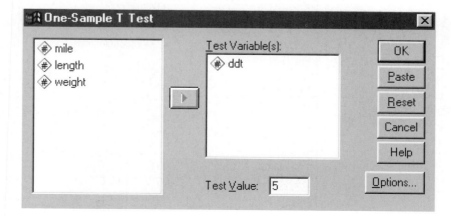

FIGURE E.17
Independent samples *t*-test for
two means dialog box

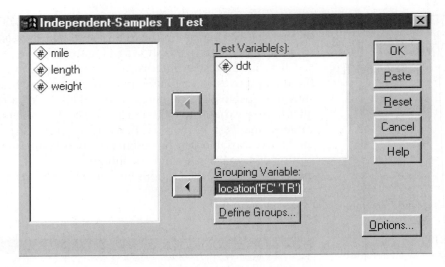

E.7
Simple Linear Regression and Multiple Regression

To conduct a regression analysis of a general linear model, click on the Analyze button on the SPSS menu bar, then click on Regression, and finally click on Linear, as shown in Figure E.18. The resulting dialog box appears as shown in Figure E.19.

Specify the quantitative dependent variable in the Dependent box and the independent variables in the Independent(s) box. Be sure to select Enter in the Method box. [*Note*: If your model includes interaction and/or squared terms, you must create and add these higher-order variables to the SPSS spreadsheet file before running a regression analysis. You can do this by clicking the Transform button on the SPSS main menu and selecting the Compute option.]

FIGURE E.18
SPSS options for regression

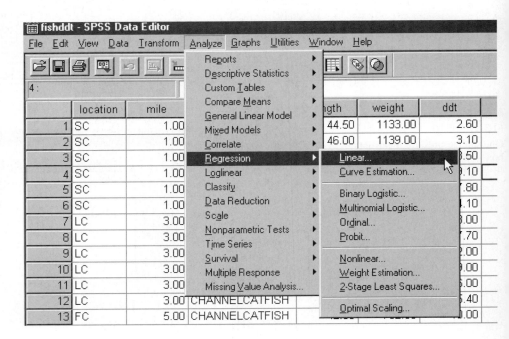

FIGURE E.19

Dialog box for linear regression

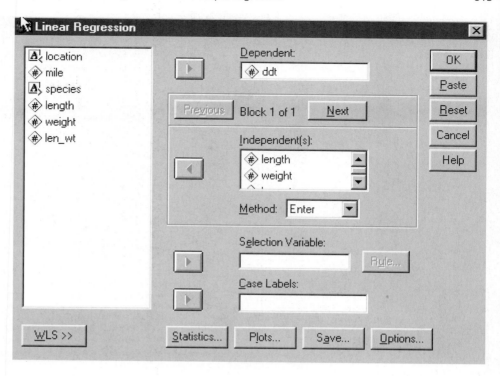

Optionally, you can get SPSS to produce confidence intervals for the model parameters and variance inflation factors by clicking the Statistics button and checking the appropriate menu items in the resulting menu list. Also, you can obtain prediction intervals by clicking the Save button and checking the appropriate item in the resulting menu list. (The prediction intervals will be added as new columns to the SPSS data spreadsheet.) Residual plots are obtained by clicking the Plots button and making the appropriate selections on the resulting menu. To return to the main Regression dialog box from any of these optional screens, click Continue. Click OK on the Regression dialog box to view the linear regression results.

As an alternative, you can fit general linear models by clicking on the Analyze button on the SPSS main menu bar, then click on General Linear Models, and finally click on Univariate, as shown in Figure E.20. The resulting dialog box appears in Figure E.21.

Specify the dependent variable in the Dependent box, the quantitative independent variables in the model in the Covariate(s) box, and the qualitative independent variables in the Fixed Factor(s) box. (*Note*: SPSS will automatically create the appropriate number of dummy variables for each qualitative variable specified.) After making the variable selections, click the Model button to view the dialog box shown in Figure E.22.

Select the Custom option, then select the terms in the model by clicking on each term. Specify interactions by highlighting the variables to be interacted and then click the Interaction button. [*Note*: Squared or higher-order terms must be previously created using the Transform button.] The model terms will appear in the Model box. Click Continue to return to the General Linear Models dialog box

SPSS options for fitting general
linear models

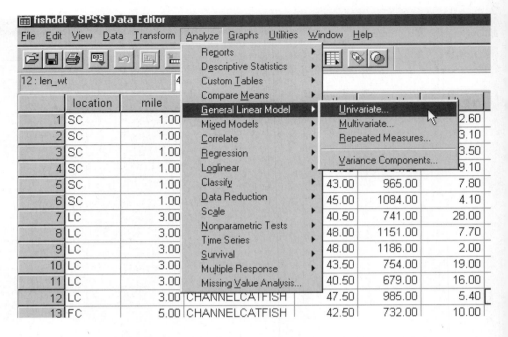

Dialog box for general linear
models

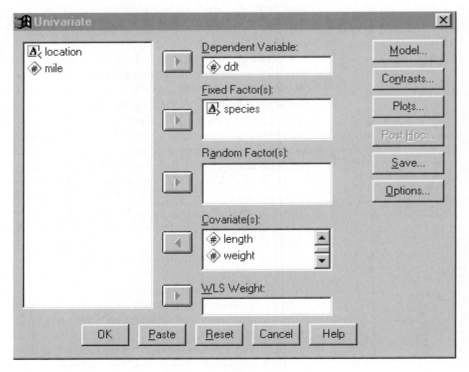

FIGURE E.22
Dialog box for selecting model
terms

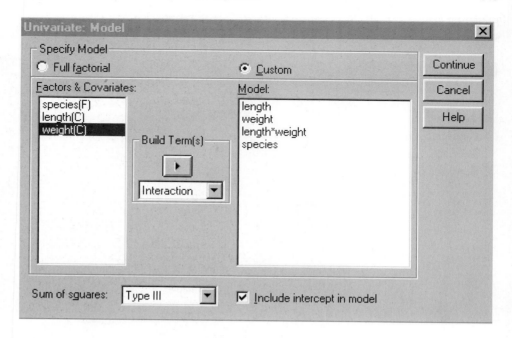

(Figure E.21). Click the Options button and check Parameter Estimates on the resulting menu to produce the estimates of the model parameters. To save predicted values and residuals, click the Save button and make the appropriate selections. When all the options you desire have been checked, click OK to view the multiple regression results.

E.8
Stepwise Regression

To conduct a stepwise regression analysis, click on the Analyze button on the main menu bar, then click on Regression, and click on Linear, as shown in Figure E.18. The resulting dialog box appears in Figure E.23. Specify the dependent variable in the Dependent box and the independent variables in the stepwise model in the Independent(s) box. Now click on the Method button and select the Stepwise option as shown in Figure E.23. [*Note*: The all-possible-regressions-selection procedure is not available in SPSS.]

As an option, you can select the value of α to use in the analysis by clicking on the Options button and specifying the value. (The default is $\alpha = .05$.) Click OK to view the stepwise regression results.

E.9
Residual Analysis and Influence Diagnostics

To conduct a residual analysis, click on the Analyze button on the SPSS main menu bar, then click on Regression, then click on Linear, as shown in Figure E.18. The resulting dialog box appears in Figure E.19. Specify the dependent variable in the Dependent box and the independent variables in the model in the Independent(s) box. To produce residual plots, click on the Plots button. The resulting menu appears as shown in Figure E.24.

FIGURE E.23
Stepwise regression menu selection

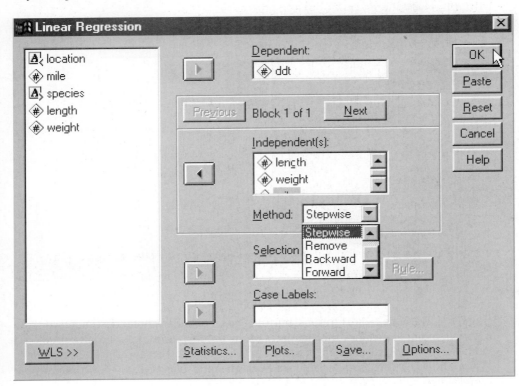

FIGURE E.24
Residual plots menu selection for multiple regression

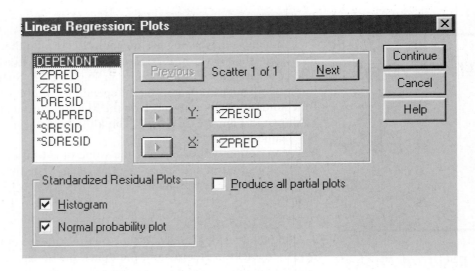

FIGURE E.25
Save menu for linear regression

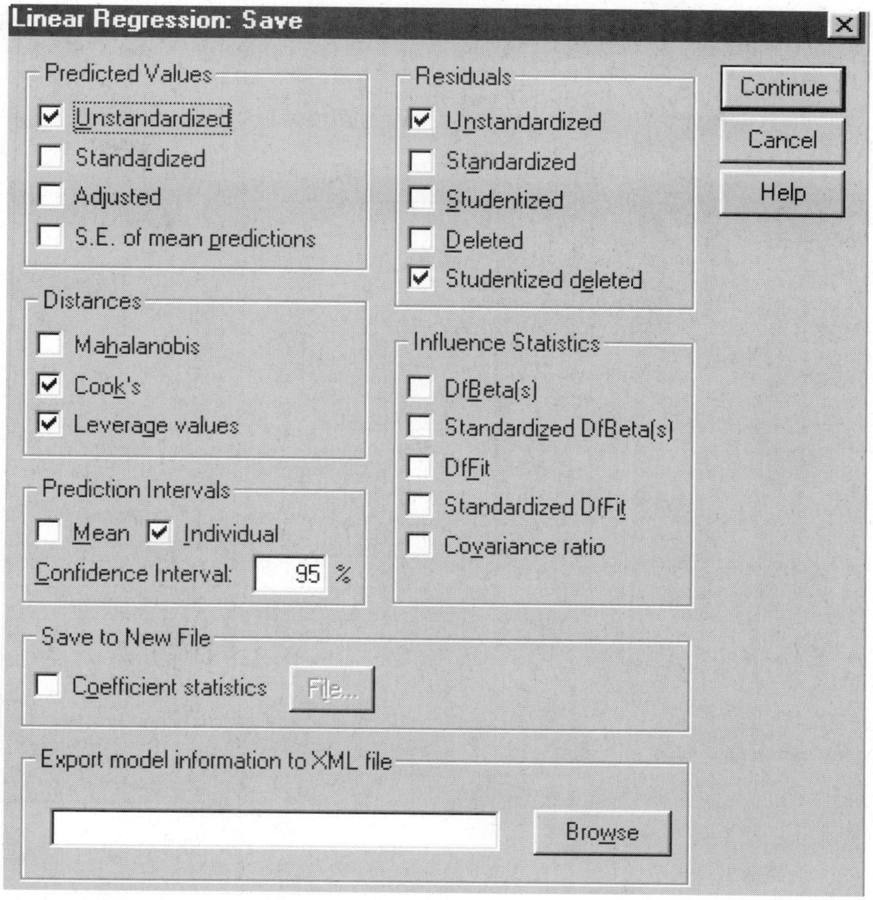

A residual scatterplot is produced by selecting the appropriate variables for the Y axis and the X axis. (The selections shown in Figure E.24 will produce a plot of residuals against predicted values.) A histogram and normal probability plot of the residuals can also be produced by checking the appropriate boxes. Click Continue to return to the Linear Regression main menu, then click OK to view the residual plots.

To produce influence diagnostics, click on the Save button in the Linear Regression dialog box (Figure E.21). The resulting menu appears as shown in Figure E.25.

Click the influence diagnostics you want to save (e.g., Cook's and Leverage), then click Continue to return to the Linear Regression dialog box. Now click OK. The influence diagnostics will appear as additional columns on the SPSS spreadsheet.

E.10
Logistic Regression

To conduct a logistic regression analysis for a two-level (bivariate) dependent qualitative variable, click on the Analyze button on the SPSS main menu bar, then click on Regression, and finally click on Binary Logistic, as shown in Figure E.26. The resulting dialog box appears in Figure E.27.

FIGURE E.26
SPSS options for logistic regression

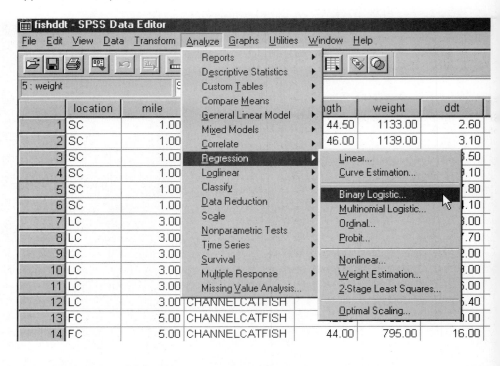

FIGURE E.27
Dialog box for logistic regression

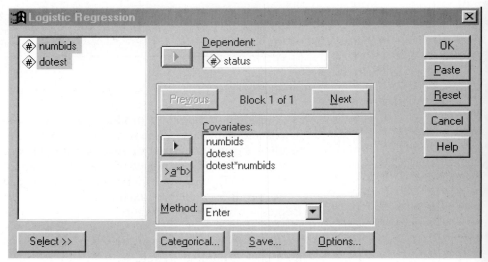

Specify the dependent variable in the Dependent box and the independent variables in the Covariates box. You can include interactions in the model by highlighting the variables to be interacted and clicking the a * b button on the menu. If any of the independent variables are qualitative, click the Categorical box and select these variables. SPSS will automatically create dummy variables for these categorical predictors. After making the variable selections, click the OK button to view the logistic regression results.

To conduct a one-way ANOVA for a completely randomized design, click on the Analyze button on the SPSS main menu bar, then click on Compare Means, and finally click on One-Way ANOVA, as shown in Figure E.28. The resulting dialog box appears in Figure E.29.

IGURE E.28

PSS options for one-way analysis
f variance

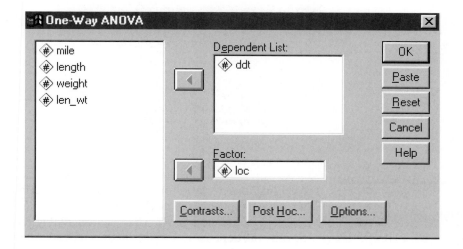

Specify the dependent variable in the Dependent List box and the qualitative variable that represents the single factor in the Factor box. To perform multiple comparisons of treatment means, click the Post Hoc button to obtain the dialog box shown in Figure E.30. On this box, select the comparison method (e.g., Bonferroni) and the comparison-wise error rate (e.g., significance level of .05). Click Continue to return to the One-Way ANOVA dialog box (Figure E.29).

IGURE E.29

ialog box for one-way ANOVA

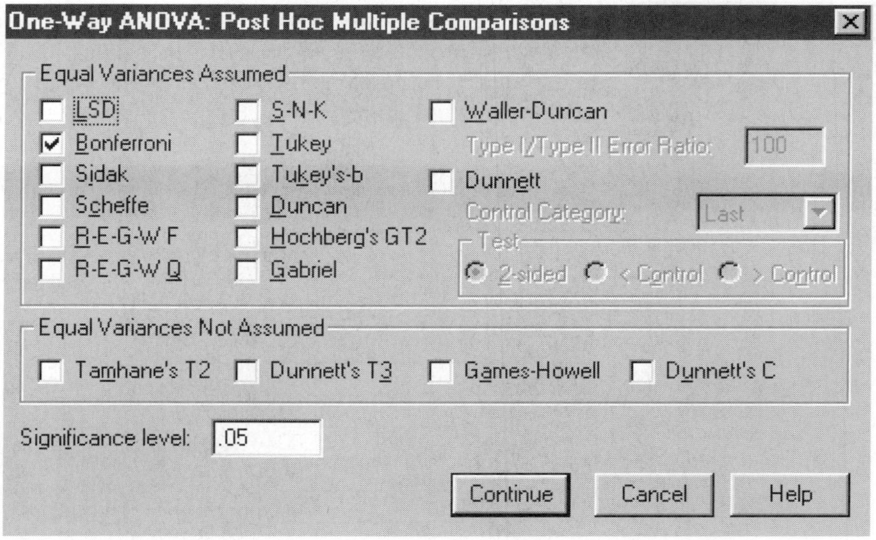

FIGURE E.30
Dialog box for multiple comparisons of means

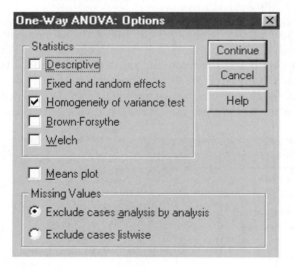

To perform a test of equality of variances, click the Options button on the One-Way ANOVA dialog box to obtain the menu shown in Figure E.31. From this menu, select Homogeneity of variance test, then click Continue to return to the One-Way ANOVA dialog box (Figure E.29). Click OK to view the ANOVA results.

FIGURE E.31
Options menu for one-way ANOVA

E.12
Analysis of Variance for Factorial Designs

To conduct an ANOVA for a factorial design, click on the Analyze button on the SPSS main menu bar, then click on General Linear Model, and finally click on Univariate (see Figure E.32). The resulting dialog box appears in Figure E.33.

Specify the dependent variable in the Dependent box and the qualitative variables that represent the factors in the Fixed Factor(s) box. To specify the factorial model

FIGURE E.32
SPSS options for factorial analysis of variance

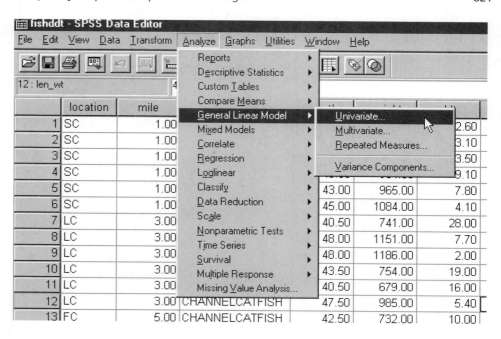

FIGURE E.33
Dialog box for factorial ANOVA

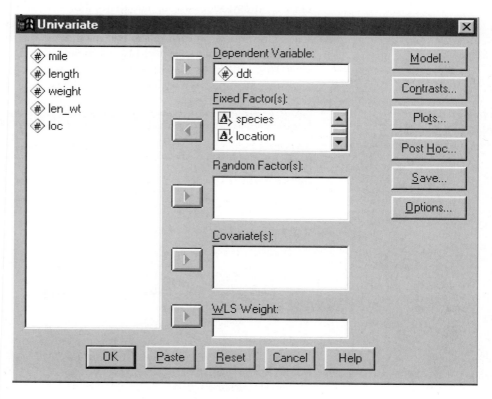

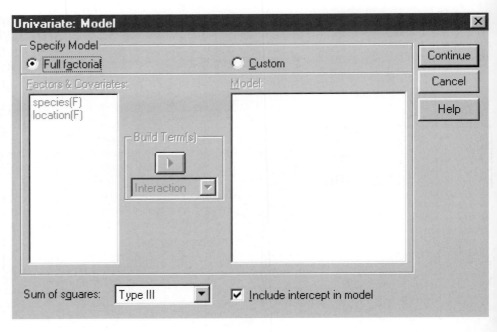

click on the Model button. The dialog box shown in Figure E.34 will appear. Select the Full Factorial option for a model with all possible main effects and interactions. Click Continue to return to the Factorial ANOVA dialog box (Figure E.33).

To run multiple comparisons of means for treatment main effects, click the Post Hoc button on the Factorial ANOVA dialog box. The dialog box shown in Figure E.35 appears. Specify the main effects to be tested in the Post Hoc Tests for

FIGURE E.35

Dialog box for multiple
comparisons of factorial means

box and select the comparison method (e.g., Bonferroni), then click Continue to return to the Factorial ANOVA dialog box (Figure E.33). [*Note*: A test of equality of variances is produced by clicking Options on the Factorial ANOVA dialog box, and selecting Homogeneity tests on the resulting menu.] Click OK on the Factorial ANOVA dialog box (Figure E.33) to run and view the ANOVA results.

E.13
Time Series Forecasting

To conduct the Durbin–Watson test for autocorrelated errors in a model for time series data, first specify the model to be fit. That is, click on the Analyze button on the SPSS main menu bar, then click on Regression, and click on Linear, as shown in Figure E.18. Specify the dependent and independent variables in the model on the resulting dialog box (see Figure E.19). Once the model has been specified, click on the Statistics button on the Linear Regression dialog box to obtain the menu shown in Figure E.36.

FIGURE E.36
Linear regression statistics menu

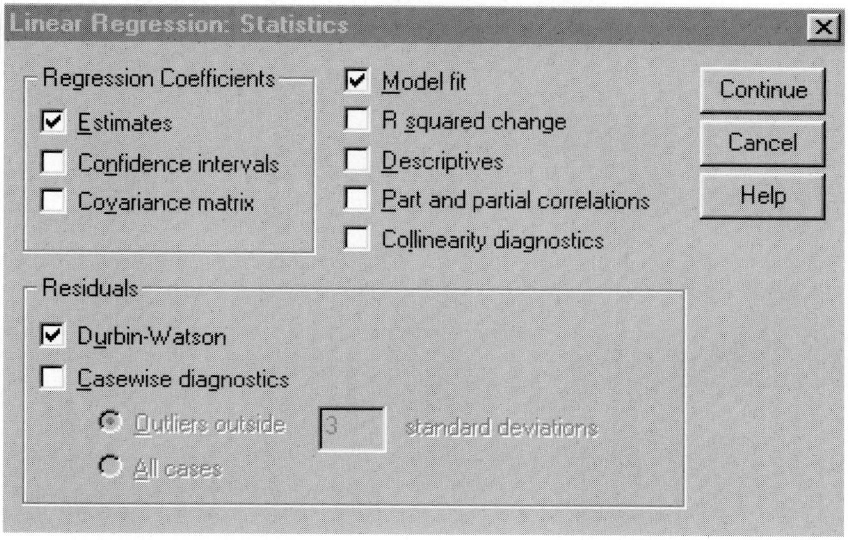

Check the Durbin–Watson box, then click Continue to return to the Linear Regression dialog box. Click OK to view the results.

To produce forecasts using exponential smoothing or the Holt–Winters method, click the Analyze button on the SPSS main menu bar, then click on Time Series, and then click on Exponential Smoothing, as shown in Figure E.37. The resulting dialog box is shown in Figure E.38.

Select the quantitative variable to be smoothed, and place it in the Variables box. For exponential smoothing, select Simple in the Model box. For the Holt–Winters method, select Holt in the Model box. To set the value of the smoothing constants, click the Parameters button and make your selections on the resulting menu screen. Click Continue to return to the Exponential Smoothing dialog box, then click OK to view the results. [*Note*: Forecasts for each time period in the data set will show up in a column on the SPSS spreadsheet screen.]

FIGURE E.37

SPSS options for exponential smoothing

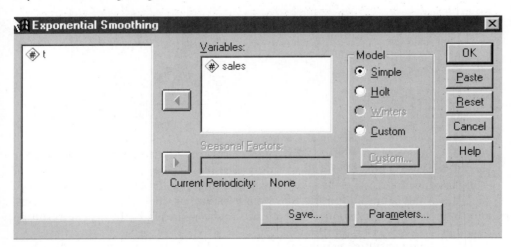

FIGURE E.38

Exponential smoothing dialog box

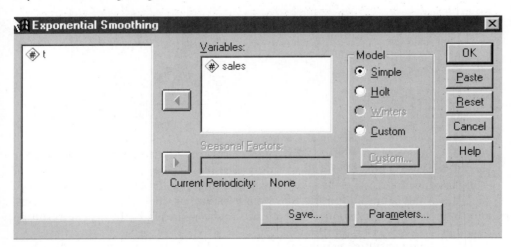

To fit a time series model with a first-order autoregressive error term, click the Analyze button on the SPSS main menu bar, then click on Time Series, and then click on Autoregression (see Figure E.37). The resulting dialog box is shown in Figure E.39.

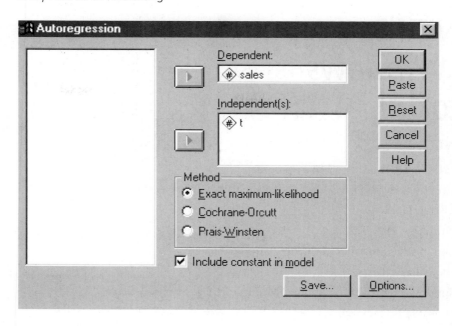

Specify the dependent time series variable in the Dependent box and the independent variables in the model in the Independent(s) box (see Figure E.39). Click Save and make the appropriate menu selections to save the forecasted values. When you return to the Autoregression dialog box, click OK to view the results.

MINITAB for Windows Tutorial

CONTENTS

F.1 MINITAB Windows Environment

Upon entering into a MINITAB session, you will see a screen similar to Figure F.1. The bottom portion of the screen is an empty spreadsheet—called a MINITAB worksheet—with columns representing variables and rows representing observations (or cases). The very top of the screen is the MINITAB main menu bar, with buttons for the different functions and procedures available in MINITAB. Once you have entered data into the spreadsheet, you can analyze the data by clicking the appropriate menu buttons. The results will appear in the Session window.

F.2 Creating a MINITAB Data Worksheet Ready for Analysis

You can create a MINITAB data file by entering data directly into the worksheet. Figure F.2 shows the length-to-width ratio data for the BONES file analyzed in Chapter 1. The variables (columns) can be named by typing in the name of each variable in the box below the column number.

If the data are saved in an external data file (e.g., the AEROBIC data set of Chapter 1), you can access it using the options available in MINITAB. Click the

FIGURE F.1

Initial screen viewed by MINITAB user

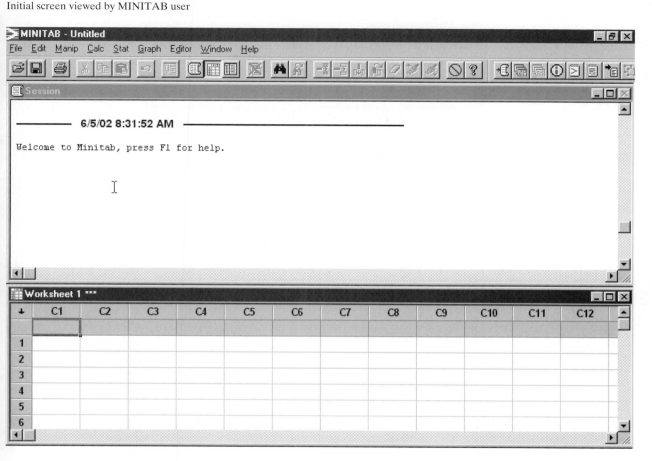

File button on the menu bar, then click Other Files, then click Import Special Text as shown in Figure F.3. The dialog box shown in Figure F.4 will appear.

Specify the variables (columns) on the data set by entering in the names C1, C2, C3, etc., in the Store data in column(s) box. Now click OK. The dialog box shown in Figure F.5 will appear.

Specify the disk drive and folder that contains the external data file, click on the file name, then click Open, as shown in Figure F.5. The MINITAB worksheet will reappear with the data from the external data file, as shown in Figure F.6.

Reminder: The variables (columns) can be named by typing in the name of each variable in the box under the column number.

F.3
Listing Data

To obtain a listing (printout) of your data, click on the Manip button on the MINITAB main menu bar, then click on Display Data (see Figure F.7). The resulting menu, or dialog box, appears as in Figure F.8. Enter the names of the variables you want to print in the Columns constants and matrices to display box

Data entered into the MINITAB
worksheet

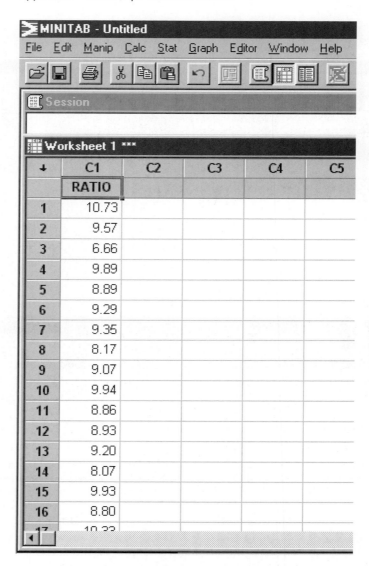

(you can do this by simply clicking on the variables), then click OK. The printout
will show up on your screen.

F.4
Graphing Data

To obtain graphical descriptions of your data (e.g., bar charts, histograms, scatterplots, etc.), click on the Graph button on the MINITAB menu bar. The resulting
menu list appears as shown in Figure F.9. Several of the options covered in this text
are (bar) Chart, (scatter) Plot, Histogram, (normal) Probability Plot, and Stem-and-
Leaf (display). Click on the graph of your choice to view the appropriate dialog
box. For example, the dialog boxes for a histogram and scatterplot are shown,
respectively, in Figures F.10 and F.11. Make the appropriate variable selections and
click OK to view the graph.

FIGURE F.3
MINITAB options for reading data from an external file

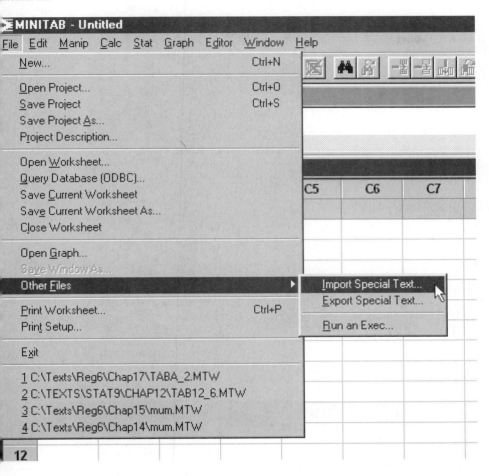

F.5
Descriptive
Statistics and
Correlations

To obtain numerical descriptive measures for a quantitative variable (e.g., mean, standard deviation, etc.), click on the Stat button on the main menu bar, then click on Display Descriptive Statistics (see Figure F.12). The resulting dialog box appears in Figure F.13.

Select the quantitative variables you want to analyze and place them in the Variables box. (As an option, you can create histograms and dot plots for the data by clicking the Graphs button and making the appropriate selections.) Click on OK to view the descriptive statistics printout.

To obtain Pearson product moment correlations for pairs of quantitative variables, click on the "Stat" button on the main menu bar, then click on Basic Statistics, and then click on "Correlation" (see Figure F.12). The resulting dialog box appears in Figure F.14. Enter the variables of interest in the Variables box, then click OK to obtain a printout of the correlations.

FIGURE F.4
Specifying the variable columns in the external
data file

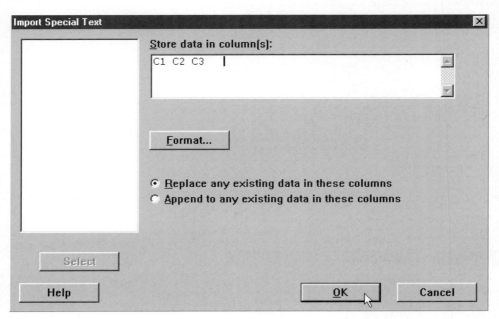

FIGURE F.5
Selecting the external data file

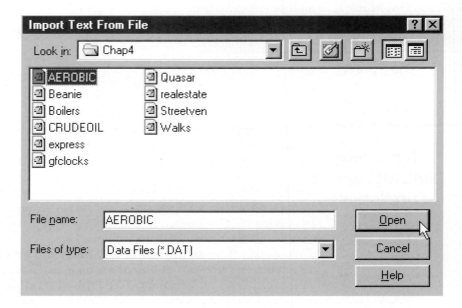

FIGURE F.6

The MINITAB worksheet with the imported data

↓	C1	C2	C3	C4	C5	C6
1	1	881	34.6			
2	2	1290	45.0			
3	3	2147	62.3			
4	4	1909	58.9			
5	5	1282	42.5			
6	6	1530	44.3			
7	7	2067	67.9			
8	8	1982	58.5			
9	9	1019	35.6			
10	10	1651	49.6			
11	11	752	33.0			
12	12	1687	52.0			
13	13	1782	61.4			
14	14	1529	50.2			
15	15	969	34.1			
16	16	1660	52.5			
17	17	2121	69.9			

Worksheet 1 ***

F.6
Hypothesis Tests

To conduct tests of hypotheses of population parameters (e.g., means, variances, proportions), click on the Stat button on the MINITAB menu bar, then click on Basic Statistics. The resulting menu appears as shown in Figure F.15. Click on the test of interest to view the appropriate dialog box, make the appropriate menu selections, and then click OK to view the results. For example, the dialog boxes for a 1-Sample t test for a population mean are shown in Figures F.16 and F.17. In Figure F.16, specify the quantitative variable to be tested in the Variables box and the null hypothesis value in the Test mean box. To specify the confidence level for a confidence interval for the mean and to select the form of the alternative hypothesis, click on Options and make the selections as shown in Figure F.17. Click OK to return to the 1-Sample t dialog box (Figure F.16), then click OK again to view the test results.

F.7
Simple Linear Regression and Multiple Regression

To conduct a regression analysis of a general linear model, click on the Stat button on the MINITAB menu bar, then click on Regression, and click on Regression again, as shown in Figure F.18. The resulting dialog box appears as shown in Figure F.19.

Specify the quantitative dependent variable in the Response box and the independent variables in the Predictors box. [*Note*: If your model includes interaction

FIGURE F.7
MINITAB menu options for
obtaining a list of your data

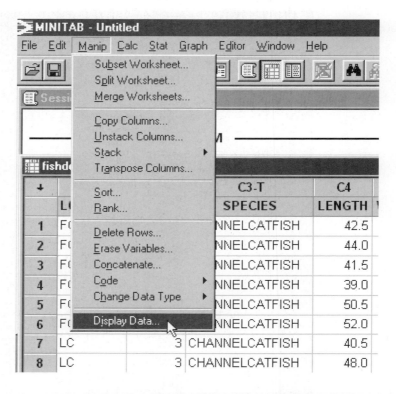

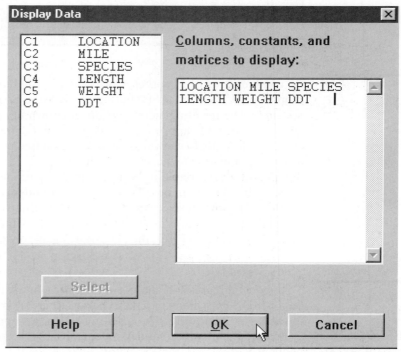

FIGURE F.9

MINITAB menu options for graphing your data

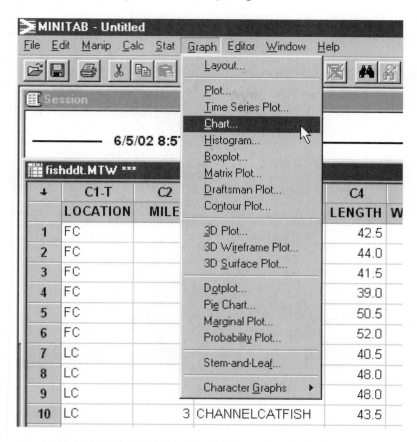

and/or squared terms, you must create and add these higher-order variables to the MINITAB worksheet before running a regression analysis. You can do this by clicking the Calc button on the MINITAB main menu and selecting the Calculator option.]

Optionally, you can get MINITAB to produce prediction intervals for future values of the dependent variable and variance inflation factors by clicking the Options button and checking the appropriate menu items in the resulting menu list. Residual plots are obtained by clicking the Graphs button and making the appropriate selections on the resulting menu. To return to the main Regression dialog box from any of these optional screens, click OK. Click OK on the Regression dialog box to view the linear regression results.

As an alternative, you can fit general linear models by clicking on the Stat button on the MINITAB main menu bar, then click on ANOVA, and finally click on General Linear Model, as shown in Figure F.20. The resulting dialog box appears in figure F.21.

Specify the dependent variable in the Responses box and the terms in the model in the Model box. [*Note*: MINITAB will automatically create the appropriate number of dummy variables for each qualitative variable specified. Also, interaction terms are indicated by placing an asterisk between the variables that are to interact, e.g., LENGTH $*$ WEIGHT. Similar notation is used for squared terms, e.g.,

FIGURE F.10
Histogram dialog box

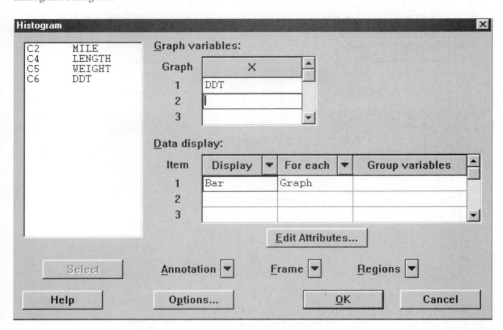

FIGURE F.11
Scatterplot dialog box

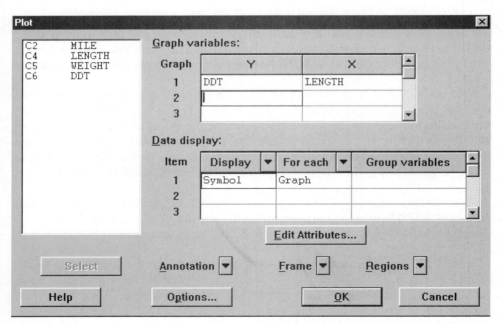

FIGURE F.12
MINITAB options for obtaining
descriptive statistics

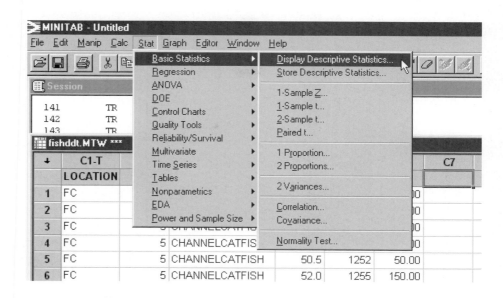

FIGURE F.13
Descriptive statistics dialog box

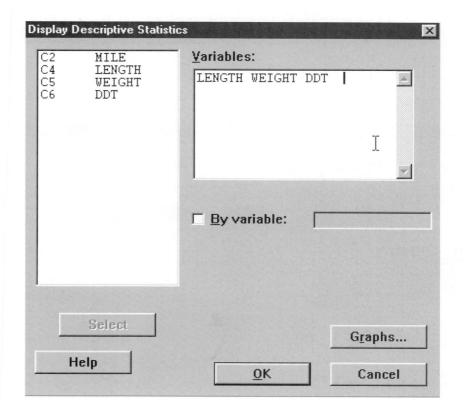

FIGURE F.14
Correlation dialog box

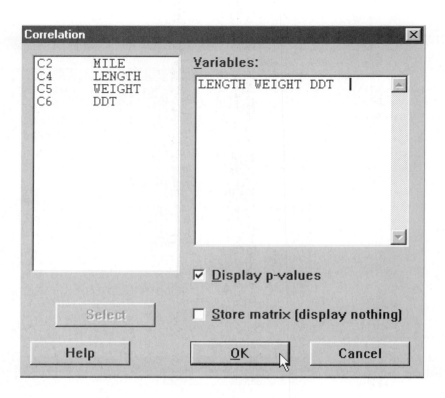

FIGURE F.15
MINITAB options for hypothesis tests

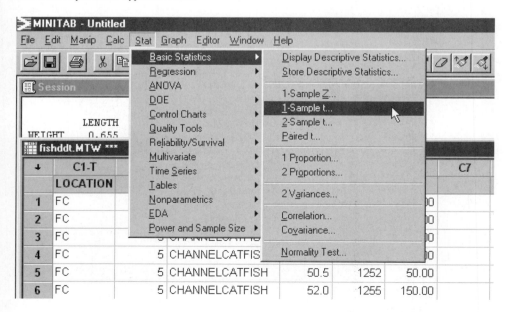

FIGURE F.16

One-sample *t*-test for mean dialog box

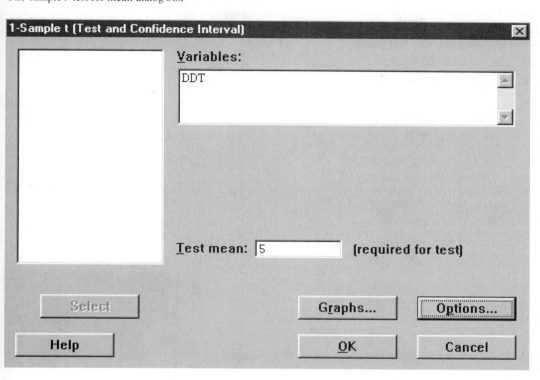

FIGURE F.17

Options dialog box for one sample
t-test

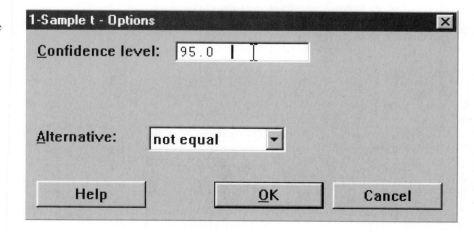

FIGURE F.18

MINITAB options for regression

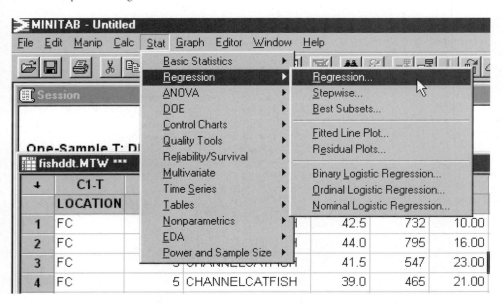

FIGURE F.19

Dialog box for regression

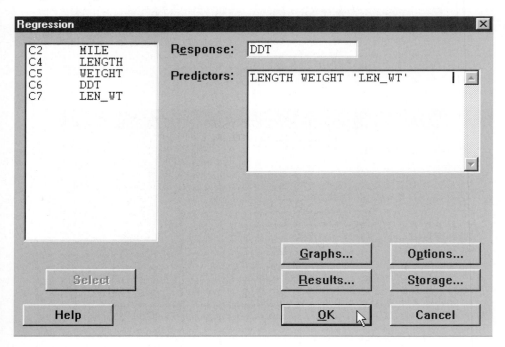

FIGURE F.20
MINITAB options for fitting general linear models

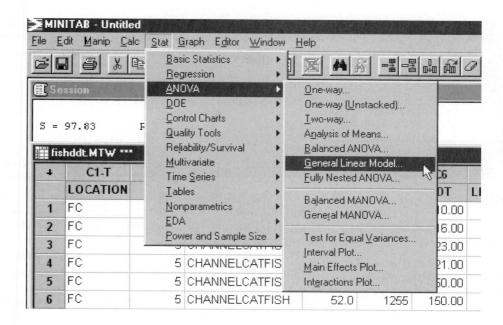

FIGURE F.21
Dialog box for general linear models

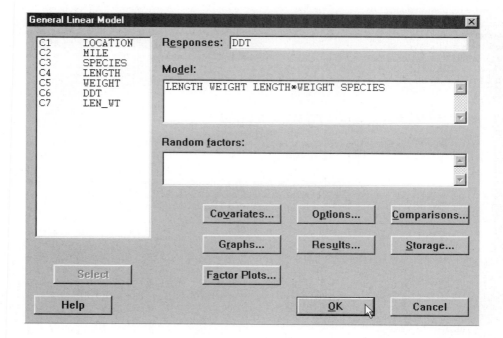

LENGTH*LENGTH.] Identify the quantitative independent variables in the model by clicking the Covariates button and placing these variables in the Covariates box.

To obtain parameter estimates for all terms in the model, click the Results button and make the appropriate selection. To save predicted values and residuals, click the Storage button and make the appropriate selections. To produce residual plots, click the Graphs button and make the your selections. When all the options you desire have been checked, click OK on the General Linear Models dialog box to view the multiple regression results.

F.8
Stepwise Regression

To conduct a stepwise regression analysis, click on the Stat button on the main menu bar, then click on Regression, and click on Stepwise (see Figure F.18). The resulting dialog box appears in Figure F.22. Specify the dependent variable in the Response box and the independent variables in the stepwise model in the Predictors box. As an option, you can select the value of α to use in the analysis by clicking on the Methods button and specifying the value. (The default is $\alpha = .15$.) Click OK to view the stepwise regression results.

To conduct an all-possible-regressions selection analysis, click on the Stat button on the main menu bar, then click on Regression, and click on Best Subsets (see Figure F.18). Specify the dependent variable in the Response box and the independent variables in the Predictors box shown on the resulting menu. Click OK to view the all-possible-regression selection results.

FIGURE F.22
Stepwise regression dialog box

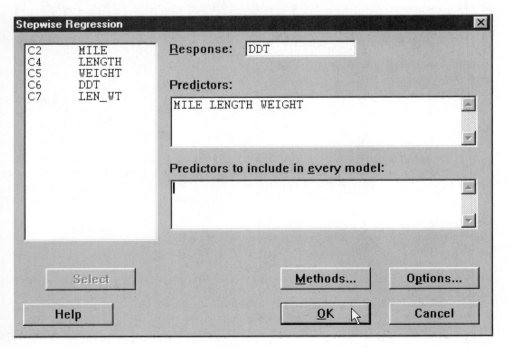

F.9
Residual Analysis and Influence Diagnostics

To conduct a residual analysis, click on the Stat button on the MINITAB main menu bar, then click on Regression, and then click on Regression again (see Figure F.18). The resulting dialog box appears in Figure F.19. Specify the dependent variable in the Response box and the independent variables in the model in the Predictors box. To produce residual plots, click on the Graphs button. The resulting menu appears as shown in Figure F.23.

To produce a histogram, select Histogram of residuals; to produce a normal probability plot, select Normal plot of residuals; to produce a plot of residuals against predicted values, select Residuals versus fits; and to produce a plot of residuals against one or more of the independent variables, put the independent variables to be plotted in the Residuals versus the variables box. Click OK to view the plots.

To produce influence diagnostics, click on the Storage button in the Regression dialog box (Figure F.19). The resulting menu appears as shown in Figure F.24.

Click the influence diagnostics you want to save [e.g., Cook's distance and Hi (leverages)], then click OK to return to the Regression dialog box. Now click OK. The influence diagnostics will appear as additional columns on the MINITAB worksheet.

FIGURE F.23
Residual plots menu selection for multiple regression

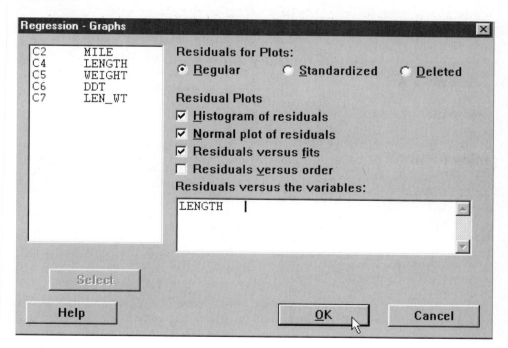

F.10
Logistic Regression

To conduct a logistic regression analysis for a two-level (binary) dependent qualitative variable, click on the Stat button on the MINITAB main menu bar, then click on Regression, and finally click on Binary Logistic Regression, as shown in Figure F.25. The resulting dialog box appears in Figure F.26.

Specify the dependent variable in the Response box and the terms in the logistic model in the Model box. [*Note:* If your model includes interaction and/or squared terms, you must create and add these higher-order variables to the MINITAB

FIGURE F.24
Storage menu options for
regression

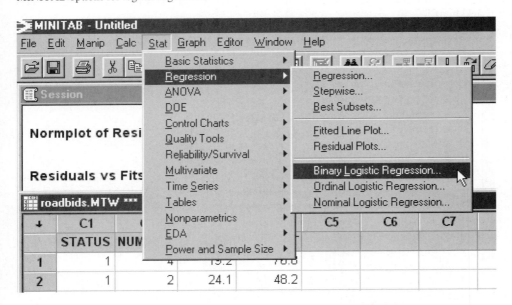

FIGURE F.25
MINITAB options for logistic regression

worksheet before running a regression analysis. You can do this by clicking the Calc button on the MINITAB main menu and selecting the Calculator option.] If any of the independent variables are qualitative, put them in the Factors box. MINITAB will automatically create dummy variables for these qualitative predictors. After making the variable selections, click the OK button to view the logistic regression results.

**F.11
One-Way Analysis
of Variance**

To conduct a one-way ANOVA for a completely randomized design, click on the Stat button on the MINITAB main menu bar, then click on ANOVA, and finally click on One-Way, as shown in Figure F.27. The resulting dialog box appears in Figure F.28.

FIGURE F.26

Dialog box for logistic regression

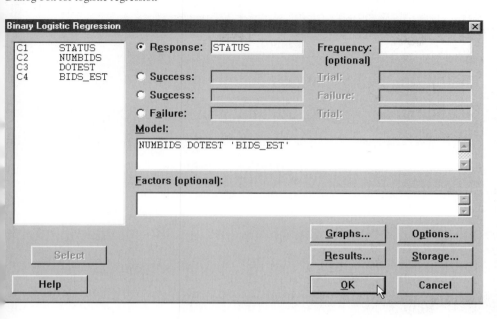

FIGURE F.27

MINITAB options for one-way analysis of variance

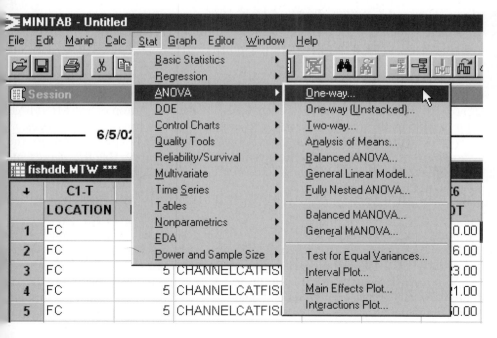

FIGURE F.28
Dialog box for one-way ANOVA

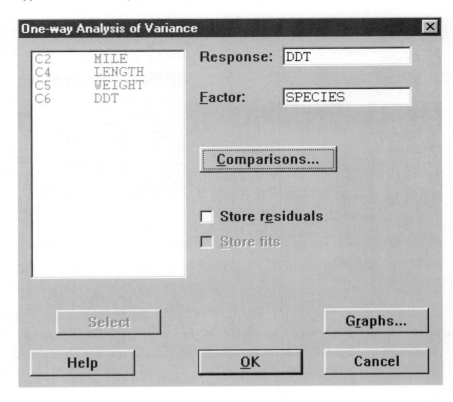

Specify the dependent variable in the Response box and the qualitative variable that represents the single factor in the Factor box. To perform multiple comparisons of treatment means, click the Comparisons button to obtain the dialog box shown in Figure F.29. On this box, select the comparison method (e.g., Tukey) and the comparison-wise error rate percentage (e.g., specify 5 for a 5% significance level). Click OK to return to the One-Way ANOVA dialog box (Figure F.28). Click OK to view the ANOVA results.

To perform a test of equality of variances, click the Stat button on the MINITAB main menu bar, then click ANOVA, and finally click Test for Equal Variances, as shown on Figure F.30. The dialog box shown in Figure F.31 appears. Specify the dependent variable in the Response box and the qualitative variable that represents the single factor in the Factors box. Specify the Confidence level, then click OK to view the test results.

F.12
Analysis of Variance for Factorial Designs

To conduct an ANOVA for a two-factor factorial design, click on the Stat button on the MINITAB main menu bar, then click on ANOVA, and finally click on Two-Way (see Figure F.32). The resulting dialog box appears in Figure F.33.

Specify the dependent variable in the Response box, the first qualitative factor in the Row Factor box, and the second qualitative factor in the Column Factor box. A model with main effects and factor interaction will automatically be fit. Click on OK to view the ANOVA results.

FIGURE F.29
Dialog box for multiple
comparisons of means

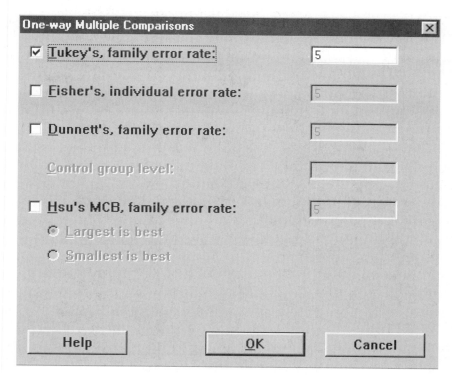

FIGURE F.30
MINITAB options testing equality of variances

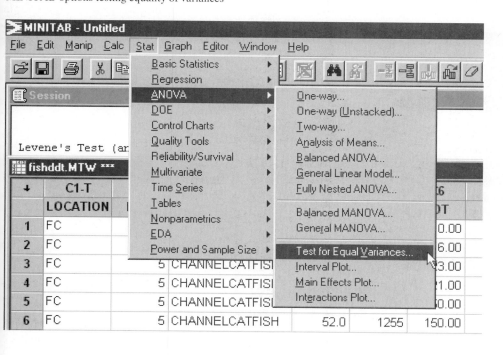

FIGURE F.31

Dialog box for testing equality of variances

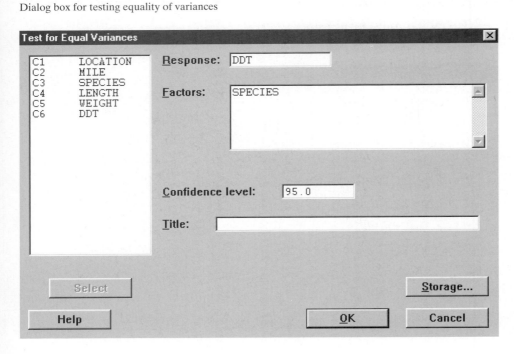

FIGURE F.32

MINITAB options for two-way (factorial)
ANOVA

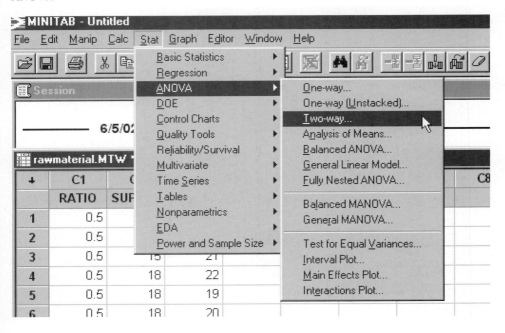

FIGURE F.33
Dialog box for two-way (factorial) ANOVA

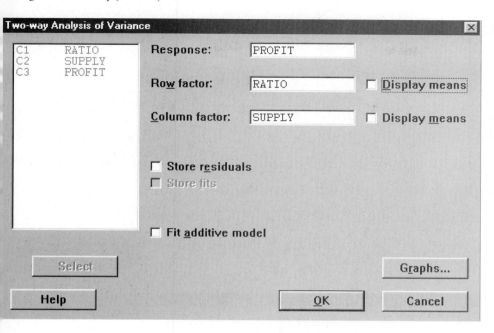

For more general factorial designs, click on the Stat button on the MINITAB main menu bar, then click on ANOVA, and finally click on General Linear Model (see Figure F.32). The resulting dialog box appears in Figure F.34.

Specify the dependent variable in the Responses box and the terms in the factorial model in the Model box. [*Note*: Interaction terms are indicated by placing an asterisk between the variables that are to be interacted, e.g., RATIO ∗ SUPPLY.]

To run multiple comparisons of means for treatment main effects, click the Comparisons button. The dialog box shown in Figure F.35 appears. Specify the main effects to be tested in the Terms box and select the comparison method (e.g., Bonferroni), then click OK to return to the Factorial ANOVA dialog box (Figure F.34). Click OK to run and view the ANOVA results.

F.13 Time Series Forecasting

To conduct the Durbin–Watson test for autocorrelated errors in a model for time series data, first specify the model to be fit. That is, click on the Stat button on the MINITAB main menu bar, then click on Regression, and click on Regression again, as shown in Figure F.18. Specify the dependent and independent variables in the model on the resulting dialog box (see Figure F.19.) Once the model has been specified, click on the Options button on the Regression dialog box to obtain the menu shown in Figure F.36.

Check the Durbin–Watson statistic box, then click OK to return to the Regression dialog box. Click OK to view the results.

To produce forecasts using moving averages, exponential smoothing or the Holt–Winters method, click the Stat button on the MINITAB main menu bar,

FIGURE F.34

Dialog box for factorial ANOVA general linear model

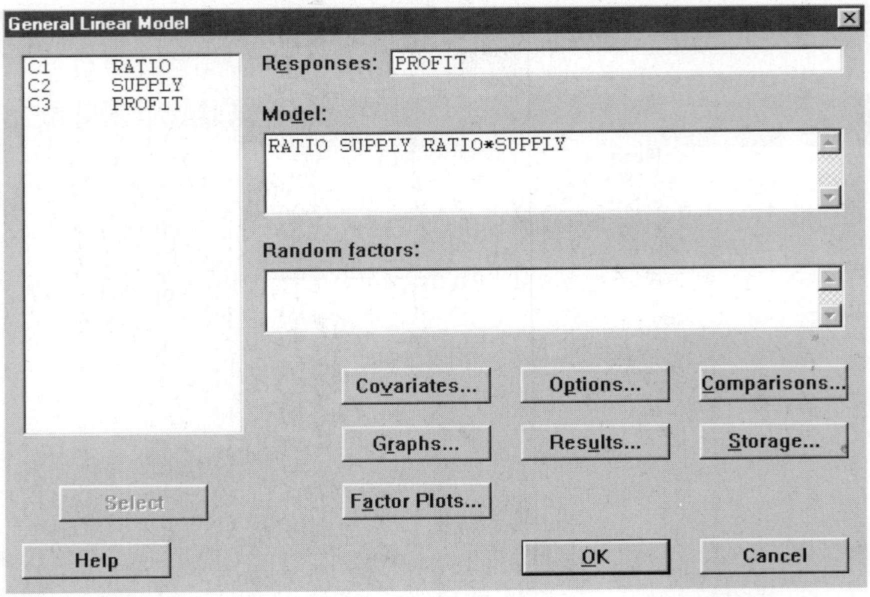

FIGURE F.35

Dialog box for multiple comparisons of factorial means

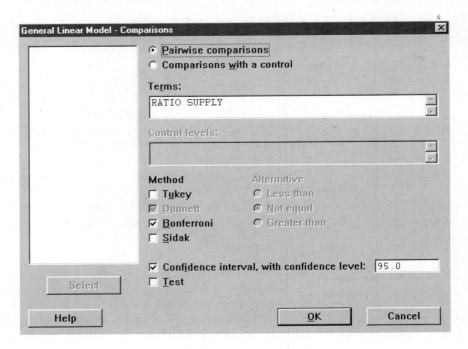

then click on Time Series. This will produce the menu list shown in Figure F.37. The options include Moving Average, Single Exp Smoothing (i.e., the exponential smoothing method covered in this text), Double Exp Smoothing (i.e., the Holt–Winter's method with trend), and Winter's Method (i.e., the Holt–Winter's method with trend and seasonality). For example, clicking Single Exp Smoothing will result in the dialog box shown in Figure F.38.

FIGURE F.36
Regression options menu

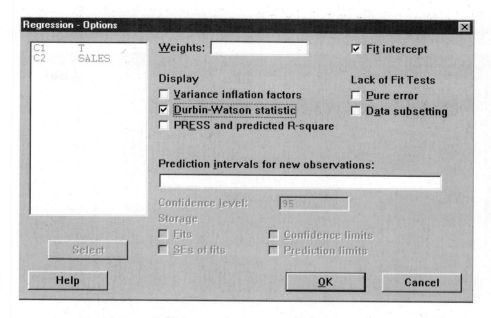

FIGURE F.37
MINITAB options for time series
analysis

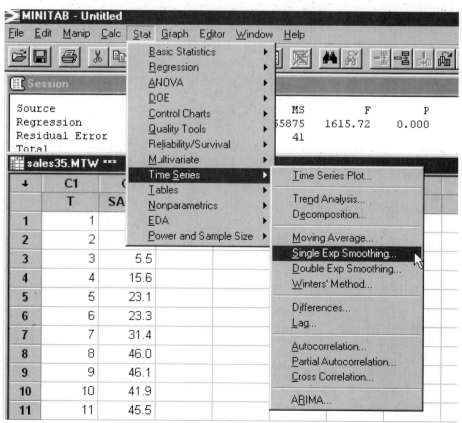

FIGURE F.38
Exponential smoothing dialog box

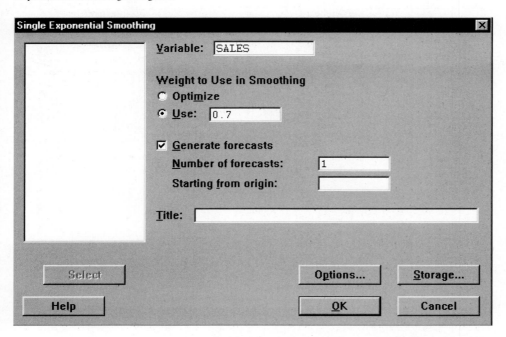

Select the quantitative variable to be smoothed and place it in the Variable box and set the value of the smoothing constant in the Weight to use in smoothing box. Click OK to view the results.

[*Note*: Time series models with a first-order autoregressive error term are not available in MINITAB.]

File Layouts
for Large Data Sets

FLAG–Sealed bid data for 235 highway construction contracts

Variable	Format	Columns	Description
CONTRACT	Numeric	1–4	Contract number
COST	Numeric	6–13	Contract cost (thousand dollars)
DOTEST	Numeric	15–22	DOT engineer's cost estimate (thousand dollars)
B2B1RAT	Numeric	24–30	Ratio of 2nd lowest bid to low bid
B3B1RAT	Numeric	32–38	Ratio of 3rd lowest bid to low bid
BHB1RAT	Numeric	40–46	Ratio of highest bid to low bid
STATUS	Numeric	51	Bid status (1 = Fixed, 0 = Competitive)
DISTRICT	Numeric	59	Location (1 = South Florida, 0 = not)
BTPRATIO	Numeric	65–71	Ratio of number of bidders to number of plan holders
DAYSEST	Numeric	75–77	DOT engineer's estimate of number of workdays required

TAMSALES7–Real estate appraisal data for 1,041 Tampa, Florida homes
in seven neighborhoods

Variable	Format	Columns	Description
SALES	Numeric	2–10	Sales price (dollars)
LANDVAL	Numeric	15–21	Land value (dollars)
IMPROVAL	Numeric	25–32	Value of improvements (dollars)
NBRHOOD	Character	35–42	Neighborhood (TAMPALMS, TEMTERR, TNC, YBOR, DAVISLE, AVILA, CWDVILL)

 TAMSALES4–Real estate appraisal data for 675 Tampa, Florida homes
in four neighborhoods

Variable	Format	Columns	Description
SALES	Numeric	2–10	Sales price (dollars)
LANDVAL	Numeric	15–21	Land value (dollars)
IMPROVAL	Numeric	25–32	Value of improvements (dollars)
NBRHOOD	Character	35–42	Neighborhood (TAMPALMS, TNC, DAVISLE, CWDVILL)

 CONDO–Sales data for 209 condominium units

Variable	Format	Columns	Description
PRICE	Numeric	7–9	Price paid (hundred dollars)
FLOORHGT	Numeric	19	Floor height (1, 2, 3, . . . , 8)
DISTELEV	Numeric	28–29	Distance from elevator (1, 2, 3, . . . , 15)
VIEW	Numeric	39	View (1 = Ocean, 0 = Bay)
ENDUNIT	Numeric	49	End unit (1 = end, 0 = not)
FURNISH	Numeric	59	Furnished (1 = Furnish, 0 = not)
METHOD	Character	61	Sales method (A = auction, F = fixed price)

SHORT ANSWERS TO ODD-NUMBERED EXERCISES

CHAPTER 1

1.1   a. Quantitative   b. Quantitative   c. Quantitative   d. Qualitative
  e. Quantitative   f. Quantitative

1.3   a. Qualitative   b. Qualitative or Quantitative   c. Qualitative   d. Quantitative

1.5   Quantitative; Qualitative; Quantitative; Qualitative; Quantitative; Qualitative; Qualitative; Quantitative; Quantitative; Quantitative

1.7   a. Amateur boxers   b. both quantitative   d. no

1.9   b. Frequency of headers and IQ   c. Quantitative; Quantitative

1.11   c. .8145; .1855

1.13   a. pie chart   b. breast cancer   c. 19%

1.15   b. no

1.19   b. yes

1.21   b. .87

1.25   a. 1.47   b. .0640   c. morning

1.27   a. $(-111, 149)$ b. $(-91, 105)$ c. SAT-Math

1.29   a. .9544   b. .1587   c. .1587   d. .8185   e. .1498   f. .9974

1.31   a. .9406   b. .9406   c. .1140

1.33   a. .2690   b. .2611   c. $(13.6, 62.2)$

1.37   a. 2.228   b. 3.747   c. -2.861 d. -1.796

1.39   a. Approximately normal; $\mu_y = 15$; $\sigma_y = 1.118$ b. ≈ 0
  c. True value of μ is highly likely to be less than 15

1.41   a. $1.13 \pm .672$ b. yes

1.43   a. $(5.89, 8.71)$

1.47   a. .025   b. .05   c. .005   d. .0985   e. .10   f. .01

1.49   a. $H_0: \mu = 15$, $H_a: \mu < 15$

1.51   a. $z = -3.72$, reject H_0

1.53   a. $z = 2.3$ b. $p = .0214$ c. $p \approx 0$

1.55 $t = 7.83$, reject H_0

1.59   a. 3.68 ± 4.277

1.61   a. $t = -7.24$, reject H_0 b. $t = -.50$, fail to reject H_0
      c. $t = -1.25$, fail to reject H_0

1.63   b. $H_0: \mu_d = 0$ c. reject H_0 at $\alpha = .05$

1.65 $t = .68$, fail to reject H_0

1.67 $F = 1.16$, fail to reject H_0

1.69 $F = 5.871$, reject H_0

1.71   a. $5, 21.5, 4.637$ b. $16.75, 36.25, 6.021$ c. $4.857, 29.81, 5.460$
      d. $4, 0, 0$

1.73   a. $z = -4$ b. $z = .5$ c. $z = 0$ d. $z = 6$

1.75   a. Quantitative   b. Quantitative   c. Quantitative   d. Qualitative
      e. Qualitative

1.77   a. $4; .635$ b. yes

1.79   c. Eclipse

1.81   a. 14.682 b. $199.974, 14.141$ c. 95% d. 90.9%
      e. Decrease; Decrease   f. $13.735, 11.960$ g. $95\%; 90.8\%$

1.83   a. $.7422$ b. $.0154$ c. no

1.85 $(.61, 2.13)$

1.87   a. 5.99 ± 3.142

1.89 $F = 1.045$, fail to reject H_0

1.91   a. $s_1, s_2 < 5.053$ c. $s_1, s_2 > 5.053$

1.93   a. $z = 1.19$, fail to reject H_0 b. $.5 \pm .690$ d. $p = .2340$

1.95 $t = -2.306$, reject H_0

CHAPTER 3

3.3   a. $\beta_0 = 2; \beta_1 = 2$ b. $\beta_0 = 4; \beta_1 = 1$ c. $\beta_0 = -2; \beta_1 = 4$ d. $\beta_0 = -4; \beta_1 = -1$

3.5   a. $\beta_1 = 2; \beta_0 = 3$ b. $\beta_1 = 1; \beta_0 = 1$ c. $\beta_1 = 3; \beta_0 = -2$ d. $\beta_1 = 5; \beta_0 = 0$
      e. $\beta_1 = -2; \beta_0 = 4$

3.7   a. $\hat{\beta}_0 = 2; \hat{\beta}_1 = -1.2$

3.9   a. $y = \beta_0 + \beta_1 x + \varepsilon$; negative   b. yes   c. no

3.11   b. $\hat{y} = -.00105 + .00321x$

3.13   a. $\hat{y} = 6.25 - .0023x$ c. 5.56

3.15 $\hat{y} = .5704 + .0264x$

3.17   a. $1.6; .5333; .730$ b. $.00068; .000052; .00723$ c. $30.767; 2.051; 1.432$
      d. $.2509; .0502; .2240$

3.19   b. $\hat{y} = 223.931 + 2,884.877x$ c. $2,496.7$ d. $11,763.43 \pm 4,993.334$ e. $11/12$

3.21   a. $t = 6.708$, reject H_0 b. $t = -5.196$, reject H_0

3.23 $(-.0039, -.0007)$

3.25   a. $t = 6.286$, reject H_0 b. $.88 \pm .23618$

3.27 $t = 2.597$, reject H_0

3.29 $t = 2.858$, reject H_0

3.31   a. .9583; .9183   b. .9487; .9000

3.33   positive

3.37   b. .1998; .0032; .3832; .0864; .9006

     c. reject H_0; fail to reject H_0; reject H_0; fail to reject H_0; reject H_0

3.39   a. $y = \beta_0 + \beta_1 x + \varepsilon$ b. $\beta_1 > 0$ d. $t = -3.12$, fail to reject H_0

3.41 $r = .5702; r^2 = .3251$

3.43   a. Negative   b. $r = -.0895$

3.45   a. 4.055; .225   b. $10.6 \pm .233$ c. $8.9 \pm .319$ d. $12.3 \pm .319$

     f. 12.3 ± 1.046

3.47   a. $.031 \pm .013$ b. $.031 \pm .004$ d. yes

3.49   a. $(1.0516, 6.6347)$ b. Narrower   c. no

3.51   a. $y = \beta_0 + \beta_1 x + \varepsilon$ b. $\hat{y} = 155.912 - 1.086x$ c. $t = -2.051$; fail to reject H_0

     d. 238.76 ± 150.32

3.53 $\hat{y} = 28.894 + .905x$; SSE $= 13{,}236.45$; $s^2 = 315.15$; $s = 17.75$; $r^2 = .0661$; $t = 1.725$;

     fail to reject H_0

3.55 $t = 3.25$, reject H_0

3.57   a. $\hat{y} = -9.2667x$ b. 12.8667; 3.2167; 1.7935   c. Yes; $t = -28.30$

     d. $-9.267 \pm .909$ e. $-9.267 \pm .909$ f. -9.267 ± 5.061

3.59   a. $\hat{y} = 5.364x$ b. Yes; $t = 25.28$ c. 18.77 ± 6.30

3.61   a. $\hat{y} = 51.18x$ b. Yes; $t = 154.56$ c. $\hat{y} = 1{,}855.35 + 47.07x$; yes, $t = 93.36$

3.63   b. $\hat{y} = 1.192 + .987x$

3.65   a. $t = 7.13$, reject H_0 b. $t = 5.98$, reject H_0 c. $t = 4.98$, reject H_0

     d. 7.0695   e. 35.86   f. 1390.08

3.67   b. $\hat{y} = 78.516 - .239x$ d. $t = -2.31$, fail to reject H_0

     e. observation #5 is an outlier   f. $\hat{y} = 139.759 - .4497x$; $t = -15.13$, reject H_0

3.69   b. $-.766$ c. $t = -4.91$, reject H_0

3.71   d. .903   e. $t = 20.83$, reject H_0 f. 4.58

3.73   a. $y = \beta_0 + \beta_1 x + \varepsilon$ b. positive   d. reject H_0

3.75   a. $\hat{y} = 14{,}012 - 1{,}783x$; no (at $\alpha = .05$), $t = -.98$

     b. $\hat{y} = 19{,}680 - 3{,}887x$; yes (at $\alpha = .05$), $t = -2.29$

     c. Predicting outside range of x $(1.52 - 4.11)$

3.71   b. $\hat{y} = 5.977 + 74.068x$ d. 11.162   e. 11.162 ± 3.739 f. 11.162 ± 1.024

3.73   a. Yes   b. Yes   d. 8.29   e. 21.15

3.75   b. Probably; however, we cannot conduct a test since n is not given

CHAPTER 4

4.1   df $= n -$ (number of independent variables $+ 1$)

4.3   b. $t = -1.00$, fail to reject H_0 c. $1.51 \pm .098$

4.5   a. $H_0: \beta_4 = 0$ vs. $H_a: \beta_4 < 0$ b. $H_0: \beta_5 = 0$ vs. $H_a: \beta_5 > 0$

4.7   estimate of β_1

4.9
a. $E(y) = \beta_0 + \beta_1 x_1 + \beta_2 x_2 + \beta_3 x_3 + \beta_4 x_4$
b. $\hat{y} = 21,087.95 + 108.45 x_1 + 557.91 x_2 - 340.17 x_3 + 85.68 x_4$
d. $t = 1.222$, fail to reject H_0

4.11
a. $E(y) = \beta_0 + \beta_1 x_1 + \beta_2 x_2 + \beta_3 x_3 + \beta_4 x_4 + \beta_5 x_5 + \beta_6 x_6 + \beta_7 x_7$
b. $\hat{y} = .998 - .0224 x_1 + .156 x_2 - .0172 x_3 - .00953 x_4$
$+ .421 x_5 + .417 x_6 - .155 x_7$

4.13 $F = 1.056$; do not reject H_0 b. .05

4.15
a. $\hat{y} = \hat{\beta}_0 + .110 x_1 + .065 x_2 + .540 x_3 - .009 x_4 - .150 x_5 - .027 x_6$
b. $H_0: \beta_1 = \beta_2 = \beta_3 = \beta_4 = \beta_5 = \beta_6 = 0$ c. $F = 32.47$, reject H_0
d. .44; .43   e. $H_0: \beta_4 = 0$ f. p-value = .860, fail to reject H_0

4.17   a. .912; .894   b. $F = 21.88$, reject H_0

4.19   a. $t = 3.88$, reject H_0 b. no   c. $F = 11.40$, reject H_0

4.21
a. 2.87   b. $t = -2.71$, reject H_0
c. Rental price, empl. growth,   d. Inflated α error
   AFDC benefits, SSI benefits

4.23   a. $H_0: \beta_3 = 0$ vs. $H_a: \beta_3 \neq 0$ c. reject H_0

4.25   a. $E(y) = \beta_0 + \beta_1 x_1 + \beta_2 x_2 + \beta_3 x_1 x_2$ b. p-value = .02, reject H_0

4.27   b. $F = 5.505$, $R^2 = .6792$, $s = .505$

4.29   b. $t = 2.69$, p-value = .031, reject H_0

4.31   a. Negative   b. $F = 1.60$, fail to reject H_0 c. $F = 1.61$, fail to reject H_0

4.33   a. $E(y) = \beta_0 + \beta_1 x_1 + \beta_2 x_1^2$ b. positive   c. no; $E(y) = \beta_0 + \beta_1 x_1$

4.35   a. (1759.7, 4275.4)   b. (2620.3, 3414.9)   c. yes

4.37   (90.69, 158.57)

4.39   (-1.233, 1.038)

4.41   a. $E(y) = \beta_0 + \beta_1 x_1 + \beta_2 x_2$ b. $E(y) = \beta_0 + \beta_1 x_1 + \beta_2 x_2 + \beta_3 x_3 + \beta_4 x_4$

4.43
a. $E(y) = \beta_0 + \beta_1 x_1$, where $x = \begin{cases} 1 & \text{if A} \\ 0 & \text{if B} \end{cases}$
b. $E(y) = \beta_0 + \beta_1 x_1 + \beta_2 x_2 + \beta_3 x_3$, where

$$x_1 = \begin{cases} 1 & \text{if A} \\ 0 & \text{if not} \end{cases} \quad x_2 = \begin{cases} 1 & \text{if B} \\ 0 & \text{if not} \end{cases} \quad x_3 = \begin{cases} 1 & \text{if C} \\ 0 & \text{if not} \end{cases}$$

$\beta_0 = \mu_D, \beta_1 = \mu_A - \mu_D, \beta_2 = \mu_B - \mu_D, \beta_3 = \mu_C - \mu_D$

4.45   c. Parallel lines

4.47   b. Second-order   c. Different shapes   d. Yes   e. Shift curves along the x_1-axis

4.49   a. $E(y) = \beta_0 + \beta_1 x_1 + \beta_2 x_2$

4.51
a. $E(y) = \beta_0 + \beta_1 x_1 + \beta_2 x_2 + \beta_3 x_1 x_2 + \beta_4 x_1^2 + \beta_5 x_2^2$
b. $\beta_4 x_1^2$ and $\beta_5 x_2^2$

4.53
a. Model 1: $t = 2.58$, reject H_0; Model 2: reject $H_0: \beta_1 = 0$ ($t = 3.32$), reject
$H_0: \beta_2 = 0$ ($t = 6.47$), reject $H_0: \beta_3 = 0$ ($t = -4.77$), do not reject $H_0: \beta_4 = 0$
($t = .24$); Model 3: reject $H_0: \beta_1 = 0$ ($t = 3.21$), reject $H_0: \beta_2 = 0$ ($t = 5.24$),
reject $H_0: \beta_3 = 0$ ($t = -4.00$), do not reject $H_0: \beta_4 = 0$ ($t = 2.28$), do not reject
$H_0 : \beta_5 = 0$ ($t = .14$)

4.55 a. $F = 3,909.25$; reject H_0 b. $s = .047$ c. $H_0: \beta_4 = 0$; $H_a: \beta_4 < 0$
e. No; since β_4 is positive

4.57 a. $E(y) = \beta_0 + \beta_1 x_1 + \beta_2 x_2$ b. $\hat{y} = .16 + .17x_1 + .17x_2$ c. $\hat{y} = .21 + .06x_1 - .03x_2$

4.59 Nested models: a and b, a and d, a and e, b and c, b and d, b and e, c and e, d and e

4.61 a. $E(y) = \beta_0 + \beta_1 x_1 + \beta_2 x_2 + \cdots + \beta_{11} x_{11}$
b. $H_0: \beta_{12} = \beta_{13} = \beta_{14} = \cdots = \beta_{35} = 0$

4.63 a. $E(y) = \beta_0 + \beta_1 x_1 + \beta_2 x_1^2 + \beta_3 x_2 + \beta_4 x_1 x_2 + \beta_5 x_1^2 x_2$
b. $H_0: \beta_2 = \beta_5 = 0$ c. $H_0: \beta_4 = \beta_5 = 0$ d. $F = .429$; .634

4.65 a. $E(y) = \beta_0 + \beta_1 x_1 + \beta_2 x_2 + \cdots + \beta_{10} x_{10}$ b. $H_0: \beta_3 = \beta_4 = \cdots = \beta_{10} = 0$
c. reject H_0 e. 14 ± 5.88 f. yes   h. $H_0: \beta_{11} = \beta_{12} = \cdots = \beta_{19} = 0$

4.67 a. $H_0: \beta_4 = \beta_5 = 0$ b. $H_0: \beta_3 = \beta_4 = \beta_5 = 0$ c. $F = .93$, do not reject H_0

4.69 a. $t = 5.96$, reject H_0 b. $t = .01$, do not reject H_0 c. $t = 1.91$, reject H_0

4.71 a. $E(y) = \beta_0 + \beta_1 x_1 + \beta_2 x_2 + \beta_3 x_3 + \beta_4 x_4 + \beta_5 x_5$ b. $F = 34.47$, reject H_0
c. $E(y) = \beta_0 + \beta_1 x_1 + \beta_2 x_2 + \cdots + \beta_7 x_7$ e. reject H_0

4.73 a. -75.51 ± 26.17 b. Do not reject H_0: $t = 1.38$ c. $R_a^2 = .25$
d. Yes; $F = 21.86$, p-value $< .01$

4.75 c. $F = 13.53$, reject H_0

4.77 a. $E(y) = \beta_0 + \beta_1 x_1 + \beta_2 x_1^2$ b. $H_0: \beta_2 = 0$ vs. $H_a: \beta_2 < 0$

4.79 a. $E(y) = \beta_0 + \beta_1 x_1 + \beta_2 x_2 + \beta_3 x_3 + \beta_4 x_4 + \beta_5 x_5$
b. $\hat{y} = 15.49 + 12.77x_1 + .71x_2 + 1.52x_3 + .32x_4 + .21x_5$
c. $R^2 = .240$, $F = 11.68$, reject H_0 d. p-value $< .025$, reject H_0

4.81 b. $t = 5$, reject H_0 c. $\hat{y} = 825$ d. supports

4.83 a. quantitative   b. quantitative   c. qualitative
d. $E(y) = \beta_0 + \beta_1 x_1 + \beta_2 x_2 + \beta_3 x_3 + \beta_4 x_4$
e. $\beta_0 = \mu_A$, $\beta_1 = \mu_B - \mu_A$, $\beta_2 = \mu_C - \mu_A$, $\beta_3 = \mu_D - \mu_A$, $\beta_4 = \mu_E - \mu_A$
f. $H_0: \beta_1 = \beta_2 = \beta_3 = \beta_4 = 0$

4.85 a. $H_0: \beta_5 = \beta_6 = \beta_7 = \beta_8 = 0$; males: $F > 2.37$; females: $F > 2.45$

CHAPTER 5

5.1 Quantitative; Quantitative; Qualitative

5.3 a. Qualitative   b. Quantitative   c. Qualitative   d. Quantitative
e. Quantitative   f. Qualitative

5.5 a. Quantitative   b. Quantitative   c. Qualitative   d. Qualitative
e. Qualitative   f. Quantitative   g. Qualitative   h. Qualitative
i. Qualitative   j. Quantitative   k. Qualitative

5.7 $E(y) = \beta_0 + \beta_1 x + \beta_2 x^2 + \beta_3 x^3$

5.9 $E(y) = \beta_0 + \beta_1 x + \beta_2 x^2$

5.11 yes

5.13 a. $E(y) = \beta_0 + \beta_1 x_1 + \beta_2 x_2 + \beta_3 x_1 x_2 + \beta_4 x_1^2 + \beta_5 x_2^2$
b. $E(y) = \beta_0 + \beta_1 x_1 + \beta_2 x_2$ c. $E(y) = \beta_0 + \beta_1 x_1 + \beta_2 x_2 + \beta_3 x_1 x_2$
d. $\beta_1 + \beta_3 x_2$ e. $\beta_2 + \beta_3 x_1$

5.15   a. $E(y) = \beta_0 + \beta_1 x_1 + \beta_2 x_1^2 + \beta_3 x_2 + \beta_4 x_2^2 + \beta_5 x_4 + \beta_6 x_4^2 + \beta_7 x_1 x_2 + \beta_8 x_1 x_4 + \beta_9 x_2 x_4$

        b. $\hat{y} = 10{,}283.2 + 276.8 x_1 + .990 x_1^2 + 3325.2 x_2 + 266.6 x_2^2 + 1301.3 x_4 + 40.22 x_4^2$
             $+ 41.98 x_1 x_2 + 15.98 x_1 x_4 + 207.4 x_2 x_4;\ F = 613.278$, reject H_0

        c. $F = 108.43$, reject H_0

5.17   a. $u = (x - 85.1)/14.81$

        b. $-.668, .446, 1.026, -1.411, -.223, 1.695, -.527, -.338$ c. $.9967$ d. $.376$

        e. $\hat{y} = 110.953 + 14.377u + 7.425u^2$

5.19   a. $E(y) = \beta_0 + \beta_1 x_1 + \beta_2 x_2 + \beta_3 x_3$

        b. $\beta_0 = \mu_4, \beta_1 = \mu_1 - \mu_4, \beta_2 = \mu_2 - \mu_4, \beta_3 = \mu_3 - \mu_4$

        c. $H_0: \beta_1 = \beta_2 = \beta_3 = 0$

5.21   a. $E(y) = \beta_0 + \beta_1 x_1 + \beta_2 x_2 + \beta_3 x_3 + \beta_4 x_4$ b. $\mu_{BA} - \mu_N$ c. $\mu_E - \mu_N$

        d. $\mu_{LAS} - \mu_N$ e. $\mu_J - \mu_N$

        f. $E(y) = (\beta_0 + \beta_5) + \beta_1 x_1 + \beta_2 x_2 + \beta_3 x_3 + \beta_4 x_4$ g. $\mu_{BA} - \mu_N$ h. $\mu_E - \mu_N$

        i. $\mu_{LAS} - \mu_N$ j. $\mu_J - \mu_N$ k. $\mu_F - \mu_M$

        l. Reject $H_0: \beta_5 = 0$; gender has an effect

5.23   a. $E(y) = \beta_0 + \beta_1 x_1 + \beta_2 x_2 + \beta_3 x_3$ b. $\beta_0 = \mu_{\text{Girls, Oldest}}, \beta_1 = \mu_{\text{Boys}} - \mu_{\text{Girls}},$

        $\beta_2 = \mu_{\text{Youngest}} - \mu_{\text{Oldest}}, \beta_3 = \mu_{\text{Middle}} - \mu_{\text{Oldest}}$

        c. $E(y) = \beta_0 + \beta_1 x_1 + \beta_2 x_2 + \beta_3 x_3 + \beta_4 x_1 x_2 + \beta_5 x_1 x_3$

        d. $.21, -.05, .06, -.03, .11, .20$ e. $H_0: \beta_4 = \beta_5 = 0$

5.25   a. $E(y) = \beta_0 + \beta_1 x_1 + \beta_2 x_2$ b. $E(y) = \beta_0 + \beta_1 x_1 + \beta_2 x_2 + \beta_3 x_1 x_2$

5.27   a. $F = 8.79$; reject H_0

        b. DF-2: 2.14; blended: 4.865; adv. timing: 7.815

5.29   a. $E(y) = \beta_0 + \beta_1 x_1 + \beta_2 x_2 + \beta_3 x_1 x_2$ b. $\beta_2 + \beta_3$ c. $F = .26$, do not reject H_0

5.31   a. Quantitative   b. Quantitative   c. Qualitative

        d. Qualitative   e. Qualitative   f. Qualitative

        g. Quantitative   h. Qualitative

5.33   a. $E(y) = \beta_0 + \beta_1 x_1 + \beta_2 x_2 + \beta_3 x_3 + \beta_4 x_1 x_2 + \beta_5 x_1 x_3$

        b. $F = 4.90$, reject H_0 (p-value $= .0394$)   c. $E(y) = \beta_0 + \beta_1 x_1 + \beta_2 x_2 + \beta_3 x_3$

        d. $F = 2.36$; no

5.35   a. $E(y) = \beta_0 + \beta_1 x_1 + \beta_2 x_2 + \beta_3 x_3 + \beta_4 x_4$

        b. $E(y) = \beta_0 + \beta_1 x_1 + \beta_2 x_2 + \beta_3 x_3 + \beta_4 x_4 + \beta_5 x_1 x_2 + \beta_6 x_1 x_3 + \beta_7 x_1 x_4$

        c. $F = 2.33$, do not reject H_0

5.37   a. $E(y) = \beta_0 + \beta_1 x_1 + \beta_2 x_2$

        b. $E(y) = \beta_0 + \beta_1 x_1 + \beta_2 x_2 + \beta_3 x_1 x_2$

        c. $E(y) = \beta_0 + \beta_1 x_1 + \beta_2 x_1^2 + \beta_3 x_2 + \beta_4 x_1 x_2 + \beta_5 x_1^2 x_2$

5.39   b. $H_0: \beta_2 = \beta_5 = 0$ c. $H_0: \beta_3 = \beta_4 = \beta_5 = 0$

5.41   a. $E(y) = \beta_0 + \beta_1 x_1 + \beta_2 x_2 + \beta_3 x_3$

        b. $E(y) = \beta_0 + \beta_1 x_1 + \beta_2 x_2 + \beta_3 x_3 + \beta_4 x_1 x_3 + \beta_5 x_2 x_3$

        c. $H_0: \beta_4 = \beta_5 = 0$

5.43   a. $E(y) = \beta_0 + \beta_1 x_1 + \beta_2 x_2 + \beta_3 x_3$ b. $H_0: \beta_3 = 0$ vs. $H_a: \beta_3 < 0$

        d. $H_0: \beta_1 = \beta_3 = \beta_4 = \beta_7 = \beta_9 = \beta_{10} = 0$

CHAPTER 6

6.1 a. x_2 b. yes

6.3 a. x_4, x_5, and x_6 b. no
c. $E(y) = \beta_0 + \beta_1 x_4 + \beta_2 x_5 + \beta_3 x_6 + \beta_4 x_4 x_5 + \beta_5 x_4 x_6 + \beta_6 x_5 x_6$

6.5 a. DOT estimate, Status, and Days

CHAPTER 7

7.7 a. no   b. no

7.9 Unable to test model adequacy since there are no degrees of freedom available for estimating σ^2 (i.e., df $= n - 3 = 0$)

7.11 a. $\hat{y} = 2.74 + .801x_1$; yes, $t = 15.918$ b. $\hat{y} = 1.66 + 12.395x_2$; yes, $t = 11.759$
c. $\hat{y} = -11.79 + 25.068x_3$; yes, $t = 2.512$

7.13 Use all three variables

7.15 Two levels each; $n \geq 4$

7.17 a. Yes, $F = 37.20$ (p-value $= .0007$)
b. x_4 possibly correlated with other x's
c. Possible multicollinearity
d. $r_{12} = .327$; $r_{13} = .231$; $r_{14} = .166$; $r_{23} = .790$; $r_{24} = .791$; $r_{34} = .881$; yes

7.19 a. .0025; no   b. .434; no   c. No
d. $\hat{y} = -45.154 + 3.097x_1 + 1.032x_2$, $F = 39,222.34$, reject
$H_0: \beta_1 = \beta_2 = 0$; $R^2 = .9998$
e. $-.8978$; high correlation f. No

7.21 df(Error) $= 0$, s^2 undefined, no test of model adequacy

CHAPTER 8

8.1 a. $\hat{y} = 2.588 + .541x$
b. $-.406, -.206, -.047, .053, .112, .212, .271, .471, .330, .230, -.411, -.611$
c. Yes; needs curvature

8.3 a. $\hat{y} = 40.35 - .207x$
b. $-4.64, -3.94, -1.83, .57, 2.58, 1.28, 4.69, 4.09, 4.39, 2.79, .50, 1.10, -6.09, -5.49$
c. Yes; model needs curvature

8.5 b. no trends   c. no trends   d. no trends   e. no trends

8.7 a. $\hat{y} = 41.153 - .262x$ d. No; no curvature

8.9 Assumption of constant error variance appears satisfied; quadratic trend suggest need for curvature

8.11 a. Yes; assumption of equal variances violated
b. Use variance-stabilizing transformation $y^* = \sqrt{y}$

8.13   a. $\hat{y} = .94 - .214x$

b. 0, .02, $-.026$, .034, .088, $-.112$, $-.058$, .002, .036, .016

c. Football shape; unequal variances

d. Use the transformation $y^* = \sin^{-1} \sqrt{y}$ and fit the model $y^* = \beta_0 + \beta_1 x + \epsilon$

e. $\hat{y}^* = 1.307 - .2496x$; possibly

8.15   Residuals are approximately normal

8.17   Residuals are approximately normal

8.19   No outliers

8.21   Observations #8 and #3 are influential

8.23   No outliers

8.25   a. No   b. Observation (child) #24 is influential

8.27   b. Model adequate at $\alpha = .05$ for all banks except bank 5

c. Reject H_0 (two-tailed at $\alpha = .05$) for banks 2 and 5; fail to reject H_o for banks 4, 6, and 8; test inconclusive for banks 1, 3, 7, and 9

8.29   a. Yes   b. $d = .173$; reject H_0 $(d_L = 1.18)$

8.31   a. $\hat{y} = 1668.44 + 105.83t$; $t_L = 2.11$, reject H_0 b. yes

c. $d_L = .845$, reject H_0

8.33   a. Misspecified model; quadratic term missing

b. Unequal variances   c. Outlier

d. Unequal variances   e. Nonnormal errors

8.35   a. No, do not reject H_0; $d_U = 1.60$ b. Yes; reject H_0; $d_L = 1.39$

8.37   Use model as specified

8.39   a. $\hat{y} = 13.875 + 2.153x$ b. Yes   c. $d = 1.061$, reject H_0

8.41   Use model as specified

CHAPTER 9

9.1   a. $E(y) = \beta_0 + \beta_1 x_1 + \beta_2 (x_1 - 15)x_2$, where $x_1 = x$ and

$$x_2 = \begin{cases} 1 & \text{if } x_1 > 15 \\ 0 & \text{if not} \end{cases}$$

b.

	y-intercept	Slope
$x \le 15$	β_0	β_1
$x > 15$	$\beta_0 - 15\beta_2$	$\beta_1 + \beta_2$

c. Test H_0: $\beta_2 = 0$

9.3   a. $E(y) = \beta_0 + \beta_1 x_1 + \beta_2 (x_1 - 320)x_2 + \beta_3 x_2$, where

$$x_1 = x, \, X_2 = \begin{cases} 1 & \text{if } x_1 > 320 \\ 0 & \text{if not} \end{cases}$$

b.

	y-intercept	slope
$x \le 320$	β_0	β_1
$x > 320$	$\beta_0 - 320\beta_2 + \beta_3$	$\beta_1 + \beta_2$

c. Test H_0: $\beta_2 = \beta_3 = 0$

9.5   a. $E(y) = \beta_0 + \beta_1 x_1 + \beta_2(x_1 - 1{,}000)x_2$, where $x_1 = x$ and

$$x_2 = \begin{cases} 1 & \text{if } x_1 > 1{,}000 \\ 0 & \text{if not} \end{cases}$$

b. $\hat{y} = 4.024 - .0021x_1 - .00139(x_1 - 1{,}000)x_2$

c. Yes; $F = 363.44$ (p-value $= .0001$)   d. $-.0021 \pm .000366$

9.7 195.7 ± 203.2

9.9   a. $\hat{y} = -2.03 + 6.06x$ b. Reject H_0: $t = 10.35$ c. $1.985 \pm .687$

9.11   a. $\hat{y} = -1.2667 + .176x$; yes, $t = 4.09$ (p-value $= .0013$)

b. Residuals: $-1.33, 6.67, -5.33, -2.33, 1.67, -6.13, 3.87, -5.13, 9.87, -1.13, -7.93,$ $14.07, -6.93, 4.07, -3.93$; appears to be violated

c.

x	Variance of Residuals ; $w_i = 1/x_i^2$
100	20.7
150	44.3
200	85.2

d. $\hat{y} = -1.5143 + .1777x$

9.15   b. $\hat{y} = -.5279 + .0750x_1 + .0747x_2 + .3912x_3$ c. Reject H_0; $F = 21.79$

d. Yes; $t = 4.01$ e. $(-.2122, .2047)$

9.17   a. Reject H_0; $\chi^2 = 20.43$ b. Yes; $\chi^2 = 4.63$ c. $(.00048, .40027)$

CHAPTER 10

10.1   b.

Year	Quarter	4-Point Moving Average
1997	1	
1997	2	
1997	3	368.53
1997	4	375.43
1998	1	382.63
1998	2	393.80
1998	3	404.25
1998	4	415.40
1999	1	417.08
1999	2	419.20
1999	3	416.63
1999	4	415.38
2000	1	415.10
2000	2	403.05
2000	3	395.18
2000	4	391.00
2001	1	392.78
2001	2	398.75
2001	3	400.83
2001	4	

c. yes   d. 87.45   e. 112.525

f. Quarter 1: 358.545 (using $M_{21} = 410$); Quarter 2: 466.979 (using $M_{22} = 415$)

10.3 a. Moving average: 1.833; Exponential smoothing: 56.05; Holt-Winters: 11.785
      b. Moving average: 2.36; Exponential smoothing: 60.22; Holt-Winters: 12.20
      c. Moving average

10.5 a. Yes
      b.

Year	5-Point Moving Average
1987	
1988	
1989	119.48
1990	124.18
1991	128.78
1992	134.46
1993	139.78
1994	145.34
1995	150.88
1996	156.12
1997	159.88
1998	163.70
1999	
2000	
2003	Forecast = 185

      c.

Year	E_t
1987	109.3
1988	111.1
1989	114.42
1990	118.972
1991	123.0232
1992	126.9339
1993	130.8804
1994	137.6482
1995	143.5489
1996	148.8894
1997	153.5336
1998	157.3202
1999	161.0321
2000	165.2193
2003	Forecast = 167.73

      d.

Year	E_t	T_t
1987		
1988	113.80	4.50
1989	118.74	4.72
1990	124.40	5.19
1991	129.39	5.09
1992	133.81	4.75
1993	137.86	4.40
1994	144.48	5.51
1995	150.95	5.99
1996	156.93	5.98
1997	161.95	5.50
1998	165.67	4.61
1999	168.81	3.88
2000	172.21	3.64
2003	Forecast = 183.13	

10.7 a.

Year	3-Point Moving Average
1971	
1972	65.89
1973	105.3733
1974	139.6367
1975	148.6333
1976	144.8333
1977	155.5333
1978	216.5333
1979	369.1033
1980	454.8133
1981	476.94
1982	424.6133
1983	394.5
1984	375.54
1985	348.4867
1986	364.6933
1987	404.57
1988	409.0167
1989	400.7367
1990	375.7967
1991	363.3333
1992	355.2333
1993	362.53
1994	375.8533
1995	385.2
1996	367.54
1997	337.69
1998	301.4133
1999	284.11
2000	276.3767
2001	

b. 1999: 284.11; 2000: 276.38; 2001: $\approx$270

c.

Year	E_t
1971	41.25
1972	55.138
1973	89.2756
1974	145.6151
1975	158.243
1976	131.4886
1977	144.9377
1978	183.7875
1979	282.9975
1980	541.4075
1981	468.7855
1982	393.1011
1983	437.8442
1984	375.8008

(continued overleaf)

(*Continued*)

Year	E_t
1985	329.0002
1986	360.096
1987	399.1472
1988	429.3734
1989	390.8427
1990	385.4245
1991	366.7729
1992	348.4106
1993	357.4981
1994	378.6996
1995	382.7719
1996	386.8024
1997	342.1765
1998	303.8273
1999	283.9495
2000	280.0779
2001	272.8476

d. 1999: 283.95; 2000: 280.08; 2001: 272.85

e.

Year	E_t	T_t
1971		
1972	58.61	17.36
1973	93.44	24.35
1974	151.32	37.76
1975	166.94	28.90
1976	139.01	6.17
1977	147.68	7.17
1978	185.77	19.54
1979	287.30	52.34
1980	552.74	137.58
1981	498.57	60.88
1982	411.23	1.59
1983	441.79	13.18
1984	379.23	−17.12
1985	326.26	−31.46
1986	353.26	−8.08
1987	396.16	12.32
1988	431.24	21.42
1989	395.50	−1.44
1990	386.07	−4.64
1991	365.97	−10.82
1992	346.09	−14.45
1993	354.14	−5.45
1994	376.94	5.85
1995	383.59	6.17
1996	388.20	5.55
1997	343.57	−14.53
1998	301.20	−25.66
1999	278.29	−24.56
2000	274.03	−16.44
2001	268.35	−12.14

1999: 278.29; 2000: 274.03; 2001: 268.35

      f. Exponential smoothing most accurate

10.9   b. $\hat{y}_t = 39.49 + 19.13t - 13.15t^2$ d. $(-31.25, 48.97)$

10.11   b. Railroads: $\hat{y}_t = 64.5 - 5.91t$; Buses: $\hat{y}_t = 35.11 - 2.42t$;
       Air Carriers: $\hat{y}_t = -3.18 + 8.61t$
    d. Railroads: -18.24; Buses: 1.23; Air Carriers: 117.36

10.13   a. $\beta_1 = \mu_{\text{post}} - \mu_{\text{pre}}$ b. μ_{pre} c. $-.55$ d. 2.53   e. 2.53

10.15   a. No, $t = -1.39$ b. 2003.48   c. No, $t = -1.61$ d. 1901.81

10.17   a. 0, 0, 0, .5, 0, 0, 0, .25, 0, 0, 0, .125, 0, 0, 0, .0625, 0, 0, 0, .03125
    b. .5, .25, .125, .0625, .03125, .0156, ...

10.19 $R_t = \phi_1 R_{t-1} + \phi_2 R_{t-2} + \phi_3 R_{t-3} + \phi_4 R_{t-4} + \varepsilon_t$

10.21   a. $E(y_t) = \beta_0 + \beta_1 x_{1t} + \beta_2 x_{2t} + \beta_3 x_{3t} + \beta_4 t$
    b. $E(y_t) = \beta_0 + \beta_1 x_{1t} + \beta_2 x_{2t} + \beta_3 x_{3t} + \beta_4 t + \beta_5 x_{1t} t + \beta_6 x_{2t} t + \beta_7 x_{3t} t$
    c. $R_t = \phi R_{t-1} + \varepsilon_t$

10.23   a. $E(y_t) = \beta_0 + \beta_1 \left[\cos\left(\frac{2\pi}{365}\right) t \right] + \beta_2 \left[\sin\left(\frac{2\pi}{365}\right) t \right]$

    c. $E(y_t) = \beta_0 + \beta_1 \left[\cos\left(\frac{2\pi}{365}\right) t \right] + \beta_2 \left[\sin\left(\frac{2\pi}{365}\right) t \right] + \beta_3 t + \beta_4 t \left[\cos\left(\frac{2\pi}{365}\right) t \right]$
       $+ \beta_5 t \left[\sin\left(\frac{2\pi}{365}\right) t \right]$

    d. No; $R_t = \phi R_{t-1} + \varepsilon_t$

10.25   a. $\hat{y}_t = \beta_0 + \beta_1 t + \phi R_{t-1} + \varepsilon_t$
    b. $\hat{y}_t = 8000 + 124.039_1 t + .5263 \hat{R}_{t-1}$
    d. $R^2 = .9957$, $s = 51.31$

10.27   a. Upward trend
    b. $y_t = \beta_0 + \beta_1 t + \beta_2 t^2 + \phi R_{t-1} + \varepsilon_t$
    c. $\hat{y}_t = 1690 - 349.57t + 21.09t^2 + .772 \hat{R}_{t-1}$; $R^2 = .9834$; $t = 7.30$, $p < .0001$, reject H_0

10.29   a. $F_{49} = 336.91$; $F_{50} = 323.41$; $F_{51} = 309.46$
    b. $Y_{49}: 336.91 \pm 6.48$; $Y_{50}: 323.41 \pm 8.34$; $Y_{51}: 309.46 \pm 9.36$

10.35   a. 2136.2   b. (1404.3, 3249.7)   c. 1944; (1301.8, 2902.9)

10.37   a.

Year	Month	3-Point Moving Average
2000	October	
2000	November	88323
2000	December	99053
2001	January	118588.3
2001	February	149737.7
2001	March	159302
2001	April	162387.7
2001	May	153476.7
2001	June	145925
2001	July	141469.3
2001	August	134834.7
2001	September	149075
2001	October	154083
2001	November	148710.3
2001	December	

    b. 152,000

c.

Year	Month	Exponential E_t
2000	October	97304
2000	November	93591.2
2000	December	83365.88
2001	January	111042.8
2001	February	134250.3
2001	March	155698.3
2001	April	157191.5
2001	May	158264
2001	June	149264
2001	July	141024.8
2001	August	143777.1
2001	September	131526.8
2001	October	159562.5
2001	November	160206.6
2001	December	128427.8

d. 115,716.3

e.

Year	Month	E_t	T_t	S_t
2000	October			
2000	November	91116.0	−6188.0	1
2000	December	79900.60	−9707.18	0.958053
2001	January	105773.77	15199.06	1.224254
2001	February	138222.33	27273.71	1.083197
2001	March	168196.62	29164.11	1.010704
2001	April	173856.49	12711.15	0.909871
2001	May	170014.46	1123.918	0.935091
2001	June	154413.75	−10583.3	0.927793
2001	July	138851.37	−14068.7	0.976094
2001	August	137280.28	−5320.36	1.060691
2001	September	126799.97	−8932.33	0.972871
2001	October	154098.86	16429.52	1.156744
2001	November	164592.95	12274.72	0.975959
2001	December	137909.54	−14996	0.867839

Forecast = 138,840.5

f. $y_t = \beta_0 + \beta_1 t + \beta_2 M_1 + \beta_3 M_2 + \beta_4 M_3 + \cdots + \beta_{12} M_{11} + \phi R_{t-1} + \varepsilon_t$

g. $\hat{y}_t = 110,203 + 5047t + 14,229m_1 + 29,514m_2 + 12,654m_3 + 8,399m_4 - 12,364m_5 - 25,141m_6 - 20,147m_7 - 46,797m_8 - 7,755m_9 - 24,704m_{10} - 63,732m_{11} - .0601\hat{R}_{t-1}$

h. (−351,710.8, 735,422.5).

10.39 a. Upward trend   b. $y_t = \beta_0 + \beta_1 t + \varepsilon_t$ c. $\hat{y}_t = 168.24 + 1.516t$

d. 181.7

e. Residuals: −20.7537, −15.2693, −11.7849, −10.3005, −4.8161, 1.6683, 5.1527, 13.6371, 12.1215, 12.6059, 11.0903, 6.5747, 6.0591, 7.5435, 4.0279, 4.5123, 4.9967, 6.4811, 6.9655, 2.4499, 1.9343, 1.4187, −3.0969, −4.6125, −7.1281, −9.6437, −11.1593, −10.6749;

Residual autocorrelation appears present.

f. $d = .132$, reject H_o

g. $y_t = \beta_0 + \beta_1 t + \phi R_{t-1} + \varepsilon_t$; $\hat{y}_t = 161.4891 + 1.7216t + .81505 R_{t-1}$

h. $(190.210, 223.185)$.

10.41 b. $y_t = \beta_0 + \beta_1 t + \phi R_{t-1} + \varepsilon_t$ d. $\hat{y}_t = 983.99 + 18.852t + .8387 R_{t-1}$

e. $(579.39, 2{,}721.49)$.

CHAPTER 11

11.1 Noise (variability) and volume (n)

11.3 a. cockatiel   b. yes   c. experimental group   d. $1, 2, 3$ e. 3

f. total consumption

g. $E(y) = \beta_0 + \beta_1 x_1 + \beta_2 x_2$, where $x_1 = \{1$ if group 1, 0 if not$\}$,

$x_2 = \{1$ if group 2, 0 if not$\}$

11.5 a. $y_{B1} = \beta_0 + \beta_2 + \beta_4 + \varepsilon_{B1}$; $y_{B2} = \beta_0 + \beta_2 + \beta_5 + \varepsilon_{B2}; \ldots; y_{B,10} = \beta_0 + \beta_2 + \varepsilon_{B,10}$;
$\overline{y}_B = \beta_0 + \beta_2 + (\beta_4 + \beta_5 + \cdots + \beta_{12})/10 + \overline{\varepsilon}_B$

b. $y_{D1} = \beta_0 + \beta_4 + \varepsilon_{D1}$; $y_{D2} = \beta_0 + \beta_5 + \varepsilon_{D2}; \ldots; y_{D,10} = \beta_0 + \varepsilon_{D,10}$; $\overline{y}_D = \beta_0 + (\beta_4 + \beta_5 + \cdots + \beta_{12})/10 + \overline{\varepsilon}_D$

11.7 a. students   b. yes; factorial design   c. class standing and type of preparation

d. class standing: low, medium, high; type of preparation: review session and practice exam

e. (low, review), (medium, review), (high, review), (low, practice), (medium, practice), (high, practice)

f. final exam score

11.9 a. No

11.11 $E(y) = \beta_0 + \beta_1 x_1 + \beta_2 x_2 + \beta_3 x_3 + \beta_4 x_4 + \beta_5 x_5 + \beta_6 x_1 x_2 + \beta_7 x_1 x_3 + \beta_8 x_1 x_4 + \beta_9 x_1 x_5 + \beta_{10} x_2 x_4 + \beta_{11} x_2 x_5 + \beta_{12} x_3 x_4 + \beta_{13} x_3 x_5 + \beta_{14} x_1 x_2 x_4 + \beta_{15} x_1 x_2 x_5 + \beta_{16} x_1 x_3 x_4 + \beta_{17} x_1 x_3 x_5$, where $x_1 = $ quantitative factor A; x_2, x_3 are dummy variables for qualitative factor B; x_4, x_5 are dummy variables for qualitative factor C

11.13 Cannot investigate factor interaction

11.15 7

11.21 8 treatments: $A_1 B_1, A_1 B_2, A_1 B_3, A_1 B_4, A_2 B_1, A_2 B_2, A_2 B_3, A_2 B_4$

11.23 $E(y) = \beta_0 + \beta_1 x_1 + \beta_2 x_2 + \beta_3 x_3 + \beta_4 x_4 + \beta_5 x_5$; 10

11.25 a. Sex and weight   b. Both qualitative   c. 4; (ML), (MH), (FL), and (FH)

CHAPTER 12

12.3 a. $E(y) = \beta_0 + \beta_1 x$, where $x = \{1$ if treatment 1, 0 if treatment 2$\}$

b. $\hat{y} = 10.667 - 1.524x$; $t = -1.775$, do not reject H_0

12.5 a. $t = -1.78$; do not reject H_0 c. Two-tailed

12.7 a. completely randomized design   b. colonies 3, 6, 9 and 12; energy expanded

c. $H_0: \mu_1 = \mu_2 = \mu_3$ d. Reject H_0

12.9　a. $H_0: \mu_1 = \mu_2 = \mu_3$　b. $E(y) = \beta_0 + \beta_1 x_1 + \beta_2 x_2$
　　　c. Reject H_0　　　　　　d. Fail to reject H_0

12.11　a.

Source	df	SS	MS	F	p
Solvent	2	3.3054	1.6527	24.51	0
Error	29	1.9553	0.0674		
Total	31	5.2607			

　　　b. $F = 24.51$, reject H_0

12.13　$F = 7.867$, reject H_0

12.15　a. 87.805　b. 53.892　c. 77.24　d. 131.132
　　　e.

Source	df	SS	MS	F
Treatment	2	53.892	26.946	13.257
Error	38	77.24	2.033	
Total	40	131.132		

　　　f. $F = 12.257$, reject H_0

12.17　a. No, do not reject H_0　b. Reject H_0, p-value = .065

12.19　a. hypothetical employees; managers　b. estimated absences
　　　c. $H_0: \mu_1 = \mu_2 = \mu_3$

12.21　a. $F = 0.61$, fail to reject H_0　b. $F = 0.61$, fail to reject H_0

12.23　No evidence of a difference among the three plant session means; $F = .019$

12.25　$F = 4.16$, reject H_0

12.27　a. Five insecticides; seven locations　b. Reject H_0, $F = 2.85$

12.29　a. factorial design
　　　b. level of coagulant (-5, 10, 20, 50, 100, and 200); Acidity level (-4, 5, 6, 7, 8, and 9); 36 combinations.

12.31　a. luckiness (lucky, unlucky, and uncertain); competition (competitive and noncompetitive)
　　　b. Interaction: $F = .72$, do not reject H_0; Luckiness: $F = 1.39$, do not reject H_0; Competition: $F = 2.84$, do not reject H_0

12.33　a.

SOURCE	df	SS	MS	F
Amount	3	104.19	34.73	20.12
Method	3	28.63	9.54	5.53
Amount × Method	9	25.13	2.79	1.62
Error	32	55.25	1.73	
Total	47	213.20		

　　　b. Do not reject H_0; $F = 1.62$

   c. Difference in mean shear strengths for any two levels of antimony amount does not depend on cooling method

   d. Amount: reject H_0, $F = 20.12$; Method: reject H_0, $F = 5.53$

12.35 b. reject H_0 c. yes

12.37 a. observational study   b. day of week and time of day; 7 and 24

   c. a =7; b =24   d. H_0: Day and Time do not interact

   e. $F = 1.22$, do not reject H_0

   f. Day: reject H_0, $F = 68.39$; Time: reject H_0, $F = 156.8$

12.39 Interaction: fail to reject H_0, $F = 1.77$; Preparation: reject H_0, $F = 14.40$; Standing: fail to reject H_0, $F = 2.17$

12.41 a. _____   b. $F = 5.21$; yes   c. $F = 2.79$; yes   d. SS(Total) $= 34.99$, $R^2 = .52$

Source	df
A	2
B	3
C	1
AB	6
AC	2
BC	3
ABC	6
Error	120
Total	143

12.43 a. $E(y) = \beta_0 + \beta_1 x_1 + \beta_2 x_1^2$ b. $E(y) = (\beta_0 + \beta_3) + (\beta_1 + \beta_6)x_1 + (\beta_2 + \beta_9)x_1^2$

   c. $E(y) = (\beta_0 + \beta_3 + \beta_4 + \beta_5) + (\beta_1 + \beta_6 + \beta_7 + \beta_8)x_1 + (\beta_2 + \beta_9 + \beta_{10} + \beta_{11})x_1^2$

   d. $\hat{y} = 31.15 + .153x_1 - .00396x_1^2 + 17.05x_2 + 1.91x_3 - 14.3x_2x_3 + .151x_1x_2 + .017x_1x_3 - .08x_1x_2x_3 - .00356x_1^2x_2 + .0006x_1^2x_3 + .0012x_1^2x_2x_3$

   e. Rolled/inconel: $\hat{y} = 53 + .241x_1 - .00572x_1^2$; Rolled/incoloy: $\hat{y} = 50.25 + .17x_1 + .00336x_1^2$; Drawn/inconel: $\hat{y} = 48.2 + .304x_1 - .00752x_1^2$; Drawn/incoloy: $\hat{y} = 31.15 + .153x_1 - .00396x_1^2$

12.45 a. $3 \times 4 \times 3 \times 3 = 108$

   b. The complete model has 108 terms, including β_0, 9 main effect terms, 30 two-way interactions, 44 three-way interactions, and 24 four-way interactions

   c. $H_0: \beta_{10} = \beta_{11} = \cdots = \beta_{107} = 0$

12.47 a. 6   b. $\mu_{12} < (\mu_3, \mu_6, \mu_9)$

12.49 a. Reject $H_0: \mu_R = \mu_P = \mu_D = \mu_A = \mu_B$ b. $(\mu_R, \mu_P) < (\mu_D, \mu_A, \mu_B)$; $\mu_D < \mu_B$

12.51 a. Evidence of accuracy $\times$ vocabulary interaction   b. Yes

   c. 75%: means for all three accuracy levels are significantly different; 87.5%: means for all three accuracy levels are significantly different; 100%: $\mu_{90\%} > (\mu_{99\%}, \mu_{95\%})$

12.53   a. Policy 1 mean differs from each of policies 3–18; 2 and 3 differ from 4–18; 4 differs from 8–18; 5, 6, and 7 differ from 9–18; 8 differs from 12–18; 9, 10, and 11 differ from 16–18

       b. Yes

12.55 $\omega = 1.82$; $(\mu_5, \mu_3, \mu_0) < (\mu_{10})$

12.57   a. yes   b. yes   c. no   d. .05

12.59   a. Reject H_0 b. $\mu_Q > (\mu_S, \mu_C)$

12.61   no significant differences

12.63   variance assumption violated

12.65   assumptions satisfied

12.67   variance assumption violated

12.69   a. Five filtering methods   b. Relevance rating

       c. Reject H_0; $\mu_1 = \mu_2 = \mu_3 = \mu_4 = \mu_5$

12.71   Reject H_0; $F = 17.66$

12.73   a. Evidence of $N \times I$ interaction; ignore tests for main effects

       b. Agree; interaction implies differences among N means depend on level of I

12.75   a.

Source	df	F
Aggressiveness (A)	1	16.43
Alcohol Cond. (C)	2	6
A × C	2	
Error	129	
Total	134	

       b. Reject H_0 c. Reject H_0

12.77   a. Factorial   b. No replications

       c. $E(y) = \beta_0 + \beta_1 x_1 + \beta_2 x_2 + \beta_3 x_1 x_2 + \beta_4 x_1^2 + \beta_5 x_2^2$

       d. Test H_0: $\beta_3 = 0$

       e. $\hat{y} = -384.75 + 3.73x_1 + 12.72x_2 - .05x_1 x_2 - .009x_1^2 - .322x_2^2$; $t = -2.05$, reject H_0 (p-value = .07)

12.79   a. (Downs, left), (Downs, right), (Normal, left), (Normal, right)

       d. reject H_0 e. $\mu_{DL} < (\mu_{DR}, \mu_{NL}, \mu_{NR})$

12.81   a.

SOURCE	df	SS	MS	F
Methods	2	.19	.10	.08
Brands	5	605.70	121.14	93.18
Error	10	13.05	1.30	
Total	17	618.94		

       c. No, do not reject H_0; $F = .08$ d. $-.233 \pm 1.469$

12.83 a. Factors (levels): accounts receivable (completed, not completed); verification (completed, not completed); treatments: CC, CN, NC, NN

      c. Yes

12.85 a. $E(y) = \beta_0 + \beta_1 x_1 + \beta_2 x_2 + \beta_3 x_1 x_2 + \beta_4 x_1^2 + \beta_5 x_2^2$

      b. $\hat{y} = 29.86 + .56x_1 - .1625x_2 - .1135x_1 x_2 - .275x_1^2 - .23125x_2^2$

      c. The two models are different   d. $R^2 = .842$ e. Yes; $F = 5.67$

12.87 a.

Source	df	F	p
Time Period	3	11.25	0.0001
Station	9		
Error	27		
Total	39		

      b. Reject H_0; $F = 11.25$

      c. Largest: 7/9-7/23 or 7/24-8/8; smallest: 8/14-8/31

12.89 Evidence of interaction ($p = .0001$)

APPENDIX A

A.1 a. $\begin{bmatrix} 6 & 3 \\ -2 & -5 \end{bmatrix}$ b. $\begin{bmatrix} 3 & 0 & 9 \\ -9 & 4 & 5 \end{bmatrix}$ c. $\begin{bmatrix} 5 & 4 \\ 1 & -4 \end{bmatrix}$

A.3 a. 3×4 b. No; the number of elements in a row of **B** does not match the number of elements in a column of **A**

A.5 a. $\begin{bmatrix} 2 & 3 \\ -9 & 0 \\ 8 & 2 \end{bmatrix}$ b. $\begin{bmatrix} 3 & 0 & 4 \end{bmatrix}$ c. $\begin{bmatrix} 14 & 7 \end{bmatrix}$

A.7 a. $\begin{bmatrix} 1 & 0 \\ 0 & 1 \end{bmatrix}$ c. $\begin{bmatrix} 1 & 0 & 0 \\ 0 & 1 & 0 \\ 0 & 0 & 1 \end{bmatrix}$

A.13 $\begin{bmatrix} 3/8 & 1/8 \\ -1/4 & 1/4 \end{bmatrix}$

A.15 a. $\mathbf{A} = \begin{bmatrix} 10 & 0 & 20 \\ 0 & 20 & 0 \\ 20 & 0 & 68 \end{bmatrix}$; $\mathbf{V} = \begin{bmatrix} v_1 \\ v_2 \\ v_3 \end{bmatrix}$; $\mathbf{G} = \begin{bmatrix} 60 \\ 60 \\ 176 \end{bmatrix}$ c. $\begin{bmatrix} 2 \\ 3 \\ 2 \end{bmatrix}$

A.17   a. $\mathbf{Y} = \begin{bmatrix} 1 \\ 2 \\ 2 \\ 3 \\ 5 \\ 5 \end{bmatrix}$; $\mathbf{X} = \begin{bmatrix} 1 & 1 \\ 1 & 2 \\ 1 & 3 \\ 1 & 4 \\ 1 & 5 \\ 1 & 6 \end{bmatrix}$ b. $\mathbf{X'X} = \begin{bmatrix} 6 & 21 \\ 21 & 91 \end{bmatrix}$; $\mathbf{X'Y} = \begin{bmatrix} 18 \\ 78 \end{bmatrix}$

       d. $\hat{\boldsymbol{\beta}} = \begin{bmatrix} 0 \\ .8571 \end{bmatrix}$ e. $\hat{y} = .8571x$

A.19 $t = -5.196$; reject H_0

A.21 $t = -2.222$; reject H_0

A.23 $(-1.1593, 2.7593)$

A.25 $(.4153, 3.0131)$

A.27 $(3.2719, 3.6581)$

A.29   a. $\mathbf{X} = \begin{bmatrix} 1 & -5 \\ 1 & -3 \\ 1 & -1 \\ 1 & 1 \\ 1 & 3 \\ 1 & 5 \end{bmatrix}$; $\mathbf{Y} = \begin{bmatrix} 1.1 \\ 1.9 \\ 3.0 \\ 3.8 \\ 5.1 \\ 6.0 \end{bmatrix}$ b. $\mathbf{X'X} = \begin{bmatrix} 6 & 0 \\ 0 & 70 \end{bmatrix}$; $\mathbf{X'Y} = \begin{bmatrix} 20.9 \\ 34.9 \end{bmatrix}$

       c. $\hat{\boldsymbol{\beta}} = \begin{bmatrix} 3.4833 \\ .4986 \end{bmatrix}$ d. $\hat{y} = 3.4833 + .4986x$

       e. SSE $= .0682$; $s^2 = .0170$ f. $t = 31.95$; yes

       g. $r^2 = .9961$ h. $(3.62, 3.85)$

A.31   a. $\mathbf{X} = \begin{bmatrix} 1 & 1 \\ 1 & 1 \\ 1 & 2 \\ 1 & 2 \\ 1 & 3 \\ 1 & 3 \\ 1 & 4 \\ 1 & 4 \\ 1 & 5 \\ 1 & 5 \\ 1 & 6 \\ 1 & 6 \end{bmatrix}$; $\mathbf{Y} = \begin{bmatrix} 1.1 \\ .5 \\ 1.8 \\ 2.0 \\ 2.9 \\ 3.8 \\ 3.4 \\ 4.1 \\ 5.0 \\ 5.0 \\ 5.8 \end{bmatrix}$ b. $\mathbf{X'X} = \begin{bmatrix} 12 & 42 \\ 42 & 182 \end{bmatrix}$; $\mathbf{X'Y} = \begin{bmatrix} 37.4 \\ 163.0 \end{bmatrix}$

       c. The elements of $\mathbf{X'X}$ are increased by a factor of 2.

       d. $(\mathbf{X'X})^{-1} = \begin{bmatrix} .4333 & -.1 \\ -.1 & .02857 \end{bmatrix}$ e. $\hat{\boldsymbol{\beta}} = \begin{bmatrix} -.09333 \\ .91714 \end{bmatrix}$;

          $\hat{y} = -.09333 + .91714x$

       f. SSE $= 1.5564$; $s^2 = .15564$;   g. $t = 13.75$; yes

       h. $r^2 = .9498$

A.33   a. 18×1 b. $\mathbf{X} = \begin{bmatrix} 1 & 1 \\ 1 & 1 \\ 1 & 1 \\ 1 & 2 \\ 1 & 2 \\ 1 & 2 \\ 1 & 3 \\ 1 & 3 \\ 1 & 3 \\ 1 & 4 \\ 1 & 4 \\ 1 & 4 \\ 1 & 5 \\ 1 & 5 \\ 1 & 5 \\ 1 & 6 \\ 1 & 6 \\ 1 & 6 \end{bmatrix}$ c. $\begin{bmatrix} 18 & 63 \\ 63 & 273 \end{bmatrix}$

d. $(\mathbf{X'X})^{-1} = \begin{bmatrix} .28889 & -.06667 \\ -.06667 & .019048 \end{bmatrix}$ e. $\mathbf{a'}(\mathbf{X'X})^{-1}\mathbf{a} = .0746$

f. Width is reduced by approximately 21%

INDEX